W0235380

Catalysis in Ionic Liquids
From Catalyst Synthesis to Application

RSC Catalysis Series

Titles in the Series:

1: Carbons and Carbon Supported Catalysts in Hydroprocessing
2: Chiral Sulfur Ligands: Asymmetric Catalysis
3: Recent Developments in Asymmetric Organocatalysis
4: Catalysis in the Refining of Fischer–Tropsch Syncrude
5: Organocatalytic Enantioselective Conjugate Addition Reactions: A Powerful Tool for the Stereocontrolled Synthesis of Complex Molecules
6: *N*-Heterocyclic Carbenes: From Laboratory Curiosities to Efficient Synthetic Tools
7: *P*-Stereogenic Ligands in Enantioselective Catalysis
8: Chemistry of the Morita–Baylis–Hillman Reaction
9: Proton-Coupled Electron Transfer: A Carrefour of Chemical Reactivity Traditions
10: Asymmetric Domino Reactions
11: C-H and C-X Bond Functionalization: Transition Metal Mediation
12: Metal Organic Frameworks as Heterogeneous Catalysts
13: Environmental Catalysis Over Gold-Based Materials
14: Computational Catalysis
15: Catalysis in Ionic Liquids: From Catalyst Synthesis to Application

How to obtain future titles on publication:
A standing order plan is available for this series. A standing order will bring delivery of each new volume immediately on publication.

For further information please contact:
Book Sales Department, Royal Society of Chemistry, Thomas Graham House, Science Park, Milton Road, Cambridge, CB4 0WF, UK
Telephone: +44 (0)1223 420066, Fax: +44 (0)1223 420247
Email: booksales@rsc.org
Visit our website at www.rsc.org/books

Catalysis in Ionic Liquids
From Catalyst Synthesis to Application

Edited by

Chris Hardacre
Queen's University Belfast, Northern Ireland
Email: c.hardacre@qub.ac.uk

Vasile Parvulescu
University of Bucharest, Romania
Email: vasile.parvulescu@g.unibuc.ro

RSC Catalysis Series No. 15

ISBN: 978-1-84973-603-9
ISSN: 1757-6725

A catalogue record for this book is available from the British Library

Published by The Royal Society of Chemistry,
Thomas Graham House, Science Park, Milton Road,
Cambridge CB4 0WF, UK

Registered Charity Number 207890

For further information see our web site at www.rsc.org

Preface

This book provides an up to date review of the state of the art of catalytic reactions in ionic liquids as well as the formation of catalytic materials using ionic liquid methods. Catalytic reactions were amongst the first to be undertaken in these neoteric solvents with electrocatalytic studies being reported in the 1960s. Thereafter, there has been an explosion in the interest in this area starting with carbon–carbon bond forming reactions utilizing ionic liquids as the catalyst as well as the solvent in Friedel–Crafts, Heck and Diels–Alder reactions. From there the field moved onto study gas–liquid reactions, asymmetric processes and the conversion of biomass. A wide range of catalysts have been utilized and modified to be compatible with ionic liquid processes including homogeneous complexes, nanoparticles, supported metal heterogeneous catalysts, supported ionic liquid based catalysts, zeolites, enzymes, electrocatalysts and photocatalysts. In the vast majority of cases, the ionic liquid based processes have been compared with analogous molecular derived systems with significant advantages being demonstrated, for example, in rate, selectivity, recycle of the catalyst or work up procedures. In a number of cases, the ionic liquid based systems have enabled new reactions to be undertaken. Due to the wide range of ionic liquids available and the ability to functionalise the cation and the anion to tailor their physical and chemical properties, the field of catalysis in ionic liquids has been transformed over the last 20 years from both the perspective of novel materials synthesis as well as reactivity-selectivity profiles. The chapters provide a perspective on how ionic liquid properties can be modified by structural changes to enable the catalytic materials and processes to be controlled. In addition, the reviews provide a summary of where our understanding lies in these systems. The complex nature of the interactions involved and the potential these systems have to change many industrial processes provide significant opportunities for future study.

RSC Catalysis Series No. 15
Catalysis in Ionic Liquids: From Catalyst Synthesis to Application
Edited by Chris Hardacre and Vasile Parvulescu

Published by the Royal Society of Chemistry, www.rsc.org

This is particularly true in the translation of the technologies under study from the bench scale to pilot and full scale industrial utilization where the recovery of the ionic liquids, their toxicity and their added value to a process, for example, are critical. We would like to thank all the authors for their hard work in reviewing the subject matter for this book and for providing their insight into the future.

Christopher Hardacre and Vasile I. Parvulescu

Contents

RSC Catalysis Series No. 15
Catalysis in Ionic Liquids: From Catalyst Synthesis to Application
Edited by Chris Hardacre and Vasile Parvulescu

Published by the Royal Society of Chemistry, www.rsc.org

CHAPTER 1

Catalytic Conversion of Biomass in Ionic Liquids

HUI WANG, LEAH E. BLOCK AND ROBIN D. ROGERS*

Center for Green Manufacturing and Department of Chemistry, The University of Alabama, Tuscaloosa, AL 35487, USA
*Email: rdrogers@as.ua.edu

1.1 Introduction

The fossil fuel-based economy is facing several problems and challenges, which involve the increasing emissions of CO_2, decreasing reserves, and increasing energy prices.[1] These challenges have driven the search for new transportation fuels and bioproducts to substitute the fossil carbon-based materials. Biomass is defined as organic matter available on a renewable basis, and it includes forest and mill residues, agricultural crops and wastes, wood and wood wastes, animal wastes, livestock operation residues, aquatic plants, and municipal and industrial wastes.[2] Biomass is deemed a sustainable and green feedstock for the production of fuels and fine chemicals, although perhaps not always in the way they are proposed to be used.

A major source of biomass is lignocellulosic biomass, which is particularly well suited for energy applications because of its large-scale availability, low cost, and environmentally benign production. Lignocelluloses are composed of cellulose, hemicellulose, lignin, extractives, and several inorganic materials, of which the first three biopolymers are the main components. The cellulose microfibrils that are present in the hemicellulose–lignin matrix are often associated in the form of bundles or macrofibrils.[3] The structure of these naturally occurring cellulose fibrils is mostly crystalline in nature and

RSC Catalysis Series No. 15
Catalysis in Ionic Liquids: From Catalyst Synthesis to Application
Edited by Chris Hardacre and Vasile Parvulescu

Published by the Royal Society of Chemistry, www.rsc.org

highly resistant to attack by enzymes. In addition, the presence of lignin also impedes enzymatic hydrolysis, as enzymes bind onto the surface of lignin and hence do not act on the cellulose chains.[4]

Usually, conversion of lignocellulosic biomass is carried out in the presence of catalysts, such as strong liquid and solid acids. Various types of lignocellulosic biomass, such as wood chips, sawdust, corncobs, and walnut shells, have been tentatively processed by liquid acid-catalyzed hydrolysis[5–8] with H_2SO_4, HCl, H_3PO_4, *etc.* Despite the relatively high catalytic activity of these liquid acids in the hydrolysis of cellulosic materials, by and large their uses are still uneconomical because the process suffers from severe corrosion, a requirement for special reactors, and costly separation and neutralization of waste acids.[9]

Recently, attention has been paid to the use of solid catalysts in the depolymerization of lignocellulosic biomass. Several types of solid acids, such as Nafion, Amberlyst, $-SO_3H$ functionalized amorphous carbon or mesoporous silica, H-form zeolites like HZSM-5, heteropolyacids, and even metal oxides (*e.g.*, γ-Al_2O_3) have been explored for their catalytic performance in the hydrolysis of lignocellulosic biomass.[10–13] It has been shown that solid Brønsted acids are efficient catalysts for the hydrolysis of lignocellulosic biomass.[14,15]

The ability of ionic liquids (ILs, now defined as salts with melting points below 100 °C[16]) to dissolve biomass provides new opportunities for the pretreatment and conversion of lignocellulosic biomass. In 2002, we reported that certain ILs, such as 1-butyl-3-methylimidazolium chloride ([C_4mim]Cl) can dissolve cellulose by as much as 25 wt% without any pretreatment.[17] Since then, increasing numbers of scientific papers, patents, and conference abstracts in this area have been published, and ILs have become one of the "hot-topics" in polysaccharide research. Up to now, ILs have been shown to be able to dissolve a number of pure biopolymers, including cellulose,[17–22] hemicellulose,[23] lignin,[24,25] chitin,[26] starch,[27] silk,[28] wool,[29] as well as a variety of raw biomass, such as wood,[30,31] bagasse,[32,33] corn stover,[34] wheat straw[35] and shrimp shell.[26] Not only is the dissolution of biomass in ILs widely studied, but also its conversion into value-enhanced products has drawn the attention of scientists.

In this chapter, the catalytic dissolution and degradation of pure cellulose, lignin (including lignin model compounds), hemicellulose, and raw lignocellulosic biomass materials in the presence of ILs will be reviewed. Several challenges in this area will also be addressed.

1.2 Catalytic Dissolution of Lignocellulosic Biomass

Lignocellulosic biomass presents a greater challenge for dissolution because of the tight, covalent, hydrogen bonded matrix of carbohydrate polymers (cellulose and hemicellulose) and phenolic polymers (lignin),[36] resulting in insolubility in common solvents. Various pretreatment methods for lignocelluloses have been developed to open the compact structure and make the conversion easier, and these methods include those that are physical (irradiation), chemical (alkali, acid, organosolv, ammonia explosion), physico-chemical (steam explosion, CO_2 explosion), or some combination of these.[37] Extreme conditions involving strong acids or bases, high temperatures, and high pressures are typically used at the expense of fragmentation of the components.

The ability of ILs to dissolve pure cellulose, lignin, and hemicellulose prompted us to study if ILs could also dissolve raw lignocellulosic biomass. In 2007, Professor Moyna, along with our group, reported that $[C_4mim]Cl$ can dissolve different sources of wood with varying hardness.[31] Later, we showed that 1-ethyl-3-methylimidazolium acetate ($[C_2mim][OAc]$) is a better solvent for wood than $[C_4mim]Cl$ under the same reaction conditions.[30] But even using $[C_2mim][OAc]$, it took 46 h to completely dissolve 0.5 g Southern yellow pine in 10 g IL at 110 °C, and the lignin content in the regenerated cellulose-rich material was 23.5%.[30] It was clear that a better separation of the lignin from the cellulose was needed.

At the time, we hypothesized that part of the difficulty in dissolving lignocellulosic biomass arose from the covalent linkages which hold lignin and the carbohydrate together *via* ether, ester, or glycoside bonds.[38] We thus sought a catalyst which would selectively cleave these bonds and facilitate the dissolution and separation of lignin and the carbohydrate. Polyoxometalates (POMs), together with O_2, had been shown to be promising systems for pulp delignification[39] and we found that when POMs ($[PV_2Mo_{10}O_{40}]^{5-}$) were present in the IL system, the dissolution of wood was greatly enhanced (*e.g.*, 0.5 g of Southern yellow pine could be dissolved in 10 g $[C_2mim][OAc]$ in 15 h *vs.* 46 h without POM).[38] The lignin content in the regenerated pulp was drastically reduced, and the lowest lignin content observed was 5.4% (*vs.* 23.5% without POM).

The form of POM (acidic POM *vs.* $[C_2mim][POM]$), POM concentration, and reaction time all affected delignification efficiency. Using $[C_2mim][POM]$, longer heating time, and higher POM loadings led to better delignification and higher lignin losses. This research indicated that the presence of an appropriate catalyst could indeed facilitate the cleavage of lignin from carbohydrate.

1.3 Catalytic Degradation of Biomass

1.3.1 Catalytic Degradation of Cellulose

Of the biopolymers in lignocellulosic biomass, cellulose is the most abundant and indeed is the most abundant renewable biodegradable biopolymer. It is a linear polysaccharide chain consisting of D-anhydroglucopyranose linked together through β-glycosidic bonds. An extensive network of inter- and intra-molecular hydrogen bonds and van der Waals forces results in a complex crystalline supramolecular structure.[40] Decrystallization and hydrolytic cleavage of cellulose polymers to other products has been a bottleneck in the path toward energy-efficient and economical utilization of cellulose and many efforts have been devoted to the depolymerization of cellulose. These include acidic hydrolysis,[41] enzymatic hydrolysis,[42] and hydrolysis in supercritical water.[43] However, progress has been limited partly due to the lack of solubility of cellulose in water and contamination of enzyme by the presence of other components. The dissolution of cellulose in ILs improves the reactivity of cellulose[44] and thus recently, more attention has been paid to the hydrolysis of cellulose in ILs.

1.3.1.1 Degradation Catalyzed by Mineral Acids

In 2007, Li and Zhao reported the hydrolysis of cellulose in [C_4mim]Cl in the presence of mineral acids, such as H_2SO_4, HCl, HNO_3, H_3PO_4.[45] The catalytic activity of H_2SO_4, HCl, and HNO_3 were similar, while H_3PO_4 showed lower catalytic activity, indicating that the acidity played an important role in the hydrolysis of cellulose in [C_4mim]Cl. Catalytic amounts of acid were sufficient to drive the hydrolysis reaction. For example, when the acid/cellulose mass ratio was 0.46, yields of total reducing sugar (TRS) and glucose were 64% and 36%, respectively, after 42 min at 100 °C.[46] However, when excess amounts of acid were loaded to the IL system, sugar yields decreased because side reactions tended to occur which consumed the hydrolysis products.

Hydrolysis of cellulose dissolved in 1-ethyl-3-methylimidazolium chloride ([C_2mim]Cl) and [C_4mim]Cl catalyzed by mineral acids can lead to the formation of glucose, cellobiose, and 5-hydroxymethylfurfural (5-HMF) as the main products.[47,48] The initial rate of glucose formation was determined to be of first order in the concentrations of dissolved glucan and acid concentration, and of zero order in the concentration of water. The independence on water concentration suggested that cleavage of the β-1,4-glycosidic bonds near chain ends is irreversible. The absence of oligosaccharides longer than cellobiose indicated that cleavage of interior glycosidic bonds is reversible due to the slow diffusional separation of cleaved chains in the highly viscous glucan–IL solution. Gradual addition of water during the glucan hydrolysis inhibited the rate of glucose dehydration to 5-HMF and the formation of humins. It was proposed that the inhibition was attributed to the stronger interactions of protons with water than the 2-OH atom of the pyranose ring of glucose, the critical step in the formation of 5-HMF. The reduction in humin formation associated with water addition was ascribed to the lowered concentration of 5-HMF, since the humins were formed through the condensation polymerization of 5-HMF with glucose.

Binder and Raines also reported the degradation of cellulose in [C_2mim]Cl with H_2SO_4 or HCl.[49] HMF was the main product, with moderate yields of glucose. The production of HMF at the expense of glucose suggested that glucose was dehydrated to HMF. To test this hypothesis, glucose was hydrolyzed in [C_2mim]Cl containing varying amounts of water. In the absence of both acid and water, glucose was recovered intact. Adding H_2SO_4 to glucose–IL solution, with little or no water, led to the rapid decay of glucose into HMF and other products. Interestingly, it was found that increasing the water content to 33 wt% in the same acidic solution enabled nearly 90% of the glucose to remain intact after 1 h, which was in accordance with Le Chatelier's principle,[50] showing that water disfavors dehydrative reactions, including glucose oligomerization and conversion into HMF. Additionally, the authors proposed that the highly nucleophilic chloride anions of the IL coordinate strongly to the carbohydrates,[51] accelerating acid-catalyzed dehydration reactions. High concentrations of water solvated chloride and thus prevented it from interacting with carbohydrates. Therefore, when water was in a large amount, the hydrolysis reaction was inhibited.

1.3.1.2 Degradation Catalyzed by Solid Acids

Rinaldi *et al.* first reported that solid acids are powerful catalysts for the hydrolysis of cellulose dissolved in ILs.[52,53] The factors responsible for the control of depolymerization of cellulose in [C_4mim]Cl using Amberlyst 15DRY as the catalyst were determined.[54] It was found that the acidic resin released H^+ into the solution, controlling the initial rate of depolymerization. The initial size of the cellulose chains was crucial in the control of initial product distribution. Long chains were preferably cleaved into shorter ones instead of producing glucose, accounting for the induction period observed for the release of glucose or total reducing sugars. Activation of cellulose towards hydrolysis requires a strong acid, which prohibits the utilization of ILs composed of a weakly basic anion, such as acetate or phosphonate. These anions could capture the available H^+ species and prevent the activation of the glycosidic bonds. Additionally, the presence of *N*-methylimidazole, an impurity in [C_4mim]Cl, decreased the catalytic performance of this system. Amberlyst 15DRY could be recycled, and after being washed with sulfuric acid, this catalyst showed the same catalytic activity as the fresh resin.

Solid acid-catalyzed hydrolysis of cellulose in IL can be substantially improved by microwave heating. H-form zeolites with a lower Si/Al molar ratio and a larger surface area showed high catalytic activity.[55] These solid catalysts exhibited better performance than styrene-based sulfonic acid resin. Compared with conventional oil bath heating, microwave irradiation at an appropriate power significantly reduced the reaction time (*e.g.*, $<$10 min at 240 W) and increased the yields of reducing sugars. A typical hydrolysis reaction with Avicel cellulose produced glucose in $\sim$37% yield within 8 min, in comparison with 7.1% by using oil bath heating at 100 °C for 10 h. Cellulose hydrolysis catalyzed by solid acids was more environmentally friendly, as it could simplify the downstream processes and circumvent waste acids and water disposal.

1.3.1.3 Degradation Catalyzed by Metal Salts

Zhao *et al.* found that $CrCl_2$ and $CrCl_3$ were efficient catalysts for the hydrolysis of cellulose in [C_2mim]Cl to HMF.[56] Later, an efficient strategy for

$CrCl_3$-mediated production of HMF in *ca.* 60% and 90% isolated yields from cellulose and glucose, respectively, in [C_4mim]Cl under microwave irradiation was reported by the same group.[57] When water was used as the solvent, glucose dehydration was essentially restrained, indicating that [C_4mim]Cl was a solvent superior to water. If H_2SO_4 was used *in lieu* of $CrCl_3$, the dehydration reaction afforded HMF in only 49% yield, and formation of insoluble humins was observed.

Dissolution of purified cellulose in a mixture of *N*,*N*-dimethylacetamide (DMAc)–LiCl and [C_2mim]Cl with the addition of $CrCl_2$ or $CrCl_3$ produced HMF in yields up to 54% within 2 h at 140 °C (with 60 wt% [C_2mim]Cl).[58] The yield compared well with results of HMF synthesis from cellulose in the patents using aqueous acid[59] or ILs.[56] Neither lithium iodide nor lithium bromide alone produced high yields of HMF because these salts in DMAc do not dissolve cellulose. However, using lithium bromide along with DMAc–LiCl did enable modest improvements in HMF yield. Likewise, using HCl as a cocatalyst also enhanced the HMF yields.

A novel catalytic system involving $CuCl_2$ (an example of a primary metal chloride catalyst) paired with a second metal chloride, such as $CrCl_2$, $PdCl_2$, $CrCl_3$, or $FeCl_3$ in [C_2mim]Cl, was found to substantially accelerate the rate of cellulose depolymerization under mild conditions.[60,61] These paired metal chlorides showed high catalytic activity for the hydrolytic cleavage of β-1,4-glycosidic bonds when compared with the rates of H_2SO_4-catalyzed hydrolysis. In contrast, single metal chlorides with the same total molar loading showed much lower activity under the same reaction conditions. Possible mechanisms involved in the paired $CuCl_2$–$PdCl_2$ catalytic system were studied experimentally in combination with theoretical calculations. Results indicated that Cu(II) was reduced during the reaction to Cu(I) only in the presence of a second metal chloride and a carbohydrate source such as cellulose in the IL system. Cu(II) generated protons by hydrolysis of water to catalyze the depolymerization step, and served to regenerate Pd(II) from Pd(0) (the added $PdCl_2$ was reduced to Pd(0) by side reactions). Pd(II) was suggested to facilitate the depolymerization step by coordinating the catalytic protons and also promoting the formation of HMF.

Considering the toxicity and environmental concerns of Cr-based catalysts, Tao and coworkers explored the catalytic activity of non-toxic and inexpensive $FeCl_2$ and $CoSO_4$ in the depolymerization of cellulose. It was found that functional acidic ILs, with the addition of $FeCl_2$ or $CoSO_4$, were an effective system for the hydrolysis of microcrystalline cellulose (MCC).[62,63] The IL, 1-(4-sulfonic acid)-butyl-3-methylimidazolium hydrogen sulfate, in combination with the metal catalyst, was found to be the most efficient system for the hydrolysis of cellulose at 150 °C, and conversion of MCC reached 84% in 300 min. The yields of HMF and furfural were up to 34% and 19%, respectively, for the $FeCl_2$ system and were shown to be 24% and 17%, respectively, for the system containing $CoSO_4$. Additionally, small amounts of levulinic acid and reducing sugars (8% and 4%, respectively) were detected. Dimers of furan compounds were the main by-products as detected by HPLC-MS, and the components of gas products, analyzed by MS, were shown to contain methane, ethane, CO, CO_2, and H_2. The IL and catalyst could be recycled by removing the solvents and reused in the hydrolysis of cellulose with favorable catalytic activity over five repeated runs.

1.3.1.4 *Degradation Catalyzed by Enzymes*

Enzymatic hydrolysis is another important method for the conversion of cellulose. However, one of the disadvantages of ILs is their strong tendency to denature enzymes. Turner *et al.*[64] studied the hydrolysis of cellulose by *Trichoderma reesei* cellulase in [C_4mim]Cl that contained 5% cellulose in 50 mM citrate buffer at 50 °C. The hydrolytic rate in this IL was poor, at least 10-fold less than that performed in aqueous buffer. Low activity in [C_4mim]Cl was attributed to the high concentration of the Cl^- ion which led to unfolding and inactivation of the enzymes, and this inactivation was shown to be irreversible. Hence, in order to preserve the activity of cellulases in the hydrolysis of cellulose, cellulose needs to be regenerated from the IL solutions and all traces of chloride-containing IL should be completely removed. This undoubtedly introduces a regeneration–separation step into the process, which would increase the overall cost and preclude the development of a single stage continuous process for conversion of lignocellulosic materials, unless more IL-tolerant enzymes can be used.

With the aim of eliminating the need to regenerate and separate cellulose, Kamiya *et al.* investigated *in situ* enzyme saccharification of cellulose in enzyme compatible IL 1-ethyl-3-methylimidazolium diethylphosphate.[65] Cellulase was directly added to the aqueous–IL mixture containing cellulose at 40 °C. The ratio of IL to water (citrate buffer with pH = 5.0) greatly affected the cellulase activity. When the volume of IL to water was greater than 3 : 2, low cellulase activity was observed. However, decreasing the volume ratio to 1 : 4 enhanced cellulase activity and over 70% of the added cellulose was converted to glucose and cellobiose.[65] A similar study by Engel *et al.*[66] demonstrated cellulase activity of up to 30% on α-cellulose in the presence of 10% (v/v) 1,3-dimethylimidazolium dimethylphosphate. Increasing viscosity and ionic strength led to decrease in enzyme activity.

1.3.1.5 *One-Step Conversion Catalyzed by Acidic Ionic Liquids*

A 100% conversion of cellulose to industrially useful chemicals was achieved in a one-step reaction by the use of cooperative IL pairs for combined dissolution and catalytic degradation.[67] One IL was selected to dissolve cellulose, while the second one was used for the catalytic conversion of the dissolved cellulose to products with low molecular weight. During the dissolution and catalytic reaction, an immiscible organic solvent, *e.g.*, hexane, was placed on top of the IL pair system to extract any small soluble organic products ($M_W < 300$ g mol^{-1}). The first IL could be [C_2mim][OAc], [C_4mim]Cl, or [C_4mim][OAc], and the second one was an acidic IL, such as

$[C_4mim][HSO_4]$, $[C_1mim][HSO_4]$, or $[C_4H_8SO_3Hmim][HSO_4]$. The acidity of the second IL was directly related to the catalytic degradation of cellulose. The strongest IL acid tested, $[C_4H_8SO_3Hmim][HSO_4]$, showed the highest catalytic performance and yielded the lowest average molecular weight products. Using the $[C_4H_8SO_3Hmim][HSO_4]$–$[C_4mim]Cl$ pair at 200 °C, 45.8 wt% of the cellulose was selectively converted to 2-(diethoxymethyl)furan.

1.3.2 Catalytic Degradation of Lignin

Lignin is a three-dimensional cross-linked amorphous phenolic polymer.[68] The composition, molecular weight, and amount of lignin differ from plant to plant, with lignin abundance generally decreasing in the order of softwoods > hardwoods > grasses.[69] Lignin fills the spaces between cellulose and hemicellulose, and acts as a resin to confer strength and rigidity to the plant.[70] Lignin is mainly composed of phenylpropane monomers that link together primarily through the C–O linkage of α- and β-ether bonds, and the β-O-4 linkage is found to be dominant.[71,72] Schematic representations of the softwood and hardwood lignin structures showing common linkages are depicted in Figure 1.1.[73] (The structures are merely pictorial and do not imply a particular sequence.) The chemical structure of lignin suggests that this polyphenolic material could potentially serve as a renewable chemical feedstock if suitable conversion chemistry is developed.

Lignin represents a vastly underutilized resource. Despite being one of the most abundant polymers on earth, ~98% of lignin is burned as a source of energy,[74,75] primarily in the pulp and paper industry, at very low efficiency. With the rapidly growing interest in conversion of lignocellulosic biomass to fuels, a large amount of lignin will be available. This will undoubtedly spur new concepts for using lignin as a resource for value-added products.

Degradation of lignin or lignin model compounds in the presence of ILs has gained more and more attention in recent years. Beech lignin was oxidatively cleaved in four ILs, including 1-ethyl-3-methylimidazolium methylsulfonate, ($[C_2mim][MeSO_3]$), 1-ethyl-3-methylimidazolium ethylsulfonate ($[C_2mim][EtSO_3]$), 1-ethyl-3-methylimidazolium trifluoromethylsulfonate ($[C_2mim][CF_3SO_3]$), and 1-methyl-3-methylimidazolium methylsulfate ($[C_1mim][MeSO_4]$), to give phenols, unsaturated propylaromatics, and aromatic aldehydes.[76] $Mn(NO_3)_2$, used as the catalyst, in combination with $[C_2mim][CF_3SO_3]$ proved to be the most effective reaction system. By adjusting the reaction conditions, the selectivity of the process could be shifted from syringaldehyde as the main product to 2,6-dimethoxy-1,4-benzoquinone (DMBQ). DMBQ could be isolated as a pure substance by a simple extraction–crystallization process,[76] and this compound is reported to be of relevance in the context of anti-tumor drugs.

Alcell and soda lignin were dissolved in the IL 1-ethyl-3-methylimidazolium diethylphosphate ($[C_2min][DEP]$) and subsequently oxidized by O_2 in the presence of several transition metal catalysts.[77] $CoCl_2 \cdot 6H_2O$ in $[C_2min][DEP]$ proved particularly effective for the oxidation. Both the Alcell

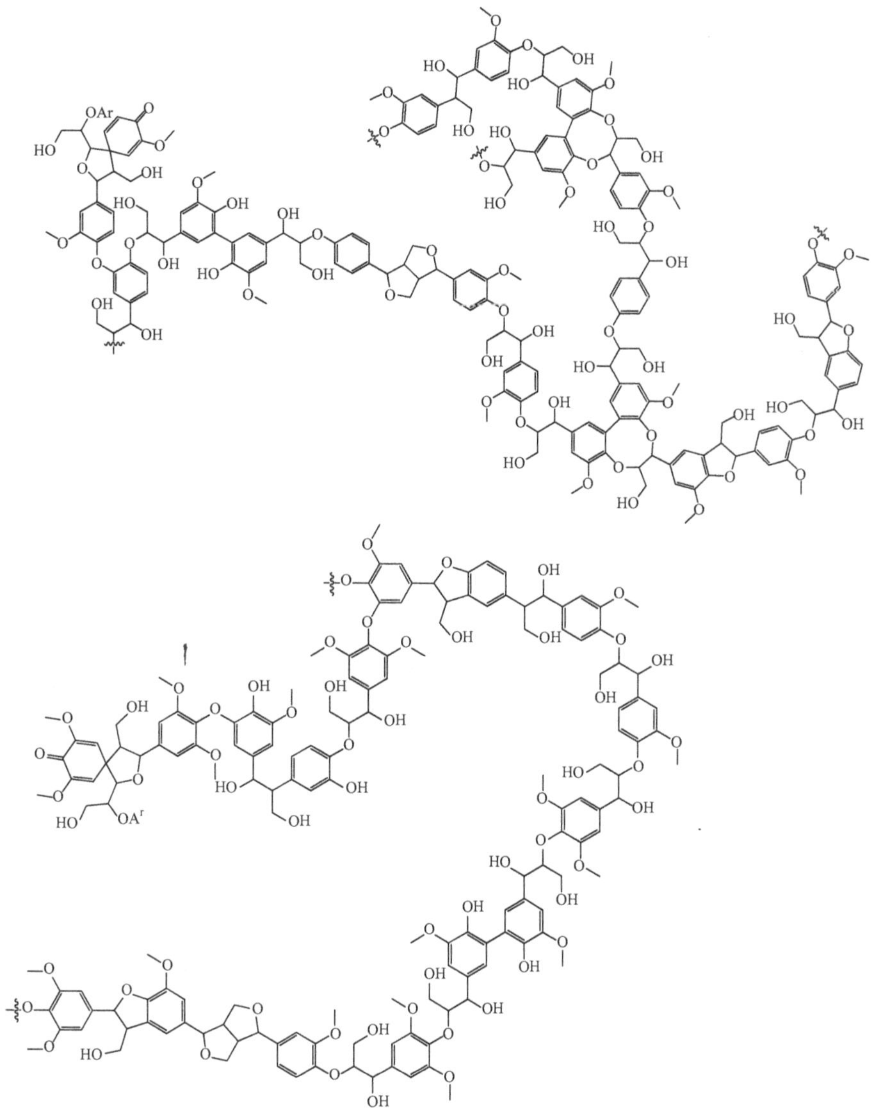

Figure 1.1 Schematic representations of softwood lignin (top) and hardwood lignin (bottom), redrawn from Ref. 73.

lignin and soda lignin readily dissolved in [C_2min][DEP] to yield dark brown solutions. Samples were taken each hour, and a small portion of each sample was used for spectroscopic analysis. Analysis of the reaction solution was conducted by first performing ethyl acetate extraction in an attempt to isolate any possible low molecular weight, monomeric products formed during the reaction; however, no monomeric products were detected by GC-MS. Attenuated total reflection infrared spectroscopy (ATR-IR) analysis of the lignin reaction samples indicated oxidation of lignin. IR peaks corresponding to alcohol and aldehyde stretches were observed, which suggested that the dissolved lignin was selectively oxidized. The absence of monomeric products, however, indicated either that the linkages in the lignin remained intact or were insufficiently disrupted to yield aromatic products of sufficiently low molecular weight for GC-MS detection.

Analysis of depolymerization of various lignin model compounds in the same system indicated that the catalyst rapidly oxidized benzyl and other alcohol functionalities in lignin, but left the phenolic functionalities and 5–5′, β-O-4, and phenylcoumaran linkages intact.[77] This system represents a potential method in a biorefinery platform to increase the oxygen functionality of lignin prior to depolymerization or functionalization of the already depolymerized lignin.

ILs based on the 1-methylimidazolium ($[Hmim]^+$) cation with chloride, bromide, hydrogen sulphate $[HSO_4]^-$, and tetrafluoroborate $[BF_4]^-$ counter ions along with $[C_4mim][HSO_4]$ were employed to degrade two lignin model compounds, guaiacylglycerol-β-guaiacyl ether (GG) and veratrylglycerol-β-guaiacyl ether (VG).[78] All the tested acidic ILs were successful in breaking down the lignin model compounds by hydrolyzing the β-O-4 ether linkages. While the acidic environment of the ILs catalyzed the hydrolysis reaction, the anions in the ILs had a significant effect on the guaiacol yield. At 150 °C, the relative guaiacol yield produced by each IL decreased in the order: [Hmim]Cl > $[C_4mim][HSO_4]$ > [Hmim]Br > $[Hmim][HSO_4]$ > $[Hmim][BF_4]$. Thus, it was assumed that the ability of the anion to hydrogen bond with the model compound was a major contributor to the ability of an acidic IL to effectively catalyze the hydrolysis of the β-O-4 ether linkage with stronger coordination leading to a chemical environment more conducive to ether bond hydrolysis.

In other work by the same group, the β-O-4 bonds of phenolic and non-phenolic lignin model compounds, GG and VG, underwent catalytic hydrolysis to guaiacol as the primary product in acidic [Hmim]Cl.[79] More than 70% of the β-O-4 bonds of both model compounds reacted with water to produce guaiacol at 150 °C. The IL could be recycled and reused without extra treatment or appreciable loss of activity.

Similarly, Binder *et al.* reported that ILs were suitable media for the Brønsted acid-catalyzed dealkylation of lignin model compounds, such as eugenol, 2-phenylethyl phenyl ether, and 4-ethylguaiacol.[80] A wide range of strong acidic catalysts enabled the formation of guaiacol in $[C_2mim]Cl$ or 1-ethyl-3-methylimidazolium triflate ($[C_2mim][OTf]$), with product yields as high as 11.6%.

Base-mediated cleavage of β-O-4 bonds in a dimeric phenolic lignin model compound GG in 1-butyl-2,3-dimethylimidazolium chloride ($[C_4C_1mim]Cl$) has also been reported.[81] Nitrogen bases of varying basicity and structures were tested at temperatures up to 150 °C. An enolether, 3-(4-hydroxy-3-methoxyphenyl)-2-(2-methoxy-phenoxy)-2-propenol, was found to be the primary product in all cases. 1,5,7-Triazabicyclo[4.4.0]dec-5-ene was shown to be the most active catalyst, leading to more than 40% of the β-O-4 bond cleaved, and its high activity was assumed to associate with the exposed nature of the nitrogen atoms, which made this catalyst function as both a base and a nucleophile.

The hydrolytic cleavage of GG and VG was also studied in [C_4mim]Cl with metal chlorides and water.[82] $FeCl_3$, $CuCl_2$, and $AlCl_3$ were found to be effective catalysts in cleaving the β-O-4 bond of GG, and it was observed that an increase in available water could lead to more β-O-4 bond cleavage of GG. After 120 min at 150 °C in the presence of $FeCl_3$ and $CuCl_2$, GG conversion reached 100%, and about 70% of the β-O-4 bonds were hydrolyzed, liberating guaiacol; while using $AlCl_3$ as the catalyst, about 80% of the β-O-4 bonds of GG were hydrolyzed with 100% GG conversion. $AlCl_3$ functioned more effectively than $FeCl_3$ and $CuCl_2$ in cleaving the β-O-4 bond of VG. About 75% of the β-O-4 bonds of VG could be hydrolyzed after 240 min at 150 °C in the presence of $AlCl_3$. The authors proposed that the catalytic activity was associated with HCl, working as the acidic catalyst, formed *in situ* by the hydrolysis of the metal chlorides.

Reichert and coworkers presented an approach of electro-oxidative cleavage of 5 wt% alkaline lignin solutions in a special protic IL, triethylammonium methanesulfonate ([$HNEt_3$][$MeSO_3$]), using electrodes coated with ruthenium–vanadium–titanium mixed oxide.[83] This protic IL provided a suitable medium for dissolution of lignin, ensured electrolysis at higher potentials, and promoted an oxidative lignin cleavage. The mixed oxide coating exhibited great oxidative stability combined with a remarkable catalytic activity for the oxidation of lignin. A wide range of aromatic fragments, such as aldehydes and ketones (benzaldehyde, 3-furaldehyde, *m*-tolualdehyde, vanillin, acetovanillone) were identified as the cleavage products as detected by GC-MS, and HPLC experiments confirmed an additional oxidative step, namely the conversion of vanillin to vanillic acid.

1.3.3 Catalytic Degradation of Hemicellulose

Hemicellulose, a mixture of polysaccharides containing xylose, arabinose, glucose, galactose, mannose, and other sugars,[84] is another major component of lignocellulosic biomass material. Studies on the degradation of hemicellulose in ILs are relatively few. Dee and Bell studied the hydrolysis of xylan, a material representing hemicellulose, in [C_2mim]Cl catalyzed by H_2SO_4.[48] They found that the rate of hemicellulose hydrolysis was approximately 1.4 times faster than that of cellulose hydrolysis. Experiments with gradual water addition produced primarily saccharide products with very low yields of dehydration products. After 30 min, the yields of xylose and furfural from xylan were 62% and 4%, respectively. The increase in furfural and xylose production from xylan prior to increasing the water concentration to 43 wt% at 60 min, resulted in the production of a black precipitate of humins. The concentration of humins continued to increase as the xylose yield decreased from 73% at 60 min to 63% at 120 min, while furfural yield only increased from 5% to 7% over the same period of time.

Using the IL [C_2mim]Cl as solvent and Brønsted acids as catalysts, Enslow and Bell investigated the kinetics of hydrolysis of xylan and subsequent dehydration/degradation reactions.[85] Xylobiose and xylose were detected as the primary products. Furfural was produced from the dehydration of xylose, and glucose as well as arabinose were also identified. Also, humins began to form and continued to accumulate throughout the duration of the experiment, which are suggested to be formed from the coupling loss reaction between xylose and furfural.[86]

The hydrolysis of xylan was compared with that of cellulose.[85] The observed initial rate constant of xylan hydrolysis was *ca.* 8 times higher on average than that of cellulose under similar reaction conditions over the temperature range from 80 to 100 °C. The lower initial rate of cellulose hydrolysis was likely due to the presence of the additional hydroxymethyl groups within cellulose. Unlike the secondary alcohols present in glucose and xylose, the primary alcohol in the methoxy group of glucose has the ability to abstract catalytic protons from solution through an equilibrium protonation–deprotonation reaction,[87] lowering the concentration of catalyst available for hydrolyzing the β-1,4-glycosidic bonds in cellulose. Thus, at the same initial catalyst loadings, fewer catalytic protons were available for cellulose hydrolysis, resulting in a lower initial rate when compared with hydrolysis of hemicellulose. This chemical process presents a viable pathway for producing sugars capable of being chemically (*via* dehydration/hydrogenation) or biologically (*via* fermentation) upgraded to potential fuel molecules.

Dehydration of hemicellulosic material, xylose, into furfural was investigated in several ILs[88] and [C_2mim]Cl as well as [C_4mim]Cl were found to be the most efficient ones. A $CrCl_2$ catalyst loading of 20 mol% with respect to xylose gave optimum conversion and yield amongst the different catalyst loadings used. For a catalyst combination of $CrCl_2$ and $CuCl_2$, the optimum mole fraction of the mixture was 0.6 : 0.4 ($CrCl_2$: $CuCl_2$), resulting in a maximum conversion of 90% and furfural yield of 49%. The addition of mineral acids had an adverse effect on furfural yield and it resulted in lower conversions when compared with reactions without the acid catalysts.

1.3.4 Catalytic Degradation of Lignocellulosic Biomass

Lignocellulosic biomass presents a more significant challenge for hydrolysis than does cellulose or hemicellulose. In addition to intractable crystalline cellulose, lignocellulosic biomass includes protective hemicellulose and lignin, heterogeneous components that are major obstacles to many biomass hydrolysis processes.[89] The dissolution of lignocellulose using ILs as the solvents offers several attractive features, and catalytic conversion of lignocelluloses in ILs has been widely studied.

Acids in ILs were demonstrated as efficient systems for hydrolysis of lignocellulosic materials with desirable TRS yields under mild conditions.[90] TRS yields could reach 66%, 74%, 81%, and 68% for hydrolysis of corn stalk, rice straw, pinewood, and bagasse, respectively, in [C_4mim]Cl in the presence of 7 wt% HCl at 100 °C under atmospheric pressure within 60 min. Analysis of hydrolyzates indicated the formation of monosaccharaides, such as glucose, xylose, arabinose, and galactose. The reactivity of other mineral acids in [C_4mim]Cl against corn stalks was also explored and it was found that hydrolysis rates decreased in the order: HCl > HNO_3 > H_2SO_4 > $HO_2CCHCHCO_2H$ > H_3PO_4. A novel outcome of this work was the discovery that acidic ILs 1-(4-sulfobutyl)-3-methylimidazolium bisulfate ([Sbmim][HSO_4]) and [C_4mim][HSO_4] could act as both catalyst and solvent for hydrolysis of corn stalk. It was found that this biomass material dissolved quickly in the acidic ILs, forming a solution with lower viscosity, implying that polysaccharide depolymerization readily occurred; however, low TRS yields (15–23%) were obtained. It was thus concluded that the strong acidic ILs might not only promote the depolymerization reaction, but also speed up the rate of sugar degradation. Corn stalk was subsequently hydrolyzed in [C_4mim]Cl, using the acidic [C_4mim][HSO_4] or [Sbmim][HSO_4] as the catalyst, and TRS yield increased to 68% and 71%, respectively.

Recently Binder and Raines[49] reported an effective process for the hydrolysis of untreated corn stover, in which water was gradually added to a catalytic system containing [C_2mim]Cl and acid catalyst (H_2SO_4 or HCl). Corn stover was hydrolyzed with [C_2mim]Cl in the presence of 10 wt% HCl at 105 °C, producing a 71% yield of xylose and 42% yield of glucose (yields were based on the xylan and cellulose content of the stover). Dilution of the reaction mixture to 70% water caused precipitation of unhydrolyzed polysaccharides and lignin. These residues were then dissolved in [C_2mim]Cl and subjected to a second-stage hydrolysis, which released additional xylose and glucose, leaving behind lignin-containing solids. After the two-stage HCl-catalyzed hydrolysis process, glucose and xylose yields reached 70% and 79%, respectively. Ion-exclusion chromatography allowed recovery of the IL and delivered sugar feedstocks that supported the vigorous growth of ethanologenic microbes. In this technique, passing the solution through a charged resin separated a mixture containing electrolyte and nonelectrolyte solutes. The charged IL was excluded from the resin, while nonelectrolytes, *e.g.*, sugars and furfural, were retained. The nonpolar HMF and furfural were adsorbed more strongly than sugars, and eluted later. Passing the corn stover hydrolyzate through a column of [C_2mim]-exchanged Dowex® 50 resin allowed laboratory-scale separation of the IL from the sugars, and recovery of the IL was higher than 95%, and those of glucose and xylose were 94% and 88%, respectively. Additionally, this chromatographic step removed inhibitory compounds such as HMF and furfural.

Miscanthus, which can be grown with lower water and soil nutrient requirements compared to other biofuel feedstocks[91] has been identified as one of several lignocellulosic feedstocks for biofuel production. The hydrolysis of raw *Miscanthus* dissolved in [C_2mim]Cl using H_2SO_4 as the catalyst was reported by Dee and Bell.[48] The kinetics of the hydrolysis of the cellulosic and hemicellulosic portions of *Miscanthus* was first order in acid concentration and zero order in water concentration. When compared with the hydrolysis of pure cellulose and xylan, it was found that rates for the hydrolysis of the cellulosic and hemicellulosic portions of *Miscanthus* were much lower, attributed to the inhibiting effects of lignin in raw *Miscanthus*.

Cleavage of the lignin–hemicellulose linkages by ethylene diamine pretreatment of *Miscanthus* increased the rate of hydrolysis of both the cellulosic and hemicellulosic materials. The conversion of *Miscanthus* to sugar products was improved by gradual addition of water to the reaction mixture, which limited the dehydration of the saccharides to furfurals and the formation of humins. Increasing the concentration of the acid catalyst increased the conversion of the cellulosic portion of *Miscanthus* to glucose but decreased the conversion of hemicellulose to xylose due to dehydration of this product to furfural and its subsequent condensation with glucose and xylose to form humins. High yields of saccharides were achieved with initial *Miscanthus* loadings of up to 9 wt%, but further increasing the initial loading, *e.g.*, 18 wt%, lowered the conversion to soluble products, most significantly for the cellulosic component.

Production of HMF and furfural from lignocellulosic biomass was also studied in $[C_4mim]Cl$ and $[C_4mim]Br$ in the presence of $CrCl_3$ under microwave irradiation.[92] Corn stalk, rice straw, and pine wood treated under typical reaction conditions produced HMF and furfural in yields of 45–52% and 23–31%, respectively, within 3 min. It was postulated that $CrCl_3$ in $[C_4mim]Cl$ formed complexes $[CrCl_{3+n}]^{n-}$ in a similar manner to $LnCl_3$.[93] These complexes would promote rapid conversion of α-anomers of glucose or xylose to β-anomers through hydrogen bonding between the Cl^- anions and the hydroxyl groups, followed by cyclic aldoses reverting to the acyclic form which combines with the chromium complex to form an enolate structure. Enolate formation would enable conversion of the aldoses to ketoses, followed by dehydration to produce HMF and furfural (Scheme 1.1). Thus, in the presence of IL, the β-1,4-glycosidic bonds were weakened at the cellulose hydrolysis step under microwave irradiation because of coordination with $[CrCl_{3+n}]^{n-}$. As a result, the β-1,4-glycosidic bonds were easily attacked by water to form glucose and oligomers. This method could

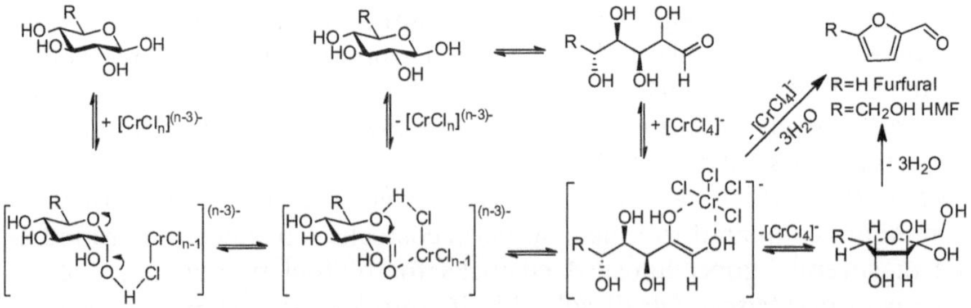

Scheme 1.1 Putative mechanism of $CrCl_3$-promoted conversion of glucose and D-xylose into HMF and furfural.[92]

facilitate energy-efficient and cost-effective conversion of biomass to biofuels and platform chemicals.

HMF could also be readily produced from untreated lignocellulosic biomass, such as corn stover and pine sawdust, under mild conditions in the mixture of DMAc–LiCl and [C_2mim]Cl with $CrCl_2$ or $CrCl_3$.[58] Yields of HMF from untreated corn stover were nearly identical to those for stover subjected to ammonia fiber expansion pretreatment. Other biomass components, such as lignin, did not interfere substantially in the process, as yields of HMF based on the cellulose content of the biomass were comparable to those from purified cellulose.

Researchers in the US Department of Energy's Joint BioEnergy Institute (JBEI) have engineered the first strains of *Escherichia coli* bacteria that can digest switchgrass biomass pretreated by ILs.[94] Both cellulolytic and hemicellulolytic strains were further engineered with three biofuel synthesis pathways to demonstrate the production of fuel substitutes or precursors suitable for gasoline, diesel, and jet engines directly from IL-treated switchgrass. This work might enable reduction in fuel production costs by consolidating two steps—depolymerizing cellulose and hemicelluloses into sugars, and fermenting the sugars into fuels—into a one-step operation, thus providing an economical route to produce advanced biofuels.

1.4 Conclusions

Ionic liquids hold the key to unlocking the bottleneck to the production of biofuels from lignocellulosic materials. By dissolving biomass materials in ILs, the hydrogen bonds that link the biopolymers together can be disrupted, which will benefit the further conversion of the biopolymers to sugars or other value-added products. Moreover, conversion of lignocellulosic biomass in the presence of ILs can be carried out under milder conditions, which can save energy; another important sustainability requirement of modern society.

However, several shortcomings of using ILs in biofuel production still exist, especially in process development and optimization. First, the choice of ILs can be a compromise between solubilizing power and catalyst compatibility. For instance, the Cl^- anion is a superior anion for biomass dissolution, but also causes many enzymes to denature. For an effective integrated process that would use enzymes *in situ* with ILs, more research on enzyme compatible ILs is required. ILs containing acetate anion are more efficient at dissolving biomass materials than the Cl-containing ILs,[23] but the acidic catalyst for the hydrolysis of biopolymers cannot coexist with basic anions, *e.g.*, acetate, in the solution.

Second, effective and energy-saving IL recycling methods should be developed for each intended process. The current method to recycle ILs in biomass processing mainly centers on evaporation of the antisolvent, which would consume a lot of energy, especially in the case of IL–water solutions. Thus, further efforts to develop effective methods to facilitate the IL recycling are needed.

Third, the price of ILs is still very high, which will make the conversion process uneconomical. This concern can be addressed by the development of new manufacturing methods, efficient scale-up technologies, and by developing better ILs. For example, better sources (*e.g.*, renewable) of raw materials, or new and effective synthetic routes could be developed.

The particular ability of some ILs to dissolve biopolymers, accompanied by a series of advantages, certainly might facilitate the conversion of raw biomass materials to value-added chemicals. However, this field is still in its infancy and much more research is needed in this area, such as optimizing large-scale pretreatment conditions, performing post-pretreatment steps in ILs, recycling and reusing ILs with reduced energy consumption and enhancing process efficiency. Moreover, the nature of lignin suggests that it might be used in the manufacture of high value chemicals, traditionally obtained from petroleum. Much of this work is now underway around the world and we look forward to the fascinating results yet to come.

References

1. J. Holm and U. Lassi, in *Ionic Liquids: Applications and Perspectives*, ed. A. Kokorin, InTech, Croatia, 1st edn, 2011, p. 545.
2. T. Vancov, A. S. Alston, T. Brown and S. McIntosh, *Renewable Energy*, 2012, **45**, 1.
3. G. Brodeur, E. Yau, K. Badal, J. Collier, K. B. Ramachandran and S. Ramakrishnan, *Enzyme Res.*, 2011, **2011**, 787532.
4. H. Palonen, F. Tjerneld, G. Zacchi and M. Tenkanen, *J. Biotechnol.*, 2004, **107**, 65.
5. B. Rivas, J. M. Dominguez, H. Dominguez and J. C. Parajo, *Enzyme Microb. Technol.*, 2002, **31**, 431.
6. A. Emmel, A. L. Mathias, F. Wypych and L. P. Ramos, *Bioresour. Technol.*, 2003, **86**, 105.
7. A. Liu, Y. K. Park, Z. Huang, B. Wang, R. O. Ankumah and P. K. Biswas, *Energy Fuels*, 2006, **20**, 446.
8. P. Lenihan, A. Orozco, E. O'Neill, M. N. M. Ahmad, D. W. Rooney and G. M. Walker, *Chem. Eng. J.*, 2010, **156**, 395.
9. C. Zhou, X. Xia, C. Lin, D. Tong and J. Beltramini, *Chem. Soc. Rev.*, 2011, **40**, 5588.
10. S. Van de Vyver, L. Peng, J. Geboers, H. Schepers, F. de Clippel, C. J. Gommes, B. Goderis, P. A. Jacobs and B. F. Sels, *Green Chem.*, 2010, **12**, 1560.
11. J. Hegner, K. C. Pereira, B. DeBoef and B. L. Lucht, *Tetrahedron Lett.*, 2010, **51**, 2356.
12. S. Suganuma, K. Nakajima, M. Kitano, D. Yamaguchi, H. Kato, S. Hayashi and M. Hara, *J. Am. Chem. Soc.*, 2008, **130**, 12787.
13. J. Tian, J. Wang, S. Zhao, C. Jiang, X. Zhang and X. Wang, *Cellulose*, 2010, **17**, 587.
14. A. Abbadi, K. F. Gotlieb and H. van Bekkum, *Starch—Stärke*, 1998, **50**, 23.

15. P. L. Dhepe, M. Ohashi, S. Inagaki, M. Ichikawa and A. Fukuoka, *Catal. Lett.*, 2005, **102**, 163.
16. R. D. Rogers and K. R. Seddon, *Science*, 2003, **302**, 792.
17. R. P. Swatloski, S. K. Spear, J. D. Holbrey and R. D. Rogers, *J. Am. Chem. Soc.*, 2002, **124**, 4974.
18. H. Zhang, J. Wu, J. Zhang and J. He, *Macromolecules*, 2005, **38**, 8272.
19. T. Heinze, K. Schwikal and S. Barthel, *Macromol. Biosci.*, 2005, **5**, 520.
20. Y. Fukaya, A. Sugimoto and H. Ohno, *Biomacromolecules*, 2006, **7**, 3295.
21. Y. Fukaya, K. Hayashi, M. Wada and H. Ohno, *Green Chem.*, 2008, **10**, 44.
22. J. Vitz, T. Erdmenger, C. Haensch and U. S. Schubert, *Green Chem.*, 2009, **11**, 417.
23. N. Sun, H. Rodriguez, M. Rahman and R. D. Rogers, *Chem. Commun.*, 2011, **47**, 1405.
24. Y. Pu, N. Jiang and A. J. Ragauskas, *J. Wood Chem. Technol.*, 2007, **27**, 23.
25. R. Mustafizur, Y. Qin, M. L. Maxim and R. D. Rogers, PCT WO 2010/056790 A1, 2008.
26. Y. Qin, X. Lu, N. Sun and R. D. Rogers, *Green Chem.*, 2010, **12**, 968.
27. K. Wilpiszewska and T. Spychaj, *Carbohydr. Polym.*, 2011, **86**, 424.
28. D. M. Phillips, L. F. Drummy, D. G. Conrady, D. M. Fox, R. R. Naik, M. O. Stone, P. C. Trulove, H. C. De Long and R. A. Mantz, *J. Am. Chem. Soc.*, 2004, **126**, 14350.
29. H. Xie, S. Li and S. Zhang, *Green Chem.*, 2005, **7**, 606.
30. N. Sun, M. Rahman, Y. Qin, M. L. Maxim, H. Rodríguez and R. D. Rogers, *Green Chem.*, 2009, **11**, 646.
31. D. A. Fort, R. C. Remsing, R. P. Swatloski, P. Moyna, G. Moyna and R. D. Rogers, *Green Chem.*, 2007, **9**, 63.
32. W. Li, N. Sun, B. Stoner, X. Jiang, X. Lu and R. D. Rogers, *Green Chem.*, 2011, **13**, 2038.
33. Z. Wang, L. Li, K. Xiao and J. Wu, *Bioresour. Technol.*, 2009, **100**, 1687.
34. Y. Cao, H. Li, Y. Zhang, J. Zhang and J. He, *J. Appl. Polym. Sci.*, 2010, **116**, 547.
35. Q. Li, Y. He, M. Xian, G. Jun, X. Xu, J. Yang and L. Li, *Bioresour. Technol.*, 2009, **100**, 3570.
36. J. Long, X. Li, B. Guo, F. Wang, Y. Yu and L. Wang, *Green Chem.*, 2012, **14**, 1935.
37. P. Kumar, D. M. Barrett, M. J. Delwiche and P. Stroeve, *Ind. Eng. Chem. Res.*, 2009, **48**, 3713.
38. N. Sun, X. Jiang, M. L. Maxim, A. Metlen and R. D. Rogers, *ChemSusChem*, 2011, **4**, 65.
39. A. R. Gaspar, J. A. F. Gamelas, D. V. Evtuguin and C. P. Neto, *Green Chem.*, 2007, **9**, 717.
40. Y. Nishiyama, J. Sugiyama, H. Chanzy and P. Langan, *J. Am. Chem. Soc.*, 2003, **125**, 14300.
41. S. Miller and R. Hester, *Chem. Eng. Commun.*, 2007, **194**, 85.

42. S. I. Mussatto, G. Dragone, M. Fernandes, A. M. F. Milagres and I. C. Roberto, *Cellulose*, 2008, **15**, 711.
43. S. Kumar and R. B. Gupta, *Ind. Eng. Chem. Res.*, 2008, **47**, 9321.
44. C. E. Wyman, B. E. Dale, R. T. Elander, M. Holtzapple, M. R. Ladisch, Y. Y. Lee, C. Mitchinson and J. N. Saddler, *Biotechnol. Prog.*, 2009, **25**, 333.
45. C. Li and Z. K. Zhao, *Adv. Synth. Catal.*, 2007, **349**, 1847.
46. H. Xie and Z. K. Zhao, *Ionic Liquids: Applications and Perspectives*, ed. A. Kokorin, InTech, Croatia, 1st edn, 2011, p. 61.
47. S. J. Dee and A. T. Bell, *ChemSusChem*, 2011, **4**, 1166.
48. S. Dee and A. T. Bell, *Green Chem.*, 2011, **13**, 1467.
49. J. B. Binder and R. T. Raines, *Proc. Natl. Acad. Sci. U. S. A.*, 2010, **107**, 4516.
50. H. L. Le Chatelier, *C. R. Seances Acad. Sci.*, 1884, **99**, 786.
51. R. C. Remsing, G. Hernandez, R. P. Swatloski, W. W. Massefski, R. D. Rogers and G. Moyna, *J. Phys. Chem. B*, 2008, **112**, 11071.
52. R. Rinaldi, R. Palkovits and F. Schüth, German Pat Appl., DE 10 2008 0 14 735.4, 2008.
53. R. Rinaldi, R. Palkovits and F. Schüth, *Angew. Chem., Int. Ed.*, 2008, **47**, 8047.
54. R. Rinaldi, N. Meine, J. vom Stein, R. Palkovits and F. Schüth, *ChemSusChem*, 2010, **3**, 266.
55. Z. Zhang and Z. K. Zhao, *Carbohydr. Res.*, 2009, **344**, 2069.
56. H. Zhao, J. E. Holladay and Z. C. Zhang, US Pat. Appl., 20080033187, 2008.
57. C. Li, Z. Zhang and Z. K. Zhao, *Tetrahedron Lett.*, 2009, **50**, 5403.
58. J. B. Binder and R. T. Raines, *J. Am. Chem. Soc.*, 2009, **131**, 1979.
59. T. Kono, H. Matsuhisa, H. Maehara, H. Horie and K. Matsuda, Jpn. Pat. Appl., 2005232116, 2005.
60. Y. Su, H. M. Brown, X. Huang, X. Zhou, J. E. Amonette and Z. C. Zhang, *Appl. Catal., A*, 2009, **361**, 117.
61. Y. Su, H. M. Brown, G. Li, X. Zhou, J. E. Amonette, J. L. Fulton, D. M. Camaioni and Z. C. Zhang, *Appl. Catal., A*, 2011, **391**, 436.
62. F. Tao, H. Song and L. Chou, *ChemSusChem*, 2010, **3**, 1298.
63. F. Tao, H. Song and L. Chou, *Carbohydr. Res.*, 2011, **346**, 58.
64. M. B. Turner, S. K. Spear, J. G. Huddleston, J. D. Holbrey and R. D. Rogers, *Green Chem.*, 2003, **5**, 443.
65. N. Kamiya, Y. Matsushita, M. Hanaki, K. Nakashima, M. Narita, M. Goto and H. Takahashi, *Biotechnol. Lett.*, 2008, **30**, 1037.
66. P. Engel, R. Mladenov, H. Wulfhorst, G. Jager and A. C. Spiess, *Green Chem.*, 2010, **12**, 1959.
67. J. Long, B. Guo, X. Li, Y. Jiang, F. Wang, S. C. Tsang, L. Wang and K. M. K. Yu, *Green Chem.*, 2011, **13**, 2334.
68. J. Zakzeski, P. C. A. Bruijnincx, A. L. Jongerius and B. M. Weckhuysen, *Chem. Rev.*, 2010, **110**, 3552.
69. T. A. D. Nguyen, K. R. Kim, S. J. Han, H. Y. Cho, J. W. Kim, S. M. Park, J. C. Park and S. J. Sim, *Bioresour. Technol.*, 2010, **101**, 7432.

70. S. K. Ritter, *Chem. Eng. News*, 2008, **86**, 57.
71. F. S. Chakar and A. J. Ragauskas, *Ind. Crops Prod.*, 2004, **20**, 131.
72. J. Gierer, *Wood Sci. Technol.*, 1980, **14**, 241.
73. J. Zakzeski, P. C. A. Bruijnincx, A. L. Jongerius and B. M. Weckhuysen, *Chem. Rev.*, 2010, **110**, 3552.
74. J. H. Lora and W. G. Glasser, *J. Polym. Environ.*, 2002, **10**, 39.
75. W. Thielemans, E. Can, S. S. Morye and R. P. Wool, *J. Appl. Polym. Sci.*, 2002, **83**, 323.
76. K. Stärk, N. Taccardi, A. Bösmann and P. Wasserscheid, *ChemSusChem*, 2010, **3**, 719.
77. J. Zakzeski, A. L. Jongerius and B. M. Weckhuysen, *Green Chem.*, 2010, **12**, 1225.
78. B. J. Cox, S. Jia, Z. C. Zhang and J. G. Ekerdt, *Polym. Degrad. Stab.*, 2011, **96**, 426.
79. S. Jia, B. J. Cox, X. Guo, Z. C. Zhang and J. G. Ekerdt, *ChemSusChem*, 2010, **3**, 1078.
80. J. B. Binder, M. J. Gray, J. F. White, Z. C. Zhang and J. E. Holladay, *Biomass Bioenergy*, 2009, **33**, 1122.
81. S. Jia, B. J. Cox, X. Guo, Z. C. Zhang and J. G. Ekerdt, *Holzforschung*, 2010, **64**, 577.
82. S. Jia, B. J. Cox, X. Guo, Z. C. Zhang and J. G. Ekerdt, *Ind. Eng. Chem. Res.*, 2011, **50**, 849.
83. E. Reichert, R. Wintringer, D. A. Volmer and R. Hempelmann, *Phys. Chem. Chem. Phys.*, 2012, **14**, 5214.
84. S. Dumitriu, ed. *Polysaccharides: Structural Diversity and Functional Versatility*, Marcel Dekker, New York, 2nd edn, 2005.
85. K. R. Enslow and A. T. Bell, *RSC Adv.*, 2012, **2**, 10028.
86. C. Sievers, I. Musin, T. Marzialetti, M. B. V. Olarte, P. K. Agrawal and C. W. Jones, *ChemSusChem*, 2009, **2**, 665.
87. T. W. G. Solomons and C. B. Fryhle, *Organic Chemistry*, John Wiley & Sons, New Jersey, 8th edn, 2004.
88. A. A. Shittu, *Catalytic Conversion of Hemicellulosic Sugars into Furfural in Ionic Liquid Media*, Master's Dissertation, The University of Toledo, 2010.
89. M. E. Himmel, S. Ding, D. K. Johnson, W. S. Adney, M. R. Nimlos, J. W. Brady and T. D. Foust, *Science*, 2007, **315**, 804.
90. C. Li, Q. Wang and Z. K. Zhao, *Green Chem.*, 2008, **10**, 177.
91. C. Somerville, H. Youngs, C. Taylor, S. C. Davis and S. P. Long, *Science*, 2010, **329**, 790.
92. Z. Zhang and Z. K. Zhao, *Bioresour. Technol.*, 2010, **101**, 1111.
93. C. C. Hines, D. B. Cordes, S. T. Griffin, S. I. Watts, V. A. Cocalia and R. D. Rogers, *New J. Chem.*, 2008, **32**, 872.
94. G. Bokinsky, P. P. Peralta-Yahya, A. George, B. M. Holmes, E. J. Steen, J. Dietrich, T. S. Lee, D. Tullman-Ercek, C. A. Voigt, B. A. Simmons and J. D. Keasling, *Proc. Natl. Acad. Sci. U. S. A.*, 2011, **108**, 19949.

CHAPTER 2

Biocatalysis in Ionic Liquids

ROGER A. SHELDON

Department of Biotechnology, Delft University of Technology, Julianalaan 136, 2628 Delft, Netherlands
Email: r.a.sheldon@tudelft.nl

2.1 Introduction

According to conventional wisdom enzymes can only be used in an aqueous medium. This constitutes a serious shortcoming as many organic substrates are sparingly soluble in water and preparative organic chemistry is generally performed in organic solvents. Sporadic reports of the use of enzymes in organic solvents date back to the beginning of the last century[1] but it was the seminal paper of Zaks and Klibanov[2] on enzymatic catalysis in organic media at 100 °C, published in 1984, that heralded the beginning of so-called "non-aqueous enzymology".[3] The observation that biocatalysis was not only feasible in non-aqueous media but that many enzymes were more thermally stable in organic solvents than in water was a revelation for organic chemists and led, in the following decades, to the broad application of biocatalysis as a mainstream synthetic tool. Today, almost three decades later, the industrial scope of biocatalysis continues to expand. Biocatalysis in non-aqueous media affords many additional benefits.[4,5] For example, certain reactions, such as esterification and amidation, are difficult to perform in aqueous media owing to equilibrium limitations and/or competing hydrolytic side reactions, but can be readily conducted in organic solvents. Other benefits are easier product recovery from low boiling solvents and elimination of microbial contamination. However, these benefits come at a cost: enzymes are able to perform as suspensions in organic solvents but their catalytic

RSC Catalysis Series No. 15
Catalysis in Ionic Liquids: From Catalyst Synthesis to Application
Edited by Chris Hardacre and Vasile Parvulescu

Published by the Royal Society of Chemistry, www.rsc.org

efficiencies are two or more orders of magnitude lower than those observed in water. We hasten to add, however, that it is difficult to make a meaningful comparison between the rate of, for example, an ester hydrolysis catalyzed by a lipase dissolved in water and the corresponding esterification catalyzed by a suspension of the same lipase in an organic solvent.

One method for improving the activity of enzymes in organic solvents is to lyophilize them in the presence of added salts,[6,7] such as potassium chloride, which was first described by Khmelnitsky and coworkers.[6] The added salt typically comprises 94–98% of the total weight. This led us to the notion that suspension of an enzyme in a room temperature ionic liquid (IL), with its salt- and water-like character, could possibly afford significant rate improvements compared to organic solvents. Ionic liquids (ILs) are substances that are composed entirely of ions and are liquid at or close to room temperature. Interest in their use as reaction media,[8] in particular for catalytic processes,[9] has increased exponentially over the last two decades. This interest stems from their negligible vapour pressure, coupled with good thermal stability and widely tunable properties, such as polarity, hydrophobicity and solvent miscibility behaviour, through manipulation of the cation and anion. Hence, they have been widely advocated as green alternatives to volatile organic solvents.

2.2 Scenarios and Definitions

Various scenarios can be envisaged for performing biotransformations in the presence of an ionic liquid (see Figure 2.1). First, we can distinguish between biocatalysis involving whole cells or isolated enzymes. Second, we can distinguish between systems involving an IL with or without water in one or two phases. In (a) the reaction is performed using whole cells in a two-phase water–IL mixture. The whole cells are suspended in the water phase and the organic substrate is dissolved in the (hydrophobic) IL phase. In (b) the free enzyme is dissolved in a water phase and the substrate in a

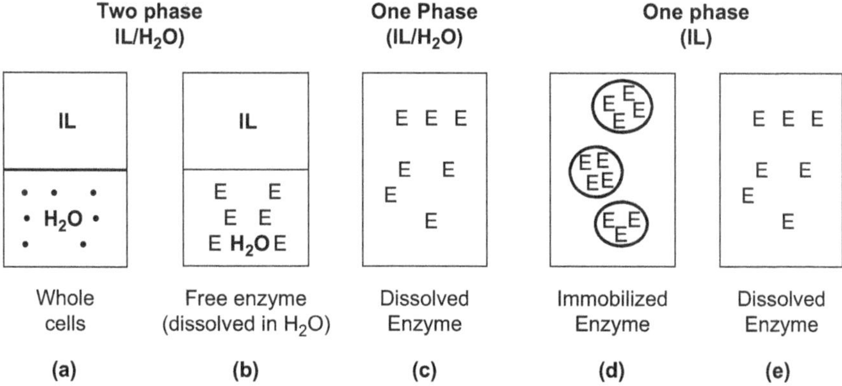

Figure 2.1 Scenarios for biocatalysis in ILs.

hydrophobic IL phase. In (c) the reaction is conducted with a free enzyme dissolved in a single phase consisting of water mixed with a hydrophilic IL. In (d) and (e) the reaction is performed in a single IL phase using an immobilized enzyme or dissolved free enzyme, respectively. Strictly speaking, the term biocatalysis in an IL refers only to a biotransformation in an IL, *in the absence of added water*. If water is added then we are concerned with an *ionic solution*, just as we would refer to a solution of sodium chloride in water or a tetraalkylammonium chloride in water as an ionic solution.

2.3 The First Examples of Biocatalysis in Ionic Liquid Media

The first example of whole-cell biocatalysis in a two-phase IL–water system was reported by Lye and Seddon and coworkers in 2000.[10] It involved the hydrolysis of 1,3-dicyanobenzene to 3-cyanobenzamide catalyzed by a nitrile hydratase contained in whole cells of *Rhodococcus* R312 (Figure 2.2).

An amidase present in the *Rhodococcus* R312 catalyzed further hydrolysis of the amide product to give 3-cyanobenzoic acid. Such reactions are generally conducted in two-phase water–organic solvent systems in order to dissolve the substrate and extract the product to minimize product inhibition. The reaction was performed in a two-phase mixture of the hydrophobic 1-butyl-3-methylimidazolium hexafluorophosphate ([bmim][PF_6]) and water. The authors noted that this system had the following benefits compared to conducting the reaction in the more conventional toluene–water system. It avoided the toxicity and flammability issues associated with the use of toluene. The cells were better dispersed and more stable towards disruption and higher activities and product yields were observed. However, although such a system has several advantages, it is not strictly speaking biocatalysis in an IL. The enzymatic reaction is taking place in the water phase and the IL is merely acting as a reservoir to dissolve the substrate and product and thereby suppress substrate and product inhibition.

In the same year Russell and coworkers[11] used [bmim][PF_6] with 5 vol% water as the reaction medium for the thermolysin catalyzed reaction of *N*-benzyloxycarbonyl-aspartic acid with L-phenylalanine methyl ester to afford *Z*-aspartame (Figure 2.3). They observed activities and yields comparable to those observed in conventional water–organic solvent systems and excellent enzyme stabilities. Furthermore, the enzyme was recycled with

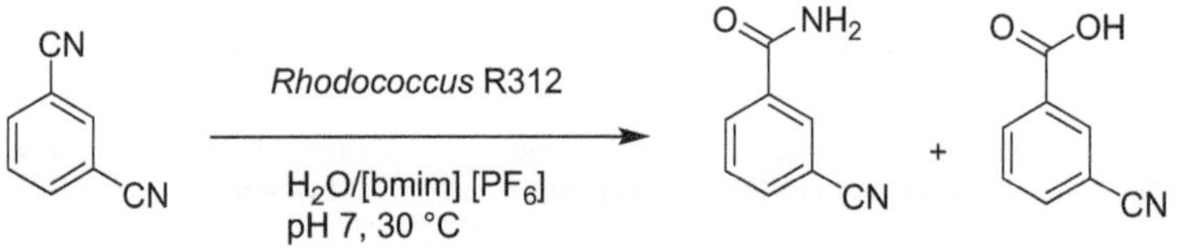

Figure 2.2 *Rhodococcus* R312 catalyzed hydrolysis of 1,3-dicyanobenzene in [bmim][PF_6]–water.

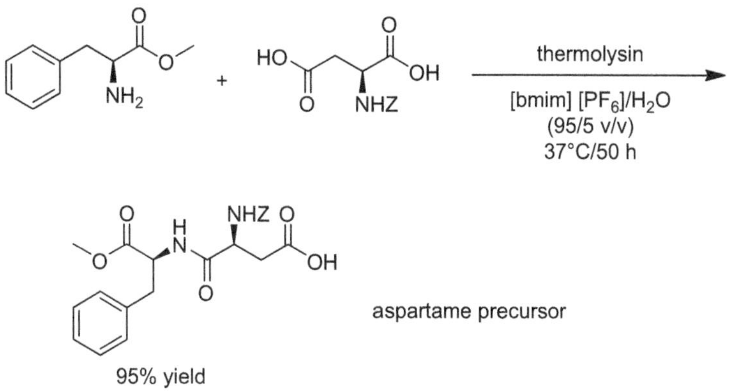

Figure 2.3 Thermolysin catalyzed synthesis of *Z*-aspartame in [bmim][PF_6].

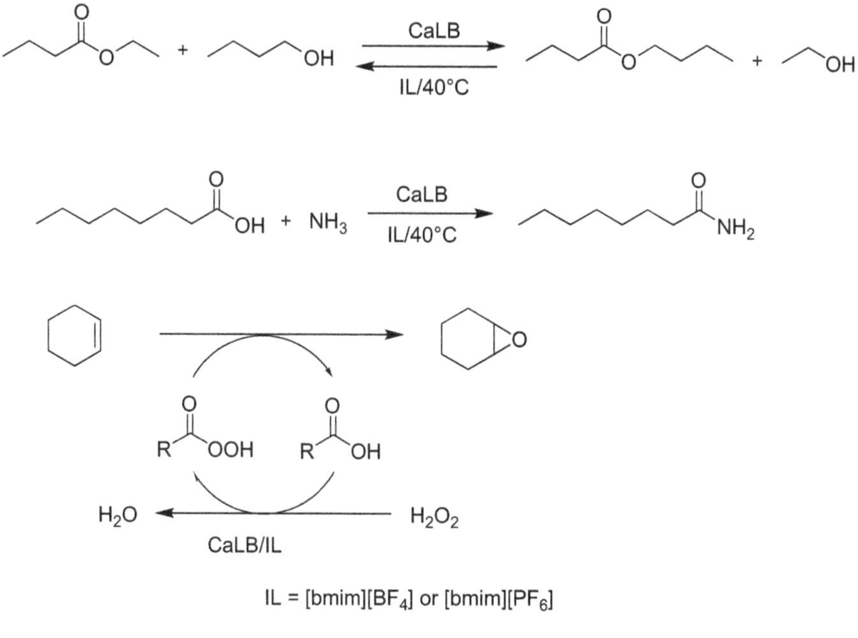

Figure 2.4 CaLB catalyzed reactions in anhydrous ILs.

no apparent loss in activity. However, as noted above this is, strictly speaking, not an example of biocatalysis *in* an ionic liquid because of the presence of 5 vol% water, which may be present as a separate phase where the enzymatic reaction takes place.

Also in the same year we[12] showed that *Candida antarctica* lipase B (CaLB) is able to catalyze a variety of reactions—transesterification, amidation and perhydrolysis—as a suspension in anhydrous [bmim][PF_6] and [bmim][BF_4] (Figure 2.4). Both the IL and the enzyme were dried over phosphorus pentoxide prior to use. On the one hand, the results were very exciting but on the other hand they were rather disappointing as rates were only marginally higher than those observed in the best organic solvents (*e.g.* *tert*-butanol, toluene and ethers). Nonetheless, this and subsequent studies showed that anhydrous ILs are eminently biocompatible with many enzymes and can afford various benefits compared to organic solvents as reaction media for biotransformations. Following this initial report in 2000, biocatalysis in anhydrous ionic liquids has been widely studied and is the subject of numerous reviews.[13–27]

2.4 The Drivers for Biotransformations in ILs

A widely heard reaction to the initial reports of biocatalysis in ILs was: biocatalysis is possible in ILs but so what. Obviously in order to have synthetic utility there have to be benefits associated with the use of enzymes in ILs. As noted above, the original motivation was the expectation that it would lead to higher activities than those observed in organic solvents. However, in practice the rates observed in ILs are at best the same as or slightly better than those observed in the best organic solvents. Another driver was the possibility to replace volatile, environmentally undesirable organic solvents with non-volatile, greener ILs. Owing to their negligible vapour pressure the risk of air pollution is largely circumvented. Other possible benefits are increased selectivity and stability (see later). Furthermore, ILs are able to dissolve large amounts of highly polar substrates, such as carbohydrates[28] and nucleosides,[29] that are sparingly soluble in common organic solvents. This provides the possibility of conducting, for example, esterifications of carbohydrates which would be impossible in water owing to equilibrium limitations and not feasible in most organic solvents owing to a lack of solubility. Moreover, biopolymers such as polysaccharides and lignocellulose exhibit good solubilities in certain ILs.[30] The current increasing interest in the use of lignocellulose as a renewable and sustainable alternative to crude oil and natural gas as a source of chemicals and transportation fuels has focused attention on the possible use of ILs as reaction media for the bio- and chemo-catalytic conversion of lignocellulosic feedstocks to *e.g.* fermentable sugars.

2.5 Properties of Ionic Liquids: How Green Are They?

Interest in conducting catalytic processes in ILs dates back to the 1960s. Initial studies were conducted typically in ILs consisting of dialkylimidazolium and alkylpyridinium cations together with chloroaluminate anions. An important shortcoming of the latter was their sensitivity towards hydrolysis and highly corrosive properties. In the 1990s these highly reactive anions were replaced by weakly coordinating anions, in particular $[BF_4]^-$

and $[PF_6]^-$, which were much more stable towards air and water and, hence, more suitable as a medium for performing catalytic reactions.[31] Consequently, initial studies of biocatalysis in ionic liquids were conducted in water miscible [bmim][BF_4] or water immiscible [bmim][PF_6]. However, these anions are not completely stable towards hydrolysis which can lead to the formation of traces of hydrofluoric acid. Hence, other weakly coordinating anions, such as trifluoroacetate, triflate, bis-triflimide, dicyanamide and methylsulfate, were introduced. In the context of biotransformations these ILs can be referred to as first generation ILs (see Figure 2.5). From both a stability and cost viewpoint it would seem more logical to have used simple anions such as chloride and acetate but their use was prohibited by the fact that such coordinating anions cause dissolution and accompanying deactivation of the enzymes (see later).

As noted above, the use of ionic liquid media for biotransformations was motivated by the possibility of replacing volatile organic solvents with non-volatile ILs with low flammability, whereby the risk of air pollution is largely circumvented. However, ILs have significant solubility in water and, hence, the environmental fate and possible ecotoxicity of ILs is a cause for concern.[32] Their high chemical and thermal stabilities suggest that problems

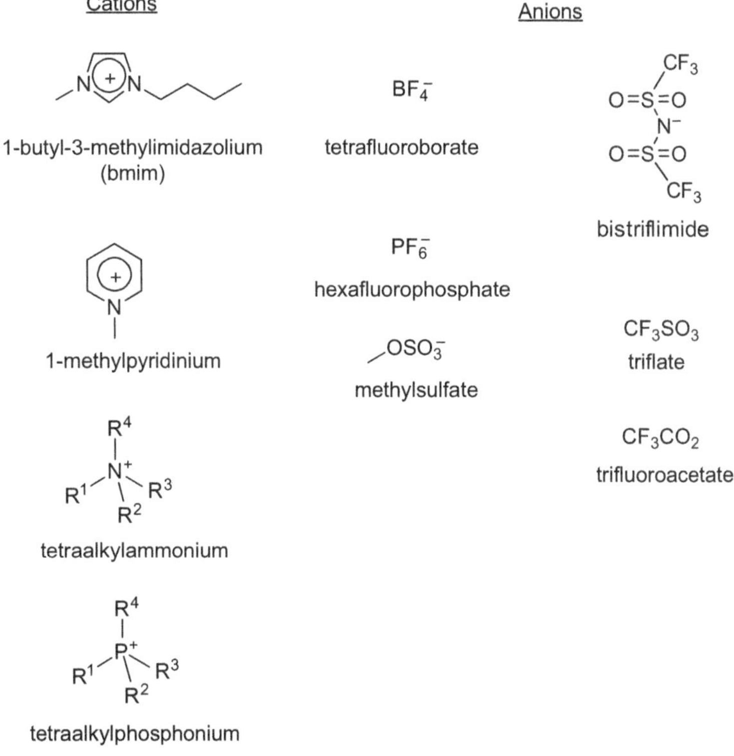

Figure 2.5 First generation ILs for biocatalysis.

with poor biodegradability, bioaccumulation and aquatic ecotoxicity can be expected. Indeed, the first generation dialkylimidazolium ILs, such as [bmim][BF_4] and [bmim][PF_6] and tetraalkylammonium based ILs exhibit acute toxicity towards a variety of aquatic organisms[33–35] and are poorly biodegradable.[36] A further shortcoming of these first generation ILs, consisting of quaternary ammonium cations in conjunction with relatively expensive anions, is that they are prohibitively expensive and many of the synthetic methods used to make them are considerably less green (high E factors) than is often claimed.[37] Consequently, second generation ILs containing more biocompatible cations and anions, often derived from more eco-friendly natural products,[38] such as carbohydrates[39,40] and amino acids,[41,42] preferably in a simple derivatization step, are emerging (see Figure 2.6 for examples).

Choline-based ILs, for example, are prepared by reaction of inexpensive choline hydroxide with a (naturally occurring) carboxylic acid affording the corresponding carboxylate salt and water as the only byproduct.[43,44] Similarly, 2-hydroxyethylammonium lactate consists of a cation closely resembling that of the natural cation, choline, and a natural, readily biodegradable anion.[45] Indeed, an important driver for the use of ILs is the fact that they lend themselves to fine tuning of their properties by an appropriate selection of cation and anion. The current trend is, therefore, towards the deliberate design of ILs that can be used for particular biotransformations while maintaining a low environmental footprint. Another recently described example is the choline based AMMOENG 110™, that is compatible with enzymes and, at the same time, is able to dissolve cellulose.[46]

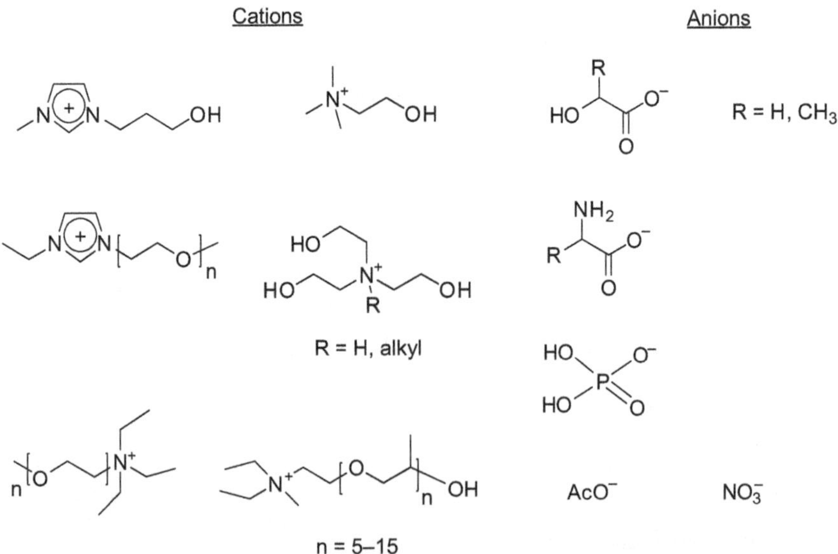

Figure 2.6 Second generation ILs for biocatalysis.

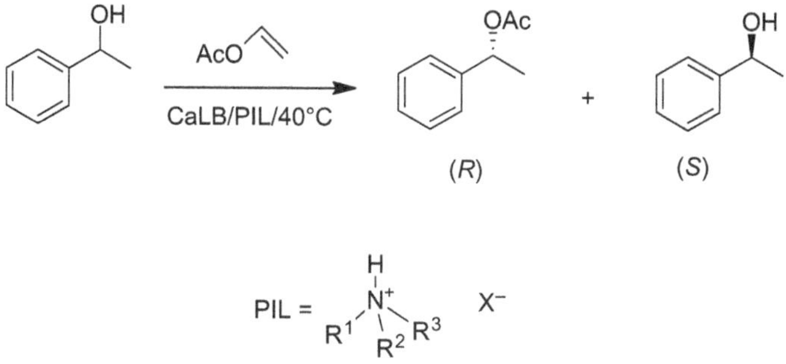

R^1	R^2	R^3	X^-	50% conv. (days)	$ee_{product}$ (%)
CH_3	CH_3	n-C_4H_9	propionate	14	95
CH_3	CH_3	n-C_4H_9	3-hexyldecanoate	2	99
i-C_4H_9	i-C_4H_9	i-C_4H_9	hexanoate	2	93
CH_3	CH_3	n-$C_{12}H_{25}$	octanoate	1	99
n-C_8H_{17}	n-C_8H_{17}	n-C_8H_{17}	acetate	4	>99
		t-BuOH		11	>99

Figure 2.7 Effective resolution of 1-phenylethanol by CaLB catalyzed resolution in PILs.

The search for inexpensive ILs that exhibit reduced ecotoxicity and improved biodegradability, and are compatible with enzymes, recently led to the use of protic ionic liquids (PILs) containing a protonated *N*-atom in the cation, as a solvent for CaLB catalyzed transesterifications (Figure 2.7).[47] PILs are exquisitely simple to prepare by simply mixing a tertiary amine with an acid, such as a carboxylic acid and are known[48,49] to exhibit better biodegradability and lower toxicity than the corresponding quaternary ammonium salts. Moreover, they have suitable hydrogen bond donating properties for interaction with and stabilization of enzymes and, when combined with alkanoate anions, they are self-buffering.

In addition to this second generation of IL media for biocatalysis another class of interesting solvents has emerged in recent years: so-called deep eutectic solvents (DES).[21,50,51] The latter are formed by mixing certain solid salts, in different proportions, with a hydrogen bond donor such as urea and glycerol. For example combining choline chloride (mp 302 °C) with urea (mp 132 °C) affords a DES which is a liquid at room temperature (mp 12 °C). Although DES are, strictly speaking, not ILs since they contain uncharged moieties, *e.g.* urea, they have properties resembling those of ILs.

2.6 Activation and Stabilization of Enzymes in ILs

In the last decade a wide variety of enzymatic transformations have been shown to be feasible in ILs. These concern mainly hydrolytic enzymes, such as lipases, esterases, proteases and glycosidases but oxido-reductases—ketoreductases, peroxidises, Baeyer–Villiger monooxygenases and laccases—and lyases, such as hydroxynitrile lyases, have also been extensively studied.[15] Most of the early examples involved reactions with dispersions of free enzymes in ILs containing weakly coordinating anions, in particular $[BF_4]^-$, $[PF_6]^-$ and $[(CF_3SO_2)N]^-$ ($[Tf_2N]^-$), which are not able to dissolve proteins.

Enzymes exhibited remarkable storage and operational stabilities in such non-coordinating ILs.[52,53] An illustrative example is the reported esterification of phthalic acids with ethanol catalyzed by the thermostable *Bacillus thermocatenulatus* lipase (BTL) in $[bmim][BF_4]$ at 120 °C.[54]

In contrast, hydrophilic ILs containing coordinating anions, such as nitrate, sulfate and chloride, are able to dissolve biopolymers by breaking the intermolecular hydrogen bonds. For example, [bmim]Cl effectively dissolves cellulose because chloride ions interact as H-acceptors with the cellulose OH groups, thereby breaking the H-bonding network of cellulose.[55,56] By the same token, proteins dissolve in ILs containing coordinating anions. Unfortunately, this can lead to deactivation of an enzyme by disruption of intramolecular hydrogen bonds that are essential for the tertiary structure and, hence, the activity of the enzyme. Thus, in a study[57] of transesterifications in a range of ILs reaction rates in $[bmim][BF_4]$, $[bmim][PF_6]$ and $[bmim][CF_3SO_3]$ were comparable to those in *tert*-BuOH. In contrast, no reaction (<5% conversion) was observed in the corresponding lactates and nitrate salts in which the enzymes were shown to dissolve. Interestingly when the solution of the inactivated enzyme in the IL was diluted with a large excess of water substantial recovery of hydrolytic activity was observed. This suggests that the enzyme unfolds and denatures on dissolving in the IL but on the addition of water it (partially) refolds into its active form. Similarly, dissolution of *Trichoderma reesei* cellulose in [bmim]Cl resulted in its deactivation, presumably owing to interactions with the strongly coordinating chloride ion.[58]

It was subsequently found that CaLB dissolves in $[Et_3MeN][MeSO_4]$ but that the dissolved enzyme was still active.[59] Seemingly, the mildly H-bond accepting $MeSO_4$ anion constitutes a borderline case and perhaps the H-bond acidity of the Et_3MeN^+ cation exerts a stabilizing effect. Using FT-IR spectroscopy, it was shown that denaturation of CaLB occurred on dissolution of CaLB in ILs in which low or no activity was observed. In contrast, the conformation of the enzyme dissolved in $[Et_3MeN][MeSO_4]$ closely resembled that of the native enzyme.

There are basically two strategies for maintaining the activity of an enzyme after dissolution in an IL: the design of enzyme compatible ILs or

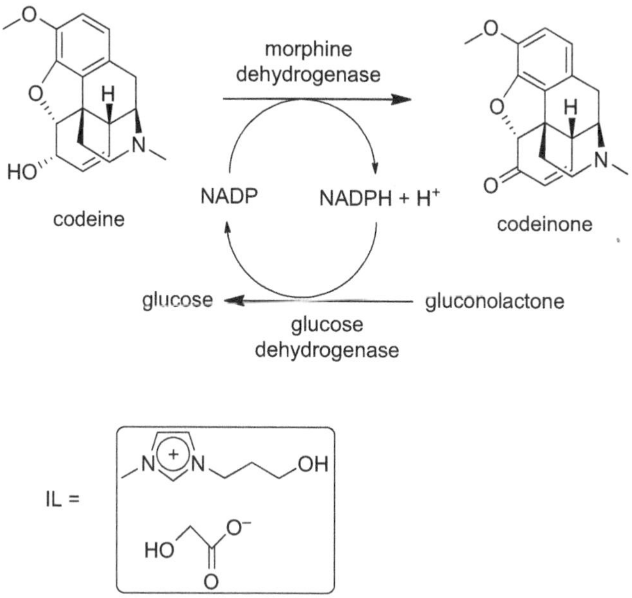

Figure 2.8 Enzymatic codeine oxidation in an ionic liquid.

modification of the enzyme to make it more resistent to denaturation by the IL. The first example of the former was, to our knowledge, reported by Bruce and Walker.[60] They studied the enzymatic oxidation of codeine to codeinone catalyzed by the NADP-dependent morphine dehydrogenase coupled with cofactor regeneration with gluconolactone–glucose dehydrogenase (Figure 2.8). The low solubility of the substrate in both water and common organic solvents led the authors to the idea of using an IL as the reaction medium. To this end they designed an IL containing a hydroxyl functionality in both the cation and the anion, based on the assumption that the enzyme would be stable in an IL that more closely resembled an aqueous environment. This proved to be the case; an IL consisting of a 3-hydroxypropyl-1-methylimidazolium cation and a glyoxylate anion dissolved the substrate, product, enzymes and cofactor and the dissolved enzyme was more active, even at a water content of 100 ppm, than as a suspension in other ILs. Moreover, the subsequent chemical conversion could be carried out in the same IL.

Similarly, Das and coworkers[61] reported that horseradish peroxidase was compatible with tetrakis (2-hydroxyethyl) ammonium triflate and Zhao and coworkers[46] designed enzyme compatible ILs, consisting of alkoxyalkyl substituted cations and acetate as the anion, that dissolve carbohydrates, including cellulose, triglycerides and amino acids (see Figure 2.9 for structures).

The second approach, modification of the enzyme, is exemplified by the cross-linked enzyme aggregates (CLEA) of *C. antarctica* lipase B (CaLB)

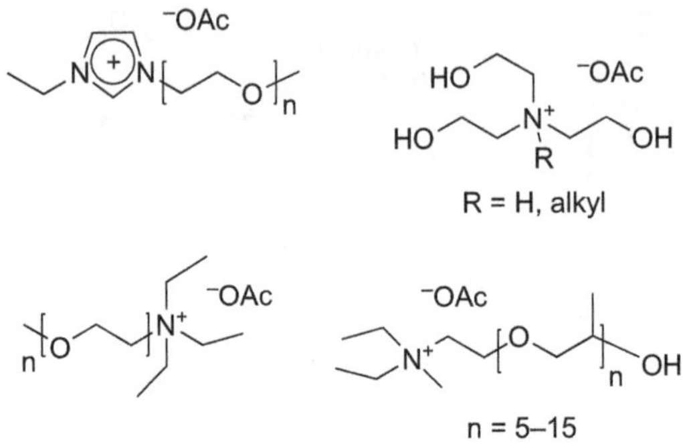

Figure 2.9 Enzyme compatible ILs.

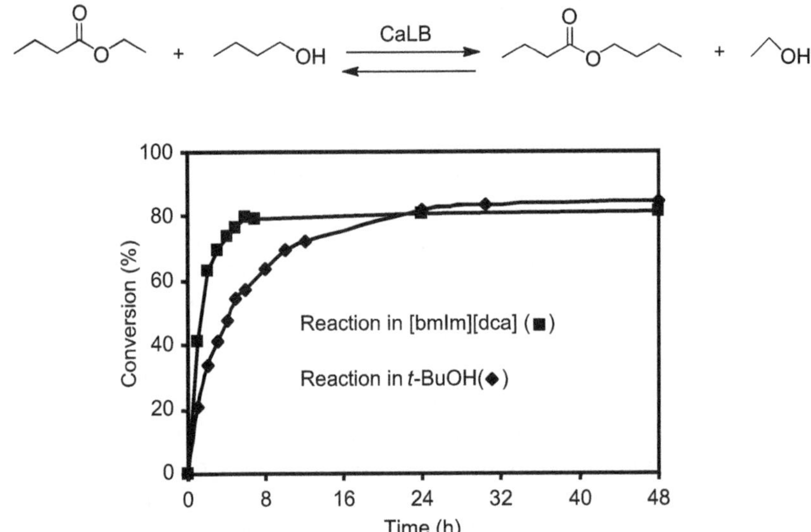

Figure 2.10 Use of suspensions of CLEAs for transesterifications in ILs.

adsorbed on microporous polypropylene (see Figure 2.10).[62] The latter was active and stable in transesterifications performed in ILs that completely deactivate the free enzyme. Vafiadi and coworkers showed that feruloyl esterase CLEAs were active and stable in the enzymatic esterification of glycerol with sinapic acid in ILs and could be recycled 5 times without loss of activity.[63] Similarly, CLEAs of *C. rugosa* lipase (CRL)[64] and *P. cepacia* lipase (PCL)[65] displayed better activities in transesterifications than the free enzymes when suspended in ILs. More recently, Yang and coworkers[66] described the use of CLEAs of *Penicillium expansum* lipase in the production of biodiesel in ILs.

2.7 Selectivity Enhancements in ILs

The use of ILs as reaction media for biotransformations can also lead to substantial enhancement of selectivity, leading to higher product purities, reduced waste and easier downstream processing. A pertinent example, where selectivity enhancement is a consequence of different solubilities of substrate and products in an IL compared to conventional solvents, was observed in the CaLB catalyzed acetylation of glucose (Figure 2.11).[67] In tetrahydrofuran (THF) as solvent the diacetylated byproduct, 3,6-O-diacetyl-D-glucose, was formed in 53% selectivity at 99% glucose conversion. In contrast, reaction in the IL afforded the desired 6-O-acetyl-D-glucose in 93% selectivity at 99% conversion. The improved chemoselectivity was rationalised on the basis of the higher solubility of glucose in the IL. Glucose is only sparingly soluble in THF and the reaction is performed with a suspension of glucose. The monoacetylated glucose is much more soluble in the organic solvents and, therefore, undergoes preferential further acetylation. In contrast, glucose is readily soluble in some ILs and the high selectivity to the 6-monoacetylated derivative is a true reflection of the inherent selectivity of the enzyme for mono- *versus* di-acylation.

Similarly, Kragl and coworkers[68] studied the enantioselective reduction of 2-octanone to 2-octanol, catalyzed by the ketoreductase from *Lactobacillus brevis,* employing an excess of isopropanol for regeneration of the NADPH cofactor in an aqueous–organic two phase system (Figure 2.12). They found that the reaction proceeded more efficiently in [bmim][(CF_3SO_2)N] compared with methyl *tert*-butyl ether (MTBE). This was attributed to the more favourable partitioning of acetone and isopropanol between the organic and the aqueous phase driving the reaction equilibrium towards product, resulting in a faster overall reaction and a higher yield.

Nov435/55°C/36 h

Solvent	Selectivity (%) I	Selectivity (%) II	Conversion (%)
[1-methyl-3-(2-methoxyethyl)imidazolium] BF_4^-	93	7	99
THF	53	47	99

Figure 2.11 CaLB catalyzed acetylation of glucose in ILs *vs.* conventional organic solvents.

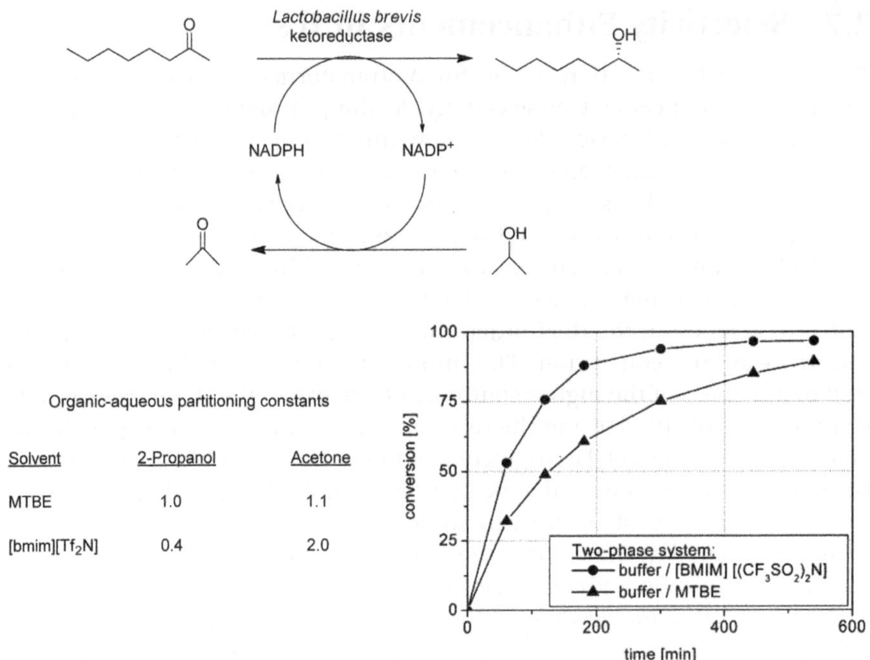

Figure 2.12 Enantioselective reduction of 2-octanone in an IL *vs.* MTBE.

Higher enantioselectivities as a result of an inherently higher enantioselectivity of the enzyme in the IL have also been observed. It is well known that the enantioselectivities of enzymes are influenced by the solvent and it is perhaps not surprising, therefore, that different enantioselectivites are observed in ILs compared to other conventional organic solvents. Two early examples of increased enantioselectivities in enzymatic kinetic resolutions of secondary alcohols, using *Burkholderia cepacia* lipase and CaLB, were reported in 2001 by the groups of Kim[69] and Kragl,[70] respectively (see Figure 2.13).

Furthermore, the enzyme could be easily recycled together with the IL.[71] In a variation on this theme, the enzyme is coated with an IL that solidifies on cooling to afford a solid coated enzyme that can be recycled. For example, Lee and Kim[72] coated *Burkholderia cepacia* lipase with molten 1-(3′-phenylpropyl)-3-methylimidazolium hexafluorophosphate, [ppmim][PF_6], which melts at 53 °C, and then cooled to room temperature. The resulting IL-coated enzyme was used in the resolution of secondary alcohols, displaying higher enantioselectivities than the native, uncoated enzyme (Figure 2.14). Similarly, Itoh and coworkers[73] used the same lipase coated with a novel IL containing a dialkylimidazolium cation and a polyoxyethylene cetyl sulfate anion as a recyclable catalyst for kinetic resolutions of 1-phenylethanol, with enhanced enantioselectivity, in hexane or di-isopropyl ether as solvent (Figure 2.14).

OH

R^1 R^2 → (AcO) OAc R^1 R^2 + OH R^1 R^2

Substrate

Burkholderia cepacia lipase

Solvent	E
MTBE	42
[bmim][Tf_2N]	200

Candida antarctica lipase B

Solvent	E
THF	26
[bmim][Tf_2N]	187

Figure 2.13 Improved enantioselectivities in enzymatic kinetic resolutions in ILs.

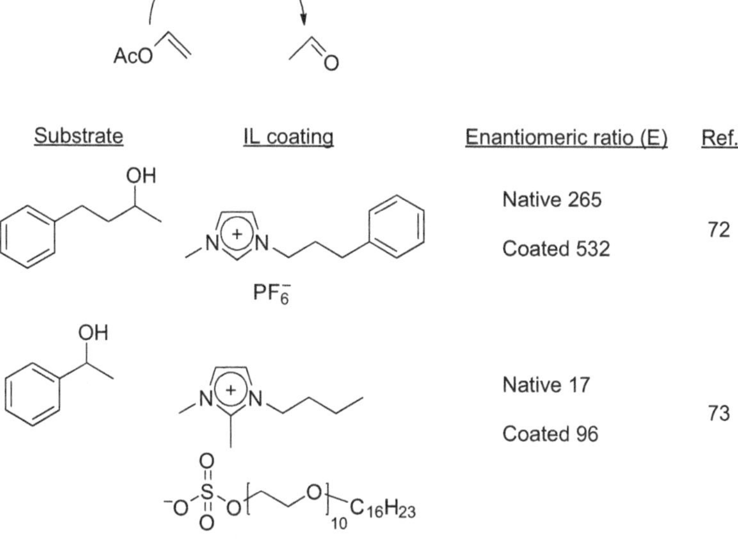

Figure 2.14 Kinetic resolution of 1-phenylethanol using an IL-coated enzyme in organic solvents.

Kim and coworkers subsequently reported[74] the dynamic kinetic resolution (DKR) of secondary alcohols using a lipase–ruthenium or protease–ruthenium combination in [bmim][PF_6]. The IL was essential for the successful performance of the DKRs at room temperature as it enhanced the activity of the ruthenium racemization catalyst.

2.8 Downstream Processing and Product Separation

Another question that is provoked by discussions of enzymatic reactions, or any other reactions for that matter, in ILs, is usually: how can the product be separated from the ionic liquid? One possibility is to extract the product with an organic solvent. However, since one of the reasons for using an IL in the first place was to avoid using volatile organic solvents, the overall gain is questionable. Nonetheless, it could lead to an improvement if a reaction was originally performed in an undesirable organic solvent, such as a chlorinated hydrocarbon, and this could be replaced by reaction in an IL followed by extraction with an environmentally attractive solvent, such as ethyl acetate.

A possibly more attractive alternative is to use supercritical carbon dioxide, $scCO_2$, as the second phase to extract the product. Brennecke and coworkers[75] already showed in 1999 that ILs, *e.g.* [bmim][PF_6], and $scCO_2$ form biphasic systems. Furthermore, although $scCO_2$ is highly soluble in the IL phase and can extract hydrophobic molecules, the IL has no measurable solubility in $scCO_2$.[76] This discovery provided the basis for the development of biphasic biocatalysis in IL–$scCO_2$ mixtures. The latter was first demonstrated in 2002 by the groups of Lozano[77] and Reetz.[78] They independently showed that $scCO_2$ could be used as a mobile phase for the extraction of products from the IL in a continuous operation mode. The product is obtained free of the IL by decompression of the $scCO_2$ and the latter can be recycled by recompression, thus providing a basis for a truly green process.[79] The concept has been applied, for example, to the CaLB catalyzed kinetic resolution of 1-phenyethanol under a wide range of conditions in high enantioselectivities (ee $> 99.9\%$) and good operational stability (15% loss of activity after 11 recycles).[75,80] Further improvements were obtained by a suitable choice of the acyl donor used.[81] Thus, by using vinyl laurate instead of vinyl acetate as acyl donor, the (*R*)-1-phenylethyl laurate product could readily be separated from the unreacted (*S*)-1-phenylethanol.

Another example is provided by the kinetic resolution of 2-octanol in an IL–$scCO_2$ mixture using succinic anhydride as the acyl donor.[82] Other enzymes have also been used, *e.g.* a cutinase from *Fusarium solari* immobilized on zeolite NaY catalyzed the kinetic resolution of 2-phenyl-1-propanol in [bmim][PF_6].[83]

More recently, the concept of using a 'miscibility switch' for performing reactions in IL–$scCO_2$ mixtures was introduced.[84] This takes advantage of the fact that, depending on the temperature and particularly the pressure, IL–$scCO_2$ mixtures can be mono- or biphasic. Hence, the reaction can be

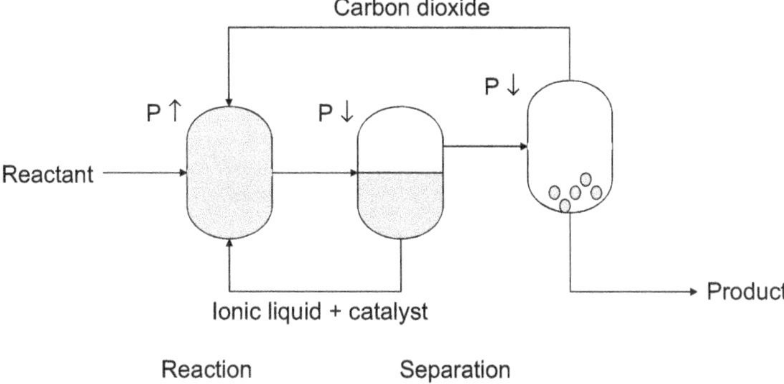

Figure 2.15 Biotransformation in an IL–$scCO_2$ mixture with a miscibility switch.

performed in a single homogeneous phase and, following adjustment of the pressure, the product separated in the $scCO_2$ phase and the IL phase recycled to the reactor (see Figure 2.15).

In another elaboration of the biphasic IL–$scCO_2$ concept, supported IL phases were prepared by covalent attachment of an IL to a polymeric support, such as cross linked polystyrene, and the resulting supported ionic liquid phase (SILP) used to absorb the free enzyme to afford a robust heterogeneous catalyst for use in $scCO_2$ as the mobile phase.[85] The reaction takes place in the SILP and the product is extracted into the mobile $scCO_2$ phase. This was used, for example, in the CaLB catalyzed kinetic resolution of 1-phenylethanol and combination of the CaLB–SILP with an acidic zeolite, also coated with the IL phase, allowed for a dynamic kinetic resolution in a one-pot procedure, affording the (*R*)-ester in 92% yield and >99% ee (Figure 2.16). Similarly, bioreactors containing CaLB–SILPs were prepared as macroporous monolithic mini-flow systems from styrene–divinylbenzene or 2-hydroxyethyl methacrylate–ethylene dimethacrylate copolymers loaded with imidazolium units. They were used for the continuous flow synthesis of citronellyl propionate in $scCO_2$ as the mobile phase.[86] These concepts offer considerable promise for the design of sustainable biotransformations by integrating efficient biocatalysis and downstream processing.

2.9 Biotransformations of Highly Polar Substrates such as Carbohydrates

Ultimately, any process conducted in an ionic liquid has to compete with the same process performed in water or an (environmentally attractive) organic solvent. However, as noted earlier some reactions, such as (trans)esterifications, cannot be performed in water owing to equilibrium limitations and/or product hydrolysis. This necessitates conducting such reactions in an organic solvent. This presents quite a challenge with highly polar substrates such as carbohydrates and nucleosides which are highly

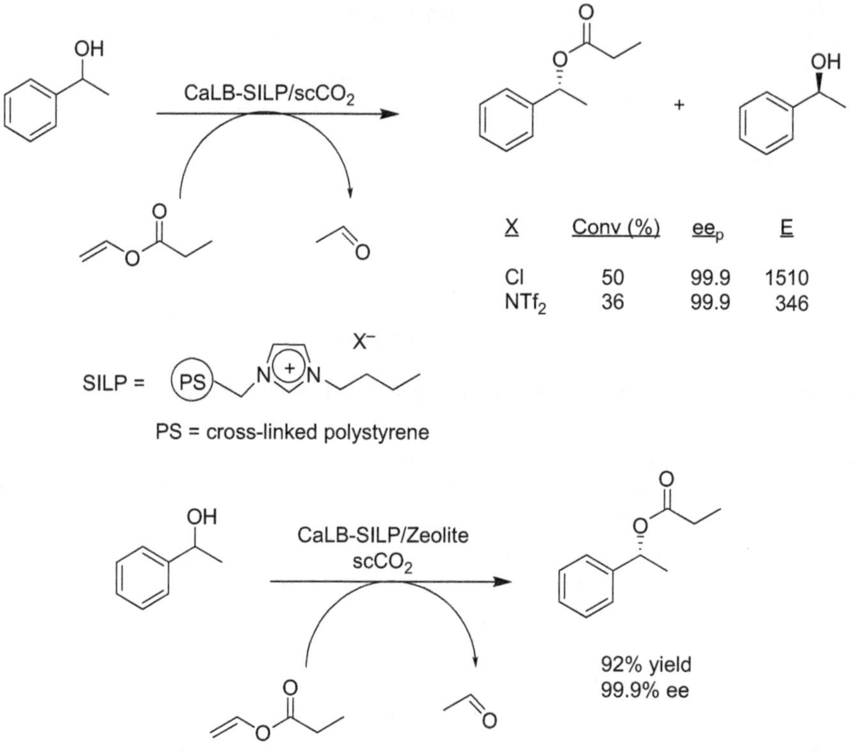

X	Conv (%)	ee_p	E
Cl	50	99.9	1510
NTf_2	36	99.9	346

Figure 2.16 CaLB–SILP catalyzed reactions in $scCO_2$.

soluble in water but have negligible solubilities in common organic solvents. The few exceptions, such as dimethyl formamide (DMF), dimethyl sulfoxide (DMSO) and pyridine, have many undesirable features and/or are incompatible with enzymes. We first pointed out the potential of using ILs as reaction media for carbohydrate transformations, in particular biocatalytic processes, in 2000.[12] This was confirmed in 2001 by Park and Kazlauskas who reported[67] the lipase catalyzed acetylation of glucose in 1-methoxyethyl-3-methylimidazolium tetrafluoroborate, [MOEMim][BF_4] (see the earlier discussion on selectivity).

In the last decade increasing interest has been focused on the design of ILs that are able to dissolve substantial amounts of mono-, oligo- and polysaccharides[87] and, hence, can be used as non-aqueous media for performing reactions of a wide variety of carbohydrates.[88] Of particular interest is the use of ILs for the biocatalytic synthesis of sugar derivatives.[89] The solubilities of carbohydrates in ILs are largely determined by the nature of the anion. Anions with strong H-bond acceptor properties, such as chloride and acetate, are able to break the intermolecular H-bonds between carbohydrate molecules and, hence, cause their dissolution in the IL. MacFarlane and coworkers[90] were the first to note that ILs containing the dicyanamide anion ($[(CN)_2N]^-$ or $[dca]^-$) dissolved glucose in concentrations in excess of 100 g L^{-1}. We subsequently showed[91] that [bmim][dca] is an excellent solvent for sucrose, lactose and β-cyclodextrin and that it could be used as a reaction medium for esterifications of sugars. Interestingly, Rogers and coworkers[92] showed that [bmim]Cl can even dissolve cellulose to the extent of 100 g L^{-1} at 100 °C. More recently, Ohno and coworkers[93] reported that ILs composed of dialkyimidazolium cations in combination with dimethyl phosphate ($[(MeO)_2PO_2]^-$) and dimethyl phosphonate ($[Me(MeO)PO_2]^-$) are able to dissolve cellulose under mild conditions. Similarly, other polysaccharides such as starch[94] and chitin[95] dissolve in ILs.

Zhao and coworkers[46] designed a series of ILs, consisting of acetate anions and an imidazolium cation containing an oligoethylene (or oligopropylene) glycol side chain (see Figure 2.9), that are both enzyme-compatible and dissolve more than 10 wt% cellulose and up to 80 wt% glucose. The length and steric bulk of the glycol chain in the cation could be designed to dissolve sufficient amounts of carbohydrates, while, at the same time, being compatible with enzymes. Free CaLB dissolved in these ILs with retention of activity, thus providing the possibility of conducting homogeneous enzymatic reactions such as the acylation of glucose and the steroid betulinic acid.[96] Koo and coworkers[97] adopted a different approach. Instead of searching for new ILs that can dissolve sugars in high concentrations they used a water mediated supersaturation method to increase the solubility of sugars in ILs. The IL is mixed with a saturated aqueous solution of the sugar and the water subsequently removed under vacuum.

Fatty acid esters of sugars are commercially important products with a wide variety of applications in food, cosmetics and pharmaceutical formulations because, in addition to being derived from renewable raw materials, they are tasteless, odourless, nontoxic, non-irritant and biodegradable.

Furthermore, their functional characteristics, such as hydrophilic–lipophilic balance (HLB) are tunable by a suitable choice of fatty acid and carbohydrate. They are generally manufactured using traditional chemical processes but there is increasing interest[98,99] in the use of enzymatic alternatives that can be conducted under milder conditions with higher selectivities and, consequently, higher product qualities.

In particular, sucrose fatty acid esters[100] find many applications in food and cosmetics, for example, as an emulsifier. They are currently manufactured by a chemical process at elevated temperatures, resulting in low selectivities and side reactions affording colored impurities. However, owing to the very low solubility of sucrose in common organic solvents, it is a challenge to find a suitable reaction medium for the (trans)esterification of sucrose with a fatty acid (ester). We already showed[101] in 1996 that Nov 435 catalyzed the selective monoacylation of sucrose with ethyl dodecanoate, giving a 1 : 1 mixture of the 6- and 6′ regioisomers (Figure 2.17). The reaction was conducted with a suspension of sucrose and Nov 435 in refluxing *tert*-butanol (82 °C), affording a 35% conversion to the 1 : 1 mixture of monoesters (and 2% di-esters) in seven days.

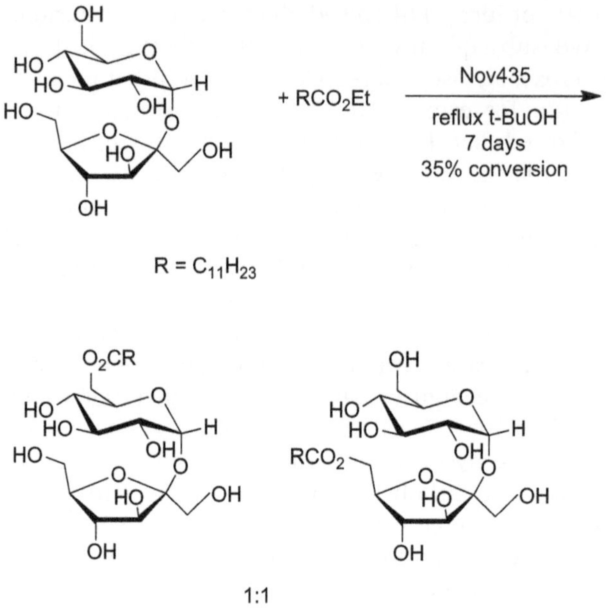

Figure 2.17 CaLB catalyzed acylation of sucrose in *tert*-butanol.

Obviously, such a slow reaction is not attractive for large scale production and, notwithstanding the substantial research effort devoted to this topic, a commercially viable process for the biocatalytic acylation of sucrose with fatty acid esters has not been forthcoming. Therefore, attention has turned to the design of ILs that can function as reaction media for the synthesis of sugar fatty acid esters in general and sucrose fatty acid esters in particular. However, most studies of sugar acylations[99] have been concentrated mainly on monosaccharides, which are relatively easy substrates compared to disaccharides such as sucrose.

Currently, emphasis is switching towards the use of ILs as reaction media for enzymatic conversion of polysaccharides, driven by, for example, the use of lignocellulose as a renewable raw material for the production of biofuels and commodity chemicals. Lignocellulose has been shown[102] to dissolve in a variety of ILs and the latter could be used as reaction media for lignocellulose pretreatment[103,104] and subsequent enzymatic hydrolysis of the cellulose.[105]

2.10 Conclusions and Future Outlook

The first papers on biocatalysis in ILs appeared just over ten years ago and were greeted with much excitement and enthusiasm regarding their novelty and potential applications. In the intervening decade considerable research has been focused on this area of non-aqueous biocatalysis. Second generation ILs based on renewable raw materials or protic ionic liquids (PILs), for example, have been developed that are less expensive, more environmentally benign and more compatible with enzymes, *i.e.* they are able to dissolve enzymes with retention of activity.

Furthermore, efficient methods have been developed for downstream processing using $scCO_2$ as a mobile phase for continuous extraction of products or a miscibility switch for performing the reaction in one phase and separating the product in a two-phase system by lowering the pressure. The use of immobilized enzymes, *e.g.* in the form of insoluble cross-linked enzyme aggregates (CLEAs), also provides for efficient separation and reuse of the enzyme.

One area where conducting a biotransformation in an IL can provide unique benefits is in reactions with highly polar substrates, such as polysaccharides and nucleosides, that have very low solubilities in common organic solvents. When particular transformations, *e.g.* esterifications, cannot be performed in water because of equilibrium limitations and/or product hydrolysis then reaction in an IL can offer a solution to the problem. Furthermore, the current interest in the enzymatic hydrolysis and oxidation of polysaccharides as renewable raw materials provides a driving force for conducting biotransformations of carbohydrates in ILs in the near future.

In short, we believe that biocatalysis in ILs still has much untapped potential for industrial scale applications in the future.

References

1. J. H. Kastle and A. S. Loevenhart, *Am. Chem. J.*, 1900, **24**, 49; E. A. Sym, *Enzymologia*, 1936, **1**, 156; F. R. Dastoli and S. Price, *Arch. Biochem. Biophys.*, 1967, **122**, 289.
2. A. Zaks and A. M. Klibanov, *Science*, 1984, **224**, 1249.
3. *Enzymatic Reactions in Organic Media*, ed. A. M. P. Koskinen and A. M. Klibanov, Blackie, Glasgow, 1996; G. Carrea and S. Riva, *Organic Synthesis with Enzymes in Non-Aqueous Media*, Wiley-VCH, Weinheim, 2008.
4. J. S. Dordick, *Enzyme Microb. Technol.*, 1989, **11**, 194.
5. J. S. Dordick, *Biotechnol. Prog.*, 1992, **8**, 259.
6. Y. L. Khmelnitsky, S. H. Welch, D. S. Clark and J. S. Dordick, *Biotechnol. Bioeng.*, 1994, **58**, 1063.
7. A. P. Borole and B. H. Davison, *Appl. Biochem. Biotechnol.*, 2008, **146**, 215.
8. *Ionic Liquids in Synthesis*, ed. P. Wasserscheid and T. Welton, Wiley-VCH, Weinheim, 2008; M. Fremantle, *Introduction to Ionic Liquids*, Royal Society of Chemistry, Cambridge, 2009; *Handbook of Green Chemistry-Green Solvents, Vol. 6, Ionic Liquids*, ed. P. Wasserscheid and A. Stark, Wiley-VCH, Weinheim, 2009; J. P. Hallett, *Chem. Rev.*, 2011, **111**, 3508.
9. R. A. Sheldon, *Chem. Commun.*, 2001, 299; V. I. Parvulescu and C. Hardacre, *Chem. Rev.*, 2007, **107**, 2615; D. Betz, P. Altmann,

M. Cokoja, W. A. Herrmann and F. E. Kuehn, *Coord. Chem. Rev.*, 2011, **255**, 1518.
10. S. G. Cull, J. D. Holbrey, V. Vargas-Mora, K. R. Seddon and G. J. Lye, *Biotechnol. Bioeng.*, 2000, **69**, 227.
11. M. Erbeldinger, A. J. Mesiano and A. J. Russell, *Biotechnol. Prog.*, 2000, **16**, 1131.
12. R. Madeira Lau, F. van Rantwijk, K. R. Seddon and R. A. Sheldon, *Org. Lett.*, 2000, **2**, 4189.
13. *Ionic Liquids in Biotransformations and Organocatalysis*, ed. P. Dominguez de Maria, Wiley, Hoboken, 2012.
14. Y. H. Moon, S. M. Lee, S. H. Ha and Y.-M. Koo, *Korean J. Chem. Eng.*, 2006, **23**, 247.
15. F. van Rantwijk and R. A. Sheldon, *Chem. Rev.*, 2007, **107**, 2757.
16. R. A. Sheldon, R. Madeira Lau, M. J. Sorgedrager, F. van Rantwijk and K. R. Seddon, *Green Chem.*, 2002, **4**, 147.
17. F. van Rantwijk, R. Madeira Lau and R. A. Sheldon, *Trends Biotechnol.*, 2003, **21**, 131.
18. S. Cantone, U. Hanefeld and A. Basso, *Green Chem.*, 2007, **9**, 954.
19. M. Sureshkumar and C.-K. Lee, *J. Mol. Catal. B: Enzym.*, 2009, **60**, 1.
20. S. Park and R. J. Kazlauskas, *Curr. Opin. Biotechnol.*, 2003, **14**, 432.
21. J. Gorke, F. Srienc and R. Kazlauskas, *Biotechnol. Bioprocess Eng.*, 2010, **15**, 40.
22. Z. Yang and W. Pan, *Enzyme Microb. Technol.*, 2005, **37**, 19.
23. N. Wehofsky, C. Wespe, V. Cerovsky, A. Pech, E. Hoess, R. Rudolph and F. Bordusa, *ChemBioChem*, 2008, **9**, 1493.
24. M. Moniruzzaman, N. Kamiya and M. Goto, *Org. Biomol. Chem.*, 2010, **8**, 2887.
25. C. Roosen, P. Muller and L. Greiner, *Appl. Microbiol. Biotechnol.*, 2008, **81**, 607.
26. F. J. Hernandez-Fernandez, A. P. De los Rios, L. J. Lozano-Blanco and C. Godinez, *J. Chem. Technol. Biotechnol.*, 2010, **85**, 1423.
27. S. H. Ha and Y-M. Koo, *Korean J. Chem. Eng.*, 2011, **28**, 2095.
28. Q. B. Liu, M. H. A. Janssen, F. Van Rantwijk and R. A. Sheldon, *Green Chem.*, 2005, **7**, 39.
29. N. Li, D. Ma and M.-H. Zong, *J. Biotechnol.*, 2008, **133**, 103.
30. R. P. Swatloski, S. K. Spear, J. D. Holbrey and R. D. Rogers, *J. Am. Chem. Soc.*, 2002, **124**, 4974.
31. R. A. Sheldon, *Chem. Commun.*, 2001, 2399.
32. T. P. T. Pham, C-W. Cho and Y-S. Yun, *Water Res.*, 2010, **44**, 352.
33. S. Stolte, M. Matzke, J. Arning, A. Böschen, W. R. Pitner, U. Welz-Biermann, B. Jastorff and J. Ranke, *Green Chem.*, 2007, **9**, 1170.
34. R. F. M. Frade and C. A. M Afonso, *Hum. Exp. Toxicol.*, 2010, **29**, 1038.
35. S. Bruzzone, C. Chiappe, S. E. Focardi, C. Pretti and M. Renzi, *Chem. Eng. J*, 2011, **175**, 17.
36. D. Coleman and N. Gathergood, *Chem. Soc. Rev.*, 2010, **39**, 600.

37. For an assessment of the greenness of ionic liquid syntheses see: M. Deetlefs and K. R. Seddon, *Green Chem.*, 2010, **12**, 17.
38. G. Imperato, B. König and C. Chiappe, *Eur. J. Org. Chem.*, 2007, 1049.
39. S. T. Handy, *Chemistry*, 2003, **9**, 2938.
40. C. Chiappe, A. Marra and A. Mele, *Top. Curr. Chem.*, 2010, **295**, 177.
41. K. Fukumoto, M. Yoshizawa and H. Ohno, *J. Am. Chem. Soc.*, 2006, **127**, 2398.
42. X. Chen, X. Li, A. Hu and F. Wang, *Tetrahedron: Asymmetry*, 2008, **19**, 1.
43. Y. Fukaya, Y. Lizuka, K. Sekikawa and H. Ohno, *Green Chem.*, 2007, **9**, 1155.
44. Y. Yu, X. Lu, Q. Zhou, K. Dong, H. Yao and S. Zhang, *Chemistry*, 2008, **14**, 11174.
45. S. Pavlovica, A. Ziemanis, E. Gziboska, M. Klavins and P. Mekss, *Green Sustainable Chem.*, 2011, **1**, 103.
46. H. Zhao, G. A. Baker, Z. Song, O. Olubajo, T. Crittle and D. Peters, *Green Chem.*, 2008, **10**, 696.
47. A. P. de los Ríos, F. van Rantwijk and R. A. Sheldon, *Green Chem.*, 2012, **14**, 1584.
48. T. L. Greaves and C. J. Drummond, *Chem. Rev.*, 2008, **108**, 206.
49. C. Pretti, C. Chiappe, I. Baldetti, S. Brunini, G. Monni and L. Intorre, *Ecotoxicol. Environ. Saf.*, 2009, **72**, 1170.
50. A. P. Abbott, D. Boothby, G. Capper, D. L. Davies and R. K. Rasheed, *J. Am. Chem. Soc.*, 2004, **126**, 9142.
51. A. P. Abbott, G. Capper, D. L. Davies, R. K. Rasheed and V. Tambyrajah, *Chem. Commun.*, 2003, 70.
52. P. Lozano, T. de Diego, J.-P. Guegan, M. Vaultier and J. L. Iborra, *Biotechnol. Bioeng.*, 2001, **75**, 563; P. Lozano, T. De Diego, D. Carrie, M. Vaultier and J. L. Iborra, *Biotechnol. Lett.*, 2001, **23**, 1529.
53. T. Frater, O. Ulbert, K. Belafi-Bako and L. Gubicza, *Commun. Agric. Appl. Biol. Sci.*, 2004, **20**, 661.
54. J. R. Martin, M. Nus, J. V. Sinisterra Gago and J. M. Sanchez-Montero, *J. Mol. Catal. B: Enzym.*, 2008, **52–53**, 162.
55. R. M. Remsing, R. P. Swatloski, R. D. Rogers and G. Moyna, *Chem. Commun.*, 2006, 1271.
56. S. Zhu, Y. Wu, Q. Chen, Z. Yu, C. Wang, S. Jin, Y. Ding and G. Wu, *Green Chem.*, 2006, **8**, 325.
57. R. A. Sheldon, R. Madeira Lau, M. J. Sorgedrager, F. Van rantwijk and K. R. Seddon, *Green Chem.*, 2002, **4**, 147.
58. M. B. Turner, S. K. Spear, J. G. Huddleston, J. D. Holbrey and R. D. Rogers, *Green Chem.*, 2003, **5**, 443.
59. R. Madeira Lau, M. J. Sorgedrager, G. Carrea, F. Van Rantwijk, F. Secundo and R. A. Sheldon, *Green Chem.*, 2004, **6**, 483.
60. A. J. Walker and N. C. Bruce, *Chem. Commun.*, 2004, 2570; A. J. Walker and N. C. Bruce, *Tetrahedron*, 2004, **60**, 561.
61. D. Das, A. Dasgupta and P. K. Das, *Tetrahedron Lett.*, 2007, **48**, 5635.

62. A. R. Toral, A. P. De los Rios, F. J. Hernandez, M. H. A. Janssen, R. Schoevaart, F. Van Rantwijk and R. A. Sheldon, *Enzyme Microb. Technol.*, 2007, **40**, 1095.
63. C. Vafiadi, E. Topakas, V. R. Nahmias, C. B. Faulds and P. Christakopoulos, *J. Biotechnol.*, 2009, **139**, 124.
64. S. Shah and M. N. Gupta, *Bioorg. Med. Chem. Lett.*, 2007, **17**, 921.
65. P. Hara, U. Hanefeld and L. T. Kanerva, *Green Chem.*, 2009, **11**, 250.
66. J.-Q Lai, Z.-L. Hub, R. A. Sheldon and Z. Yang, *Process Biochem.*, 2012, **47**, 2058.
67. S. Park and R. J. Kazlauskas, *J. Org. Chem.*, 2001, **66**, 8395.
68. M. Eckstein, M. Villela Filho, A. Liese and U. Kragl, *Chem. Commun.*, 2004, 1084.
69. K. W. Kim, B. Song, M. Y. Choi and M. J. Kim, *Org. Lett.*, 2001, **3**, 1507.
70. S. H. Schöfer, N. Kaftzik, P. Wasserscheid and U. Kragl, *Chem. Commun.*, 2001, 425.
71. T. Itoh, E. Akasaki, K. Kudo and S. Shirakami, *Chem. Lett.*, 2001, **30**, 262.
72. J. K. Lee and M. J. Kim, *J. Org. Chem.*, 2002, **67**, 6845.
73. T. Itoh, S. Han, Y. Matsushita and S. Hayase, *Green Chem.*, 2004, **6**, 437.
74. M.-J. Kim, H. M. Kim, D. Kim, Y. Ahn and J. Park, *Green Chem.*, 2004, **6**, 471.
75. L. A. Blanchard, D. Hancu, E. J. Beckman and J. F. Brennecke, *Nature*, 1999, **399**, 28.
76. C. Cadena, J. L. Anthony, J. K. Shah, T. I. Morrow, J. F. Brenecke and E. J. Maginn, *J. Am. Chem. Soc.*, 2004, **126**, 5300.
77. P. Lozano, T. De Diego, D. Carrié, M. Vaultier and J. L. Iborra, *Chem. Commun.*, 2002, 692.
78. M. T. Reetz, W. Wiesenhofer, G. Francio and W. Leitner, *Chem. Commun.*, 2002, 992.
79. For a review of recent developments see P. Lozano and E. Garcia-Verdugo, in *Ionic liquids in biotransformations and organocatalysis*, ed. P. Dominguez de Maria, Wiley, New York, 2012, ch. 4.
80. P. Lozano, T. De Diego, D. E. Carrié, M. Vaultier and J. L. Iborra, *Biotechnol. Prog.*, 2003, **19**, 380.
81. M. T. Reetz, W. Wiesenhofer, G. Francio and W. Leitner, *Adv. Synth. Catal.*, 2003, **345**, 1221.
82. R. Bogel-Lukasik, V. Najdanovic-Visak, S. Barreiros and M. Nunes da Ponte, *Ind. Eng. Chem. Res.*, 2008, **47**, 4473.
83. S. Garcia, N. M. T. Lourenco, D. Lousa, A. F. Sequeira, P. Mimoso, J. M. S. Cabral, C. A. M. Afonso and S. Barreiros, *Green Chem.*, 2004, **6**, 466.
84. M. D. Bermejo, A. J. Kotlewska, L. J. Florusse, M. J. Cocero, F. van Rantwijk and C. J. Peters, *Green Chem.*, 2008, **10**, 1049.
85. P. Lozano, E. Garcıa-Verdugo, N. Karbass, K. Montague, T. De Diego, M. I. Burgueteb and S. V. Luis, *Green Chem.*, 2010, **12**, 1803.

86. P. Lozano, E. Garcia-Verdugo, R. Piamtongkam, N. Karbass, T. De Diego, M. I. Burguete, S. V. Luis and J. L. Iborra, *Adv. Synth. Catal.*, 2007, **349**, 1077.
87. For a review see M. E. Zakrzewska, E. Bogel-Lukasik and R. Bogel-Lukasik, *Energy Fuels*, 2010, **24**, 737.
88. S. Murugesan and R. J. Linhardt, *Curr. Org. Synth.*, 2005, **2**, 437.
89. For a recent review see N. Galonde, K. Nott, A. Debuigne, M. Deleu, C. Jerome, M. Paquot and J.-P. Wathelet, *J. Chem. Technol. Biotechnol.*, 2012, **87**, 451.
90. D. R. MacFarlane, J. Golding, S. Forsyth, M. Forsyth and G. B. Deacon, *Chem. Commun.*, 2001, 1430.
91. Q. Liu, M. H. A. Janssen, F. Van Rantwijk and R. A. Sheldon, *Green Chem.*, 2005, **7**, 39.
92. R. P. Swatloski, S. K. Spear, J. D. Holbrey and R. D. Rogers, *J. Am. Chem. Soc.*, 2002, **124**, 4974.
93. Y. Fukaya, K. Hayashi, M. Wada and H. Ohno, *Green Chem.*, 2008, **10**, 44.
94. Q. Xu, J. F. Kennedy and L. Liu, *Carbohydr. Polym.*, 2008, **72**, 113.
95. Y. Wu, T. Sasaki, S. Irie and K. Sakurai, *Polymer*, 2008, **49**, 2321.
96. H. Zhao, C. L. Jones and J. V. Cowins, *Green Chem.*, 2009, **11**, 1128.
97. S. H. Lee, D. T. Dang, S. H. Ha, W.-J. Chang and Y.-M. Koo, *Biotechnol. Bioeng.*, 2008, **99**, 1.
98. A. M. Gumel, M. S. M. Annuara, T. Heidelberg and Y. Chisti, *Process Biochem.*, 2011, **46**, 2079.
99. Z. Yang and Z.-L. Huang, *Catal. Sci. Technol.*, 2012, **2**, 1767.
100. Y.-G. Shi, J.-R. Li and Y.-H. Chu, *J. Chem. Technol. Biotechnol.*, 2011, **86**, 1457.
101. M. Woudenberg-van Oosterom, F. van Rantwijk and R. A. Sheldon, *Biotechnol. Bioeng.*, 1996, **49**, 328.
102. M. Zavrel, D. Bross, M. Funke, J. Büchs and A. C. Spiess, *Bioresour. Technol.*, 2009, **100**, 2580.
103. T. Vancov, A.-S. Alston, T. Brown and S. McIntosh, *Renewable Energy*, 2012, **45**, 1.
104. J. Long, X. Li, B. Guo, F. Wang, Y. Yu and L. Wang, *Green Chem.*, 2012, **14**, 1935.
105. Z. Qiu, G. M. Aita and M. S. Walker, *Bioresour. Technol.*, 2012, **117**, 251.

CHAPTER 3

Homogeneous Catalysis in Ionic Liquids

SIMON DOHERTY

NUCAT, School of Chemistry, Bedson Building, Newcastle University, Newcastle upon Tyne, NE1 7RU, UK
Email: simon.doherty@ncl.ac.uk

3.1 Concepts, Principles and Practise

3.1.1 General Introduction

The aim of this chapter is to describe the principles, concepts and strategies underlying the use of ionic liquids in homogenous catalysis as well as current challenges, to highlight the current level of understanding in this area and emphasise potential applications. The first section will be purposefully instructive in nature as it aims to provide readers inexperienced in the use of ionic liquids with sufficient working knowledge and understanding to make an informed decision about the possible benefits of applying ionic liquids to their research.

There are a number of relevant and highly informative specialist reviews that the interested reader is directed to, which either complement or supplement the subject matter presented herein; these include Catalysis in Ionic Liquids,[1] Ionic Liquids and Catalysis: Recent Progress from Knowledge to Applications,[2] Homogeneous Catalysis in Ionic Liquids,[3] Transition Metal-Catalysed Reactions in Non-Conventional Media,[4] Room-Temperature Ionic Liquids: Solvents for Synthesis and Catalysis,[5] Catalysts with Ionic Tags and their Uses in Ionic Liquids,[6] Functionalised Imidazolium Salts for

RSC Catalysis Series No. 15
Catalysis in Ionic Liquids: From Catalyst Synthesis to Application
Edited by Chris Hardacre and Vasile Parvulescu

Published by the Royal Society of Chemistry, www.rsc.org

Task-Specific Ionic Liquids and their Applications,[7] Recent Advances in Ionic Liquid Catalysis,[8] Olefin Metathesis in Ionic Liquids,[9] Recent Advances in Oxidation Catalysis using Ionic Liquids as Solvents,[10] Ionic Liquids as Solvents for Catalysed Oxidations of Organic Compounds,[11] Multi-component Reactions and Ionic Liquids: A Perfect Synergy for Eco-Compatible Heterocyclic Synthesis,[12] Ionic Liquid-Supported (ILS) Catalysts for Asymmetric Organic Synthesis,[13] Immobilisation of Molecular Catalysts in Supported Ionic Liquid Phases,[14] Are Ionic Liquids Suitable Media for Organocatalytic Reactions?,[15] Applications of Ionic Liquids in the Chemical Industry,[16] Transition Metal Nanoparticle Catalysis in Ionic Liquids,[17] Poly(ionic) Liquids: An Update.[18] Details of more 'matter specific' reviews will be provided in the relevant sections. For ease of navigation the chapter has been divided into five main sections "Carbonylation, Hydroformylation and Cross-Coupling", "Oxidation and Hydrogenation", "Lewis and Brønsted Acid Catalysis", "Organocatalysis, Metathesis, Ring Opening Polymerisation and Dimerisation/Oligomerisation" and "Biomass Transformations and Catalysis with Transition Metal Nanoparticles".

For the purpose of this chapter an ionic liquid will be taken to mean a salt with a melting point typically <100 °C, although others have used the looser and more inclusive definition of a salt that is low melting. In contrast to conventional solvents, one of the most remarkable features of ionic liquids is their diversity as well as potential for functionalisation. In this regard, an immense range of anion and cation combinations is possible through which properties such as polarity, stability, viscosity, density, conductivity, electrochemical window, hydrophilicity/hydrophobicity and solvent–solute interactions can be modified and fine-tuned, for instance, to optimise catalyst performance (stability, activity and selectivity) and to facilitate product isolation and catalyst recycling by efficient immobilisation and retention in the ionic liquid phase. A selection of the main cation architectures and anions in common use in catalysis is shown in Figure 3.1.1. A variety of ionic liquids containing task-specific functionalities such as Brønsted acids or bases, pendent coordinating heteroatom donors, oligoethers, nitriles and

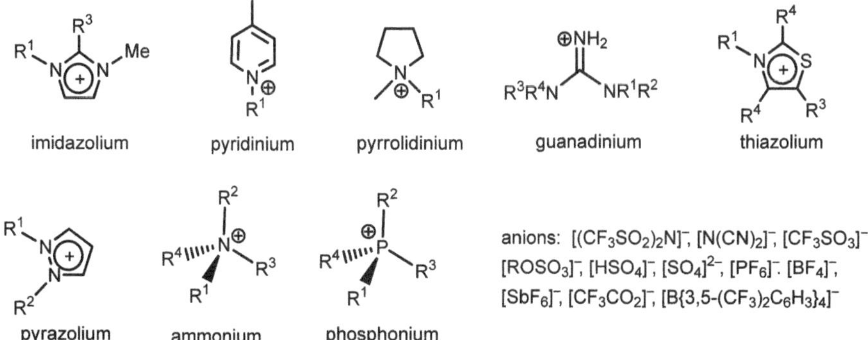

Figure 3.1.1 A selection of commonly used cations and anions for ionic liquids.

Lewis acids covalently attached to either the cation or anion have also been developed to modify and optimise catalyst properties and catalytic processes; other applications of task-specific ionic liquids include selective metal ion extraction,[19a,b] carbon dioxide capture[20a,b] and the synthesis of new ion conductive materials.[21]

3.1.2 Catalyst Solubility and Ligand Substitution/Abstraction in Ionic Liquids

Catalyst precursors are often either soluble in ionic liquids at the concentrations typically used for catalysis or can be converted into an 'active' soluble cationic Lewis acid complex by virtue of replacing an anionic ligand such as a halide with a non-coordinating anion that is 'compatible' with or the same as that of the ionic liquid solvent. In the case of halide abstraction a dichloromethane solution of the precatalyst is typically treated with a stoichiometric amount of a silver salt of a non-coordinating anion then filtered to remove the silver halide by-product before adding the ionic liquid and removing the volatile components to leave an ionic liquid solution of the cationic Lewis acid complex. However, the use of a silver salt is not always necessary as the halide may dissociate in an ionic liquid/solvent mixture and be effectively "diluted" in the pool of the ionic liquid. In this regard, and on the basis that ligand substitution (often of an anionic ligand with a neutral molecule/substrate) is a key step in every homogeneous catalytic process, Shaughnessy and co-workers have shown that ionic liquids promote substitution of anionic ligands by pyridine derivatives to form charge separated adducts and that the extent of this effect depends on the nature of the ionic liquid.[22] This dissolution-based exchange has several advantages as it avoids the use of a silver salt as well as the associated filtration/purification step and, moreover, it allows a single precursor to be used and the counterion varied (and thus catalyst optimised) through a judicious choice of ionic liquid. Several studies have reported that comparable activities and selectivities are obtained using the halide abstraction and dissolution-exchange approach, which strongly suggests that both methods generate the same active catalyst. For example, a mixture of $CuCl_2$ and bis(oxazoline) in $[C_2mim][NTf_2]$ gave the same ee, dr and yield for the cyclopropanation of styrene as $Cu(OTf)_2$ and bis(oxazoline) in dichloromethane[23a,b] while $[IrCl(COD)]_2$ and phosphinooxazoline in $[C_6mim][PF_6]$ gave the same ee and yield for the asymmetric hydrogenation of *N*-(1-phenylethylidene)aniline as a mixture of $[IrCl(COD)]_2$, $Na[B\{3,5\text{-}(CF_3)_2C_6H_3\}_4]$ and phosphinooxazoline in dichloromethane prior to conducting the hydrogenation in $[C_6mim][PF_6]$.[24] The ionic liquid was critical for both these dissolution based exchanges as $CuCl_2$ and bis(oxazoline) in dichloromethane gave much lower yields and ee's and a mixture of $[IrCl(COD)]_2$ and phosphinooxazoline in dichloromethane did not form the active catalyst.

In cases where the catalyst is only poorly soluble a ligand may be modified by introducing a polar or ionic functionality to both enhance the solubility and/or improve retention of the catalyst in the ionic liquid during extraction and isolation of the product, which will also facilitate recycling. To this end, the polar or ionic tag should be introduced remote from the active site to limit the effect on catalyst performance unless it is designed to have a dual role of improving solubility/retention and modifying catalyst efficacy, for instance, through ionic and/or hydrogen-bonding interactions with the substrate to assist coordination or control regio or stereochemistry through transition state effects. However, as a note of caution, even though ligand modification is now a well-developed concept for solubilising and immobilising catalysts, metal and/or ligand leaching may well still occur during work-up and extraction if the ligand reversibly dissociates from the metal, which will be manifested in a drop in conversion and/or selectivity. Thus, the additional expense required to synthesise such ionic liquid-compatible ligands must be justifiable in terms of the value-added benefit provided by improved catalyst stability/longevity, activity, selectivity and/or recyclability (*vide infra*). The use of ionic liquids can also reduce catalyst generation times quite considerably as reactions that rely on either ligand association or exchange to generate the active catalysts occur much more efficiently than in conventional solvents. For example, high ee's and good reproducibility were obtained for Diels–Alder reactions conducted in ionic liquids using catalyst generated from $Cu(OTf)_2$ and (*S,S*)-*t*-Bu-bis(oxazoline) with short metal–ligand complexation times (<5 min)[25a] whereas catalyst aging times >3 h were required to reach high ee's in dichloromethane.[25b]

3.1.3 Synthesis, Purification and Stability

One of the principal reasons/drivers underlying the application of ionic liquids for synthesis and catalysis has been their 'green credentials' as alternatives to volatile organic solvents (VOS), principally on the basis that they have low volatility, are non-flammable and have good thermal and chemical stability, which simplifies distillative isolation of volatile products, reduces atmospheric pollution and improves the safety of a process, particularly for reactions which could form explosive mixtures in conventional solvents *e.g.* oxidations. However, the synthesis of ionic liquids does somewhat limit their 'green credentials' as the most popular pathway involves a quaternisation–metathesis sequence [Equation (3.1.1)]; this necessarily generates stoichiometric quantities of by-product which can be difficult, time consuming and costly to remove. Santini, Chauvin and co-workers have recently addressed many of the problems associated with the synthesis of water miscible ionic liquids by eliminating the use of silver salts (metathesis step) and water by performing the metathesis reaction in the melt of the onium-halide salt (70 °C for imidazolium and 120 °C for pyrrolidinium and phosphonium) in the presence of the lithium or sodium salt of the desired anion, then extracting the ionic liquid into either CH_2Cl_2 or THF. This approach has been successfully applied to the synthesis of a series of imidazolium, pyrrolidinium and phosphonium ionic liquids of

$[N(CN)_2]^-$, $[BF_4]^-$ and $[NTf_2]^-$ in good yield and high purity (>99.5%).[26] Other routes include neutralisation of a base with a Brønsted acid [Equation (3.1.2)], direct alkylation of an alkylimidazole or a pyridine with an electrophilic alkylating reagent of a non-coordinating anion [Equation (3.1.3)], and alkylation coupled with decarboxylation using dimethylcarbonate (DMC) as a 'green' alkylating agent for the halide-free synthesis of imidazolium, methylammonium and methylphosphonium-based ionic liquids as the methyl carbonate salt [Equation (3.1.4)]; the anion can subsequently be exchanged by reaction with either a Brønsted acid or water to afford a variety of ionic liquids with different physiochemical properties [Equation (3.1.5)].[27] Finally, microwave activation has also been used for the efficient two-step one-pot solvent free synthesis of imidazolium and pyridinium ionic liquids based on an S_N2-metathesis sequence followed by extraction or precipitation. In all cases, the use of microwave heating gave much higher yields (89–98%) than conventional heating (34–73%) under otherwise identical conditions.[28] The synthesis of task-specific ionic liquids, *i.e.* those with functionality (*vide infra*) is more specialised and generally requires a lengthy multi-step route.

R¹-imidazole(R²) —R³-X→ [R¹,R²,R³-imidazolium]⁺[X]⁻ —MA→ [R¹,R²,R³-imidazolium]⁺[A]⁻ (3.1.1)

R¹-imidazole(R²) —HA→ [R¹,R²,H-imidazolium]⁺[A]⁻ (3.1.2)

R¹-imidazole(R²) —R³-X→ [R¹,R²,R³-imidazolium]⁺[X]⁻ X ≠ Halide (3.1.3)

$$(C_8H_{13})_3P \xrightarrow{MeOC(O)OMe} [(C_8H_{13})_3PMe]^+[OC(O)OMe]^- \quad (3.1.4)$$

$$[(C_8H_{13})_3PMe]^+ \xrightarrow{HA} [(C_8H_{13})_3PMe][A] + HOC(O)OMe\ (CO_2 + MeOH) \quad (3.1.5)$$

The purity of ionic liquids can be critical for applications in catalysis as even small amounts of impurity, such as inorganic halides, unreacted alkylating agents, water, other protic reagents and amines, can modify their physical and chemical properties and influence catalyst performance. For example, halide anions are responsible for deactivation of the arene hydrogenation precatalyst $[H_4Ru_4(\eta^6\text{-}C_6H_6)_4][BF_4]$[29] and Cu(II)-*bis*(oxazoline)-based Lewis acid catalysts. However, while impurities most often act to poison a catalyst, water has been shown to play a critical role in the activation of chloride-based precatalysts by facilitating solvation of the dissociated chloride; for instance, solvation of chloride was proposed to be integral to generation of the active hydride species from [(*p*-cymene)-RuCl(dppm)] in the biphasic ruthenium-catalysed hydrogenation of styrene as the precatalyst remained inactive in 'dry' ionic liquid. In contrast, in 'dried' ionic liquid activation occurs *via* a completely different pathway (section 3.2.2.2.4).[30a,b] A number of analytical protocols have been applied to quantify impurities in ionic liquids including ^1H NMR spectroscopy,[31a,b] electrochemical methods[32] and X-ray photoelectron spectroscopy.[33]

The first room temperature ionic liquids to be used in homogeneous transition metal catalysis were weakly acidic imidazolium-based chloroaluminates for ethylene polymerisation with a Zeigler–Natta catalyst and propene dimerisation using $[NiCl_2(P^iPr_3)_2]$ or $[Ni_2Br_2(\text{2-methylallyl})_2]$ as catalyst. However, chloroaluminate-based ionic liquids are highly moisture sensitive and 'spontaneously' hydrolyse to HCl and aluminium oxides, which was a major drawback. Replacement of the chloroaluminate with tetrafluoroborate or nitrate improved the hydrolytic stability of ionic liquids and enabled their applications in synthesis and catalysis to be more widely explored and developed. Recent studies have extended the range of available anions that can be used to modify and optimise catalyst performance. While modern *i.e.* second and later generation ionic liquids are generally considered stable and chemically inert, those based on tetrafluoroborate and hexafluorophosphate have been shown to be sensitive to hydrolysis under quite mild conditions and within the timescale of a reaction. Major problems associated with the generation of HF from these ionic liquids include partial loss of ionic liquid, corrosion, and catalyst deactivation through irreversible coordination of fluoride. A host of alternative anions that are both more stable with respect to hydrolysis and non-coordinating have since been developed and one of the most popular is bis(trifluoromethanesulfonyl)amide $[NTf_2]^-$, which also has a number of other favourable properties including low viscosity and high thermal stability; however, the high price of such ionic liquids may limit their practical applications. Even ionic liquids that are hydrolytically stable can accumulate significant concentrations of water if they are not protected from the atmosphere and handled under anhydrous conditions. To this end, the hydrophilicity of an ionic liquid is determined largely by the anion, with more basic anions leading to higher solubility of water. Even hydrolytically sensitive solutes have shown unexpectedly high stability in wet ionic liquids which has been attributed to the interaction between water and the ionic liquid anion reducing its nucleophilicity and, thereby, its hydrolytic activity/power.[34]

3.1.4 Coordination and Non-Innocent Behaviour of Ionic Liquids

The anion of an ionic liquid used for transition metal-based catalysis is typically chosen to be non-coordinating. However, there is now clear evidence that anions previously assumed to be non-coordinating such as $[NTf_2]^-$ can interact with a metal in the absence of a more strongly coordinating ligand or substrate. A variety of coordination modes have been proposed but the most common appear to be monodentate κ(*O*), found in $[Ti(O^iPr)_2(^iPrOH)_2(NTf_2)_2]$[35] and κ(*N*) coordination in $[CpFe(CO)_2(NTf_2)]$[36] and a vast array of linear gold(I) complexes of the type $PR_3Au(NTf_2)$.[37] However, even though the $[NTf_2]^-$ anion interacts with the metal, complexes of this anion are markedly more electrophilic than their corresponding chlorides and, in this regard, the choice of anion has close parallels in the use of non-coordinating anions in coordination chemistry and Lewis acid catalysis by electrophilic metal complexes.[38] In cases where anion coordination appears to inhibit activity, for example, the nickel(II)-dimine/methylaluminoxane-catalysed oligomerisation of ethylene it is possible to use a designer non-coordinating anion such as $[B\{3,5\text{-}(CF_3)_2C_6H_3\}_4]^-$ ($[BAr_f]$); such anions have successfully been used in a host of transition metal-catalysed reactions conducted in conventional solvents.[39] Although it is generally accepted that ionic liquids with 'non-coordinating' anions are the preferred choice for use in catalysis, there have been a number of reports in which the anion of an ionic liquid such as $[NBu_4][X]$ ($X = Cl^-$, OAc^-) has been proposed to play an integral role in a catalytic transformation such as formation of a more active anionic palladium species $[ArPdX_3]^{2-}$ ($X = Br^-$, AcO^-) that facilitates oxidative addition of challenging substrates,[40a,b] or by providing nucleophilic assistance in the rate determining transmetalation step of the Stille cross-coupling.[41a,b] Moreover, this 'assistance' may not be restricted to nucleophilic anions as a positive influence of pyrrolidinium-based ionic liquids in the Stille coupling has also been attributed to a 'nucleophilic assistance' effect of the $[NTf_2]^-$ anion on the transmetalation step. In this case, the reduced cation–anion interaction increases the competitive ability of even the non-coordinating $[NTf_2]^-$ to interact with dissolved species.[42] Thus, as the nucleophilicity of the ionic liquid anion is to some extent determined by the strength of the anion–cation interaction it may be possible to design an ionic liquid with a bifunctional role as both solvent and 'participating' anion to improve catalyst efficiency.

The cation of imidazolium-based ionic liquids is also a potentially non-innocent fragment as it has been shown to react with transition metal catalysts and catalyst precursors to form derived metal carbene complexes; this can occur either *via* direct oxidative addition of the acidic C(2)–H bond to an electron-rich low valent complex or *via* deprotonation of the imidazolium salt and association of the resulting carbene [Equation (3.1.6) and (3.1.7)].

$$[\text{Me-N}(+)\text{N-Bu}]X^- + L_nPd(0) \longrightarrow \left[L_nPd{-}H(\text{Me-N}\widehat{\ }\text{N-Bu})\right]^+ X^- \quad (3.1.6)$$

$$[\text{Me-N}(+)\text{N-Bu}]X^- + Pd(OAc)_2 \xrightarrow{\text{Base}} \left[(OAc)Pd(\text{Me-N}\widehat{\ }\text{N-Bu})\right]^+ X^- \quad (3.1.7)$$

As carbenes are a highly versatile class of ligand that have been used across a host of transition metal-catalysed transformations it is not surprising that there is strong evidence that imidazolium-based ionic liquid-derived transition metal carbene complexes can themselves be good catalysts. For example, mixed phosphine-imidazolylidene palladium complexes of the type $[(PPh_3)_2Pd(C_4mimy)X]X$ ($X = Cl^-, Br^-$) have been identified in the $[C_4mim][BF_4]$-based palladium-catalysed Suzuki–Miyaura cross-coupling between aryl halides and aryl boronic acids; strong support was provided for the generation of these species either from $[Pd(PPh_3)_2(Ph)Br]$ or $[(PPh_3)_2PdCl_2]$, *via* acid–base chemistry rather than oxidative addition. Convincing evidence has also been presented for reversible oxidative addition and reductive elimination of 1,3-dialkyl-2-arylimidazolium salts at palladium [Equation (3.1.8)] as the palladium-catalysed cross-coupling between bromobenzene and tolylboronic acid with $[Pd(PPh_3)_4]$ as precatalyst generated the 1-butyl-2-phenyl-3-methylimidazolium cation ($[C_4(C_6H_5)mim]^+$), which is entirely consistent with reductive elimination from a palladium aryl-imidazolylidene intermediate, while the palladium-catalysed coupling between 4-bromotoluene and tolylboronic acid in the presence of $[C_4(C_6H_5)mim][BF4]$, gave both 4,4′-dimethylbiphenyl (42.6%) and 4-methylbiphenyl (14.4%); the latter of which must have been derived from coupling of the phenyl moiety derived from $[C_4(C_6H_5)mim]^+$ with the tolylboronic acid which also strongly supports a pathway involving oxidative addition of the 2-aryl-imidazolium cation to Pd(0).[43]

$$[\text{Me-N}(+)\text{N-Bu}, \text{2-Ar}]X^- + L_nPd(0) \rightleftharpoons \left[L_nPd{-}Ar(\text{Me-N}\widehat{\ }\text{N-Bu})\right]^+ X^- \quad (3.1.8)$$

A cationic nickel carbene-hydride generated by oxidative addition of $[C_4mim][NTf_2]$ to $Ni(COD)_2$ catalyses the oligomerisation of ethylene but ultimately deactivates *via* reductive elimination of the 2-ethyl-substituted imidazolium salt from the corresponding ethyl-carbene intermediate (Scheme 3.1.1).[39] Stable carbene-hydride complexes of palladium and platinum have also been prepared by reaction of an imidazolium salt with a suitable precursor, $L_2Pd(0)$ (L = NHC) and $[Pt(PPh_3)_4]$, respectively.[44]

Ionic liquid-derived carbenes have also been shown to deactivate catalysts, for example, inhibition of $Pd(OAc)_2$/tppts-catalysed butadiene–methanol

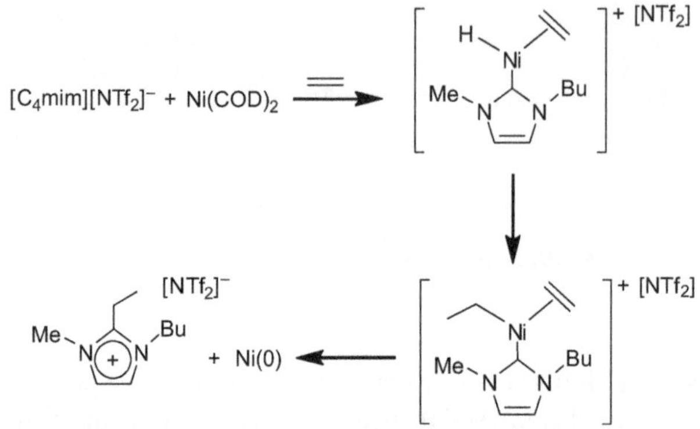

Scheme 3.1.1 Formation of an *N*-heterocyclic carbene during activation of a Ni(0) complex.

telomerisation in the presence of [C_4mim][NTf_2] as co-solvent was attributed to the formation of a stable Pd-imidazolylidene. The formation of such species was avoided by the use of a dialkylimidazolium-based ionic liquid substituted at the C2-position; the resulting systems based on [$C_4C_1{}^2$mim][X] (X = BF_4^-, NTf_2^-) as co-solvent were highly active and selective for the telomerisation of butadiene with methanol.[45] Deuterium exchange reactions at the C2-position of the imidazolium cation of d_3-[C_4mim][NTf_2] in the [Rh(acac)(CO)$_2$]/xantphos-catalysed hydroformylation of 1-octene are also consistent with the formation of metal-carbene species. These carbene species do not cause any significant changes in the catalytic activity or selectivity of the hydroformylation reaction.[46] While substitution of an imidazolium cation at the C2 position is the most obvious and straightforward approach to protect imidazolium-based ionic liquids it is not always sufficient to prevent formation of a carbene as C-H activation at the 4- and 5-positions has also been described, albeit, there are far fewer examples.[47] Thus, there is a clear need to be aware of the potential non-innocent nature of imidazolium-derived ionic liquids for use in catalysis and allied applications and whether carbene formation is responsible for the outcome of a reaction, be it in the form of an active catalyst or a catalyst deactivation pathway. The involvement of a carbene may be identified by isolation/spectroscopic characterisation or may be probed and inferred by either a change in activity and/or product distribution as a function of time, a comparison of catalyst performance in ionic liquids that can't form carbenes or varying the supporting ligands as a difference in activity may indicate that a metal carbene does not form or has only a minor influence.[48]

Interestingly, a catalyst deactivation process has recently been reported to occur *via* different pathways in ionic liquid compared with conventional organic solvent. In conventional solvents deactivation of the alkyne cyclotrimerisation catalyst [Cp(1,4-σ-C_4Ph_4)PPh_3] occurs *via* a rate-limiting dissociation of triphenylphosphine followed by reductive elimination to afford [CpCo(η^4-C_4Ph_4)] whereas in ionic liquid $\Delta V^\ddagger$ and $\Delta S^\ddagger$ values are more consistent with direct reductive elimination; this was attributed to electrostatic stabilisation of the charge-separated transition state for reductive elimination by the highly polar ionic liquid.[49]

3.1.5 Strategies in Ionic Liquid Catalysis

Heterogenisation of a well-defined active and selective homogeneous catalyst by covalent attachment to a support material was popularised as a strategy to facilitate product isolation, catalyst recovery and reuse. However, this approach has several limitations as the immobilised catalyst is often less active and/or less selective than its homogeneous counterpart. Immobilisation of a catalyst in an ionic liquid presents a plethora of alternative strategies and advantages as the fundamental properties of the catalyst are retained, a wide range of catalyst–ionic liquid combinations can be tested in a combinatorial manner to rapidly identify an optimum system, catalyst performance (rate and selectivity) is often enhanced compared with reactions in conventional solvents, the catalyst often exhibits improved stability and longevity, the product is readily isolated by either extraction, decantation or *via* a fully integrated solventless process.

3.1.5.1 Homogeneous and Liquid–Liquid Biphasic Catalysis and Catalyst Immobilisation

Catalysis in ionic liquids greatly simplifies product isolation compared with reactions conducted in conventional solvents and, in this regard, combines the benefits of homogeneous (a well-defined optimised catalyst) with heterogeneous catalysis (ease of product isolation, catalyst recovery and recycling). Reactions are either conducted under homogeneous conditions and the product isolated by extraction or distillation or more commonly under liquid–liquid biphasic conditions. In an ideal liquid–liquid biphasic system the ionic liquid is able to dissolve the active species while also being partially miscible with the substrate and the resulting product should have limited solubility in the ionic phase to allow it to be isolated by decantation, however, in reality a subsequent extraction step is often required to optimise the yield. Liquid–liquid biphasic catalysis has also been shown to change reaction selectivity and modify the product distribution as a result of differential solubility of products; examples include hydrogenation of dienes, olefin oligomerisation and metathesis. Thus, ionic liquids are ideally suited to liquid–liquid biphasic catalysis because the physiochemical properties of the ionic components can be modified and tuned in a rational manner to optimise performance (catalyst retention, activity, selectivity and stability). When an extraction step is necessary it is critical to limit the extent of leaching of both catalyst and ionic liquid to avoid contamination of the product as well as the loss of expensive solvent, metal and ligand. Moreover, efficient retention of the catalyst in the ionic liquid layer also enables the system to be recycled, provided the catalyst remains unchanged and is stable. While charged catalysts and catalyst precursors are generally highly soluble and efficiently immobilised in ionic liquids by virtue of coulombic interactions, neutral species are often less soluble and/or may not be efficiently retained and could be extracted during work-up or leach during biphasic catalysis. Moreover, even though the catalytic species may be charged, the ligand and/or metal may still leach if the catalyst forms an equilibrium with the free ligand and uncomplexed metal as has been reported for copper(II)-*bis*(oxazoline)-catalysed Diels–Alder reactions and cyclopropanations.[50] In cases where the solubility of the catalyst/precursor is low and/or retention in the ionic liquid is poor, the catalyst can be modified by introducing an inert ionic tag onto the ligand remote from the active site so that it does not affect performance but improves solubility and/or retention in the ionic liquid phase. However, the use of a charge-tagged ligand may not prevent metal leaching if the active species exists in dissociative equilibrium with the free metal; in such cases it may be possible to restore activity by replenishing the metal with a suitable precursor provided the ligand is retained in the ionic liquid phase.

In addition to simplifying product separation, the use of ionic liquids for catalysis also enables the ionic liquid–catalyst mixture to be recycled. Thus, after the product has been decanted, extracted or distilled the ionic liquid can either be reused directly or washed in a purification step, dried and recharged with further substrate; the same ionic liquid–catalyst mixture could ideally be recycled and reused with a range of different substrates and/or used to catalyse an entirely different reaction with no cross contamination or loss in activity/selectivity. This is an important benefit since the catalyst and ionic liquid are often by far the most expensive components of a reaction; the former is even more valuable/precious when it requires a time consuming, lengthy multi-step synthesis, as is the case for an elaborate chiral ligand. In many cases a drop in conversion, and for asymmetric reactions a drop in enantioselectivity, is often observed after several runs; the main cause of this has been attributed to leaching of the catalyst/ligand during the extraction process or during purification of the ionic liquid–catalyst mixture between successive cycles. While incorporation of an ionic tag to improve catalyst immobilisation is a simple idea, the additional expense and effort associated with performing a lengthy multi-step synthesis to modify an optimum ligand for a particular reaction and substrate combination will have to be compensated for in terms of efficiency of immobilisation, recycling and ultimately catalyst longevity and performance as measured by the turnover number (TON). Since the inception of this concept in 1996 a wide range of charge-tagged achiral and chiral ligands have been developed for a host a transition metal-catalysed transformations; more recently modification of organocatalysts with remote imidazolium and pyrrolidinium tags has resulted in a significant improvement in performance compared with the corresponding unmodified catalyst. In the vast majority of cases the ionic tag incorporated onto the ligand scaffold is often designed to be compatible with or complement the corresponding anion or cation of the ionic liquid; the most common cationic tags are imidazolium, ammonium, phosphonium and pyridinium introduced *via* quaternisation of the corresponding imidazole, amine, phosphine or pyridine with the appropriate ligand-derived electrophile

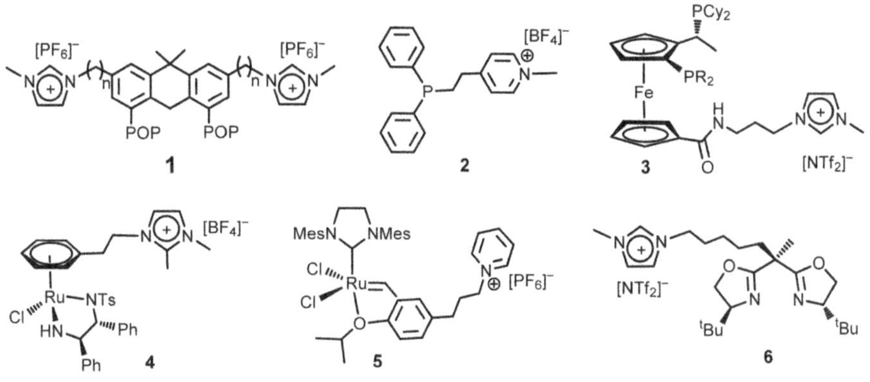

Figure 3.1.2 A selection of imidazolium and pyridinium-modified ligands and precatalysts.

(Figure 3.1.2, **1–6**). This strategy has been applied to imidazolium-, guanidinium- and pyridinium-tagged monodentate and/or bidentate phosphines for rhodium-catalysed hydroformylation; imidazolium- and pyridinium-tagged Grubbs–Hoveyda catalysts for olefin metathesis, imidazolium-tagged pyridines for palladium-catalysed Heck reactions, pyridinium-tagged nitriles for palladium-catalysed Suzuki–Miyaura cross-coupling, imidazolium-tagged arenes for ruthenium-catalysed transfer hydrogenation with half-sandwich complexes, imidazolium-tagged Schiff bases, bipyridines and porphyrins for oxidation catalysis, imidazolium-tagged *bis*(oxazolines) for asymmetric Diels–Alder cycloaddition and imidazolium-tagged chiral diphosphines for rhodium-catalysed asymmetric hydrogenation. A comprehensive review article entitled *Catalysts with Ionic Tags and their Use in Ionic Liquids* covers developments in this area up to 2008 and more recent examples will be discussed later in dedicated sections.[6]

3.1.5.2 Thermoregulated Reversible Ionic Liquid Biphasic Systems

A number of thermoregulated ionic liquid-based reversible liquid–liquid biphasic systems have been developed that are homogeneous at high temperatures and biphasic at 'low temperature'. Such systems combine the advantages of efficient substrate–catalyst mixing under the homogeneous conditions of the reaction with ease of separation by decanting the product-containing organic phase at low temperatures. The earliest examples were based on a fluorous-tagged ionic liquid–toluene combination and an ionic liquid–water system for hydrosilylation[51] and hydrogenation, respectively;[52] more recently polyether,[53] heteropolyanion[54] and dicationic Brønsted acid[55] based ionic liquids have all shown thermoregulated biphasic properties and been used in a range of catalytic transformations. For example, rhodium NPs dispersed in $[Et_3N(CH_2CH_2O)_{16}Me][MeSO_3]$ ($Rh@IL_{PEG750}$) and having a mean diameter of 2.1 ± 0.3 nm formed a thermoregulated biphasic system with a toluene/heptane mixture that catalysed the hydrogenation of cyclohexene with TOFs in excess of 1160 h^{-1} at 60 °C and 0.5 MPa H_2. The system recycled efficiently across 9 runs, rhodium leaching was below the detection limits and there was no change in morphology; the high stability of these RhNPs was attributed to complexation with the polyether.[53b] The same RhNP@IL_{PEG750}/organic biphasic system also catalysed the hydroformylation of 1-octene and gave good conversions at 90 °C under 5 MPa CO/H_2 (1 : 1) with a catalyst loading of only 0.05 mol%.[53d] Palladium nanoparticles stabilised in the same thermoregulated ionic liquid/organic system composed efficiently catalysed the Heck reaction between aryl iodides and system or methyl acrylate at 80–100 °C. High yields were obtained and the system showed good long term stability and recycled without any loss in activity.[53d]

3.1.5.3 Ionic Liquid–$scCO_2$ Biphasic Catalysis

While conventional solvents are most commonly used to extract reaction product(s) and unreacted reagents from the ionic liquid–catalyst phase their use diminishes the green credentials of a process and can lead to contamination of the ionic liquid. As an alternative to organic solvents, $scCO_2$ has a number of beneficial properties as a solvent because it is non-toxic, easy to recover and can extract a wide range of substrates and products from ionic liquids with remarkable efficacy. Compared with conventional liquid–liquid biphasic homogeneous catalysis which requires downstream separation/processing, ionic liquid–$scCO_2$ systems combine rapid mass transfer with quantitative phase separation which enables reaction and separation to be run in a single unit operation. This concept was first successfully demonstrated under batch operation for the rhodium-catalysed hydrogenation of 1-decene in [C_4mim][PF_6][56] and the ruthenium-catalysed asymmetric hydrogenation of tiglic acid in 'wet' [C_4mim][PF_6][57] and in both cases the product was extracted efficiently using $scCO_2$ and the ionic liquid–catalyst solution reused. In addition to ease of processing, the use of [C_2mim][PF_6]–$scCO_2$ for the iridium-catalysed asymmetric hydrogenation of imines was shown to increase catalyst stability and enhance the rate of reaction compared with conventional solvents; the increase in rate was attributed to enhanced H_2 availability in the ionic liquid phase.[24]

In early pioneering studies, Cole-Hamilton and co-workers developed and demonstrated the concept of a continuous flow process based on a biphasic ionic liquid–$scCO_2$ system with the catalyst immobilised in an ionic liquid and $scCO_2$ acting as the transport vector for the substrate(s) and product(s). The high solubility of $scCO_2$ in ionic liquids is an additional benefit as it decreases the ionic liquid viscosity which can facilitate mass transfer during catalysis. The rhodium-catalysed hydroformylation of 1-octene was used to demonstrate this principle with $scCO_2$ transporting the substrate into the reactor containing the ionic liquid–catalyst mixture [Rh(acac)$(CO)_2$]–[Ph_2P(3-$C_6H_4SO_3$)][C_4mim]–[C_4mim][PF_6]; the product was also removed with $scCO_2$ and collected by decompressing into a second autoclave held at low temperature (Figure 3.1.3). However, high reaction rates could only be achieved with alkenes that were highly soluble in the ionic liquid and under high

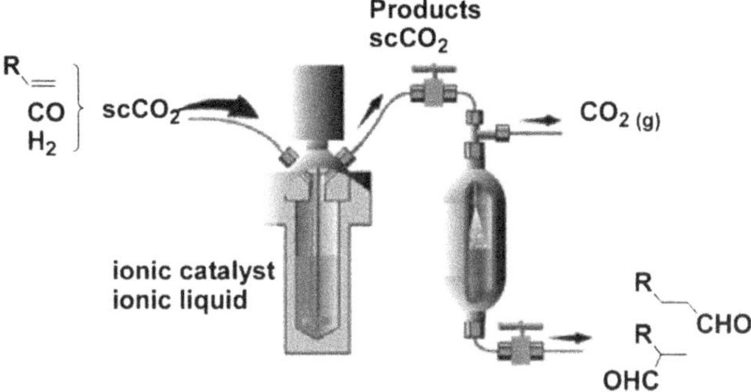

Figure 3.1.3 Schematic of continuous flow homogeneous catalysis using a $scCO_2$–ionic liquid biphasic system.
Reproduced from reference 59.

pressures to make $scCO_2$ a good solvent for the products. A subsequent study on the hydroformylation of 1-octene showed that conversions increased as the length of the alkyl chain on the imidazolium cation increased and $[NTf_2]^-$-based salts gave better rates than their corresponding $[PF_6]^-$ counterparts, both effects were attributed to increased solubility of the alkene in the ionic liquid.[58,59] A biphasic IL–$scCO_2$ system has also been developed for Ru/BINAP-catalysed asymmetric hydrogenation of methyl propionylacetate, however, an acidic ionic liquid additive was required to achieve/maintain high catalyst activity and stability but this was at the expense of ee which dropped from 97–98% in organic solvent to 80–82% in ionic liquid. The continuous flow process was reasoned to be more resource efficient and required less man power for operation than the batch process.[60]

3.1.5.4 *Supported Ionic Liquid Phase Catalysis*

Ionic liquids are designer solvents by virtue of their synthesis and/or high cost and the large volumes typically required for liquid–liquid biphasic catalysis can therefore be prohibitively expensive while their high viscosity can cause mass transfer limitations if the reaction is fast, as only a minor amount of the active species will be accessible to catalyse the reaction. Supported ionic liquid phase (SILP) catalysis was developed to address these issues. In this concept a thin film of ionic liquid containing the homogeneous catalyst is immobilised on a high surface area porous support material such as silica to afford a heterogenised type of homogeneous ionic liquid catalyst [Figure 3.1.4(a)]. This approach is conceptually more straightforward than covalent immobilisation of a catalyst to a support but the use of a liquid reaction phase requires the catalyst to be efficiently immobilised in the ionic liquid, the ionic liquid to be completely insoluble in the reaction mixture and the material to have good mechanical stability in order to avoid leaching of the catalyst and/or ionic liquid layer both under continuous operation or recycling of a batch process. To this end, the most efficient mode of operation for a SILP catalyst is in gas phase applications.[61]

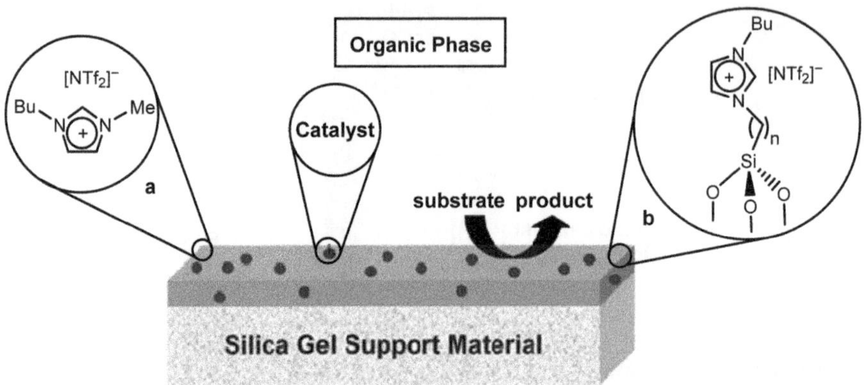

Figure 3.1.4 A supported ionic liquid phase catalyst. (a) The ionic liquid phase $[C_4mim][NTf_2]$ containing the active catalyst is immobilised as a thin film on the surface of the support material. (b) The ionic liquid–catalyst mixture is impregnated on a support modified with a monolayer of covalently anchored ionic liquid.

The use of a highly dispersed thin film of ionic liquid–catalyst mixture can reduce cost quite significantly because only a small volume of ionic liquid is required to cover the support and efficiency is improved because the entire catalyst is accessible for reaction by virtue of being confined within a thin layer of ionic liquid so that the substrate can readily diffuse to the catalyst. The former point is critical to the economic viability of a process while the latter point ensures that reactions are not mass transfer limited. In addition to combining the favourable properties of homogeneous and heterogeneous catalysis, the SILP concept can also be applied to a continuous flow-operated fixed-bed system.

SILP catalysts are generally prepared by impregnation which involves dissolving a metal complex and ligand in the minimum amount of organic solvent, such as dichloromethane, acetone or methanol, to generate the precatalyst. Ionic liquid is then added and the resulting mixture poured onto a support material such as a porous silica gel and after stirring the volatile component is removed by evaporation under reduced pressure to leave the catalyst as free-flowing powder. Mechanical integrity can be improved and leaching reduced further by modifying the surface of the support material with a monolayer of covalently anchored ionic liquid-like fragments to enhance the interaction between the free ionic liquid–catalyst mixture and the surface [Figure 3.1.4(b)].[62] The concept of SILP catalysis and related configurations has been thoroughly reviewed[63,64] and the interested reader is encouraged to refer to a particularly instructive perspective entitled *Immobilisation of Molecular Catalysts in Supported Ionic Liquid Phases*.[65]

Acidic chloroaluminate-based ionic liquids were among the first catalysts to be immobilised by impregnation on a pre-dried support; in this system chloroaluminate anions were anchored to a silica support though covalent bonds between aluminium and surface OH groups.[66] Improved activity and selectivity was achieved for the alkylation of benzene with decene compared with the free ionic liquid and leaching studies showed that immobilisation was highly efficient with negligible loss of active species. However, conversions dropped on repeated use due to the moisture sensitivity of the active chloroaluminate.[67] In an alternative approach, an imidazolium cation was anchored to the surface of a support and subsequently treated with two equivalents of $AlCl_3$ to generate a supported Lewis acid which was also highly active and selective for the alkylation of benzene [Equation (3.1.9)].

$$\text{silica surface} \ldots Cl^- \xrightarrow{2\,AlCl_3} \ldots [Al_2Cl_7]^- \tag{3.1.9}$$

The concept of SILP has also been applied to a host of transition metal-catalysed transformations: the most thoroughly studied are hydroformylation, hydrogenation, oxidation and various carbonylations. The hydroformylation of 1-hexene demonstrated some of the advantages and potential drawbacks of SILP catalysis. A SILP catalyst prepared by modifying a silica surface with a monolayer of imidazolium cations immobilised through covalent Si–O bonds and impregnated with an ionic liquid–catalyst mixture composed of $[C_4mim][PF_6]$ and catalyst generated from $[Rh(CO)_2(acac)]$ and tri(*m*-sulfonyl)triphenyl phosphine *tris*(1-butyl-3-methyl-imidazolium) salt (tppti) (Rh : P ratio of 1 : 10) gave a turnover frequency (TOF) of 65 min^{-1}, nearly three times higher than that of 23 min^{-1} obtained under ionic liquid biphasic conditions, and with comparable *n*/*i* selectivity. However, the ionic liquid partially dissolved in the organic phase at high aldehyde concentrations which resulted in leaching of the rhodium complex and depleted the ionic layer. The improved activity was attributed to the high concentration of active catalyst at the reaction interface compared with the biphasic system and leaching of the ionic liquid and rhodium could be reduced to an acceptable level at aldehyde concentration less than 50 wt%; rhodium leaching could be further suppressed by increasing the phosphine ligand concentration.[68] A similar enhancement in activity compared with the corresponding homogeneous and liquid–liquid biphasic systems was obtained for a SILP-based hydrogenation catalyst generated by immobilisation of $[Rh(PPh_3)_2(NBD)][PF_6]$ in $[C_4mim][PF_6]$ on a silica gel support; the system also exhibited remarkable stability by recycling 18 times with no loss in activity.[69a] In a more recent development, a SILP-type system based on phosphine-free palladium ($PdCl_2$) immobilised on imidazolium-modified silica (Aerosil 300) has been reported to efficiently catalyse the alkoxycarbonylation, phenoxycarbonylation and aminocarbonylation of a range of aryl iodides; high yields were obtained at 80 °C under 0.5 MPa CO in reasonable reactions times for a range of substrate combinations and the system recycled 4 times with only a minor reduction in yield.[69b]

The first successful application of a SILP–$scCO_2$ continuous flow system gave high rates (up to 800 h^{-1}) for the rhodium-catalysed hydroformylation of 1-octene using $scCO_2$ as the transport vector over a fixed-bed SILP catalyst composed of $[C_3mim][Ph_2P(3\text{-}C_6H_4SO_3)]/[Rh(acac)(CO)_2]$ in $[C_8mim][NTf_2]$ on silica gel; the high rates were due to a combination of efficient diffusion of the substrates and gases to the catalyst surface, high solubility of the substrates and gases within the SIL and extraction of heavy product which would otherwise foul the catalyst by blocking the pore.[70] The same research group also applied $scCO_2$–ionic liquid-based continuous flow technology to olefin metathesis with built-in catalyst separation and demonstrated its use in self-metathesis and cross-metathesis of methyl oleate.[71] A fully integrated continuous flow hydrogenation of dimethylitaconate using a SILP catalyst comprised of Rh/(S_a,R_c-1-naphthyl-QUINAPHOS and $[C_2mim][NTf_2]$ with $scCO_2$ as the mobile phase gave ee's >99%, TONs in excess of 100 000 and a space time yield of 0.7 kg L^{-1} h^{-1} and presents a viable option for the production of chiral final chemicals and pharmaceuticals.[72]

3.1.5.5 *Polymer Immobilised Ionic Liquid Phase Catalysis*

The concept of supported ionic liquid phase catalysis has been extended to include the use of polymer immobilised ionic liquids as the support with the idea of combining the advantages of SILP with the tuneable microstructure, ionic microenvironment, charge density and distribution, functionality and hydrophilicity/hydrophobicity of polymers. A ROMP-derived linear cation-decorated polymer has been combined with the Venturello peroxophosphotungstate $[PO_4\{(O(O_2)_2\}_4]^{3-}$ to afford the derived polymer immobilised ionic liquid phase (PIILP) catalyst **7** as a free-flowing powder. The system catalyses the epoxidation of allylic alcohols and alkenes and can be recovered in an operationally straightforward procedure and reused with only a minor reduction in performance.[73] Copper(II)-bis(oxazoline)-derived PIILP systems based on pyrrolidinium decorated cross-linked polystyrenes catalysed the asymmetric Diels–Alder reaction between *N*-acryloyloxazolidinone and cyclopentadiene to give good conversions, *endo*/*exo* ratios as high as 95 : 5 and excellent ee's (>99%) in short reaction times at room temperature. However, catalyst performance varied as a function of the ionic polymer which highlights the complex nature of these PIILP systems. The successful employment of IPs suggests that tailoring of the correct ionic environment on a support and not necessarily ionic solvation of these metal triflate complexes in a bulk IL or under SILP conditions could be sufficient for catalyst stabilisation.[74a] A PIILP Lewis acid system based on gadolinium triflate immobilised on an imidazolium-decorated polystyrene catalyses the Michael addition of amines and thiols to α,β-unsaturated esters and acrylonitrile and gives good yields under mild conditions with a catalyst loading of 0.5 mol%. The system operates most efficiently under solvent free conditions as reactions conducted in acetonitrile resulted in high levels of leaching (<600 ppm).[74b]

Palladium immobilised on a gel-supported ionic liquid-like phase (g-SIILP) based on an imidazolium-decorated polystyrene divinylbenzene (PS-DVB) resin **8** acts as a reservoir to release and capture active palladium species for the Heck arylation. The palladium was supported on the *g*-SILLP in the form of an *N*-heterocyclic carbene and the resulting material was highly stable against air and moisture. Activity was found to depend on the nature and loading of the *g*-SILLP and the optimum system recycled efficiently to give constant yields and kinetic parameters over five runs with very low levels of palladium leaching. The release is controlled by the ionic microenvironment of the palladium complexes and the resulting soluble nanoclusters are stabilised with respect to further aggregation by the additional imidazolium cations decorating the surface. A high charge density (ionic loading) led to a lower activity but efficient recycling, strongly coordinating anions reduced activity, while weakly coordinating anions such as $[SbF_6]^-$ and $[BF_4]^-$ led to high activity but poor recyclability.[75] The thermal stability, hydrophilicity, swelling properties and polarity of these materials can be modified and tuned in a rational manner by varying the anion and cation and adjusting the loading and morphology, in particular the microenvironment of these polymers provides a similar medium in terms of polarity as the bulk molecular ionic liquid.[76] A multitask two-component supported ionic liquid-like phase blend acts both as a source of active soluble palladium species for the Heck, Suzuki-Miyaura and Sonogoshira cross-coupling.[77] The catalyst is easy to prepare, highly active, stable and recycled with remarkable efficiency across eight runs with different reactions and a very low level of palladium leaching (<0.2% of the initial palladium). A continuous flow Heck arylation based on a packed-bed of the polymer blend and using *sc*CO_2 as the solvent and transport vector operated over 60 min and gave a constant yield of 80%. A macroporous imidazolium-decorated cross-linked polystyrene has been used as a support to engineer a PIILP bioreactor based on CALB (*Candida antarctica* lipase B) for continuous flow transesterification. The most hydrophilic monoliths were the most efficient and operated over 10 h with no loss in activity, outperforming the corresponding silica-supported CALB system under the same conditions.[78]

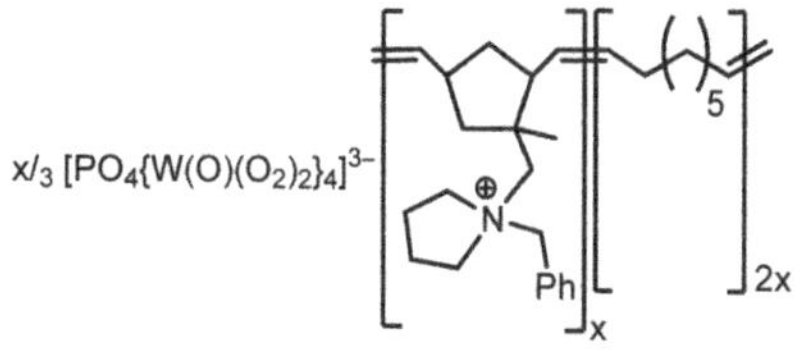

7 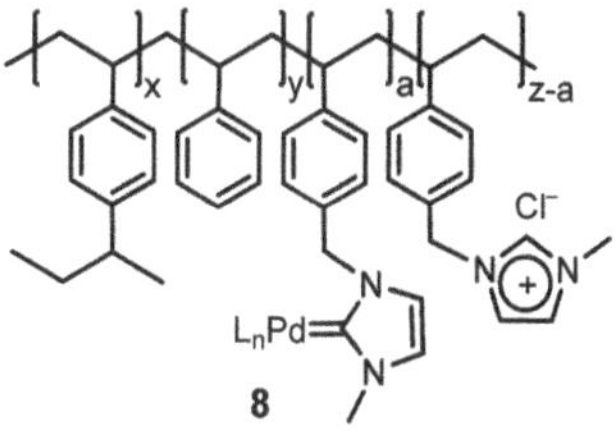 **8**

A dual catalyst system that comprises Pd nanoparticles (NPs) stabilised by water soluble sulfonated ionic liquid-decorated polymers **9** or **10** and a heteropolyacid (HPA) catalyses the hydrogenation of phenol with remarkable selectivity (>99%) and gives cyclohexanone in quantitative yield. Lower yields and/or selectivities were obtained with Pt, Ru and Rh NPs under similar conditions and with Pd/C or when the catalysis was conducted in the absence of acid. The ionic polymer was suggested to stabilise the PdNPs and control selectivity by creating a macromolecular reaction cavity decorated with ion pairs and sulfonic acid groups which concentrates the phenol through favourable hydrogen-bonding interactions and displaces the cyclohexanone before it can react further. The HPA was proposed to enhance the activity by acting as a super Brønsted acid and forming a phenol dication. In this system, the sulfonated ionic liquid-decorated polymer and the HPA act synergistically to enhance both selectivity and reaction rates. The design of reaction specific cavities to control selectivity (and activity) is a potentially useful concept which could be applied to a range of reaction types.[79] A series of carboxyl and hydroxyl-functionalised ionic liquids immobilised on cross-linked polymers have recently been reported to catalyse the fixation of CO_2 with epoxides, the most efficient of which appear to be carboxyl-based systems; full details and references are provided in section 3.4.1.7.

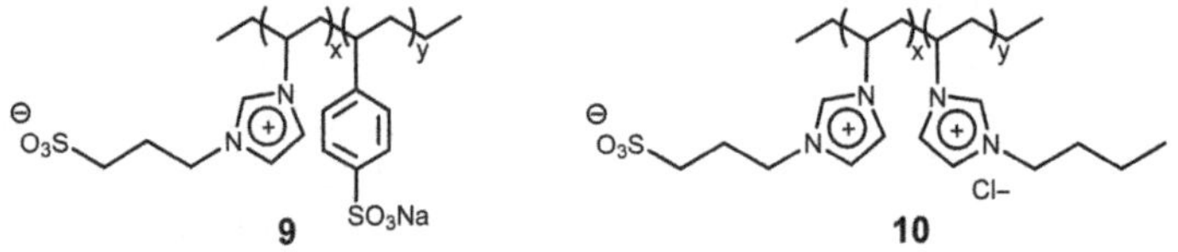

3.1.6 Ionic Liquid Effects

3.1.6.1 Enhancement in Rate

While immobilisation, catalyst stabilisation, facile separation of the product and efficient recycling are often the primary motives for conducting reactions in ionic liquids there have also been numerous reports of substantial enhancements in rate as well as significant changes in selectivity for Lewis acid metal-catalysed carbon–carbon bond forming reactions conducted in ionic liquid compared with conventional organic solvents.[80] For example, the $Sc(OTf)_3$-catalysed Diels–Alder reaction between 2,3-dimethylbutadiene and 1,4-naphthoquinone and the cyanosilylation of benzaldehyde both gave good conversions in short reaction at low catalyst loadings (0.1–0.2 mol%) in $[C_4mim][X]$, ($X = PF_6^-$, SbF_6^-) whereas the same reactions in dichloromethane were extremely sluggish; interestingly an enhancement in rate was also obtained even in the presence of only one equivalent of ionic liquid.[81,82] The effect was even more pronounced for the $Sc(OTf)_3$-catalysed Friedel–Crafts alkylation of benzene with 1-hexene as a near quantitative conversion was obtained in $[C_4mim][X]$ ($X = PF_6^-$, SbF_6^-) after 12 h at room temperature while there was no evidence for reaction in conventional organic solvents.[83] A similar enhancement in activity was achieved for the alkenylation of benzene with 1-phenyl-1-propyne in $[C_4mim][X]$ ($X = PF_6^-$, SbF_6^-) as 10 mol% $Sc(OTf)_3$ or $Hf(OTf)_4$ gave excellent yields of *Z*-alkenylation product in 4 h whereas a yield of only 27% was obtained after 96 h in the absence of ionic liquid [Equation (3.1.10)].[84,85] The increase in activity was proposed to be associated with an exchange process that increased the Lewis acidity of the ionic liquid-immobilised $Sc(OT)_3$ catalyst [Equation (3.1.11)]. This claim was supported by anion exchange studies between Merrifield resin-bound imidazolium hexafluoroantimonate and $Sc(OTf)_3$ which formed highly electrophilic superacidic species of the type $Sc(OTf)_{3-x}(SbF_6)_x$.[86] The idea that the Lewis acidity of a catalyst can be fine-tuned by a judicious choice of counterion was further reinforced by conducting the trimethylsilyltriflate-catalysed alkenylation of *p*-xylene with 1-phenyl-1-propyne in the presence of a range of quaternary ammonium salts with different anions; catalyst efficiency, as measured by the yield of product, increased in the order $[SbF_6]^- > [PF_6]^- > [BF_4]^- > [OTf]^-$.

Benzene + Ph—≡—Me —(10 mol% $Sc(OTf)_3$, 85 °C)→ PhC(Ar)=CHMe (3.1.10)

$[C_4mim][SbF_6]$: 91% (4h)
without ionic liquid: 27% (96h)

$$M(OTf)_n + x\,[C_4mim][SbF_6] \xrightarrow{\text{anion exchange}} [M(OTf)_{n-x}][SbF_6]_x + x\,[C_4mim][OTf] \tag{3.1.11}$$

more electrophilic Lewis acidic catalyst

The super acidic Lewis acid ionic liquid $[C_4mim][Sb_2F_{11}]$ was sufficiently active to enable the kinetic *E*-isomeric arene alkenylation adduct to be obtained with high selectivity (95 : 5) at low temperature while the thermodynamic *Z*-isomer was generated in 99 : 1 selectivity at 90 °C.[87] In addition to the formation of super electrophilic Lewis acids, there is also convincing evidence that stabilisation of reactive intermediates such as vinyl and arenium cations contributes to the rate acceleration obtained for metal triflate-catalysed Friedel–Crafts reactions conducted in ionic liquids.[88] Activation of the arene and stabilisation of the resulting arenium intermediate by the Lewis acid ionic liquid $[C_4mim]Cl\text{-}(AlCl_3)_x$ ($x = 0.67$, where x = mole fraction of $AlCl_3$) were suggested to be the principal factors responsible for facilitating the complete hydrogenation of benzene using Pd/C as the system was completely inactive in the absence of ionic liquid and with a conventional Lewis acid such as $AlCl_3$. Exceptional enhancements in activity as well as substantial increases in enantioselectivity have also been obtained for Cu(II)-*bis*(oxazoline)-catalysed Diels–Alder reactions[25a,89,90] and the Mukaiyama aldol reaction[91] conducted in ionic liquid compared with dichloromethane; reactions in ionic liquid are typically complete within 2 min whereas much longer reaction times (60–120 min) are required in dichloromethane. In some cases, the enhancement in rate enabled the catalyst loading to be

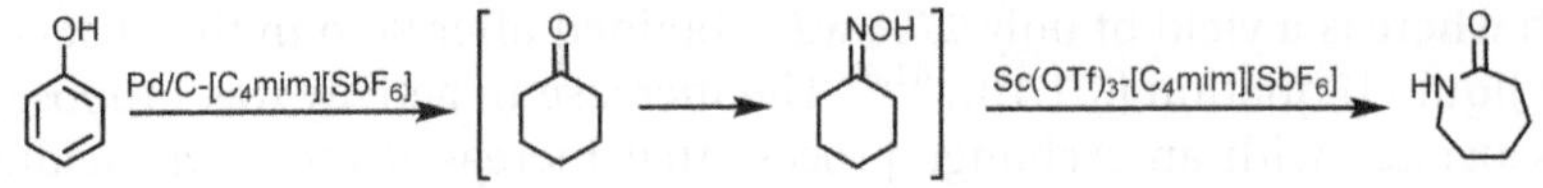

Pd/C-[C4mim][SbF6]
Sc(OTf)3-[C4mim][SbF6]

Scheme 3.1.2 A multifunctional Pd/Sc(OTf)$_3$/ionic liquid catalyst for the tandem one-pot conversion of phenol to ε-caprolactam.

reduced such that near quantitative conversions could still be obtained with Cu(OTf)$_2$–bis(oxazoline) loadings as low as 0.5–0.6 mol%, which represents a significant saving as bis(oxazolines) are commercially available but relatively expensive. In addition to shorter reaction times and lower catalyst loadings, such a large increase in rate for reactions conducted in ionic liquid could also be beneficial for challenging less reactive substrate combinations.

A facile Sc(OTf)$_3$ in ionic liquid-catalysed conversion of cyclohexanone to cyclohexanone oxime and subsequent Beckmann rearrangement is a key step in the tandem one-pot conversion of phenol to ε-caprolactam (Scheme 3.1.2) with the multicomponent Pd/Sc(OTf)$_3$/[C$_4$mim][SbF$_6$] catalyst. The Sc(OTf)$_3$ and ionic liquid were shown to cooperate in the final Beckmann rearrangement as the reaction between cyclohexanone and hydroxylamine in 1,2-dichloroethane only gave oxime. The ionic liquid was proposed to enhance the rate of this transformation by increasing the Lewis acidity of Sc(OTf)$_3$ *via* the anion exchange process described above [Equation (3.1.11)], and stabilising the charged intermediate in the Beckmann rearrangement.[92]

A four-fold increase in the rate of the Diels–Alder reaction between 1,3-cyclohexadiene and *N*-benzylmaleimide was obtained in [C$_n$mim]Cl ($n=$ 8–14) based ionic liquid–aqueous micellar systems compared to pure water. Relative rate constants k_{IL}/k_W increased in the order C8 < C10 < C12 < C14 and for each system tested the maximum rate occurred above the CMC of the respective amphiphilic ionic liquid (Figure 3.1.5). Reaction rates were further increased by addition of 0.1 equivalents of indium(III) chloride to generate the Lewis acid ionic liquid [C$_{12}$mim]Cl–InCl$_3$ ($\chi = 0.09$). Micellar catalysis in ionic liquid–aqueous systems is a promising new concept as it should be possible to modify the size and shape of the micelles in a rational manner through the cation–cation contact domains and to tune the CMC in order to influence reaction rates and selectivities.[93] An increase in rate (and *endo*/*exo* selectivity) for the Diels–Alder reaction between cyclopentadiene with methyl acrylate has been correlated with the hydrogen bond-donating capacity of imidazolium-based ionic liquids and the greatest selectivities were predicted to occur in ionic liquids with the strongest hydrogen-bond donor cation coupled with the weakest hydrogen-bond accepting anion.[94]

The superhydrophobic environment of cross-linked polystyrene-grafted imidazolium and pyridinium-based ionic liquids is responsible for the enhancement in transesterification rate compared with conventional catalysts such as SBA-15-[C$_1$vim][SO$_3$CF$_3$] and Amberlyst 15. The increase in rate was attributed to the much better wettability of the reactant than the product which increases the concentration of the reactant in the catalyst, in much the same manner as reactant enrichment in carbon nanotube catalysts.[95]

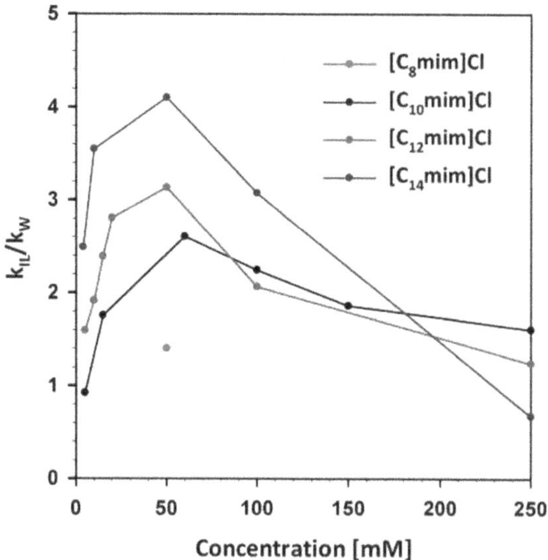

Figure 3.1.5 Relative reaction rate k_{IL}/k_W of a Diels–Alder reaction in aqueous solutions of the amphiphilic ionic liquids $[C_n\text{mim}]Cl$ ($n = 8$–14). Reproduced from reference 93.

Enhancements in the rate of C–C coupling reactions such as Heck, Stille and Suzuki–Miyaura have been associated with the formation of an electron-rich anionic species of the type $[L_nPdX]^-$ which is highly active towards oxidative addition, as proposed by Amatore, Jutand *et al.*,[40b] and/or for the latter two reactions a nucleophilic assistance effect in the transmetalation step whereby the anion promotes exchange by expanding the coordination sphere of the Sn (or B) reagent.[41] This latter effect can either manifest itself as a direct function of the nucleophilicity of a poorly solvated anion or may be masked as a cation effect by virtue of the strength of the cation–anion interaction *i.e.* reducing the cation–anion interaction can increase the competitive ability of the anion to interact with dissolved species. In this regard, both Br^- and $[NTf_2]^-$ have been proposed to facilitate transmetalation. While these effects appear to be associated with catalysis by NPs it is conceivable that either an anion–cation pair or a task-specific ionic liquid could also be designed to optimise catalysis by well-defined molecular complexes.

3.1.6.2 Enhancement in Selectivity

One of the most profound and dramatic effects on selectivity for reactions conducted in ionic liquid compared with conventional solvents has been the regioselectivity of the Heck arylation of electron-rich 1-olefins. Terminal olefins bearing electron-withdrawing groups generally give high selectivity for the β-regioisomer whereas electron-rich olefins often give a mixture of regioisomers with very low selectivity [Equation (3.1.12)].

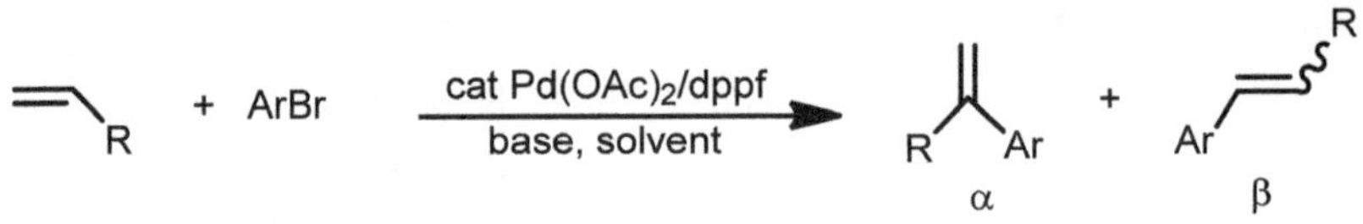

(3.1.12)

High selectivity for α-arylation can be obtained when aryl triflates are used as the coupling partner or in the presence of a stoichiometric quantity of a thallium or silver salt,[96–98] but both these approaches have obvious drawbacks. A dramatic and substantial improvement in α-selectivity for the Pd/dppp-catalysed Heck arylation of electron-rich olefins was achieved in $[C_4mim][BF_4]$ compared with the same catalyst in conventional solvents such as acetonitrile, DMF, dioxane, toluene and DMSO.[99] For example, the Heck arylation between butyl vinyl ether and a range of aryl bromides catalysed by $Pd(OAc)_2$/dppp in $[C_4mim][BF_4]$ gave the corresponding α-arylated vinyl ether as the sole product whereas the same reaction in DMF gave $\alpha : \beta$ ratios from 46 : 54 to 69 : 71 [Equation (3.1.12)]. A tentative explanation reasoned that the ionic liquid facilitates dissociation of halide to favour the ionic pathway, which is known to be highly α-selective as a result of electrostatic and frontier orbital interactions.

A remarkable improvement in the regioselectivity of the Heck arylation between 4-bromotoluene and *trans*-ethyl cinnamate was also achieved in TBAB with TBAA as base which gave >99 : 1 selectivity for the *trans* isomer **11** compared with reactions in conventional solvents such as DMF and DMA which gave moderate selectivity,[100] or in TBAB with other bases such as sodium dicarbonate, potassium phosphate, DABCO, sodium acetate and tributylamine which lacked stereoselectivity (Table 3.1.1). This enhancement in regioselectivity was attributed to rapid intramolecular neutralisation of an olefin-coordinated Pd–H species by metal bound acetate.[101]

Table 3.1.1 Base effect on the stereoselectivity of the Heck reaction.[a]

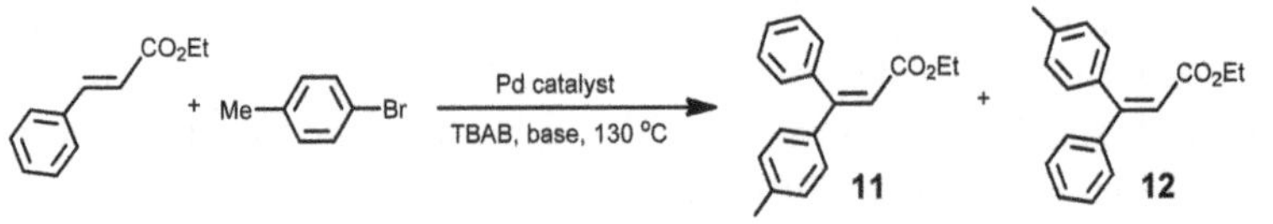

Base	*Solvent*	*t (h)*	*Conv. (%)*	***11* : *12* ratio**
NaHCO	TBAB	7	100	59 : 41
K_3PO_4	TBAB	4	75	61 : 39
DABCO	TBAB	5	85	62 : 38
Na_2CO_3	TBAB	5	74	59 : 41
Bu_3N	TBAB	8	94	60 : 40
$NaHCO_3$	$[C_4mim]Br$	24	22	65 : 35
TABB	TBAB	1	100	>99 : 1

[a]Reprinted with permission from reference 101. Copyright 2003 American Chemical Society.

Several studies have reported a marked enhancement in enantioselectivity for Lewis acid-catalysed transformations such as the Diels–Alder reaction and the Mukiayama aldol reaction in ionic liquids compared with conventional solvents. Meracz and Oh were first to disclose an ionic liquid-induced enhancement in ee for the $Cu(OTf)_2$/((*S,S*)-*t*-Box)-catalysed Diels–Alder cycloaddition between crotonyl oxazolidinone and cyclopentadiene [Equation (3.1.13)] at room temperature which increased from 52% in dichloromethane to 92% in $[C_4C_4im][BF_4]$; a much higher yield of adduct **13** and *endo*/*exo* ratio was also obtained in $[C_4C_4im][BF_4]$ (65%, 97 : 3) compared with dichloromethane (4%, 79 : 21).[89] This was follwed by numerous additional reports of significant enhancements in enantioselectivity ($\Delta ee > 20\%$) for $Cu(OTf)_2$/bis(oxazoline) and Pt/diphosphine-catalysed Diels–Alder reactions. Doherty *et al.* reported an increase in ee from 78% to 95% for the $Cu(OTf)_2$–((*S,S*)-*t*-box)-catalysed Diels–Alder reaction between *N*-acryloyl oxazolidinone and cyclopentadiene [Equation (3.1.13)] at room temperature in $[C_2mim][NTf_2]$[25a] and Kim and co-workers reported increases from 68% to 94% and 85% for the same reaction catalysed by $Cu(OTf)_2$/indabox in $[C_4mim][SbF_6]$ and $[C_4mim][PF_6]$, respectively, at 3 °C whereas reactions in hydrophilic ionic liquids such as $[C_4mim][BF_4]$ and $[C_4mim][OTf]$ gave racemic product.[90] However, ee's approaching those obtained at room temperature in ionic liquid can be obtained at low temperature in dichloromethane (−78 °C) but much longer reactions times are typically required. Doherty and Hardacre and co-workers also reported a significant enhancement in ee (as well as reaction rate) for the same Diels–Alder reaction catalysed by Lewis acid platinum complexes of BINAP and conformationally flexible NUPHOS-type diphosphines in ionic liquid compared with dichloromethane. A comparison between BINAP/Pt and NUPHOS/Pt catalysts in ionic liquid and dichloromethane indicated that the enhancement in ee was associated with a decrease in the rate of racemisation of the NUPHOS/Pt catalysts in ionic liquid compared with dichloromethane as well as an intrinsic solvent effect.[102] In addition to the advantage of an increase in rate, ionic liquids also allow high ee's to be achieved without the need to recourse to the low temperatures often required in conventional solvents.

$Cu(OTf)_2$ / bis(oxazoline)

R = H, Me

(*S*)-*endo*-**13** (*S*)-*exo*-**13**

(3.1.13)

Baiker has recently demonstrated that the chemoselectivity of nitrobenzene hydrogenation catalysed by ionic liquid-supported platinum nanoparticles can be tuned through the reaction conditions (acid or base)

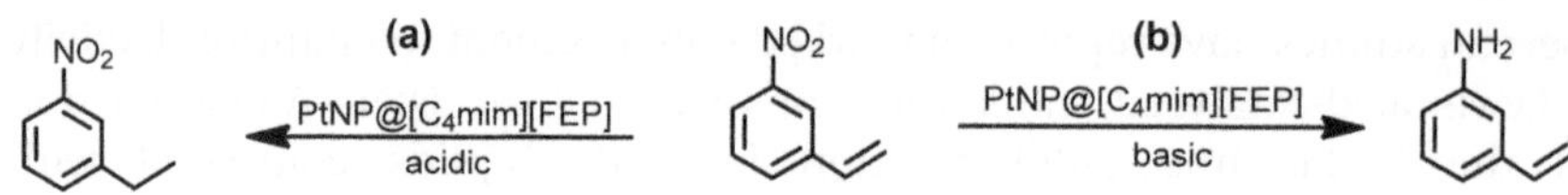

Scheme 3.1.3 Tuning the chemoselective hydrogenation of 2-nitrostyrene catalysed by ionic liquid-supported platinum nanoparticles.

and that the presence of ionic liquid was critical to achieving chemoselective reduction of the nitro group. High selectivity (95%) for reduction of the C=C double bond [Scheme 3.1.3(a)] can be achieved with Pt@[C_4mim][FEP] in the presence of bipyridine as co-stabiliser under acidic conditions (CF_3CO_2H) while selective reduction of the nitro group (89%) was achieved under basic conditions (NEt_3) with Pt@[C_4mim][FEP] (FEP = tris(pentafluoroethyl)trifluorophosphate) stabilised with quinoline and supported on carbon nanotubes [Scheme 3.1.3(b)]. A comparison with ionic liquid-free catalyst systems revealed that ionic liquid was mandatory to achieve selective reduction of the nitro group.[103]

3.1.7 Catalysis in Microstructured Reactors and Supports

An insightful review entitled *Microstructured Reactors and Supports for Ionic Liquids* addresses the achievements, challenges and potential uses of ionic liquids in microstructured reactors and/or microstructured catalyst supports with an emphasis on improving reaction selectivity and yield.[104] A continuous flow low pressure microflow system gave markedly higher selectivities and/or yields for the palladium-catalysed carbonylative Sonogoshira coupling and the aminocarbonylation of aryl iodides compared with a conventional batch system.[105] The microstructured reactor comprising two T-shaped micromixers and a stainless steel tube reactor mixes ionic liquid containing the palladium catalyst with CO and the substrates, successively, in separate Xmicromixers and then passes the resulting multiphase through a heated capillary tube reactor to give the desired product. The improvement of the microflow system over the batch reactor was attributed to the occurrence of a plug flow which provided a larger specific interfacial area between the CO and liquid phase and facilitates the diffusion of CO into the thin ionic liquid plugs. An automated continuous flow microreactor system based on an *N*-heterocyclic carbene palladium catalyst immobilised in [C_4mim][NTf_2] was continuously recycled and gave 115.3 g (10 g h^{-1}) of butyl cinnamate from the Mizoroki–Heck coupling between iodobenzene with butyl acrylate corresponding to an overall yield of 80%.[106]

Supported ionic liquid phase catalysis with microstructured supports (SSILP) combines the advantages of the use of ionic liquids as solvents for homogeneous catalysis with the benefits of a structured heterogeneous catalyst and has been applied to a range of hydrogenation reactions. For example, a thin film of [C_4mim][BF_4] dispersed over a high surface area structured support consisting of sintered metal fibres (SMFs) coated with a layer of carbon nanofibres (CNF) ensures efficient use of the transition metal catalyst without mass transfer limitations to give a TOF of 250 h^{-1} for the rhodium-catalysed hydrogenation of 1,3-cyclohexadiene with 96% selectivity for cyclohexene.[107] Supported ionic liquid phase palladium nanoparticles on a CNF–SMF composite catalyse the hydrogenation of acetylene to ethane with high efficacy and excellent selectivity in a continuous flow tubular reactor; the catalyst also exhibited exceptional long term stability with no deactivation even after 6 h on stream and showed promise for industrial application.[108] A similar structured supported ionic liquid phase catalyst based on PdNPs immobilised on an activated carbon cloth (ACC) was used for the hydrogenation of citral. Initial rates and TOFs based on citral conversion were up to three times greater than those obtained with palladium on ACC in the absence of ionic liquid, but the ionic liquid system deactivated more rapidly due to accumulation of hydrogenation products and impurities in the ionic liquid. Within the series of ionic liquids investigated, both anion and cation had a dramatic effect on activity and selectivity.[109,110] Efficient mass transfer for the rhodium-catalysed hydrogenation of propene has been achieved using corrugated porous silica filled polyethylene composites (PESC) to support a thin layer of $[Rh(nbd)(PPh_3)_2][PF_6]$/dppb in $[C_4mim][PF_6]$. The increase in mass transfer was due to the greater contact area provided by the distribution of the ionic liquid–catalyst solution within the pores of the PESC allowing more efficient contact between the catalyst and reagents. Good conversions were maintained over hundreds of hours of operation and activity was maintained even after storage for 3 months in air.[111]

3.1.8 Nomenclature

As the number and type of available ionic liquids has escalated different abbreviations to represent their ions have evolved. While abbreviations derived from the name of the ion such as BMIM for 1-butyl-3-methylimidazolium are common place this approach does introduce an element of ambiguity and as such this chapter will use an alphanumeric system similar to that adopted in the recent literature and relevant review articles. In this manner 1-alkyl-3-methylimidazolium cations are denoted by $[C_n\text{mim}]^+$, 1-alkyl-2,3-dimethylimidazolium cations as $[C_nC_1{}^2\text{mim}]^+$, 1-alkyl-3-alkyl'imidazolium cations as $[C_nC'_n\text{im}]^+$, *N*-alkylpyridinium cations as $[C_n\text{pyr}]^+$, 4-methyl-*N*-alkylpyridinium cations as $[C_n\text{mpyr}]^+$, *N*-alkyl,*N*-methyl pyrrolidinium cations as $[C_n\text{mpyrr}]^+$, picolinium $[\text{pic}]^+$, tetraalkylphosphonium ions as $[P_{n,n',n'',n'''}]^+$, and tetraalkyammonium ions as $[N_{n,n',n'',n'''}]^+$ where *n* indicates the length of the alkyl chain attached to the P and N atoms, respectively. This method of abbreviation will be extend in a logical and rational manner to indicate the type and location of a functional group for instance, $[N{\equiv}CC_4\text{mim}]^+$ indicates the presence of a cyano group on the terminal carbon atom of the butyl chain of the imidazolium cation while $[SO_3NaC_4\text{mim}]^+$ and $[SO_3HC_4\text{mim}]^+$ indicates a butyl chain terminated with sodium sulfonate and a sulfonic acid, respectively. Common anions such as bis(trifluoromethylsulfonyl)imide ($N\{CF_3SO_2\}_2$) and triflate (CF_3SO_3) are denoted as $[NTf_2]^-$ and $[OTf]^-$, respectively, while the same abbreviation system will be used to describe anions bearing alkyl chain such as $[C_4OSO_3]^-$ which denotes butylsulfate.

3.2 Applications of Ionic Liquids in Homogeneous Carbonylation, Hydroformylation and Transition Metal-Catalysed Cross-Coupling Catalysis

3.2.1 Carbonylations Involving Electrophiles

The first examples of a palladium-catalysed carbonylation in ionic liquid reported by Tanaka and co-workers demonstrated a marked enhancement in rate for the alkoxycarbonylation of aryl halides compared with the same reactions conducted in alcohol.[112] In a comparative study, yields between 83–100% were obtained when the alkoxycarbonylation of iodobenzene with 2-propanol was carried out in $[C_4mim][X]$ ($X=PF_6^-$, BF_4^-) with 0.5 mol% $Pd(OAc)_2/PPh_3$ and 150 kg cm^{-2} CO at 150 °C, whereas reaction in neat 2-propanol gave a yield of only 12% under the same conditions. Interestingly, the use of $[C_4mim][PF_6]$ also reduced the selectivity for double carbonylation. The rate enhancement in ionic liquid was attributed to stabilisation of a polar or ionic transition state such as an acyl palladium species attacked by an alcohol or hydroxide ion. A rate enhancement was also obtained for the hydroxycarbonylation of iodobenzene in $[C_4mim][X]$ ($X=PF_6^-$, BF_4^-) and Aliquat® 336 compared with the same reaction in water-immiscible solvents such as dichloromethane and benzene, however, the TOFs in these ionic liquids (170–180 h^{-1}) were slightly lower than those in water miscible-solvents such as THF and DMF (200–260 h^{-1}). Even though the TOFs were slightly lower in IL, the high activity of this two phase IL system, facile isolation of product by extraction of its ammonium carboxylate salt into water and efficient recycling in an operationally straightforward manner, with only a minor decrease in yield over four runs, enabled a high total TON to be achieved.[113a,b] (*E*)- and (*Z*) mixtures of *β*-vinyl bromides (>5 : 1) can be stereoselectively carbonylated in $[C_4mim][PF_6]$ at 100 °C under 20 bar CO for 10 h using 5 mol% $[Pd(PPh_3)_2Cl_2]$ to afford the corresponding (*E*)-α,β-unsaturated carboxylic acids in moderate yield and with *E*/*Z* mole ratios up to 99 : 1 [Equation (3.2.1)]. Yields improved in the second and third runs but dropped in runs four and five while the *E*/*Z* ratios remained high. However, the carbonylation of a 59 : 41 *E*/*Z* mixture of (2-bromoprop-1-en-1-yl)benzene under the same conditions gave a 67 : 33 more ratio of α,β-unsaturated carboxylic acids.[114]

$$\underset{E/Z = 88:12}{ArCH{=}CHBr} \xrightarrow[{[C_4mim][PF_6]}]{Pd(PPh_3)_2Cl_2} \underset{E/Z \text{ up to } 99:1}{Ar\text{–}CH{=}CH\text{–}CO_2H} \tag{3.2.1}$$

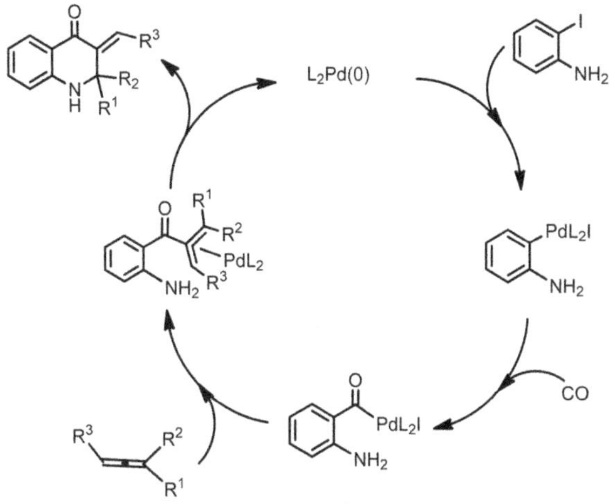

Scheme 3.2.1 A possible mechanism for the cyclocarbonylation of *o*-iodoanilines with allenes.

Ionic liquids have been used as both solvent and promoter to enhance the efficacy of a multicomponent cyclocarbonylation between *o*-iodo-aniline and allenes to afford 3-methylene-2,3-dihydro-1H-quinolin-4-ones [Equation (3.2.2)]. The beneficial effect of using an ionic liquid as the solvent was clearly demonstrated by a comparison of the same reaction in benzene. When the reaction was conducted in $[C_4mim][PF_6]$ using $Pd_2(dba)_3$/dppb as catalyst, N^iPrEt_2 as base at 90 °C under 5 atm CO the product was isolated in 74% yield, which was a marked improvement on the 37% obtained in benzene. The ionic liquid–catalyst mixture recycled with reasonable efficiency provided additional dppb was added to the reaction mixture. The authors proposed a mechanism based on oxidative addition of palladium to *o*-iodoaniline, CO insertion to afford a palladium-acyl intermediate, regioselective 1,2-insertion of the allene into the Pd-acyl bond to afford a π-allylpalladium species which undergoes intramolecular nucleophilic attack to liberate the product (Scheme 3.2.1).[115] The same Pd/dppb–ionic liquid system also catalyses the cyclocarbonylation of 2-allylphenols and anilines, 2-vinylphenols, and 2-aminostyrenes to give the corresponding lactones and lactams in high yield and good to excellent selectivity for one isomer. The ionic liquid containing catalyst recycled with no loss in activity over several runs but the product distribution changed quite dramatically; this change was suggested to be associated with the formation of a palladium carbene-based catalyst that promoted cyclocarbonylation with a markedly different selectivity.[48] The carbonylation of 3-alkynyl-1-ols and 1-alkyn-4-ols catalysed by $Pd(OAc)_2$/2-pyPPh$_2$ occurs quantitatively and selectively under biphasic conditions in $[C_4mim][BF_4]$ to afford *exo*-methylene γ- and δ-lactones, respectively. The ionic liquid–catalyst mixture recycled after the product was removed by distillation but the activity dropped quite dramatically after the first run.[116]

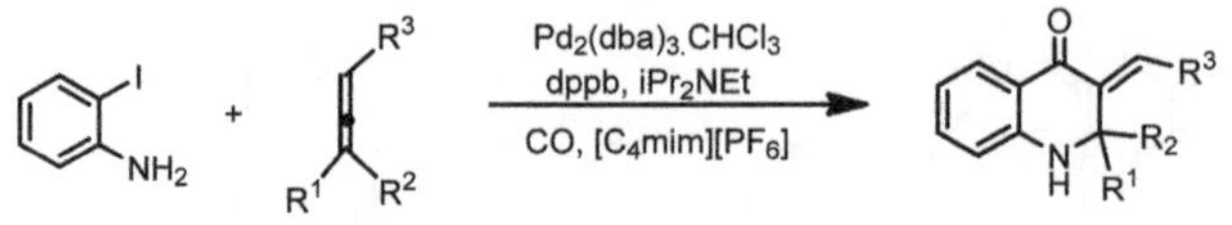

(3.2.2)

Ionic liquids have a dramatic effect on the palladium-catalysed cyclocarbonylation of enynol with thiols compared with THF as reactions in $[C_4mim][PF_6]$ occur with high selectivity for monocarbonylation to afford thioether-substituted 6-membered α,β-unsaturated lactone **14** as the dominant product while in THF double carbonylation to afford thioester-containing 6-membered ring lactone **15** is favoured (Scheme 3.2.2). For example, a $[C_4mim][PF_6]$ solution containing 2 mol% $Pd(OAc)_2/PPh_3$ catalysed the cyclocarbonylation of 3-phenyl-2-penten-4-yn-1-ol and thiophenol at 110 °C under 500 psi CO to afford the monocarbonylated product **14** in 91% yield whereas the same reaction in THF gave the dicarbonylated adduct **15** as the sole product in 62% yield.[117a,b]

Although the overwhelming majority of catalysis performed in ionic liquids has been conducted in nitrogen-based systems, phosphonium salt ionic liquids (PSILs) have recently been investigated as solvents for palladium-catalysed processes because their high stability towards thermal and chemical degradation and their exceptionally low-volatile renders them ideal for high temperature reactions or for processes in which the product can be removed by distillation. In this regard, McNulty and Alper and co-workers have pioneered the use of phosphonium-based ionic liquids in palladium-based catalysis and in many cases have demonstrated a beneficial effect on efficiency. Alper and co-workers have shown that trihexyl(tetradecyl)phosphonium hexafluorophosphate $[P_{6,6,6,14}][PF_6]$ is an excellent solvent for the palladium-catalysed thiocarbonylation of aryl iodides with thiols using 5 mol% $Pd(OAc)_2/PPh_3$ and triethylamine as base under 200 psi CO at 100 °C. Under these conditions, the thiocarbonylation of iodobenzene with thiophenol gave the corresponding thioester in 91% yield, which was markedly higher than the yield of 28% obtained in THF. The catalyst also recycled efficiently after extraction of the product with hexane, although the reuse was limited to a single run.[118] The use of $[P_{6,6,6,14}]Br$ as solvent also dramatically enhanced the efficiency of the palladium-catalysed ligand-free

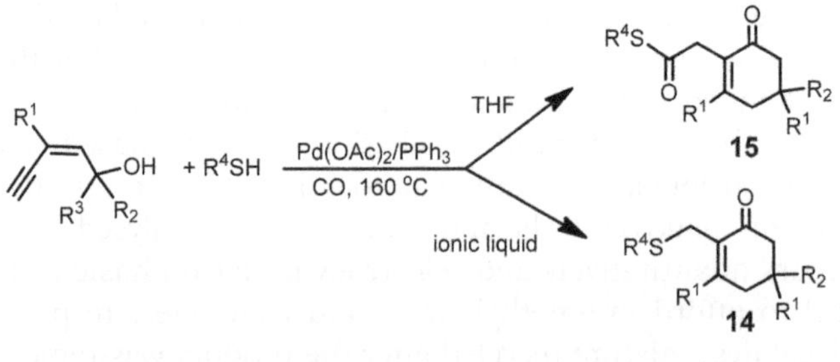

Scheme 3.2.2 Effect of ionic liquid on the chemoselectivity of the palladium-catalysed cyclocarbonylation of enynols with thiols.

cyclocarbonylation of *o*-iodophenols with terminal acetylenes [Equation (3.2.3)]. Good to excellent yields of a diverse range of chromones were obtained using 5 mol% $PdCl_2$ under 1 atm of CO at 100 °C whereas reactions conducted under the same conditions in conventional organic solvent did not give any product. A survey of ionic liquids $[P_{6,6,6,14}][X]$ (X = Br^-, Cl^-, NTf_2^-) revealed a pronounced anion effect with both yields and selectivity increasing in the order $[Br]^- > [Cl]^- > [NTf_2]^-$. The ionic liquid–catalyst system was only recycled once but a good yield of the desired chromone was obtained.[119]

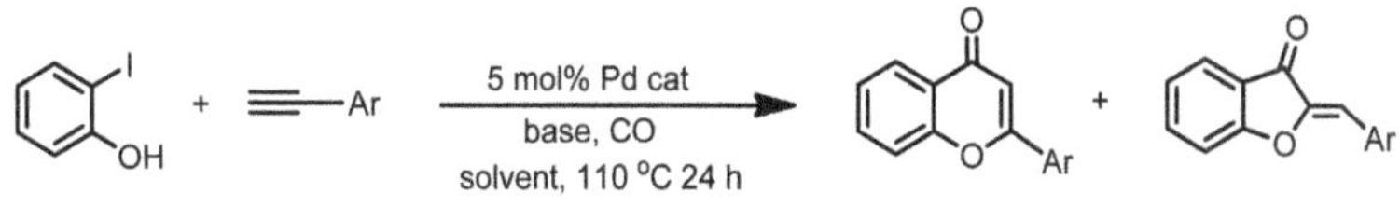

5 mol% Pd cat; base, CO; solvent, 110 °C 24 h (3.2.3)

Phosphonium-based ionic liquids also enhance the efficacy of a palladium-catalysed carbonylation-hydroamination sequence involving 1-halo-2-alkynylbenzene and amines to afford substituted 3-methylene-isoindolin-1-ones in good yields and high selectivity in favour of the *Z*-isomers (Scheme 3.2.3). Under optimum conditions 5 mol% $[PdCl_2(PPh_3)_2]$ in $[P_{6,6,6,14}]Br$ catalysed the carbonylation–hydroamination between 1-bromo-2-(phenylethynyl)benzene and benzylamine at 100 °C under 1 atm of CO in the presence of DBU as base to afford the corresponding 3-methylene-isoindolin-1-ones in 84% yield; comparable yields were obtained for a host of other substrate combinations. The beneficial effect of the ionic liquid on catalyst performance was underpinned by comparative reactions in THF which gave markedly lower yields in the same time.[120] The target isoindolin-1-ones could also be synthesised in a one-pot Sonogoshira coupling-carbonylation-hydroaminatino sequence involving an 2-bromo-ioodbenzene, a terminal alkyne and an amine. A positive 'bromide' effect was identified for the palladium-catalysed alkoxycarbonylation of 4-iodotoluene with 4 mol% $Pd(OAc)_2$/dppf in a series of phosphonium-based ionic liquids $[P_{6,6,6,14}][X]$ (X = OTs^-, Cl^-, $C(CN)_2^-$, NTf_2^-, Br^-, PF_6^-, BF_4^-, phosphate, sulphate, decanoate) as a much higher isolated yield was obtained in $[P_{6,6,6,14}]Br$ than all other ionic liquids tested. The authors reasoned that the bromide anion acted as a nucleophile by intercepting the acyl-palladium intermediate to liberate a reactive acid bromide that subsequently reacted with a protic alcohol or amine to afford the observed ester or amide, respectively.[121]

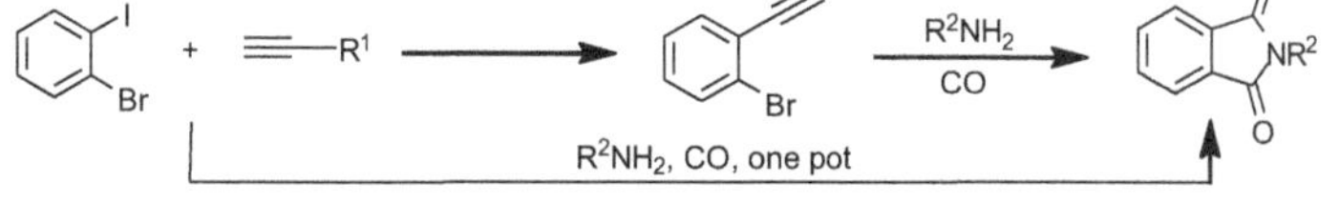

R^2NH_2; CO; R^2NH_2, CO, one pot

Scheme 3.2.3 Multi-step and one-pot synthesis of substituted 3-methyleneisoindolin-1-ones.

High rates and high selectivity have been achieved for the palladium-catalysed alkoxycarbonylation of iodobenzene in [C_4mim][*p*-$MeC_6H_4SO_3$] using $Pd(OAc)_2$ and water soluble phosphines. A survey of sulfonated phosphines revealed that catalyst efficiency, as measured by the TOF, decreased in the order tppts (1938 h^{-1}) > tppds (1470 h^{-1}) > tppms (1156 h^{-1}). Hydrophilic imidazolium-based ionic liquids proved superior to their hydrophobic counterparts as TOFs decreased in the order [*p*-$MeC_6H_4SO_3$]$^-$ (1938 h^{-1}) > [BF_4]$^-$ (1660 h^{-1}) > [PF_6]$^-$ (1430 h^{-1}); the TOF of 1930 h^{-1} and 1692 h^{-1} obtained in dioxane and THF, respectively, compared favourably with those in [C_4mim][*p*-$MeC_6H_4SO_3$] and [C_4mim][BF_4], respectively. The higher TOFs obtained in hydrophilic ionic liquids were attributed to the higher solubility of the palladium catalyst in [C_4mim][*p*-$MeC_6H_4SO_3$] and [C_4mim][*p*-$MeC_6H_4SO_3$] compared with [C_4mim][PF_6]. Even though high TOFs could be obtained in dioxane and THF separation of the product from the catalyst was non-trivial for both solvent systems whereas the product was extracted from the ionic liquid–catalyst mixture with cyclohexane in an operatically straightforward procedure and the immobilised catalyst recycled 10 times without any significant decrease in activity or selectivity.[122] A tandem catalytic sequence involving Pd(0)-catalysed deallylation-Pd(II)-catalysed carbonylative heterocyclisation of 1-(2-allyloxyphenyl)-2-yn-1-ols to afford benzofurans was conducted in ionic liquids; the highest yields were obtained in [C_4mim][BF_4] under 30 atm of CO at 100 °C using PdI_2/KI/PPh_3/H_2O. The system recycled efficiently over seven runs by extracting the product with diethyl ether and adding a freshly prepared solution of substrate, water and methanol.[123] An efficient ionic liquid-based three-component synthesis of substituted endocyclic enol lactones has been developed which involves a highly regioselective palladium-catalysed carbonylative coupling of alkynes and 1,3-diketones [Equation (3.2.4)]. The model reaction between phenylacetylene and 2,4-pentandione identified [C_4mim][NTf_2] as the most appropriate solvent and [$PdCl_2(PPh_3)_2$]/dppp as the most efficient catalyst under 200 psi of CO at 100 °C. Under these conditions, moderate to good yields of a range of substituted endocyclic enol lactones were obtained. The catalyst could be recycled five times by extracting the product with diethyl ether at the end of each run. This approach provides an alternative synthesis of endocyclic enol lactones that complements the palladium-catalysed cyclisation of α,ω-alkynyloic acids, which generally gives mixtures of exocyclic and endocyclic product.[124]

$$R^1C(O)CH_2C(O)R^2 + R^3{-}{\equiv}{-}R^4 \xrightarrow[\text{[C}_4\text{mim][NTf}_2\text{], 110 °C}]{\text{Pd(PPh}_3\text{)Cl}_2\text{/dppp}} \text{lactone product} \quad (3.2.4)$$

Phosphonium-based ionic liquids are also effective solvents for the high temperature reductive carbonylation of a range of electron-rich and electron-poor mono and dinitroarenes to afford urethanes **16** in good yield and high selectivity [Equation (3.2.5)]. Reactions conducted in [$P_{6,6,6,14}$][PF_6] using

5 mol% catalyst generated from $PdCl_2$ and phenanthroline at 135 °C under 200 psi of CO gave the urethane as the sole product in near quantitative yield; other catalytic systems were less selective and gave a mixture of the urethane **16** and the corresponding amine **17**. A marked improvement in conversion and selectivity for the reductive carbonylation of nitrobenzene catalysed by $[PdCl_2(phen)]$ was obtained by addition of ionic liquid co-catalyst; conversions increased from 67% to 91% while selectivity for ethyl phenyl carbamate increased from 81% to 95%.[125] A further improvement was achieved when Brønsted acid-functionalised imidazolium-based ionic liquids were used; the highest selectivities of 98.8% and 97.9% were obtained using $[CO_2HC_1C_4im][BF_4]$ and $[CO_2HC_1C_4im][PF_6]$ as co-catalyst, respectively, at 150 °C under 6 MPa of CO. Conversions and selectivities were markedly lower with the corresponding bromide and chloride-based Brønsted acid-functionalised imidazolium-based ionic liquids and decreased with increasing number of carbon atoms between the imidazolium cation and the acid group. The efficiency of the catalyst system was reasoned to be due to facile exchange of the chloride in $[PdCl_2(phen)]$ with $[PF_6]^-$ and $[BF_4]^-$ to form $[Pd(phen)][PF_6]_2$ and $[Pd(phen)][BF_4]_2$, which are coordinatively unsaturated, electrophilic, highly Lewis acidic and very reactive, as well as to carboxyl group-assisted substitution.[126]

(3.2.5)

A homogeneous solution of $[PdCl_2(phen)]$ in $[C_4mim][X]$ ($X = BF_4^-$, Cl^-, PF_6^-, $FeCl_4^-$) catalyses the carbonylation of aniline to give carbamates and ureas under 5 MPa of CO at 170 °C. The anion has a dramatic effect on catalyst performance with $[BF_4]^-$ and $[FeCl_4]^-$ giving markedly higher TOFs (4540 and 4420 h^{-1}, respectively) than $[Cl]^-$ and $[PF_6]^-$-based ionic liquids (2680 and 228 h^{-1}, respectively). The exceptionally poor performance of the latter was attributed to the low stability of $[C_4mim][PF_6]$ under the reaction conditions as significant decomposition of this ionic liquid was noted at the end of the reaction. Low yields were obtained with $Pd(OAc)_2$ and $PdCl_2$ in the absence of phenathroline and poor conversions were also obtained in the absence of ionic liquid.[127]

Acrylamides are an important class of compound that is used widely in organic reactions and polymer synthesis. The most common route to these derivatives is hydration of an acrylonitrile, although there is considerable interest in the palladium-catalysed aminocarbonylation of alkynes as an atom economical and clean alternative. However, the need for an acid component limits the potential of this approach somewhat as the acid renders the process corrosive. In this regard, the use of an ionic liquid as both reaction solvent and promoter for the regioselective synthesis of 2-substituted acrylamides *via* the palladium-catalysed aminocarbonylation of alkynes under mild conditions without the need for an acid additive was an important discovery [Equation (3.2.6)]. Optimisation of the benchmark reaction between 1-octyne and diethylamine identified [C_4mim][NTf_2] and Pd$(OAc)_2$/dppp to be the most efficient ionic liquid–catalyst combination which gave the corresponding acrylamide in 66% isolated yield and a regioselectivity of 99 : 1 in favour of **18**. For comparison, only a trace amount of product was obtained when the reaction was conducted in either THF or DMF under the same conditions. Good yields were obtained for a range of alkyne–amine combinations and the catalyst recycled five times with no loss in activity.[128a] The methoxycarbonylation of ethylene to afford methyl propionate is a technologically important process as the product is an intermediate in the production of methyl methacrylate. However, the process requires the use of a strong Brønsted acid to promote the reaction but corrosion and rapid phosphine alkylation are major drawbacks. To this end, a range of imidazolium, pyridinium, phosphonium and ammonium-based Brønsted acid ionic liquids have recently been employed as replacements for methane sulfonic acid in the palladium-catalysed methoxycarbonylation of ethylene; advantages of these systems included phase separation of the methyl propionate and facile recovery of the ionic liquid as well as stabilisation of palladium intermediates and a reduction in the formation of palladium black. The most efficient system recycled 15 times without any significant loss in activity or selectivity.[128b]

$$R^1\text{—}\equiv \;+\; R^1R^2NH_2 \xrightarrow[\text{[C}_4\text{mim][NTf}_2\text{]}]{\text{Pd(OAc)}_2\text{/dppp, CO}} \mathbf{18} + \mathbf{19} \tag{3.2.6}$$

Substituted ureas are important intermediates in the synthesis of biologically active compounds, pesticides, pigments and herbicides and their synthesis *via* the carbonylation of amines using CO_2 as the carbonyl source is of particular interest as an alternative to the use of phosgene and carbon monoxide. In this regard, symmetric urea derivatives have been prepared in good yield from amines and CO_2 in a recyclable catalytic system comprising an ionic liquid and CsOH as catalyst; importantly this system also avoids the need for a stoichiometric amount of dehydrating agent. The cation and anion have a profound effect on the formation of the urea derivatives; the highest yield of 98% was obtained in [C_4mim]Cl after 4 h whereas [C_4mim][BF_4] and [C_4mim][PF_6] gave yields of 84% and 56.6%, respectively, in the same time. Ultimately, this technology could be used to synthesize isocyanates from CO_2 without the need for stoichiometric quantities of dehydrating agent.[129] An investigation into the role/efficiency of ionic liquids as catalyst in the synthesis of 1,3-substituted ureas by the carboxylation of amines [Equation (3.2.7)] has shown that significantly higher conversions are obtained in [C_4mim]-based salts of strongly basic anions such as chloride, dimethylphosphate and acetate, with chloride proving to be the most efficient; in comparison much lower conversions were obtained in their $[BF_4]^-$, $[PF_6]^-$, $[MeSO_4]^-$ and $[CH_3SO_3]^-$ counterparts. Interestingly, the halide salts of the tetrabutylphosphonium cation, $[P_{4,4,4,4}]$ gave yields that matched those obtained with [C_4mim]Cl whereas the corresponding tetrabutylammonium halides gave a moderate yield of 1,3-DBU as did pyrrolidinium, piperidinium and pyridinium halides. Computational studies on the mechanism of the carboxylation indicated a three-step process involving (i) interaction of three molecules of RNH_2 with CO_2 to ultimately generate carbamic acid or the carbonate salt, (ii) attack of amine on the nitrogen-bonded carbon atom of the carbamic acid to generate $(RNH)_2C(OH)_2$ and (iiii) proton transfer and dehydration. The use of [C_2mim]Cl significantly reduces the activation energies for all of the transition states by up to 10 kcal mol^{-1} as well as the relative energies of the intermediate products compared to reaction in the absence of ionic liquid. Although dehydration is accepted as the rate determining step for the synthesis of ureas from CO_2 and amines stabilisation of the transition state for this step in ionic liquids renders addition of the second amine rate determining. The activation energies of the transition states and intermediate ionic species are lowered through a relatively strong interaction between the carbonyl oxygen and the acidic C(2)-H of the $[C_2mim]^+$ cation as well as by a network of multiple hydrogen-bonding interactions involving free amine, carbonyl-bonded amine and the chloride anions.[130]

$$2\,RNH_2 + CO_2 \xrightarrow[\text{[C}_4\text{mim][X]}]{\text{Cat}} RHN\text{–C(=O)–}NHR + H_2O \quad (3.2.7)$$

R = alkyl, cycloalkyl

$$2\,RNH_2 + CO + {}^1/{}_2O_2 \xrightarrow[\text{[C}_4\text{mim][Se(O)}_2\text{OMe]}]{\text{Cat}} RHN\text{–C(=O)–}NHR + H_2O \quad (3.2.8)$$

Room temperature ionic liquids of the type [C_2mim][$Se(O)_2OMe$], consisting of imidazolium cations and selenium-based anions, are remarkably active catalysts for the oxidative carbonylation of aniline to phenyl carbamate and diphenyl urea [Equation (3.2.8)]; the TOFs of 93–98 h^{-1} at 40 °C and up to 3680 h^{-1} at 120 °C are a significant improvement on that of 18 h^{-1} obtained with [$KSe(O)_2OMe$]. The ionic liquid–catalyst mixture recycled efficiently with no significant reduction in activity or selectivity over five runs. Selenium contamination of the urea product is a major problem with the use of conventional alkali metal-containing selenium catalysts, however, the diphenyl urea obtained using these ionic liquid–catalyst systems has a selenium content of only 2.4 ppm after washing with methanol which is significantly lower than the 13.2 ppm obtained under similar conditions with [$KSe(O)_2OMe$].[131] The corresponding tetraalkylphosphonium methylselenites [R_4P][$Se(O)_2OMe$] (R = Et, *n*-Bu) also catalyse the oxidative carbonylation of aniline to afford ureas although the TOF of 73 and 76 h^{-1} are slightly lower than those obtained with their imidazolium counterparts. ^{13}C NMR labelling studies on $[C_2mim][Se(O)_2(O^{13}CH_3)]$ shed light on the mechanism of oxidative carbonylation which appears to involve exchange of the methoxy group with amido, uptake and insertion of CO into the resulting Se–N bond to generate a selenium carbamoyl complex which undergoes a nucleophilic abstraction-type process with a second molecule of amine to liberate the *N,N′*-dialkyl urea together with [Se(O)(OH)], which is ultimately reoxidised to $[Se(O)_2(OH)]$.[132]

Hydrogen carbonate-based ionic liquids are highly efficient catalysts for the carbonylation of cyclohexylamine and afford cyclohexylformamide in 83–88% yield after 4 h at 140 °C under 4 MPa of CO. The remarkable efficacy of this system was evident in a comparison with $[C_4mim]^+$-based salts of other anions which gave much lower conversions (15–40%) under the same conditions. Catalytic activity appears to correlate with the basicity of the amine and increases in the order $[HCO_3]^- > [MeCO_2]^- > [MePHO_3]^- > [MePO_4]^- > [MeSO_4]^- > [MeSO_3]^- > [BF_4]^-$. Computational studies suggest that the $[C_2mim][HCO_3]$-catalysed carbonylation occurs in a concerted manner involving activation of amine by $[HCO_3]^-$, C–N bond formation between the activated amine and CO to generate a carbamoyl species and hydrogen transfer from $[HCO_3]^-$ to the carbon atom of the carbamoyl. The high activity of $[C_2mim][HCO_3]$ was attributed to the ability of the anion to act as a bifunctional hydrogen atom acceptor and donor in the transition state.[133]

Highly dispersed ionic liquid catalysts (8–53 wt%) have been prepared by physical confinement or encapsulation of ionic liquids of the type $[C_nmim][BF_4]$ ($n = 2$, 4, 10, 16) in a silica gel matrix by traditional sol–gel processing in the presence of $[Rh(PPh_3)_3Cl]$ or $[Pd(PPh_3)_2Cl_2]$ to afford a mesoporous silica gel-supported ionic liquid nanocatalyst. In this design concept, the ionic liquid acts as both the catalyst and reaction medium while the solid matrix is the reactor and as such the pore size should be large enough to contain the ionic liquid, while the channel size should be small enough to prevent leaching of the ionic liquid–catalyst mixture but large enough to allow free transportation of the reactant and product (Figure 3.2.1). A study of the stability of the ionic liquid-confined silica gel revealed that those with longer alkyl chains ($n = 10$, 16) remained confined in the silica gel pores whereas smaller ones ($n = 2$, 4) were washed out under vigorous reflux in acetone. N_2 adsorption measurements conducted on the $[C_4mim][BF_4]$ template silica gel matrix remaining after removal of the ionic liquid by washing showed that the BET surface area decreased from 933 to 322 $m^2\,g^{-1}$ but the average pore volume and average pore diameter increased from 0.298 to 1.35 $cm^3\,g^{-1}$ and 29 to 109 Å, respectively, when the amount of confined ionic liquid was increased from 0 to 53 wt%. A marked enhancement in rate was obtained when $[Rh(PPh_3)_3Cl]$-derived silica gel-confined ionic liquid ($[Rh(PPh_3)_3Cl]$–$[C_{10}mim][BF_4]$–silica gel) was used for the carbonylation of nitrobenzene, compared with the reactions conducted in bulk ionic liquid or as a SILP system on silica. The best catalytic performance for the

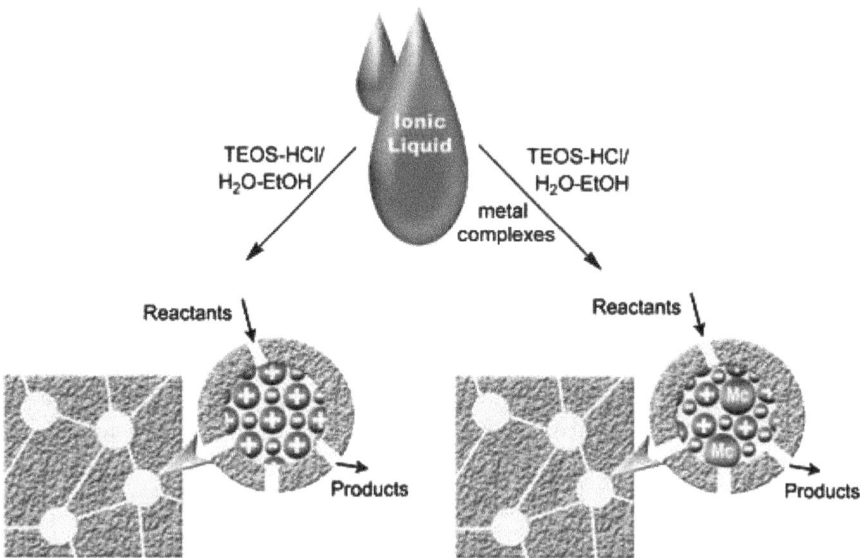

Figure 3.2.1 Illustration of the silica gel-confined ionic liquid with and without metal complex.
Reproduced with permission from reference 134. Copyright 2005 Wiley VCH.

carbonylation of nitrobenzene was achieved over the $[Rh(PPh_3)_3Cl]$-based silica gel-confined ionic liquid catalyst (Rh–$[C_{10}mim][BF_4]$–silica gel) which gave a TOF of 11 548 h^{-1} compared to 7463 h^{-1} for a physical mixture of $[Rh(PPh_3)_3Cl]$, $[C_{10}mim][BF_4]$ and silica gel. The enhancement in catalyst activity was attributed to the high concentration of ionic liquid-containing metal complex by confinement in the pores or cavities of the silica gel matrix. A marked enhancement in rate was also obtained for the oxidative carbonylation of aniline to phenyl carbamate with rhodium- and palladium-based silica gel-confined ionic liquid catalysts; the TOFs of 2650 h^{-1} and 4900 h^{-1} were markedly higher than those of 494 h^{-1} and 752 h^{-1}, respectively in neat ionic liquid. In both cases the amount of ionic liquid used in the silica gel-confined ionic liquid was much lower than in either the physical mixtures or the SILP catalyst systems.[134]

3.2.2 Hydroformylation

Rhodium-catalysed hydroformylation is the largest scale industrial process involving a homogeneous catalyst. However, efficient separation of the product(s) from the catalyst remains a challenge particularly for long chain substrates. Separation can be accomplished by distillation of the product but only for C3–C5 feedstock as the catalyst decomposes at the temperatures required to distil higher aldehydes. The industrial water–organic biphasic process developed by Ruhrchemie/Rhône-Pulenc overcomes this problem but is limited to propene as higher feedstocks are not sufficiently soluble and transport of the olefin to the catalytic site is rate determining. As hydroformylation of higher olefins is of particular industrial interest there is still a need to identify and develop new processes that allow efficient HF of longer chain olefins and separation of the catalyst. In this regard, ionic liquids have been investigated for liquid–liquid biphasic hydroformylation as olefins are much more soluble in ionic liquids than water and they are compatible with water-sensitive phosphite-based ligands. Moreover, the use of ionic liquids for rhodium-catalysed liquid–liquid biphasic hydroformylation has long been the standard benchmark reaction with which to evaluate new immobilisation concepts. A review entitled *Hydroformylation in Room Temperature Ionic Liquids (RTILs): Catalyst and Process Developments* provides an informative account of the development of room temperature ionic liquids as alternative solvents for biphasic catalysis and covers the literature up to 2007.[135]

3.2.2.1 Biphasic Hydroformylation with Charge-Tagged Catalysts

Chauvin *et al.* were first to describe the use of ionic liquids for the biphasic hydroformylation of 1-pentene with $[Rh(CO)_2(acac)]$/triarylphosphine in $[C_4mim][X]$ ($X = BF_4^-$, PF_6^-). The highest TOF of 333 h^{-1} obtained using triphenylphosphine as ligand and $[C_4mim][PF_6]$ as solvent competed with that of 297 h^{-1} obtained in toluene. However, as the active rhodium catalyst leached into the organic phase it was necessary to immobilise the precursor in the ionic liquid by using the sodium salts of monosulfonated or trisulfonated triphenylphosphine. While the TOF of 59 h^{-1} and 103 h^{-1} obtained with $[Rh(acac)(CO)_2]$/tppms and $[Rh(acac)(CO)_2]$/tppts, respectively, were significantly lower than their triphenylphosphine counterparts extraction of rhodium catalyst was completely prevented.[136] A concentration-time profile for the Rh/tppts-catalysed hydroformylation of 1-octene in a biphasic mixture of $[C_4mim][PF_6]$ and decane revealed that *n*-nonal and a mixture of three branched aldehydes formed and that isomerisation of 1-octene was significant in the early stages of reaction but reached a maximum after 1 h; hydroformylation of the resulting *iso*-octenes led to a decrease in the *n*/*iso* ratio from 2.6 to 0.8 at longer reaction times. The initial rate showed a first order dependence on the concentration of catalyst, a fractional dependence (0.75) with respect to olefin concentration and a fractional order with respect to H_2 pressure (0.46). The reaction rate also increases with increasing CO partial pressure up to 20 bar and then decreases with a further increase in CO pressure; this is typical for hydroformylation reactions conducted under homogeneous and biphasic conditions. The activation energy of 107.9 kJ mol^{-1} is within the range observed for hydroformylation reactions.[137]

Efficient immobilisation of rhodium hydroformylation catalysts in $[C_4mim][PF_6]$ was achieved by modifying neutral phosphines with the phenylguanidinium cation. For example, the combination of [Rh(acac)$(CO)_2$] and guanidinium-modified triphenylphosphine **20** catalysed the hydroformylation of 1-octene in [C_4mim][PF_6] with a TOF of 276 h^{-1} and an $n:i$ ratio of 2.7 compared to a TOF of 680 h^{-1} and an $n:i$ ratio of 2.8 for the corresponding PPh_3-based system. However, the activity of the PPh_3 system dropped to 100 h^{-1} after only two cycles due to substantial leaching of the rhodium whereas its guanidinium-modified triphenylphosphine counterpart retained its activity; for comparison catalyst based on sodium triphenylphosphine monosulfonate showed no evidence of leaching but was less active (TOF = 80 h^{-1}). The concept was also applied to the guanidinium-modified xantphos diphosphine **21** as this ligand architecture affords much higher regioselectivities in the hydroformylation of terminal olefins; the xanthane class of diphosphine was developed by van Leeuwen and co-workers and gave up to 98% selectivity for linear aldehyde in the hydroformylation of 1-octene.[138,139] The biphasic hydroformylation of 1-octene in [C_4mim][PF_6] using [Rh(acac)$(CO)_2$] in combination with phenylguanidinium-modified xantphos gave high selectivity for linear aldehyde ($n:i$ = 18.0–21.3) and a total TON of 3500 after ten recycles with less than 0.07% rhodium leaching into the organic layer. An increase in activity after each recycle up to the fourth run was attributed to a catalyst preforming time as well as the presence of 3-iodophenylguanidine impurity which was gradually removed over the first few runs. Thus, modification of a phosphine ligand with the guanidinium group appears to be a straightforward and highly effective approach to immobilise a transition metal catalyst which does not modify its steric or electronic properties or influence/effectiveness as a ligand.[140]

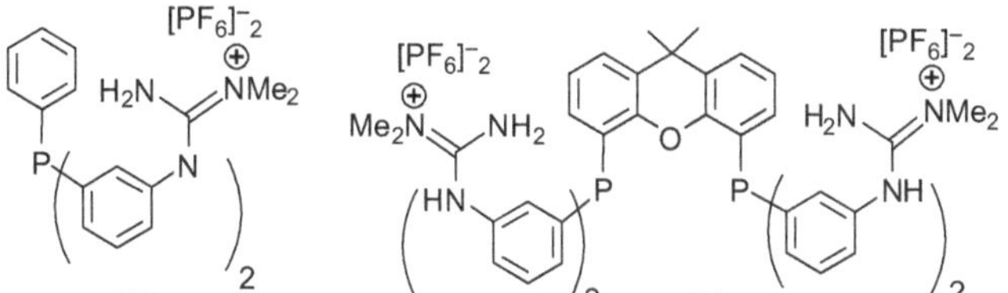

$[PF_6]^-_2$ H_2N NMe_2 P N 2 **20** $[PF_6]^-_2$ Me_2N NH_2 HN P O P 2 **21** $[PF_6]^-_2$ H_2N NMe_2 NH 2

The catalytic activity of xantphos-based catalysts improves quite significantly by replacing the diphenylphosphino groups with phenoxophosphines (POP).[141,142] This also proved to be a successful strategy for improving the efficacy of the biphasic rhodium-catalysed hydroformylation of 1-octene in [C_4mim][PF_6]. In order to ensure efficient immobilisation of the catalyst in the ionic liquid a phenoxaphosphino-modified xantphos-type diphosphine was modified with two 1-methyl-3-pentylimidazolium hexafluorophosphate groups so that it closely resembled the reaction medium. Catalyst prepared by reaction of **22** with [Rh$(CO)_2$(acac)] was highly active and selective for the hydroformylation of 1-octene in [C_4mim][PF_6] at 100 °C and 17 bar H_2-CO (1 : 1). Under these conditions $n:i$ ratios up to 44 were obtained which improved to 55 as the hydrogen pressure was increased, however, the activity was moderate but increased to 112 h^{-1} on successive runs and ultimately exceeded 300 h^{-1} when the pH_2 was increased to 40 bar. Further optimisation studies revealed that the TOF increased dramatically as the rhodium concentration was reduced and the stirring speed increased; the latter indicates that the system operates under mass transfer limitations. An unprecedentedly high activity (TOF > 6200 h^{-1}) and selectivity ($n : i$ ratio > 41) for the hydroformylation of 1-octene was obtained under optimised conditions at 100 °C, a high hydrogen pressure (CO : H_2, 1 : 9, 60 bar), a catalyst substrate : ratio of 3823, a low rhodium concentration (1.7 mM) and a stirring speed of 900 rpm in [C_4mim][PF_6]. The ionic liquid–catalyst mixture also recycled efficiently without any loss in activity or selectivity over 10 runs and only a trace amount of rhodium leaching (0.07–0.08% of the initial rhodium intake).[143,144]

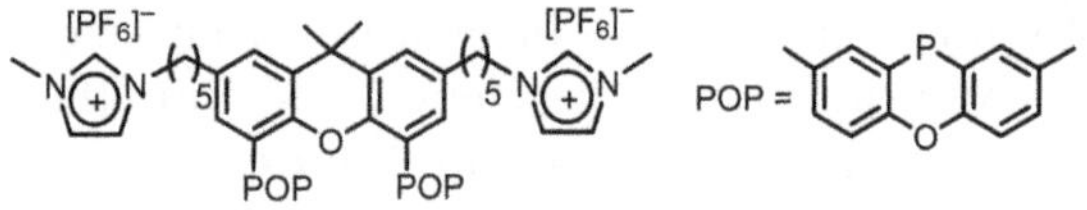

22

As for catalyst based on triphenylphosphine compared to its sulfonated derivatives, catalyst generated from sulfonated-xantphos (xantphosSulf) and [Rh(acac)(CO)$_2$] was less active than its unmodified counterpart for the biphasic hydroformylation of 1-octene in [C_4mim][PF_6], at 80 °C under 50 bar CO–H_2 (1 : 1) and 0.1 mol% rhodium. However, while the TOF of 245 h^{-1} (n : i = 2.1) obtained with Rh/xantphos dropped dramatically on recycle due to extraction of the majority of the rhodium into the organic phase, Rh/xantphosSulf recycled efficiently over four runs with no loss in activity, albeit it with a much lower TOF (41 h^{-1}) and an $n : i$ selectivity of 1.7. The $n : i$ selectivity could be improved to 4.1 by increasing the phosphine/Rh ratio to five but at the expense of the TOF which dropped to 16 h^{-1}; the selectivity was ultimately optimised at 13.1 under 15 bar H_2–CO (1 : 1) at 100 °C.[145] High linear selectivity for the rhodium-catalysed hydroformylation of 1-octene was also obtained under liquid–liquid biphasic conditions using Rh/sulfoxantphos in [C_4pyr][BF_4]–heptane under 41.4 bar syngas and 373 K. An $n : i$ ratio of 30 and a TOF of *ca.* 40 h^{-1} was maintained over several cycles with no evidence for catalyst leaching. Kinetic studies in the temperature range 353–373 K gave a first order dependence of the rate on the catalyst and 1-octene concentration, an increase in rate with H_2 pressure (fractional order) and inhibition at high partial pressures of CO due to formation of an inactive rhodium dicarbonyl complex; two models that accurately fit the initial kinetic data were identified.[146] Similarly, ionic liquid-like phosphines modified with pyridinium, pyrrolidinium and imidazolium groups and weakly coordinating anions have also been used to immobilise rhodium-hydroformylation catalysts on the basis that these groups are architecturally similar to the cations of the most common ionic liquids. In each case the catalyst was efficiently immobilised in the ionic liquid, but activity and regioselectivity were similar to the corresponding triphenylphosphine system. Interestingly, tetrabutyl ammonium salts of sulfonated triarylphosphites are hydrolytically stable in ionic liquid and afford higher $n : i$ ratio than their tppms counterparts.[147,148]

Several thermoregulated ionic liquid biphasic systems that are homogeneous at the temperature of reaction and biphasic at low temperature have been developed; the majority of these comprise imidazolium-based ionic liquids.[51,52,149] On the basis that quaternary ammonium ionic liquids are less toxic and safer than their imidazolium counterparts, Jiang developed a series of quaternary ammonium alkane sulfonates $[Et_3N(OCH_2CH_2)_nMe]$ (IL_{PEGX}, X = 350, 550, 750) that form a thermoregulated biphasic ternary system with *n*-pentane and toluene and applied it to the Rh/tppts-catalysed hydroformylation of 1-decene. A mixture of IL_{PEG750} containing Rh/tppts and heptane–toluene containing 1-decene is homogeneous at 108 °C and biphasic at room temperature which allowed the product to be separated and the catalyst phase to be recovered and reused. Optimisation studies led to high yields of aldehyde using 0.6 mol% Rh/tppts at 110 °C under 5 MPa CO–H_2 (1 : 1) after 5 h, corresponding to a TOF of 295 h^{-1}. The system recycled with exceptional efficiency and no loss of activity over eight runs, provided additional heptane was added to maintain the T_m of the system; Rh leaching was <1% in each recycle.[53]

3.2.2.2 Continuous Flow Hydroformylation using Ionic Liquid–$scCO_2$

Cole-Hamilton and others exploited the gas–liquid properties of $scCO_2$ to develop the first system for the continuous flow biphasic hydroformylation of low volatility olefins.[150] This process operates by dissolving an ionic catalyst in an ionic liquid and using the $scCO_2$ as the transport vector for the substrate, gaseous reactants and the aldehyde product.[58] The success of this approach relies on the high solubility of CO_2 in ionic liquids, the capacity for $scCO_2$ to extract the product from an ionic liquid without removing any ionic liquid and the liquid-like ability of $scCO_2$ to dissolve organic compounds of low to medium polarity. The $scCO_2$–IL biphasic system enabled continuous operation and unlike conventional liquid–liquid biphasic systems allowed separation of the product from the catalyst and reaction solvent to be conducted under a single set of conditions. Moreover, decompression of the gaseous mixture downstream gave product free from reaction solvent and catalyst thereby eliminating the need for further purification by distillation. The use of triphenylphosphite for the rhodium-catalysed hydroformylation of 1-hexene in $[C_4mim][PF_6]$–$scCO_2$ under semi-continuous operation resulted in a gradual decrease in selectivity and a severe drop in activity after the third cycle due to hydrolysis of $[PF_6]^-$ and subsequent attack of the released HF on the phosphite ligand. While catalyst generated from the sodium salt of triphenylphosphine monosulfonate gave low yields as a result of poor solubility in the ionic liquid, exchange of the sodium cation for

$[C_3mim]^+$ resulted in a significant improvement in reactivity and commercially acceptable rates (TOF > 500 h^{-1}) were achieved under continuous flow using Rh/$[C_3mim][Ph_2P(3\text{-}C_6H_4SO_3)]$ in $[C_8mim][NTf_2]$–*sc*CO_2 at 200 bar CO_2 and 100 °C and a high 1-octene flow rate (0.4 mL min^{-1}). Under certain process conditions, the ionic liquid–catalyst mixture exhibited good long term operational stability and gave an average mass balance of 88% and a total production of 330 mL of aldehyde from 12 mL of a 0.015 M solution of catalyst over 41.6 h. The presence of adventitious oxygen resulted in formation of phosphine oxide together with $[Rh_4(CO)_{12}]$ which forms an efficient catalyst for alkene isomerisation as well as hydroformylation, but with poor selectivity for the linear product; furthermore the high solubility of this complex in *sc*CO_2 resulted in increased levels of leaching which was more problematic in systems that used low grade CO_2. Rhodium leaching into the product stream was reduced to 0.012 ppm at high CO–H_2 partial pressures under steady state operation which corresponds to 1 g of rhodium in approximately 40 tonnes of product. A comparison with commercial hydroformylation processes revealed a number of favourable features, however, the selectivity of 75% was slightly lower than required for commercialisation and a high pressure was required, although this is comparable to that used in the unmodified cobalt system. Rates varied with the length of the imidazolium alkyl group and within a series of ionic liquids the lowest rates were obtained in $[C_4mim][PF_6]$ due to mass transfer limitations associated with the poor solubility of the substrate in the ionic liquid. Rates increased quite dramatically as the length of the alkyl chain on the imidazolium cation increased, consistent with an increase in the solubility of alkene in more hydrophobic ionic liquids.[151] Although the above study demonstrated that it was possible to achieve commercially acceptable rates, good long term stability and very low levels of rhodium leaching under continuous flow operation for up to 72 h, the $n:i$ ratios of 2–3 were unacceptably low. Taking a lead from the work of van Leeuwen and others,[143,144] nixantphos **25** was modified in a straightforward two-step procedure involving deprotonation of **23**, reaction of the amide with 1-bromo-3-chloropropane followed by quaternisation of the resulting *N*-(chloropropyl)nixantphos **24** with *N*-methylimidazole to afford imidazolium-tagged **25** (Scheme 3.2.4). Batch reactions for the hydroformylation of 1-octene run at 40 bar CO–H_2 in the presence of CO_2

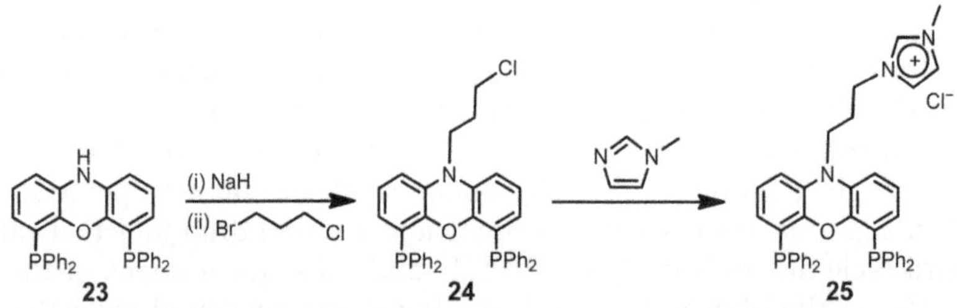

Scheme 3.2.4 Route for the synthesis of imidazolium-modified ligand **25** from nixantphos **23**.

(200 bar) gave high selectivity for linear aldehyde, with an *n* : *i* ratio of 37, very low isomerisation activity and minimal rhodium leaching (170–220 ppb). Continuous flow operation of Rh/**25** in $[C_8mim][NTf_2]$-*sc*CO_2 gave a constant TOF of 272 h^{-1} for the first eight hours, after which it decreased quite substantially; during this time the *n* : *i* ratio also remained constant at 40 : 1 and the degree of isomerisation was only *ca.* 2%. After 12 h on line the *n* : *i* ratio dropped dramatically and the isomerisation activity increased; these changes appear to be associated with complete oxidation of the phosphine and catalysis by an unliganded rhodium species. In this regard, unliganded rhodium is (i) highly active for hydroformylation of olefins in *sc*CO_2, (ii) easily extracted which explains the level of rhodium leaching, and (iii) gives poor l : b ratios with high isomerisation activity. The introduction of $[C_3mim]$[tppms] as a sacrificial oxygen scavenger to ameliorate the problem associated with oxidation gave a constant rate over 30 h of continuous operation but with a reduced *n* : *i* ratio of 16 which eventually dropped to six.[59,152]

3.2.2.3 *Supported Ionic Liquid Phase Hydroformylation*

The concept of supported ionic liquid phase catalysis evolved to address the drawbacks associated with the amount of expensive ionic liquid required for liquid–liquid biphasic catalysis and the often intrinsically high viscosity which leads to slow mass transport which becomes problematic when the reaction is fast. SILP materials are based on a thin film of ionic liquid–catalyst mixture physiosorbed onto a high surface area porous support (section 3.1.5.4) which minimises diffusion problems as the film is typically within the nanometre range; thus in such a configuration all of the catalyst is involved in the reaction. The negligible vapour pressure, large liquid range and high thermal stability of ionic liquids ensures that the solvent is retained on the support in its fluid state at common reaction temperatures which renders SILP catalysis ideally suited to continuous processing. To this end, the SILP concept has been successfully applied to the continuous gas-phase hydroformylation of propene and 1-butene using a Rh/sulfoxantphos catalyst; the former achieved a TOF of 501 h^{-1} at 140 °C while the latter reached 564 h^{-1} at 120 °C and both systems showed good long term stability. A slight deactivation of the former system over time was not due to catalyst decomposition as there was no change in selectivity; this was more likely to be due to accumulation of high boiling by-products such as 2-ethyl-hexenal and 2-ethyl-hexanol which dissolve in the ionic liquid layer and lowers the effective concentration of the catalyst. However, unfortunaetly Rh-sulfoxantphos is not capable of converting industrial mixed technical C4-feedstock comprising 1-butene and 2-butenes into linear pentenal *via* tandem isomerisation–hydroformylation.[153–155] Reasoning that rhodium-phosphite complexes are known to catalyse selective isomerisation–hydroformylation and convert internal alkenes to linear aldehydes,[156] Wasserscheid and co-workers employed the elaborate biaryl phosphite **26** in a rhodium-based

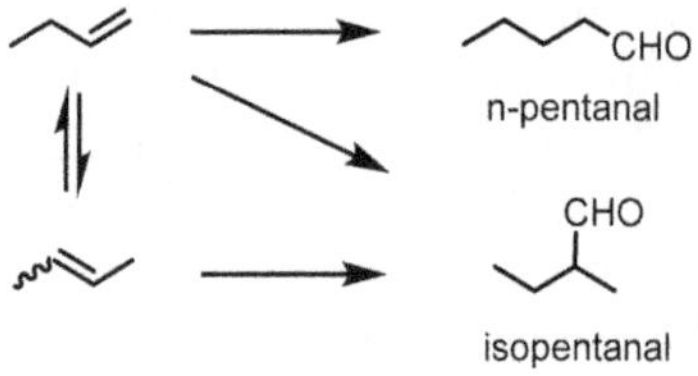

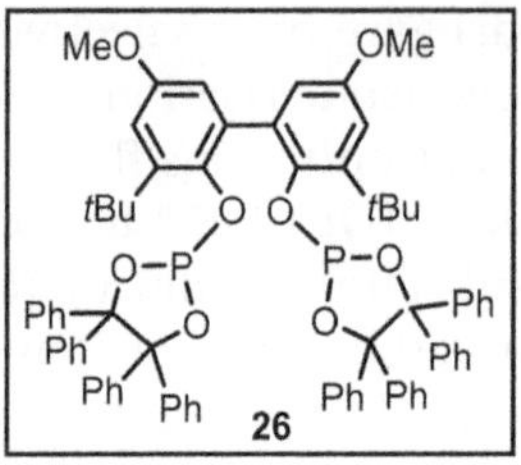

Scheme 3.2.5 Left: possible reaction pathways for hydroformylation of mixed C4 feed to yield predominantly *n*-pentanal; Right: the novel diphosphite ligand **26**.

SILP catalyst for the continuous gas-phase hydroformylation of industrial mixed C4-feedstock (raffinate 1) and obtained exceptional regioselectivity for *n*-pentanal (99.5%); this is quite remarkable given the complex composition of the feedstock which comprises 43.1% isobutene, 25.6% 1-butene, 9.1% *trans*-2-butene, 7.0% *cis*-2-butene, 14.9% butanes (Scheme 3.2.5).[157] Thus, $Rh(acac)(CO)_2$/**26** SILP is an efficient isomerisation catalyst that adds CO–H_2 to 1-butene exclusively but leaves 2-butenes and isobutene untouched. Good activity (TOF = 330 h^{-1}) was obtained at low conversion but the system was sensitive to hydrolysis and rapidly deactivated. The use of a well-dried feed gas and addition of *bis*(2,2,6,6-tetramethyl-4-piperidyl)sebacate to the supported catalyst as acid scavenger improved the stability of the system such that it operated for 800 h and maintained the same initial conversion of 25% with no loss in selectivity; this corresponds to an average activity of 410 h^{-1} and an accumulative TON of 350 000. Catalyst performance was further optimised at 120 °C and a total pressure of 25 bar to give a TOF of 3600 h^{-1} and a space-time yield of 850 kg *n*-pentanal per cubic meter per hour which could be maintained for 10 h with no loss in activity. In contrast to the vast array of ligands that have been modified with an ionic tag to effect retention of the catalyst in the ionic liquid phase, this system uses a neutral phosphite and exhibits good long term stability with no rhodium leaching in a continuous gas-phase SILP process.

A system for the continuous flow hydroformylation of 1-octene based on a supported ionic liquid phase catalyst prepared from $Rh(acac)(CO)_2$/ $[C_3mim][Ph_2P(3\text{-}C_6H_4SO_3)]$ immobilised in $[C_8mim][NTf_2]$ and supported on microporous silica has been developed using compressed CO_2 to transport the substrate, reacting gases and product over the catalyst bed.[158] The reaction rate was not influenced by the thickness of the ionic liquid film or the syn gas : substrate ratio. The phase behaviour had a profound influence on the reaction rate, which increased with increasing pressure up to 100 bar, but then decreased as the pressure was further increased. Rates were low at low CO_2 pressure due to poor gas diffusion to the catalytic sites in the SILP while the optimum rate was obtained in a CO_2-expanded liquid phase due to enhanced gas solubility and reduced interfacial tension and viscosity, which enables better transfer of gases to the catalyst. At high pressure gas availability is no longer rate limiting and the reaction rate decreases as the pressure increases because the flowing phase becomes a better solvent for the substrate reducing its partitioning into the ionic liquid and thus its availability to the active catalyst. Under optimum conditions the catalyst was stable for 40 h of uninterrupted operation and gave steady production of aldehyde at a TOF of 500 h^{-1} and 42% conversion of 1-octene.

3.2.2.4 Spectroscopic Studies

Even though interface processes such as substrate diffusion into the catalyst phase, the reaction rate at the phase boundary and product diffusion back to the organic phase are critical to the performance of a multiphase catalytic system, there have been relatively few experimental studies to investigate the chemical nature of the liquid surface of such systems. In this regard, the surface composition of ionic catalyst solutions containing $[Rh(acac)(CO)_2]$ and *tris*(3-sodium sulfonatophenyl)phosphine (tppts) as applied in ionic liquid biphasic hydroformylation has been studied by angle-resolved X-ray photoelectron spectroscopy (ARXPS), a technique that allows the surface of an ionic liquid to be probed at different depths.[159] Studies on a saturated solution of $[Rh(CO)_2(acac)]$ in $[C_4mim][C_8OSO_3]$ in the absence of tppts indicated that the top layer of the ionic liquid was virtually free of rhodium; this was attributed to the long alkyl chains of the octylsulfate groups forming a highly apolar surface region which largely excludes the rhodium. A similar study on a tppts-saturated solution of $[C_2mim][C_2OSO_3]$ revealed that the surface was dominated by tppts in an orientated manner such that the nonpolar phenyl groups were directed towards the vacuum and the P atoms and the *meta*-SO_3^- groups were directed towards the bulk of the ionic liquid; the outer surface was also found to be depleted in imidazolium ions due to the enrichment and preferential orientation of the tppts ligand. ARXPS spectra of the rhodium-containing saturated solution of tppts revealed surface enrichment of the *in situ* formed rhodium complex, again with a similar orientation of the *meta*-SO_3^- and P atom towards the bulk of the ionic liquid. Thus, in terms of surface composition, this catalyst solution has a significant and even slightly preferred presence of the Rh/tppts complex in the top 1–1.5 nm of the catalytic ionic liquid. This study demonstrated that the nature of the applied ligand influences the preferred position of the catalyst complex in either the bulk phase or at the interface of a multiphase catalytic system and provided spectroscopic support for earlier molecular dynamics studies which claimed significant interface activity of Rh/tppts complexes in organic–ionic liquid biphasic systems.[160,161] Such studies could ultimately be used to maximise the catalyst concentration at the phase boundary and thus optimise reactions, (*e.g.* reduce mass transfer limitations) by manipulating the surface activity of transition metal complexes in multiphase catalytic systems.

The activity, selectivity and stability of rhodium-based SILP catalysts for the hydroformylation of propene were shown to be highly sensitive to the ligand composition, ligand/Rh ratio, ionic liquid composition, ionic liquid loading and temperature of the silica pre-treatment.[162] Preliminary optimisation studies established that the highest activity is achieved using sulfoxantphos as the ligand and $[C_4mim][C_8OSO_3]$ as the ionic liquid immobilised on silica pre-treated at 100 °C with a Rh/sulfoxantphos ratio of 0.1 and an ionic liquid loading corresponding to $\alpha = 0.2$ (where α is defined as the ratio between the volume of ionic liquid used and the pore volume of the support). *In situ* FT-IR and solid state ^{31}P and ^{29}Si MAS NMR spectroscopic studies suggest that the active form of the SILP catalyst $[HRh(CO)_2(sulfoxantphos)]$ is stabilised by the support through interaction of the sulfonate groups of the sulfoxantphos with the surface silanol groups *via* modes A and B shown in Figure 3.2.2. Silanol groups on the surface of the support are required to achieve high activity and stability as a reduction in the surface concentration of silanol groups by dehydration led to a severe reduction in activity as the fraction of the sulfoxantphos ligands interacting with the support through the phosphino group increased which rendered them unavailable for coordination with the rhodium cation. At low sulfoxantphos/Rh ratios (*ca.* 3) a high fraction of the ligand binds to the surface through the phosphino group and the concentration of active complex stabilised by the support through interactions of the sulfonate groups of the sulfoxantphos ligand is low. Above the optimum sulfoxantphos/Rh ratio of 10, complexes of the type $[HRh(sulfoxantphos)_2]$ form which are less active propene hydroformylation catalysts. A high loading of $[C_4mim][C_8OSO_3]$ was also shown to reduce activity and stability as the ionic liquid can interact with the surface silanol groups and interfere with or inhibit formation of species B resulting in the formation of $[Rh(CO)(\mu\text{-}CO)(sulfoxantphos)]_2$ which is inactive for hydroformylation.

The rhodium-catalysed hydroformylation of 1-octene in novel *N*-alkyl caprolactam-based ionic liquid crystals $[C_n\text{-}CP][X]$ ($n = 12$, 16, 18, $X = MeSO_4^-$, *p*-TSA) and $[C_{16}mim][X]$ ($X = MeSO_4^-$, BF_4^-) has been compared because the regiochemistry and chemoselectivity of organic reactions in liquid crystalline solvents are often different from those in conventional organic solvents and in common ionic liquids.[163] For reactions conducted in $[C_n\text{-}CP][p\text{-}TSA]$, regioselectivity increased with an increase in alkyl chain

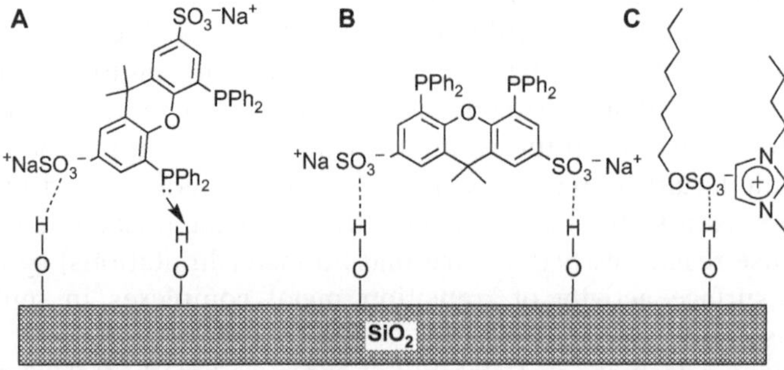

Figure 3.2.2 Possible interactions of sulfoxantphos (**A** and **B**) and $[C_4mim][C_8OSO_3]$ (**C**) on the surface of silica.

length from 1.2 ($n = 6$) to 2.9 ($n = 18$). The anion had a profound effect on regioselectivity and the *n*/*iso* ratio dropped dramatically to 0.3 in [C_{16}-CP][$MeSO_4$]. Regioselectivities were higher in imidazolium-based ionic liquids as [C_{16}mim][OMs] gave an *n*/*iso* ratio of 3.7 and a TOF of 74 h^{-1} compared with an *n*/*iso* ratio of 0.3 and a TOF of 285 h^{-1} in [C_{16}-CP][OMs]. Although *n*/*iso* ratios in caprolactam-based ionic liquid crystals were lower than in imidazolium-based systems, conversions were higher in the former. A survey of reaction parameters revealed that temperature and pressure affected the *n*/*iso* ratio and TOF in caprolactam and imidazolium-based ionic liquids in quite different ways. In the case of caprolactam-based ionic liquid crystals, high pressures favour formation of linear aldehyde in the mesophase (105 °C) whereas at higher temperatures (135 °C) in the isotropic phase a lower pressure favours the branched product. In contrast, in [C_{16}mim][$MeSO_4$] the highest *n*/*iso* ratios were obtained at low pressures in the mesophase.[164]

The activity of the [$HRh(CO)(PPh_3)_3$]-catalysed hydroformylation of ethylene in imidazolium-based ionic liquids decreased with increasing length of the imidazolium alkyl chain and the TOF of 10 627 h^{-1} obtained in the optimum system Rh/[C_4mim][BF_4] is higher than that of 9856 h^{-1} obtained in toluene under the same conditions. Within a series of [C_4mim]$^+$-based ionic liquids activity depended markedly on the anion and a slightly lower TOF was obtained in [C_4mim][BF_4] while the catalyst was completely inactive in their [OTs]$^-$, [OAc]$^-$, [SCN]$^-$ and [HSO_4]$^-$ based counterparts. ESI-MS analysis of the hydroformylation mixtures after the reaction identified an equilibrium between [$HRh(PPh_3)_2(CO)$], [$HRh(PPh_3)_3$] and [$HRh(PPh_3)_2$] in toluene whereas the major rhodium-containing species in [C_4mim][BF_4] was identified as the *N*-heterocyclic carbene complex [$Rh(CO)(PPh_3)_2$(1-butyl-3-methylimidazolylidene)]$^+$ which the authors claimed to be a new stable and active catalyst for hydroformylation which does not form inactive clusters; however, identification by ESI-MS alone is not sufficient evidence for its role as a catalyst. The Rh/[C_4mim][BF_4] system recycled six times with no loss in activity and ESI-MS identified the same rhodium complex after the sixth run.[165]

3.2.3 Transition Metal-Catalysed Cross-Coupling in Ionic Liquid

3.2.3.1 The Heck Reaction

The Heck reaction has evolved into a powerful tool for C–C bond formation that has been applied to the synthesis of a wide variety of useful intermediates and products. While ionic liquids are perhaps an obvious choice as solvent for the Heck reaction on the basis that they have tunable polarity and are high boiling, their use has also enhanced catalyst stability, improved activity and regioselectivity, altered reaction pathways and enabled straightforward catalyst separation and recycling protocols to be developed. There is an expanse of literature covering this area as well as several useful and instructive review articles to which the interested reader is directed.[166,167] To this end, this section will not be exhaustive in terms of the literature but will draw upon a variety of articles to emphasise historical studies, useful principles, observations and practises and state-of-the-art discoveries.

3.2.3.1.1 Early Developments and the Influence of the Ionic Liquid Composition. The first well-documented use of an ionic liquid as the reaction medium for the Heck reaction was a ligand-free system based on either $Pd(OAc)_2$ or $PdCl_2$ in $[P_{4,4,4,16}]Br$ which coupled *trans* cinnamic acid butyl-ester with bromobenzene. The catalyst remained in the ionic liquid melt after extraction of the product and could be used for two additional cycles.[168] Reasoning that Heck reactions are typically conducted in polar solvents at high temperature and that salt additives can activate and stabilise the catalyst, non-aqueous ionic liquids (NAILs) were shown to be suitable solvents for the reaction between aryl bromides and styrene catalysed by palladacycle **27** in $[NBu_4]Br$ with NaOAc as base; yields compared favourably with those obtained in DMAc or DMF and TONs of 1 000 000 were obtained with activated aryl bromides. The use of NAILs for the Heck reaction of deactivated aryl chlorides gave a significant enhancement in activity compared with the same reaction conducted in molecular solvents.[169] A subsequent study demonstrated that halide-based ionic liquids improve the efficiency of a wide range of catalysts for the Heck arylation between styrenes and aryl chlorides compared with their performance in conventional organic solvents. A comparison of solvent effects revealed that a coordinating halide anion was necessary to achieve high yields, as salts with non-coordinating anions were poor solvents as were those based on the imidazolium cation, and that NBu_4Br gave both higher rates and longer term stability than DMF for a range of precatalysts and palladium precursor-ligand combinations. A survey of substrates also revealed that catalyst efficiency depends on steric and electronic factors. The authors considered several potential explanations for the observed effects of NAILs including (i) stabilisation of palladium colloids generated from ligand-free Pd(II) salts (ii) an *anion effect* that generates an active $[BrPd(ligand)]^-$ anion and (iii) a possible Pd(II)/Pd(IV) catalytic cycle but concluded that each catalyst should be considered individually. A recycle experiment using 1 mol% of palladacycle precatalyst **27** and $[NBu_4][OAc]$ as base for the arylation of styrene with bromobenzene gave near quantitative yields across 12 runs.[170]

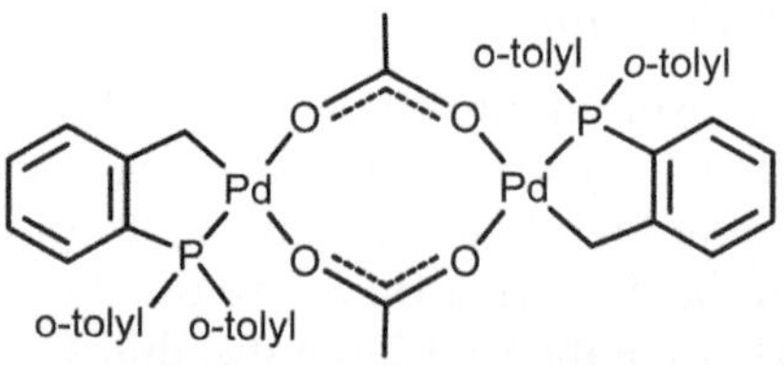

o-tolyl o-tolyl
O O P
Pd Pd
P O O
o-tolyl o-tolyl

27

The effect of ionic liquid and phosphine additive on the palladium-catalysed Heck reaction between iodobenzene and ethyl acrylate revealed a marked anion effect as reactions in ionic liquids with coordinating anions such as [C_6pyr]Cl were more efficient than those in [C_6pyr][X] ($X = BF_4^-$, PF_6^-); an excellent yield was obtained in the former after 24 h at 40 °C whereas the latter required high temperatures and longer reaction times to achieve comparable conversions. Although the addition of phosphine to [C_6pyr]Cl/Pd$(OAc)_2$ decreased the activity, near quantitative conversions could still be obtained at higher temperatures. In contrast, a study of ligand effects on the more challenging reaction between 4-bromoanisole and ethyl acrylate in [C_4mim][PF_6] revealed that addition of triphenylphosphine and tri(*o*-tolyl)phosphine gave a marked enhancement in catalyst activity at 100 °C compared with the ligand-free system under similar conditions. The difference in activity between reactions in pyridinium and imidazolium-based ionic liquids was tentatively attributed to formation of an imidazolylidene-palladium carbene complex *via* activation of the C2-H proton while the effect of added phosphine on activity was considered to reflect a different mechanistic pathway.[171] Moderate to good yields have also been obtained for the Heck arylation of a range of allylic alcohols in molten [Bu_4N]Br at 80–120 °C under ligand-free conditions using 10 mol% $PdCl_2$ and sodium hydrogen carbonate as base. Even though recycle experiments led to a dramatic decrease in activity, the authors claimed that the ammonium salt stabilised the active catalyst.[172] A molten mixture of [Bu_4N][OAc]–[*n*-Bu_4N]Br (2 : 15) is a highly effective combination solvent for the palladium-catalysed Heck arylation of methyl cinnamate with a variety of electron-rich and electron-poor electrophiles [Equation (3.2.9), R = Me]; the corresponding diaryl acrylates were obtained in good to excellent yield and with high *E*-selectivity. Under the same conditions the arylation of methyl-3-acrylates occurred with high *Z*-selectivity (*E*/*Z*, 4 : 96 to 1 : 99).[173] The benzothiazole-derived carbene [bis(2,3-dihydro-3-methylbenzothiazole-2-ylidene)PdI_2] also forms a highly efficient catalyst for the Heck arylation of butyl acrylate with aryl bromides in molten tetrabutylammonium bromide (TBAB). Good yields were obtained with 1 mol% Pd at 130 °C using sodium formate as the reducing agent and sodium bicarbonate as base to limit Hofmann elimination of the tetrabutylammonium salt. The same conditions were also highly effective for the Heck arylation of *trans*-cinnamates with electron-rich and electron-poor aryl bromides [Equation (3.2.9)] as well as the double Heck arylation of butyl acrylate. For each reaction, consistently better results were obtained in TBAB than in conventional organic solvents such as DMF or DMA. The authors tentatively attributed the high efficiency of this system to small TBAB-stabilised palladium nanoparticles, as reported in the pioneering studies of Reetz,[174] and provided tentative evidence that the poorly solvated bromide anion of TBAB acts as a nucleophile to form the activated anionic 16-electron palladium complex $[L_2PdBr]^-[NR_4]^+$ and reasoned that the large cation impeded aggregation of the palladium by imposing a Coulombic barrier to collision as well as electrostatically assisting in the removal of the bromide from the derived oxidative addition adduct to facilitate olefin coordination and insertion.[175,176]

Ph–CH=CH–CO_2R + Ar-X $\xrightarrow[\text{TBAB}]{\text{Pd catalyst}}$ Ph(Ar)C=CH–CO_2Et + Ph–CH=C(Ar)–CO_2Et

R = Me, Et 3-5%

(3.2.9)

Highly efficient Heck reactions between butyl vinyl ether with a range of bromides have been conducted in $[C_4mim][PF_6]$ under microwave heating. Reactions were highly regioselective for β-arylation and good yields of the corresponding (*E*)-cinnamate esters were obtained after short reaction times (5–45 min) at temperatures between 180 °C and 220 °C with 4 mol% $PdCl_2/P(o\text{-}tolyl)_3$ as catalyst. Interestingly the same article reported that the $Pd(OAc)_2$/dppp-catalysed arylation of butyl vinyl ether with 2-bromonaphthalene in $[C_4mim][PF_6]$ under microwave heating occurred with high regioselectivity (>99 : 1) for the α-arylated product.[177] Bis(imidazoles) are also highly efficient ligands for the palladium-catalysed Heck reaction between *n*-butyl acrylate and aryl iodides in $[C_4mim][PF_6]$ as near quantitative yields of (*E*)-cinnamate ester were obtained after heating at 120 °C for only 1 h and the catalyst recycled five times with no loss in activity.[178] High α-selectivity (>9 : 1) was obtained for the arylation of butyl vinyl ether with 4-iodoanisole, which compared favourably to that reported later by Xiao and discussed in section 3.2.3.1.2.

In situ XAFS investigations have shown that palladium clusters 0.8–1.6 nm in diameter are the main species present during the Heck reaction in $[C_4mim][X]$ and $[C_4pyr][X]$ ($X = PF_6^-$, BF_4^-) and $[N_{6,2,2,2}][NTf_2]$, both in the absence and presence of PPh_3 as well as reagents. Changing the anion to chloride prevented the formation of metal and in $[C_6mim]Cl$ data was consistent with formation of a bis(carbene) complex *via* activation of the C(2)-H bond of the imidazolium whereas $[PdCl_4]^{2-}$ was formed in $[C_4pyr]Cl$ and $[C_6C_1{}^2mim]Cl$. An induction period for reactions conducted in $[C_4mim][PF_6]$ below 70 °C was attributed to the gradual formation of palladium metal; this induction period decreased in the absence of reagents as well as at higher temperatures but increased in the presence of PPh_3.[179]

3.2.3.1.2 Evidence for Catalysis by *N*-Heterocyclic Carbene Complexes. Early definitive evidence for the generation of an active palladium *N*-heterocyclic carbene species during Heck reactions conducted in imidazolium-based ionic liquids was provided by Xiao and co-workers.[180] Having shown that the Heck reaction between aryl halides and acrylates or styrene proceeds more efficiently in $[C_4mim]Br$ than in its tetrafluoroborate counterpart, the nature of the palladium species was investigated. A mixture of dimeric $[PdBr(\mu\text{-}Br)(bmiy)]_2$ (**28**) and monomeric $[PdBr_2(bmiy)_2]$ (**29**) complexes was formed by heating a $[C_4mim]Br$ solution of $Pd(OAc)_2$ in the absence of substrate under conditions similar to those used for the catalysis. Prolonged reaction times transformed **28** into *trans*-**29** which subsequently isomerised to *cis*-**29**, unequivocally demonstrating that **28** is an intermediate. Both complexes were identified to be present in the catalytic Heck reaction mixture.

Isolated *trans*-[$PdBr_2$(bmiy)$_2$] was shown to form an active catalyst for the olefination of iodobenzene and 4-bromobenzaldehyde by acrylates in [C_4mim]Br, although yields were slightly lower than those obtained with Pd(OAc)$_2$–[C_4mim]Br. Interestingly, and in contrast, the same isolated bis(carbene) complex only gave poor conversions in [C_4mim][BF_4], which appeared to be associated with transformation into a less active species by the solvent.[180]

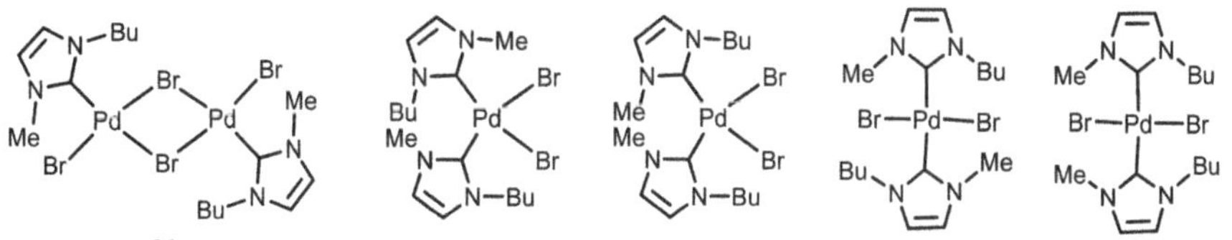

28 *cis-anti*-**29** *cis-syn*-**29** *trans-anti*-**29** *trans-syn*-**29**

Shreeve and co-workers have also provided convincing evidence that the active catalyst for the Heck reaction in bis(imidazolium)-based ionic liquids is a bis(carbene)-palladium species as the yields obtained from the reaction between butyl acrylate and aryl halides in [1,1′-dibutyl-3,3′-methylenebis-(imidazolium)][NTf_2]$_2$ using 1 mol% of the derived palladium carbene precatalyst [1,1′-methylene-3,3′-dibutylbis(imidazolin-2,2′-diylidene]palladium dichloride **30** matched those obtained with $PdCl_2$ in the same ionic liquids.[181]

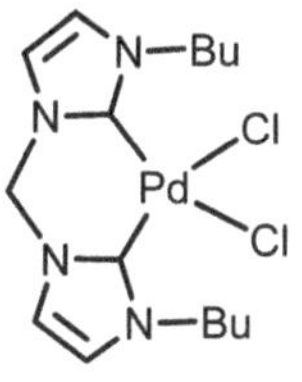

30

While rate enhancements for catalytic transformations conducted in ionic liquids compared with conventional solvents have often been accounted for by formation of a palladium-carbene-based catalyst, reaction profiles for the Mizoroki–Heck reaction between *n*-butyl acrylate and bromobenzene in [C_4mim][NTf_2] and DMF using 5 mol% *trans*-[$PdCl_2$(IBuMe)$_2$], [$PdCl_2$(MeCN)$_2$] and [$PdCl_2$(MeCN)$_2$]–AgCl(IBuMe) showed that for all precatalysts the rate of reaction in ionic liquid was greater than that in DMF. The latter two systems both performed markedly better in ionic liquid compared with DMF and *trans*-[$PdCl_2$(IBuMe)$_2$] gave much higher conversions in [C_4mim][NTf_2] compared with DMF in the same time (Table 3.2.1). The authors concluded that the increase in reaction rate could not be solely attributed to the formation of a carbene but is likely to be due to the nature of the solvent and/or stabilisation of a [NTf_2]$^-$ complex.[182]

3.2.3.1.3 Control of Regioselectivity and Chemoselectivity. High regioselectivity (>99 : 1) for the α-arylation of electron-rich olefins has been

Table 3.2.1 Extent of the Heck reaction between bromobenzene and n-butyl acrylate.

Precatalyst	*[C_4mim][NTf_2] (30 min)*	*DMF (30 min)*
trans-$PdCl_2(IBuMe)_2$	$(99 \pm 7)\%$	$(20 \pm 2)\%$
$PdCl_2(MeCN)_2$	$(27 \pm 5)\%$	$(7 \pm 5)\%$
$PdCl_2(MeCN)_2$ + [AgCl(IBuMe)]	$(99 \pm 7)\%$	$(13 \pm 2)\%$

Table 3.2.2 Solvent effect on the Heck arylation between n-butyl vinyl ether and 4-bromobenzaldehyde.[a]

Solvent	*Conversion*	*α/β*	*E/Z*
[C_4mim][PF_6]	100	>99 : 1	—
Toluene	18	47 : 53	68 : 32
Dioxane	26	35 : 65	82 : 18
Acetonitrile	33	45 : 55	63 : 37
DMAc	98	24 : 76	72 : 26
DMF	100	47 : 53	80 : 20
DMSO	100	86 : 14	79 : 22

[a]Adapted with permission from reference 185. Copyright 2001 American Chemical Society.

obtained by using a Pd/diphosphine-based catalyst and either an aryl triflate or an aryl halide in combination with a stoichiometric quantity of a silver or thallium salt as halide scavenger. However, the use of aryl triflates and/or the need for an additive are significant drawbacks. As the Heck reaction is generally accepted to occur either *via* a neutral pathway that leads mainly to the linear β-substituted olefin or *via* an ionic pathway that yields the branched or α-product as the predominant isomer, Xiao and co-workers reasoned that the use of an ionic liquid as solvent for the Heck reaction would favour the ionic pathway and give selective α-arylation by stabilising the catonic palladium-olefin intermediate.[184] To this end, early studies by Amatore and Jutand[40b,183] as well as Milstein[185] support the participation of ionic species in polar solvents. The Heck arylation of butyl vinyl ether with 4-bromobenzaldyde catalysed by 2.5 mol% $Pd(OAc)_2$/dppp demonstrated the stark and disparate difference in selectivity for reactions conducted in ionic liquid compared with conventional organic solvents, details of which are presented in Table 3.2.2. A comparison of the Heck reaction in [C_4mim][BF_4] and DMF emphasised the marked influence of the solvent on selectivity; the former gave >99 : 1 α/β selectivity while the latter gave a mixture of products with α/β ratios ranging from 46 : 54 to 69 : 31. Excellent levels of regiocontrol for the α-adduct were obtained for a wide range of electron-rich and electron-poor aryl bromides with a host of vinyl ethers, allyltrimethylsilane and enamides to afford α-arylated vinyl ethers and enamides, aryl ketones and

β-arylated allylic silanes, respectively. In addition to modifying the α/β selectivity the ionic liquid also stabilises the catalyst with respect to formation of palladium black. A survey of ligand effects revealed that activity and selectivity correlate with the bite angle; good conversions were achieved with wide bite angle phosphines while high α-selectivity required a medium bite angle phosphine and dppp ultimately proved to be the optimum compromise. Increasing the ligand bite angle was suggested to increase steric interactions between the inserting olefin and the phenyl rings at phosphorus which promoted α-selectivity by destabilising the transition state to the β-arylated product; however, if the bite angle was too large the ligand would inhibit rotation of the olefin to the in-plane position required for insertion which would reduce α-selectivity.[99,185]

A pronounced change in regioselectivity for the palladium-catalysed Heck reaction in $[C_4mim][BF_4]$ compared with conventional solvents has also been reported for the arylation of electron-rich 5-hexen-2-one with aryl bromides. Reactions conducted in $[C_4mim][BF_4]$ gave γ-arylated ketone **31** as the dominant product whereas (*Z*)- and (*E*)-δ-arylated γ,δ-unsaturated ketone **32** was the major product in toluene, dioxane, acetonitrile, DMF and DMSO [Equation (3.2.10)]. Rather interestingly, the regioselectivity of arylation in $[C_4mim][BF_4]$ was also highly sensitive to the nature of the ligand and replacement of dppp with PPh_3, dppe, BINAP, xantphos or dppb resulted in a dramatic switch in selectivity to afford the linear δ-arylated adduct **32**, predominantly as the *E*-isomer. The authors claimed that the high regioselectivity for α-arylation was also consistent with the ionic pathway and discussed the possibility of chelation-assisted substitution of bromide involving the carbonyl group in the substrate.[186]

+ Ar-Br —[$Pd(OAc)_2$/dppp, 115 °C, NH^iPr_2, solvent]→ **31** + **32**

(3.2.10)

The high regioselectivity for palladium-catalysed α-arylation in ionic liquids has been used to synthesise 5- and 7-membered cyclic ketals by arylating hydroxyalkyl vinyl ethers. In a typical reaction, $Pd(OAc)_2$/dppp in $[C_4mim][BF_4]$ gave moderate to good yields of the cyclic ketal with outstanding regioselectivity for α-arylation (>99 : 1) after 12–24 h at 115 °C, without the need for a halide scavenger. For comparison the same coupling in DMF requires a stoichiometric quantity of thallium acetate (as halide scavenger) to achieve high α-regioselectivity and reactions were much slower.[187] The first examples of highly regioselective internal α-arylation of unsaturated alcohols with aryl bromides were also achieved with the $Pd(OAc)_2$/dppp combination, again without the need for a halide scavenger [Equation (3.2.11)]. The arylation of a wide range of unsaturated alcohols occurred with exceptionally high regioselectivity (**33** : **34**; $n=0$, >99 : 1; $n=1 \sim 80$: 20, $n=2$, ~ 85 : 15) in $[C_4mim][BF_4]$/DMSO (4 mol% $Pd(OAc)_2$/

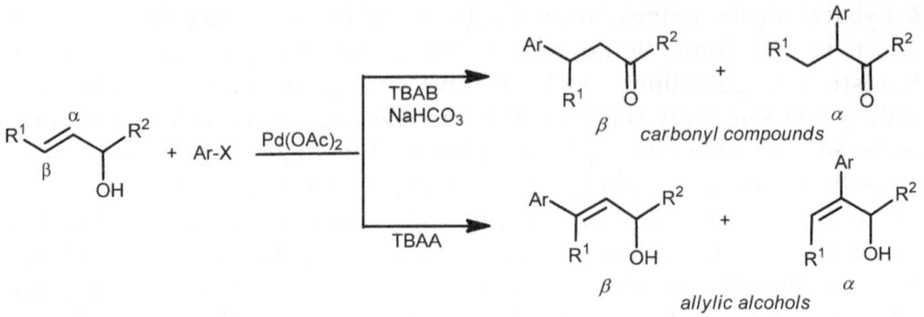

Scheme 3.2.6 labels: R^1, R^2, α, β, OH + Ar-X, Pd(OAc)$_2$; TBAB, NaHCO$_3$ → carbonyl compounds (β and α); TBAA → allylic alcohols (β and α)

Scheme 3.2.6 Effect of base on the selective formation of aromatic carbonyl compounds or aromatic conjugated alcohols in the Heck arylation of allylic alcohols.

dppp at 115 °C) to give good yields of the α-arylated alcohol **33** with no evidence for isomerisation to the carbonyl isomer.[188]

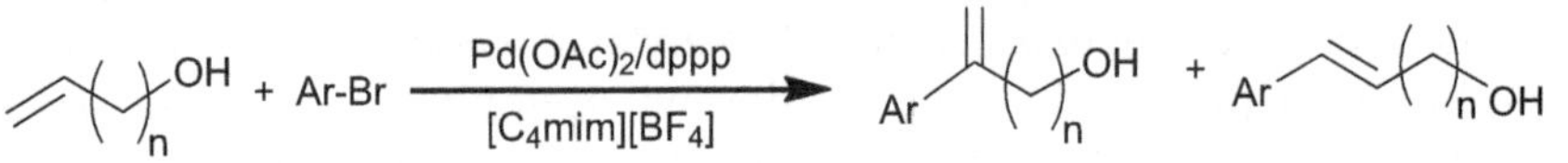

$$\text{alkenol} + \text{Ar-Br} \xrightarrow[\text{[C}_4\text{mim][BF}_4]]{\text{Pd(OAc)}_2\text{/dppp}} \mathbf{33} + \mathbf{34} \quad (3.2.11)$$

The selectivity of the Pd/nanocolloid-catalysed Heck arylation between allylic alcohols and aryl bromides and iodides depends quite markedly on the nature of the base as well as the ionic liquid such that either the *aromatic carbonyl product* or the *allylic alcohol* can be obtained with exceptional selectivity through a judicious choice of both (Scheme 3.2.6). The arylation of terminal alcohols in tetrabutylammonium bromide (TBAB) affords *β*-arylation as the predominant product in the form of the *arylated carbonyl isomer* for all bases tested with the exception of tetrabutylammonium acetate (TBAA) which was remarkably selective for the *β-arylated allylic alcohol*. The use of TBAA as both solvent and base results in much higher activity such that near quantitative conversions to the allylic alcohol are achieved at reaction temperatures as low as 60 °C; for comparison the same reactions conducted in organic solvent with TBAA as base were much slower and less selective. Two pathways were proposed to account for the marked difference in selectivity. In TBAB, pathway A is dominant which involves slow coordination of the allylic alcohol, sterically controlled migratory insertion to place the aryl group at the *β*-position and reversible *β*-hydride elimination to ultimately afford the thermodynamically favoured carbonyl product. In TBAA, the allylic alcohol coordinates to palladium only after displacement of the acetate to afford a cationic π-complex, the subsequent migratory insertion also favours the *β*-arylated product but the *β*-OH chelates to the cationic palladium centre rendering *β*-hydride elimination of the benzylic hydrogen to afford the allylic alcohol more favourable (Scheme 3.2.7).[189]

3.2.3.1.4 Catalysts Modified with Ionic Tags and Task-Specific Functional Groups. Shreeve and co-workers designed monoimidazolium salt **35** as both

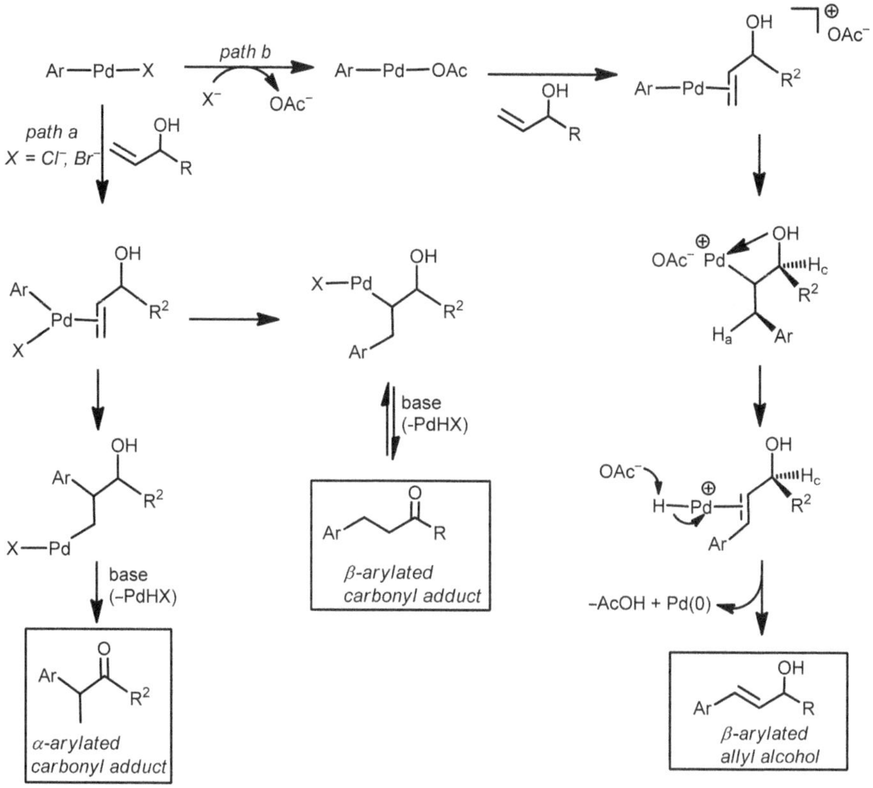

Scheme 3.2.7 Mechanism to account for the effect of base on the selectivity of the Heck arylation of allylic alcohols.

ionic liquid solvent and ligand for the palladium-catalysed Heck reaction on the basis that an ionic liquid-coordinated catalyst would be firmly immobilised in and retained by the ionic liquid during extraction of the product. Excellent yields of (*E*)-methyl cinnamate were obtained for the arylation of methyl acrylate with iodobenzene after heating at 100 °C for 4 h, using **35** as solvent, 2 mol% of the *trans*-[$PdCl_2$(**35**)] as precatalyst and Na_2CO_3 as base. The ionic liquid–catalyst mixture was recovered and recycled four times with no loss in activity and then subsequently used for two different substrate combinations both of which also recycled with remarkable efficiency.[190]

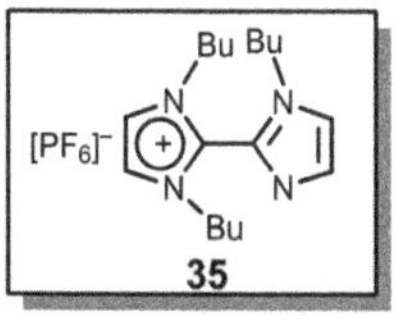

35

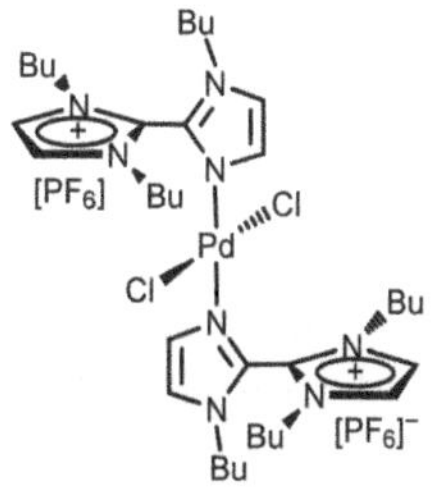

trans-$PdCl_2$(**35**)$_2$

Table 3.2.3 Recycle Heck reaction using pyrazolyl-functionalised 2-methylimidazolium-based ionic liquids **36a–b**.

		Cycle no.								
IL	*Catalyst*	*1*	*2*	*3*	*4*	*5*	*6*	*7*	*8*	*9*
36a	**37a**	>99	>99	>99	>99	>99	>99	>99	>99	>99
$[C_4C_1{}^2mim][NTf_2]$	**37a**	>99	>99	>99	>99	>99	>99	>99	>99	>99
$[C_4C_1{}^2mim][NTf_2]$	$PdCl_2$	>99	86	45	12	3				
36b	$PdCl_2$	>99	>99	>99	>99	>99	>99	>99	>99	>99

Similarly, Shreeve also prepared imidazolium-tagged pyrazoles **36a–b** that could act as both a ligand through the basic nitrogen atom of the pyrazole and as the ionic liquid solvent,[191] reasoning that the imidazolium tag would limit leaching of the catalyst and/or ligand and on the basis that pyrazolyl-based palladium(II) complexes form efficient catalysts.[192] Imidazolium-modified pyrazolyl complexes **37a–b** [Equation (3.2.12)] form highly active and stable catalysts for the Heck arylation of *n*-butyl acrylate with iodobenzene in **36** and $[C_4C_2{}^1mim][NTf_2]$ to give near quantitative yields of *n*-butyl (*E*)-cinnamate after 3 h at 110 °C with 2 mol% of **37**; both systems also recycled with remarkable efficiency over nine runs with no loss in activity. For comparison, 2 mol% of $PdCl_2$ in **36** gave similar yields and recycled with the same efficacy suggesting that the same catalyst is generated *in situ* whereas $PdCl_2/[C_4mim][NTf_2]$ rapidly deposited palladium black under the conditions of catalysis (Table 3.2.3). The same ionic liquid–precatalyst combination also gave good conversions in the Suzuki–Miyaura cross-coupling of aryl bromides with phenyl boronic acid in the presence of water at 100 °C; the system is remarkably stable and remains active after 14 recycles with no loss in activity.

(3.2.12)

A parallel study showed that palladium complexes of pyridyl-functionalised imidazolium-based ionic liquids **38** are also highly efficient catalysts for the arylation of *n*-butyl acrylate with iodobenzene; excellent yields (>99%) were obtained after 3 h at 120 °C and the system recycled with outstanding efficiency over 10 runs. Coordination of pyridine was proposed to prevent the formation of palladium black which occurred rapidly with $PdCl_2$/[1-butyl-2-phenyl-3-methylimidazolium][NTf_2] ($PdCl_2$/**39**) under the same conditions.[193]

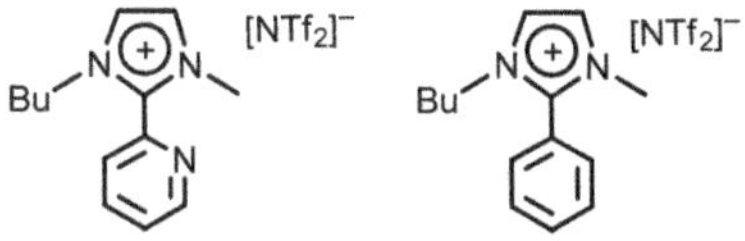

38 **39**

The ionic liquid-like hemilabile hybrid P,N-donor **40** combines with $PdCl_2$ to form an efficient and recyclable catalyst for the Heck reaction between ethyl acrylate and a range of aryl iodides in [C_4mim][BF_4]. Good conversions to the corresponding *E*-ethyl cinnamate were obtained after heating at 110 °C for 2 h whereas slightly lower conversions were obtained in DMF. Comparative recycle experiments demonstrated the beneficial effect on stability and activity of combining the phosphino-imidazolium salt and the piperidine as $PdCl_2$/**40** recycled seven times with no loss in activity or selectivity and no palladium leaching (>0.1 μg g^{-1}); for comparison $PdCl_2$/**42** gave a good conversion in the first run but gradually deactivated with formation of palladium black whereas $PdCl_2$/**41** was more stable but slightly less active than its PN counterpart.[194]

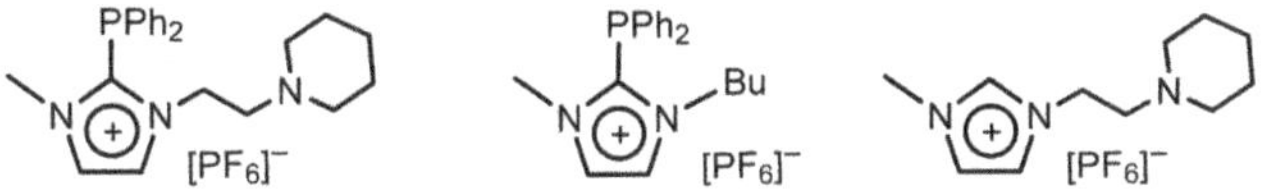

40 **41** **42**

Pyrazolyl ring-modified heterocyclic carbenes have been used for the palladium-catalysed Heck reaction in ionic liquids on the basis that the pyrazolyl ring is more weakly coordinating than a pyridyl ring and therefore the derived complexes should be hemilabile and highly efficient catalysts. To this end, the resulting hemilabile complex **43** forms a highly efficient and recyclable catalyst for the arylation of *n*-butyl acrylate with a host of electronically diverse aryl bromides and iodides in [C_4mim][PF_6], giving good yields of the corresponding *n*-butyl (*E*)-cinnamate in good yield after 8 h at 120 °C. The ionic liquid–catalyst mixture recycled five times with different substrate combinations and showed no loss in activity. The potential role and benefits of the mixed pyrazolyl-*N*-heterocyclic carbene were discussed.[195]

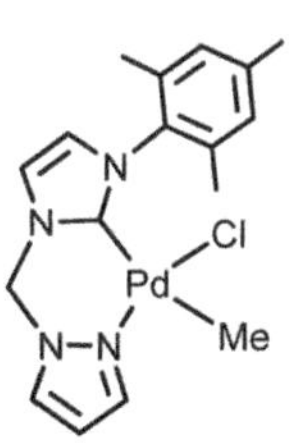

43

Hydrophilic halogen-free ionic liquids such as [$C_n$mim][p-$CH_3C_6H_4SO_3$] ($n = 1, 2, 4, 6, 8, 12$) are effective solvents for the $PdCl_2$/tppts-catalysed Heck arylation of styrene with aryl bromides and good conversions were obtained with 1 mol% $PdCl_2$ at 110 °C. A combination of TEM and mercury poisoning tests provided a strong indication that palladium nanoparticles (5–10 nm) were responsible for the catalysis. A survey of the ionic liquid revealed that yields increased with the length of the alkyl chain on the imidazolium cation from 14.8% ($n = 1$) to 99.6% ($n = 4$) but decreased to 15.3% with a further increase in chain length ($n = 12$). The increase in yield up to $n = 4$ is consistent with an increase in the solubility of the hydrophobic substrate with increasing length of the imidazolium alkyl chain *i.e.* mass transfer limitation. Above $n = 4$ conversions decreased due to a decrease in the solubility of the hydrophilic catalyst as the ionic liquid became more hydrophobic. Addition of water to the ionic liquid also improved yields due to an increase in the solubility of the catalyst; the optimum water content was *ca.* 20 % and above this yields decreased due to mass transfer limitations of the hydrophobic substrate.[196] The combination of [C_4mim][tppms], [C_4mim][OAc] and $PdCl_2$ forms an efficient and recyclable system for the palladium-catalysed Heck arylation of ethyl acrylate with a range of aryl bromides. Good conversions were typically obtained after 3 h at 140 °C using 1 mol% $PdCl_2$/[C_4mim][tppms]. For comparison, while comparable conversions were obtained with PPh_3 in place of [C_4mim][tppms] palladium black precipitated and the catalyst mixture could not be recycled. In contrast, $PdCl_2$/[C_4mim][tppms] in [C_4mim][OAc] was highly stable and recycled 10 times without any decrease in activity or leaching of the palladium. The authors proposed a multiple role for [C_4mim][OAc] as both base and solvent as well as a source of *N*-heterocyclic carbene to form a highly active and stable palladium catalyst. Rapid precipitation of palladium black and a marked decrease in yield when [$C_4C_1{}^2$mim][OAc] was used as base lent some support to the role of a stable *N*-heterocyclic carbene/Pd catalyst.[197]

Biphasic Heck arylations conducted in ethanolamine and *N,N*-diethylethanolamine-derived ionic liquids at 100 °C with 1.5 mol% $PdCl_2$ are more efficient than the same reactions in organic solvent; the improvement was attributed to the high solubility of the precatalyst and an increase in the stability and life-time of the catalyst. A straightforward recycle protocol involving extraction of the product with diethyl ether–hexane allowed the catalyst to be reused five times with only a minor drop in activity.[198] Amino-functionalised imidazolium-based ionic liquids of the type [NH_2C_3mim][X] ($X = Cl^-$, BF_4^-, PF_6^-) that act both as ligand and base are excellent co-solvents/additives for the Heck arylation between *n*-butyl acrylate and aryl iodides and bromides in [C_4mim][PF_6] using 1 mol% palladium submicron powder as catalyst. Good yields of product were obtained after 24 h at 120 °C and activity increased in the order $[PF_6]^- > [BF_4]^- > [Cl]^-$. No catalytic activity was obtained in the absence of the amino-functionalised ionic liquid and only a trace amount of product was obtained in [C_4mim][BF_4] with added K_2CO_3. While the use of $Pd(OAc)_2$ gave comparable conversions recovery and reuse was more problematic than the palladium submicron powder.[199]

3.2.3.1.5 Rate Enhancements. A substantial enhancement in rate has been obtained for the ultrasound-assisted palladium-catalysed Heck arylation between methyl acrylate or styrene and iodobenzene in $[C_4C_4im][X]$ ($X = Br^-$, BF_4^-) compared with conventional thermal heating. Good yields of product were obtained in the presence of 1 mol% $Pd(OAc)_2$ at 30 °C in short reaction times (1.5–3.0 h) whereas no reaction occurred under similar sonication conditions in conventional organic solvents such as DMF as NMP. Ultrasound irradiation of a $[C_4C_4im][X]$ solution of $Pd(OAc)_2$ or $PdCl_2$ and sodium acetate generated the bis(dibutylimidazolylidene)-palladium complex *trans*-$[Pd(C_4C_4imy)_2X_2]$ which undergoes sonochemical conversion into highly stabilised clusters of palladium nanoparticles under the conditions used for the ultrasound assisted Heck reaction.[200]

A remarkable enhancement in the rate of $Pd(OAc)_2$/dppp-catalysed Heck reactions in $[C_4mim][BF_4]$ has been obtained by addition of hydrogen bond-donating ammonium salts such as $[NEt_3H][BF_4]$ to reactions conducted in either ionic liquid or conventional solvents. As the presence of bromide in $[C_4mim][BF_4]$ inhibits the palladium-catalysed Heck arylation, $[NEt_3H][BF_4]$ was added as a hydrogen-bond donor to trap and effectively remove bromide. The positive influence of added hydrogen-bond donor was demonstrated in a comparison of the palladium-catalysed arylation of *n*-butyl vinyl ether with bromoacetophenone in the presence of 1.5 equivalents of $[NBu_4]Br$ and $[HNEt_3][BF_4]$, which gave conversions of 2% and 78%, respectively. Remarkably the arylation of a range of electron-rich olefins, including hydroxyalkyl vinyl ethers, vinyl ethers and enamides, with aryl bromides in $[C_4mim][BF_4]$ in the presence of $[NEt_3H][BF_4]$ occurred with high regioselectivity (>99 : 1) in favour of the α-adduct and complete conversion in less than 3 h; compared with 36 h in the absence of $[NEt_3H][BF_4]$. Catalyst loading was lowered to 0.1 mol% to give a TOF and TON of 83 h^{-1} and 1000, respectively. An acceleration in rate was also obtained in conventional organic solvents such that reactions conducted in a DMF/$[NEt_3H][BF_4]$ mixture gave the α-arylated product exclusively with good conversion after only 1 h; for comparison reaction was much slower and less selective in the absence of $[NEt_3H][BF_4]$ suggesting that the salt additive exerts a significant promoting effect on the ionic pathway, although a detailed mechanism of this rate acceleration was not fully elucidated.[201]

Interested in the positive effect of ionic liquids in homogeneous catalysis, Sierra and co-workers obtained a marked improvement in TOF from 290 to 840–1000 for the $Pd(0)/P(OPh)_3$-catalysed Heck arylation of styrene with one of the more challenging electrophilic partners (3-chlorobromobenzene) in a DMF–ionic liquid mixture compared with the same reaction in neat DMF. Moreover, a TON of 27 000 and TOF of 1350 h^{-1} was obtained using 0.001 mol% $Pd_2(dba)_3$ and neat *N*-octylpicolinium hexafluorophosphate $[C_8pic][PF_6]$ as the solvent. The increase in TON for methyl methacrylate was much more modest due to the lower solubility of MMA in the DMF–IL mixture and in neat ionic liquid, however, a TON of 33 000 and TOF of 1650 h^{-1} was obtained in $[C_8pic][PF_6]$.[202]

A dramatic increase in conversion using ionic liquid as additive has also been reported for the $Pd(OAc)_2$-catalysed Heck arylation between 2,3-dihydrofuran and iodobenzene in either DMF or a DMF–water mixture. Addition of $[C_4pyr][BF_4]$ (IL/Pd = 80) to neat DMF increased conversions from 23% to 91% whereas imidazolium salts appeared to deactivate the catalyst. In contrast, selected imidazolium-based ionic liquids gave a small increase in conversion for reactions conducted in 1 : 1 DMF–water up to a IL : Pd ratio of 20 although conversions were highly sensitive to the concentration of ionic liquid and dropped rapidly to <5% at an IL : Pd ratio above 30. The use of morpholinium-based ionic liquids as additive revealed a strong anion effect; a marked improvement in yield was obtained with carboxylate anions such as benzyloxypropionate and methylnitromethyloxocyclopentaneacetate compared with neat DMF whereas yields with morpholinium salts of non-coordinating anions such as $[BF_4]^-$ and $[PF_6]^-$ were much lower. This was explained by the operation of a halide-free pathway involving a highly active palladium-carboxylate as proposed earlier by Amatore and Jutand.[40b] The possible involvement of palladium nanoparticles with the ionic liquid cation stabilising the negatively charged surface was also considered as a possible explanation for the dramatic effect of the cation on catalyst performance in both DMF and DMF–water. An ee of 10% was achieved for the Heck reaction between 2,3-dihydrofuran and iodobenzene using a morpholinium salt with a chiral anion as the only source of asymmetry.[203]

3.2.3.1.6 Alternative Coupling Partners. The first dehydrative palladium-catalysed Heck arylation using a secondary aryl alcohol as the source of styrene occurred efficiently in ionic liquid under microwave heating [Equation (3.2.13)]. A systematic optimisation identified $[C_6mim]Br$ as the best solvent, $[PdCl_2(PPh_3)_2]$ as the precatalyst of choice, sodium formate as base and LiCl as additive; no reaction occurred in DMF which highlights the critical role of the ionic liquid. Under these conditions a diverse range of stilbenes can be obtained in good yields in 15–30 min with 4 mol% $[PdCl_2(PPh_3)_2]$ under microwave heating (150 °C, 120 W). Primary alcohols were much more reluctant to undergo this dehydration–Heck coupling sequence. This approach presents a number of advantages over conventional Heck coupling including the use of readily available inexpensive secondary alcohols, a broad substrate scope and *in situ* generation of polymerisation sensitive styrenes.[204]

$$\text{ArCH(OH)CH}_3 + \text{Ar-X} \xrightarrow[\text{C}_6\text{mim][Br]}]{\text{Pd(OAc)}_2} \text{Ar-CH=CH-Ar}^1 \quad (3.2.13)$$

A marked effect of the anion on the efficiency of the Heck arylation between aza-endocyclic acrylates and arenediazonium salts catalysed by $Pd(OAc)_2$ as the source of palladium manifested itself in the yields when ionic liquids were used as either solvent or additive. Poor conversions were obtained in

$[C_4mim][X]$ ($X = Br^-$, BF_4^-) while reactions in $[C_4mim][X]$ ($X = PF_6^-$, NTf_2^-) gave good to excellent conversions at the same temperature. Aryl iodides are less reactive coupling partners for aza-endocyclic acrylates and good conversions could only be obtained in the presence of Ag_2CO_3 after much longer reaction times at 100–120 °C. With the aim of applying this reaction to the asymmetric synthesis of paroxetine a range of chiral imidazolium and pyridinium-based ionic liquids derived from *N*-α-methylbenzylamine, phenyl glycinol and nicotine were used as solvent or additive each of which gave good yields but no asymmetric induction.[205] Arene diazonium salts have also been used as the electrophilic partner in the palladium-catalysed Matsuda–Heck arylation of styrene and its derivatives in $[C_4mim][X]$ ($X = BF_4^-$, PF_6^-) to give good yields of *trans*-stilbene after heating for 5–8 h at 80 °C. The diazonium salts can also be generated *in situ* in a two-step diazotisation–arylation sequence but isolated yields are slightly lower.[206]

3.2.3.1.7 Catalysis of the Heck Reaction in Immobilised/Encapsulated Ionic Liquids. Ionic liquids have been immobilised in silica matrices by conventional sol–gel processing and the resulting ionogels have attracted interest as supports for transition metal catalysis because they retain liquid-like dynamics while only using a small amount of ionic liquid and as such combine the advantages of homogeneous and heterogeneous catalysis. Two types of palladium-encapsulated ionogels (Figure 3.2.3), one ligand free the other with added PPh_3, were prepared by adding a dichloromethane solution of $Pd(OAc)_2$ (or $Pd(OAc)_2/PPh_3$) to a mixture of $[C_4mim]Br$ (0.5 equivalents) and formic acid (2 equivalents) and stirring for 10 min before adding tetramethoxysilane (1 equivalent). Both types of ionogel catalysed the arylation of ethyl acrylate with iodobenzene at 100 °C; the ligand-free Pd-ionogel experienced an initiation period of ~50 min whereas the initial rate achieved with the Pd/PPh_3-based ionogel was higher than that under homogeneous conditions in toluene and also markedly higher than its phosphine free counterpart. Hot filtration experiments and analysis of the solvent for ionic liquid and palladium both

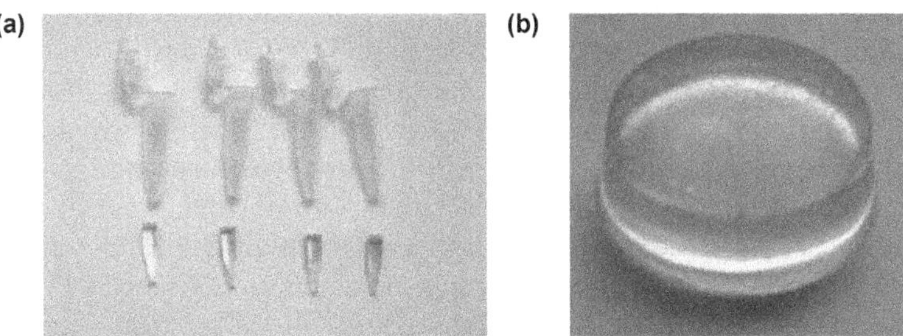

Figure 3.2.3 (a) Pd-doped ionogel pellets (b) Pd-doped cone ionocats. Reproduced from reference 207 with permission from the Centre National de la Recherché Scientifique (CNRS) and The Royal Society of Chemistry.

indicated that catalysis occurred in the confined ionic liquid phase. The tetraethylammonim bromide by-product was retained in the encapsulated ionic liquid which simplified the separation and purification procedure as the product could be isolated by decantation when reactions were quantitative, however, the accumulation of this by-product could ultimately limit the recyclability of these ionogels which remains to be investigated.[207]

3.2.3.2 *Suzuki–Miyaura Cross-Coupling*

3.2.3.2.1 Early Studies, Rate Enhancements and the Involvement of *N*-Heterocyclic Carbenes. The first examples of palladium-catalysed Suzuki–Miyaura cross-couplings in an ionic liquid were performed in $[C_4mim][BF_4]$ using an '*improved catalyst aging protocol*' and demonstrated a number of benefits over reactions in conventional solvents including (i) a marked enhancement in rate at reduced catalyst concentration, particularly for non-activated substrates, (ii) formation of high purity product not contaminated with homo-coupled biaryl by-product, (iii) no loss in yield for reactions conducted in air and (iv) a straightforward recycling procedure with no loss in activity. For example, the TOF of 465 h^{-1} for the reaction between bromobenzene and phenyl broronic acid was 90 times higher than that obtained under conventional conditions and an even more marked enhancement from 2 h^{-1} to 400 h^{-1} was achieved for 4-bromoanisole.[208] Welton and co-workers later identified and subsequently isolated the mixed phosphine-imidazolylidene palladium complex $[(PPh_3)_2Pd(bmimy)Cl][BF_4]$ during the Suzuki–Miayaura cross-coupling between bromobenzene and 3-bromopyridine in $[C_4mim][BF_4]$, using $Pd(PPh_3)_4$ as the precatalyst and NaCl as the halide source to prevent decomposition of the catalyst. The complex was shown to form *via* the oxidative addition adduct $[Pd(PPh_3)_2(Ar)Br]$ ($X=Cl^-$, Br^-) upon addition of aqueous $NaHCO_3$; the same product could also be generated from PdX_2 ($X=Cl^-$, OAc^-) and PPh_3. While enhancement in catalyst performance for reactions conducted in ionic liquid compared with conventional organic solvents had previously been attributed to the unique 'ionic environment' the identification of a mixed phosphine-imidazolylidene palladium complex under the conditions of catalysis provided the first definitive evidence for the non-innocent nature of imidazolium-based ionic liquids as $[(PPh_3)_2Pd(bmimy)Cl][BF_4]$ gave a TOF of 930 h^{-1} when used for the Suzuki–Miyaura coupling of phenyl boronic acid with bromobenzene.[43a] A subsequent and more detailed survey of the effect of changing the palladium precatalyst and the ionic liquid components on the palladium-catalysed Suzuki–Miyaura cross-coupling revealed that (i) both Pd(0) and Pd(II) precatalysts could be used as the source of palladium, (ii) a mixed phosphine–imidazolylidene complex of the type $[(PPh_3)_2Pd(bmimy)X][BF_4]$ was present in all highly active and stable catalyst solutions and (iii) reactions conducted in the C2-protected ionic liquid $[C_4C_1{}^2mim][BF_4]$ led to poor TONs and rapid catalyst deactivation, all of which lend support but not definitive proof for the possible involvement of an active mixed phosphine–NHC complex. Variation of the anion revealed a dramatic effect on performance with catalyst solutions in $[C_4mim][X]$ ($X = BF_4^-$, OTf^-, NTf_2^-) giving high TON and no evidence for decomposition whereas poor conversions were obtained in $[C_4mim][X]$ ($X = PF_6^-$, SbF_6^-) and a solution of $[C_4mim]Cl$ was completely inactive. Detection of the 1-butyl-2-phenyl-3-methylimidazolium cation during an attempt to identify a palladium complex containing an NHC provided additional support for the involvement of an NHC-based species in the catalytic Suzuki–Miyaura reaction (Section 3.1.4).[43b] A marked improvement in yield was obtained for the $[Pd(PPh_3)_4]$-catalysed Suzuki–Miyaura cross-coupling of *N*-heterocyclic chlorides with naphthalene boronic acid in $[C_2mim][BF_4]$ compared with toluene.[209] Good conversions were also obtained with $Pd(PPh_3)_2Cl_2$ and $Pd(OAc)_2/2PPh_3$ but poor yields were obtained for reactions catalysed by $Pd(OAc)_2$ in the absence of phosphine; this was taken to support the formation of an active mixed phosphine–imidazolylidene palladium cation, as previously described by Welton and co-workers.[43a,209] Good yields of biaryl can be obtained in short reaction times (20–90 min) from the ultrasound-promoted palladium-catalysed Suzuki–Miyaura cross-coupling between aryl halides and phenyl boronic acid in $[C_4C_4im][BF_4]$–MeOH using 0.2 mol% $Pd(OAc)_2$ as a source of palladium and sodium acetate as base. However, under these conditions inactive palladium black precipitated which prevented recycling. A comparison showed that much lower conversions were obtained at the same temperature in the absence of ultrasound even after extended reaction times.[210] Nitrogen donor-based palladium complexes generated *in situ* by reaction of $[PdCl_2(MeCN)_2]$ with four equivalents of 1-methyl imidazole (mim), catalyse the Suzuki–Miyaura cross-coupling of bromobenzene with 4-tolylboronic acid.[211] A thermal pre-treatment of the $[PdCl_2(MeCN)_2]$–mim-ionic liquid mixture at 110 °C was necessary to achieve efficient catalysis otherwise significant catalyst decomposition led to poor yields; a similar thermal aging process has previously been reported for a palladium/phosphine-ionic liquid-based catalyst. While higher TOFs were obtained in dioxane (239 h^{-1}), toluene (225 h^{-1}), and water (232 h^{-1}) than in $[C_4mim][NTf_2]$ (106 h^{-1}) the catalyst decomposed in conventional solvents and could not be recycled whereas the ionic liquid-based system was highly stable and recycled five times with no loss in activity. The highest TOFs were obtained in $[C_4C_1{}^2mim][BF_4]$ (206 h^{-1}) and $[C_4C_1pyrr][NTf_2]$ (209 h^{-1}) but in both cases the catalyst decomposed; it is worth noting that neither of these ionic liquids can generate an imidazolylidene fragment. As the most stable catalysts are formed in imidazolium ionic liquids that can form imidazolylidene ligands it is tempting to attribute the catalyst stability to a mixed imidazole–imidazolylidene complex analogous to the mixed phosphine–imidazolylidene palladium catalysts reported earlier and described above.[43,208]

The use of quaternary phosphonium-based ionic liquids as solvents for catalysis has received much less attention than their imidazolium, pyridinium and *N*-alkylammonim counterparts. To this end, high conversions of aryl iodides, bromides and electron deficient chlorides with aryl boronic acids have been obtained in tetradecyltrihexylphosphonium chloride, $[P_{6,6,6,14}]Cl$. Reactions involving aryl iodides were typically complete within 1 h at 50 °C using 1 mol% $Pd_2(dba)_3$ while aryl bromides required slightly longer to reach completion (3 h). Even though reactions of aryl chlorides were much slower good conversions could be obtained after heating at 70 °C for 30 h; for comparison the same cross-couplings with aryl chlorides did not proceed in imidazolium-based ionic liquids even at temperatures as high as 110 °C. The ionic liquid catalyst mixture was recovered and recycled five times with no loss in activity by adding water and hexane to generate a triphasic system with the catalyst immobilised in the central ionic liquid layer, the product in the top organic layer and the inorganic salts in the lower aqueous layer.[212]

Ionic liquids have also been reported to exert a marked rate accelerating effect on the solid phase Suzuki–Miyaura cross-coupling between aryl iodide-based Wang resin and phenyl boronic acid compared with the same reaction in DMF. A series of parallel experiments in neat DMF and a 1 : 1 mixture of DMF–$[C_4mim][BF_4]$ with 5 mol% $Pd(PPh_3)_4$ and Na_2CO_3 as base highlighted the beneficial effect of the ionic liquid as the former achieved a yield of 31% after 30 min which is much lower than the corresponding yield of 59% in DMF–$[C_4mim][BF_4]$. The co-solvent was proposed to be necessary to ensure swelling as no product was detected when neat ionic liquid was used as the solvent. Moreover, the ratio of ionic liquid to DMF could be lowered to 1 : 9 with no loss in activity, which would render this approach more cost effective.[213]

3.2.3.2.2 Catalysts Modified with an Ionic Tag or a Task-Specific Functional Group.

Ether/polyether-functionalised imidazolium-based ionic liquids were introduced as solvents for the biphasic palladium-catalysed Suzuki–Miyaura cross-coupling of aryl iodides/bromides on the basis that the ether donor would stabilise the active catalytic species *via* weak co-ordination and the ionic liquid would provide effective immobilisation for recycling. In general, higher yields were obtained in $[MeOC_2mim][NTf_2]$ and $[MeOC_2pyr][NTf_2]$ than in their unfunctionalised counterparts $[C_4mim][NTf_2]$ and $[C_4pyr][NTf_2]$; the former also recycled efficiently over five runs with no loss in activity. The superior yields obtained in ether-functionalised ionic liquids were proposed to result from weak interactions between the oxygen atom and the active palladium species that stabilise the catalyst with respect to decomposition and aggregation. Pyridinium-based polyether-functionalised ionic liquids were the preferred solvents as their imidazolium-based counterparts form *N*-heterocyclic carbenes which appear to terminate the catalysis. Thus, polyether-based ionic liquids could therefore be potential nonvolatile and reusable alternatives to the more volatile and flammable donor solvents such as ether, tetrahydrofuran and dioxane for transformations in which the ether group plays a central role in catalyst stabilisation.[214]

In principle, a thermoregulated biphasic ionic liquid-based system combines the advantage of being homogeneous at the elevated temperature used for catalysis with the ease of product isolation associated with a heterogeneous catalyst as the system is biphasic at room temperature. Such a multiphase protocol has been developed for the palladium-catalysed Suzuki–Miyaura cross-coupling of aryl bromides with aryl boronic acids using the triethylammonium-tagged diphenylalkylphosphine, $[Et_3N(CH_2)_3PPh_2][NTf_2]$, to confine/immobilise the catalyst in the ionic liquid layer. Good conversions were typically obtained after 2 h at 65 °C in $[C_4C_1pyrr][NTf_2]$–H_2O (2 : 1) using K_3PO_4 as base and 1 mol% of *trans*-$[PdCl_2\{Ph_2P(CH_2)_3NEt_3\}][NTf_2]_2$ as precatalyst. Under optimum conditions 0.01 mol% $[PdCl_2\{Ph_2P(CH_2)_3NEt_3\}][NTf_2]_2$ gave a TON and TOF of 7800 and 5200 h^{-1}, respectively, for the coupling between phenyl boronic acid and 2-bromotoluene. Good yields were also obtained in $[C_4mim][NTf_2]$ but no reaction occurred in $[1\text{-CNMepyr}][NTf_2]$ or $[P_{6,6,6,14}]Cl$. For comparison, reactions conducted in THF and MeCN gave yields of 0% and 12%, respectively, under the same conditions. The catalyst containing ionic liquid phase recycled six times with no loss in activity with less than 10 ppb of palladium leaching into the product during recycle.[215,216]

Even though ionic liquids have been shown to stabilise metal NPs through electrostatic interactions their tendency to aggregate in simple unfunctionalised imidazolium-based ionic liquids during reaction can result in loss of activity and poor recyclability. To this end, a range of heteroatom-functionalised ionic liquids have been employed as both the reaction media and stabiliser for palladium NPs with varying levels of success; these studies suggest that a stable catalyst requires the correct balance of coordination and electrostatic interaction. A family of tuneable coordinating click ionic liquids **45** [Equation (3.2.14)] have been synthesised by reaction of *N*-propargyl-imidazolium salts **44** with organic azides followed by metathesis with MX ($M = Li^+, K^+, X = PF_6^-, NTf_2^-$), their physiochemical properties as a function of anion and substituent on the imidazolium and triazole ring investigated by TGA and DSC and the charge and orbital distributions determined using computational methods. A solution of 2 mol% $PdCl_2$ in **45** ($R = Me$, $R^1 = Me$, $R^2 = n\text{-Bu}$) forms a catalyst that couples aryl iodides with aryl boronic acids to afford near quantitative yields of biaryl product after heating at 100 °C for 5 h. The catalyst mixture recycled with no loss in activity after seven runs indicating that the palladium species is immobilised effectively in the click-derived ionic liquid; TEM and XPS analysis of the catalyst mixture after reaction confirmed the presence of PdNPs with a mean diameter of 3.77 ± 0.77 nm. The authors proposed that the NPs are immobilised and stabilised by the synergistic effect of coordination and electrostatic interaction of the triazolyl and imidazolium units, respectively.[217]

[X]⁻ **44** —(R^2N_3; $CuSO_4.5H_2O$; MeOH/H_2O)→ [X]⁻ **45** (3.2.14)

Inexpensive, easy-to-prepare and stable formate-based hydroxyethylammonium and morpholinium ionic liquids have been used as the reducing agent for Pd(II) and as base and solvent for the phosphine-free palladium-catalysed homocoupling of aryl bromides and iodides; good yields of the corresponding biaryls and heterobiaryls were obtained using 3 mol% $PdCl_2$ at 100 °C. The phosphorylated (2-hydroxyethyl)ammonium formate $[HCO_2][NH_3CH_2CH_2OPPh_2]$ also formed an efficient system for homocoupling provided base was present. Recycle experiments showed that this phosphinite-based ionic liquid system recycled with remarkable efficiency and could be reused 10 times with only a minor drop in activity.[218]

3.2.3.3 Stille and Other Cross-Couplings

An efficient and recyclable system for the Stille cross-coupling between α – iodoenones or aryl iodides and phenyl or vinyltributylstannane based on 5 mol% $[PdCl_2(PhCN)_2]/AsPh_3$ and CuI in $[C_4mim][BF_4]$ gave moderate to good yields at 80 °C. While the same multicomponent catalyst gave poor yields for the Stille coupling of bromobenzene, $Pd(PPh_3)_4$ gave yields in excess of 90% after heating at 80 °C for 18 h. Even though reactions conducted in $[C_4mim][BF_4]$ were slower than those in NMP under the same conditions, both systems recycled five times with no loss in activity.[219] A survey of the efficacy of ionic liquids as solvent for the $Pd_2(dba)_3/AsPh_3$-catalysed Stille cross-coupling between iodobenzene and vinyltributylstannane revealed $[C_4mim][C_8OSO_3]$ to be the solvent of choice as near quantitative yields were obtained with 5 mol% palladium after only 1 h at 80 °C; for comparison nucleophilic ionic liquids such as $[C_4mim]Br$ and $[C_2mim][OTs]$ gave much lower yields of product. The ligandless Stille cross-coupling between iodobenzene and vinyltributylstannane with $Pd(OAc)_2$ was significantly slower and required longer to reach comparable conversions; the highest conversions were obtained in $[C_4mim][NTf_2]$, $[C_4C_1{}^2mim][NTf_2]$, $[Hpyr][NTf_2]$ and $[C_4C_1pyrr][NTf_2]$. The efficacy of $[NTf_2]^-$-based ionic liquids was attributed to nucleophilic assistance in the transmetalation step, which would necessarily be a function of the strength of the cation–anion interaction. In this regard, it is not surprising that the ability of the anion to interact with the tin is influenced by the cation *i.e.* a strong cation–anion interaction might inhibit/hamper the ability of the anion to assist the transmetalation.[41a] Biaryls and aromatic alkynes can be prepared by the Stille cross-coupling of aryl halides with aryltributylstannane and alkynyltributylstannane, respectively, in $[C_4mim][PF_6]$; good yields were obtained with a variety of coupling partners after heating at 80 °C for 12–20 h in the presence of 5 mol% $Pd(PPh_3)_4$. The product was readily isolated by extraction and the recovered ionic liquid–catalyst mixture reused five times with no loss in activity.[220]

While $[P_{6,6,6,14}]Cl$ is the ionic liquid of choice for a host of Suzuki–Miyaura cross-couplings and Heck arylations[221,222] very low conversions were obtained when the Buchwald–Hartwig amination between 4-bromobiphenyl and the weakly nucleophilic diphenylamine was attempted in this ionic liquid as well as in its bromide and dicyanamide counterparts. A near quantitative yield was obtained in $[P_{6,6,6,14}][NTf_2]$ after 2 h at 104 °C with 4 mol% catalyst generated from $Pd_2(dba)_3$ and 1-isobutyl-2,2,6,6-tetramethylphosphorinane. A survey of nine different phosphonium-based ionic liquids underpinned the superiority of $[P_{6,6,6,14}][NTf_2]$ as reaction was complete within 2 h whereas the same reaction in $[P_{6,6,6,14}][BF_4]$ and $[P_{6,6,6,14}]$[decanoate] only reached 59% and 57% conversion, respectively, even after 24 h, while its closest rival $[P_{6,6,6,14}]$[saccharide] gave 90% conversion after 24 h; in general nucleophilic counter-anions are detrimental to this amination. The authors reasoned that while added halide may facilitate the oxidative addition step through an intermediate of the type $[L_2PdX]^-$, as previously described by Amatore and Jutand[40b] a high concentration of halide would hinder subsequent ligand exchange steps. Thus, the role of the polar $[NTf_2]^-$-based ionic liquid would be to promote ionisation of the halide from the oxidative addition adduct to afford $[L_2PdAr]^+$ and thereby facilitate coordination of the weakly nucleophilic NPh_2H; deprotonation followed by reductive elimination would then liberate the product and complete the catalytic cycle.[223]

The organoaluminium adduct bis(trimethylaluminium)-1,4-diazobicyclo[2.2.2]octane **46** (DABAL-Me_3) is a proficient air-stable methyl transfer agent for the palladium-catalysed methylation of aryl and vinyl halides. A recyclable biphasic ionic liquid protocol using THF and either $[C_4C_1{}^2mim][BF_4]$ or $[C_6DABCO][BF_4]$ with 2 mol% $[PdCl_2(MeCN)_2]$ and 4 mol% XPhos or its sulfonated counterpart (XPhos-SO_3H) gave good conversions and isolated yields after 2 h at 65 °C. While XPhos readily leached into the organic phase XPhos-SO_3H was immobilised more effectively and retained by the ionic liquid; encouraging recycle experiments were reported for $[C_4C_1{}^2mim][BF_4]$–THF whereas $[C_4mim][BF_4]$–THF recycled poorly, indicating that the formation of an inactive imidazolium-derived palladium-imidazolylidene complex may be a catalyst decomposition pathway.[224]

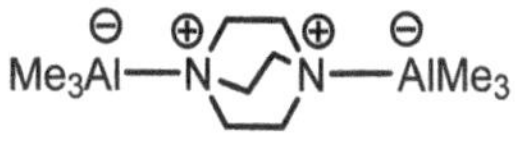

$$Me_3\overset{\ominus}{Al}—\overset{\oplus}{N}(CH_2CH_2)_3\overset{\oplus}{N}—\overset{\ominus}{Al}Me_3$$

46 (DABAL-Me_3)

Ionic liquids doped with pyrrolidinium-tagged π-acidic alkenes based on the chalcone and benzylidene acetone framework are excellent media for palladium-catalysed allyl–aryl coupling as the product can be separated in an operationally straightforward manner because the alkene is efficiently immobilised in the ionic liquid. Allyl–aryl coupling between cyclohexyl-carbonate and an aryl siloxane occurrs in high yield in $[C_5C_1pyrr][NTf_2]$ doped with 8 mol% of the phase-tagged chalcone **47a** (Z = H, OMe, NO_2; X = PF_6^-, NTf_2^-, Br^-) or benzylidene acetone **47b** (X = PF_6^-, NTf_2^-, Br^-) with 4 mol% $Pd(OAc)_2$ and two equivalents of $[NBu_4][F]$. For comparison, only trace amounts of product were obtained under the same conditions in the absence of immobilised alkene. The author speculated that the size of the palladium nanoparticles formed during the reaction and thus the efficiency of the cross-coupling appeared to be a function of the electronic property of the alkene with an electron-releasing OMe group promoting much faster reactions than an electron withdrawing group.[225]

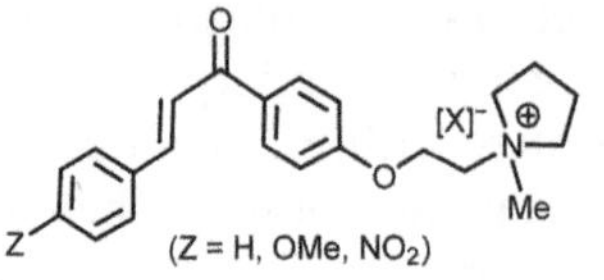

3.3 Applications of Ionic Liquids in Homogeneous Oxidation and Hydrogenation Catalysis

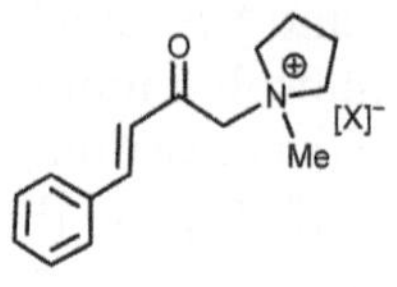

3.3.1 Oxidations

A wide range of oxidations have been conducted in ionic liquids and in the vast majority of cases the ionic liquid acts as the solvent to immobilise and stabilise the catalyst as well as facilitate its recovery and reuse, however, they can also function as extractant or as the catalyst. In addition, marked and substantial enhancements in rate as well as improvements in selectivity have been reported for oxidation reactions conducted in ionic liquids compared with conventional organic solvent. In this regard, the overview provided below will attempt to introduce the applications and potential benefits of using ionic liquids in oxidation catalysis. A review article published in 2011 entitled *Recent Advances in Oxidation Catalysis using Ionic Liquids as Solvents*[10] makes for a highly informative read and a number of additional general and subject specific review articles also provide excellent coverage of this rapidly expanding area.[11,226]

3.3.1.1 Epoxidation and Dihydroxylation

3.3.1.1.1 Methyl Trioxorhenium. Hermann first demonstrated that methyl trioxorhenium (MTO) in combination with urea hydrogen peroxide (UHT) or H_2O_2 was an effective catalyst for the epoxidation of olefins[227a] and the applications of this system has been the subject of a recent comprehensive review article.[227b] With the aim of overcoming the poor solubility of UHP in organic solvents the epoxidation of a range of olefins was conducted in [C_4mim][BF_4] using 2 mol% MTO as catalyst; the solution was homogeneous, yields of epoxide compared favourably with those obtained in organic solvent and high selectivity for epoxide was obtained as the system was essentially water free which prevented ring opening formation of diol.[228] Oxygen atom transfer from the catalytically active diperoxorhenium (dpRe) and monoperoxorhenium (mpRe) complexes to various olefin substrates in a range of ionic liquids has been studied using UV-vis spectroscopy and ^{2}H NMR spectroscopy. Oxygen atom transfer from the dpRe was shown to be markedly more rapid than transfer from the mpRe in a range of ionic liquids ($k_4 \approx 4.5 \times k_3$) whereas the mpRe is in general more reactive towards olefins than the dpRe in 1 : 1 CH_3CN–H_2O (Scheme 3.3.1). Both rate constants k_3 and k_4 were largely unaffected by the nature of the cation but were highly sensitive to the nature of the anion.[229]

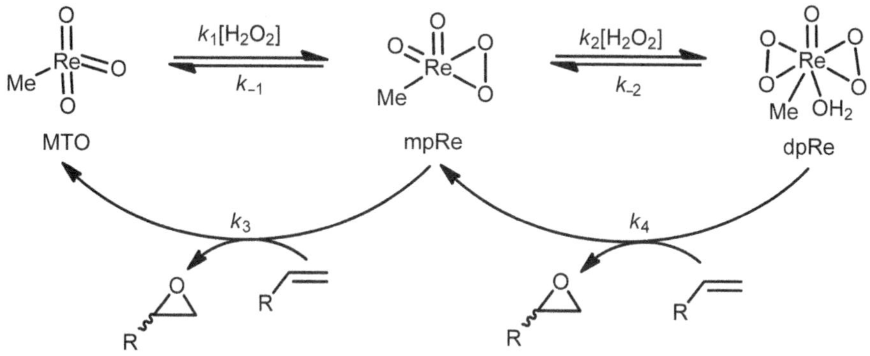

Scheme 3.3.1 Reaction of MTO with hydrogen peroxide to form two peroxorhenium complexes.

3.3.1.1.2 Catalysis with Porphyrin, Schiff Base and Salen-Derived Metal Complexes. High turnover numbers and good recyclability have been obtained for the epoxidation of alkenes in an ionic liquid biphasic system based on 1 mol% of an iron(III) complex of the sodium salt of tetrakis(2′,6′-dichloro-3′-sulfonatophenyl)porphyrin immobilised in [C_4mim]Br. Markedly higher yields were obtained in this system compared with the same catalyst in MeCN–water (1 : 1, v/v) and the immobilised catalyst recycled five times with no loss in activity.[230] A cationic manganese(III) complex of a pyridinium-tagged porphyrin [Mn^{III}(tetrakis-(*N*-methyl-4-pyridinium)-porphyrin][PF_6]$_5$ immobilised in [C_4pyr][BF_4] catalyses the epoxidation of styrene under mild conditions with PhIO as oxidant. The conversion of 96% obtained after 1 h with 0.4 mol% catalyst matched that in MeCN while the 90% selectivity for styrene epoxide was slightly higher than that of 80% in MeCN; the major by-product was phenylacetaldeyhde. For comparison, the corresponding neutral manganese porphyrin catalyst was less active and selective in [C_4pyr][BF_4] and gave a conversion of 48% with 73% selectivity, whereas the conversion and selectivity obtained in MeCN matched its pyridinium-tagged counterpart. The pyridinium-tagged catalyst recycled efficiently and gave good conversions over five runs while the activity of its neutral counterpart dropped dramatically after the first run. However, the selectivity gradually changed on successive recycles and phenylacetaldehyde was obtained as the sole product in the fifth run; this was attributed to the formation of other Mn(III, IV, V)-oxo species.[231]

There are numerous reports of the catalytic asymmetric epoxidation of alkenes in ionic liquids using metal complexes coordinated by salen or salen-type ligands. High ee's were obtained for the asymmetric epoxidation of olefins catalysed by Jacobsen's chiral (salen)Mn^{III} complex [(*R*,*R*)-*N*,*N*′-bis(3,5-di-*tert*-butylsalicylidene)-1,2-cyclohexanediamine]manganese(III) chloride in a 1 : 4 mixture of [C_4mim][PF_6]–CH_2Cl_2 with NaOCl as the terminal oxidant. A marked enhancement in activity was also obtained in the presence of an ionic liquid compared with the same reaction in neat dichloromethane; for example, epoxidation of 2,2-dimethylchromene was complete within 2 h in [C_4mim][PF_6]–CH_2Cl_2 whereas the same reaction in neat dichloromethane required 6 h to reach a comparable conversion. The catalyst could be recovered but conversions dropped from 86% to 53% over five consecutive cycles while ee's dropped from 96% to 88%.[232] Manganese(III) complexes of amino acid-derived chiral salen-like Schiff bases **48** catalyse the asymmetric epoxidation of chromenes in L-1-ethyl-3-(1′-hydroxy-2′-propanyl)imidazolium bromide with NaClO as oxidant and pyridine *N*-oxide as the axial base. The highest ee's (94%) were obtained with catalysts formed from Schiff bases derived from the more bulky amino acids ($R = Me < i\text{-}Pr < CH_2Ph$).[233] Dinuclear Mn(III)-salen complexes **49–50** derived from 1,2-diphenylethylenediamine or 1,2-diaminocyclohexane and bearing *N,N*-dibutylamine substituents at the 5,5′-positions to enhance their water solubility and inherent phase transfer capability catalyse the asymmetric epoxidation of alkenes in ionic liquid–dichloromethane mixtures of $[C_4mim][PF_6]$ and [L-1-ethyl-3-(1′-hydroxy-2′-propanyl)imidazolium]Br using NaOCl as the terminal oxidant and pyridine-*N*-oxide as co-catalyst. Reaction times in both ionic liquid systems were shorter than in neat dichloromethane and the 1,2-diaminocyclohexane-derived catalyst generally gave higher ee's than its 1,2-diphenylethylenediamine counterpart, across the range of substrates examined. Both catalyst–ionic liquid combinations gave poor to moderate ee's for styrene, α-methyl styrene, stilbene and indene. Markedly higher ee's (up to 94%) were obtained for chromene-based substrates but there was limited evidence for a beneficial effect of the ionic liquid as the ee's obtained in neat dichloromethane either compared favourably with or were higher than those in the ionic liquid–dichloromethane systems. However, both ionic liquid–catalyst systems recycled efficiently with only a minor loss in activity and enantioselectivity over six runs.[234] Ionic liquid tagged Mn(III) complexes of salen ligands modified with imidazolium tetrafluoroborate either at the 5 or 5 and 5′- positions of the aromatic rings catalyse the asymmetric epoxidation of styrene at 0 °C with *m*-CPBA as oxidant and pyridine *N*-oxide as axial base. Even though ee's were moderate (34–40%) they matched those obtained with unmodified catalyst with the advantage that the catalyst could be recovered and used for 10 successive runs with no loss in activity.[235]

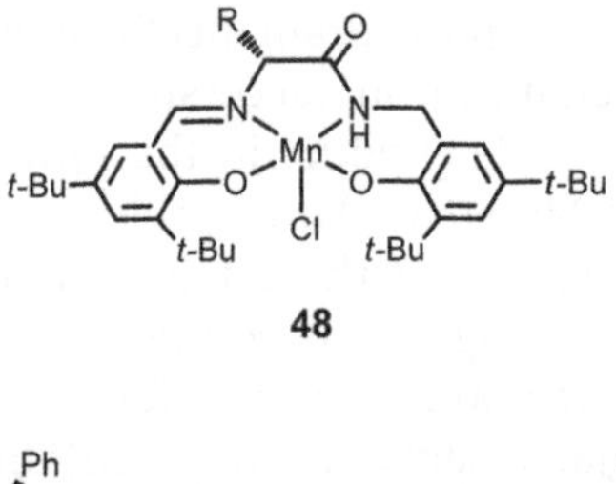

48

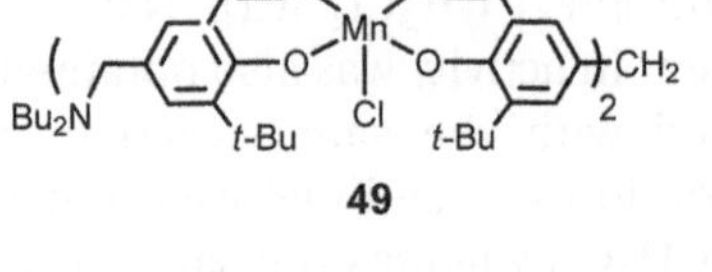

49 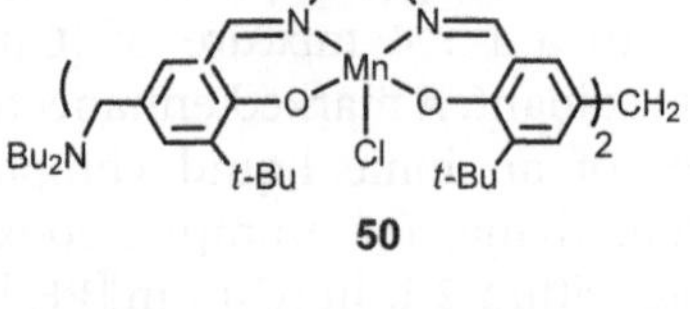 **50**

3.3.1.1.3 Metal Oxo-Peroxo-Based Catalysts. A host of dioxomolybdenum(VI) and rhenium(VI) complexes supported by nitrogen donor ligands have been reported to catalyse the epoxidation of olefins in ionic liquids. Neutral diazabutadiene-supported dioxomolybdenum(VI) complexes of the type $[MoO_2X_2\{p\text{-tolyl-}(CH_3\text{-DAB})\}]$ (X = Cl, Me) catalyse the ionic liquid-based biphasic epoxidation of cyclooctene with 100% selectivity using *tert*-butyl hydroperoxide (TBHP) as oxidant. However, higher TOFs were obtained in the absence of solvent than in either chlorinated solvent or ionic liquid. In contrast, the initial TOF obtained with the cationic terdentate amine-supported dioxomolybdenum(VI) complex $[MoO_2Cl(Bu_3Me_3\text{-tame})][BF_4]$ (Bu_3Me_3-tame = *N*,*N*′,*N*″-tribenzyl-1,1,1-tris(methylaminomethyl)ethane) matched that obtained in the absence of solvent, but the selectivity of 85% was lower than that obtained with the neutral dioxomolybdenum complexes due to ring opening formation of the 1,2-cyclooctanediol. However, interestingly selectivity increased on recycle and eventually reached 100% after five runs.[236] A systematic comparison of the efficiency of the dioxomolybdenum(VI) complex $[MoO_2Cl_2(2,2'\text{-bipy})]$ and Schiff/Lewis base adducts of $MeReO_3$ as catalysts for the epoxidation of cyclooctene as a function of the ionic liquid revealed that the MTO. adduct–H_2O combinations were more active than $[MoO_2Cl_2(2,2'\text{-bipy})]$–TBHP as the former gave an optimum TOF of 479 h^{-1} in $[C_4mim][PF_6]$ whereas the latter gave a maximum TOF of 113 h^{-1} in $[C_4mim][NTf_2]$. Moreover, MTO-based catalysts gave better yields in all ionic liquids tested than under solventless conditions whereas the $[MoO_2Cl_2(2,2'\text{-bipy})]$ was more active under solventless conditions. The water content of the ionic liquid was considered to be crucial as $[MoO_2Cl_2(2,2'\text{-bipy})]$ is hydrolytically sensitive.[237] In stark contrast, the ionic liquid-based biphasic epoxidation of cyclooctene catalysed by 0.1 mol% $[MoO_2Cl_2(L_2)]$ (L = 4,4′-bis-alkoxycarbonyl-2,2′-bipyridine) with TBHP as oxidant gave markedly higher TOF than under solventless conditions. The highest TOFs of 3420–8090 h^{-1} were obtained in $[C_4mim][NTf_2]$ whereas those under solventless conditions were much lower and ranged from 1880–1950 h^{-1}. The catalyst was also highly selective with no evidence for the formation of diol and phase separation of the ionic liquid and product after the reaction facilitated isolation of the product and recovery of the ionic liquid–catalyst mixture which recycled without any noticeable loss in activity.[238] Commercially available inexpensive molybdenum precursors such as MoO_3 and $[Mo_2O_7][NH_4]_2$ form active catalysts (presumed to be an oxodiperoxomolybdenum species) for the epoxidation of olefins in $[C_4mim][PF_6]$ with UHP as oxidant. Good conversions (>90%) were obtained for a range of substrates after 18 h at 60 °C with 2.5 mol% catalyst, whereas reactions in conventional solvents such as methanol, water and dichloromethane gave <5% conversion under the same conditions.[239,240] However, the reaction times were longer and the temperatures higher compared with the conditions required for the same epoxidations catalysed by $[PPh_4][MoO(O_2)_2(QO)]$ (QOH = 8-quinolinol) in MeCN.[241]

Monomeric and dimeric dioxomolybdenum complexes supported by bis(oxazolines) and oxazolinyl-pyridines catalyse the epoxidation of olefins and give good conversions to the epoxide with 100% selectivity in ionic liquid–dichloromethane mixtures as well as in neat dichloromethane, with TBHP as oxidant; $[C_4C_1pyrr][NTf_2]$ was identified as the optimum ionic liquid. The monomeric catalysts gave slightly higher conversions in dichloromethane than in ionic liquid due to phase transfer limitations as cyclooctene is immiscible with these ionic liquids while diol formed rapidly with $[MoO_2Cl_2(dme)]$ in $[C_4C_1pyrr][NTf_2]$–CH_2Cl_2 *i.e.* the absence of an oxazoline-based ligand causes rapid ring opening hydrolysis of epoxide. Recycle experiments with a dimolybdenum oxazolinyl-pyridine-based catalyst in $[C_4C_1pyrr][NTf_2]$ resulted in a loss in activity after three runs due to catalyst decomposition whereas reactions run in $[C_4mim][PF_6]$ recycled five times before activity dropped which was tentatively considered to be due to formation of a less active molybdenum *N*-heterocyclic carbene. Epoxidation of (*R*)-limonene in $[C_4C_1pyrr][NTf_2]$ catalysed by a dimeric dioxomolybdenum catalyst gave low yields of epoxide but with excellent diastereocontrol (100%) whereas the same catalyst was more active in dichloromethane but less selective due to epoxide ring opening. In contrast, monomeric dioxomolybdenum catalysts were more active in $[C_4C_1pyrr][NTf_2]$ than their dimeric counterpart but they gave poor diastereoselectivity.[242]

The $[MoO(O_2)_2(H_2O)_n]$-catalysed hydrogen peroxide-mediated epoxidation of cyclooctene occurs more rapidly in ionic liquid media than in chloroform and the addition of pyrazole and 3,5-dimethylpyrazole results in a marked increase in activity for reactions conducted in ionic liquid but not for those conducted in chloroform. Under optimum conditions, 2.5 mol% $[MoO(O_2)_2(H_2O)_n]$ and 10 mol% 3,5-dimethylpyrazole in $[C_{12}mim][PF_6]$ gave complete conversion to the epoxide after heating at 60 °C for 2 h whereas the same catalyst only reached 23% conversion in chloroform. Stoichiometric studies with $[MoO(O_2)_2(L)]$ (L = pyrazole, 3,5-dimethylpyrazole) revealed that oxodiperoxomolybdenum complexes only oxidise one equivalent of alkene even though they contain two peroxo fragments.[243] A detailed study on the influence of base and ionic liquid on this oxodiperoxomolybdenum-based system revealed that (i) selectivity for epoxide improved with increasing chain length of the *N*-alkyl imidazolium cation in the order C_4 (0%) < C_8 (22%) < C_{12} (100%), due to the reduced water miscibility inhibiting ring opening hydrolysis, (ii) the addition of pyridine as additive improved the selectivity described above to give C_4 (64%) < C_8 (100%) = C_{12} (100%) but did not interfere with conversions, (iii) bipyridines inhibit catalysis and (iv) the addition of a pyrazole base increased catalyst activity in the order pyrazole < 3-methylpyrazole < 3,5-dimethylpyrazole. The ionic liquid catalyst mixture retained its activity over 10 runs provided additional 3,5-dimethylpyrazole was added after each cycle whereas activity dropped gradually on successive runs in the absence of additional base due to its depletion during extraction of the product.[244]

Sulfonated terdentate Schiff bases such as the sodium salt of sulfonated-*N*-salicylidene-2-aminophenolate (NaH_2sSAP) have been used to improve both the solubility of dioxomolybdenum(VI) precatalysts in ionic liquid and increase their Lewis acidity and therefore activity as catalysts. The biphasic epoxidation of cyclooctene in [C_4mim][NTf_2] catalysed by 1 mol% [MoO_2(sSAP)][Na] at 55 °C using TBHP in decane as oxidant gave 63% yield of the epoxide in 271 min whereas the same reaction under solventless conditions only achieved 46% yield after 450 min. Low yields were obtained in [C_4mim][CF_3SO_3] and [C_4mim][*p*-$CH_3C_6H_4SO_3$] which was attributed to poor catalyst solubility and the high viscosity of the solvent.[245]

The oxidation of cyclooctene and sulfides in ionic liquids is catalysed by oxo- and dioxovanadium complexes supported by Schiff base, salen, salophen and hydrazone ligands. Poor yields were obtained for the epoxidation of cyclooctene in [C_4mim][PF_6] at room temperature using a catalyst loading of 1 mol%, however, the same catalysts gave good conversions and were highly selective for the oxidation of thioanisole to its sulfoxide, although higher conversion were obtained in acetonitrile and trifluoroethanol. Quantitative conversions were obtained within 2 h for the oxidation of *p*-tolyl thioether in [C_4mim][PF_6] under the same conditions but the reaction was less selective and gave a mixture of sulfoxide and sulfone. The use of microwave activation (2 W, 32 °C) for the vanadium-catalysed oxidation of *p*-tolyl thioether in [C_4mim][PF_6] led to quantitative conversions in much shorter reaction times (20 s) with 100% selectivity for the sulfoxide, even when the catalyst loading was reduced to 0.03 mol%.[246]

3.3.1.1.4 Epoxidations Catalysed by Peroxometalates. Peroxophosphotungstates are an efficient class of catalyst for the epoxidation of olefins with hydrogen peroxide and their use in imidazolium-based ionic liquids has been widely explored. Ionic liquids can act as solvent and activator for the [$PO_4\{WO(O_2)_2\}_4$]-catalysed epoxidation of *cis*-cyclooctene resulting in a significant and substantial enhancement in TON and selectivity compared with dichloromethane. Activity and selectivity depends quite markedly on the ionic liquid anion and [C_4mim][PF_6] proved to be the ionic liquid of choice with a TON of 289 and 99% selectivity for epoxide, compared with a TON of 1 and 75% selectivity in dichloromethane. For comparison, the use of [C_4mim][BF_4] was detrimental as the TON dropped below 1 with a selectivity of only 12%. ^{31}P NMR spectroscopic studies on the formation of the active Venturello catalyst and a kinetic profile of the epoxidation indicated that the enhancement in catalyst efficiency was associated with rapid generation of the active catalyst in [C_4mim][PF_6] as the epoxidation in acetonitrile experienced a significant induction period due to much slower formation of the active catalyst in this solvent.[247] The ionic liquid immobilised imidazolium-decatungstate [C_4mim]$_4$[$W_{10}O_{23}$] is also an efficient catalyst for the epoxidation of olefins under mild conditions. Good yields were obtained for a variety of substrates and the ionic liquid–catalyst mixture recycled efficiently with only a minor loss in activity after five runs.[248]

Recent studies have demonstrated that guanidinium-based ionic liquids in combination with either $[PW_{12}O_{40}][Y]_3$ or Venturello's peroxometalate $[PO_4\{WO(O_2)_2\}_4][Y]_3$ ($Y = [NR_4]^+$, guanidinium, $[C_4mim]^+$) and H_2O_2 form efficient catalysts for the epoxidation of olefins.[249] For reactions catalysed by $[PO_4\{W(O)_2)_2\}_3][NMe(Oct)_3]_3$ conversions varied with the ionic liquid anion as well as the substitution pattern of the guanidinium cation and ranged from 13–79%; the highest conversions were obtained in water soluble triflate-based guanidinium ionic liquids. Guanadinium-based phosphotungstates gave higher conversions and/or better recyclability for epoxidation than their tetralkylammonium counterparts and their performance compared favourably with their imidazolium-based counterpart $[C_4mim]_3[PW_{12}O_{40}]$ reported earlier by Zhao and co-workers and described above.[247]

Highly efficient, self-separating thermoregulated peroxometalate-based room temperature ionic liquids **51a–b** catalyse the epoxidation of cyclooctene to give complete conversion with 99% selectivity after 4 h at 60 °C. The ionic liquid catalyst was retrieved by extracting the product with cyclohexane and reused a further five times with no loss in activity. Good conversions were also obtained for the epoxidation of styrene and cyclohexene under the same conditions but with slightly lower selectivity as a result of acid hydrolysis of the epoxide. In a survey of co-solvents, ethyl acetate was found to improve selectivity quite substantially such that **51b** catalysed the epoxidation of cyclohexene to give 90% conversion and 97% selectivity, which is a marked improvement on the selectivity of 23% obtained in the absence of solvent; a slight improvement in selectivity was also obtained for the epoxidation of styrene. Ionic liquid **51a** always gave lower conversions and selectivity than **51b** which was attributed to its greater hydrophilicity. Under the conditions of catalysis **51b** forms a homogeneous solution but on cooling the ionic liquid–catalyst phase separates and can be recovered and reused with remarkable efficiency.[250]

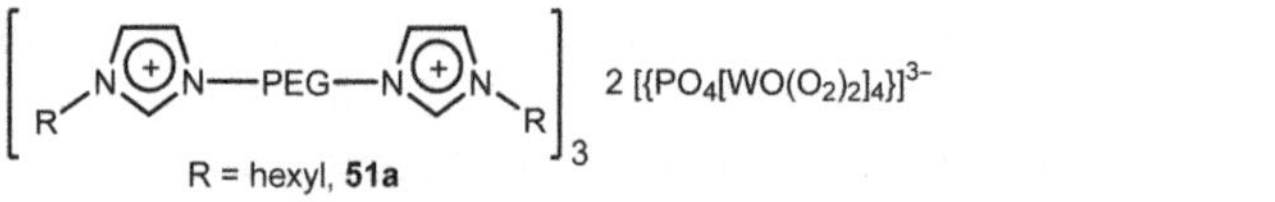

[R–N⊕N—PEG—N⊕N–R]$_3$ 2 $[\{PO_4[WO(O_2)_2]_4\}]^{3-}$

R = hexyl, **51a**
R = dodecyl, **51b**

A homogeneous H_2O_2-based peroxometalate-catalysed epoxidation of lipophilic olefins combines the efficiency of a homogeneous system with the recovery/recyclability of a heterogeneous catalyst. The design concept was based on obtaining an ionic liquid soluble polyoxometalate salt through a judicious choice of cation and tuning the reaction environment of the ionic liquid solvent to accommodate the lipophilicity of the alkene substrate as well as the hydrophilicity of aqueous H_2O_2. After optimisation, a conversion of 88% and a selectivity of 95% was obtained for the epoxidation of cyclooctene after 1 h at 80 °C using 1 mol% $[C_{12}pyr]_3[PW_{12}O_{40}]$ in a 1:1 mixture of $[C_{12}pyr][BF_4]$ and $[C_4pyr][BF_4]$ as amphiphatic solvent and H_2O_2 as oxidant; for comparison the same reaction in DMF only gave 69% conversion albeit with 100% selectivity. The ionic liquid–catalyst solution recycled with high conversions and good epoxide selectivity over three runs but accumulation of water in the ionic liquid–catalyst mixture resulted in a biphasic system and partial precipitation of the catalyst, however, activity could be recovered by decanting the water and replenishing the ionic liquid. The system recycled more efficiently by removing the by-product after each run.[251]

A broad range of alkenes have been oxidised with high selectivity using the peroxotungstate $[\{W(=O)(O_2)_2(H_2O)\}_2(\mu\text{-}O)]$ immobilised on dihydroimidazolium-based ionic liquid-modified silica; reactions were conducted at 333 K and good to excellent yields were obtained using 1 mol% catalyst. The performance of this SILP catalyst compares favourably with the corresponding homogeneous system $[n\text{-}N_{1,1,1,12}]_2[\{W(=O)(O_2)_2(H_2O)\}_2(\mu\text{-}O)]$ in acetonitrile.[252] Linear pyrrolidinium-decorated polymers with tuneable surface properties and microstructure have been prepared by ring opening metathesis polymerisation and used to support peroxophosphotungstate $[PO_4\{W(O)_2)_2\}_3]^{3-}$. The resulting polymer immobilised ionic liquid phase (PIILP) peroxometalates are highly efficient catalysts for the epoxidation of allylic alcohols and alkenes that can be recovered in an operationally straightforward filtration and reused with only a minor reduction in performance.[73]

3.3.1.1.5 Ruthenium-Carbene-Based Epoxidation. The first example of a ruthenium-catalysed epoxidation of olefins in ionic liquids used 1 mol% *trans*-$[Ru(CN\text{-}Me)(trpy)(H_2O)][PF_6]_2$ (CN-Me = 3-methyl-1-(pyridin-2-yl)-imidazolylidene and trpy = 2,2′:6′,2″-terpyridine) in a 0.8 : 1.2 mixture of $[C_4mim][PF_6]$ and dichloromethane with $PhI(OAc)_2$ as the terminal oxidant. Although the conversions obtained in the ionic liquid–dichloromethane mixture were slightly lower than those in neat dichloromethane the ionic liquid–catalyst mixture recycled 10 times for the epoxidation of cyclooctene without any loss in activity (TON ~ 1000) whereas the corresponding reaction in dichloromethane recycled poorly as the conversion dropped to 35% in the second run.[253]

An ionic liquid-based lipase driven epoxidation of olefins that uses enzyme-catalysed H_2O_2-mediated perhydrolysis of octanoic acid to generate the peracid *in situ* avoids the transportation and handling of large quantities of organic peracid. Epoxidations were conducted at room temperatures and at 50 °C in a range of hydrogen bond-donating ionic liquids and $[C_4mim][X]$ ($X = NO_3^-$, PF_6^-, BF_4^-) using Novozym 435 as catalyst, peroctanoic acid as oxidant and cyclohexene, cyclooctene and styrene as substrates. For each substrate the highest yields were obtained in $[HOC_3mim][NO_3]$ while those in $[C_4mim][BF_4]$ and $[C_4mim][PF_6]$ were up to 23% lower. Similarly, $[HOC_3mim][NO_3]$ was also the ionic liquid of choice for the lipase-mediated Baeyer–Villiger oxidation of cyclic ketones, aliphatic aldehydes and ketones. Reactions were typically much faster than the corresponding transformations in organic solvent and good conversions were obtained after 5 h at 50 °C. The authors reasoned that the hydrogen bond-donating ionic liquid facilitated the rate determining rearrangement of the Criegee-intermediate. In addition to facilitating the oxygen transfer by peracid, the ionic liquid–enzyme mixtures are stable which could allow for recycling and/or a continuous flow process to be developed.[254]

3.3.1.2 *Wacker-Type Oxidations*

The Wacker reaction is a versatile method for the conversion of terminal olefins to methyl ketones using a Pd(II) catalyst and either O_2 or H_2O_2 as the oxidising agent, but efficiency of H_2O_2 utilisation and moderate conversions have been identified as major drawbacks. A marked enhancement in conversion was obtained for the palladium-catalysed oxidation of styrene with H_2O_2 [Equation (3.3.1)] in the presence of a catalytic quantity (2.5 wt%) of either [C_4mim][BF_4] or [C_4mim][PF_6], compared with the same reaction conducted in the absence of ionic liquid. Under optimum conditions 0.5 mol% $PdCl_2$ and 2.5 wt % [C_4mim][X] ($X = BF_4^-$, PF_6^-) gave quantitative conversion and 91–92% selectivity for acetophenone, together with 4–5% benzaldehyde, compared with a conversion of 60% and a selectivity of 70% in the absence of ionic liquid. The authors reasoned that this enhancement resulted from the formation of an oxaziridinium ion-type intermediate involving the imidazolium cation.[255] The presence of water is crucial to achieving good conversions and high ketone/aldehyde selectivity for the $PdCl_2$/CuCl-catalysed Wacker oxidation of styrenes and aliphatic olefins [Equation (3.3.1)] in ionic liquid as reactions are immeasurably slow in neat [C_4mim][BF_4]. The use of [C_4mim][BF_4]–H_2O (2 : 1, v/v) gave good to excellent conversions for a range of substrates. Recycle experiments with 4-methyl styrene gave consistent conversions over eight runs with only a minor loss in activity whereas the selectivity for ketone gradually increased from 80% to 96% in the final run.[256]

PdCl$_2$-CuCl/[O], ionic liquid (styrene → acetophenone + benzaldehyde) (3.3.1)

The nature of the ionic liquid has a profound and dramatic effect on oxidation *versus* dimerisation selectivity as well as the efficiency of the palladium-catalysed oxidation of styrene with H_2O_2. Reactions conducted in hydrophilic ionic liquids such as [C_2mim][$N(CN)_2$], [C_2mim][BF_4], [C_1C_1pip][X] ($X = CF_3CO_2^-$, NO_3^-), [C_1C_1pyrr][X] ($X = NO_3^-$, HSO_4^-, HCO_2^-, ClO_4^-) and catalysed by 1 mol% $PdCl_2$ with H_2O_2 as oxidant at 60 °C are highly selective for acetophenone (90–100%); a minor amount of benzaldehyde (5–10%) was generated and only a trace quantity of the anti-Markovnikov product was detected. In stark contrast, high conversions were also obtained in hydrophobic ionic liquids but selectivity for acetophenone

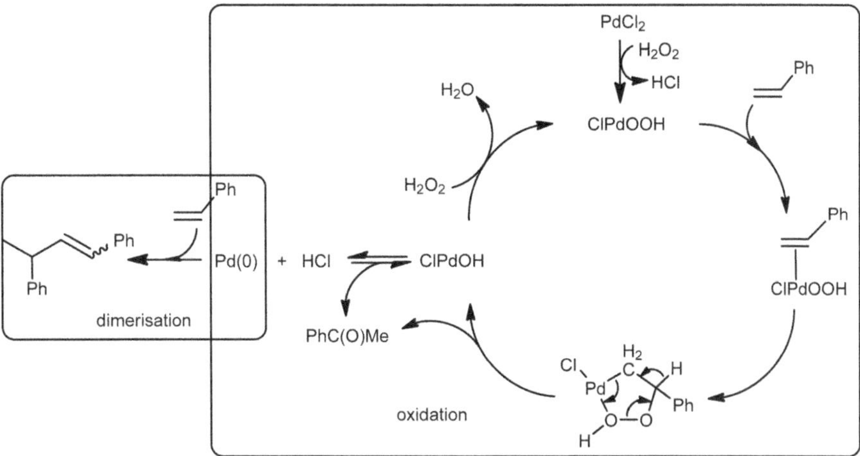

Scheme 3.3.2 Proposed mechanism for competitive palladium-catalysed styrene oxidation and dimerisation.
Reproduced from reference 258.

varied between 20–75% with the styrene dimer 1,3-diphenyl-1-butene as the major by-product. The authors reasoned that dimerisation of styrene would compete with oxidation under conditions in which regeneration of the active palladium hydroperoxide (XPd-OOH) from XPd-OH was disfavoured as it is in equilibrium with Pd(0) which would coordinate styrene and promote the dimerisation pathway, as previously reported.[257] Thus, hydrophilic ionic liquids give high selectivity for oxidation because the concentration of H_2O_2 is high and regeneration of Pd-OOH is favoured, whereas the biphasic system obtained with hydrophobic ionic liquids and their relatively high viscosity disfavours regeneration of the active hydroperoxide species (Scheme 3.3.2). However, hydrophilicity alone was not the sole requirement for efficient oxidation as only low conversions were obtained in ionic liquids composed of basic or amphoteric anions whereas both high conversions and selectivities were obtained in $[C_1C_1pip][CF_3CO_2]$ and $[C_1C_1pyrr][X]$ ($X = CF_3CO_2^-$, NO_3^-).[258]

3.3.1.3 Oxidation of Alcohols

3.3.1.3.1 TEMPO-Catalysed Oxidations. Ionic liquids have been shown to be suitable solvents for the selective oxidation of primary alcohols catalysed by TEMPO (2,2,6,6-tetramethyl-piperidyl-1-oxy) and a stoichiometric quantity of oxidant. A mild, effective and straightforward TEMPO/CuCl-catalysed selective oxidation of primary and secondary alcohols to the desired aldehydes and ketones in $[C_4mim][PF_6]$ with oxygen as the terminal oxidant gave good to excellent conversions at 65 °C for a range of substrates with no evidence for over oxidation. The system also recycled eight times with only a slight loss in activity.[259] The selective oxidation of primary alcohols to aldehydes has been achieved in high yields and under mild conditions in ionic liquids using a three-component catalyst consisting of acetamido-TEMPO, a copper(II) salt and 4-dimethylaminopyridine (DMAP) with oxygen as the terminal oxidant. A survey of ionic liquids and copper salts identified $Cu(ClO_4)_2$ and $[C_4pyr][PF_6]$ to be the optimum combination to obtain good isolated yields with a range of benzylic, allylic and aliphatic alcohols. The efficient recyclability of this system was attributed to the high solubility of the acetamido-TEMPO and DMAP in the ionic liquid. The DMAP was proposed to have a dual role as both base to deprotonate the alcohol and as an N-donor to coordinate to the copper. The catalyst is easy to handle and recycled efficiently over five runs with only a minor loss in activity. Inactivity towards secondary alcohols was attributed to hindered access of TEMPO to the sterically congested coordination sphere of copper.[260] In a related study, base was also shown to accelerate the aerobic CuCl–TEMPO-catalysed oxidation of alcohols to aldehydes and ketones in $[C_4mim][PF_6]$. For example, the oxidation of *p*-nitrobenzyl alcohol catalysed by 5 mol% TEMPO/CuCl and 10 mol% pyridine reached completion after 3 h at 65 °C whereas in the absence of base the same reaction required 24 h to reach 97% completion. The authors proposed that as CuCl/TEMPO/pyridine catalysed the oxidation of both primary and secondary alcohols reaction most likely occurs *via* a less-hindered copper species which may have only one molecule of coordinated pyridine (compare with DMAP above).[261] A three-component system with 3 mol% TEMPO, 6 mol% HBr and H_2O_2 in $[C_4mim][PF_6]$ catalysed the selective oxidation of benzylic alcohols to the corresponding aldehyde to afford good yields in 2 h at 40 °C. Although the TEMPO could not be recycled, its ionic liquid soluble–ether insoluble counterpart acetamido-TEMPO gave similar conversions and selectivities and could be recovered and reused three times with no loss in activity; the recovered catalyst was also used for the oxidation of a second different substrate with no cross contamination.[262]

The first imidazolium-modified TEMPO **52** (TEMPO-IL) used for the oxidation of allylic, benzylic and aliphatic alcohols in $[C_4mim][PF_6]$ gave conversions and selectivities that matched those obtained with unmodified TEMPO but with the advantage that it could be recovered and recycled in an operationally straightforward procedure.[263] Molecular sieves were found to enhance the rate of the CuCl/TEMPO–IL (**52**, $X = PF_6^-$) catalysed oxidation of benzylic and allylic alcohols in $[C_4mim][PF_6]$ by acting as a heterogeneous Brønsted base rather than desiccant. Reaction times were reduced quite considerably and the oxidation of benzylic alcohols catalysed by 5 mol% CuCl and 5 mol% of an imidazolium-modified TEMPO in $[C_4mim][PF_6]$ typically reached near quantitative conversion in 3–5 h at 80 °C compared with 8–16 h for the same reactions in the absence of sieves.[264] A highly efficient three-component metal-free system comprising imidazolium-modified TEMPO–IL **52** ($X = Cl^-$, BF_4^-), imidazolium-modified carboxylic acid **53** and sodium nitrite catalyses the aerobic oxidation of benzylic, allylic and heterocyclic alcohols to afford the corresponding carbonyl compounds under mild conditions with high activity and exceptional selectivity (>99%);

a TON of 5000 was achieved for the oxidation of benzyl alcohol with 0.001 mol% **52** ($X = Cl^-$). Recycle studies gave good conversions across four runs with only a slight loss in activity provided that the $NaNO_2$ was replenished after each cycle.[265] A mixture of imidazolium-modified TEMPO–IL **52** combined with NaBr and *N*-chlorosuccinimide (NCS) in $[C_4mim][X]$ ($X = BF_4^-$, PF_6^-) also catalyses the chemoselective oxidation of diols containing a benzylic alcohol and a primary alcohol as well as a mixture of a primary and secondary benzylic alcohols or a benzyl alcohol and a primary aliphatic alcohol. High selectivity for the primary benzyl alcohol was obtained in each case. The ionic liquid–catalyst mixture could be recycled 10 times with no loss in activity by extracting the product with ether and washing the ionic liquid phase with water.[266]

$[X]^-$ $[X]^- = PF_6^-, BF_4^-, Cl^-$ TEMPO-IL **52**

CO_2H $[X]^- = BF_4^-, Cl^-$ **53**

Copper-TEMPO SILP catalyst based on either a bimodal pore structure silica or an ionogel as the support material and $[C_4mim][X]$ ($X = PF_6^-$, BF_4^-, $C_8OSO_3^-$) as the catalyst-philic phase dispersed in the form of a film, oxidises benzylic alcohols in diethyl ether under 1 atm of O_2 at 65 °C (Figure 3.3.1). The ionogel-based/$[C_4mim][C_8OSO_3]$ SILP catalyst was the most active with a TON of 3920 after 7 h (0.11 g catalyst, 2.5 mol% $CuCl_2$) which increased to 6533 in the absence of solvent. In contrast its $[PF_6]^-$ counterpart was much less active and gave a TON of 225 under the same conditions. For comparison the corresponding SILP catalyst generated from the bimodal pore structure silica was less active and gave a TON of 1020

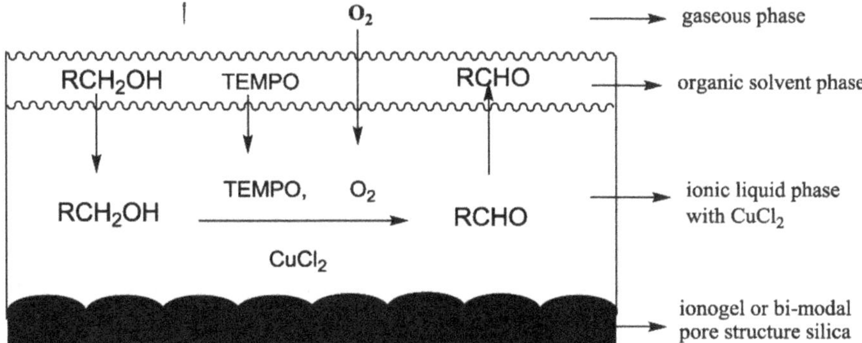

Figure 3.3.1 Copper TEMPO-catalysed aerobic oxidation of alcohols in the presence of SILP catalysts.
Reproduced with permission from reference 267. Copyright 2010 Elsevier.

compared with 1920 for the ionogel-based SILP system under similar conditions (5 mol% TEMPO, 5 mol% $CuCl_2$, 7 h). Recycle experiments with benzyl alcohol as substrate showed that both catalysts could be recovered by filtration and reused seven times with no loss of activity provided a fresh portion of TEMPO was added in each cycle.[267]

3.3.1.3.2 Ruthenium and Tungsten Catalysed Oxidations. Ionic liquids have been used as an alternative solvent for the highly selective perruthenate-catalysed oxidation of alcohols to aldehydes and ketones. The $[^nPr_4N][RuO_4]$ (TPAP)-catalysed oxidation of primary and secondary alcohols in $[C_4mim][BF_4]$ gave moderate to good conversions after only 2 h at room temperature with 5 mol% TPAP and *N*-methyl morpholine oxide as co-oxidant. Slightly lower yields were obtained with unactivated primary alcohols due to the biphasic nature of the system and the poor mass transport arising from the low miscibility of the substrate with $[C_4mim][BF_4]$; however, conversions could be improved by sonication. In comparison to TPAP/NMO, the combination of TPAP/O_2/CuCl/2-aminopyridine in $[C_4mim][NTf_2]$ is almost entirely selective towards benzylic alcohols but was essentially inactive towards aliphatic and secondary alcohols.[268] The selective aerobic oxidation of aliphatic and aromatic alcohols to their corresponding aldehydes and ketones has been catalysed by $RuCl_3$, $[(p\text{-cymene})RuCl_2]_2$ and $[RuCl_2(PPh_3)_3]$ in various ammonium salts without the need for a co-catalyst. No reaction occurred in $[C_4mim][X]$ ($X = BF_4^-$, PF_6^-) and only very low yields were obtained in $[C_4mim]Cl$, even in the presence of added base. High conversions were obtained in $[nBu_4N][OH].5H_2O$ and $[N_{1,8,8,8}]Cl$ but the product had to be isolated by distillation from the latter due to the high solubility of the ionic liquid in organic solvents typically used for extraction. The activity of these systems compared favourably with the $[RuCl_2(PPh_3)_3]$/TEMPO combination reported earlier.[269]

The tris(imidazolium)-based peroxophosphotungstate $[C_4mim]_3[PO_4\{WO(O_2)_2\}_4]$ also catalyses the oxidation of a range of secondary alcohols in $[C_4mim][BF_4]$ to give the corresponding ketones in good yield after 2–4 h at 90 °C; the product was extracted with ether and the ionic liquid–catalyst mixture recycled three times with only a slight decrease in activity. The same catalyst also oxidised primary alcohols such as benzyl alcohol but gave a mixture of aldehyde and carboxylic acid.[270]

3.3.1.3.3 Palladium and Nickel-Catalysed Oxidations. Looking to improve on the palladium(II) ethanoate-catalysed liquid phase homogeneous aerobic oxidation of primary and secondary alcohols in dimethyl sulfoxide developed by Peterson and Larock,[271a] Seddon and co-workers explored the same transformation in ionic liquids. Following an earlier reported method, the TOF of 1.2 obtained for the oxidation of benzyl alcohol in $[C_4C_1{}^2mim][BF_4]$ (5 mol% $Pd(O_2CMe)_2$, 1 atm O_2, 80 °C) was a marked improvement on the corresponding TOF of 0.4 in DMSO. Reactions in $[C_6mim]Cl$ were not selective and gave dibenzyl ether and benzoic acid; the former was attributed to the generation of $[PdCl_4]^{2-}$ due to the presence of a small amount of chloride impurity in the ionic liquid, while the benzoic acid was formed *via* over oxidation brought about by water in the ionic liquid. A study of the influence of water content on activity and selectivity revealed that the amount of water influences the selectivity and that the ionic liquid–catalyst mixtures remain highly selective for benzaldehyde with a water content as high as 25 vol%, above which benzoic acid was the dominant product, and that the rate of oxidation increases with increasing water content. The high tolerance of this catalyst system towards water suggested that the ionic liquid deactivates the water and prevents it from further reaction to benzoic acid. Recycle experiments demonstrated that the ionic liquid–catalyst mixture could be reused five times but it was necessary to remove water that had accumulated either as by-product or through adsorption from the atmosphere in order to retain high selectivity.[271b]

The miscibility of 2-octanol and 2-octanone as well as 1-phenyl ethanol and acetophenone with a variety of ionic liquids has been determined in order to identify ionic liquids that undergo thermoregulated phase separation to simplify product isolation (Figure 3.3.2). Of the four ionic liquids that met the solubility criteria and that were also oxygen resistant, $[C_4mim][BF_4]$ gave the highest yield for the palladium-catalysed aerobic oxidation of 2-octanol at 120 °C with 2.5 mol% $Pd(OAc)_2$ under 10 bar of O_2 and an ionic liquid/substrate ratio of one. Much poorer conversions (10%) were obtained at 90 °C due to the low solubility of 2-octanol in the ionic liquid but yields improved to 26% when the ionic liquid/substrate ratio was reduced to 0.1. Good conversions were also obtained for 2-hexanol (48%) and 2-decanol (87%) while the low yield for 2-dodecanol (21%) was attributed to poor miscibility with the ionic liquid at the reaction temperature. The ionic liquid–catalyst phase can be recovered by decanting the product after allowing the reaction mixture to cool and phase separate; palladium leaching into the product phase was determined to be less than 0.1%. The ionic liquid–catalyst layer recycled efficiently with no cross contamination from different substrates.[272]

The oxidation of primary and secondary benzylic and aliphatic alcohols in $[C_2mim][X]$ has also been catalysed by nickel(II) complexes supported by a

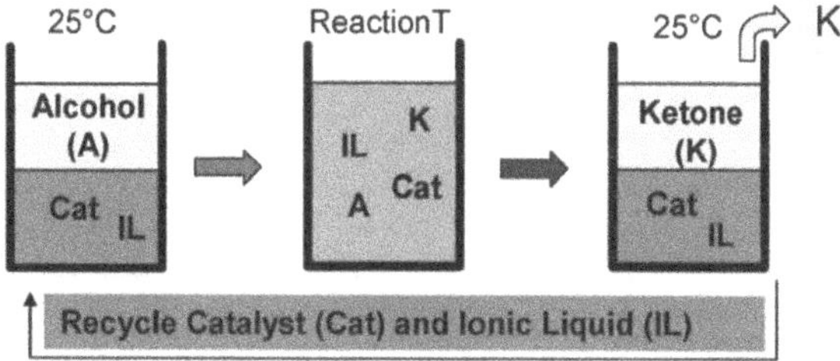

Figure 3.3.2 A recyclable catalytic ionic liquid system using the concept of temperature-dependent miscibility of ionic liquids and organic components. Reproduced from reference 272 with permission from the PCCP Owner Societies.

N-(2-pyridyl)-*N′*-(5-substituted-salicylidene)hydrazine and triphenylphosphine with NaOCl as terminal oxidant. Good yields were obtained in short reaction times (15–30 min) with 2 mol% catalyst; reactions were carried out in air and the catalyst recycled with remarkable efficiency over ten runs.[273]

3.3.1.4 Dihydroxylation

Osmium-catalysed dihydroxylation of olefins has found wide-spread use in organic synthesis but the high cost of the osmium catalyst, and the ligands required for asymmetric dihydroxylation, as well as the high toxicity and volatility of OsO_4 are drawbacks. The first attempt to address these limitations by immobilising the osmium catalyst in an ionic liquid used standard Upjohn conditions (2 mol% OsO_4, 1.2 equivalents NMO in *t*-BuOH–H_2O) in the presence of [C_4mim][PF_6] and gave diol in good yields but with poor recyclability. Reasoning that reversible binding of the amine to OsO_4 was responsible for the leaching and poor recyclability, reactions were conducted with 4-(dimethylamino)pyridine on the basis that this would bind more strongly and improve catalyst retention (Figure 3.3.3). To this end the use of 2.4 mol% DMAP gave a dramatic improvement in catalyst recyclability and a variety of substrates were dihydroxylated over six runs with only a minor reduction in activity in each case.[274] A parallel study reported that the dihydroxylation of olefins in [C_2mim][BF_4] is efficiently catalysed by 5 mol% OsO_4 with NMO.H_2O as co-oxidant to give good yields across a range of substrates after 18 h at room temperature. The system recycled efficiently and consistent conversions were obtained across five runs.[275]

A similar protocol proved successful for the asymmetric dihydroxylation of olefins either under biphasic conditions in [C_4mim][PF_6]–H_2O or under homogeneous conditions in [C_4mim][PF_6]–*tert*-BuOH–water with 0.5 mol% $K_2OsO_2(OH)_4$, three equivalents of $K_3Fe(CN)_6$ as oxidant and 1 mol% 1,4-bis(9-*O*-dihydroquininyl)phthalazine or 1,4-bis(9-*O*-dihydroquininyl)biphenylpyrimidine as the chiral ligand. Yields and ee's either rivalled or exceeded those obtained in *tert*-BuOH–H_2O for a range of substrates and both systems

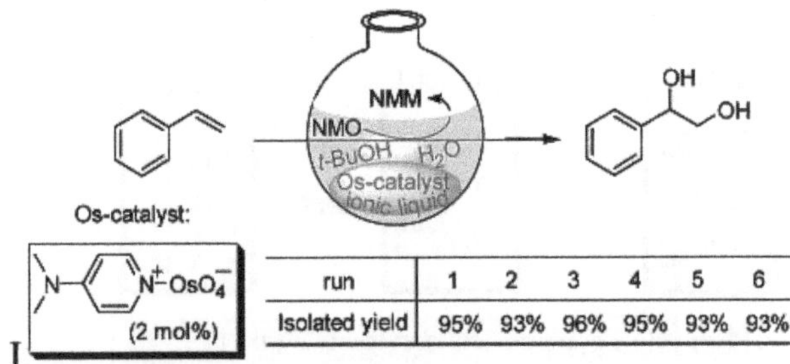

run	1	2	3	4	5	6
Isolated yield	95%	93%	96%	95%	93%	93%

Figure 3.3.3 A recyclable and reusable DMAP-based catalytic system for olefin dihydroxylation.
Reprinted with permission from reference 274. Copyright 2002 American Chemical Society.

were highly stable and recycled with remarkable efficiency as yields and ee's only dropped after runs 10 and 12 for the biphasic and homogeneous systems, respectively. The catalyst was efficiently immobilised in the ionic liquid phase which retained greater than 90% of the osmium from the previous cycle and the product phase was contaminated with less than 3% osmium (7 ppb), which is significantly lower than that for reactions in *tert*-BuOH–H_2O. The above recycle experiments corresponded to total TONs of 1334 and 1720 for the biphasic and homogenous systems, respectively.[276a,b] While 1 mol% OsO_4 in combination with 2.5 mol% 1,4-bis(9-*O*-dihydroquininyl)phthalazine [$(DHQD)_2$PHAL] and NMO as co-oxidant gave good yields and high ee's for the asymmetric dihydroxylation of olefins in either [C_4mim][BF_4] or [C_4mim][PF_6] and acetone–water, both systems recycled poorly due to excessive leaching of both osmium and the $(DHQD)_2$PHAL. A marked improvement in recyclability was achieved by replacing (DHQD)PHAL with the new bis-cinchona alkaloid $(QN)_2$PHAL, **54**, which is efficiently retained in the ionic liquid after *in situ* bis(dihydroxylation) to the more polar alcohol **55**. Recycle experiments with 0.1 mol% OsO_4 and 5 mol% $(QN)_2$PHAL gave a total TON of 2370, however, activity gradually decreased on successive reuses and longer reaction times became necessary.[277]

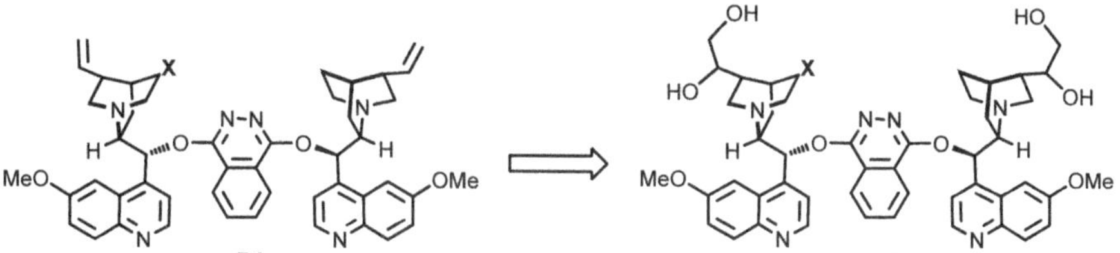

A highly efficient, robust and clean osmium-catalysed asymmetric dihydroxylation of olefins has been developed by using an ionic liquid as the reaction medium/immobilisation phase and *sc*CO_2 to recover/extract the product (Figure 3.3.4). Optimisation studies on 1-hexene and styrene with 0.5 mol% $K_2OsO_2(OH)_4$, 1 mol% $(DHQD)_2$PHAL and $K_3[Fe(CN)_6]$ or NMO as co-oxidant identified [C_4mim][NTf_2] and [$C_4C_1{}^2$mim][NTf_2] as the ionic liquids of choice and 100 bar of CO_2 for the extraction. Under these conditions conversions and ee's matched those obtained in organic solvent–water and ionic liquid–co-solvent systems with the added benefit that slow addition of the substrate was NOT necessary to achieve high ee's due to the poor solubility of the olefin in the ionic liquid. Recycle experiments gave greater than 80% conversion for the first six runs, ee's in excess of 90% for the first seven runs and an extremely low level of catalyst leaching (0.21–0.34%).[278]

A high level of asymmetric induction has been achieved for the dihydroxylation of olefins using a chiral ionic liquid as the sole source of chirality *i.e.* in the absence of a Sharpless chiral ligand. For example, the asymmetric dihydroxylation of 1-hexene in the guanidinium-based ionic liquids [$(di\text{-}h)_2$dmg][L-lactic] (**56**) and [$(di\text{-}h)_2$dmg][quinic] (**57**) catalysed by 0.5 mol% $K_2OsO_2(OH)_4$ and NMO as co-oxidant gave (*R*)-1,2-hexanediol in 85% and

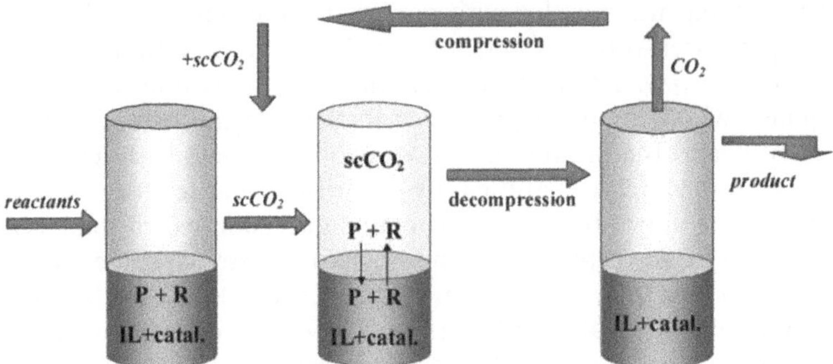

Figure 3.3.4 Schematic of a *sc*CO_2–IL biphasic system for ionic liquid–catalyst phase recycle and product recovery.
Reprinted with permission from reference 279. Copyright 2011 American Chemical Society.

81% ee, respectively, with yields in excess of 90%. These yields and ee's compared favourably with those obtained in *tert*-BuOH–water with the chiral ligand $(DHQD)_2$PHAL. The amount of chiral ionic liquid could be reduced to a 1 : 1 mixture of [C_8mim][BF_4] and [(di-h)$_2$dmg][quinic] with only a slight reduction in ee from 85% to 79% and, moreover, an ee of 64% was still obtained even when the ratio of [C_8mim][BF_4] to chiral ionic liquid was reduced to 9 : 1. Recycle experiments using either *tert*-butyl methyl ether or hexane as the extractant resulted in a marked reduction in ee to 37% and 25%, respectively, by run 3 due to leaching of the chiral ionic liquid. Although the leaching was not quantified it would be reasonable to assume that the majority of ionic liquid was extracted in the first cycle on the basis that an ee of 61% was obtained with a 10% solution of [(di-h)$_2$dmg][quinic] in [C_8mim][BF_4]. Efficient recycling was achieved using [C_{10}mim][BF_4] as the catalyst immobilisation phase and *sc*CO_2 as the extractant at 40 °C and 120 bar (Figure 3.3.4). Under these conditions ee's and yields remained constant over five runs and osmium leaching was below the detection limit of the analysis methods (0.03–0.08%).[279]

(di-h)$_2$dmg **56** (L-lactic) **57** (quinic)

3.3.1.5 *Catalytic Oxidative Desulfurization*

The use of oxidation coupled with extraction into ionic liquids (extraction and catalytic oxidative desulfurization) to remove sulfur containing compounds such as dibenzothiophene (DBT) and its alkyl substituted derivatives from fuel is attracting interest as an alternative to hydrodesulfurization (HDS). In this concept an oil immiscible ionic liquid containing a suitable

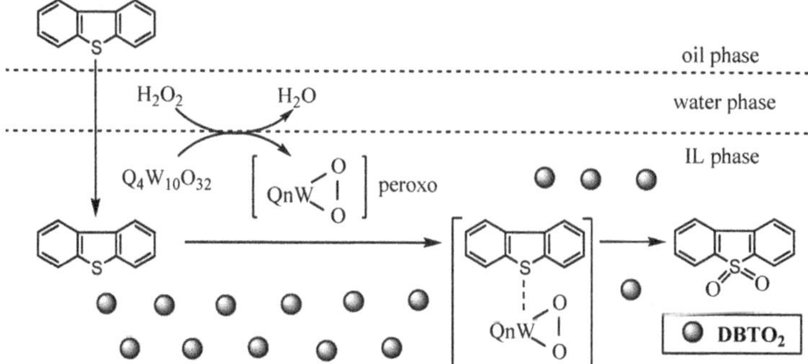

Figure 3.3.5 Schematic of the concept of biphasic ionic liquid-based extractive catalytic oxidative desulfurization.
Reprinted with permission from reference 283. Copyright 2009 American Chemical Society.

oxidation catalyst extracts heteroaromatic sulfides which are subsequently oxidised to the corresponding sulfone and efficiently retained by the ionic liquid phase due to their high polarity; the purified oil can then be recovered by decanting the ionic liquid–catalyst phase (Figure 3.3.5).

Extraction and catalytic oxidative desulfurization (ECODS) of model crude oil (1000 ppm substrate in *n*-octane) with 1 mol% phosphotungstic acid and 3 equivalents of H_2O_2 in $[C_4mim][BF_4]$ removes 98.2% of dibenzothiophene (DBT) after 1 h at 30 °C; a similar efficacy of extraction was achieved with $[C_4mim][PF_6]$ whereas $[C_8mim][BF_4]$ and $[C_8mim][PF_6]$ were much less efficient and removed 64.7% and 63.5% of the DBT, respectively. Under these conditions the efficiency of this oxidative extraction decreases in the order dibenzothiophene (98.2%) > 4,6-dimethylbenzothiophene (74.6%) > benzothiophene (65.7%), however, complete removal of each substrate was achieved after 3 h at 70 °C. The ionic liquid–catalyst mixture also recycled five times with only a minor drop in efficiency from 98.2 to 96.1%.[280] The combination of catalytic oxidation and liquid–liquid extraction in an ionic liquid-based biphasic system is markedly more efficient than simple extraction into ionic liquid. For example, an optimised system containing 5 mol% Na_2MoO_4 and 4 equivalents of 30% H_2O_2 in $[C_4mim][BF_4]$ removed 99% of the DBT from model crude oil (1000 ppm DBT in octane) after 3 h at 70 °C compared with only 13.6% by ionic liquid extraction and 4.1% by oxidation in the absence of ionic liquid. Increasing the alkyl chain length of the imidazolium cation or changing the anion to $[PF_6]^-$ or $[CF_3CO_2]^-$ reduced the extraction efficiency. Under the same conditions removal of benzothiophene (BT) and 4,6-dimethyldibenzothiophene (4,6-DMDBT) only reached 61% and 89.5%, respectively. Extraction efficacy for removal of DBT remained close to 98% across five recycles with <0.9% of the catalyst leaching into the oil phase.[281a] A similar efficiency of oxidative desulfurization of model crude oil was achieved using 1 mol% $[Q_3][PO_4\{MoO(O_2)_2\}_4]$ ($Q = NEt_4^+$, $[N_{1,1,14}]^+$) in combination with two equivalents of H_2O_2 and

$[C_4mim][BF_4]$ as extractant. Under optimum conditions 97.3% of DBT was removed within 3 h at 70 °C compared with only 16.8% in the absence of ionic liquid.[281b] A marked improvement in the efficiency of oxidative extraction was obtained using an ionic liquid-catalyst mixture based on the decatungstates $[Q]_4[W_{10}O_{32}]$ ($Q = [NBu_4]^+$, $[NMe_4]^+$, $[NEt_3(C_7H_7)]^+$). After optimisation 1 mol% $[NBu_4]_4[W_{10}O_{32}]$ in $[C_4mim][BF_4]$ removed 98.0% of the DBT in model crude oil after only 30 min at 60 °C, whereas only 36.0% of the DBT was removed in the absence of ionic liquid. A similar level of extraction was obtained with $[NMe_4]_4[W_{10}O_{32}]$–$[C_4mim][BF_4]$ (96.0%) while $[NEt_3(C_7H_7][W_{10}O_{32}]$–$[C_4mim][BF_4]$ was much less efficient and only removed 66.0% of the DBT. The optimum system recycled efficiently although reaction times had to be extended to 1 h after the third run.[282] Peroxovanadium species generated by reaction of V_2O_5 with H_2O_2 catalyse the ionic liquid-based biphasic oxidative desulfurization of model crude oil. Up to 98.7% removal of DBT was achieved using 5 mol% V_2O_5 with six equivalents of H_2O_2 after 4 h at 30 °C, compared with only 19.3% in the same ionic liquid without V_2O_5. A study of the influence of temperature on conversion showed that the highest conversions were obtained at 30 °C for reaction times >1.5 h and although initial extraction rates were more rapid at higher temperatures self-decomposition of the H_2O_2 at these temperatures limited the final conversion to ~90%. Systems based on other water immiscible ionic liquids including $[C_8mim][BF_4]$ (28.9%), $[C_4mim][PF_6]$ (22.1%) and $[C_8mim][PF_6]$ (70.1%) were much less efficient which was attributed to the formation of a triphasic system comprising a lower ionic liquid layer, a middle H_2O_2 layer and an upper oil layer.[283] The heteropolytungstate $[SO_3HC_3pyr]_3[PW_{12}O_{40}].2H_2O$ forms an exceptionally efficient extraction and catalytic desulfurization system in $[C_8mim][PF_6]$ which removes up to 99.5% of the DBT in model crude oil after only 1 h at 30 °C with 1 mol% catalyst and four equivalents of H_2O_2 as oxidant. While $[C_4mim][PF_6]$–$[SO_3HC_3pyr]_3[PW_{12}O_{40}]\cdot 2H_2O$ also gave a high level of extraction efficiency, those based on $[C_nmim][BF_4]$ ($n = 4, 8$) were much less effective and only removed 77.1% and 81.5% of the DBT, respectively, *i.e.* the anion exerts a marked influence on the performance of these ECODS systems. A similar level of extraction (98.8%) was obtained with 4,6-DMDBT in the same time while BT only reached 69.9% removal. This system is remarkably robust as it recycled nine times with no loss in activity and also reduced the sulfur level of FCC (Fluid Catalyst Cracking) gasoline from 360 ppm to 70 ppm after only 3 h at 30 °C.[284] Hexatungstate-based aqueous H_2O_2-in-ionic liquid emulsions generated from $[R_3NMe][W_6O_{19}]$ ($R = C_4H_9, C_8H_{17}, C_{12}H_{13}$) and $[C_8mim][PF_6]$ are also efficient catalysts for the oxidative desulfurisation of model oil and remove up to 98% of the DBT after 1 h at 60 °C with a catalyst : DBT ratio of 1 : 60. The oxidation was proposed to occur in the begnin microenvironment of the emulsion droplets and phase separatation of the system after reaction facilitated recovery of the clean oil and recycling of the catalyst up to 15 times with no loss in effeciency.[285a] Similarly, the catalytic H_2O_2-in-ionic liquid emulsion generated from the self-emulsifiable task-specific ionic liquid $[P_{6,6,6,14}][W_6O_{19}]$ also removes up to 98% DBT from model oil *via* oxidative desulfurisation at 60 °C by providing an optimum microenvironment for the oxidation, efficiently extracting the resulting sulfone and acting as the peroxo-based catalyst.[285b]

Ionic liquid solutions of the Lewis acidic vanadium complexes V(O)(X-acac)$_2$ (where X corresponds to a substituent on the central carbon of the acetylacetonate) catalyse the desulfurization of model crude oil, either as a biphasic or triphasic system depending on the miscibility of the ionic liquid with water. The highest conversions of DBT to its sulfone were achieved in either [C$_4$mim][NTf$_2$] or [C$_4$mim][PF$_6$] with 5–6 mol% V(O)(X-acac)$_2$ and 4 equivalents of H_2O_2 which gave conversions of 95–96% at room temperature in 30 min; for comparsion, conversions in [C$_4$mim][BF$_4$] or [C$_4$mim][OTf] were moderate.[286] Redox active ionic liquids based on $FeCl_3$ have also acted as both catalyst and extractant for the removal of DBT from model crude oil. The highest extraction efficiency was achieved with [C$_4$mim]Cl/FeCl$_3$, a model oil/ionic liquid mass ratio of three and a substrate/H_2O_2 ratio of three which removed 99.2% of BDT, 90.3% of 4,6-DMDBT and 75.0% of the BT from model crude oil samples after 30 min at 30 °C. Sulfur removal only dropped from 99.2% to 98.6% over nine recycles provided that the ionic liquid was washed with carbon tetrachloride to remove sulfone that accumulated. This system was suggested to operate *via* Fenton-like chemistry involving iron(III)-peroxo intermediates which generates Fe^{2+} and reactive peroxo radicals and/or the ferryl ion and $OH^{\bullet}$ radicals.[287a] Fenton-like ionic liquids of the type [R$_3$MeN]Cl/[FeCl$_3$] (R = C_4H_9, C_8H_{17}, $C_{10}H_{21}$) catalyse the deep oxidative desulfurisation of model oil with remarkable efficiency and remove 96–99% of the DBT after 1 h at room temperature with a H_2O_2/DBT ratio of 14. For comparison, the corresponding $CuCl_2$, $SnCl_2$ and $ZnCl_2$ based ionic liquids are much less effective as desulfurisation only reached 19–26% under the same conditions. The optimum system recycled six times with no significant decrease in desulfurisation activity.[287b]

The protic-at-nitrogen Brønsted acidic ionic liquid [HC$_1$pyrr][BF$_4$] has been used as both catalyst and extractant for the oxidative desulfurization of dibenzothiophene from model crude oil. The authors proposed that the oxidation was caused by hydroxyl radicals generated by decomposition of the H_2O_2. Kinetic studies showed that oxidative desulfurization of a system comprising 10 mL octane, 1550 μg mL^{-1} of DBT, 10 mL [HC$_1$pyrr][BF$_4$] and a H_2O_2/DBT ratio of 3 was complete after 1 h at 60 °C and followed first order kinetics with an apparent rate constant of 0.0816 min^{-1}.[288a,b] The Brønsted acid ionic liquid [Hmim][BF$_4$] has a similar bifunctional role for the oxidative desulfurization of crude oil. However, a relatively high temperature (90 °C), a H_2O_2/substrate ratio of 10 and a large volume of ionic liquid ($V_{\text{model oil}}/V_{\text{IL}} = 3.2/5$) was required to achieve efficient extraction. Under these conditions 93% of the BDT in a sample of model crude oil (1000 μg mL^{-1}) was removed in 6 h while removal of sulfur from commercial crude oil reached 50% and increased to 73% when the H_2O_2/sulfur ratio was increased to 40.[289] High levels of sulfur removal have also been achieved under mild conditions in relatively short reaction times using the protic-at-anion Brønsted acidic ionic liquid [C$_4$mim][HSO$_4$] as both catalyst and extractant. Up to 99.6% of the sulfur was removed at room temperature in 90 min from octane containing 1000 ppm DBT with a system containing a $V_{model\ oil}/V_{IL}$ ratio of 2 and a H_2O_2/DBT molar ratio of 5; this was markedly higher than the 32.8% extracted with $[C_4pyr][BF_4]$ under the same conditions. A decrease in desulfurization efficiency with increasing temperature (for reaction times > 15 min) was explained by a change in viscosity influencing the rate of decomposition of hydrogen peroxide. In a comparative study the efficiency of sulfur removal followed the order DBT (99.6%), BT (94.2%), thiophene (86.6%) > 4,6-DMDBT (85.2%).[290] Task-specific imidazolium-based ionic liquids with a carboxylic acid tethered to the cation are also efficient catalysts and extractants for deep oxidative desulfurization. Within a series of related Brønsted acidic ionic liquids, extraction efficiency paralleled the rate of formation of the peroxycarboxylic acid and decreased in the order $[(CO_2H)C_2mim][HSO_4]$ > $[(CO_2H)C_2mim]$ $[H_2PO_4]$ > $[(CO_2H)C_2mim]$ $[Cl]$ > $[(CO_2H)C_1mim]$ $[HSO_4]$ > $[(CO_2H)$ $C_1mim][H_2PO_4]$. Removal of DBT and 4,6-DMDBT from a 10 mL sample of model crude oil (*n*-tetradecane containing 500 μg of substrate g^{-1}) reached 96.7% and 95.1%, respectively, with 2.5 mmol of $[(CO_2H)C_2mim][HSO_4]$ and 1 mL of 30% H_2O_2. Desulfurization was proposed to occur by extraction of the sulfur compound from the diesel into the ionic liquid layer where it is oxidised by the *in situ* generated ionic liquid-based peroxycarboxylic acid.[291a] The BAIL $[SO_3HC_4C_2im][NTf_2]$ acts as both extractant and catalyst to reduce the sulphur content of DBT in model oil from 1600 ppm to less than 20 ppm, whereas the extraction efficiency of conventional non-protic ionic liquids such as $[C_4pyr][NTf_2]$, $[C_4mim][NTf_2]$ and $[C_4mim][PF_6]$ were much lower and gave sulfur contents between 1200 and 900 ppm under comparable conditions. Desulfurisations conducted with NaClO reached equilibrium much faster than those using H_2O_2 as oxidant as the former was complete after only 5 min at 30 °C whereas the latter required significantly longer (> 60 min) even at 60 °C. The disparate extraction efficiencies of H_2O_2 and NaClO were suggested to reflect the different mechanisms; the BAIL was proposed to decompose the H_2O_2 to $OH^{\bullet}$ radicals which oxidise the DBT after it has been extracted into the ionic liquid phase whereas the NaClO oxidises the DBT at the oil-water interface allowing the resulting sulfone to be rapidly extracted into the IL. The optimum hydrogen peroxide-based system recycled with no loss in desulfurisation efficiency provided the ionic liquid was back-extracted with water to remove the sulfone.[291b]

Lanthanide-containing polyoxometalates $Na_7H_2LnW_{10}O_{36} \cdot 32H_2O$ (Ln = Eu, La) in combination with $[C_4mim][BF_4]$ as extractant and H_2O_2 as oxidant are particularly efficient systems for deep desulfurization of model oil containing BDT, BT and 4,6-DMDBT. Under optimum conditions with a H_2O_2/DBT/$La_7H_2W_{10}O_{36}$ ratio of 500/100/5 complete removal of DBT only required 25 min at 30 °C and the system recycled 10 times with only a minor drop in efficiency. This lanthanide-based oxidative desulfurization system is markedly more efficient than other peroxometalate catalysts, gives high conversions under mild conditions in short reactions times, has a simple work-up procedure and recycles with remarkable efficiency.[292]

The amphiphilic lanthanide-containing peroxometalates (dodecyltrimethylammonium)$_9$LaW$_{10}$O$_{36}\cdot 4H_2O$, (trimethylstearylammonium)$_9$LaW$_{10}$O$_{36}$ and (dimethyldioctylammonium)$_9$LaW$_{10}$O$_{36}$ encapsulated by surfactants of varying length form emulsions with [C$_8$mim][PF$_6$] that are even more efficient than the Na$_7$H$_2$LnW$_{10}$O$_{36}\cdot 32H_2O$–[C$_4$mim][BF$_4$] biphasic system. Extraction efficiency varied with the ionic liquid and increased in the order [C$_8$mim][PF$_6$] > [C$_4$mim][BF$_4$] > [C$_4$mim][PF$_6$] > [C$_8$mim][BF$_4$] at 30 °C and with the length of the surfactant alkyl chain such that (DDA)$_9$LaW$_{10}$O$_{36}$ reaches 100% removal of DBT after 14 min whereas (TSA)$_9$LaW$_{10}$O$_{36}$ and (DODA)$_9$LaW$_{10}$O$_{36}$ require 20 min and 45 min, respectively. For comparison, Na$_7$H$_2$LnW$_{10}$O$_{36}\cdot 32H_2O$–[C$_4$mim][BF$_4$] requires 40 min to completely remove the DBT under similar conditions. The long alkyl chains on the POM surface were proposed to function by absorbing and trapping the weakly polar sulfide through hydrophobic–hydrophobic interactions, which is then oxidised to the sulfone by the active POM species, while also providing access to the H_2O_2 oxidising agent to reform the active tungsten peroxo catalyst. The efficiency of this system allows 100% sulfur removal from model oil with an S content of 50 ppm within 1 min.[293]

3.3.1.6 *Oxidative Condensation*

Benzothiazoles have been prepared *via* ruthenium(III)-catalysed oxidative condensation–cyclisation between 2-aminothiophenol and aldehydes with air as the oxidant. Yields were higher and reaction times shorter in ionic liquids compared with conventional solvents and reaction in [C$_4$mim][PF$_6$] gave the highest yield of 83% after 0.5 h at 80 °C compared with toluene which only reached 50% after 3 h. The rate enhancement was attributed to a hydrogen bond interaction between the acidic imidazolium C2-H and the aldehyde carbonyl activating it towards electrophilic condensation and a C(2)—H—N(imine) hydrogen bond facilitating the intramolecular nucleophilic cyclisation step to afford the dihydrobenzothiazole intermediate which is ultimately oxidised to the corresponding thiazole by ruthenium(III) (Scheme 3.3.3). This protocol was also successfully applied to the synthesis of pyrimidine nucleoside-benzothiazole hybrids.[294] The same transformation between 2-aminothiophenol and benzaldehyde catalysed by $FeCl_3\cdot 6H_2O$

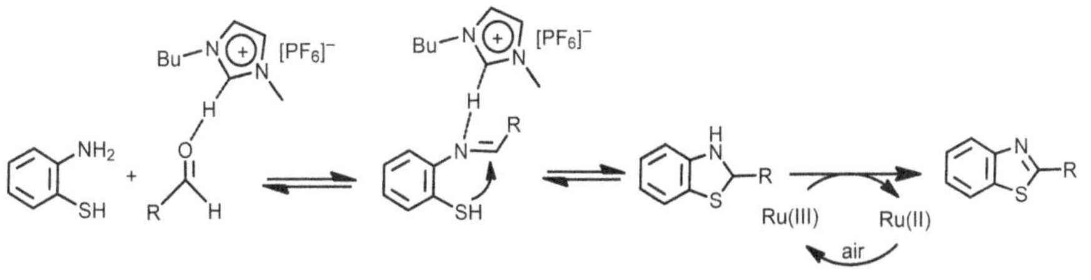

Scheme 3.3.3 Proposed mechanism for the $RuCl_3$-catalysed oxidative formation of 2-substituted benzothiazole.

in $[C_4mim][BF_4]$ also gave the corresponding 2-benzylbenzothiazole, which gradually transformed into 2-benzoylbenzothiazole in a subsequent Fe(III)-catalysed benzylic oxidation. The 2-benzoylbenzothiazole was obtained as the sole product after heating the reaction mixture at 80 °C for 12 h, whereas a mixture of 2-benzylbenzothiazole and 2-benzoylbenzothiazole was obtained under the same conditions in conventional organic solvents. In addition to the ionic liquid acting as both solvent and co-catalyst the $FeCl_3$ may also function as a Lewis acid to activate the aldehyde and imine.[295]

3.3.1.7 Oxidation of Sulfides

A modified electron deficient flavin selectively catalyses the hydrogen peroxide-mediated oxidation of a range of electron-deficient and electron-rich sulfides to their sulfoxides in $[C_4mim][PF_6]$–MeOH to give good yields within 3–5 h. Recycle experiments on several substrates showed that the ionic liquid–catalyst mixture retained its activity over six runs and was sufficiently stable to be stored for several days between runs with no loss in activity.[296] Highly efficient and selective oxidation of sulfides to sulfoxides has been catalysed by a bifunctional ionic liquid catalyst comprising a Brønsted acidic cation, $[SO_3HC_3pyr]^+$, and the hexafluorotitanate anion. In a systematic comparison, ~7 mol% of $[SO_3HC_3pyr][TiF_6]$ catalysed the oxidation of diphenylsulfide at room temperature to give 84% conversion with 87% selectivity to the sulfoxide whereas $[SO_3HC_3pyr]_2[SO_4]$ was less active and gave 51% conversion with 94% selectivity for the sulfide and $[C_4pyr][TiF_6]$ only gave 15% conversion but with 85% selectivity, which shows that the cooperative effect manifests itself in the conversions but that selectivity remains high regardless of the composition of the ionic liquid catalyst. High yields of sulfoxide were obtained for a wide range of substrates in short reaction times (10 min–2 h) using 7 mol% $[SO_3HC_3pyr][TiF_6]$ and one equivalent of 30% H_2O_2 at 25 °C. The efficiency of $[SO_3HC_3pyr][TiF_6]$ was proposed to be due to an interaction between the SO_3H proton and a fluoride in $[TiF_6]^{2-}$ facilitating fluoride–peroxide exchange to afford the active peroxo-titanium species responsible for transferring the oxygen atom to the sulfide (Scheme 3.3.4).[297]

Peroxotungstate $[W_2O_3(O_2)_4]^{2-}$ immobilised on imidazolium and pyridinium-modified silica [Figure 3.3.6(a)] catalyses the selective oxidation of sulfides to sulfoxides. Good yields of sulfoxide and only trace amounts of sulfone could be obtained after 2.5–4 h at 8–12 °C with 1.5 mol% catalyst and H_2O_2 as oxidant. Good yields of sulfone were obtained by increasing the catalyst loading to 2 mol%, using 2.5 equivalents of H_2O_2 and increasing the reaction time.[298] The same peroxotungstate immobilised on a multilayer ionic liquid-modified silica [Figure 3.3.6(b)] is a more efficient catalyst for the selective oxidation of sulfides than the corresponding monolayer ionic liquid system as 1.5 mol% of the former gave yields in excess of 90% after 1.5 h whereas the latter required 2.5 h to reach a similar conversion. The improvement in performance was attributed to the three-dimensional environment of the multilayer system which more closely resembles the liquid

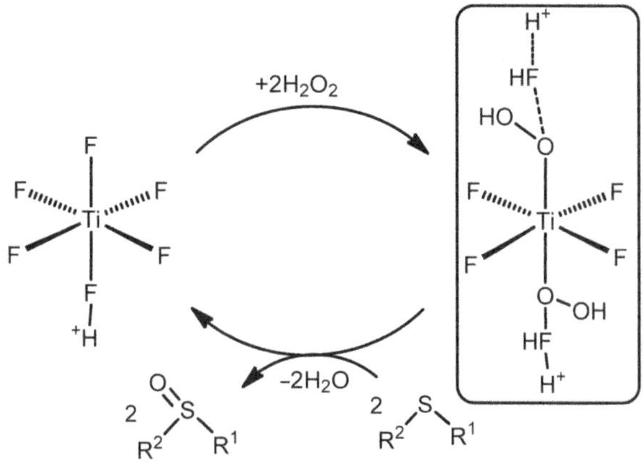

Scheme 3.3.4 Proposed mechanism for sulfoxidation catalysed by $[SO_3HC_3pyr]$-$[TiF_6]$ using H_2O_2.

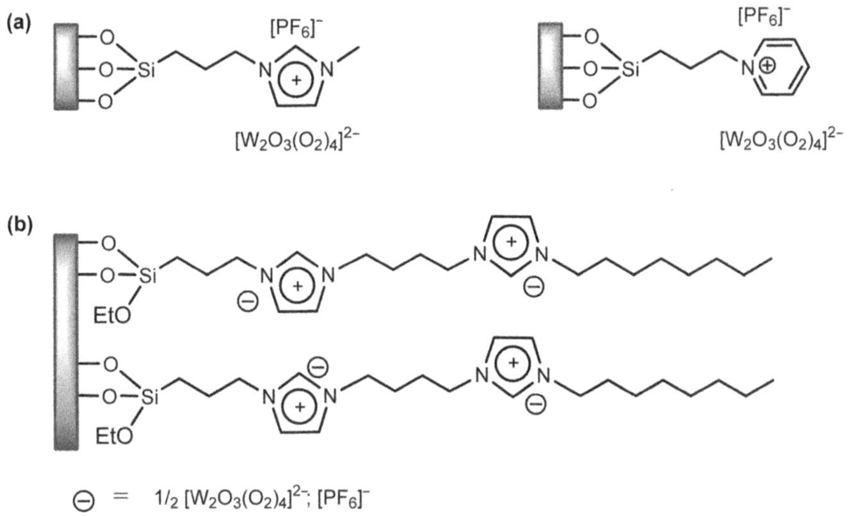

Figure 3.3.6 Peroxotungstate immobilised on (a) ionic liquid-modified silica and (b) multilayer ionic liquid-modified silica.

phase character of the ionic liquid. A direct comparison with the performance of the same peroxometalate in the corresponding ionic liquid under homogeneous conditions would have been instructive and informative. Conversions varied with the length of the *N*-alkyl chain attached to the terminal imidazolium cation, albeit it only by ± 5%, with the *N*-octyl system giving the highest conversion; presumably by providing the appropriate balance of hydrophobicity and hydrophilicity for effective substrate access and product departure. Catalyst recovery was straightforward and the system recycled efficiently with only a minor reduction in activity after eight runs. As for the monolayer system described above high selectivity for the sulfone was obtained by increasing the catalyst loading and the reaction time.[299]

The peroxotitanate-catalysed hydrogen peroxide-mediated oxidation of sulfides to their sulfones occurs with slightly faster rates in $[C_4mim][BF_4]$–MeOH at room temperature than in neat methanol. Good conversions were obtained in short reaction times for a range of aryl and alkyl sulfides.[300] The electrocatalytic selective oxidation of sulfides to sulfoxides based on electroactivation of urea hydrogen peroxide in $[C_4mim][BF_4]$ gave yields that matched those obtained by acid catalysis with 1*R*-(−)-10-camphor sulfonic acid. The involvement of electrogenerated BF_3 as a Lewis acid catalyst from $[C_4mim][BF_4]$ was implicated.[301]

3.3.1.8 Bayer–Villiger Oxidation

Bernini, Fabrizi and co-workers reported the first examples of the Baeyer–Villiger oxidation of cyclic ketones to be catalysed by $MeReO_3$–H_2O_2 in $[C_4mim][BF_4]$ and obtained reasonable to good yields with a variety of substrates. The product was isolated by extraction with diethyl ether and the ionic liquid–catalyst mixture recycled five times, indicating that the catalyst was both stable and efficiently immobilised.[302] The platinum(II)-catalysed Baeyer–Villiger oxidation of cyclohexanone in ionic liquids gave significantly higher yields than in halogenated solvents and in shorter reaction times. Reactions conducted with $[Pt_2(\mu\text{-}OH)_2(dppb)_2][BF_4]_2$ under biphasic conditions *i.e.* in hydrophobic ionic liquids such as $[C_4mim][PF_6]$ and $[C_nmim][NTf_2]$ ($n = 4, 6, 8$) gave better yields than reactions conducted under homogeneous conditions in hydrophilic ionic liquids $[C_4mim][X]$ ($X = BF_4^-$, OTf^-). Yields varied as a function of the anion and cation and $[C_4mim][NTf_2]$ was identified as the optimum solvent.[303] Significantly higher yields were obtained in ionic liquids compared with dichloromethane for the Lewis acid-catalysed Baeyer–Villiger oxidation of cyclic ketones with bis(trimethylsilyl)peroxide (BTSP) as oxidant. The oxidation of cyclopentanone and cyclohexanone catalysed by $BF_3.OEt_2$ in $[C_4mim][NTf_2]$ gave δ-valerolactone and ε-caprolactone in 95% and 88% yield, respectively, compared with yields of 73% and 60%, respectively, in dichloromethane. Interestingly, $[C_4mim][OTf]$ acted as both solvent and catalyst in the Baeyer–Villiger oxidation of cyclic ketones and gave good yields in short reaction times. The triflate anion was shown to be crucial for the catalysis and was proposed to facilitate formation of the Criegee adduct **58** while the liberated trimethylsilyl triflate by-product was involved in the subsequent rearrangement of **59** to the final ester, according to Scheme 3.3.5.[304]

Surface-immobilised Brønsted acidic ionic liquids prepared by anchoring 1-methyl-3-(triethoxysilylpropyl)imidazolium hydrogen sulphate to a multimodal pore structure silica catalyses the Baeyer–Villiger oxidation of cyclic ketones at 50 °C to give the corresponding lactones in excellent yield in reasonable reaction times (5–20 h).[305] The Baeyer–Villger oxidation of cyclic ketones in $[C_4mim][NTf_2]$ with molecular oxygen and benzaldehyde as

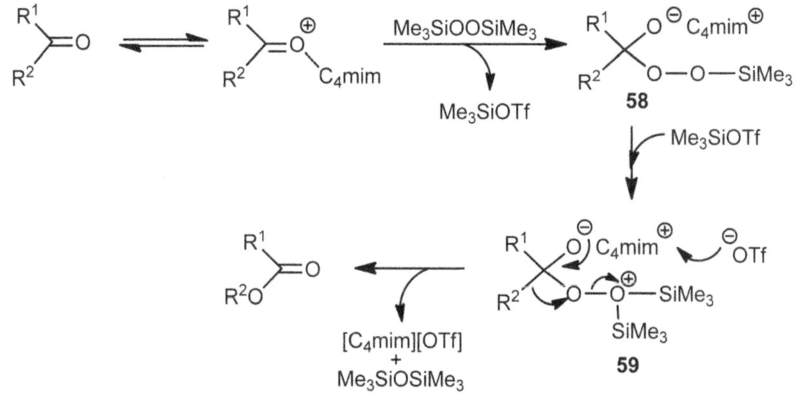

Scheme 3.3.5 Proposed mechanism for the Baeyer–Villiger oxidation of ketones by BTSP in the presence of [C_4mim][OTf] as the solvent and catalyst.

oxidant shows a significant enhancement in rate at 90 °C in the presence of a catalytic amount of 1,1′-azobis(cyclohexanecarbonitrile) (ACHN) as radical initiator. Two radical pathways were considered both involving an acylperoxy radical intermediate.[306]

3.3.2 Hydrogenation

The solubility of hydrogen in a range of ionic liquids has been determined by high pressure ^{1}H NMR measurements and is low compared to conventional organic solvents and similar to that of water. Thus, not unreasonably one might anticipate that the low solubility of dihydrogen in ionic liquids and the potential mass transfer problems would limit their use as solvents for reactions involving hydrogen. Dyson *et al.* studied the hydrogenation of benzene in eleven different ionic liquids and, surprisingly, within experimental error there was no difference in the rates of hydrogenation across all eleven ionic liquids; this was attributed to the higher solubility of H_2 in the substrate and substrate–IL mixture than in the neat ionic liquid which therefore has less influence on the reaction rate. As a corollary, in an ideal biphasic process the ratio of substrate to ionic liquid should be high to decrease the effect of mass transfer. Moreover the fast diffusion of hydrogen into ionic liquids results in high hydrogen transfer rates into the catalyst layer such that the consumed hydrogen, albeit in low concentration, is replenished rapidly.[307]

3.3.2.1 Arene Hydrogenation

A remarkable enhancement in the rate of hydrogenation of arenes has been obtained in ionic liquids compared with conventional solvents. A [C_4mim][BF_4] solution of [Ru(η^6-*p*-cymene)(η^2-TRIPHOS)Cl][PF_6] catalyses the hydrogenation of benzene and toluene with TOFs of 476 h^{-1} and 205 h^{-1}, respectively, whereas the corresponding TOFs in dichloromethane

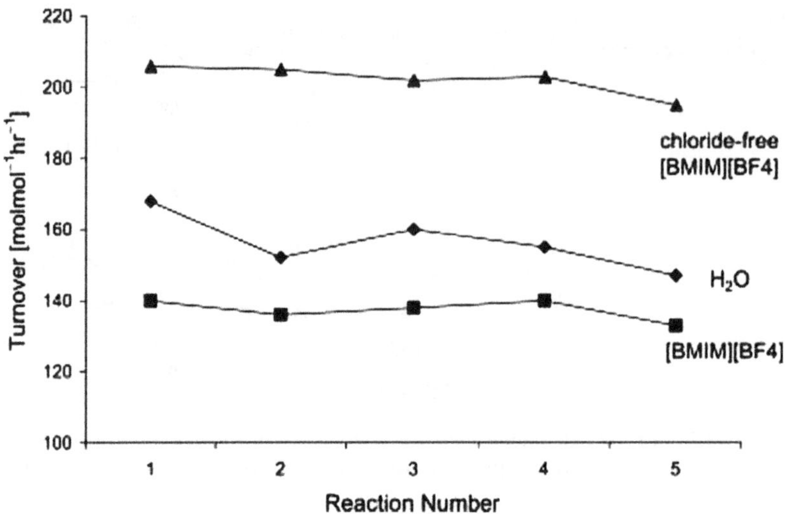

Figure 3.3.7 Reuse of [(η^6-*p*-cymene)Ru(1,3,5-triaza-7-phosphaadamantane)Cl_2] for the hydrogenation of benzene to cyclohexane in water, [bmim][BF_4] and halide-free [bmim][BF_4] illustrating the differences in TOFs. Reproduced with permission from reference 29. Copyright 2003 Wiley.

were only 242 h^{-1} and 74 h^{-1}, respectively. The same system was essentially inactive towards the hydrogenation of styrene and its derivatives but selectively hydrogenated allylbenzene to allyl cyclohexane with a high TOF. The authors favoured a slippage mechanism for this arene hydrogenation on the basis of isolating recovered catalyst, albeit with benzene in place of *p*-cymene, and isolation of a single isomer of $C_6H_6D_6$ from the hydrogenation of C_6D_6; both are consistent with addition of dihydrogen to the arene from one face of a coordinated arene. The pendent phosphino group was shown to be essential for arene hydrogenation and was thought to operate by promoting η^6 to η^4-ring slippage *via* an arm-on/off mechanism to effect hydrogenation of the uncoordinated double bond.[308] The same authors have also highlighted the detrimental effect of halide impurities on the ionic liquid biphasic ruthenium-catalysed hydrogenation of arenes. The use of [(η^6-*p*-cymene)Ru(1,3,5-triaza-7-phosphaadamantane)Cl_2] for the hydrogenation of benzene in [C_4mim][BF_4] under 60 bar of H_2 at 90 °C gave a TOF of 124 mol mol^{-1} h^{-1}, which was significantly lower than that of 170 mol mol^{-1} h^{-1} obtained under aqueous phase conditions (Figure 3.3.7). The same hydrogenation conducted in 'halide free' [C_4mim][BF_4] prepared by methylation of 1-butylimidazole with [Me_3O][BF_4] resulted in a substantial increase in TOF to 206 mol mol^{-1} h^{-1}; similar improvements were obtained for the hydrogenation of toluene, ethylbenzene and chlorobenzene for reactions conducted in halide-free ionic liquid compared with ionic liquid prepared by metathesis and contaminated with chloride. Both the ionic liquid and aqueous phase systems showed good stability over five recycles with negligible loss in activity.[29]

On the basis that the high solubility of charged clusters in ionic liquids should render them suitable for applications in catalysis the efficiency of $[H_4Ru_4(\eta^6\text{-}C_6H_6)_4][BF_4]_2$–$[C_4mim][BF_4]$ for the biphasic hydrogenation of benzene, toluene and cumene was investigated and compared against the corresponding aqueous phase system. The TOF of 364 h^{-1} obtained for the hydrogenation of benzene at 90 °C under 60 atm of H_2 was similar to that of 352 h^{-1} in water suggesting that the active species could be the same in both systems, possibly the oxidative addition adduct $[H_6Ru_4(\eta^6\text{-}C_6H_6)_4][BF_4]_2$. Moreover, these TOFs compared favourably with that of 246 h^{-1} for $[(\eta^6\text{-}C_6H_6)Ru(\mu\text{-}Cl)Cl]_2$ at 50 °C under 50 atm H_2. The main advantage of this ionic liquid–organic system over its aqueous counterpart lies in the ease of separating the catalyst from the product/starting material stream and subsequent purification of the solvent; in this regard, the same batch of ionic liquid could be used for the catalytic hydrogenation of several different substrates.[309]

The first highly enantioselective phosphine-free ruthenium-catalysed asymmetric hydrogenation of 2-methyl quinoline was achieved with [(*p*-cymene)Ru(TsDPEN)X] ($X = Cl^-$, OTf^-) in $[C_4mim][PF_6]$ under slightly acidic conditions [Equation (3.3.2)]. Under optimum conditions 1 mol% Ru/TsDPEN gave quantitative conversion of 2-methylquinoline to 1,2,3,4-tetrahydroquinoline in 99% ee, which was a significant improvement on that of 94% obtained in methanol. In addition to the enhancement in enantioselectivity the ionic liquid provided effective stabilisation of the catalyst which was reused with no loss in activity or enantioselectivity after exposure to air for 30 days; for comparison, under the same conditions the catalyst decomposed within one week in methanol. Such a dramatic enhancement in catalyst stability by $[C_4mim][PF_6]$ was attributed to a combination of solvation of the cationic ruthenium catalyst and the low solubility of oxygen in the ionic liquid. The ionic liquid–catalyst mixture also recycled efficiently with no loss in enantioselectivity over eight runs and only a minor reduction in activity after the sixth run. Excellent ee's (97–99%) and conversions were obtained for a range of substituted quinolines. The authors proposed that this hydrogenation occurs *via* a different mechanism to that of ketones in methanol and suggested a pathway involving deprotonation of a dihydrogen adduct followed by a series of proton and hydride transfer steps (Scheme 3.3.6).[310]

[(p-cymene)RuX(TsDPEN)]
H_2, $[C_4mim][PF_6]$ (3.3.2)

3.3.2.2 *Alkene Hydrogenation*

3.3.2.2.1 Olefins and Dienes. The earliest report of an ionic liquid biphasic hydrogenation was conducted in $[C_4mim]^+$-based salts with

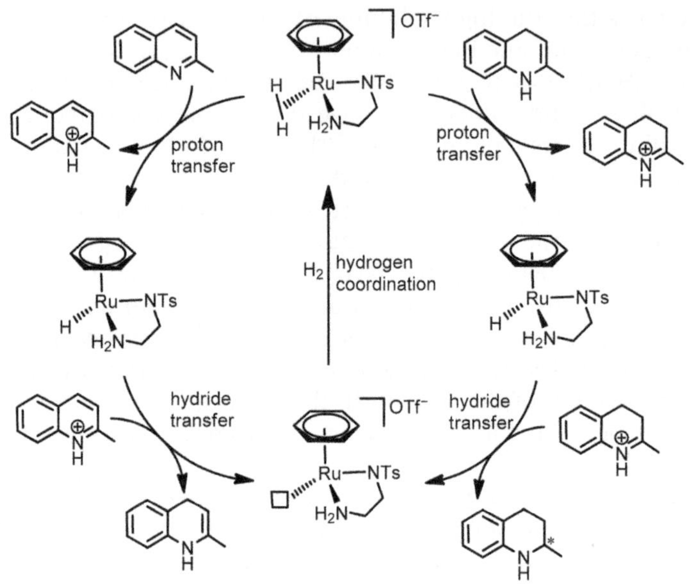

Scheme 3.3.6 Proposed mechanism for the hydrogenation of 2-methylquinoline catalysed by [(*p* cymene)Ru(TsDPEN)OTf] in an ionic liquid.

0.27 mol% $[Rh(nbd)(PPh_3)_2][PF_6]$ for the rhodium-catalysed hydrogenation of pent-1-ene at 30 °C (0.1 MPa H_2) and under these conditions rates were five times higher than the corresponding homogeneous system in acetone. The TOF of 2.54 min^{-1} in $[C_4mim][SbF_6]$ was approximately twice that of 1.72 min^{-1} in $[C_4mim][PF_6]$ which was tentatively attributed to the higher solubility of substrate in the former. A low conversion obtained in $[C_4mim][BF_4]$ was attributed to the presence of impurity chloride ions deactivating the catalyst by coordination.[136] The vast majority of hydrogenations involving ionic liquid-immobilised catalysts are now conducted under biphasic conditions. Solutions of $[Rh(PPh_3)_3Cl]$ and $[Rh(cod)_2][BF_4]$ in $[C_4mim][X]$ ($X = BF_4^-$, PF_6^-) efficiently catalyse the hydrogenation of cyclohexene under 10 atm of H_2 at room temperature with substrate : catalyst ratios of 15 000 for the former and 6550 for the latter. High turnover frequencies have also been obtained for the hydrogenation of 1-hexene, cyclohexene and butadiene using $[RuCl_2(PPh_3)_3]$ immobilised in $[C_4mim][BF_4]$ under 25 atm of H_2 at room temperature. Under these conditions TOFs of 537 h^{-1} and 170 h^{-1} were obtained for the hydrogenation of 1-hexene and cyclohexene, respectively, while butadiene gave a mixture of butenes and butane with moderate selectivity. The ruthenium catalyst was almost completely retained in the ionic liquid after the product was decanted.[311a,b]

Carbonyl cluster anions have also been explored as catalyst/catalyst precursors for hydrogenation of olefins on the basis that these highly charged compounds would be soluble and efficiently retained in an ionic liquid and that the polar non-nucleophilic environment of ionic liquids would increase catalyst lifetime by effective solvation and protection. Four cluster anions $[Fe_3(CO_{11}]^-$, $[HWOs_3(CO)_{14}]^-$, $[H_3Os_4(CO)_{12}]^-$ and $[RuC(CO)_{16}]^{2-}$ were evaluated as catalysts for the hydrogenation of styrene in $[C_4mim][BF_4]$, octane and methanol at 100 °C under 50.7 bar of H_2. The activity of $[Fe_3(CO_{11}]^-$ and $[RuC(CO)_{16}]^{2-}$ were essentially the same in ionic liquid and molecular solvents; the former cluster is almost inactive whereas the latter gave good conversions, however, mercury poising strongly indicated that the active catalyst was in fact ruthenium colloid/nanoparticles and the minor difference in activity was most likely due to different particle sizes/distributions. Both osmium clusters are markedly more active in ionic liquid than methanol or octane which appears to be due to an increase in the stability of the catalyst in ionic liquid compared with molecular solvents. The TOF for the hydrogenation of styrene using $[H_3Os_4(CO)_{12}]^-$ varied from 392 to 718 h^{-1} across a range of ionic liquids; the lowest was obtained in $[C_4C_1{}^2mim][PF_6]$ which is the most viscous ionic liquid while the highest was obtained in $[C_8mpyr][BF_4]$ which formed a single phase with the substrate and was therefore not subjected to mass transfer resistance.[312]

3.3.2.2.2 Asymmetric Hydrogenation of Dehydroamino Acids and Unsaturated Acids/Amides

3.3.2.2.2.1 Asymmetric Hydrogenation with Unmodified Ligands. Catalytic asymmetric hydrogenation is an exceptionally powerful method for the synthesis of non-racemic compounds but platinum group metal precursors and/or the chiral ligands are rather expensive and as such there has been considerable interest in developing methods for separating and recycling these catalysts. In this regard, one of the principle drivers for conducting asymmetric transformations in ionic liquids under liquid–liquid biphasic conditions has been the stabilisation and immobilisation of expensive enantiopure catalysts to facilitate their separation and reuse. The first report of an asymmetric hydrogenation conducted in an ionic liquid appeared in 1995 in which 1 mol% $[Rh(COD)\{(-)diop\}][PF_6]$ was immobilised in $[C_4mim][SbF_6]$ and used for the hydrogenation of α-acetamidocinnamic acid to give (*S*)-phenylalanine in 64% *ee*.[136] Shortly after, $[RuCl_2\{(S)\text{-BINAP}\}]_2.NEt_3$ immobilised in $[C_4mim][BF_4]$ with the aid of alcohol as co-solvent catalysed the asymmetric hydrogenation of 1-arylacrylic acids and gave ee's that matched those in neat alcohol. While the methanol-based system formed a homogeneous solution at the end of the reaction *i*-PrOH–$[C_4mim][BF_4]$ formed a two phase mixture which allowed the product to be separated and the recovered ionic liquid–catalyst phase to be reused several times without significant loss in enantioselectivity or activity.[313] The concentration of hydrogen in solution has been shown to influence both the activity and enantioselectivity. Molecular hydrogen is approximately four times more soluble in $[C_4mim][BF_4]$ than in

$[C_4mim][PF_6]$ under the same pressure and as a consequence 1 mol% [{(−)-1,2-bis(2*R*,5*R*)-2,5-diethylphospholano)benzene}Rh(COD)][OTf] gave 73% conversion and 93% ee for the asymmetric hydrogenation of 2-α-acetamindo cinnamic acid in $[C_4mim][BF_4]$ whereas the same catalyst only gave 26% conversion and 81% ee in $[C_4mim][PF_6]$; this was related to the availability of hydrogen at the catalyst site, *i.e.* conversion and enantioselectivity appear to increase when the solubility of hydrogen in the ionic liquid increases. The influence on enantioselectivity of the hydrogen concentration in solution was clearly evident as a higher hydrogen pressure led to a marked improvement in ee which reached 90% at 100 atm. The kinetic resolution of (±)-methyl-3-hydroxy-2-methylenebutanoate catalysed by $[RuCl_2\{(S)\text{-tol-BINAP}\}]_2 \cdot NEt_3$ immobilised in $[C_4mim][BF_4]$–*iso*-PrOH also showed a dramatic dependence on hydrogen pressure and the ee of 98% obtained under 40 atm of H_2 was markedly higher than that of 75% at 20 atm and closer to the 97% obtained under homogeneous conditions in neat methanol.[314] The ee's obtained for the asymmetric hydrogenation of enamides catalysed by [Rh{(*R*,*R*)-Me-DUPHOS}(COD)] in $[C_4mim][PF_6]$ matched those obtained under homogeneous conditions, with the added benefit of improving the stability of the highly oxygen-sensitive catalyst and facilitating efficient recycling. The stability provided by the ionic liquid was clearly evident in comparative recycle experiments under homogenous conditions in *iso*-PrOH and under biphasic conditions in $[C_4mim][BF_4]$; although quantitative conversion and an ee of 99% was obtained under homogeneous conditions for reactions conducted with rigorous exclusion of air, the ee dropped to 57% with only 5% conversion when the catalyst was transferred to the reactor in the presence of air. In contrast, catalyst immobilised in ionic liquid gave consistently high ee's across five runs (96–94%) and although the activity dropped after the first run, conversions remained constant for the next four runs, the last of which was performed after leaving the catalyst solution exposed to air for 24 h.[315]

The solubility of hydrogen also has a dramatic influence on the efficiency of the ruthenium-catalysed asymmetric hydrogenation of class I and class II α,β-unsaturated carboxylic acids (Figure 3.3.8) in ionic liquid and CO_2-expanded ionic liquid; class I substrates are hydrogenated in high ee at high hydrogen concentrations while class II substrates give high ee's at low hydrogen concentrations. The hydrogenation of tiglic acid (class II) catalysed by $[Ru(O_2CMe)\{(R)\text{-tol-BINAP}\}]$ gave high ee's in all ionic liquids tested where low H_2 solubility is beneficial. The authors argued that a higher ee was

Figure 3.3.8 Structures of typical class I and class II substrates.

obtained in [C_4mim][PF_6] (93%) than in [C_4mim][BF_4] (88%) because the latter is less viscous, has a lower surface tension and is better able to dissolve hydrogen which would result in a greater concentration of H_2, which is unfavourable for the hydrogenation of class II substrates. Similarly, hydrogenation in the mixed solvent [C_4mim][PF_6]–*iso*-PrOH was less selective than in neat ionic liquid, as evidenced by ee's of 93% and 76%, respectively, which was again attributed to the lower viscosity and more efficient mass transfer of hydrogen in the mixed solvent system. In contrast, hydrogenation of adipic acid (class I) requires a high concentration of hydrogen and high mass transfer rates to achieve high enantioselectivity. The highest ee of 92% was obtained in methanol under 50 bar of hydrogen whereas much lower ee's were obtained in ionic liquid, varying between 15 and 32%; intermediate ee's were obtained in ionic liquid–alcohol systems consistent with the addition of alcohol lowering the viscosity and improving mass transfer. The variation in enantioselectivity observed across a range of ionic liquids could not be explained by differences in hydrogen solubility and mass transfer alone and it is likely that other solvent parameters such as polarity, coordinating ability and hydrophobicity may be important. The use of CO_2-expanded ionic liquids for the hydrogenation of adipic acid supported this interpretation as an ee of 32% was achieved in [C_4mim][PF_6] which increased to 57% in CO_2-expanded [C_4mim][PF_6]; a similar increase was obtained in [C_4mim][PF_6]–alcohol mixture which is consistent with greater H_2 solubility and mass transfer rates. Conversely, and as expected, the ee's obtained for the hydrogenation of tiglic acid in CO_2-expanded ionic liquid were significantly lower than those obtained in the neat ionic liquid. Thus, it is possible to achieve high ee's across a range of substrates with a judicious choice of ionic liquid or ionic liquid–co-solvent system and conditions.[57,316]

Enhanced reaction rates and/or ee's were obtained for the [Rh(NBD)$_2$][BF_4]/Taniaphos-catalysed asymmetric hydrogenation of methyl α-acetamidoacrylate in wet ionic liquids such as [C_8mim][BF_4]–H_2O and [C_4mim][[NTf_2]–H_2O compared with either the pure ionic liquid, ionic liquid–alcohol mixture or neat alcohol. A ternary combination based on ionic liquid, water and toluene gave the highest ee's for hydrogenation of the more lipophilic substrate methyl α-acetamidocinnamate. Interestingly, the positive effect of wet ionic liquid was ligand dependent; all ferrocenyl diphosphines investigated gave higher ee's and conversions in ionic liquid–water than in methanol whereas bppm, Prophos and Me-Duphos tended to give inferior results. Recycle experiments showed that Rh/Taniaphos in [C_4mim][BF_4]–H_2O gave near complete conversion and >99% ee for the asymmetric hydrogenation of methyl α-acetamidoacrylate over five consecutive runs with only a slight drop in activity in run six which was attributed to catalyst deactivation by trace air during work-up rather than leaching as the rhodium content in the aqueous phase was only 0.5%. Industrially relevant turnover numbers in excess of 10 000 were also achieved for the hydrogenation of methyl α-acetamidoacrylate by using 0.0001 mol%

Rh/Taniaphos in [C_4mim][BF_4]–H_2O under 10 bar of H_2.[317] The Rh/Et-DuPHOS and Ru/BINAP-catalysed asymmetric hydrogenation of methyl α-acetamidoacrylate and methyl acetoacetate, respectively, cannot be performed in neat ionic liquid as both catalysts were completely inactive in [C_4mim][PF_6] and [C_4mim][BF_4] but high TOFs and good ee's could be achieved in the presence of a co-solvent either under homogeneous conditions or as a liquid–liquid biphasic system. For example, the TOF of 3012 h^{-1} and ee of 97% for the Rh/Et-DuPHOS-catalysed hydrogenation of α-acetamidoacrylate in [C_4mim][PF_6]–MeOH matched that of 3225 h^{-1} and 97% obtained in neat methanol under the same conditions; similarly, the TOF and ee obtained for the hydrogenation of methyl acetoacetate with Ru/BINAP in [C_4mim][PF_6]–MeOH was also close to that obtained in neat methanol or ethanol. Reactions conducted in [C_4mim][BF_4]–methanol gave slightly lower TOF and ee's for both substrate–catalyst combinations than in neat alcohol while the use of [C_4mim][BF_4]–H_2O resulted in a dramatic reduction in performance. The use of biphasic conditions involving [C_4mim][PF_6] and either hexane, diethyl ether or 2-propanol led to much lower TOFs than those obtained under monophasic conditions and interestingly water had a beneficial influence relative to the other co-solvents in the Rh/Et-DuPHOS-catalysed hydrogenation of α-acetamidoacrylate, this was attributed to efficient mixing.[318]

Fluorinated BINOL-derived phosphites gave poor yields but moderate to good ee's for the rhodium-catalysed asymmetric hydrogenation of dimethyl itaconate and methyl-2-acetamidoacrylate in [C_4mim][PF_6] under 1 bar of H_2 at 40 °C with catalyst generated *in situ* from 2.5 mol% [Rh(COD)$_2$][BF_4] and 5 mol% phosphite. Ee's decreased quite dramatically as the hydrogen pressure increased with no noticeable difference in activity and while the higher hydrogen solubility in [C_4mim][PF_6]–*sc*CO_2 resulted in a substantial increase in activity, the enantioselectivity decreased. Recycle experiments gave consistent conversions over 10 runs but racemic product was obtained after the first run due to ligand leaching.[319]

3.3.2.2.2.2 Asymmetric Hydrogenation with Charge-Tagged Ligands. Modification of a phosphine by attachment of an inert ionic tag as a strategy to immobilise a catalyst, improve its retention and prevent/limit leaching as well as to enhance its stability and facilitate reuse has been widely investigated. The imidazolium cation is the most commonly used tag as it closely resembles the cation of the ionic liquid and is generally relatively straightforward to introduce at a site remote to the coordinating heterotom(s). Numerous papers report the use of this strategy for the immobilisation of rhodium and ruthenium alkene hydrogenation catalysts based on diphosphines, phosphites and phosphoramidates, a selection of which are shown in Figure 3.3.9.

Imidazolium-tagged diphosphine **60** combines with [Rh(COD)$_2$][BF_4] to form a highly efficient catalyst for the biphasic asymmetric hydrogenation of *N*-acetylphenylethenamine in [C_4mim][SbF_6]–*iso*-PrOH under 1 atm of H_2

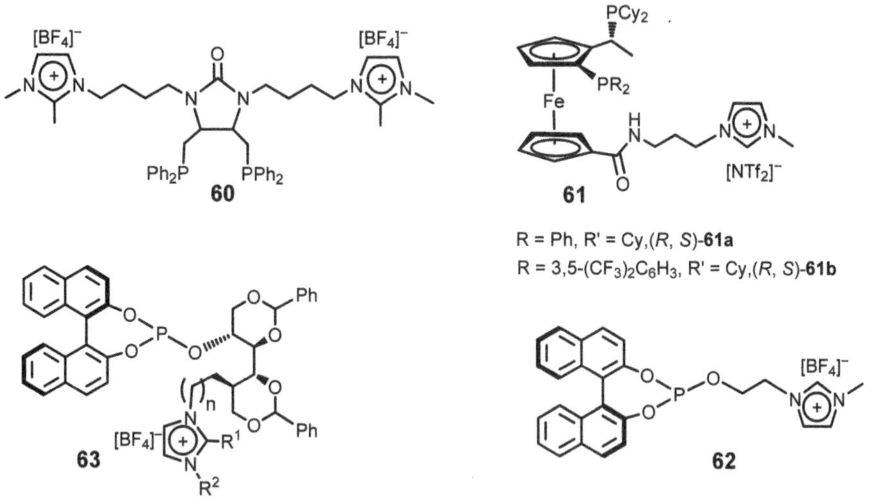

R = Ph, R' = Cy,(*R*, *S*)-**61a**
R = 3,5-$(CF_3)_2C_6H_3$, R' = Cy,(*R*, *S*)-**61b**

Figure 3.3.9 A selection of imidazolium-tagged diphosphines and phosphites.

which recycles more efficiently than its unmodified counterpart. Excellent ee's were obtained with the imidazolium-tagged catalyst (95.4–97.0%) as well as its unmodified counterpart (91.4–95.8%), however, the former recycled efficiently and was reused three times without any loss in activity; catalyst activity decreased slightly in the fourth run but good conversions were obtained after longer reaction times. In contrast, the conversions obtained with the unmodified catalyst dropped quite significantly after only one recycle. The tagged catalyst showed no evidence for either rhodium or diphosphine leaching after the first run whereas 2% of the rhodium and 6% of the phosphorus of the unmodified catalyst leached to the *iso*-PrOH layer during the first run. However, leaching alone did not account for the activity–recycle profile of these two systems, which the authors suggested reflected the greater stability of the imidazolium-tagged catalyst.[320] Josiphos-type diphosphines modified with an imidazolium tag are also remarkably efficient ligands for the rhodium-catalysed asymmetric hydrogenation of methyl acetamidoacrylate and dimethyl itaconate. The conversions (100%) and ee's (97–99%) obtained with 0.5 mol% $[Rh(NBD)_2][BF_4]$ and either **61a–b** under biphasic conditions in $[C_4mim][BF_4]$–TBME, $[C_4mim][BF_4]$–*iso*-PrOH or $[C_4mim][BF_4]$–toluene paralleled those obtained with their unmodified counterparts in organic solvent as well as under the same biphasic conditions. However, recycle studies showed that the tagged catalyst had much better reusability in $[C_4mim][BF_4]$–TBME than the corresponding unmodified catalyst as the former could be recycled eight times with only a minor drop in TOF (15%) whereas the activity of the latter dropped much more dramatically. Even though ICP-AES analysis of the TBME layer showed that 12 ppm of rhodium (2.3%) leached from the unmodified system compared with only 0.4 ppm (0.1%) from the tagged catalyst this alone did not account for the poor catalyst reusability. In addition to more efficient immobilisation the authors reasoned that the high affinity of the imidazolium-modified catalyst for the ionic liquid provided better protection with respect to deactivation by oxygen introduced into the system during work-up and recycle; conversely the unmodified catalyst would be 'more exposed' to the oxygen in the TMBE layer.[321] Monodentate BINOL-derived phosphites and phosphoramidates have evolved into highly versatile ligands for a host of asymmetric transformations and as such their modification and use in ionic liquids has been explored. The first example of an imidazolium-tagged phosphite was prepared by Gavrilov, Reetz and co-workers for the rhodium-catalysed asymmetric hydrogenation of dimethyl itaconate. The ee of 79% obtained with 0.01 mol% [Rh(COD)$_2$][BF$_4$]/**62** in [C$_4$C$_1{}^2$mim][BF$_4$] under 25 bar of H_2 at 22 °C was slightly lower than that of 94% obtained for the same catalyst in dichloromethane. The imidazolium-tagged precatalyst [Rh(COD)(**62**)$_2$][BF$_4$] was also anchored to phosphotungstic acid-modified silica and the resulting material shown to be an efficient heterogeneous catalyst for the asymmetric hydrogenation of dimethyl itaconate; under these conditions 1 mol% rhodium gave complete conversion and 88% ee over three runs before showing any reduction in performance.[322] Carbohydrate-based monophosphites are also effective ligands for asymmetric hydrogenation and as such have been modified with imidazolium tags to afford **63** ($n=4$, 6, 12; $R^1=H$, Me, $R^2=Me$, *n*-Bu), which are also highly efficient ligands for asymmetric hydrogenation. The derived rhodium precatalysts [Rh(COD)(**63**)$_2$][BF$_4$] gave quantitative conversions and ee's in excess of 99% for the asymmetric hydrogenation of *N*-acetylphenylethenamime in dichloromethane, under 10 atm of H_2 at room temperature. Excellent ee's (92–99%) were also obtained for the asymmetric hydrogenation of α-dehydroamino acid esters in dichloromethane under the same conditions. Recycle experiments under biphasic conditions in [C$_4$mim][BF$_4$]–toluene showed that each catalyst gave good conversions (97–100%) and excellent ee's (97–99%) for the asymmetric hydrogenation of *N*-acetylphenylethenamine in the first run but that reusability depends on the structure of the imidazolium tag and the ionic liquid. As rhodium leaching was determined to be less than 0.1% for all catalysts tested the erosion in performance on recycle was proposed to be due to deactivation of the catalyst by oxygen contaminant introduced during work-up and recycle; those ligands with the highest affinity for the ionic liquid were the most efficiently protected. The most efficient combination of [C$_4$mim][BF$_4$] and [Rh(COD)(**63**)$_2$][BF$_4$] ($R^1=H$, $R^2=Me$) recycled 10 times with negligible loss in ee or conversion.[323] The efficiency and recyclability of BINAP tagged at the 4,4′- and 5,5′-positions with methylammonium groups (4,4′- and 5,5′-*diam*BINAP **64** and **65**, respectively) in the ruthenium-catalysed asymmetric hydrogenation of ethyl acetoacetate showed a marked dependence on the nature of the ionic liquid. Catalysts prepared *in situ* from the bromide salt of 4,4′- or 5,5′-*diam*BINAP (**64** or **65**) and [Ru(η^3-2-methylallyl)$_2$(η^2-COD)] gave quantitative conversions in [C$_4$mim][X] ($X=PF_6{}^-$, $NTf_2{}^-$, $BF_4{}^-$) with ee's ranging from 71 to 86% under 40 bar of H_2 at 50 °C whereas poor conversions (30–80%) and pitifully low ee's were obtained in the phosphonium-based ionic liquids $[C_{14}PCy_3][X]$ ($X = Cl^-$, NTf_2^-). For comparison, the ee of 95% obtained with Ru/4,4′-*diam*BINAP in water was higher than that in any ionic liquid. Only a single recycle run was conducted with 4,4′-*diam*BINAP/Ru in $[C_4mim][BF_4]$ and $[C_4pyr][NTf_2]$ but in both cases ee's improved after the first run.[324]

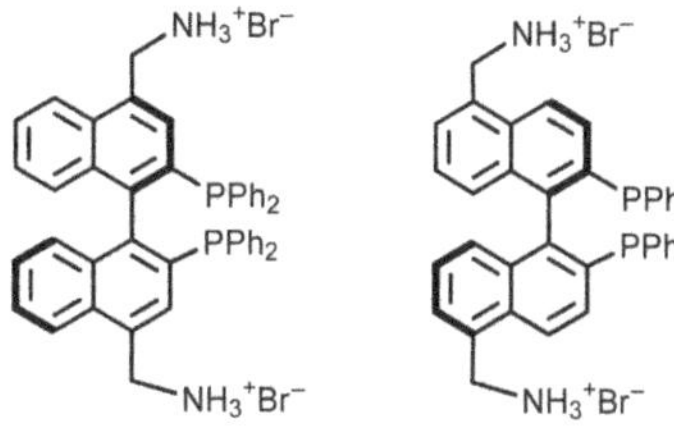

$NH_3^+Br^-$ PPh2 PPh2 $NH_3^+Br^-$

64

$NH_3^+Br^-$ PPh2 PPh2 $NH_3^+Br^-$

65

As pyrrolidinodiphosphines BPPM and PhCAPP are highly efficient and versatile ligands for rhodium-catalysed asymmetric hydrogenation, imidazolium and amino acid-modified versions **66** and **67**, respectively, have been prepared with the aim of developing ionic liquid immobilised systems. Catalyst generated from **66** and $[Rh(COD)_2][BF_4]$ gave the highest ee's of 95% for the asymmetric hydrogenation of methyl (*Z*)-2-acetamidocinnamate in methanol with a TOF of 24 000 h^{-1} while the combination of **67** and $[Rh(COD)Cl]_2$ gave a similar ee (94%) but a much lower TOF (<5000 h^{-1}). However, the same catalyst combinations gave moderate ee's and low TOFs in $[C_4mim][BF_4]$. A number of factors were considered to be responsible including (i) the high viscosity of the ionic liquid and low H_2 solubility resulting in poor mass transfer, (ii) high viscosity of the ionic liquid disfavouring formation of the induced-fit coordination environment for the flexible non-rigid 7-membered chelate, (iii) unfavourable dissociation of chloride due to low solvation enthalpy as reported by Dyson *et al.*,[30] (iv) the presence of chloride impurity (poisoning) and (v) lack of solvent involvement in the catalytic cycle. Markedly higher ee's and near quantitative conversions were obtained under homogeneous conditions with 1 mol% $[Rh(COD)_2][BF_4]$/**66** in a 1 : 5 (vol/vol) mixture of $[C_4mim][BF_4]$ and MeOH under 1 atm of H_2 at 25 °C; the ee' s obtained under these conditions match those in neat methanol. Under these conditions, the tagged catalyst combination $[Rh(COD)_2][BF_4]$/**66** recycled more efficiently than its unmodified counterpart; high activities and enantioselectivities were maintained over eight runs whereas the corresponding unmodified catalyst only recycled three times before conversions dropped. The loss of activity and enantioselectivity during recycle was attributed to leaching (3% rhodium loss in first cycle) and catalyst deactivation due to ligand oxidation; for comparison the tagged catalyst only suffered 0.1 mol% rhodium loss on recycle and the phosphine was markedly more stable with respect to oxidation.[325]

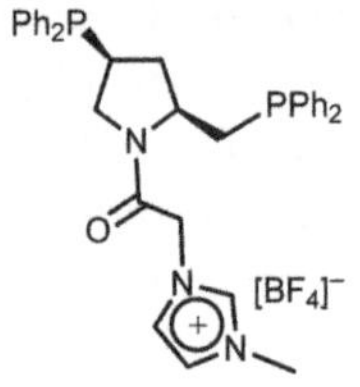

Ph2P PPh2 N O N [BF4]− N 66 Ph2P PPh2 N O H2N CO2H 67

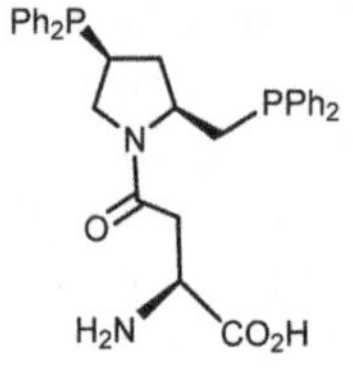

Cationic dendritic pyrophos-rhodium complexes based on poly(propylene imine) and poly(amido amine) dendrimers bearing between 4 and 64 $[Rh(NBD)]^+$ units on the periphery and hyperbranched poly(ethylene imines) with 9 to 139 active sites in the polymer structure have been immobilised in the thermomorphic $[C_4mim][BF_4]$–*iso*-PrOH mixture and used for the rhodium-catalysed asymmetric hydrogenation of *Z*-methyl α-acetamidocinnamate with the aim of developing a recyclable system. Catalytic reactions were conducted at 55 °C as the solvent mixture is thermomorphic and homogeneous at this temperature. Both types of dendrimer exhibited a negative 'dendritic effect' with increasing size with respect to activity, recyclability and enantioselectivity, which is more pronounced than in methanol. The hyperbranched poly(ethylene imine)-based catalysts also showed a strong negative 'dendrimer effect' with increasing size of the molecular support but this was less pronounced than the poly(propylene imine) and poly(amido amine)-based systems; poly(propylene imine) bound pyrophos-Rh(I) complexes of generation one and two displayed good reusability with no loss in activity or selectivity after three cycles. Loss in catalyst activity for the large dendrimers was associated with catalyst degradation rather than leaching.[326]

3.3.2.2.2.3 Asymmetric Hydrogenation by Activation and Deactivation in Chiral Ionic Liquids. Tropos diphosphines have been used successfully for asymmetric catalysis by either on metal resolution protocols or *via* asymmetric activation/chiral poisoning.[327a–c] To this end a significant ee (67%) was obtained for the rhodium-catalysed asymmetric hydrogenation of 2-acetamidoacrylate using the tropos diphosphine 5,5′-disulfonated 2,2′-bis(diphenylphosphino)-1,1′-biphenyl **68** in combination with (*S*)-proline methyl ester-derived chiral ionic liquid $[MeProl][NTf_2]$ as the sole source of chirality (Scheme 3.3.7).[328]

The same chiral ionic liquid induces a high level of enantioselectivity in the asymmetric hydrogenation of methyl *N*-acetamido acrylate and dimethyl itaconate when combined with Rh/*rac*-BINAP (Figure 3.3.10). An identical level of asymmetric induction was obtained by conducting the asymmetric hydrogenation of dimethyl itaconate in a 5 : 1 mixture of dichloromethane and $[MeProl][NTf_2]$ using catalyst generated *in situ* from [Rh(acac)(COD)] and *rac*-BINAP as that obtained with catalyst generated from the same precursor and enantiopure BINAP (Table 3.3.1). Table 3.3.1 also shows that the

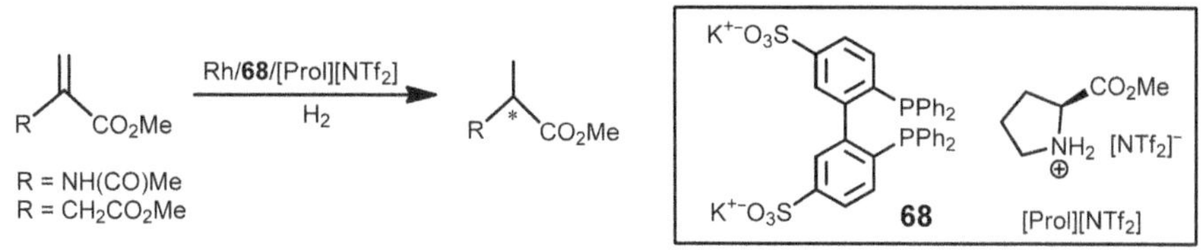

Scheme 3.3.7 Homogeneous rhodium-catalysed hydrogenation of benchmark substrates in chiral ionic liquid using catalyst derived from the tropos ligand 5,5′-disulfonato-2,2′-bis(diphenylphosphino)-1,1′-biphenyl.

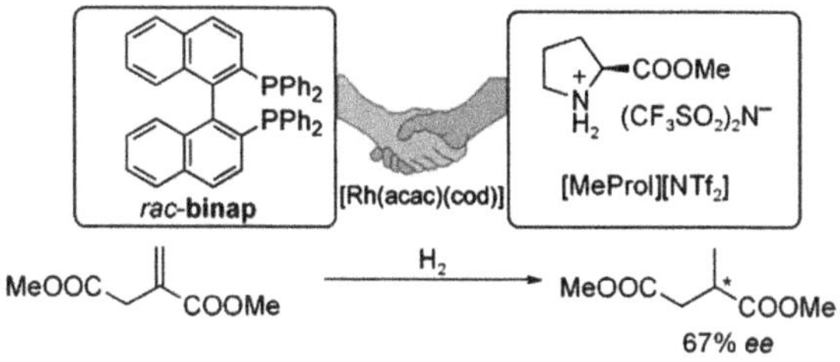

Figure 3.3.10 Enantioselective hydrogenation with *rac*-BINAP in the presence of a (*S*)-proline methyl ester-derived chiral ionic liquid.
Reproduced with permission from reference 329. Copyright 2008 Wiley.

Table 3.3.1 Rhodium-catalysed hydrogenation of dimethyl itaconate in the presence of the chiral ionic liquid [MeProl][NTf_2].

Ligand	*[Rh]*	*ee (%)*
Rac-BINAP	[Rh(acac)(COD)]	67 (*S*) 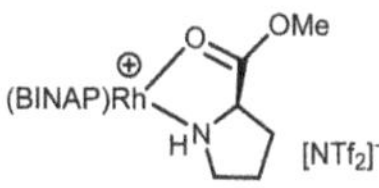
(*R*)-BINAP	[Rh(acac)(COD)]	71 (*R*)
(*S*)-BINAP	[Rh(acac)(COD)]	64 (*S*)

presence of the chiral ionic liquid did not affect the sense of asymmetric induction of the chiral ligand and that the product is formed primarily by an (*S*)-BINAP/Rh-based catalyst with *rac*-BINAP/Rh in [MeProl][NTf_2]. Hydrogen uptake experiments with enantiopure catalyst in [MeProl][NTf_2] confirmed that (*S*)-BINAP/Rh was markedly more active than its (*R*)-BINAP counterpart with initial rates estimated to be 6.7 : 1 in favour of (*S*)-BINAP/Rh. Kinetic and spectroscopic studies supported a pathway that involved effective chiral poisoning of the (*R*)-BINAP/Rh combination to be the principal mechanism of differentiating the two enantiomeric catalysts as the (*S*)-methyl ester of proline coordinates to rhodium to afford a 2.5 : 1 mixture of [{(*R*)-BINAP}Rh{(*S*)-MeProl}]$^+$ and [{(*S*)-BINAP}Rh{(*S*)-MeProl}]$^+$; the former is more stable and less reactive than its (*S*)-BINAP diastereoisomeric counterpart. In contrast, the same ionic liquid changed the sense of asymmetric induction in the (*R*)-BINAP/Rh(I)-catalysed hydrogenation of *N*-acetamido acrylate. Product with (*S*)-configuration was obtained in 25% ee in dichloromethane but this decreased when increasing amounts of [MeProl][NTf_2] were added until the sense of asymmetric induction switched to give product with (*R*)-configuration in a 1 : 1 mixture of dichloromethane and [MeProl][NTf_2]; an ee of 41% was obtained in neat [MeProl][NTf_2].[329]

3.3.2.2.2.4 Continuous Flow Asymmetric Hydrogenation. A highly efficient continuous flow process for the asymmetric hydrogenation of dimethylitaconate has been demonstrated by combining a supported ionic liquid phase catalyst based on [Rh(COD){(R_a,R_c)-1-naphthyl-QUINAPHOS}][NTf_2] immobilised in a thin layer of [C_2mim][NTf_2] on silica gel 100 and featuring fully integrated product separation (Figure 3.3.11). In dichloromethane, dimethylitaconate was reduced in >99% ee with a maximum TOF of 45 000 h^{-1} whereas this dropped to 4350 h^{-1} in [C_2mim][NTf_2] under 100 bar CO_2 due to lower H_2 solubility and higher solvent viscosity. High ee's (>99%) a minimum TOF of 3800 h^{-1} and a TON of 30 000 were obtained under continuous flow catalysis in a biphasic IL–*sc*CO_2 system but ee's dropped as the time on stream increased such that a total TON of 60 000 was achieved after 30 h but with a final ee below 50%. In contrast, the high initial activity (3800 h^{-1}) obtained with a continuous flow SILP–*sc*CO_2 system was maintained over 30 h on stream with stable single pass conversions >90%, corresponding to a total TON in excess of 100 000 and a space time yield of 750 g L^{-1} h^{-1}. Analytically pure product was isolated without the use of an organic solvent with productivity greater than 100 kg product per gram of rhodium. A slight decrease in ee (to 70%) after 20 h on stream was attributed to the formation of a less selective rhodium species. The stability of the SILP–*sc*CO_2 system is sensitive to the surface

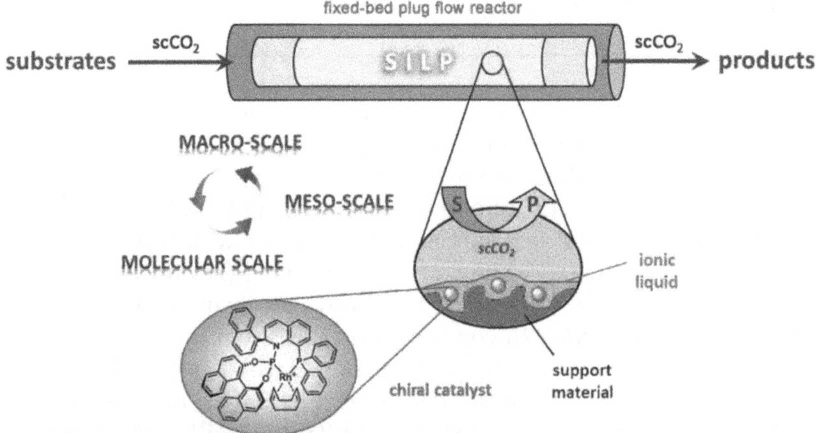

Figure 3.3.11 Concept of continuous flow supported ionic liquid phase catalysis with *sc*CO_2.
Reproduced with permission from reference 72. Copyright 2013 Elsevier.

properties of the support material and the best long term stability was achieved either with hydrophobic fluorinated materials or by introducing a water scavenger into the processing unit. Under these conditions an ee of 98–99% was obtained over 10 h on stream which dropped slightly but remained > 90% over 30 h with a TON between 70 000–100 000. This system is quite remarkable as it delivers high selectivity, excellent productivity, good long term stability, fully integrated product separation and no detectable catalyst leaching based on an 'unmodified' catalyst.[72]

3.3.2.3 Asymmetric Hydrogenation and Transfer Hydrogenation of Ketones, Ketoesters and Imine

3.3.2.3.1 Ketones. The first examples of asymmetric transfer hydrogenation in ionic liquids were conducted in tetralkyl and tetraarylphosphonium tosylates, which are solid at room temperature but liquid at the reaction temperature. In an unoptimised study, the combination of $[Rh_2(OAc)_4]$ and (−)-DIOP in $[E_{8,8,8,2}][OTs]$ (E = N, P) catalysed the asymmetric hydrogenation of acetophenone in 92% ee but only 50% conversion while the use of (1*S*, *2R)-cis*-aminoindanol led to poor conversion and negligible ee's.[330]

While modification of a ligand with an ionic tag is a popular strategy to improve catalyst solubility and retention in ionic liquids, the introduction of such a tag should ideally not influence or modify catalyst performance. However, even ionic tags introduced remote from the active site have been shown to have a profound influence on performance relative to that of the unmodified catalyst. For example, modification of the η^6-arene amino alcohol and TSDPEN-based transfer hydrogenation catalysts with an imidazolium tag led to efficient immobilisation for the biphasic ionic liquid-based ruthenium-catalysed transfer hydrogenation of ketones (Figure 3.3.12), but

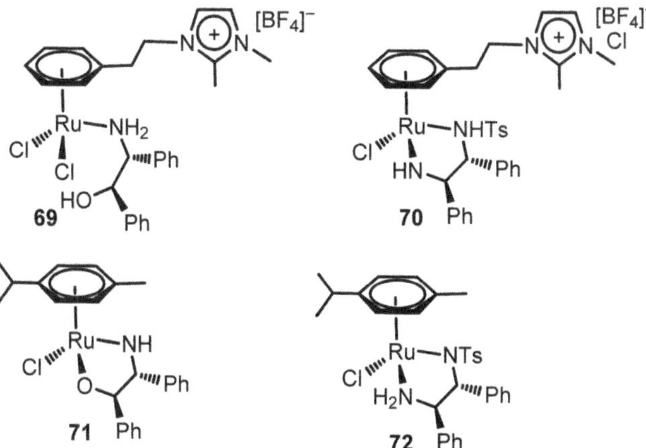

Figure 3.3.12 Imidazolium-modified η^6-arene ruthenium amino alcohol (**69**) and Ts-DPEN (**70**) transfer hydrogen precatalysts and their 'corresponding' unmodified counterparts, **71** and **72**, respectively.

unfortunately the ee obtained with amino alcohol **69** for the transfer hydrogenation of acetophenone in 2-propanol–KOH was markedly lower than its unmodified *p*-cymene counterpart **71.** In [C_4mim][PF_6] both **69** and **70** are found exclusively in the ionic liquid layer whereas their neutral counterparts are partitioned between both phases, however, the tagged catalysts were also found to leach during the course of the reaction probably due to base-induced catalyst degradation. The degree of leaching could be reduced by lowering the amount of base and under these conditions catalyst loss was 10 times lower for the tagged systems relative to their neutral counterparts. While **69** and **70** still rapidly deactivated under these conditions **70** was stable for 72 h and recycled four times although conversions did drop from 81% in the first run to 21% in the fourth. Surprisingly, the unmodified diamino-based catalyst **72** recycled much more efficiently than its tagged counterpart when formic acid was used as the source of hydrogen and although this was attributed to the aqueous wash after the product extraction step the extent of catalyst leaching was not quantified.[331]

The Ts-DPEN-derived ruthenium *p*-cymene transfer hydrogenation catalyst **73** has also been modified by introducing an imidazolium tag at the 4-position of the *N*-tosyl aryl ring to improve its recyclability. The ee's and conversions obtained for the asymmetric transfer hydrogenation of various aryl-alkyl ketones using the resulting catalyst matched those obtained with its neutral counterpart while the benefit of incorporating an imidazolium group was manifested in recycle experiments. Thus, while the unmodified catalyst gave good ee's and high conversions up to the third cycle its activity decreased thereafter, whereas, the imidazolium-tagged catalyst maintained high conversions over a greater number of runs.[332] A similar approach was adopted by Zhou *et al.* in which imidazolium-tagged aminosulfonamide **74** was combined with [(*p*-cymene)RuCl_2]$_2$ in [C_4mim][PF_6] to form an active catalyst for the HCO_2H–NEt_3 transfer hydrogenation of a range of aromatic ketones; high conversions and good to excellent ee's were obtained across a range of electron-rich and electron-poor substrates with a S : C ratio of 100. The catalyst recycled after extraction of the product with diethyl ether and although activity dropped dramatically good conversions (and high ee's) were still obtained after longer reaction times.[333]

The sodium salt of disulfonated (1*S*,2*S*)-1,2-diphenyl-1,2-ethylene diamine **75** (1*S*,2*S*-DPENDS) has been used as a chiral modifier for the ruthenium-catalysed asymmetric hydrogenation of ketones in hydrophilic ionic liquids. Quantitative conversions were obtained for the asymmetric hydrogenation of acetophenone using catalyst generated from [RuCl_2(tppts)$_2$]$_2$ and (1*S*,2*S*)-DPENDS and KOH in hydrophilic [$C_n$mim][*p*-$CH_3C_6H_4SO_3$] ($n=2$, 4, 8, 12) under 5MPa of H_2 at 50 °C while negligible conversions were obtained in [C_4mim][X] (X = PF_6^-, BF_4^-). Within the series [$C_n$mim][*p*-$CH_3C_6H_4SO_3$] ee's dropped steadily as the length of the imidazolium alkyl chain increased, from 79.2% in [C_4mim][*p*-$CH_3C_6H_4SO_3$] to 59.4% in [C_{12}mim][*p*-$CH_3C_6H_4SO_3$]. Catalyst performance also depended on the concentration of KOH and ee's varied between 72.5–79.2% for concentrations between

4.0 and 36.0 mol L^{-1} whereas there was no conversion in the absence of KOH. Good conversions and reasonable ee's were obtained for the asymmetric hydrogenation of a range of ketones under these optimised conditions. Recycle experiments demonstrated that the $[C_4mim][p\text{-}CH_3C_6H_4SO_3]$-catalyst mixture could be reused and conversions remained high for the first five runs but then gradually dropped, however, activity could be restored by the addition of KOH; ee's dropped steadily after the first run but remained close to 70% even after the ninth run.[334]

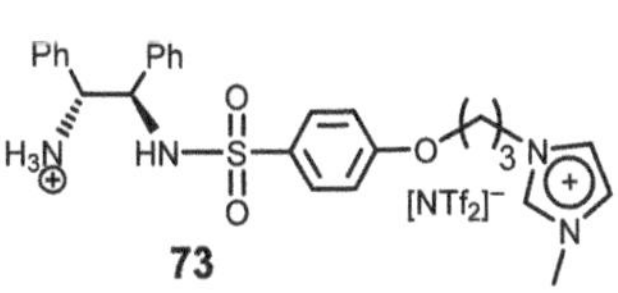

A rhodium/BINAP-based catalyst derived from the sacrificial alkene-containing rhodacarborane precursor [*closo*-1,3-{μ-(η^2-3-$CH_2=CHCH_2CH_2$)}-3-H-3-PPh_3-3,1,2-$RhC_2B_9H_{10}$] and (*R*)-BINAP catalyses the asymmetric hydrogenation of acetophenone and ethyl benzoylformate in $[C_8mim][PF_6]$, $[C_4mim][PF_6]$, the 1-carbadodecaborate-based ionic liquid $[C_4pyr][CB_{11}H_{12}]$ and in THF. Higher rates, as measured by the TOF, and ee's were obtained for both substrates in all of the ionic liquids compared with THF and within the limited range of the ionic liquids studied the highest ee's and TOFs were obtained in $[C_4pyr][CB_{11}H_{12}]$. Conversions, ee's and TOF all decreased quite severely when the rhodacarborane precursor was replaced by $[Rh(COD)Cl]_2$, which was tentatively suggested to reflect a different mechanism for hydrogenation.[335]

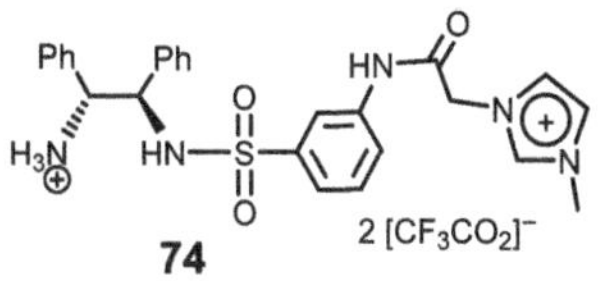

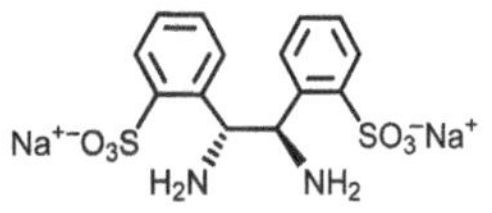

3.3.2.3.2 Ketoesters. The hydrogenation of *β*-ketoesters is an important transformation as the resulting *β*-aryl *β*-hydroxy acids and their esters are useful building blocks as well as precursors to a host of important drugs. Remarkably high enantioselectivities (up to 99.5%) have been achieved for the asymmetric hydrogenation of ethyl benzoylacetate at room temperature under 1400 psi of H_2 using 1 mol% [$RuCl_2${4,4′-substituted-(*R*)-BINAP}] whereas an ee of only 85% was obtained with unmodified (*R*)-BINAP. Having identified an efficient catalyst and optimum conditions the asymmetric hydrogenation of a range of *β*-ketoesters was conducted under homogeneous conditions in methanol and under liquid–liquid biphasic conditions in $[C_4mim][BF_4]$–MeOH using ruthenium catalysts derived from **76** and **77**. The former gave slightly lower ee's on average in $[C_4mim][BF_4]$–MeOH compared with neat MeOH whereas the latter consistently performed better in the ionic liquid–methanol mixture than in neat methanol. Both catalysts recycled with only a moderate loss in activity but a significant reduction in ee after each recycle. As ruthenium leaching was determined to be less than 0.02–0.04% this deterioration was attributed to the instability of the active ruthenium-hydride species.[336]

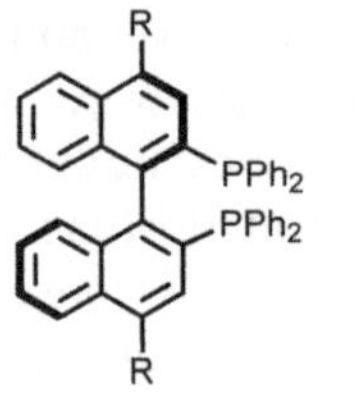

76, R = $P(O)(OH)_2$
77, R = $SiMe_3$

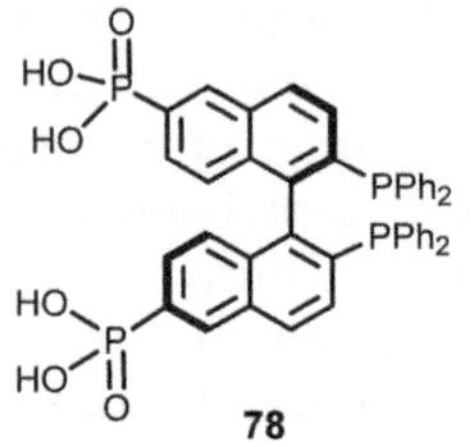

78

Polar phosphoric acid-derived ruthenium-BINAP complexes have been designed to facilitate separation of the ionic liquid–catalyst from the organic product. To this end catalyst formed from (*R*)-2,2′-bis(diphenylphosphino)-1,1′-binaphthyl-6,6′-bis(phosphonic acid) **78** is highly active for the monophasic asymmetric hydrogenation of a range of *β*-ketoesters in $[C_4mim][BF_4]$, $[C_4mim][PF_6]$ and $[C_3C_1{}^2mim][NTf_2]$ with equal volumes of methanol as co-solvent under 1500 psi of H_2 at room temperature. Enantioselectivities were sensitive to the nature of the ionic liquid and $[C_4mim][BF_4]$ gave the highest ee's which were either comparable to or better than those obtained in neat methanol. The catalyst phase was recycled up to four times with no loss in activity or enantioselectivity.[337]

A marked and significant enhancement in activity was obtained for the asymmetric hydrogenation of methyl acetoacetate conducted in ionic liquid–methanol mixtures compared with the same reaction in neat methanol using 1 mol% catalyst generated from $[Ru(2\text{-methylallyl})_2(COD)]$ and two equivalents of phenyl-4,5-dihydro-3H-dinaphtho[2,1-c;1′2′-e)phosphepine. A detailed survey revealed that activity and enantioselectivity varied with the nature of the cation for a range of triflimide-based ionic liquids. For all imidazolium-based ionic liquids examined, yields and ee's were lower in the ionic liquid–methanol mixture than in neat methanol and the authors related this to slow or incomplete precatalyst activation caused by the low solvation enthalpy of the chloride in the imidazolium salt, as reported earlier by Dyson.[30] Similarly lower yields and poorer ee's were obtained for reactions conducted in $[N_{8,8,8,1}][NTf_2]$–methanol compared with neat methanol. In contrast, although there was a significant induction period for reactions conducted in $[NMe_4][NTf_2]$–methanol, which appeared to correspond to a delay in the formation of the active species, yields and ee's eventually approached those obtained in neat methanol. The enhancement in activity was unique to ammonium cations bearing alcohol functionalities and appeared to be associated with rapid and facile activation of the precatalyst. Thus, reactions conducted in ionic liquid–methanol mixtures based on $[HOC_2NMe_3][NTf_2]$, $[HOC_2NH_3][NTf_2]$ and $[(HOC_2)_2NMe_2][NTf_2]$ all gave higher conversions than those in neat methanol with ee's between 86–93%. A study of enantioselectivity with time revealed that ee's were high in the early stages of the reaction, which is also consistent with rapid formation of the most active and selective species. Each of the hydroxyl-functionalised ionic liquid–methanol catalyst systems recycled three times with only a minor loss in activity and very little change in ee whereas catalyst in $[C_2mim][NTf_2]$-methanol recycled much less efficiently.[338] The influence of trace impurities in ionic liquids on the enantioselectivity of asymmetric hydrogenation has been systematically investigated in the reversibly biphasic mixture $[C_4mim][PF_6]$-ethanol, using methyl acetoacetate as substrate and [(*p*-cymene)RuCl{(*R*)-BINAP}] as the precatalyst. In house prepared $[C_4mim][PF_6]$ was subjected to a series of normalised purification steps based on a theoretical semi-mechanistic model which provided direct evidence on the number and duration of necessary purification steps together with an effective value for each. Hydrogenations conducted in non-purified $[C_4mim][PF_6]$-ethanol gave good conversions to product but ee's varied quite dramatically from 25–85% depending on the purity of the supply while reproducible ee's of 97% were obtained with $[C_4mim][PF_6]$ that had been subjected to the optimised set of extraction steps. This study demonstrates the potential for acquiring unreliable and misleading data in ionic liquids and highlights the impact of purification.[339]

Since its inception, the concept of supporting a transition metal catalyst on a thin film of ionic liquid immobilised on a high surface area porous support material (SILP) has also been applied to a host of reactions including hydroformylation, methanol carbonylation, hydroamination and alkene hydrogenation. SILP catalysis has been applied to the continuous flow gas phase asymmetric hydrogenation of methyl acetoacetate (MAA) by using an elaborate precatalyst, [Ru{3-(2,5-(2*R*,5*R*)-dimethylphospholanyl-1)-4-di-*o*-tolylphosphino-2,5-dimethyl-thiophene}Br_2], immobilised in a thin film of $[C_2mim][NTf_2]$ on silica gel 30. The use of an ionic liquid film on the support was essential to achieve activity and a high degree of pore filling α ($\alpha =$ ionic liquid volume/pore volume) with the ionic liquid was beneficial, which may well be to ensure efficient protection of the catalyst from unfavourable interactions with the support. Catalyst activation times were shortened from 27 h to 20 h by using dried non-calcinated silica and hydrogen pre-treatment which was thought to be associated with accelerated formation of the active hydride species. A conversion *versus* time on stream (TOS) profile for the SILP-catalysed hydrogenation of MAA (125 °C, 10 bar of H_2, 5000 rpm, 0.125 g h^{-1} MMA) showed an initial period of activation during which conversion and ee increased steadily, followed by a period of stable catalyst operation ($X_{max} = 32.4\%$) and a period of catalyst deactivation in which conversion and ee both dropped. The period of stable catalyst operation increased at lower temperatures from 10 h TOS at 125 °C to 35 h TOS at 105 °C, although a lower maximum conversion was obtained. The conversion profile improved ($X_{max} = 87.3\%$) when a continuous feed of gaseous methanol was introduced into the reactor, although the period of stable operation was not extended. Thermal degradation of the SILP catalyst and build-up of high boiling side-products that fill and block the open pore volume of the SILP catalyst were identified as the major threats to catalyst stability. The use of a plug flow reactor improved catalyst stability and full catalytic activation was achieved after 35 h TOS and stable catalytic operation was maintained for a further 70 h at 105 °C to give 70% conversion and 75–80% ee. The use of a SILP catalyst for continuous gas phase hydrogenation allows simple catalyst/product separation and could also be applied to a host of well-known catalytic systems without the need for further ligand modification.[340] The continuous flow asymmetric hydrogenation of methyl propionylacetate in a biphasic ionic liquid–*sc*CO_2 system based on Ru/BINAP immobilised in $[C_3C_1{}^2mim][NTf_2]$ with *sc*CO_2 as the mobile phase and $[SO_3HC_4C_4im][NTf_2]$ as acid additive gave high reaction rates with good catalyst stability but slightly reduced ee's compared with reactions conducted in the absence of acid. For comparison, continuous flow hydrogenation in the absence of acid gave 96% ee but reached a maximum conversion of only 18% after 5 h which dropped slowly thereafter, as did the ee. In contrast, in the presence of acid high single pass conversions (>95%) and good enantioselectivity (80–82%) were maintained over the first 140 h to give a total TON of 16 000 and a space time yield of 138 g L^{-1} h^{-1} with no evidence for metal or ligand leaching. A comparison between batch and continuous flow systems for the production of 100 kg of methyl 3-hydroxypentanoate demonstrated that the latter is more resource efficient (Table 3.3.2) even though reactions are conducted at much higher pressure as this is offset by the reduced scale of the equipment. The major drawback associated with this process is the reduced enantioselectivity (80–82 *versus* 97–98%) in the presence of acid, particularly for applications that demand optically pure products such as pharmaceuticals.[60] The continuous flow gas phase hydrogenation of methyl pyruvate to methyl lactate with a SILP-based catalyst comprising a solution of Ru/BINAP in $[HOC_3py][NTf_2]$ immobilised on calcined silica gel 100 gave stable catalyst performance over 50 h on stream with good yields (80–84%) but low ee's (26–30%).[341]

3.3.2.3.3 Imines. There have been relatively few reports of the asymmetric hydrogenation of imines in ionic liquids which is surprising since it leads to synthetically valuable chiral amines that are found in pharmaceutical and agrochemical products. The first such study compared the asymmetric hydrogenation of trimethylindolenine catalysed by

Table 3.3.2 Calculated figures for the production of 100 kg of methyl 3-hydroxypentanoate through batch and continuous flow hydrogenation.

	Batch	*Continuous flow*
Number of batches	4	1
Reactor unit	50 L	5.4 L
Reaction solvent	100 L	2.7 L
Catalyst A	350 g	25 g
Reaction time	4×6 h	96 h
Processing time	1–2 weeks	<1 week
Pressure	4 bar H_2	250 bar H_2–CO_2
Temperature	100 °C	60 °C

iridium/XYLIPHOS in a range of ionic liquids against the same reaction in organic solvent. A slightly higher temperature was required for reactions conducted in ionic liquid to overcome the solvent viscosity and mass transfer limitations. Good reproducibility was ultimately achieved after allowing the $[IrCl(COD)]_2$, XYLIPHOS, tetrabutylammonium iodide and trifluoroacetic acid precatalyst mixture to age for 1 h as shorter aging times led to irreproducible data. Under optimum conditions the quantitative conversion and 86% ee obtained after 15 h in $[C_{10}mim][BF_4]$, using 0.4 mol% Ir/XYLIPHOS at 40 bar H_2 and 50 °C, matched that achieved in toluene under the same conditions. While short exposure of a toluene solution of the catalyst to air had a severe detrimental effect on conversion an ionic liquid solution of the catalyst was much more tolerant/robust and gave 98% conversion and 84% ee after transfer to the autoclave in air. Such a marked improvement in the stability/tolerance of an ionic liquid-immobilised catalyst with respect to oxygen could simplify transfer protocols as well as the manipulation of catalyst solutions and recycle procedures.[342] While an Ir/phosphinooxazoline combination has been shown to be an efficient catalyst for the asymmetric hydrogenation of imines under homogeneous conditions in *sc*CO_2, immobilisation and recycle was hampered by significant deactivation during the recycle procedure.[343] Leitner and co-workers combined ionic liquids and *sc*CO_2 to overcome these problems and developed an efficient, stable and recyclable biphasic system with a number of additional benefits. Hydrogenation of *N*-(1-phenylethylidene)aniline carried out in imidazolium and pyridinium-based ionic liquids revealed the beneficial effect of biphasic IL–*sc*CO_2 systems as quantitative yields of product were obtained at a hydrogen partial pressure of 30 bar in the presence of CO_2 whereas a much higher pressure (100 bar) was required to achieve good conversions in the absence of CO_2. This apparent 'activation' of the catalyst system by CO_2 was related to improved H_2 availability in the ionic liquid due to the increased hydrogen solubility and reduced viscosity in the presence of CO_2. Facile anion exchange between the precatalyst [Ir(phosphinooxazoline)(COD)][X] and various ionic liquids revealed a remarkably strong influence on enantioselectivity of the anion with ee's varying from 30% in $[C_4mim][BF_4]$ to 78% in $[C_4mim][BAr_f]$. Facile anion exchange allowed the catalyst to be generated/activated directly either by mixing $[IrCl(COD)]_2$ and the phosphinooxazoline in dichloromethane, removing the solvent and redissolving the resulting chloride-iridium complex in an appropriate ionic liquid or by mixing the metal precursor and ligand in a mixture of dichloromethane and ionic liquid and then removing the organic solvent immediately prior to catalysis; near identical results that matched those obtained with preformed catalyst were obtained using both protocols. The catalyst recycled efficiently after extraction of the product by *sc*CO_2 and stable activity and selectivity was obtained over seven runs, which was a marked improvement on the corresponding CO_2-immobilised system. Ionic liquid solutions of the catalyst were also shown to be significantly less air-sensitive than solutions of the same catalyst in organic solvent, even on prolonged exposure to air (75% conversions and 56% ee after 20 h in air under standard conditions) *i.e.* the ionic liquid environment stabilises catalytically active intermediates. Thus, the use of a biphasic IL–*sc*CO_2 system allows catalyst activation, tuning and immobilisation that would not be possible under conventional homogeneous conditions.[24]

3.3.2.4 Catalyst Activation Studies

The high nucleophilicity of the chloride anion in imidazolium-based ionic liquids compared with water has been rationalised on the basis of weak interaction between an imidazolium cation and the chloride anion, determined to be *ca.* 15 kJ mol^{-1}. A further potential consequence is that chloride dissociation from a transition metal complex will be thermodynamically less favoured in ionic liquid than in aqueous solution and as a result metal complexes that must dissociate chloride to form an active species may be completely inactive in ionic liquid. Daguenet and Dyson first reported that $[(\eta^6$-*p*-cymene)Ru(dppm)Cl]Cl formed an active catalyst for the hydrogenation of styrene in water (TOF 1000 h^{-1}) but was inactive in $[C_4mim][OTf]$ whereas the acetonitrile adduct $[(\eta^6$-*p*-cymene)Ru(dppm)(MeCN)]$[OTf]_2$ gave near quantitative conversions after 15 h under biphasic conditions in $[C_4mim][OTf]$ under 50 bar of H_2 at 100 °C; this was proposed to reflect the difference in the thermodynamics of the ligand dissociation step *i.e.* chloride dissociation *versus* acetonitrile dissociation. Addition of water as a co-solvent to an ionic liquid solution of $[(\eta^6$-*p*-cymene)Ru(dppm)Cl]Cl appeared to render chloride dissociation more favourable and the system catalysed the hydrogenation of styrene and gave conversions that matched those obtained with its acetonitrile adduct. Thus, it is clearly important to consider the consequences of performing catalysis in ionic liquids where chloride dissociation is either essential to activation of a precatalyst or is an integral step of a catalytic cycle; approaches to overcome these problems include design of a suitably 'labile' precursor or addition of a co-solvent that can solvate the dissociated chloride. While this explanation appears eminently reasonable and plausible, and the concept may still have some validity, subsequent studies shed further light and a more detailed understanding on the observed effects. A more detailed analysis of this system revealed that generation of the active catalyst was dependent on both the cation and the anion of the ionic liquid and that the active species is different from that generated in water. Interestingly, while $[(\eta^6$-*p*-cymene)Ru(dppm)Cl]Cl was shown to form an active catalyst for the hydrogenation of styrene in $[C_4mim][NTf_2]$, $[(\eta^6$-*p*-cymene)Ru(dppm)Cl]$[BF_4]$ was completely inactive and further experiments confirmed that the presence of chloride was necessary for hydrogenation activity. The activity of $[(\eta^6$-*p*-cymene)Ru(dppm)Cl]Cl in $[C_4mim][X]$ was shown to depend strongly on the anion of the ionic liquid and decreased in the order $[NTf_2]^- >> [CF_3SO_3]^- \sim [CF_3CO_2]^- > [SnCl_3]^- \sim [BF_4]^- \sim [PF_6]^-$. The high activity obtained in $[C_4mim][NTf_2]$ was proposed to be associated with low/poor chloride solvation (as measured by the empirical solvent polarity scale), the low molar conductivity (poor diffusion of $[C_4mim]^+$ cation) and poor ion pairing with the cationic Ru(II) precursor. The rate of hydrogenation also showed a marked dependence on the ionic liquid cation which was again attributed to solvation effects as efficient desolvation of the chloride is required to form the active species. On the basis of kinetic studies the formation of $[(dppm)Ru(\mu\text{-}Cl)_3Ru(dppm)]^+$ was proposed to be the rate determining step in the generation of the active catalyst. This study highlighted that both the anion and cation of an ionic liquid can influence the formation of the active species which appears to be the result of differences in solvent–solute interactions that take place in each ionic liquid. In this case, the influence of the ionic liquid on the rate of hydrogenation was related to the different solvation capability of each ionic liquid towards chloride ions with higher catalytic activities being obtained in ionic liquids that provided poor chloride solvation and thus favoured formation of the postulated active catalyst $[(dppm)Ru(\mu\text{-}Cl)_3Ru(dppm)]^+$.[30a]

The influence of the nature of the ionic liquid on the $[Rh(COD)(PPh_3)_2][NTf_2]$-catalysed selective hydrogenation of cyclohexadiene (CYD) revealed that the initial rate in $[C_4mim][NTf_2]$ is approximately twice that in $[C_4C_1{}^2mim][NTf_2]$. A study of the two key steps, ligand exchange to afford a $[Rh(CYD)(PPh_3)_2][NTf_2]$ and the subsequent hydrogenation, revealed that both steps are slower in $[C_4C_1{}^2mim][NTf_2]$ than in $[C_4mim][NTf_2]$ and that the latter step is rate determining. The solubility of CYD in both ionic liquids and the excess molar enthalpy of mixing for the binary mixtures of CYD with $[C_4mim][NTf_2]$ and $[C_4C_1{}^2mim][NTf_2]$ were quantified and even though both factors indicated a more favourable interaction between CYD and the $[C_4mim]^+$ cation (higher solubility and smaller positive enthalpy of mixing) than $[C_4C_1{}^2mim]^+$, these factors alone do not account for the different rates of hydrogenation. On the basis that the higher viscosity of $[C_4C_1{}^2mim][NTf_2]$ induced a higher diffusion coefficient of CYD in $[C_4mim][NTf_2]$ (187.7 $\mu m^2\ s^{-1}$) compared with $[C_4C_1{}^2mim][NTf_2]$ (97 $\mu m^2\ s^{-1}$), by a factor of 1.9, the authors reasoned that the difference in the rate of both ligand exchange and hydrogenation of CYD with $[Rh(COD)(PPh_3)_2][NTf_2]$ in $[C_4C_1{}^2mim][NTf_2]$ and $[C_4mim][NTf_2]$ was associated with the difference in viscosity of the ionic liquids and hence the mobility of molecules in solution.[344] A subsequent study found a linear relationship between the initial rates of reaction and the inverse of viscosity *i.e.* the reactions kinetics are controlled by mass transfer and in more viscous systems the mobility of molecules is lower and a lower reactivity is observed. The influence of stirring speed, hydrogen pressure and catalyst loading all reinforced the nature of mass transfer control in this system.[345]

3.3.2.5 *Hydrosilylation*

The hydrosilylation of C–C double bonds is a technologically important reaction for the synthesis of organomodified polydimethylsiloxanes. Reactions are typically conducted under homogeneous conditions, however, removal of the catalyst can be a time consuming and costly process. Ionic liquids have been investigated as a catalyst immobilisation phase for hydrosilylation in order to reduce contamination of the product with the catalyst and to enable recycling of expensive precious metal catalyst. The first example of an ionic liquid-based hydrosilylation used the perfluorinated Wilkinson-type catalyst $[RhCl(PArf)_3]$ (Arf = $-C_6H_4\{SiMe_2CH_2CH_2C_6F_{13}\}$-*p*) for the anti-Markovnikov addition of $PhMe_2SiH$ to 1-octene in $[C_4mim][B\{C_6H_4(SiMe_2CH_2CH_2C_6F_{13})$-*p*$\}_4]$; the TOF of 400 h^{-1} obtained under homogeneous conditions at 100 °C was the same as that obtained with $[RhCl(PPh_3)_3]$ in $[C_4mim][BF_4]$ but both were lower than the TOF of 1800 h^{-1} obtained with $[RhCl(PPh_3)_3]$ in benzene. However, phase separation at 0 °C enabled the ionic liquid–catalyst mixture to be recovered and recycled 15 times to give a total TON of 4000 (92% catalyst activity retained per cycle); this is significantly higher than that obtained with conventional catalysts under monophasic conditions. The minor drop in conversion on successive cycles was due to leaching of the rhodium (4%) and phosphine (2%) per cycle.[51] The platinum-catalysed hydrosilylation (and hydroboration) of terminal alkynes is catalysed by Spiers catalyst, $[H_2PtCl_6]$, in $[N_{8,8,8,1}][NTf_2]$ with high selectivity for the corresponding β-*E*-vinylsilane; the product was readily isolated from the ionic liquid by bulb-to-bulb distillation and the recovered ionic liquid–catalyst solution recycled without loss in activity.[346]

The use of ionic liquids to immobilise an unmodified hydrosilylation catalyst with the aim of developing a heterogeneous process has been explored by researchers at Degussa. A range of Si–H-functionalised polydimethylsiloxane, olefin and ionic liquid combinations were explored with an emphasis on achieving efficient separation of the ionic liquid–catalyst mixture from the product. The authors concluded that successful recovery of the catalyst and its reusability required an appropriate combination of a catalyst and an ionic liquid, which had to be harmonised with the hydrophilicity/hydrophobicity of the product. Suitable catalyst–ionic liquid combinations were H_2PtCl_6–$[C_1C_1{}^2mim][MeSO_4]$, H_2PtCl_6–$[4\text{-}MeC_4pyr][BF_4]$ and H_2PtCl_6–$[3\text{-}MeC_4pyr]Cl$ for the range of olefins (400EO, 540, 400PO) and Si–H functional polydimethylsiloxanes ($n = 18$, 28, 78) examined. The failure of hydrosilylations conducted in 1,3-dialkylimidazolium salts was attributed to the formation of *N*-heterocyclic carbenes which coordinated to and deactivated the catalyst.[347]

A more detailed and systematic study of ionic liquid-based biphasic platinum-catalysed hydrosilylation between 1-hexadecene and polydimethylmethylhydrogen-polysiloxane [Equation (3.3.3)] identified *N*-butyl-4-picolinium tetrafluoroborate and *N*-butyl pyridinium tetrafluoroborate to be the optimum ionic liquids as they combined good conversions and efficient immobilisation for H_2PtCl_6 and $Pt(PPh_3)_4$. Systematic activity and stability studies in $[C_4pyr][BF_4]$ demonstrated that K_2PtCl_4 and $[PtCl_2(C_6H_{10})_2]$ formed the most active catalyst while molecular precatalysts such as $PtCl_2(PPh_3)_2$ and $Pt(PPh_3)_4$ were somewhat more stable and could be recycled 10 times to give total TONs in excess of 1 500 000. Highly active nanoparticles formed rapidly under the conditions of catalysis with K_2PtCl_4, $[PtCl_2(C_6H_{10})]_2$ and $PtCl_2(COD)$ whereas $PtCl_2(PPh_3)_2$ and $Pt(PPh_3)_4$ were more stable and the active species could not be unequivocally identified. The authors concluded that several different active species *i.e.* soluble nanoparticles and molecular catalysts most likely operate at the same time.[348]

$$\mathrm{Me{-}SiMe_2{-}[O{-}SiMe_2]_n{-}[O{-}SiMe(H)]_m{-}O{-}SiMe_3} + \mathrm{CH_2{=}CH{-}C_4H_{14}} \xrightarrow[\text{ionic liquid}]{\text{Pt-catalyst}} \mathrm{Me{-}SiMe_2{-}[O{-}SiMe_2]_n{-}[O{-}SiMe(R)]_m{-}O{-}SiMe_3} \quad (3.3.3)$$

Following the success of $[C_1C_1{}^2mim][MeSO_4]$ as a solvent for platinum-catalysed hydrosilylation, the same ionic liquid was used to immobilise $[Rh(COD)(\mu\text{-}OSiMe_3)]_2$, $[Rh(COD)(PCy_3)(OSiMe_3)]_2$ and $[Rh(\mu\text{-}OSiMe_3)(tfb)]_2$ which were all highly efficient catalysts for the liquid–liquid biphasic hydrosilylation of allyl glycidyl ether, allyl phenyl ether, allyl butyl ether and dimethoxyallylbenzene using heptamethyltrisiloxane as a model for polysiloxanes. The ionic liquid–catalyst mixture $[C_1C_1{}^2mim][MeSO_4]$/ $[Rh(COD)(\mu\text{-}OSiMe_3)]_2$ recycled efficiently and consistent conversions were maintained over 14 recycles. Reassuringly, good conversions of each substrate were also obtained using poly(dimethyl-hydromethyl)siloxane in $[C_1C_1{}^2mim][MeSO_4]$ at 120 °C with 5 ppm $[Rh(COD)(\mu\text{-}OSiMe_3)]_2$. A survey of several different phosphonium-based ionic liquids identified propoxymethyltrihexylphosphonium saccharinate in combination with $[Rh(COD)(\mu\text{-}OSiMe_3)]_2$ to be as efficient as its $[C_1C_1{}^2mim][MeSO_4]$ counterpart.[349] Quaternary ammonium-based ionic liquids bearing allyl of vinyl groups are also effective solvents for the biphasic rhodium- and platinum-catalysed hydrosilylation of 1-octene with 1,1,1,3,5,5,5-heptamethyltrisiloxane. The most efficient metal/ionic liquid combinations were $[Rh(COD)(\mu\text{-}OSiMe_3)]_2$ with diallyldimethylammonium methylsulfonate or trimethylvinylammonium ethanesulfonate; both are considerably more active than platinum-based systems as well as Wilkinson's complex and within a series of platinum precatalysts activity decreased in the order Pt(0) > Pt(II) > Pt(IV) which parallels the order of olefin complexation. The same optimum systems also catalysed the hydrosilylation of the more hydrophilic substrate allyl glycidyl ether and recycled with remarkable efficiency over 10 runs; this was a marked improvement on reactions carried out in other ionic liquids devoid of unsaturated groups which experienced leaching and a reduction in activity on successive runs. The effective retention in allyl and vinyl-containing ionic liquids may be associated with coordination of the olefin group to the catalyst, possibly in a bidentate manner for the most efficient diallyldimethylammonium-based ionic liquid.[350]

The platinum-catalysed hydrosilylation of allyl chloride is an industrially important multi-tonne transformation as the product is an additive for polymers and an intermediate to other surface active organosilanes. This

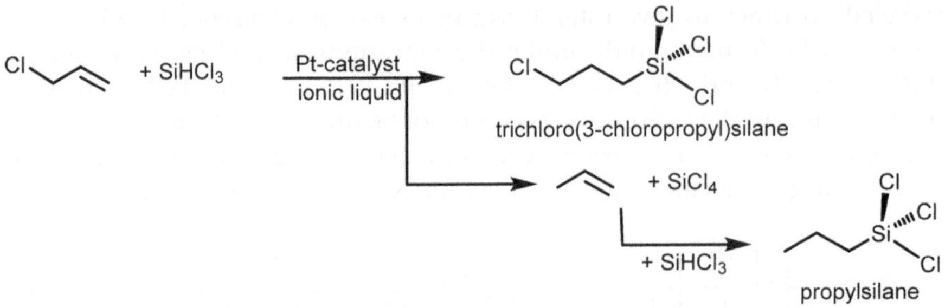

Scheme 3.3.8 Hydrosilylation of allyl chloride with trichlorosilane to form (3-chloropropyl)trichlorosilane; main reaction and relevant side reactions forming tetrachlorosilane, propene and propyltrichlorosilane.

transformation presents a number of technical challenges as trichlorosilane is flammable, highly volatile, easily hydrolysed and chemically aggressive and the reaction shows a complex selectivity profile with propyltrichlorosilane forming as the major undesired by-product (Scheme 3.3.8). A study of this reaction under liquid–liquid biphasic conditions revealed that $[NTf_2]$-based ionic liquids are suitable solvents for $SiHCl_3$ whereas $[C_2mim][EtOSO_3]$, $[C_4mim][BF_4]$, $[C_2mim][OTs]$, $[C_2mim][Et_2PO_4]$ and $[C_2mim][CF_3SO_3]$ all resulted in either rapid and complete or partial decomposition of the trichorosilane; a preliminary screening identified $[C_4C_1{}^2mim][NTf_2]$ to be the optimum ionic liquid for phase separation and performance and $PtCl_4$ to be the most practical source of platinum. Experiments in semi-batch mode gave a TOF of 3780 h^{-1} and 80.9% selectivity for (3-chloropropyl)trichlorosilane at 100 °C (300 ppm $PtCl_4$) and recycle studies showed no loss in selectivity over nine runs with platinum leaching below the detection limit of the spectrometer and no evidence for leaching of the ionic liquid. In a continuous loop reactor a $[C_2C_1{}^2mim][NTf_2]$ solution of 400 ppm $PtCl_4$ operated over 48 h with stable conversion and selectivity to give a total organic throughput of 14 kg with an average conversion greater than 50% and total selectivity for (3-chloropropyl)trichlorosilane of 61–72%.[351] A recyclable homogeneous monophasic platinum-catalysed system for the hydrosilylation of allyl chloride with trichlorosilane was subsequently developed by adding the fully miscible ionic liquid $[C_{10}C_1{}^2mim][NTf_2]$ to the organic reaction mixture to stabilise the catalyst during thermal distillation of the product (17–20 mbar, 100 °C). In this manner, the homogeneous system recycled ten times with no loss in activity with a final TON in excess of 14 200. The stabilisation was attributed to a tight interaction between the platinum nanoparticles and the ionic liquid *i.e.* the micelle nanocolloid stabilisation effect. The selectivities and productivities compared favourably with those obtained in the actual state-of-the-art process, but with the distinct advantage that no off-line catalyst make-up is required to recover and recycle the catalyst which would improve the process economy and the technology could be implemented without the need for modification of existing plants.[352]

Ionic liquids are also suitable solvents for palladium-catalysed hydrostannylation of α-heteroalkynes and alkynyl esters. Reactions in [C_4mim][PF_6] catalysed by 2 mol% $Pd(PPh_3)_4$ at 60 °C gave good yields, high regioselectivity (>95%) and complete *E*-selectivity. The ionic liquid-immobilised catalyst was recovered and reused five times after which the activity dropped quite significantly.[353]

3.4 Lewis and Brønsted Acid Catalysis

3.4.1 Lewis Acid Catalysis

3.4.1.1 Lewis Acid-Catalysed Diels–Alder Reactions

Scandium triflate-catalysed carbon–carbon bond forming reactions in organic solvents are often limited by low TONs (10–20) which somewhat restricts their applications, particularly for large scale reactions. To this end, the use of ionic liquids as solvent for $Sc(OTf)_3$-catalysed Diels–Alder reactions results in a remarkable acceleration in rate, an improvement in selectivity as well as easier product isolation, catalyst recovery and recycling compared with dichloromethane. For example, the reaction between 1,4-naphthoquinone and 1,3-dimethylbutadiene in [C_4mim][PF_6] with 10 mol% $Sc(OTf)_3$ was extremely rapid and reached completion in seconds at room temperatures. Although longer reaction times were required when the catalyst loading was reduced to 0.2 mol% quantitative conversions were still achieved within 2 h; for comparison the same reaction in dichloromethane gave 22% yield in the same time. The kinetic study reproduced in Figure 3.4.1

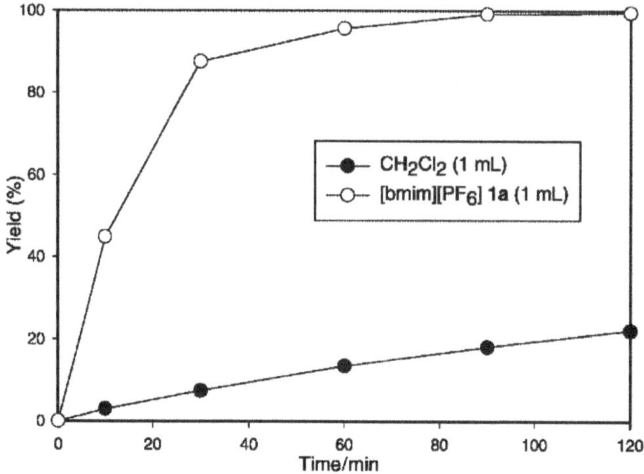

Figure 3.4.1 Kinetic study of the Diels–Alder reaction between 1,4-naphthoquinone and 2,3-dimethylbuta-1,3-diene with 0.2 mol% $Sc(OTf)_3$ illustrating the marked difference between the rates in ionic liquid and dichloromethane.
Reproduced from reference 81.

clearly shows the remarkable difference between the rates in these two solvents. A similar acceleration in rate was achieved for reactions conducted in dichloromethane with only one equivalent of $[C_4mim][PF_6]$ which means that the amount of ionic liquid required can be reduced. A significant improvement in the *endo*/*exo* selectivity was also obtained for reactions conducted in $[C_4mim][PF_6]$ (>99 : 1) compared with dichloromethane (94 : 6). The ionic liquid–catalyst mixture also recycled with exceptional efficiency and could be reused 11 times with only a minor reduction in activity.[81]

A significant enhancement in rate and enantioselectivity for $Cu(OTf)_2$/bis(oxazoline)-catalysed Diels–Alder reaction between *N*-acryloyl or *N*-crotonyloxazolidinone with cyclopentadiene or 1,3-cyclohexadiene was achieved in $[C_4mim][BF_4]$ compared with dichloromethane. Complete conversion and 95% ee was obtained within 2 min for the reaction between *N*-acryloyloxazolidinone and cyclopentadiene in $[C_4mim][BF_4]$ while the same reaction in dichloromethane required 60 min to reach completion with an ee of 78%. The enhancement in rate in ionic liquid enable the catalyst loading to be reduced to 0.5 mol% while retaining the same level of enantioselectivity. Even the more challenging combination of *N*-crotonyloxazolidinone and cyclopentadiene gave a similar rate enhancement in $[C_4mim][BF_4]$ and reached completion with 97% enantioselectivity for the *endo* diastereoisomer within 60 min; for comparison there was no reaction in dichloromethane even after 2 h. While excellent ee's and high diastereoselectivities can also be obtained with $[Cu\{(S,S)\text{-}tert\text{-butyl-box})][OTf]_2$ in dichloromethane, the rate enhancement in ionic liquid is particularly beneficial as it enables lower catalyst loadings and more challenging less reactive substrate combinations to be reacted. An additional benefit for reactions conducted in ionic liquid is the significantly shorter complexation times required to achieve reproducible selectivity compared with dichloromethane; the former requires a catalyst aging time of 5 min whereas a complexation time of 3 h is typically required in dichloromethane. On the basis of the encouraging performance of Cu/bis(oxazoline) catalysts, the corresponding imidazolium bis(triflimide)-tagged bis(oxazoline) **79** was prepared and its $Cu(OTf)_2$-derived catalyst was shown to recycle ten times without any loss in activity or enantioselectivity and to have a much higher affinity for the ionic liquid phase during recycle than the unmodified catalyst.[25a] An enhancement in rate was also reported for the MgX_2/bis(oxazoline) ($X = ClO_4^-$, NTf_2^-, OTf^-) catalysed Diels–Alder reaction between *N*-acryloyloxazolidinone and cyclopentadiene conducted in $[C_2mim][NTf_2]$ compared with dichloromethane. Interestingly, the use of ionic liquid effected a reversal of the configuration for reactions catalysed by Mg(II) complexes of the less hindered Box-Ph compared with dichloromethane; this was attributed to the geometry of the catalyst determining the sense of asymmetric induction. Evidence was presented for $[NTf_2]^-$ acting as a chelating ligand to afford octahedral coordination in $[C_2mim][NTf_2]$ whereas tetrahedral coordination is favoured in dichloromethane.[354] Following the lead of Doherty and co-workers,[25a] a range of related C_2-symmetric and unsymmetrical imidazolium-tagged bis(oxazolines) were prepared with a variety of anions and their performance in the Diels–Alder reaction between *N*-acryloyloxazolidinone and 1,3-cyclohexadiene evaluated. Not surprisingly given the precedent, good conversions, high *endo*/*exo* selectivity and ee's up to 97% were obtained in $[C_4mim][NTf_2]$ with 10 mol% $Cu(OTf)_2$ in combination with **80**, (*S*,*S*)-*tert*-butyl-box modified with two $[C_4C_1{}^2mim][OTs]$ tags. Recycle studies demonstrated that this catalyst could be reused 20 times with no loss in performance whereas its neutral counterpart could only be used seven times before activity and enantioselectivity dropped quite severely. The tagged bis(oxazolines) were more toxic than traditional, bis(oxazolines) but were considered environmentally more friendly because of their efficient recyclability.[355]

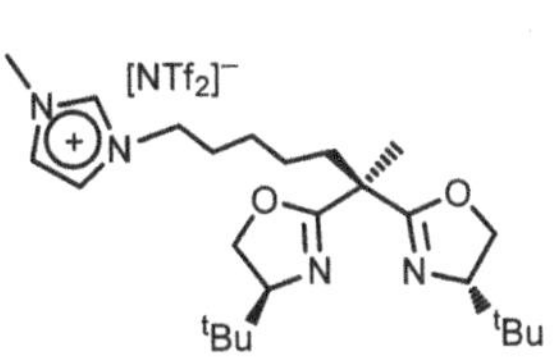

$[NTf_2]^-$ ^tBu ^tBu **79**

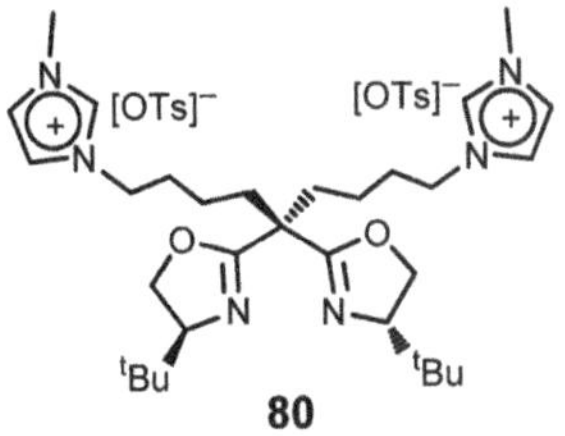

$[OTs]^-$ $[OTs]^-$ ^tBu ^tBu **80**

A significant enhancement in rate as well as a marked improvement in enantioselectivity (Δee ~20%) has also been obtained for ionic liquid and biphasic ionic liquid–diethyl ether-based Diels–Alder reaction between cyclopentadiene and *N*-acryloyloxazolidinone catalysed by enantiopure Lewis acid platinum complexes of the type $[Pt(diphosphine)][SbF_6]$ supported by BINAP and atropisomeric NUPHOS-type diphosphines. Reactions conducted in ionic liquid–diethyl ether mixtures were slightly slower but gave higher ee's than the same reaction in neat ionic liquid and, moreover, allowed high ee's to be obtained without the need to recourse to low temperatures. Both systems recycled efficiently by extraction of the product into diethyl ether with only a minor decrease in conversion and no change in ee on successive reaction.[102] Enantiopure Lewis acid platinum complexes of atropisomeric NUPHOS diphosphines also catalyse the carbonyl-ene reaction between unsymmetrical 1,1′-disubstituted alkenes and phenyl glyoxal or ethyl glyoxylate and for several substrate combinations the ee's obtained in $[C_4mim][NTf_2]$ were markedly higher than those obtained in dichloromethane (Δee ~ 5–50%); this was attributed to a combination of slower racemisation of the catalyst in $[C_4mim][NTf_2]$ compared with dichloromethane and an intrinsic solvent effect.[356]

3.4.1.2 Lewis Acid-Catalysed Cyclisations and Additions

Gold-catalysed cyclisations initiated by coordination of a multiple bond to a Lewis acid gold(I) complex has evolved into a powerful strategy for the synthesis of elaborate carbocyclic and heterocyclic motifs. As such transformations are typically catalysed by cationic metal complexes of the type

[LAu][X] ionic liquids are expected to be ideal solvents because coulombic interactions should immobilise the catalyst with respect to leaching during product isolation and thereby facilitate recycling. To this end, highly substituted furans have been prepared by cyclisation of 2-(1-alkynyl)-2-alken-1-ones in the presence of various nucleophiles using $[Bu_4N][AuCl_4]$ as catalyst in $[C_4mim][BF_4]$; the system recycled six times without loss in activity [Equation (3.4.1)].[357] Similarly, 2-substituted indoles have been prepared by Au(III)-catalysed cyclisation of 2-alkynylanilines in $[C_4mim][BF_4]$ [Equation (3.4.2)].[358]

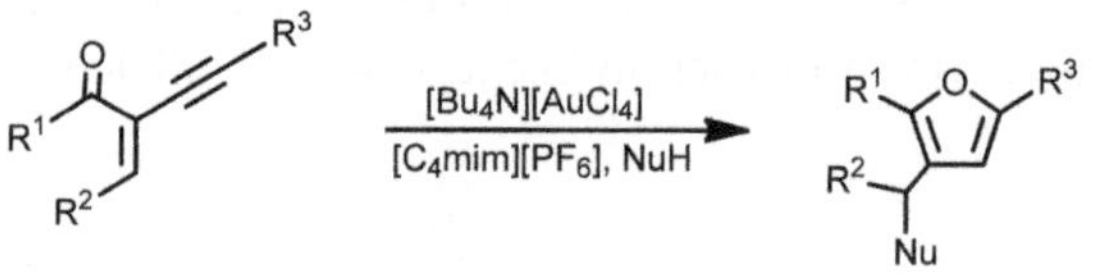

$[Bu_4N][AuCl_4]$; $[C_4mim][PF_6]$, NuH (3.4.1)

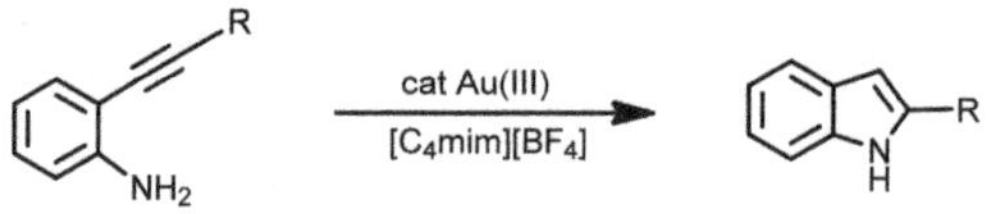

cat Au(III); $[C_4mim][BF_4]$ (3.4.2)

The gold(I)-catalysed cycloisomerisation of α-hydroxyallenes **81** occurs with complete axis to centre chirality transfer in $[C_4mim][PF_6]$ using 1 mol% $AuBr_3$ [Equation (3.4.3)] and quantitative conversion to the corresponding 2,5-dihydrofurans **82** was obtained after only 10 min at room temperature. The ionic liquid–catalyst mixture was air-stable and recycled efficiently although reaction times increased after the first run for selected substrates. After five runs the gold content in the extracts was just 0.03% (650 ppb in 2200 ppm) confirmed that the catalyst was efficiently retained in the ionic liquid.[359]

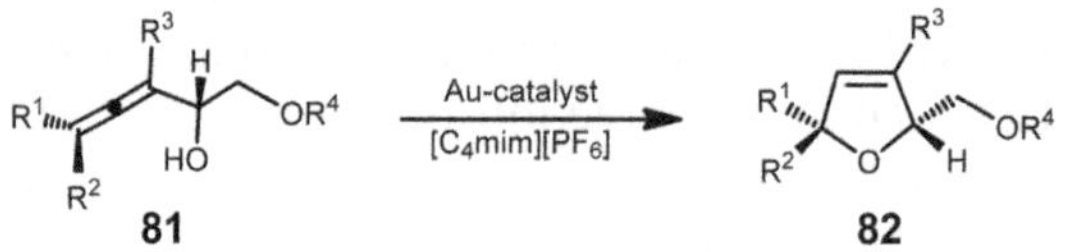

81 → Au-catalyst; $[C_4mim][PF_6]$ → **82** (3.4.3)

The selectivity of the AuCl and $AuCl_3$-catalysed cycloisomerisation of dienyne **83** in $[C_4mim][X]$ ($X = NTf_2^-$, PF_6^-, BF_4^-) showed a marked dependence on the anion and in some cases was more rapid and selective than the same reaction in dichloromethane (Scheme 3.4.1). Reactions conducted in hydrophobic ionic liquids such as $[C_4mim][NTf_2]$ and $[C_4mim][PF_6]$ gave improved selectivity for **84** and required much shorter reaction times than the same reaction in dichloromethane; whereas reactions conducted in the

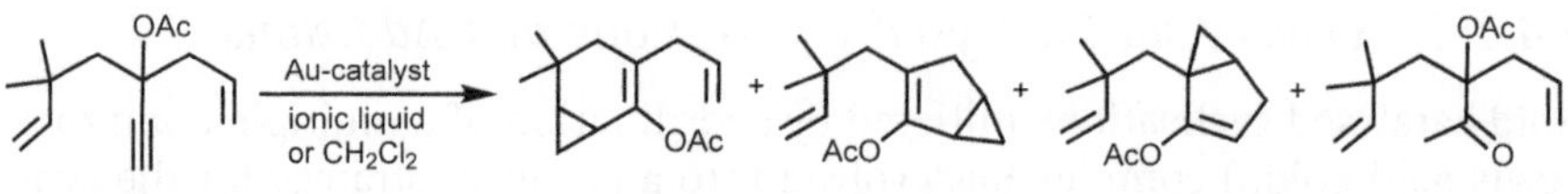

83 → Au-catalyst; ionic liquid or CH_2Cl_2 → **84** + **85** + **86** + **87**

Scheme 3.4.1 Distribution of products from the gold-catalysed cycloisomerization of enyne **83** carried out in various ionic liquids.

more hydrophilic [C_4mim][BF_4] required longer, were less selective and gave a significant amount of hydration product **87**. In contrast, cationic gold(I) catalysts generated by abstraction of chloride from Au(PPh_3)Cl were poor catalysts for the cycloisomerisation and gave **87** as the major component.[360] The gold-catalysed hydrative cyclisation of a range of 1,6-diynes gave good yields of cyclohexanone with 1–5 mol% [Au(PPh_3)][NO_3], methanol as co-solvent and methane sulfonic acid as promoter. The system recycled efficiently over six runs after extracting the product with diethyl ether.[361]

The lanthanide triflate-catalysed intramolecular hydroalkoxylation/cyclisation of unactivated primary/secondary and aliphatic/aromatic alkenols [Equation (3.4.4)] to afford furan, pyran, spirobicyclic pyran/furan, benzofuran and isochroman derivatives in room temperature ionic liquids showed a marked dependence on the anion of the ionic liquid as well as the lanthanide. A marked enhancement in rate was achieved with all ionic liquid–catalyst combinations compared with reaction in nitromethane and Yb(OTf)$_3$/[C_2mim][OTf] proved to be the optimum combination giving a TOF of 6.37 h^{-1} (at 120 °C) compared with only 0.1 h^{-1} (at 100 °C) in nitromethane. This 70-fold increase in TOF was explained by the relative Lewis acidity of Ln^{3+} in ionic liquid compared with nitromethane as the latter is polar, aprotic and moderately cation coordinating which decreases the Lewis acidity of Ln^{3+} by solvation. The cyclisation of a range of alkenols occurs with Markovnikov selectivity to give the corresponding five- and six-membered heterocycles in good yield. For substrates with terminal alkenes the relative ordering of catalyst activity was $Yb^{3+}>Sm^{3+}>La^{3+}$ whereas La(OTf)$_3$ with a large ionic radius was a more efficient catalyst for the cyclisation of sterically more challenging alkenols. Mechanistic studies support a pathway involving rate limiting hydroxyl coordination to the electron deficient Ln^{3+} centre, intramolecular H^+ transfer followed by ring closing nucleophilic attack of the alkoxide.[362a,b]

HO–C(R¹)(R¹)–alkenyl–R² → (1 mol% Ln(OTf)$_3$, [C_2mim][OTf], 120 °C) → 2,2-R¹,R¹-tetrahydrofuran with R$_2$ (3.4.4)

The cyclisation of 1,6-enynes catalysed by $PtCl_2$, $AuCl_3$ an [$RuCl_2(CO)_3$]$_2$ in ionic liquids occurs at much lower temperatures and in shorter reaction times than required in conventional organic solvents [Equation (3.4.5)]. For example, the cyclisation of **88** catalysed by 4 mol% $PtCl_2$ in toluene gave the corresponding vinyl cyclopentene **89** in 86% yield after 3 h at 80 °C whereas the same reaction in [C_4mim][PF_6] reached 79% yield after only 15 min at the same temperature and 81% yield after 5 h at 40 °C.[363]

88 → (1 mol% Ln(OTf)$_3$, [C_2mim][OTf], Temp) → **89** (3.4.5)

3.4.1.3 Lewis Acid-Catalysed Friedel–Crafts Alkylation and Acylation

3.4.1.3.1 Reactions Catalysed by Lewis Acid Ionic Liquids. The principal criterion for efficient Friedel–Crafts activity in ionic liquids is their Lewis acidity. In this regard, the first Friedel–Crafts reactions conducted in ionic liquid employed $[C_2mim]Cl/AlCl_3$ as both catalyst and solvent and reaction only occurred when the ionic liquid was Lewis acidic *i.e.* when $AlCl_3$ was in excess over $[C_2mim]Cl$ such that the Lewis acid $[Al_2Cl_7]^-$ was present. For example, a Lewis acid ionic liquid prepared by adding two mole equivalents of $AlCl_3$ to $[C_2mim]Cl$ has an apparent $AlCl_3$ mole fraction, N, of 0.67 and thus when $N<0.5$ the ionic liquid will necessarily be basic. Under these Lewis acidic conditions the alkylation of benzene with alkyl chlorides gave monoalkylated and polyalkylated products together with evidence that primary carbonium ions rearrange to their more stable secondary counterparts. Friedel–Crafts acylations are also catalysed by $[C_2mim]Cl–AlCl_3$ mixtures and the rate of acetylation increased with increasing Lewis acidity of the ionic liquid as the initial rates correlate with the calculated concentration of $[Al_2Cl_7]^-$. Turnover experiments demonstrated that the system was catalytic in $[Al_2Cl_7]^-$ even though it was present in larger excess by virtue of being the solvent.[364] The same Lewis acid ionic liquid mixtures also catalyse Friedel–Crafts acylations involving naphthalene, toluene, anisole, chlorobenzene and indane derivatives at 0 °C to give excellent yields of product with high regiocontrol. Under the same conditions the acylation of anthracene was shown to be reversible as the initial product 9-acetylanthracene slowly disproportionates to form anthracene and 1,5-diacetylanthracene. Evidence was presented for a proton-catalysed deacetylation pathway.[365] Chloroaluminate-based Lewis acid ionic liquid mixtures generated from $[Et_xNH_{4-x}]Cl$ ($x=1$–3) and $AlCl_3$ are also efficient solvents and catalysts for the Friedel–Crafts alkylation of 2-methylnaphthalene with long chain alkenes. After optimisation, high conversions of olefin and excellent selectivity for monoalkylation were obtained with $[Et_3NH]Cl/AlCl_3$ ($N=0.67$–0.7.5) whereas reactions catalysed by $AlCl_3$ were complicated and gave a plethora of side reactions including isomerization, disproportionation, hydrogenation and polyalkylation. In comparison, chloroaluminate ionic liquids were markedly more efficient than zeolites as they operated at much lower temperatures (353 K compared with 453–523 K) and the authors attributed this efficiency to highly dispersed superacid protons in the ionic liquid. Within a limited series of catalysts surveyed activity increased in the order $[EtNH_3]Cl/AlCl_3<[Et_2NH_2]Cl/AlCl_3<[Et_3NH]Cl/AlCl_3$ and optimum conversions were obtained with a naphthalene to alkene ratio of four after 30 min at 313 K.[366a,b] A survey of Lewis acid ionic liquids of the type $[C_nmim][Al_2Cl_6X]$ ($n=4$, 8, 10; $X=Cl^-$, Br^-, I^-) as catalysts for the Friedel–Crafts alkylation between benzene and 1-dodecene revealed $[C_4mim][Al_2Cl_6Br]$ to be the optimum combination and a catalyst loading of 0.5 mol% gave 91.8% conversion and 38.0 selectivity for 2-dodecylbenzene at 305 K. Conversions dropped quite dramatically after the first recycle while selectivity remained constant; the drop in activity was associated with a reduction in Lewis acidity due to loss of $[Al_2Cl_5Br]^-$.[367] High selectivity and good yields for the Friedel–Crafts phosphorylation of benzene to give dichlorophenylphosphine have been obtained using an ionic liquid–catalyst mixture with the composition $[NEt_3H]Cl/AlCl_3$ ($N = 0.60–0.71$). Optimum yields and selectivity were obtained when $N = 0.71$ with a PCl_3 : benzene : IL reactant composition of 30 : 10 : 1 after 16 h at reflux. Yields dropped rapidly on recycle from 64.4% in the first run to 8.3% in run three. The authors proposed a mechanism involving generation of the dichlorophosphinium ion from an adduct between PCl_3 and $[Al_2Cl_7]^-$, electrophilic addition of $[PCl_2]^+$ to benzene and proton abstraction by $[AlCl_4]^-$ to liberate HCl and regenerate $[Al_2Cl_7]^-$.[368] Similarly, the sulfonation of benzene and its derivatives with *p*-toluenesulfonylchloride in $[C_4mim]Cl/FeCl_3$ ($N = 0.6$) as both solvent and catalyst gave good yields of the sulfone after 2–8 h at a reaction temperature of 30–50 °C. Comparable yields could be obtained when the catalyst loading was reduced to 5–10 mol% either by raising the temperature to 80 °C or by using microwave heating (300 W) for much shorter reaction times (5–30 min).[369]

The Lewis acidic ionic liquid mixture $[C_4mim]Cl/FeCl_3$ catalyses the regioselective Friedel-Crafts benzylation of arenes and heteroarenes with 1-phenylethyl acetate under mild conditions to afford diarylmethanes, a common motif in a host of biologically active compounds. Optimisation on the benzylation of *o*-xylene with 10 mol% $[C_4mim]Cl/FeCl_3$ (1 : 1.2, $N = 0.54$) at 80 °C gave 1,2-dimethyl-4-(1-phenylethyl)benzene in near quantitative yield and high regioselectivity (96%). Although $FeCl_3$ also catalysed this reaction and gave comparable yields it was highly air and moisture-sensitive and did not easily recycle. Interestingly, $[C_4mim][BF_4]/FeCl_3$ gave a conversion of only 4% which strongly suggests that $FeCl_3$ is unlikely to be the active species and that there may be a cooperative role between the $[C_4mim]^+$ cation and $[FeCl_4]^-$. The ionic liquid–catalyst mixture could be recycled five times in an operationally straightforward procedure with no change in activity or selectivity. This protocol offers a number of practical benefits as it is operationally simple, reactions are conducted under mild conditions, it is inexpensive, easy/safe to handle and recycles efficiently.[370]

A systematic study of the Lewis acid ionic liquid-catalysed alkylation of isobutene with 2-butene as a function of the cation and anion as well as the $[C_nmim][X]/AlCl_3$ ratio (Lewis acidity) and temperature identified $[C_8mim]Br-AlCl_3$ ($N = 0.58$) to be the optimum catalyst at an operating temperature of 80 °C. Conversions increased as the chain length of the imidazolium alkyl group lengthened due to the higher solubility of the reactants. The superior performance of $[C_8mim]Br/AlCl_3$ was attributed to the higher inherent acidity of bromide-based aluminates compared with their chloride and iodide counterparts. Within the series $[C_8mim][X]/AlCl_3$ the trimethylpentanes/dimethylhexanes (TMPs/DMHs) selectivity decreased in the order $Br^- > Cl^- > I^-$. Activity and TMP selectivity increased with increasing anion ratio due to formation of the strongly acidic $[Al_2Cl_6Br]^-$ which was proposed to generate HBr by abstraction of the C2-H proton from the imidazolium cation. Although more active, this system was highly moisture sensitive and gradually deactivated with time on stream. The maximum olefin consumption and TMP selectivity was obtained at 80 °C and at higher temperatures decomposition of the imidazolium cation reduced the concentration of active Brønsted acid sites. A comparison of $[C_8mim]Br/AlCl_3$ with sulfuric acid revealed that the former was more active over the entire reaction time due its high Lewis acidity but gave lower TMP selectivity, consequently an ionic liquid with a higher concentration of Brønsted acid sites is likely to be an effective alkylation catalyst.[371]

An *in situ* IR spectroscopic study conclusively confirmed that the mechanism of the Friedel–Crafts acylation of benzene is the same in ionic liquid and in 1,2-dichloroethane *i.e.* acetyl chloride interacts with the Lewis acid MCl_3 (M = Fe, Al) to form various adducts such as $CH_3CClO\text{-}MCl_3$ and the 2 : 1 adduct $CH_3ClCO\text{-}2MCl_3$ (M = Al, $\nu(CO) = 1644\ cm^{-1}$, $1571\ cm^{-1}$; M = Fe, $\nu(CO) = 1675\ cm^{-1}$) in equilibrium with the key intermediate acylium cation $[CH_3CO][MCl_4]$ (M = Al, $\nu(CO) = 2300\ cm^{-1}$; M = Fe, $\nu(CO) = 2292\ cm^{-1}$) which attacks benzene to afford the MCl_3 adduct of acetophenone. Mössbauer measurements on a mixture of $FeCl_3$ and $[C_4mim]Cl$ identified an equilibrium involving $[FeCl_3]$, $[C_4mim][Fe_2Cl_7]$ and $[Fe_2Cl_6]$ and/or $[C_4mim][FeCl_4]$ depending on the mole ratio of $FeCl_3$ to ionic liquid; theoretical calculations lend some support to the above mechanism.[372a,b]

Isomerisation of alkanes to highly branched hydrocarbons of the same carbon number for use as fuel additives has been catalysed by solid acids *via* high temperature hydroisomerisation as well as by superacid systems consisting of a mixture of a strong Lewis acid such as SbF_5, TaF_5, NbF_5 and a Brønsted acid such as CF_3COOH or CF_3SO_3H at temperatures as low as 298 K; however, these systems are highly toxic and strongly corrosive and there is a clear need to identify a more benign alternative. In this regard, superacidic ionic liquids of the type $[cation]Cl/AlCl_3/H_2SO_4$ also catalyse the isomerisation of *n*-octane under mild conditions (283–323 K). The optimum composition based on a $[C_4mim]Cl/AlCl_3$ ratio of 0.5 and a $H_2SO_4/AlCl_3$ ratio of 0.18 gave an *n*-octane conversion of 77.2% and a yield of branched C4–C8 alkane of 54.2%, of which 32.4% were branched C6–C8 alkanes. The H_2SO_4/$AlCl_3$ ratio of 0.18 required to achieve the optimum conversion corresponds closely to the predicted ideal composition of 3/16 (or 0.1875), according to the equilibria present in the ionic liquid system to form superacid 'naked' protons [Equation (3.4.6) and (3.4.7)]. This system is easier and safer to handle than that based on HCl, operates under much milder conditions and avoids the formation of coke; the low operating temperatures also favour formation of the thermodynamically more stable branched isomer.[373]

$$2[Al_2Cl_7]^- + 3H_2SO_4 \rightarrow Al_2(SO_4)_{3(s)} + 6HCl + 2[AlCl_4]^- \quad (3.4.6)$$

$$6[Al_2Cl_7]^- + 6HCl \rightleftharpoons 12[AlCl_4]^- + 6\,\text{“}H^+\text{”} \quad (3.4.7)$$

3.4.1.3.2 Reactions Catalysed by Lewis Acids in Ionic Liquids. The Friedel–Crafts alkylation between benzene with 1-bromopropane, 1-chlorobutane or benzylchloride in pyridinium-based ionic liquids $[C_2pyr][X]$ ($X = BF_4^-$, $CF_3CO_2^-$) is more efficiently promoted by $FeCl_3$ than the environmentally more hazardous $AlCl_3$. The highest conversions were obtained in $[C_2pyr][BF_4]/FeCl_3$ at 50 °C with a benzene : electrophile : catalyst : IL ratio of 2 : 1 : 2 : 1.[374] The same pyridinium-based ionic liquids are also suitable solvents for the Friedel–Crafts acylation of aromatic compounds with acetic anhydride. Each combination of $[C_2pyr][X]$ ($X = BF_4^-$, $CF_3CO_2^-$) and MCl_3 (M = Al, Fe) gave good conversions and high selectivity at room temperature, 50 °C and 75 °C with the optimum arene : anhydride : IL : MCl_3 ratio of 2 : 1 : 1 : 2. While similar yields were obtained for each catalyst in their respective ionic liquids those in $[C_2pyr][CF_3CO_2]/FeCl_3$ were marginally better. The product was easily separated from the biphasic reaction mixture and the catalyst reused with only a slight drop in activity over three runs.[375]

Scandium(III) triflate immobilised in hydrophobic ionic liquids catalyses the Friedel–Crafts alkylation of benzene, phenol and anisole with alkenes with remarkable efficiency. Good to excellent yields of monoalkylated product were obtained with 20 mol% $Sc(OTf)_3$ after 12 h at 20 °C in $[C_nmim][PF_6]$ ($n = 4$–6) and $[C_nmim][SbF_6]$ ($n = 2$, 4). The product can be readily separated from the biphasic reaction mixture and the recovered catalyst recycled. In stark contrast, there was no reaction in conventional organic solvents such as dichloromethane, acetonitrile, nitromethane, $PhNO_2$ or in hydrophilic ionic liquids such as $[C_nmim][X]$ ($n = 2$, 4; $X = OTf^-$, BF_4^-).[83] The Friedel–Crafts alkenylation of aromatic compounds [Equation (3.4.8)] presents a number of challenges such as competing polymerisation of terminal alkynes and slow reactions rates for internal alkynes. To this end, the use of hydrophobic ionic liquids for the metal triflate-catalysed Friedel–Crafts alkenylation of arenes gave a substantial and significant enhancement in rate as well as a reduction in the formation of undesired by-product, compared with reactions conducted in the absence of ionic liquid. Under optimum conditions, 10 mol% $Sc(OTf)_3$ catalysed the alkenylation of benzene with phenylpropyne in $[C_4mim][X]$ ($X = PF_6^-$, SbF_6^-) to give quantitative conversion after only 4 h at 85 °C; for comparison the same reaction conducted in the absence of ionic liquid only reached 27% in 96 h. Other metal triflates including $In(OTf)_3$, $Hf(OTf)_4$ and $Y(OTf)_3$ were also efficient catalysts for this transformation in $[C_4mim][SbF_6]$. The rate acceleration was attributed to stabilisation of the unstable vinyl cation intermediate in the highly polar ionic liquid. High yields were obtained in short reaction times for the Friedel–Crafts alkenylation of a range of substrate combinations in $[C_4mim][SbF_6]$. The product was decanted by cooling the reaction mixture to −40 °C and the recovered ionic liquid–catalyst phase used up to eight times with no loss in conversion. This protocol was extended to an intramolecular Friedel–Crafts alkenylation to construct the courmarin and 2(1*H*)-quinolinone motifs [Equation (3.4.9)].[84,85]

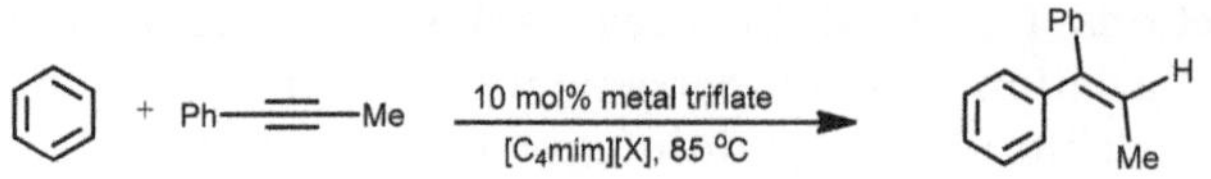

(3.4.8)

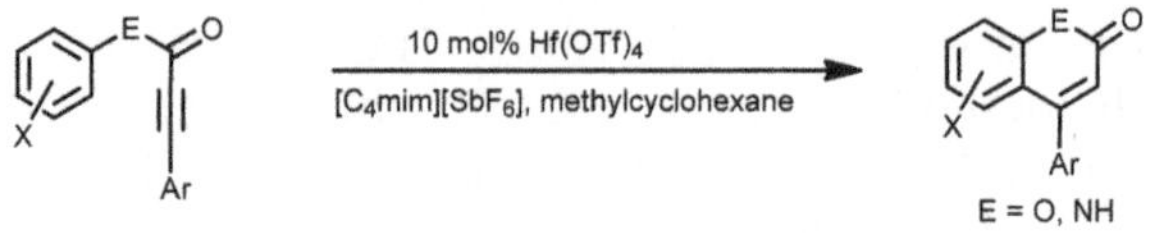

(3.4.9)

Ferrocene is efficiently and selectively acylated in [C_4mim][NTf_2] with 5 mol% Sc$(OTf)_3$ as catalyst, acetic anhydride as the acylating agent and microwave (MW) heating (5–40 W; 65–110 °C); under these conditions quantitative yields of acetylferrocene were obtained in less than 2 min. For comparison, much longer reaction times (2.5–4 h) were required to achieve acceptable conversions under conventional heating at 50 °C. Hydrophilic ionic liquids such as [C_4mim][OTf] are extremely poor solvents for this reaction and gave <1% yield after 96 h. This IL–MW–Sc$(OTf)_3$ combination is among the most efficient protocols for the acylation of ferrocene.[376]

A survey of the effect of the metal triflate and solvent on the Friedel–Crafts benzoylation of anisole with benzoyl chloride revealed that quantitative conversion and high *ortho*/*para* selectivity (4 : 96) could be achieved with 10 mol% Cu$(OTf)_2$ in [C_4mim][BF_4] after 1 h at 80 °C whereas conversions and selectivities were lower with Zn$(OTf)_2$, Sn$(OTf)_2$ or Sc$(OTf)_3$ and/or when reactions were conducted in acetonitrile or 1,2-dichloroethane. Reactions using acetic anhydride in place of the acid chloride were much slower due to precipitation of Cu$(OAc)_2$.H_2O arising from trace amounts of water in the ionic liquid, which lowered the concentration of active Cu(II) species.[377] The Friedel–Crafts benzoylation of aromatic compounds is also catalysed by a host of main group and transition metal bis{(trifluoromethyl)sulfonyl}amides in [C_4mim][NTf_2] at 60–160 °C, using either the acid chloride, carboxylic acid or anhydride as the acylating agent. Moreover, catalysts generated *in situ* by dissolving the metal halide in [C_4mim][NTf_2] gave conversion that matched those obtained with the preformed metal amide.[378] A detailed kinetic study of the Friedel–Crafts benzoylation of anisole with benzoic anhydride in [C_4mim][NTf_2] with a range of metal triflates provided convincing support for the generation of free acid catalyst ($HNTf_2$ in the case of [C_4mim][NTf_2]) by a complex exchange of ligands between the metal salt, benzoic anhydride and the ionic liquid with formation of the mixed anhydride. Deactivation of the catalyst with increasing conversions was attributed to a H-bonding interaction between the carbonyl group of 4-methoxybenzophenone and $HNTf_2$. The authors modified the two cycle mechanism originally proposed by Dumeunier and Markov to account for the catalyst deactivation as well as an equilibrium between the ionic liquid and the acid catalyst which increased the concentration of the more active $HNTf_2$ relative to HOTf (Scheme 3.4.2).[379] Hardacre *et al.* subsequently identified optimum reaction conditions and separation methods

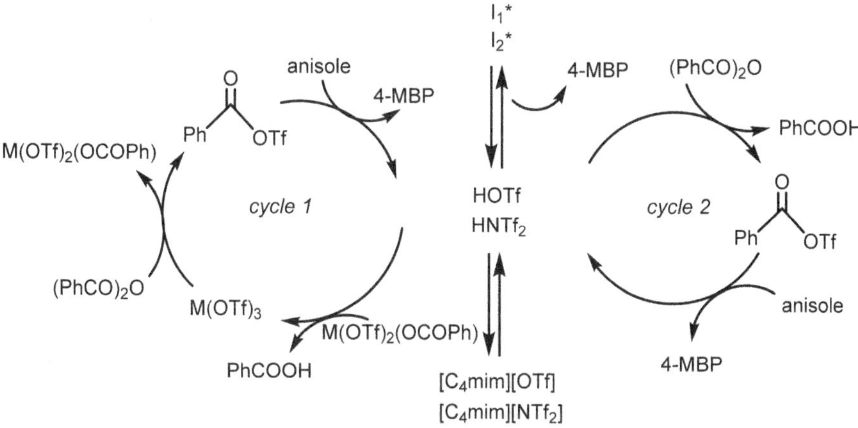

Scheme 3.4.2 Proposed catalytic cycle for the benzoylation of anisole in ionic liquids.

for the large scale benzoylation of anisole in ionic liquids after a detailed economic evaluation and a system based on zeolites and $[C_2mim][NTf_2]$ was identified to be the most cost effective. The predominant catalytically active species was shown to be $HNTf_2$ generated *via* exchange of surface acidic protons on the zeolite with the ionic liquid cation. Of the separation strategies explored only vacuum/steam distillation gave product of suitable quality (99.5% purity and 94.7% selectivity to the 4-isomer).[380]

A dramatic increase in activity was obtained for the bismuth(III)-catalysed Friedel–Crafts acylation of aromatic substrates with benzoyl chloride in ionic liquids compared with the same reactions conducted under solventless conditions. For example, the benzoylation of anisole in $[C_2mim][NTf_2]$ catalysed by 10 mol% $Bi(OTf)_3$ or Bi_2O_3 at 80 °C was complete after 5 min ($o/m/p = 4:0:96$) whereas the corresponding reaction in neat anisole required 3 h heating at 110 °C to achieve a yield of 92% ($o/m/p = 8:0:92$); moreover, good yields were also obtained after only 4 h when the catalyst loading was decreased to 1 mol%. The benzoylation of more challenging substrates required higher temperatures (150 °C) and in a limited comparison higher conversions were obtained with $Bi(OTf)_3/[C_2mim][NTf_2]$ than $B_2O_3/[C_2mim][NTf_2]$. In a survey of ionic liquids, catalyst activity was shown to depend on the nature of the cation and anion; good conversions were obtained in hydrophobic imidazolium-based ionic liquids $[C_nmim][X]$ ($n = 2, 4$; $X = NTf_2^-, PF_6^-$), while slightly lower conversions were obtained in tetraalkylammonium and phosphonium ionic liquids of $[PF_6]^-$ and $[NTf_2]^-$ and only poor yields were obtained in hydrophilic ionic liquids based on $[BF_4]^-$ and $[OTf]^-$.[381] More recently bismuth triflate was reported to be an effective catalyst for the Friedel–Crafts acylation of various electron-rich aromatics with the anhydride or acid chloride in ionic liquid; good yields were obtained with 5 mol% $Bi(OTf)_3$ immobilised in $[C_4mim][PF_6]$ after 30 min under conventional heating at 80 °C or within 5 min using microware heating (60 W).[382]

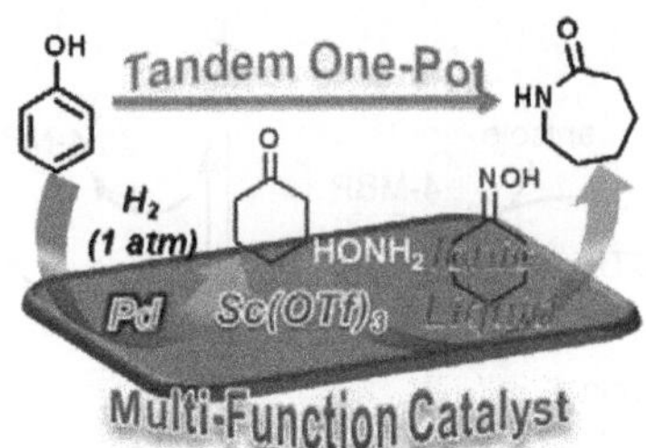

Figure 3.4.2 A multifunctional Pd–Sc(OTf)$_3$–ionic liquid catalyst system for the tandem one-pot conversion of phenol to ε-caprolactam.
Reprinted with permission from reference 92. Copyright 2013 American Chemical Society.

3.4.1.4 Lewis Acid-Catalysed Tandem Reactions

A multifunctional Pd/C–Sc(OTf)$_3$–ionic liquid system catalyses the tandem and highly selective one-pot conversion of phenol to ε-caprolactam (Figure 3.4.2).[92] The Pd/C and Sc(OTf)$_3$ cooperate to catalyse the quantitative and highly selective (>99.9%) hydrogenation of phenol to cyclohexanone[383] while the Sc(OTf)$_3$ and ionic liquid cooperate to catalyse the tandem transformation of cyclohexanone into its oxime and subsequent Beckmann rearrangement to afford ε-caprolactam. The Sc(OTf)$_3$ is crucial for the final rearrangement as only oxime was obtained from the reaction between cyclohexanone and hydroxylamine hydrochloride with 10 mol% Sc(OTf)$_3$ in dichloroethane whereas the same reaction in a mixture of dichloroethane and [C$_4$mim][SbF$_6$] (5 : 1, v/v) gave ε-caprolactam in excellent yield. In this system, the Lewis acid Sc(OTf)$_3$ appears to accelerate the hydrogenation of phenol to cyclohexanone and suppress reduction to cyclohexanol while the ionic liquid increases the Lewis acidity of Sc(OTf)$_3$ *via* anionic exchange and lowers the activation energy for the Beckmann rearrangement by stabilising the charged intermediate. Such cooperative action in multicomponent catalytic systems could be applied to the design of other tandem reaction sequences.

Ionic liquid-coordinated ytterbium(III) sulfonate [SO$_3$C$_3$mim]$_3$Yb[BF$_4$]$_3$ is a recoverable and recyclable Lewis acid that catalyses the Michael addition of indoles to *β*-nitrostyrenes in ethanol at 70 °C to give good yields of adduct **90** in short reaction times. The product was subsequently converted into the 1,2,3,4-tetrahydro-*β*-carboline **92** *via* **91** by sodium formate reduction of the nitro group followed by a Pictet–Spengler cyclisation with benzaldehyde using [Ph$_3$PC$_3$SO$_3$H][*p*-TSA]/[C$_4$mim][BF$_4$] as catalyst and solvent according to Scheme 3.4.3.[384]

A marked enhancement in yield of the Cu(0)/phenanthroline-catalysed tandem reaction involving 2-iodophenols and isothiocyanate [Equation (3.4.10)] was obtained in [C$_4$mim][PF$_6$] compared with toluene. In a representative example, the reaction between 4-chloro-2-iodophenol and phenylisothiocyanate in [C$_4$mim][PF$_6$] at 90 °C reached 83% conversion after 24 h whereas only 52% yield was obtained after 40 h in toluene. The ionic

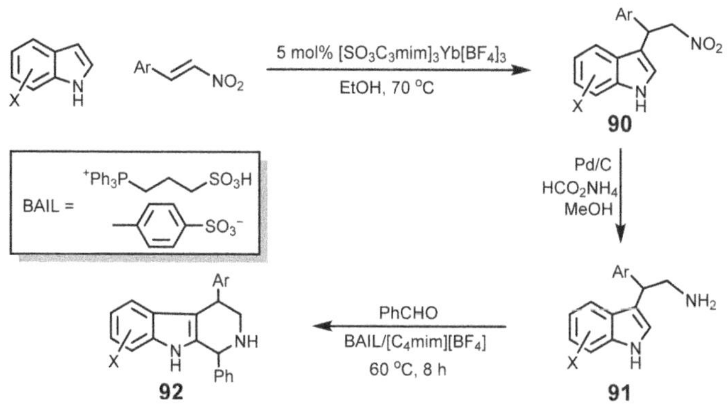

Scheme 3.4.3 Lewis acid ionic liquid-catalysed Michael addition of indoles to *β*-nitrostyrenes and transformation of adduct **90** into 1,2,3,4-tetrahydro-*β*-carboline **92**.

liquid-immobilised catalyst is reasonably robust and stable and recycled several times with no loss in activity, even after being stored in air for several weeks.[385]

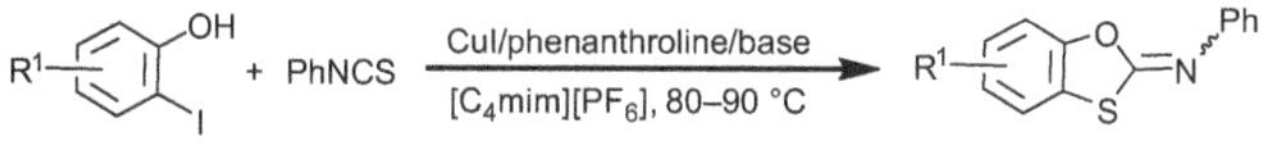

(3.4.10)

Silica aerogels prepared in a two-step acid–base-catalysed sol–gel process have been used as a support for ionic liquid solutions of Cp_2ZrCl_2. The resulting ASILP Lewis acids catalyse the condensation between ketones and *p*-phenylenediamine to afford 1,5-benzodiazepines in good yield (84–87%) after 8–9 h whereas in the absence of aerogel Cp_2ZrCl_2 only achieved yields of 48–51% after 24 h. The aerogel-confined ionic liquid containing catalyst is firmly immobilised in the porous network of the aerogel as it recycled efficiently with only a minor reduction in yield over four runs and <3.6% leaching of the Lewis acid.[386]

3.4.1.5 Lewis Acid-Catalysed Nitrene and Carbene Transfer Reactions

The aziridination of alkenes [Scheme 3.4.4(a)] with chloramine-T as the nitrene transfer agent and 10 mol% *N*-bromosuccinimide as catalyst gave higher conversions in [C_4mim][X] (X = BF_4^-, PF_6^-) compared with acetonitrile and nitromethane; the enhanced activity was attributed to increased polarisation of the N–Br bond in the more polar ionic liquid facilitating formation of the bromonium ion. Recycle experiments gave consistent yields (88–90%) over three runs.[387] The [{tris(pyrazolyl)methane}Cu(MeCN)][BF_4]-catalysed aziridination of styrene [Scheme 3.4.4(a)] and the C–H bond amidation of tetrahydrofuran [Scheme (3.4.5b)] in [C_4mim][PF_6] with chloramine-T (*N*-chloro 4-methylbenzenesulfonamide, sodium salt;

Scheme 3.4.4 Nitrene transfer reactions (a) aziridation and (b) C–H insertion of nitrene.

Na[4-MeC$_6$H$_4$SO$_2$NCl]) as the nitrene source gave good yields of product in 24 h and 3 h, respectively, at room temperature. The immobilised catalyst recycled after extraction of the product with diethyl ether and addition of more chloramine-T and substrate to the ionic liquid; conversions dropped after the fourth (aziridination) and sixth (amidation) runs. The system appeared to be tolerant with respect to accumulated sodium chloride which may well be associated with the equilibrium between [{tris(pyrazolyl)methane}CuCl] and [{tris(pyrazolyl)methane}Cu]$^+$ favouring the active cationic species in ionic liquid *i.e.* the neutral complex acts a reservoir for the active catalyst through a dissociative equilibrium.[388]

The same pyrazolylmethane-based copper complexes also catalyse carbene transfer reactions from ethyl diazoacetate to a range of unsaturated (cyclopropanation) and saturated (C–H functionalisation) substrates under biphasic conditions in [C$_4$mim][PF$_6$]–hexane to give conversions that matched those obtained under homogeneous conditions in dichloromethane.[389,390] The ionic liquid-immobilised copper catalyst was recovered and gave consistent conversions over five recycles with no loss in activity, which is a marked improvement on previously reported systems.[23b,50,391] The electrophilic character (and presumably the reactivity) of the copper complex was quantified using the ν(CO) values of the carbonyl adducts [Tp*Cu(CO)] and [{HC(3,5-Me$_2$pz)$_3$}Cu(CO)][BF$_4$] as the stretching frequency generally correlates with the catalytic activity in such a way that the higher the ν(CO) value the higher the catalytic activity. The values of 2066 cm^{-1} and 2113 cm^{-1} for [Tp*Cu(CO)] and [{HC(3,5-Me$_2$pz)$_3$}Cu(CO)][BF$_4$] respectively, indicate that the latter is more electrophilic by virtue of having a high ν(CO) due to less efficient backbonding; the former does not catalyse this transformation which is entirely consistent with its lower ν(CO) value.

The efficiency of ionic liquid immobilised copper(II) complexes of bis(oxazolines) as catalysts for the cyclopropanation of styrene with ethyl diazoacetate in ionic liquids depends on the cation and anion, the presence of water, the IL/copper salt ratio as well as the purity of the ionic liquid. Ee's obtained in [C$_2$mim][NTf$_2$] or [C$_2$mim][OTf] matched those in dichloromethane provided the former were dried over P$_2$O$_5$ prior to use. However, ee's dropped on recycle which appeared to be due to a shift of the equilibrium towards the ligand free achiral catalyst due to extraction of uncoordinated bis(oxazoline) into the hexane phase. Lower yields and ee's were also obtained in [C$_4$mim][OTf] which was again attributed to a shift of the equilibrium towards ligand free copper due to the reduction in polarity; the reduction in ee and activity was even more pronounced in $[Oct_3NMe][OTf]$. The highest ee's were obtained with a high $[C_2mim][OTf]/CuCl_2$-bis(oxazoline) ratio which was interpreted in terms of the equilibrium between the chloride and triflate forms of the chiral copper complex favouring the more active and selective triflate.[23b] One of the main problems encountered with the use of copper complexes of chiral bis(oxazolines) for asymmetric catalysis in ionic liquids is the equilibrium between the coordinated and free ligand as the latter is extracted during the product isolation step and the ligand-free copper remaining in the ionic liquid causes a reduction in enantioselectivity; this effect is compounded on each recycle. Rather than add additional chiral ligand to solve this problem, Majoral and co-workers prepared an aza-bis(oxazoline) on the basis that it was estimated to shift the equilibrium in favour of the copper complex by 5.7 kcal mol^{-1} and should therefore reduce ligand leaching. Enantioselectivities remained constant at 90–92% for the *trans* isomer over eight runs and although conversions dropped after the first cycle they remained constant for the next seven; for comparison the corresponding CuCl/bis(oxazoline) system recycled poorly and ee's dropped gradually after the first cycle falling from ~85% to ~60% in the fifth run. Moreover, the ionic liquid immobilised azabis(oxazoline)/CuCl system catalysed the cyclopropanation of a series of four different alkenes with remarkable efficacy and in each case gave the product in high ee and purity.[50]

3.4.1.6 *Lewis Acid-Catalysed Cyanosilylation and Azide–Alkyne Coupling*

A remarkable increase in the efficiency of the scandium triflate-catalysed cyanosilylation of benzaldehyde [Equation (3.4.11)] was achieved in $[C_4mim][SbF_6]$ compared with dichloromethane. The magnitude of this ionic liquid enhancement in activity was patently evident in the conversions as 0.1 mol% catalyst gave >99% conversion after only 5 min in $[C_4mim][SbF_6]$ whereas only 7% conversion was obtained in dichloromethane. Moreover, a quantitative conversion was also obtained after only 30 min when the catalyst loading was reduced to 0.01 mol%, corresponding to a TOF of 20 000 h^{-1}. Moreover, a recycle study with 0.01 mol% $Sc(OTf)_3$ showed no loss in activity after 10 runs and a total TON close to 100 000 indicating that the catalyst was efficiently immobilised and stabilised in the ionic liquid. Other metal triflates including $Ho(OTf)_3$, $Gd(OTf)_3$, $Er(OTf)_3$, $Tm(OTf)_3$, $Y(OTf)_3$ and $In(OTf)_3$ also gave good conversions with a catalyst loading of only 0.1 mol%. The high activity was suggested to reflect the increased Lewis acidity of the catalyst by virtue of anion exchange with the ionic liquid, as discussed in section 3.1.6.1. Excellent conversions were obtained for the cyanosilylation of a range of electronically and functionally diverse aldehydes and ketones using 0.1 mol% $Sc(OTf)_3$ in $[C_4mim][SbF_6]$ after reaction times of 5 min and 2 h, respectively.[392]

$$PhCHO + TMSCN \xrightarrow[[C_4mim][X]]{Sc(OTf)_3} PhCH(OSiMe_3)CN \qquad (3.4.11)$$

The first copper-catalysed azide–alkyne click reactions between the sugar azide, methyl 6-azidoglucopyranoside and an alkyne, ethynyl *C*-galactoside, in ionic liquid gave the corresponding triazole-linked disaccharides reaching 95% yield in $[C_8dabco][N(CN)_2]$; for comparison the same reaction conducted in toluene and DMF gave 96% and 70% yield, respectively. However, although work-up of the reaction mixture was non-trivial as small amounts of ionic liquid extracted into the organic solvent the system still recycled with reasonable efficiency.[393]

3.4.1.7 *Lewis Acid-Catalysed Ring Opening Transformations of Epoxides*

3.4.1.7.1 Asymmetric Ring Opening. The efficiency of the Cr(salen)-catalysed asymmetric ring opening of *meso* epoxides with $TMSN_3$ in $[C_4mim][X]$ ($X = SbF_6^-$, PF_6^-, BF_4^-, OTf^-) is highly anion dependent. The azido silyl ethers were obtained in good yields and high ee's in hydrophobic ionic liquid ($X = SbF_6^-$, 94%; PF_6^-, 87%) with 3 mol% $[Cr(salen)N_3]$ whereas hydrophilic ionic liquids gave poor yields and negligible ee's. Interestingly a 5 : 1 mixture (v/v) of $[C_4mim][PF_6]$ and $[C_4mim][OTf]$ proved to be the optimum solvent for immobilisation of the catalyst and recycle studies gave ee's of 93–94% across five runs with no loss in activity.[394]

3.4.1.7.2 Fixation of Carbon Dioxide (Cycloaddition). Ionic liquids have been used as both solvent and catalyst for the chemical fixation of carbon dioxide with epoxides to afford cyclic carbonates [Equation (3.4.12)]. The first cycloadditions between CO_2 and propylene oxide (R = Me) in an ionic liquid were conducted in $[C_4mim][X]$ ($X = Cl^-$, PF_6^-, BF_4^-) or $[C_4pyr][BF_4]$ and showed that both the anion and cation influenced activity. Reactions were performed at 110 °C and a $p(CO_2)$ of 2.0–2.5 MPa and under these conditions $[C_4mim][BF_4]$ proved to be the most efficient by quite some margin giving a TON of 60.1 which increased to 449 at 140 °C. The $[C_4mim]^+$ cation was tentatively suggested to be the main site for activation of the CO_2-propylene oxide complex as the anion is non-nucleophilic.[395]

$$\text{R-epoxide} + CO_2 \xrightarrow[\text{ionic liquid}]{\text{cat}} \text{R-cyclic carbonate} \quad (3.4.12)$$

The tetrabutylammonium halides are also effective solvents and catalysts for the fixation of carbon dioxide with oxiranes. Reactions conducted either in a 1 : 1 mixture of $[Bu_4N]Br$ and $[Bu_4N][I]$ at 120 °C or in $[Bu_4N][I]$ at 60 °C at atmospheric pressure gave good to excellent yields of the corresponding cyclic carbonate with a wide range of substrates including polymerisation sensitive epoxides. The products can be isolated either by distillation or extraction. Evidence was presented to support a mechanism involving nucleophilic ring opening of the epoxide by the halide ion to afford a reactive

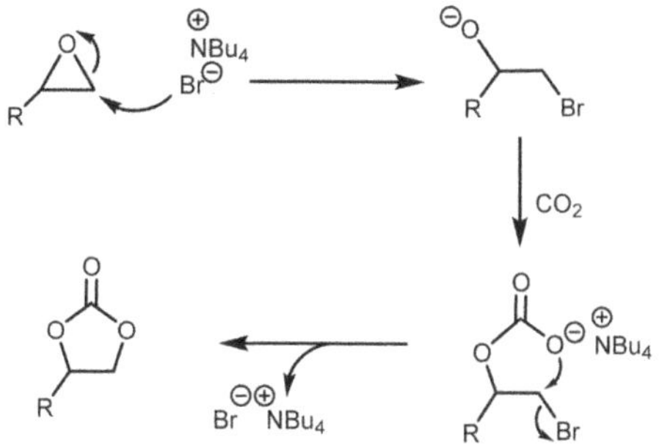

Scheme 3.4.5 Proposed mechanism for the ring opening of the epoxide.

oxyanion which attacks CO_2 according to Scheme 3.4.5. The role of the cation was investigated using [C_4mim][I] and as it is a very poor solvent/catalyst for this transformation the authors reasoned that this reflected the greater nucleophilicity of the iodide in [Bu_4N][I] due to poor/weak association with the bulky $[Bu_4N]^+$ cation relative to the planar imidazolium cation which would associate with the anion more tightly and render it less nucleophilic for the ring opening step.[396]

Remarkably high TOFs were obtained for the synthesis of propylene carbonate from carbon dioxide and propylene oxide under supercritical conditions (14 MPa, 100 °C) in the presence of ionic liquid. Within a series of imidazolium-based ionic liquids the highest activity was obtained in [C_4mim][BF_4] and conversions increased as the alkyl chain length on the imidazolium cation increased such that [C_8mim][BF_4]–*sc*CO_2 gave quantitative yields and complete selectivity for propylene carbonate in 5 min, corresponding to a TOF of 517 h^{-1}. This increase was attributed to the higher solubility of CO_2 and PO in ionic liquids with long alkyl chains. Quantitative conversions were also achieved at 60 °C after 120 min and good conversions were obtained for a selection of substrates.[397] A dramatic increase in the activity of imidazolium halides for the synthesis of cyclic carbonates was obtained by substituting the halide anion with zinc tetrahalide. A comparative study revealed the magnitude of this enhancement as the [C_4mim]Br-catalysed cycloaddition between carbon dioxide and ethylene oxide gave a TOF of 78 h^{-1} at 100 °C under 3.5 MPa CO_2 with a catalyst/substrate ratio of 5000, while the combination of [C_4mim]Cl and $ZnBr_2$ gave a TOF of 2112 h^{-1} which increased to 3545 h^{-1} for a mixture of [C_4mim]Br and $ZnBr_2$. Similar activities were achieved with the corresponding preformed imidazolium zinc tetrahalides $[C_4mim]_2[ZnX_2Y_2]$. The use of imidazolium zinc tetrahalides also gave a comparable rate enhancement for the cycloaddition between carbon dioxide and propylene oxide. Activity was found to depend strongly on the halide and decreased in the order $[ZnBr_4]^- > [ZnBr_2Cl_2]^- >>> [ZnCl_4]^-$ but showed limited variation with the imidazolium cation. The authors related the high activity to a complementary combination of Lewis acidity (presumably by virtue of halide dissociation from $[ZnX_2Y_2]^-$) to activate the epoxide and the Lewis basicity of the halide ion to ring open the coordinated epoxide.[398] Arai and co-workers also reported [C_4mim]Cl/$ZnBr_2$ to be a highly active combination for the cycloaddition of styrene epoxide with carbon dioxide as it gave styrene carbonate in 94% yield after 1 h at 80 °C under 14 MPa CO_2. Catalyst efficiency varied with the metal halide as well as the anion of the ionic liquid and to a lesser extent the length of the alkyl chain on the imidazolium cation. Poor conversions and low selectivities were obtained with $ZnBr_2$ in [C_4mim][X] ($X = BF_4^-$ and PF_6^-) while the combinations of [C_4mim]Cl with $FeBr_2$, $FeBr_3$, $MgBr_2$, ZnI_2 or LiBr all gave moderate to good yields. The highest yields were obtained with [C_8mim]Cl/$ZnBr_2$, which is consistent with the earlier work of Kim *et al.* described above.[398] The efficiency of [$C_n$mim]Cl/$ZnBr_2$ ($n = 4$, 8) was suggested to be the result of cooperative action of the Lewis acidic zinc and the nucleophilic halide on the key activation and ring opening steps.[399] The combination of a Brønsted acid-modified imidazolium-based ionic liquid and $ZnBr_2$ catalysed the coupling of carbon dioxide with propylene oxide to afford propylene carbonate. Several Brønsted acid ionic liquid–metal halide combinations were examined and **93**/$ZnBr_2$ was identified to be the optimum system reaching a TOF of 4248 h^{-1} at 110 °C under 1.5 MPa CO_2. A survey of metal salts revealed that activity decreased in the order $Zn^{2+} > Fe^{3+} > Cu^{2+} \sim Cu^+$ and for a series of zinc salts activity appeared to be a compromise between the nucleophilicity and leaving ability of the anion such that $Br^- > I^- > Cl^- > SO_4^{2-}$. Activity was also sensitive to the nature of the Brønsted acid ionic liquid and decreased in the order **93**>**94**∼**95**>**96**∼**97**>**98** (Figure 3.4.3). The authors proposed a mechanism involving dual or synergistic activation of the epoxide through coordination to the Lewis acid and hydrogen-bond formation with the carboxylate group of the Brønsted acid ionic liquid followed by nucleophilic ring opening of the epoxide by the X^- anion, insertion of CO_2 into the Zr-O σ-bond and ring closing *via* nucleophilic displacement of bromide by the oxygen atom of the carbonate (Scheme 3.4.6).[400a] Brønsted acid modified halide-based ionic liquids have also been reported to act as acid based bifunctional catalyst for the *solvent and metal-free* fixation of CO_2 with propylene oxide. A structure-efficiency survey identified [{$C_3(CO_2H)$}$_2$im][Br] to be the most efficient as good yields of propylene carbonate were obtained at 398 K under 2MPa CO_2 over five runs. The catalysts are straightforward to prepare, environmentally benign, stable and activate a range of epoxides *via* carboxyl-halide synergism *i.e.* activation of the epoxide by hydrogen bonding with the carboxyl residue and nucleophilic activation of the carbon atom by the Lewis basic halide.[400b] Subsequently, imidazolium bromide-based ionic liquids immobilised on

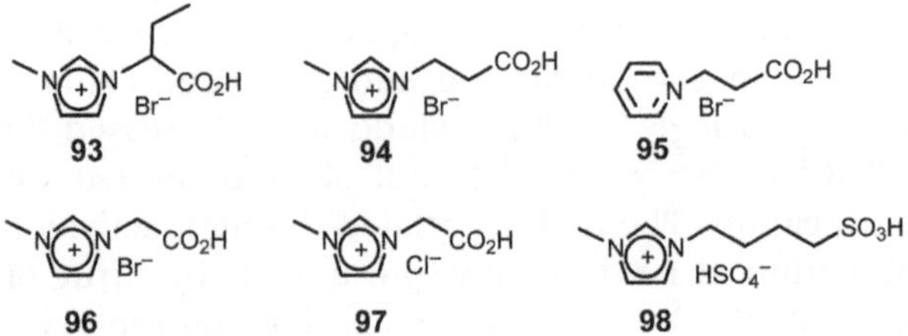

Figure 3.4.3 Acid-functionalised ionic liquids.

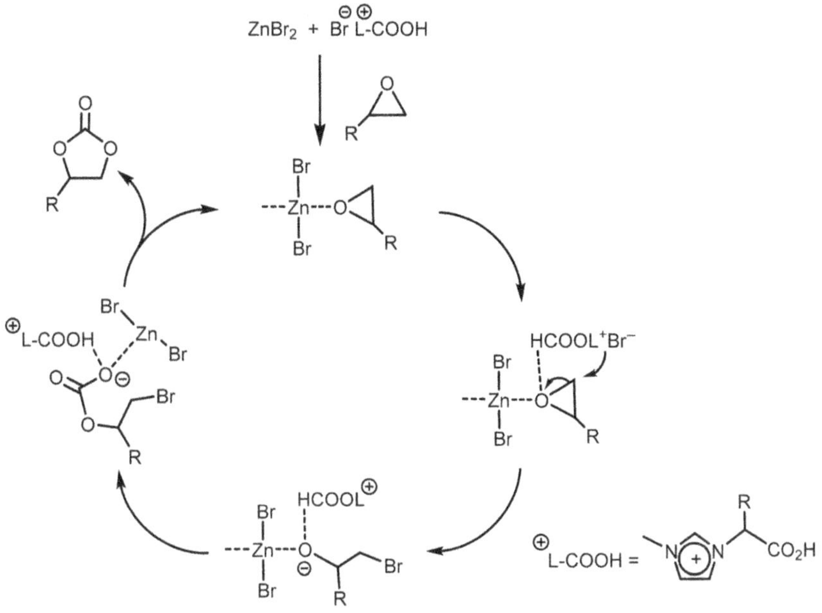

Scheme 3.4.6 Possible mechanism for the Brønsted acid ionic liquid–metal halide-catalysed coupling of carbon dioxide with an epoxide.

silica and combined with zinc salts such as $ZnCl_2$, $ZnBr_2$, $Zn(OAc)_2$ and $ZnSO_4$ were reported to be efficient heterogeneous catalysts for the fixation of CO2 with a range of epoxies giving high conversions (TOF = 2398–2714 h^{-1}) at 100 °C and 1.5 MPa CO_2 and excellent selectivity (>98%); in contrast, combinations based on other metals salts were much less effective.[400c]

Hydroxyl-functionalised SBA-15-supported 1,2,4-triazolium-based ionic liquids are efficient and recyclable catalysts for the fixation of CO_2 with epoxides. Optimisation of reaction parameters gave high yields (80–99%) with excellent selectivity for the cyclic carbonate (97–99%) under mild conditions (110 °C) and in reasonable reaction times. The system was proposed to operate by activating the epoxide through hydrogen bonding to the hydroxyl group and the CO_2 by interaction with the tertiary amine of the triazolium cation; the former promotes ring opening by the ionic liquid anion while the latter facilitates addition of the ring-opened intermediate to CO_2 to afford the carbonate anion which ultimately undergoes ring closing nucleophile displacement to generate the cyclic carbonate.[401a] Cross-linked divinylbenzene polymer has also been used to immobilise hydroxethyl-functionalised ionic liquid 3-(2-hydroxyl-ethyl)-1-(3-amino-propyl)imidazolium bromide ([HEMIM][Br]) and the derived material is an efficient and selective heterogeneous catalyst for the solventless coupling of CO_2 and epoxide to cyclic carbonates.[401b] A structure efficiency study demonstrated that the hydroxyethyl group and bromide acts synergistically to facilitate the coupling. The catalyst loading of 0.44 mol% was significantly lower than that of 1.6 mol% required for the closely related 1-(2-hydroxyl-ethyl)-imidazolium bromide on polystyrene resin reported by Sun *et al.*[401c] Both systems are

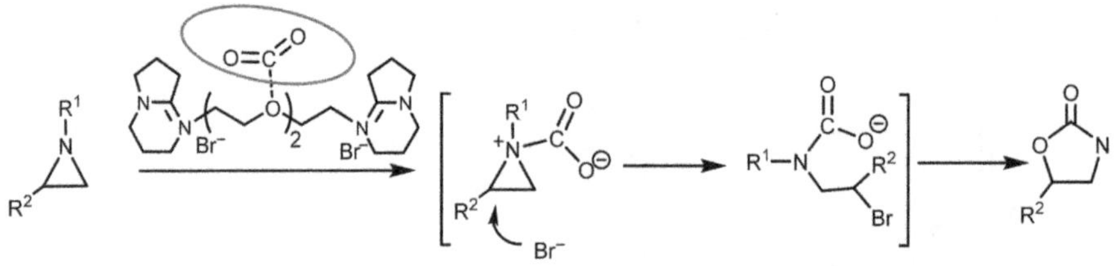

Scheme 3.4.7 Proposed mechanism for the BrDBNPEG$_{150}$DBNBr-catalysed cycloaddition between CO_2 and aziridines.

attractive in terms of their simplicity in addition to their high stable and gave good recyclability. Reasoning that a vicinal diol should impart efficient activation of an epoxide by hydrogen-bonding chelation to the oxygen atom, Bhanage demonstrated that polymer-immobilised diol-modified imidazolium-based ionic liquids were more efficient than either their unsupported counterparts or the corresponding mono-hydroxyl-based systems. Good yields and excellent selectivities were obtained at 130 °C under 2 MPa CO_2 after 3 h and the system retained its performance over four cycles.[401d] The same polymer-immobilised diol-modified imidazolium-based ionic liquids also catalysed the regioselective coupling of aziridines with CO_2 and gave good yields of the corresponding 5-aryl 2-oxazolidinones in short reactions times at room temperature using 1.5 mol% catalysts.[401e]

Reasoning PEG to be a CO_2-philic material, He and co-workers prepared a series of PEG-derived ionic liquids to catalyse transformations of CO_2. An early study reported that the PEG$_{6000}$-supported ammonium salts PEG$_{6000}$(NBu$_3$Br)$_2$ efficiently catalysed the cycloaddition of aziridine with CO_2 to afford the corresponding 5-aryl-2-oxazolidinones [Equation (3.4.13)] in good yield and excellent regioselectivity (82–99%).[402] Slightly higher regioselectivities (94–98%) were later obtained with BrDBNPEG$_{150}$DBNBr (**99**). Both catalysts could be recovered, either by centrifugation or by distillation of the product, and recycled with remarkable efficiency. These catalysts were proposed to activate CO_2 by interaction with the ether link of the PEG while the aziridine simultaneously formed an adduct with CO_2 (IR evidence) prior to regioselective nucleophilic ring opening by attack of the Br^- anion at the most substituted carbon atom to generate the corresponding carbamate salt stabilised by the [BrDBNPEG$_{150}$DBN]$^+$ cation (Scheme 3.4.7). Replacement of the PEG linker with an alkyl group resulted in a much lower activity which was taken as evidence for a CO_2 activating interaction with the ether.[403]

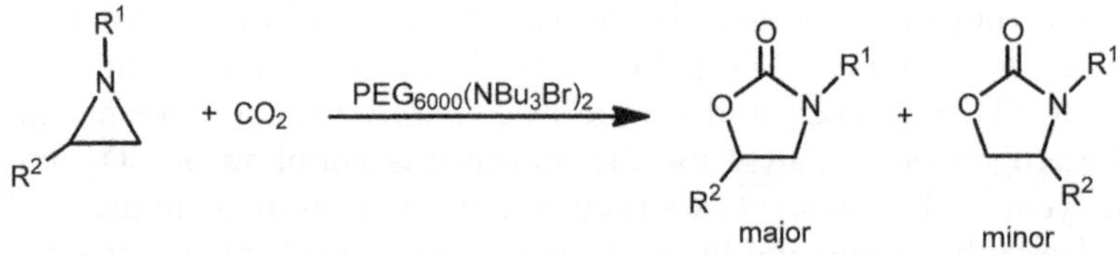

(3.4.13)

An architecturally similar task-specific ionic liquid BrTBDPEG$_{150}$TBDBr (**100**) is also an efficient catalyst for the synthesis of cyclic carbonates from

CO_2 and epoxides. In this case the catalyst was proposed to operate by interaction of the secondary amine with CO_2 to form the corresponding carbamic acid while the epoxide is activated through hydrogen bonding either with the carbamic acid or the secondary amine in **100**. Nucleophilic ring opening of the epoxide by bromide occurs with high regioselectivity on the less sterically hindered β-carbon atom and the resulting bromoalkyl alcohol species $HOCHRCH_2Br$ attacks the carbamic acid to generate the corresponding carbonate that undergoes intramolecular ring closure.[404]

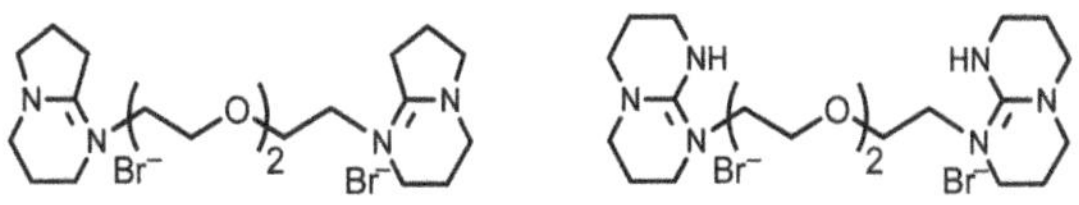

BrDBNPEG$_{150}$DBNBr (**99**) BrTBDPEG$_{150}$TBDBr (**100**)

DBU, DABCO, methylimidazole, and pyridine-derived protic ionic liquids are inexpensive, easy to prepare, robust and environmentally benign catalysts for the fixation of CO_2 with aziridines and give good yields of the 5-aryl-2-oxazolidone with high regioselectivity under mild conditions. ^{1}H NMR and *in situ* FT-IR studies further support hydrogen bond activation of the aziridine.[405]

Until recently the vast majority of ionic liquids employed as solvent and/or catalyst for the fixation of CO_2 with epoxides have been based on ammonium, phosphonium, pyridinium and imidazolium cations. Recently a series of easy-to-prepare air- and water-stable bifunctional Lewis basic ionic liquids derived from $[HDBU]^+$, $[C_nDABCO]^+$, $[HTBD]^+$ and $[HHMTA]^+$ have been reported to catalyse the cycloaddition between CO_2 and propylene oxide with remarkable efficiency. The design concept was based on a bifunctional ionic liquid containing a Lewis basic tertiary amine that would interact with CO_2 to form a carbonate and an N–H that would activate the epoxide through an O—H–N hydrogen bond. Under optimum conditions (140 °C under 1 MPa CO_2) catalyst efficiency varied with the cation and decreased in the order $[HDBU]^+ > [HTBD]^+ > [C_8mim]^+ > [C_4DABCO]^+ > [C_8DABCO]^+ > [C_4mim]^+ > [HHMTA]^+$ as well as with the anion with $[OH]^-$, $[Br]^-$, $[Cl]^-$ and $[OAc]^-$ giving good conversions while $[NTf_2]^-$, $[PF_6]^-$ and $[BF_4]^-$ were ineffective. The optimum catalyst [HDBU]Cl gave good yields across a variety of substrates and recycled over five runs with no loss in activity. The proposed mechanism involves activation of the epoxide by a hydrogen bond interaction with the N–H proton and simultaneous nucleophilic attack of the chloride at the less hindered carbon atom of the epoxide to afford the ring-opened adduct which attacks the carbamate salt formed by coordination of the tertiary nitrogen atom of the catalyst to the CO_2; the product is formed in a final intramolecular ring closing step (Scheme 3.4.8).[406] The Lewis basic DABCO-derived ionic liquid $[C_nDABCO][X]$ ($X = Br^-$; $n = 3, 7, 11$) also catalyses the cycloaddition between aziridine and CO_2 and gives good yields with high chemo- and regioselectivity for the 5-aryl-2-oxazolidinone under mild conditions. IR spectroscopic studies support a CO_2-activating interaction with the nucleophilic tertiary nitrogen atom of the ionic liquid, as described above.[407]

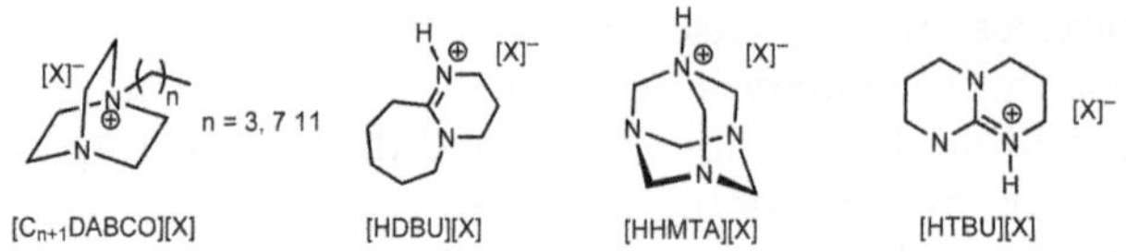

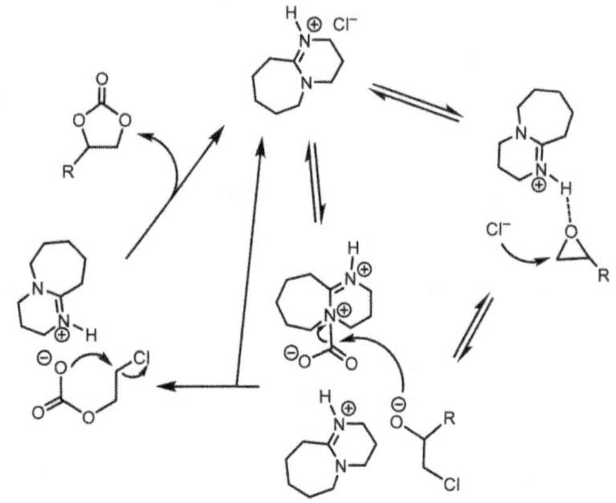

Scheme 3.4.8 Proposed mechanism for the cycloaddition between CO_2 and epoxides using [HDBU]Cl as a catalyst.
Adapted with permission from reference 406. Copyright 2010 Wiley.

The cycloaddition of CO_2 with styrene epoxide catalysed by [*N*,*N*-dibutylprolinium]Br under 1 atm of CO_2 at 90 °C gave >99% selectivity for styrene epoxide albeit in only 19% yield. The use of triethylamine as co-catalyst improved the yield from 19% after 24 h to 77% in the same time; a similar enhancement in activity was obtained with other bases including DBU, DMAP and NBu_3. Although the role of the NEt_3 remained speculative the corresponding ammonium salt was considered to activate the epoxide by hydrogen bonding to the oxygen atom and/or facilitating carbonate forming cyclisation by interaction with the proton in the carboxylic acid of the ionic liquid.[408]

The synthesis of cyclic carbonates from CO_2 and epoxides has been conducted under much milder conditions (50 °C) than previously reported in the reversible room temperature ionic liquid generated *in situ* from (*E*)-*N*′-hexyl-*N*,*N*-dimethylacetamidine ($R^1 = -(CH_2)_6H$; C6-amidine), hexylamine and CO_2 [Equation (3.4.14)] with 2 mol% C6-amidine.HBr as catalyst. Under these conditions good yields (91–94%) of carbonate were obtained with a selection of epoxides. Other catalysts including LiBr, NaBr, $ZnBr_2$, $ZnCl_2$ and NEt_4Br gave lower yields under the same conditions and in the same time. Yields increased with increasing temperature as a result of the lower viscosity of the ionic liquid as well as an increase in the intrinsic rate. A lower yield obtained in the corresponding amidinium dithiocarbamate-based ionic liquid was attributed to its higher viscosity. Ionic liquid–catalyst mixtures based on C6-hexylamine-CO_2 and either amidine.HBr or LiBr both recycled efficiently over three runs with only a minor reduction in conversion.[409]

$$R^1{-}N{=}C(CH_3){-}NMe_2 + R^2NH_2 \underset{N_2}{\overset{+CO_2}{\rightleftharpoons}} R^1{-}NH^{\oplus}{=}C(CH_3){-}NMe_2 \; {}^{\ominus}O_2C{-}NHR^2 \tag{3.4.14}$$

The chromium-salen-catalysed cycloaddition of styrene epoxide with carbon dioxide in $[C_4mim][PF_6]$ gave modest yields (25–48%) and reasonable selectivity (71%) after 6 h at 80 °C under 100 bar CO_2 using a substrate to catalyst ratio of 875; the major by-product was identified as phenylglycol resulting from Lewis acid-catalysed hydrolysis of styrene epoxide. Several shortcomings of this system were identified including lower yields than in dichloromethane, the need for exhaustive liquid–liquid extraction to recycle and gradual leaching of the chromium catalyst into the organic phase after extensive reuse.[410]

Ionic liquids anchored to a cross-linked styrene-based polymer matrix **101a** are highly efficient catalysts for the heterogeneous cycloaddition of CO_2 with epoxides. Good yields of cyclic carbonate (97.4%) were obtained after 6 h at 383 K under 6 MPa CO_2; the catalyst exhibited excellent stability and recycled five times without any loss in activity. The polystyrene–ionic liquid was more active than either $[C_4mim]Cl$ or [vinylmim]Cl because the poor miscibility of the monomers with the substrate caused interphase mass transfer limitations whereas the PSIL microparticles were well-dispersed in the reaction mixture under efficient stirring. The ionic polymer catalyst was readily separated from the product by filtration, washed with THF, acetone and methanol, dried and reused five times with no loss in activity.[411a] More recently, the first polymer-immobilised carboxyl-functionalised ionic liquids (**101b**) have been prepared and reported to catalyse the cycloaddition of CO_2 with epoxides, reaching high conversions after 4 h under 2MPa CO_2 at 140 °C. The system was proposed to operate by Lewis acid activation of the epoxide *via* interaction with the hydrogen-bond donating carboxyl residue (and possibly through electrostatic interactions with the imidazolium ring), ring opening nucleophilic attack by the Lewis basic bromide, addition of the resulting b-bromoalkoxide to CO_2 and ring closing displacement of bromide to regenerate the catalyst and liberate the cyclic carbonate.[411b] Within in a series of polymer-grafted bis(imidazolium) based ionic liquids (**101c**) that functionalised with a carboxyl group was more active for the cycloaddition of CO_2 with epoxide than those containing hydroxyl or amino groups. High yields of cyclic carbonate were obtained with catalyst loadings as low as 0.3 mol% after 4 h at 120 °C under 2 MPa CO_2. The efficiency of this system was attributed to activation of the epoxide *via* hydrogen-bond interactions and electrostatic interactions with the double imidazolium cation as well as nucleophilic activation of the halide anion. Recycle studies showed that the system was stable and could be reused five times with no loss in activity of selectivity.[411c] The efficiency of carboxyl-functionalised ionic systems as catalysts for the cycloaddition of epoxides with CO_2 was clearly evident in a comparative study of the efficiency of silica-immobilised carboxyl-functionalised imidazolium-based ionic liquids against their corresponding hydroxyl and alkyl counterparts, which revealed that the former were more active and selective, probably due to activation of the epoxide by the carboxyl group. Activity increased in the order of nucleophilicity $I^- > Br^- > Cl^-$, optimum conditions were identified as 3 h, 1.82 MPa CO_2 at 110 °C and the system recycled with remarkable efficiency.[411d]

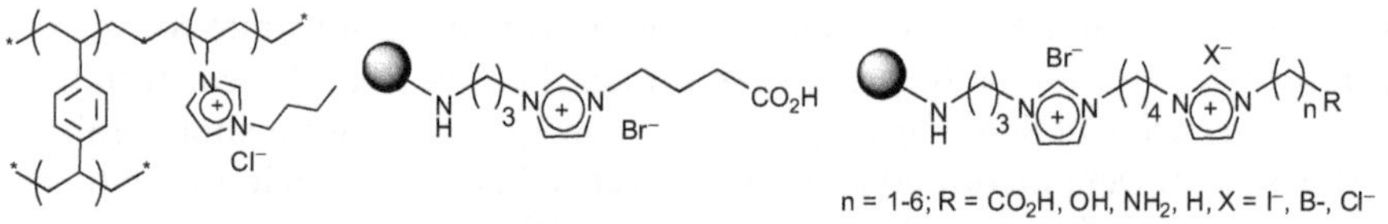

n = 1-6; R = CO_2H, OH, NH_2, H, X = I^-, B-, Cl^-

101a **101b** **101c**

3.4.1.7.3 Synthesis of Dimethylcarbonate. Bifunctional ionic liquids **102a** containing a quaternary ammonium site and a remote tertiary amine catalyse the one-pot synthesis of dimethylcarbonate from ethylene oxide and CO_2 in methanol to give good yields with reasonable selectivity after 8 h at 150 °C under 2 MPa CO_2. Catalyst efficiency as measured by TON increased in the order $[N_{114,6N11}]I > [N_{114,6N11}]Br > [N_{114,6N11}]Cl > [N_{1,1,4,6N1,1}][BF_4] > [N_{114,6N11}][NTf_2]$ which paralleled the anion nucleophilicity. A much lower yield and selectivity was obtained with the bis(quaternary) ammonium salt $[N_{114}\text{-}C_6\text{-}N_{114}][Br]_2$ (**102b**), which provided a strong indication that both the amine and quaternary ammonium groups are required for efficient catalysis. Under optimum conditions $[N_{114,6N11}][I]$ gave 99% ethylene oxide conversion and 74% dimethyl carbonate selectivity which was maintained over six runs after which yields of DMC decreased quite dramatically.[412]

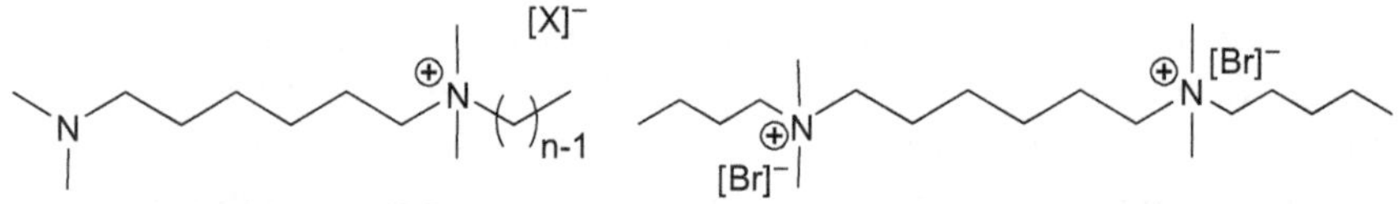

102a, $[N_{11n,6N11}][X]$ **102b**, $[N_{114}\text{-}C_6\text{-}N_{114}][Br]_2$

A survey of the reactivity and selectivity of the transesterification of ethylene carbonate (EC) with methanol to afford dimethylcarbonate (DMC) as a function of the ionic liquid identified 1,3-dimethylimidazolium-2-carboxylate to be a particularly efficient catalyst as it gave DMC in 81% yield and 99% selectivity after heating at 110 °C for 80 min; in contrast, selectivity was much lower below 80 °C due to formation of 2-hydroxyethyl methyl carbonate. The catalyst was proposed to operate by hydrogen-bond activation of methanol, attack of the resulting methoxide at the carbonyl carbon atom of EC to generate the monoester, reaction with another equivalent of methoxide to liberate DMC and the ethylene glycol anion which then picks up a proton to complete the cycle (Scheme 3.4.9). A polystyrene immobilised 1,3-dimethylimidazolium-2-carboxylate was operated in a continuous flow fixed-bed reactor at 110 °C and 0.3 MPa to give 67% conversion of EC and 95% selectivity for DMC over 200 h of operation with no loss in activity.[413]

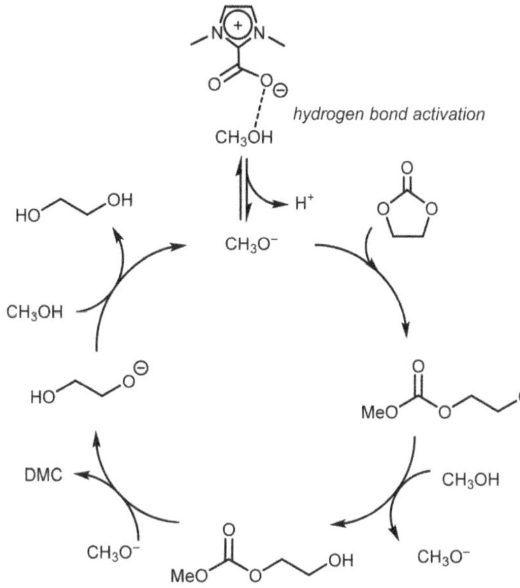

Scheme 3.4.9 Proposed mechanism for the 1,3-dimethyl-imidazolium-2-carboxylate-catalysed synthesis of dimethylcarbonate *via* the transesterification of ethylene carbonate with methanol.

3.4.1.8 Lewis Acid-Catalysed Prins Cyclisation

Prins cyclisation of homoallylic alcohols, amines and thiols with aldehydes in neat $NEt_4F \cdot 5HF$ occurs with exceptional efficiency to afford 4-fluorinated tetrahydropyrans, 4-fluorothiacyclohexanes and 4-fluoropiperidines in good yields and with very high selectivity for the *cis* product. The high selectivity was attributed to the formation of an oxonium ion chair-like transition state which avoids unfavourable 1,3-diaxial interactions. The ionic liquid recycles efficiently to afford yields in excess of 90% across five runs.[414a,b] The aza-Prins cyclisation between aliphatic/aromatic aldehydes and *N*-tosyl-3-butenylamine under liquid–liquid biphasic conditions in either $[C_4mim][BF_4]$-benzotrifluoride using $FeCl_3$ as promoter or in a mixture of $[C_4mim][FeCl_4]$ and benzotrifluoride (BTF) gave good yields of the corresponding piperidine with high stereoselectivity. Yields were markedly higher than those previously obtained in dichloromethane and the product-containing BFT phase could be readily isolated by liquid–liquid phase separation.[415]

3.4.2 Catalysis Involving Brønsted Acid Ionic Liquids

Brønsted acid ionic liquids (BAILs) are attracting interest as potential surrogates for conventional mineral acids and solid acids in chemicals manufacture as the former generate large amounts of waste during neutralisation, are highly corrosive and difficult to separate and reuse while the latter can suffer from limited accessibility of their active sites, have high molecular weight/active site ratios and are prone to deactivation. Brønsted acid ionic

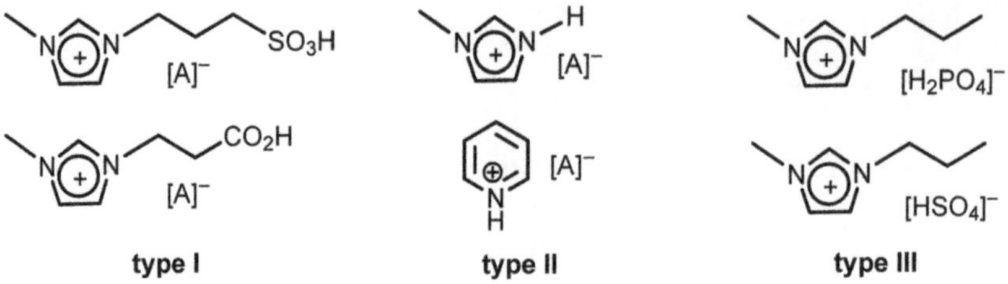

Figure 3.4.4 Classes of Brønsted acid ionic liquids.

liquids potentially combine the advantages of solid acids such as low volatility, non-corrosiveness and ease of separation with liquid-like mobility, greater effective surface area and high activity. The acidity of ionic liquids can be centred on the cation either in the form of a covalently linked acid (type I, $-SO_3H$, $-CO_2H$, $-PO_3H_2$) or directly attached to a quaternary nitrogen/phosphorus atom (type II) or have available protons on the anion (type III, HSO_4^-, $H_2PO_4^-$) and this section will present examples of catalysis with each (Figure 3.4.4). A recent microreview entitled *Structural Effects on the Physiochemical and Catalytic Properties of Acidic Ionic Liquids* provides a detailed and highly informative overview of the synthesis, properties and applications of this class of ionic liquid.[416]

The first strong Brønsted acid ionic liquids were developed by Davis and co-workers and used as dual solvents–catalysts for Fischer esterification, alcohol dehydrodimerisation and the pinacol rearrangement. The acids are prepared in a 100% atom economical synthesis and their behaviour was fully consistent with the donor acid (CF_3SO_3H and *p*-TSA.H_2O) being fully incorporated into their respective ionic liquid structures, rather than remaining as a mixture of added strong acid and dissolved zwitterion. For each reaction tested, the new BAILs gave good product selectivities as well as a balance between yields achievable with a homogeneous acid catalyst and the ease of product isolation and catalyst recycling for a heterogeneous system. The tuneability of such task-specific ionic liquids means that it should be possible to design an optimum catalyst for a range of acid-catalysed transformations.[417]

3.4.2.1 *Brønsted Acid-Catalysed Selective Alkylation of Arenes and Olefins*

The alkylation of *p*-cresol with *tert*-butyl alcohol (TBA) to give 2-*tert*-butyl-*p*-cresol (2-TBC) and 2,6-di-*tert*-butyl-*p*-cresol (DTBC) has been catalysed by environmentally benign Brønsted acid ionic liquids [Equation (3.4.15)]. A comparison of the activity of *N*-methyl imidazolium, pyridinium and triethylammonium-based Brønsted acid ionic liquid catalysts against other solid acid catalysts revealed that high conversions (80.8%) and selectivity for 2-TBC (90%) are obtained at temperatures as low as 70 °C with either $[Et_3NC_4SO_3H][HSO_4]$ or $[SO_3HC_4mim][HSO_4]$ acting as both catalyst and solvent. For comparison conventional solid acid catalysts required higher temperatures (90–130 °C) to reach similar conversions and often gave much lower selectivities. The ionic liquids recycled five times with no significant loss in activity or selectivity.[418] The same BAILs can also be used in place of traditional acids to catalyse the alkylation of *m*-cresol with *tert*-butyl alcohol [Equation (3.4.16)] to give good yields (81%) and excellent selectivity (96%) for 2-*tert*-butyl-5-methyl phenol under mild conditions (363 K). In addition to matching or outperforming conventional solid and mineral acids $[Et_3NC_4SO_3H][HSO_4]$ also recycled with remarkable efficiency.[419] The same researchers also reported that the alkylation of *p*-cresol with TBA can be conducted in $[SO_3HC_4pyr][HSO_4]$ as both catalyst and solvent to give 2-TBC in good yield (79%) and high selectivity (up to 92%) under much milder conditions (343 K) than those required with conventional solid acid catalysts such as WO_x/ZrO_2 and TPA/TiO_2 (403 K).[420]

OH + *tert*-BuOH —BAIL→ 2-TBC (3.4.15)

OH + *tert*-BuOH —BAIL→ (3.4.16)

Ionic liquid–super acid mixtures based on $[C_n mim][CHF_2CF_2SO_3]$ ($n = 2, 4, 12$) and $CHF_2CF_2SO_3H$ have been used to catalyse the alkylation of *p*-xylene with 1-dodecene. The system is homogeneous in the early stages of the reaction but readily phase separates into product and ionic liquid–acid layers as the reaction progresses, which allows the product to be isolated in a straightforward procedure. Optimum conditions (ionic liquid/super acid ratio of 3 : 1) gave excellent conversions after 60 min at 100 °C (95–99%) and the catalyst could be recovered by decantation to afford product contaminated with less than 1% of the acid and ionic liquid. Similarly, the acid-catalysed polymerisation of 1,3-propanediol in $[C_2mim][CHF_2CF_2SO_3]$–$CHF_2CF_2SO_3H$ is homogeneous at the reaction temperature of 160 °C but separates into two phases as the polymer forms; the upper layer contains polyol ($M_n = 2907$) and the lower layer contains the IL and acid which can be decanted and recycled. Such phase separation combines the advantages of a homogeneous and heterogeneous catalyst and could be used to improve a range of chemical processes.[421]

Mixtures of acidic ionic liquid such as $[SO_3HC_n mim][X]$ ($X = NTf_2^-$, OTf^-) or $[C_n mim][HSO_4]$ ($n = 2, 4, 6$) and strong protic acids such as H_2SO_4 and CF_3SO_3H catalyse the alkylation of *iso*-butane with 1-butene [Equation (3.4.17)] to give conversions, C8-selectivity and trimethylpentane/dimethylhexane (TMP/DMH) selectivity that either match or are better than that obtained in either the pure protic acid or the pure acidic ionic liquid. The optimum binary combination of $[C_8mim][HSO_4]$ with 76.4 wt% CF_3SO_3H

gave a markedly higher C_8-selectivity (76.4%) and TMP/DMH ratio (6.8) than the 41.1% and 1.4, respectively, obtained in neat CF_3SO_3H. The addition of [C_8mim][HSO_4] was proposed to tune the acidity of strong acids as well as improve *iso*-butene solubility in the catalyst mixture. The [C_8mim][HSO_4]–CF_3SO_3H mixture, recycled over 25 runs with remarkable efficacy, was not prone to the cracking reactions typical of neat acid and inhibited the formation of heavies and the ensuing deactivation. Having demonstrated the viability of this concept, it should be possible to optimise ionic liquid acidity, reactant solubility and mass transfer properties for use in other acid-catalysed reactions.[422]

BAIL

DMH TMP

(3.4.17)

Protic-at-nitrogen ammonium-based ionic liquids (AMILs) also enhance the efficiency of the acid-catalysed alkylation of *iso*-butane with 1-butene. Under optimum conditions triflic acid combined with [NEt_3H][HSO_4] (3 : 1, v/v) gave 85.1% TMP selectivity and a research octane number (RON) of 98, which is an improvement on the 65% selectivity obtained with commercial H_2SO_4 and pure TfOH. The high C8 selectivity appears to be associated with the *iso*-butane/olefin (I/O) solubility ratio as a higher I/O ratio feed favours alkylation of olefin over polymerisation. Within a series of $[HSO_4]^-$-based ionic liquids C8 selectivity and RON showed only a minor variation with the nature of the ammonium cation whereas C8 selectivity varied as a function of the ionic liquid to TfOH ratio as well as the anion, with $[HSO_4]^-$-based systems outperforming their $[OTf]^-$ counterparts in terms of activity and selectivity. The addition of AMIL was proposed to increase the I/O ratio in the system and adjust the acidity to favour alkylation over polymerisation both of which render alkylation more favourable and lead to high TMP selectivity.[423]

3.4.2.2 *Brønsted Acid-Catalysed Esterifications*

An initial study found that yields increased during recycling of the BAIL-catalysed Fischer esterification of acetic acid with [$Ph_3PC_3SO_3H$][OTf] up to the third run after which conversions dropped. A reinvestigation into this system established that this increase in yield was due to retention of by-products, in particular water. Control experiments showed that addition of 30% water (w/w) to rigorously dried ionic liquid gave a 22% increase in yield compared to the same reaction in the absence of water.[424] Brønsted acid ionic liquids based on imidazolium, pyridinium and ammonium cations functionalised with sulfonic acid groups are markedly more active and selective catalysts for the esterification of alcohols with carboxylic acids than sulfuric acid. Under optimum conditions **103–105** each catalysed the esterification of ethanol with acetic acid to give conversions in excess of 90% with 100% selectivity, whereas concentrated sulfuric acid gave 60.3% conversion and 98.2% selectivity. This protocol has several additional advantages over the use of conventional mineral acids as water does not need to be removed, liquid esters are immiscible with the ionic liquid and can be decanted at the end of the reaction and the ionic liquid catalyst can be recovered and reused after drying.[425]

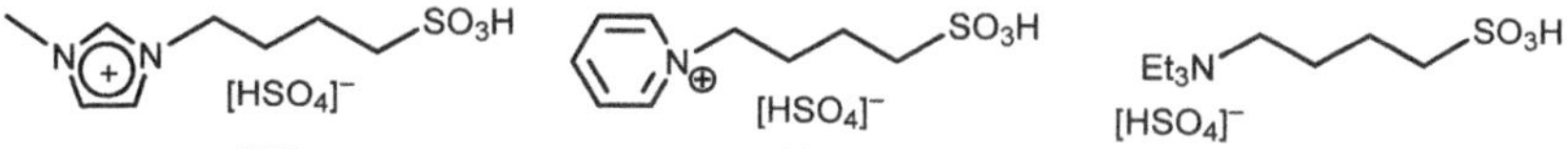

SO_3H $[HSO_4]^-$ **103** SO_3H $[HSO_4]^-$ **104** Et_3N SO_3H $[HSO_4]^-$ **105**

The efficiency of the BAIL-catalysed esterification of benzoic acid with $[SO_3HC_3pyr][X]$ ($X = BF_4^-$, HSO_4^-, $H_2PO_4^-$, *p*-TSA) varies with the anion such that activity increases with increasing Brønsted acidity of the ionic liquid and optimisation studies identified $[SO_3HC_3pyr][HSO_4]$ to be the most active. The reaction mixture phase separates as the esterification progresses which shifts the equilibrium in favour of the product, enables the product to be isolated by decantation and allows the recovered catalyst to be reused. Thus, high catalyst efficiency, ease of product isolation and recyclability can be achieved by balancing acidity and phase behaviour.[426] The efficiency of trialkylammonium-based sulfonic acid-functionalised ionic liquids $[SO_3HC_3NR_3][X]$ ($X = HSO_4^-$, $H_2PO_2^-$) as catalysts for the esterification of acetic acid, metacetonic acid and benzoic acid depends on the length of the *N*-alkyl chain such that yields of ester decrease in the order Me > Et > Bu. This was related to changes in the lipophilicity and polarity as the reagents are more soluble in the less lipophilic ionic liquids while the ester is poorly miscibile. For a given cation, higher yields were obtained with BAILs of the more acidic anion $[HSO_4]^-$ than with $[H_2PO_4]^-$. Reactions went to completion without removing the produced water as the product phase separated and the ionic liquid catalyst could be recovered and reused several times with no loss in activity.[427] Brønsted acid imidazolium- and pyridinium-based ionic liquids **106–107** bearing aromatic sulfonic acids groups ($X = SO_3H$) are highly efficient catalysts for the esterification of long chain aliphatic acids with methanol and ethanol. Good yields (86–99%) of ester are obtained after 3.5 h at room temperature whereas reactions conducted in the corresponding neutral ionic liquid ($X = H$) gave lower yields of ester (7–52%) after much longer reaction times (5 h). The mixture became biphasic as the reaction progressed allowing the product to be decanted and the catalyst to be recovered and reused. Yields of product decreased with increasing chain length of the aliphatic acid due to their low solubility in the BAIL and the more facile emulsification of long chain esters which made removal of the water more difficult.[428]

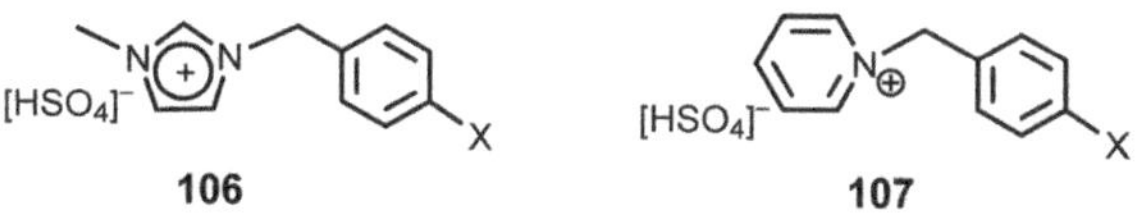

$[HSO_4]^-$ X **106** $[HSO_4]^-$ X **107**

The heteropolyanion-based Brønsted acid ionic liquid $[SO_3HC_3\text{-}mim]_3[PW_{12}O_{40}]$ is a reaction induced self-separating catalyst for

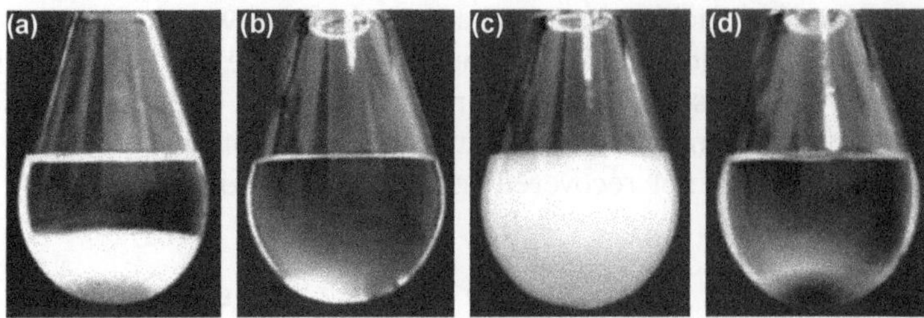

Figure 3.4.5 Photographs of the esterification of citric acid with *n*-butanol over $[SO_3HC_3mim]_3[PW_{12}O_{40}]$. (a) $[SO_3HC_3mim]_3[PW_{12}O_{40}]$ (light grey solid at bottom), citric acid (white solid in the middle), and alcohol (liquid in the upper level) before mixing; (b) homogeneous mixture during the reaction; (c) heterogeneous mixture near completion of the reaction; (d) at the end of the reaction; the catalyst has precipitated. Reproduced with permission from reference 54. Copyright 2009 Wiley.

esterification. In this concept, the reaction mixture is homogeneous at the early stages of the esterification but as the reactants are consumed the system becomes heterogeneous which induces spontaneous separation of the catalyst that can be filtered, recovered and reused (Figure 3.4.5). The success of this system relies on one of the reactants being a polycarboxylic acid (*e.g.* citric acid, succinic acid, lactic acid) or polyol (ethylene glycol, glycerol) in which the catalyst is soluble; as the ester forms the catalyst becomes immiscible and the reaction switches from homogeneous to heterogeneous. Good yields are obtained at 100–130 °C for a range of substrate combinations and in the majority of cases the catalyst can be recovered by a conventional filtration. High yields were even obtained for substrates which remained heterogeneous. Pure $[H_3PW_{12}O_{40}]$ gave comparable yields under the same conditions but remained soluble throughout and isolation was problematic. Similarly, $[SO_3HC_3\text{-}pyr][PW_{12}O_{40}]$ and $[SO_3HC_3NEt_3][PW_{12}O_{40}]$ also acted as self-separating catalysts but were slightly less active than $[SO_3HC_3mim]_3[PW_{12}O_{40}]$.[54]

While Brønsted acid ionic liquids present a number of operational and environmental advantages over conventional acids and solid acids they are relatively expensive and even though they can be separated and recycled even small losses can ultimately be costly. This drawback could be limited if the catalyst loading could be reduced and, in this regard, Xia and co-workers have shown that $[SO_3HC_4mim][OTf]$, $[SO_3HC_4mim][HSO_4]$, $[SO_3HC_4C_{10}im][OTf]$ and $[SO_3HC_4pyr][OTf]$ are all highly efficient catalysts for the esterification of *n*-butyric acid with methanol and give excellent yields (85–92%) after heating at 80 °C for 2 h at very low catalyst loadings (0.3–0.5 mol%); for comparison, yields obtained with conventional acids were lower. The ionic liquid–catalyst was recovered by distillation of the methyl butyrate and methanol and reused ten times with only a minor reduction in activity. An acidity–activity relationship (Hammett acidity function) indicated that the order of activity parallels the acidity.[429] In addition to the above, numerous other studies have reported that the catalytic activity of alkylsulfonic acid-based BAILs depends on the anion and the most efficient are generally those supported by $[HSO_4]^-$ and $[BF_4]^-$ while those with *p*-TSA and $[H_2PO_4]^-$ are much less active. A study of the acidity of a series of sulfonic acid-functionalised pyridinium-based ionic liquids [SO_3HC_4pyr][X] (X = BF_4^-, $H_2PO_4^-$, HSO_4^-, *p*-TSA) using the Hammett method showed that acidity depended strongly on the anion which for this series decreases in the order BF_4^- > HSO_4^-, > *p*-TSA > $H_2PO_4^-$ with associated H_0 values of −3.6. −3.3, −2.1, −1.2, respectively; this order also parallels the activity order for these BAILs in a range of acid-catalysed reactions. The H–O bond distances in these BAILs increases in the reverse order of the acidity BF_4^- < HSO_4^-, < *p*-TSA < $H_2PO_4^-$, due to the increasing strength of the interaction between the anion and the sulfonic acid proton and is the reverse order of their acidities (and activities).[430]

Introduction of a perfluoroalkyl tether into hydrogensulfate-based imidazolium-tagged Brønsted acid ionic liquids was shown to reduce their efficiency as esterification catalysts relative to their protio-counterparts. Poor miscibility of the fluorinated ionic liquid with *n*-propanol, reagent and product solubility as well as differences in viscosity were considered to be responsible for the low activity.[431]

3.4.2.3 Brønsted Acid-Catalysed Oligomerisation

Silica-supported imidazolium-based Brønsted acid ionic liquids [$SO_3HC_4C_n$im][X] (n = 1, 4; X = HSO_4^-, $CF_3SO_3^-$) are efficient catalysts for the oligomerisation of isobutene. The product distribution varies with the length of the alkyl chain on the imidazolium cation, the nature of the anion, the silica pre-treatment temperature and the reaction temperature. The TONs (~57) and TOFs (~11.5 h^{-1}) of the SILP catalyst are an order of magnitude higher than those of their respective ionic liquids, which were typically 5.6 and 1.14 h^{-1}, respectively. A C8-selectivity of 87% was obtained with [$SO_3HC_4C_4$im][CF_3SO_3] at 60 °C while the best C12-selectivity of 84% was obtained with [SO_3HC_4mim][CF_3SO_3] at 100 °C. Good recyclability was obtained across 6–8 runs with total leaching between 2 and 4.8%.[432]

3.4.2.4 Brønsted Acid-Catalysed Beckmann Rearrangements

The combination of a triflate-based BAIL and $ZnCl_2$ forms an efficient catalyst for the Beckmann rearrangement of ketone oximes into their amides [Equation (3.4.18)], whereas no reaction occurs with the individual components. The optimum combination based on 5 mol% each of the 'gemini' bis(imidazolium) dication **108** (n = 1, 6, 10) and $ZnCl_2$ gave 99% conversion of acetophenone oxime to its amide after heating at 80 °C for 5 h; slightly lower conversions were obtained with [SO_3HC_4mim][OTf]/$ZnCl_2$ (92–93%) while systems based on [SO_3HC_4mim][H_2PO_4] and [SO_3HC_4mim][CF_3CO_2] gave poor yields (5–47%).[433a,b]

(3.4.18)

3.4.2.5 Brønsted Acid-Catalysed Michael Additions and other Carbon–Heteroatom Bond Forming Reactions

Acidic-at-nitrogen imidazolium salts [Hmim][X] (X = HSO_4^-, OTs^-, BF_4^-) catalyse the Michael addition of nitrogen, sulfur and oxygen-based nucleophiles to α,β-unsaturated ketones more efficiently than their sulfonic acid-functionalised counterparts (Scheme 3.4.10). In a comparative survey using the addition of ethyl carbamate to cyclohexanone as the benchmark, reactions conducted in [Hmim][X] gave conversions between 70–89% while those in [C_4mim][HSO_4] and [SO_3HC_4mim][HSO_4] only reached conversions of 62% and 50%, respectively, and there was no reaction in [C_4mim][BF_4], [C_4mim][PF_6] or under solventless conditions. Under optimum conditions 30 mol% [Hmim][OTs] also catalysed Michael addition reactions involving less reactive nucleophiles such as alcohols and thiols as well as a host of heterocyclic nitrogen nucleophiles.[434] In a related study Brønsted acid ionic liquid [SO_3HC_4pyr][*p*-$CH_3C_6H_4SO_3$] catalyses the conjugate addition between indoles and α,β-unsaturated ketones to give good yields of the β-indolylketone for a range of substrate combinations and recycled with reasonable efficiency over three runs.[435]

Oxadiazoles can be synthesised *via* acid-catalysed C–N bond forming condensation between substituted 1,2,4-triazoles and a primary organoamine [Equation (3.4.19)] but the reaction is highly sensitive to steric hindrance and the use of bulky amines typically requires harsh conditions and results in low yields. Protic ionic liquids have been used as both catalyst and solvent for this transformation and good yields were obtained in shorter reaction times and at lower temperatures, particularly for sterically demanding substrate combinations, than with traditional acid catalysts. Conditions were optimised for the condensation of 2-phenyl-1,3,4-oxadiazole with arylamines and [pyrH][X] (X = OAc^-, $CF_3CO_2^-$) was identified as the most efficient solvent/catalyst as it reached 84% yield after 20 min at 110 °C, a marked improvement on the 1–13 h required to reach moderate conversions with other protic ionic liquids such as [Hmim][X] (X = OAc^-, OTs^-, $CF_3CO_2^-$, HSO_4^-) and [HDABCO][OAc]. The activating role of the protic ionic liquid was further underpinned as a neutral ionic liquid with added acid such as [C_4pyr]Br/100 mol% CF_3CO_2H, only gave 46% yield after 4 h.

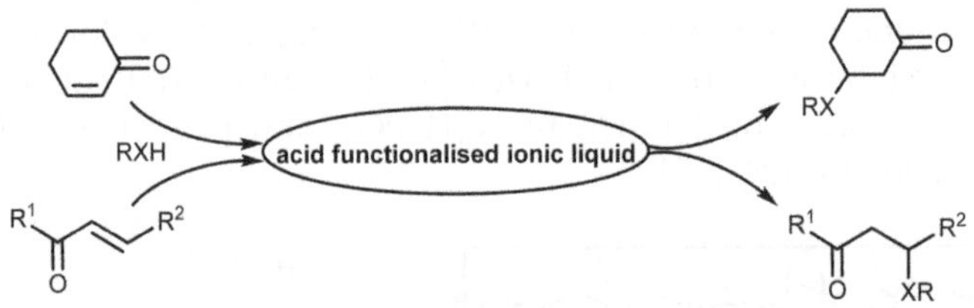

Scheme 3.4.10 Brønsted acid ionic liquid–catalysed Hetero-Michael reactions of nitrogen, sulfur, oxygen nucleophiles with α,β-unsaturated enones.

Interestingly, [pyrH][OAc] was more efficient than its $[CF_3CO_2]^-$ counterpart for reactions involving alkyl amines. The authors presented a modified mechanism for this transformation and proposed that the ionic liquid acts in both its acid and base forms to account for the efficiency of [pyrH][X] as well as the catalyst dependent performance of aryl and alkyl amines.[436]

Ph–(oxadiazole) + RNH_2 $\xrightarrow{\text{BAIL}}$ Ph–(triazole, N–R) (3.4.19)

The direct amination of alcohols has been catalysed by a variety of Brønsted acids but in most cases there are drawbacks such as poor selectivity, low catalyst activity, narrow substrate scope and difficulty of product isolation. Several of these limitations have been addressed as a range of sulfonic acid-functionalised imidazolium, phosphonium, guanidinium, pyridinium, pyrrolidinium and morpholinium-based ionic liquids catalyse the amination of alcohols with high selectivity. The most efficient catalyst, $[SO_3HC_4C_{14}im][OTf]$, for the amination of carbinol with *p*-toluene sulfonamide gave *N*-benzhydryl-4-methylbenzenesulfonamide as the sole product in 92% yield after heating at 80 °C for 3 h. For comparison, triflic acid was much less selective and under the same conditions gave 56% yield of the desired amide together with 42% of bis(diphenylmethyl)ether as unwanted by-product. Good yields were also obtained with 10 mol% catalyst in 1,4-dioxane as co-solvent. The catalyst has broad substrate scope, gives good yields with benzylic, propargylic and aliphatic alcohols, sulfonamides, carbamates, aromatic amines and *N*-heterocycles and recycles with consistent yields over six runs after aqueous work-up.[437]

The synthesis of carbamates from the carbonylation of primary aliphatic amines with dimethylcarbonate [Equation (3.4.20)] can be catalysed by the imidazolium-based Brønsted acid-functionalised ionic liquids $[CO_2HC_1C_n\text{-}im][X]$ ($X = BF_4^-$, PF_6^-, $n = 2, 3, 5, 7$) and $[SO_3HC_nmim][OTf]$ ($n = 2, 4, 6, 8$). Both classes of BAIL gave quantitative conversions under optimum conditions (80 °C/4 h) but the SO_3-functionalised systems were much more selective (91–95%) than their carboxylic acid counterparts (~54–71%). For comparison $[C_4mim][X]$ ($X = BF_4^-$, PF_6^-) both gave poor conversions and dismal selectivities under the same conditions while $[C_4mim]Cl$ gave quantitative conversion but a moderate selectivity (68%). The benefits of this BAIL-catalysed protocol include good catalyst stability, ease of recovery, efficient recyclability, mild reaction conditions and a solvent free synthesis.[438] The *N*-formylation of amines catalysed by *N*-protic imidazolium-based ionic liquids affords good yields of the corresponding *N*-formyl amine in a solvent free process that is functional group compatible and operationally straightforward. The BAIL activates the carbonyl of formic acid towards nucleophilic attack which enables less reactive nucleophiles such as primary alcohols to undergo addition.[439]

$$RNH_2 + MeOC(O)OMe \xrightarrow[80\,^{\circ}C/4\,h]{BAIL} RHNC(O)OMe + MeOH \quad (3.4.20)$$

The cyclotrimerisation of aldehdyes to afford 1,3,5-trioxanes catalysed by 1 mol% sulfonic acid-functionalised pyridinium and 4,4-bipyridinium ionic liquids gave excellent yields (98%) and remarkably high selectivity (99%) after 60 min at 25 °C. However, the authors' claim that this is the most efficient catalyst may be flawed since it was not clear whether the comparisons were conducted with the same number of acid equivalents.[440]

3.4.2.6 Brønsted Acid-Catalysed Multicomponent Reactions

A temperature-dependent biphasic system comprising the PEG-1000-based dication acidic ionic liquid (DAIL) **109** (PEG_{1000}-DAIL) and toluene catalysed the three-component condensation between malononitrile, 5,5-dimethyl-1,3-cyclohexanedione and a range of aromatic aldehydes [Equation (3.4.21)] and gave good yields of the corresponding benzopyran (85–93%). A slightly lower yield of 78% was obtained when neat **109** was used as both solvent and catalyst which was reduced to 46% with sulfuric acid as catalyst. Reactions were conducted at 80 °C as the system is homogeneous at this temperature. Upon cooling to room temperature the ionic liquid separates from the product containing toluene phase which can be isolated by decantation (Figure 3.4.6). The ionic liquid phase was charged with additional substrate and toluene and used a further nine times with no apparent loss in activity and only a 7.5% loss in weight.[55]

ArCHO + NCCH₂R + 5,5-dimethylcyclohexane-1,3-dione → (PEG_{1000}-DAIL/toluene) 2-amino-4-aryl-3-R-7,7-dimethyl-5-oxo-tetrahydrobenzopyran (3.4.21)

HO_3S–(CH₂)₄–imidazolium–(CH₂)₂–(O–CH₂CH₂)$_n$–O–(CH₂)₂–imidazolium–(CH₂)₄–SO_3H $[HSO_4]^-$ $[HSO_4]^-$ **109**

The three-component condensation between β-naphthol, an aromatic aldehyde and an amide derivative to afford 1-amidoalkyl-2-naphthols [Equation (3.4.22)] is catalysed by a host of Brønsted acids and Lewis acids but all present drawbacks such as low yields, long reaction times, high catalyst loadings, toxic/expensive catalysts and poor substrate scope. Sulfonic acid-functionalised imidazolium-based ionic liquids including 3-methyl-1-sulfonic acid imidazolium chloride, 1,3-disulfonic acid imidazolium chloride and 3-methyl-1-sulfonic acid imidazolium tetrachloroaluminate all catalyse this transformation with remarkable efficiency such that near

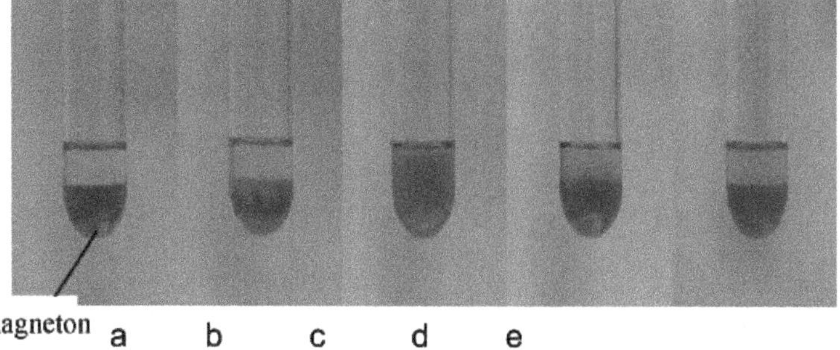

Figure 3.4.6 Reaction in PEG_{1000}-DAIL/toluene (a) PEG_{1000}-DAIL/toluene at room temperature; (b) after addition of substrates; (c) homogeneous phase at 80 °C; (d) phase separation with cooling; (e) complete phase separation after cooling to RT.
Reproduced from reference 55.

quantitative yields can be obtained in short reaction times (1–40 min) at 120 °C with 10 mol% catalyst. The efficacy of these Brønsted acidic ionic liquids as measured by TOF (4.8–6.4 min^{-1}) are a significant improvement on those achieved with $[H_3PW_{12}O_{40}]$ (0.55 min^{-1}), $[Fe(HSO_4)_3]$ (1.7 min^{-1}), and $HClO_4/SiO_2$ (3.7 min^{-1}). Additional advantages of this protocol include substrate generality, short reaction times, a clean reaction profile, simplicity and ease of preparation of the catalyst.[441] The same one-pot three-component condensation is also efficiently catalysed by the PEG-based bis(imidazolium) sulfonic acid-functionalised ionic liquid, PEG_{1000}-DAIL **109**, at 80 °C to give good yields of the 1-amidoalkyl-2-naphthol in short reaction times. The catalyst was recovered in an operationally straightforward procedure and recycled eight times with only a minor reduction in yield and 6.4% weight loss. A comparative survey revealed that PEG_{1000}-DAIL gave higher yields in much shorter reaction times at significantly lower temperatures and with lower catalyst loadings than conventional acid catalysts.[442]

OH + R^1CHO + $R^2C(O)NH_2$ —BAIL→ R^1, $NHC)O)R^2$, OH

(3.4.22)

The carbonylation of formaldehyde is a technologically important transformation as the product methyl glyoxylate (MG) is an intermediate to ethylene glycol. The reaction has been catalysed by concentrated sulfuric acid under extremely harsh conditions which leads to corrosion. Imidazolium-based Brønsted acid ionic liquids catalyse the one-pot two-step carbonylation of HCHO and esterification of the resulting glycolic acid with methanol under milder conditions [Equation (3.4.23) and (3.4.24)]. The best yield (97.6%) and highest selectivity for MG (98.3%) was obtained with $[SO_3HC_4mim][OTf]$ at 170 °C under 5 MPa of CO while slightly lower yields and selectivities were obtained with the corresponding $[HSO_4]^-$ and *p*-TSA-based ionic liquids. The optimum system recycled eight times with no loss in activity.[443]

$$HCHO + CO + H_2O \xrightarrow{BAIL} HOCH_2COOH \tag{3.4.23}$$

$$HOCH_2COOH + MeOH \xrightarrow{BAIL} HOCH_2COOMe + H_2O \tag{3.4.24}$$

The efficiency of sulfonic acid-functionalised benzimidazolium-based ionic liquid catalysts **110** ($X^- = HSO_4^-$, $H_2PO_4^-$, *p*-TSA, BF_4^-, OTf^-) for the one-pot aldol condensation between aromatic aldehydes and 4-hydroxycoumarin correlates with the relative acidity. The order of acidity based on the Hammett function H_0 $[OTf]^-$ (−0.09) > $[HSO_4]^-$ (0.05) > $[BF_4]^-$ (0.19) > *p*-TSA (0.42) > $[H_2PO_4]^-$ (0.64) parallels the activity as measured by the yield obtained after 2 h at 70 °C with 10 mol% catalyst. *Ab initio* geometry optimisations revealed that the ionic liquids with the shortest H–O bonds were the most acidic and active for the synthesis of biscoumarins.[444]

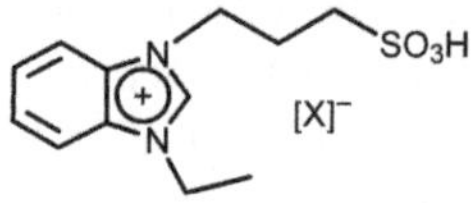

SO$_3$H [X]$^-$

110

3.4.2.7 Brønsted Acid-Catalysed Protection Reactions

The chemoselective trimethylsilylation of hydroxyl groups by hexamethyldisilazane (HMDS) is efficiently catalysed by 2 mol% 2-methyl-1-sulfonic acid imidazolium hydrogen sulphate, $[SO_3Hmim][HSO_4]$, under solvent free conditions to give good yields in extremely short reaction times (1–8 min). The protocol has been successfully applied to a range of acid sensitive, electron-donating and electron-withdrawing primary and secondary alcohols, hindered secondary and tertiary alcohols, diols, acyloins, phenols and naphthols. Primary alcohols were protected with 100% selectivity in the presence of secondary alcohols but selectivity was lower for a mixture of secondary and tertiary alcohols.[445] An efficient, mild and practical protocol for the chemoselective *N*-*tert*-butoxycarbonylation of amines uses 10 mol% of 1,1,3,3-tetramethyl guanidinium acetate as catalyst. The protocol applies to the protection of electron-rich and poor aromatic amines, aliphatic amines, amino alcohols, *N*-heterocycles and hydrazine and reactions were typically complete within 5–30 min at room temperature. The products can be extracted into diethyl ether and the catalyst recovered and reused with only a minor drop in activity. The catalyst was proposed to operate *via* electrophilic activation of the $(Boc)_2O$ making the carbonyl group susceptible to nucleophilic attack by the amine.[446]

Aprotic imidazolium-derived ionic liquids **111** (E = OMe, $N(CH_2)_4$, $A^- = BF_4^-$, PF_6^-, NTf_2^-, $C_8OSO_3^-$), bearing either an ester or amide group, catalyse the acetalization of aromatic aldehydes in the presence of a protic

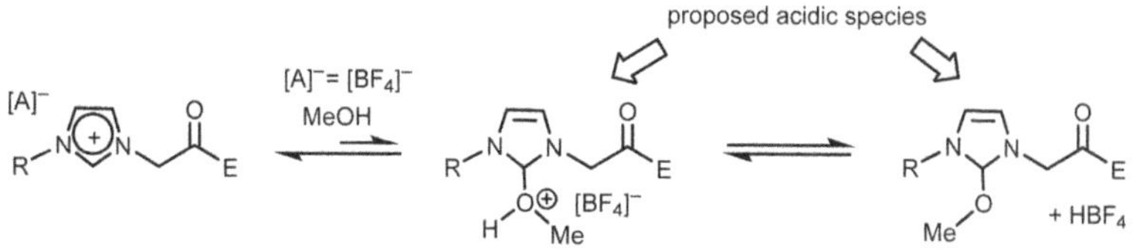

Scheme 3.4.11 Proposed mode of action of catalytic aprotic imidazolium ions involving addition of the protic additive (*e.g.* MeOH) to the C-2 position of the imidazolium ion to generate the acidic species **112a** and **112b**.

additive such as methanol by acting as a Brønsted acid catalyst.[447] A survey of ionic liquids revealed that the anion has a strong influence on efficacy and $[CO_2MeC_1mim][BF_4]$ was identified as the optimum catalyst. Good to excellent yields were obtained for a range of electron-rich, electron-poor and hindered aldehydes after 24 h at room temperature with a low catalyst loading. The catalyst recycled 15 times without any loss in activity. The catalyst was proposed to operate by addition of methanol to the C2-position of the imidazolium cation to generate the active Brønsted acid **112a–b** (Scheme 3.4.11) in much the same manner that related pyridinium-derived ionic liquids were reported to act as Brønsted acid catalysts.[448] These aprotic ionic liquid catalysts were determined to have low antibacterial and antifungal toxicity and as such should have a reduced environmental impact and could represent a template for the design of new greener catalysts.

Pentaerythritol diacetals and diketals have been prepared in good yield (87–95%) by catalytic acetalization of aldehydes and ketones with pentaerythritol at 100 °C using SO_3H-functionalised pyridinium, imidazolium and ammonium ionic liquids as catalyst. The acetalization starts as a homogeneous reaction and becomes biphasic as the reaction progresses which shifts the equilibrium and facilitates recovery and reuse of the catalyst. The yields are markedly higher than those obtained in concentrated sulfuric acid, which has been reported to be the most efficient catalyst for this transformation. The acidity of the ionic liquid, as measured by the Hammett function, H_0, shows a reasonable correlation with the activity and minimum energy geometries revealed that ionic liquids with the shortest H–O bonds were the most acidic and active for acetalization. This system addresses many of the drawbacks associated with the preparation of pentaerythritol diacetals such as harsh reactions conditions, low yields, the use of volatile hazardous organic solvents and cumbersome and lengthy isolation procedures.[449]

3.4.2.8 Brønsted Acid-Catalysed Intra- and Intermolecular Hydroalkoxylation

The intramolecular hydroalkoxylation of (±)-6-methyl-5-hepten-2-ol catalysed by SO_3H-tethered imidazolium and triazolium-based ionic liquids occurs with remarkable efficacy (>95%) and complete regioselectivity for the corresponding pyran. The protocol has been applied to the synthesis of a range of five- and six-membered oxygen heterocycles and although catalyst loadings between 10–100 mol% are required it is a marked improvement over the use of triflic acid which is highly corrosive, toxic and typically used in over stoichiometric amounts. Moreover, the products can be isolated by extraction with diethyl ether and the catalyst recycled three times with only a minor drop in yield.[450] The hydration of alkynes to ketones is efficiently catalysed by an ionic liquid system based on a six-carbon-linked bis(pyrrolidinium) hydrogen sulphate and a small amount of sulfuric acid catalyst (0.5–8 equivalents). Complete hydration of phenyl acetylene on a 1.0 mmol scale was achieved in 30 min under mild conditions (40 °C) in a mixture of ionic liquid (7.0 mmol) and H_2SO_4 (8.0 mmol). Excellent conversions were also obtained on scale-up and the ionic liquid–catalyst mixture was highly robust and could be reused ten times with no loss in activity. The efficiency of this system was attributed to the high chemical activity of the proton in the ionic liquid. One of the major benefits of this system is the small quantity of H_2SO_4 required (8 equivalents) compared with conventional acid catalysis which requires >1500 equivalents.[451]

3.5 Organocatalysis, Metathesis, Ring Opening Polymerisation and Dimerisation/Oligomerisation

3.5.1 Organocatalysis

Asymmetric organocatalysis is evolving into a powerful and complementary methodology in organic synthesis. The three most versatile classes of catalyst are those based on L-proline and its derivatives which react by formation of enamines or iminium salts with aldehydes and ketones, chiral urea/thiourea derivatives which activate by forming H-bonding interactions and chiral Brønsted acids. As the vast majority of organocatalysed transformations are conducted in polar high boiling solvents and often involve charged transition states the use of ionic liquids was perhaps a predictable and natural extension to develop improved product isolation and purification protocols as well as to recycle the catalyst. In this regard, ionic liquids are potentially ideal solvents for organocatalysis because they have tuneable polarity, miscibility and viscosity and can be modified with a task-specific functionality and/or a protic group to optimise performance. As there are two comprehensively referenced review articles that survey the use of ionic liquids in organocatalysis up to 2009 the majority of this section will be dedicated to more recent developments, after a brief overview of some of the early seminal studies.[13,15]

3.5.1.1 Organocatalysis in Ionic Liquids: Aldol, Michael, Mannich and Diels–Alder Reactions

Early studies showed that imidazolium, guanidinium, phosphonium and pyrrolidinium-based ionic liquids are suitable solvents for proline (**113**)[452a,b] and prolinamide (**114**)[453] catalysed asymmetric aldol reactions between aromatic aldehydes and ketones [Equation (3.5.1)] as well as cross-aldol reaction involving aliphatic aldehydes [Equation (3.5.2)].[454] In many cases the ee's (and de's) as well as yields either competed with or were better than those obtained in organic solvents and in some cases the use of an ionic liquid enabled the loading of (*S*)-proline to be reduced to 5 mol% from the 30 mol% typically required in organic solvents. Some researchers have also claimed substantial enhancements in yield and ee for (*S*)-proline-catalysed aldol reactions in ionic liquid compared with organic solvents. For example, the (*S*)-proline-catalysed aldol reaction of aldehydes with acetone as both solvent and reagent gave the corresponding *β*-hydroxyketone in 6% yield and 70% ee with 20 mol% catalyst whereas the same reaction in guanidinium-based ionic liquids reached 46% yield and 98% ee with only 10 mol% catalyst while selected substrates gave ee's in excess of 99%.[455] Similarly, a significant enhancement in rate was also reported for the Michael addition of ketone and aldehyde donors to *β*-nitrostyrenes [Equation (3.5.3)] as reactions conducted with 5 mol% (*S*)-proline in [C_4mim][PF_6] reached good conversions after 24 h at room temperature compared with the 2–4 days and 15–20 mol% catalyst required in organic solvent; a comparable enhancement in rate was also obtained with a 1 : 1 mixture of [C_4mim][PF_6] and chloroform.[456]

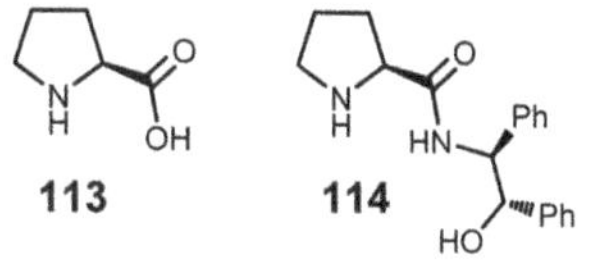

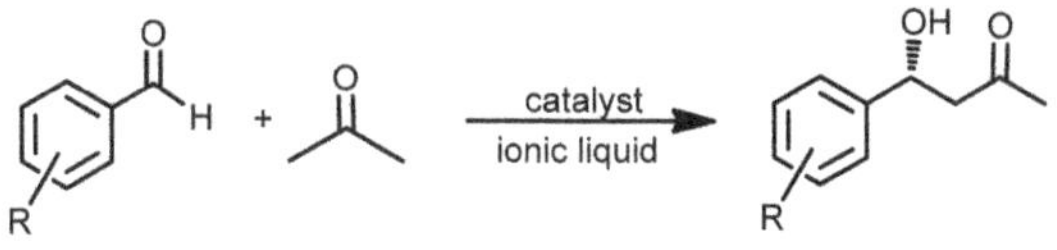

113 114

catalyst / ionic liquid (3.5.1)

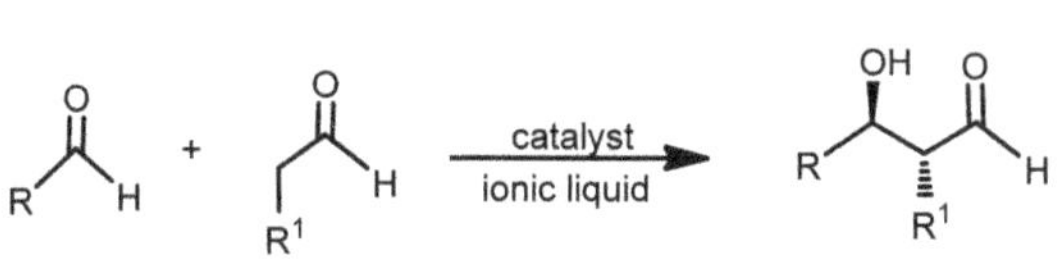

catalyst / ionic liquid (3.5.2)

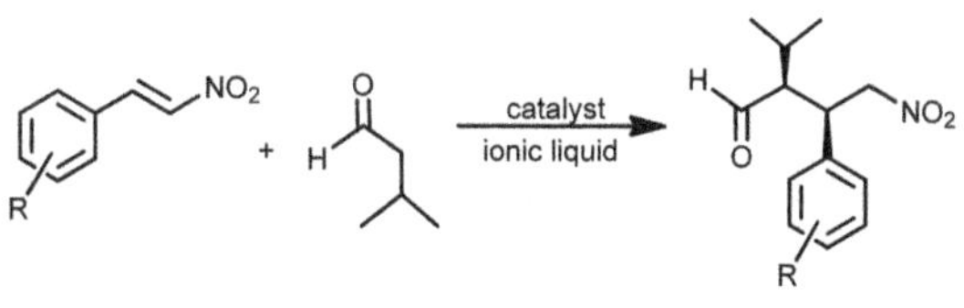

catalyst / ionic liquid (3.5.3)

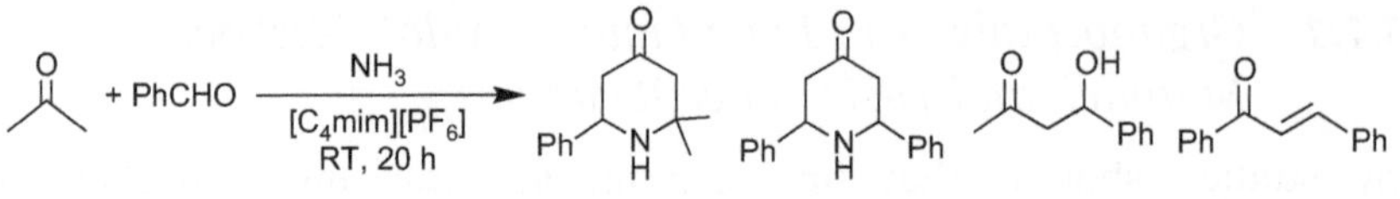

Scheme 3.5.1 Product distribution for the L-proline-catalysed tandem Mannich reaction of ammonia, aldehydes and acetone.

The L-proline-catalysed Michael addition of cyclohexanone to β-nitrostyrene identified methoxyethyl-substituted [MeOC_2mim][OMs] to be the solvent of choice as it gave the γ-nitroketone in 75% ee, which is a marked and substantial improvement on the ee's of 20% and 50% obtained in DMSO and MeOH, respectively; other ionic liquids such as [C_4mim][BF_4], [C_4mim]Cl and [C_4mim][PF_6] also gave much lower ee's.[457] While the overwhelming majority of studies have been restricted to imidazolium-based ionic liquids with $[BF_4]^-$, $[PF_6]^-$ or $[Cl]^-$ anions, those that have surveyed the influence of the ionic liquid have revealed that performance shows a marked dependence on the nature of the anion.[458] The influence of the ionic liquid anion on catalyst efficiency became clearly evident in the (*S*)-proline-catalysed tandem Mannich reaction between ammonia, an aldehyde and acetone (Scheme 3.5.1). High chemoselectivity (up to 93%) was obtained in [C_4mim][PF_6] at room temperature to afford 2,2-dimethyl-6-substituted-4-piperidones **115** as the major product while much lower chemoselectivity was obtained in [C_4mim]Cl, [C_4mim][BF_4] and [C_2mim][NTf_2] as well as in ethanol and ethylene glycol; Brønsted acids and Lewis acids in [C_4mim][PF_6] were also less efficient than proline.[459]

Highly enantioselective Michael addition of 2-oxindoles to vinyl selenone in ionic liquid has been catalysed by a thiourea-based *Cinchona* alkaloid designed to optimise hydrogen bond interactions between the chiral catalyst and substrate in order to generate a tighter chiral ion pair intermediate. The anion has a dramatic influence on catalyst performance with tetrafluoroborate-based pyridinium and imidazolium ionic liquids [$C_n$pyr][BF_4] ($n = 4, 6, 8$) and [C_4mim][BF_4], respectively, giving ee's in excess of 90% whereas [C_4pyr][NO_3] and [C_4pyr][OTf] gave low yields and/or poor ee's. Thiourea-based *Cinchona* alkaloid **119** was proposed to act as a bifunctional organocatalyst and a transition state model involving dual nucleophilic activation of the oxindole and electrophilic activation of the vinyl selenone was proposed to account for the high catalyst efficiency and sense of asymmetric indication.[460]

An enhancement in the *endo*/*exo* selectivity and ee for the Diels–Alder cycloaddition of dienes with α,β-unsaturated aldehydes catalysed by the MacMillan iminium salt of imidazolidin-4-one **120** was obtained in ionic liquid/water (3 : 1, v/v) under homogeneous conditions compared with previous reports. The highest ee of 93% and dr ratio of 93 : 5 was obtained in [C_1pyr][OTf] with 33% water and 3 mol% catalyst. The efficacy of this system was proposed to be due to a weak interaction between the ionic liquid cation and anion which favoured a significant interaction of the pyridinium cation with the transition state resulting in more pronounced shielding of the *Re*-face of the dienophile and consequently greater stereocontrol in the approach.[461]

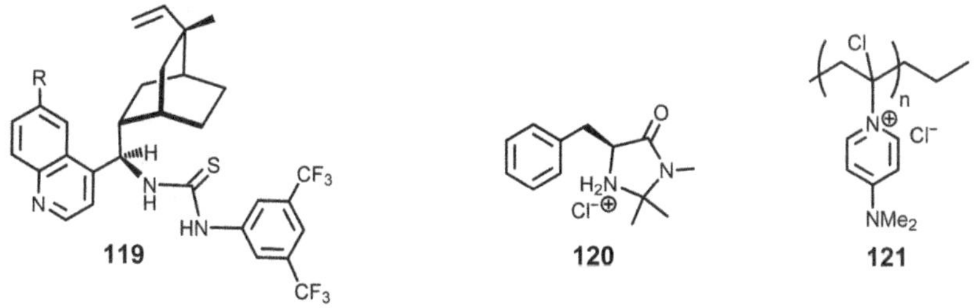

Polyvinylidene chloride supported ionic liquid derived from 4-dimethylaminopyridine **121** in combination with L-proline efficiently catalysed the aldol reaction between cyclohexanone and a variety of aromatic aldehydes in water at room temperature to give the *anti*-adduct in high dr (up to 6 : 94, *syn*/*anti*) and excellent ee (up to 98%); the system also recycled efficiently with no loss in ee over six runs.[462]

3.5.1.2 Organocatalysts with Ionic Tags

Although the use of ionic liquids for organocatalysed reactions has led to high ee's and enabled catalyst loadings to be reduced leaching of the catalyst during extraction of the product often results in poor recyclability. One potential solution involves the attachment of an ionic tag such as an imidazolium or pyridinium group to the proline in order to improve its retention in the ionic liquid during work-up.

In this regard, substantial enhancements in rate (5 fold) and enantioselectivity (Δee 20–30%) have been obtained for the cross-aldol reaction between aromatic aldehydes and acetone conducted in [C_4mim][NTf_2] and catalysed by imidazolium- or ammonium-tagged proline (**122–123**), compared with the same reaction in DMSO. The enhancement in rate was attributed to stabilisation of the charged transition state from reactants to the iminium intermediate while a tighter transition state in the rate determining step caused by the high affinity of the ion pair organocatalyst for the ionic liquid was considered as a possible explanation for the higher enantioselectivity.[463] Zhou and Wang have also reported a marked enhancement in rate and ee for the same transformation catalysed by the ether-linked imidazolium-tagged (*S*)-proline **124** in [C_4mim][BF_4] compared with that in acetone or DMSO; moreover, the ionic liquid-immobilised catalyst recycled with remarkable efficiency without any significant loss in enantioselectivity and only a minor reduction in yield after six runs.[464]

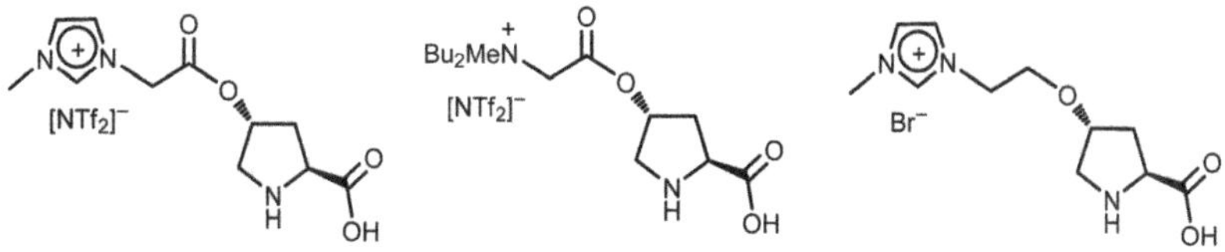

Amphiphilic long-chain alkyl-functionalised imidazolium-tagged (*S*)-proline organocatalyst **125** ($x=11, y=4$, $[A]^- = PF_6^-$) also gave excellent ee's (up to 99%) and disastereoselectivities (up to 97 : 3, *anti*/*syn*) for the direct asymmetric aldol reaction in water and recycled with exceptional efficiency, however, as its short-chain alkyl-functionalised counterpart ($x=1$, $y=1$, $[A]^- = BF_4^-$) was completely inactive under aqueous conditions the efficiency of the former may well be associated with its hydrophobicity and its ability to solubilise the reactants to promote reaction.[465] Interestingly, while imidazolium-modified chiral pyrrolidines **126** ($n=3$, $[A]^- = Br^-$, BF_4^-, PF_6^-; $n=8$, $[A]^- = Br^-$) only gave modest enantioselectivities and poor diastereoselectivities for the direct aldol reaction, even in the presence of acid additives[466] the same catalyst, in combination with CF_3CO_2H as additive, catalysed the Michael addition of a range of cyclic and acyclic ketones to *β*-nitrostyrenes with exceptional efficiency to give quantitative conversions in 8 h, ee's > 99% and *syn*/*anti* ratios up to 99 : 1. This is also a marked improvement on the ee's and de's reported earlier for the L-proline-catalysed Michael addition in $[MeOC_2mim][OMs]$.[457] The catalyst was precipitated, recovered addition of diethyl ether and recycled with only a minor drop in activity and a slight variation in ee. The ionic tag clearly has a marked influence on catalyst efficiency as the corresponding imidazole-modified pyrrolidine conjugate required 18 h to reach a similar conversion and with a lower ee (91%). The efficient stereocontrol was explained by a model in which the ionic liquid tag shields the *Si* face of the enamine double bond in the ketone donor such that reaction occurs through *Re*–*Re* approach.[467] A DFT study on this pyrrolidine-catalysed Michael addition showed that added acid changes the imine formation mechanism from a stepwise to a concerted process, lowers the reaction barrier by enhancing the electrophilicity of the carbonyl carbon and makes the enamine formation highly exothermic. Under acid conditions the Br^- anion of the imidazolium-modified pyrrolidine acts as a proton acceptor to facilitate the imine–enamine tautomerisation while the imidazolium cation stabilises the developing negative charge in the transition state of the C–C bond formation step *i.e.* this study rationalises the role of acid additive and the imidazolium tag and could provide a basis for the rational design of more efficient catalysts.[468]

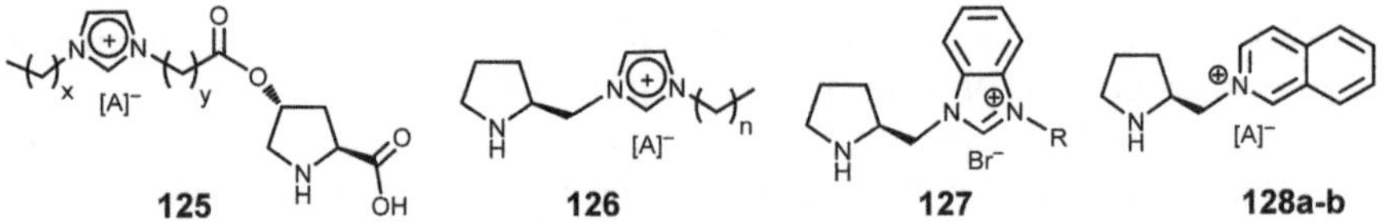

125 **126** **127** **128a-b**

Miao and Chan demonstrated that the imidazolium-modified (2*S*,3*R*)-4-hydroxyproline **125** ($x=0$, $y=1$, $[A]^- = BF_4^-$) gave consistently higher ee's (Δee = 10–30%) for the aldol reaction than unmodified (*S*)-proline when reactions were conducted in neat acetone whereas the two catalysts gave comparable ee's in DMSO; thus it appears that the imidazolium tag is not just an anchor that immobilises the catalyst but it also has a beneficial effect on performance even when reactions are not conducted in an ionic liquid.[469]

The enantioselective desymmetrisation of 4-substituted cyclohexanones *via* asymmetric Michael addition to *β*-nitrostyrenes is also catalysed by a benzimidazolium-tagged pyrrolidine **127**, in combination with trifluoroacetic acid as additive to afford the corresponding Michael adducts bearing three stereocentres with ee's up to 99%.[470] The same benzimidazolium-tagged pyrrolidine is also an efficient organocatalyst for the highly enantioselective S_N1 type alkylation of cyclic ketones with a range of alcohols in chlorinated solvents at room temperature [Equation (3.5.4)]. For comparison, other imidazolium-modified pyrrolidines were much less efficient while conventional proline organocatalysts either gave lower yields and/or ee's or formed by-products. Exceptional ee's and de's were also obtained for the desymmetrisation of *para*-substituted cyclohexanones. The stereochemical outcome was rationalised by a transition state model with the *Si*-face of the enamine blocked by the benzimidazolium cation as well as steric repulsions between the incoming carbocation and the *para*-substituent of the enamine; electrostatic interactions were also suggested to play a role in the stereocontrol.[471]

(3.5.4)

Even though some of the catalysis described above was not performed in an ionic liquid the terminology adopted by the authors describing these organocatalysts as functionalised chiral ionic liquids merits their inclusion in this section. Interestingly, onium-functionalised organocatalysts in either water, water–acid or in the presence of acid appear to outperform their neutral counterparts in ionic liquid, which suggests that the phase tag may have a dual function *i.e.* to immobilise the catalyst and facilitate recycling as well as influence the efficiency of stereocontrol. As such, it may not be necessary to use bulk ionic liquid as a solvent which would reduce cost quite substantially, especially if the system recycled efficiently.

The isoquinolinium cation is also an effective tag for pyrrolidine-based ionic liquid organocatalysis as 20 mol% **128a** ($[A]^- = Br^-$) and **128b** ($[A]^- = PF_6^-$) gave yields and ee's of 99% and 93% ee, respectively, for the Michael addition of cyclohexanone to β-nitrostyrene at room temperature in $[C_4mim][BF_4]$, which were a marked enhancement on those of 0–59% obtained in organic solvents such as *i*-PrOH, MeOH, MeCN, THF, DMSO and CH_2Cl_2. Catalyst efficiency depends strongly on the nature of the phase tag as the corresponding pyridinium, quinolinium and acridinium-modified organocatalysts gave poor to moderate ee's for the same transformation. The optimum organocatalyst gave high conversions and good to excellent ee's for a range Michael acceptors after 20 h at room temperatures and the system recycled efficiently to give high ee's and conversions across five runs. The sense of asymmetric induction was accounted for by employing an earlier model developed by Seebach in which the *Si*-face of the enamine intermediate was shielded by the isoquinolinium cation such that the Michael acceptor approaches the less hindered *Re*-face. The enhancement in rate was suggested to be due to stabilisation of the iminium ion transition state by favourable electrostatic interactions in the ionic liquid.[472] In contrast, the imidazolium-tagged pyrrolidine **126** ($X = BF_4^-$) did not show a solvent dependent performance for the same reaction as there was no significant variation between reactions conducted in $[C_4mim][PF_6]$, DMSO, DMF or *i*-PrOH.[473] Pyrrolidine **129** modified by a click-derived imidazolium tag is also a remarkably efficient catalyst for the Michael addition of cyclic and acyclic ketones to *β*-nitrostyrenes and gave good yields, high ee's end excellent dr's in DSMO, chloroform and water in the presence of 5 mol% TFA co-catalyst; other acids gave slightly lower yields and/or ee's and only mediocre yields were obtained in the absence of acid. The ionic liquid catalyst recycled efficiently over four runs with no loss in activity or selectivity.[474]

The ammonium-tagged L-prolinamides **130** (R=Et, *n*-Bu) in combination with 5 mol% TFA as additive catalyse the Michael addition of aldehydes to *β*-nitrostyrenes in either THF or dichloromethane to give the corresponding γ-nitroketones in moderate to high diastereoselectivity (67 : 33–90 : 10 *syn*/*anti*) and good enantioselectivity (72–88%). The ionic liquid catalyst was proposed to operate *via* an enamine pathway and the bulky ammonium cation and a hydrogen bond between the amide NH and the nitro group of the acceptor were considered to be responsible for the high activity and stereoselectivity while an ionic interaction between the nitro group and the ammonium was thought to stabilise the intermediate.[475]

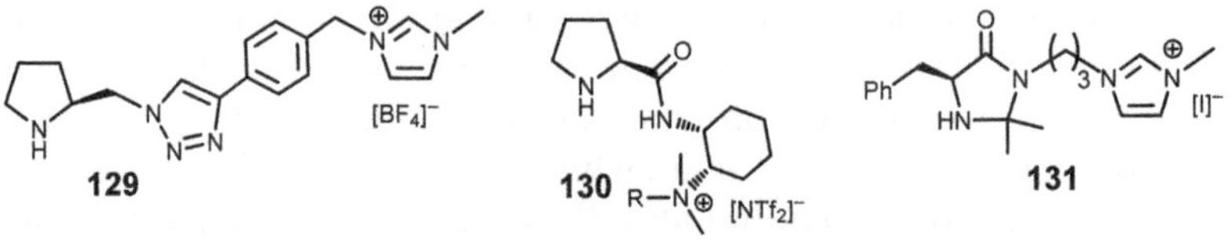

Imidazolium-modified imidazolidinone **131** catalyses the dipolar cycloaddition between nitrones and α,β-unsaturated aldehydes with remarkable efficiency and gives the corresponding isoxazolidines in good to excellent *endo*-enantioselectivity at −20 °C in CH_3NO_2–water with HBF_4 as co-catalyst; the catalyst can be recovered and recycled five times without any major reduction in yield or ee.[476]

Substitution of the (1*R*,2*R*)-cyclohexyldiamine-derived tether in the C_2-symmetric imidazolium-modified bis(prolinamide) **132**[477] with an achiral tether simplifies the synthesis and renders the resulting catalysts **133** more cost effective without unduly compromising their efficiency for the aldol reaction *i.e.* the linker does not appear to alter the stereodifferentiation.

However, the corresponding *o*-phenylene-bridged bis(prolinamide) without an imidazolium tag was much less efficient, which further underpins the important role of hydrophobic interactions between the catalyst, reactants and water. The *p*-phenylene-bridged imidazolium-modified catalyst recycled with exceptional efficiency over 15 runs.[478]

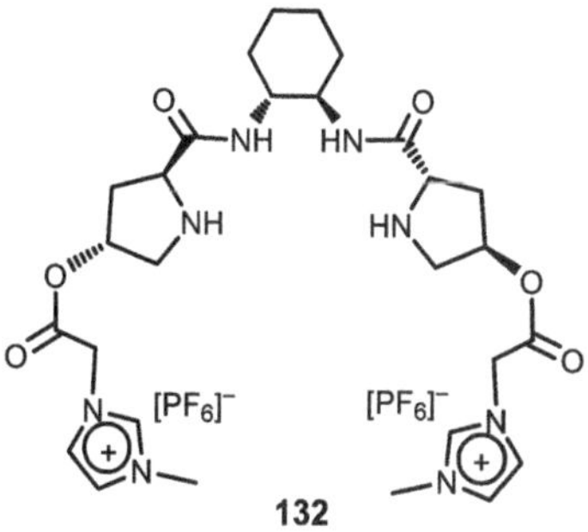

132 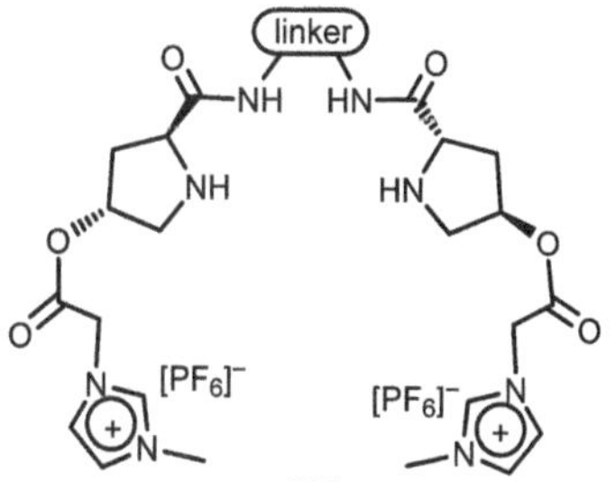 **133**

Imidazolium-modified (*S*)-proline sulfonamides (**134–135**) are highly efficient and recyclable catalysts for the asymmetric Michael addition of ketones and aldehydes to *β*-nitrostyrenes. The design concept for this class of functionalised chiral ionic liquid was a motif based on a proline and a sulfonamide with an acidic NH that could form hydrogen bonds. High diastereoselectivities (89 : 11–93 : 8, *syn*/*anti*) and reasonable to good enantioselectivities (65–82%) were obtained for the Michael addition between aldehydes and β-nitrostyrene at 4 °C in either diethyl ether or methanol but yields were moderate even after 6 days. The addition of 5 mol% TFA to reactions conducted in methanol resulted in a dramatic increase in rate and a slight improvement in ee. However, even in the presence of acid co-catalyst, yields remained moderate for reactions involving cyclohexanone.[479a,b] Markedly higher yields and ee's were obtained in shorter reaction times (16–36 h) using the more acidic (*S*)-pyrrolidine sulfonamide (**136**) and, in addition the high activity of this catalyst also eliminated the need for additional acid co-catalyst. The improved activity and selectivity was attributed to the higher acidity of the N–H which formed stronger hydrogen bonds in the transition state as well as the close proximity of the sterically demanding imidazolium cation to the catalytic site. The catalyst was recycled after precipitation with ethyl acetate and reused five times with no significant loss in enantioselectivity and only a minor drop in conversion.[480]

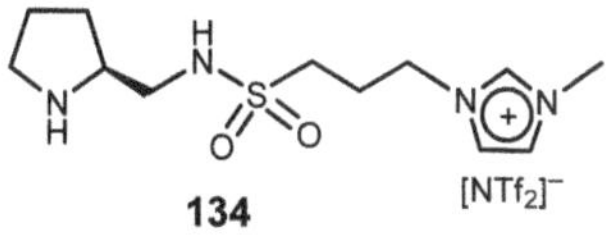

134 $[NTf_2]^-$ **135** $[NTf_2]^-$ **136** $[BF_4]^-$

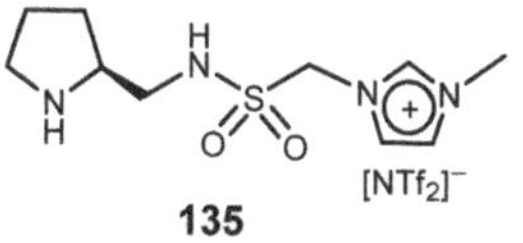

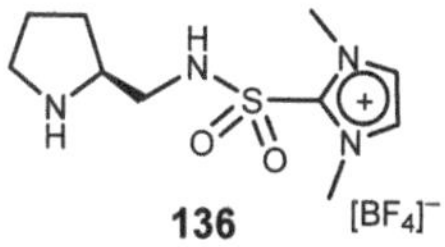

While modification of organocatalysts with a cationic group such as an imidazolium or pyridinium has led to some quite remarkable enhancements in efficiency for the asymmetric aldol reaction compared with the corresponding unmodified catalyst either in ionic liquid or conventional organic solvents, there are relatively few examples in which a chiral anion-derived organocatalyst has been used to effect this transformation. The first of these based on [C_2mim][Pro] (**137**), an imidazolium salt with L-proline as the anion, catalyses the direct aldol reaction between aromatic aldehydes and cyclic ketones in [C_4mim][BF_4] to give the corresponding aldol products in good isolated yield, excellent anti-diastereoselectivity (up to 3 : 97, *syn*/*anti*) and excellent enantioselectivity (up to 99%). While similar yields were obtained in conventional solvents such as DMSO, DMF, methanol, toluene, dichloromethane and water, the ee's and dr's were generally pitifully low.[481] In an elegant five-step synthesis starting from *trans*-hydroxyproline Gauchot and Schmitzer prepared 1-butyl-3-methylimidazolium (3*R*,5*S*)-5-(methoxycarbonyl)-pyrrolidin-3-yl sulfate (**138**) which catalyses the aldol reaction between aromatic aldehydes and cyclic ketones in [C_4mim][NTf_2] to give excellent yields and good selectivities (up to 99 : 1 dr and 89% ee). However, less reactive aldehydes such as 2-naphthaldeyhde and 2-thiophenecarboxaldehyde required activation by the hydrogen bond forming organocatalyst **139** which gave moderate yields and excellent ee's and de's. Both the ester and amide catalysts gave the same sense of asymmetric induction which was proposed to arise from the close proximity of the imidazolium cation and sulfonate group which hinders approach of the aldehyde on this side and favours formation of the *anti* diastereoisomer.[482]

137 **138** **139**

Motivated by replacing the trifluoroacetic acid required in organocatalysis with (*S*)-1-[(pyrrolidin-2-yl)methyl]pyrrolidine **140** and reasoning that the incorporation of a bulky quaternary ammonium group could enhance chiral induction *via* electrophilic activation (Figure 3.5.1), even in the absence of hydrogen bonding, proline-derived basic chiral ionic liquids **141** ($n = 1$, 4, 8,

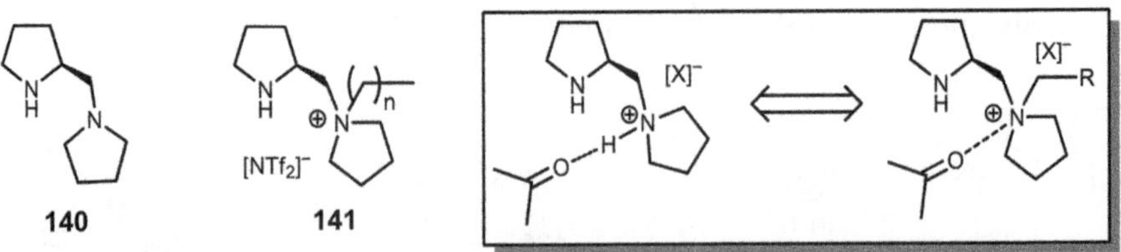

Figure 3.5.1 Strategy for replacement of acid used in (*S*)-1-[(pyrrolidin-2-yl)methyl]pyrrolidine-based organocatalysis with a bulky alkylated pyrrolidinium ionic liquid.

12) containing a bulky alkylated pyrrolidinium cation were developed and shown to catalyse the asymmetric aldol reaction between 4-nitrobenzaldehyde and acetone in good yield (67%) and up to 75% ee in the absence of additional acid. As this compares favourably with the 45% yield and 67% ee obtained with **140**/CF_3CO_2H as additive it appears that electrophilic activation by a quaternary ammonium group can replace the hydrogen-bond activation of a protonated ammonium. This strategy will reduce the use of toxic and corrosive acid as well as broaden the substrate scope to include acid sensitive compounds.[483]

3.5.1.3 Ionic Liquid Organocatalysts

The imidazolium-based ionic liquid [C_5mim]Br is a remarkably efficient catalyst for the three-component synthesis of dithiocarbamates from an amine, carbon disulfide and an activated alkene, dichloromethane or an epoxide (Scheme 3.5.2). Reactions are exceptionally rapid compared to those in organic solvents and typically complete in 10–30 min at room temperature. The imidazolium cation activates the CS_2 towards nucleophilic attack by the amine to afford the dithiocarbamate anion which then undergoes a subsequent Michael-type addition to the conjugated alkene.[484]

(+ve) ESI and MALDI-TOF-TOF MS and MS-MS studies on the ionic liquid-catalysed aza-Michael reaction identified supramolecular assemblies between [C_4mim][$MeOSO_3$], an α,β-unsaturated ketone and an aniline that involve a relay of cooperative hydrogen bonds and charge–charge interactions. Reasoning that the catalytic power of the ionic liquid is governed by its ability to form such supramolecular structures, the authors demonstrated that it was possible to obtain a measure of the catalytic efficiency by using mass spectroscopic techniques to estimate the abundance/concentration of these supramolecular assemblies during the course of the reaction. The ionic liquid appears to have a dual role of electrophilic–nucleophilic activation in which the carbonyl oxygen atom of the α,β-unsaturated ketone forms a hydrogen bond (electrophilic activation) with the C2-hydrogen of the imidazolium cation while the nitrogen lone pair forms an electrostatic interaction with the quaternary nitrogen atom of the [C_4mim]$^+$ cation and facilitates hydrogen bond formation with the oxygen atom of one of the S=O groups on the [$MeOSO_3$]$^-$ anion (nucleophilic activation); the resulting six-membered chair-like cyclic structure (**A**) orientates/positions the nitrogen atom of the aniline for nucleophilic addition to the β-carbon of the α,β-unsaturated ketone (Scheme 3.5.3). Such a model could provide the basis for

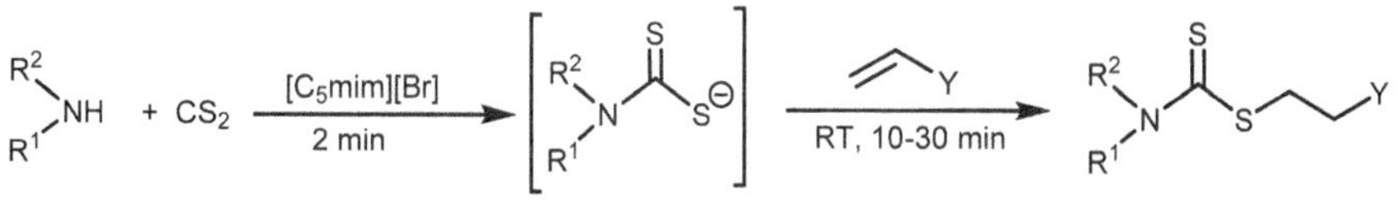

$$R^1R^2NH + CS_2 \xrightarrow[\text{2 min}]{[C_5mim][Br]} \left[R^1R^2N{-}C(=S){-}S^{\ominus} \right] \xrightarrow[\text{RT, 10-30 min}]{CH_2{=}CH{-}Y} R^1R^2N{-}C(=S){-}S{-}CH_2CH_2{-}Y$$

Scheme 3.5.2 Three-component synthesis of dithiocarbamates catalysed by [C_5mim]Br.

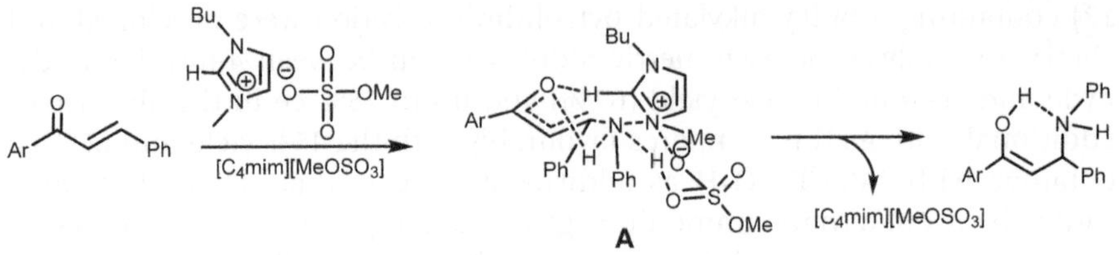

Scheme 3.5.3 Role of [bmim][$MeSO_4$] in the aza-Michael reaction between aniline and an α,β-unsaturated ketone.

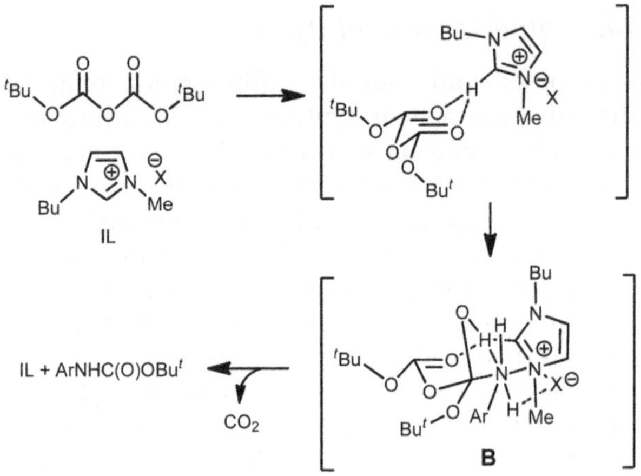

Scheme 3.5.4 Role of [C_4mim][X] as catalyst in the chemoselective *N-tert*-butyloxycarbonylation of amines.

rational design and optimisation of new organocatalysts and enable selectivities to be predicted and/or rationalised.[485]

Imidazolium-based ionic liquids [C_4mim][X] ($X = NTf_2^-$, BF_4^-, OAc^-, HSO_4^-, $MeOSO_3^-$, ClO_4^-, HCO_2^-, N_3^-, PF_6^-) are efficient organocatalysts for the highly chemoselective *N-tert*-butyloxycarbonylation of a range of aromatic, heteroaromatic, aryl alkyl, heteroaryl alkyl and alkyl amines and give good yields in very short reaction times (1–45 min). The ionic liquid was proposed to operate *via* a dual electrophilic–nucleophilic activation by forming a cooperative hydrogen-bond network **B** in much the same manner as for the aza-Michael addition described above (Scheme 3.5.4). Electrophilic activation of the $(Boc)_2O$ occurs through a bifurcated hydrogen bond between the C2-H of the [C_4mim]$^+$ cation and the carbonyl oxygen atom(s) of $(Boc)_2O$ and a hydrogen bond between the ionic liquid anion and one of the hydrogen atoms of the NH_2 group induces nucleophilic activation of the nitrogen while also orientating it in close proximity to the 'activated' carbonyl group of $(Boc)_2O$ to facilitate nucleophilic attack. The differential reactivity of aromatic and aliphatic amines enables selective *N-t*-Boc formation for inter- and intramolecular competition experiments. The superior performance of [C_4mim][NTf_2] was demonstrated in a comparison with a host of

Lewis acid catalysts, all of which gave markedly lower yields under the same conditions.[486] This dual electrophilic–nucleophilic activation *via* a relay of cooperative hydrogen bonds and charge–charge interactions was also identified earlier for the *O*-*tert*-butoxycarbonylation of 2-naphthol.[487]

Perosa and co-workers have demonstrated cooperative nucleophilic and electrophilic organocatalysis by ionic liquids for the solvent free Baylis–Hilmann dimerisation of cyclohexenone by comparing the performance of a series of ionic liquids that allowed the contributions of the anionic and cationic components to be discriminated. Even though the basicities of the anion in $[P_{8,8,8,1}][OCO_2Me]$ and $[P_{8,8,8,1}][OCO_2H]$ are far lower than phosphazene P_1-*t*Bu and DBU, they catalyse the dimerisation at similar or greater rates which was taken as evidence for an electrophilic catalytic role involving the $[P_{8,8,8,1}]^+$ cation (*i.e.* electrophilic activation of the carbonyl by the phosphonium ion). Further support was provided as the rate of dimerisation in a mixture of $[P_{8,8,8,1}]$Br and DBU increased with increasing concentration of the electrophilic component whereas $[P_{8,8,8,1}]$Br is inactive as the anion is not a strong enough nucleophile.[488]

Imidazolium-based ionic liquids act as solvent and catalyst for the three-component one-pot aminomethylation of electron-rich aromatics. Good yields have been obtained at room temperature with a range of different ionic liquids the most efficient of which was $[C_8mim][BF_4]$.[489] *N*,*N*-Dimethylethanolammonium-based ionic liquids catalyse the Knoevenagel reaction by acting in a bifunctional manner, both as a hydrogen bond donor through the alcohol group of the cation to activate the carbonyl and as a base to deprotonate the active methylene compound. The activity of a series of ionic liquids appears to parallel the hydrogen-bond acceptor parameter β indicating its potential importance in determining activity.[490]

The basic imidazolide-derived imidazolium ionic liquid $[C_4mim][Im]$ catalyses the aza-Markovnikov addition of imidazole to vinyl acetate to give the expected product in high yield with a catalyst loading of only 2 mol% (Scheme 3.5.5). Comparable yields were obtained under similar conditions with $[C_4mim][Im]$ and [TBAB][Im] which indicates that hydrogen bonding between the C2-H and the vinyl ester is not necessary for catalysis. The proposed mechanism involved the imidazolide anion reacting with vinyl acetate to form 1-(1*H*-imidazol-1-yl)ethanone **142** and ethenolate, the latter of which deprotonates the imidazole to form acetaldehyde. Finally, the 1-(1*H*-imidazol-1-yl)ethanolate **143** formed by addition of imidazolide to acetaldehyde reacts with the **142** to generate the Markovnikov adduct **144** and liberate the catalyst (Scheme 3.5.5).[491]

The basic ionic liquid $[C_2mim][OAc]$ efficiently catalyses the benzoin condensation of aromatic aldehydes as well as oxidation of the benzil adduct and subsequent hydroacylation of the resulting ketone (Scheme 3.5.6); previous reports and experimental studies were consistent with CO_2 as the oxidant for the benzil oxidation step while computational studies suggest that the benzaldehyde for the hydroacylation step results from a reverse benzoin condensation.[492]

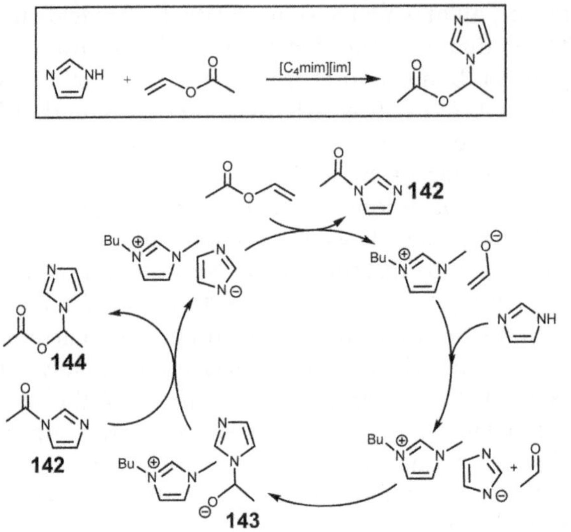

Scheme 3.5.5 Proposed mechanism for the [Bmim][Im]-catalysed aza-Markovnikov addition.

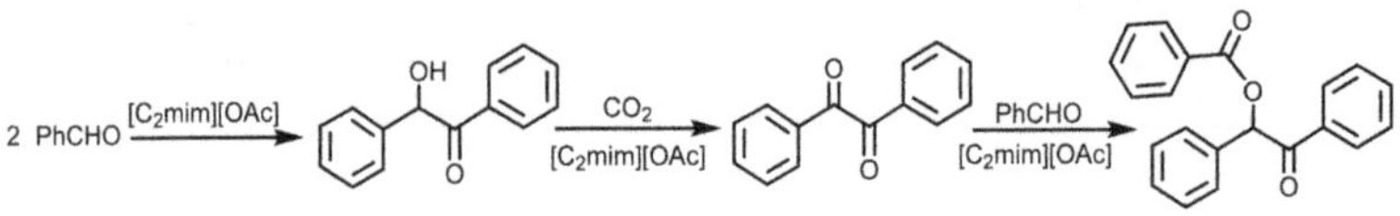

Scheme 3.5.6 Benzoin condensation, oxidation by air (CO_2) and subsequent hydroacylation, catalysed by [C_4mim][OAc].

Basic phosphonium ionic liquids with methylcarbonate or bicarbonate as the anion are efficient catalysts for the nitroaldol reaction of aldehydes and cyclic ketones with nitromethane and nitroethane. Good yields of adduct are obtained at room temperature or 50 °C using 1 mol% [$P_{8,8,8,1}$][CO_2OMe] or [$P_{8,8,8,1}$][CO_2OH] and the efficiency of both catalysts compared favourably with that of non-nucleophilic sterically hindered bases such as DBU. The catalyst was proposed to operate *via* an initial acid–base reaction between the phosphonium salt and nitromethane to generate the nucleophilic nitronate anion and methylcarbonic acid which decomposes to CO_2 and methanol.[493] The bicarbonate exchanged phosphonium salt [$P_{8,8,8,1}$][CO_2OH] also catalysed the Michael addition of various donors to α,β-unsaturated ketones under solventless conditions between 4–40 °C in the presence of 0.4–2 mol% catalyst and conversions matched or exceeded those obtained with the phosphazene base P_1-*t*Bu or DBU. A comparison of the catalyst activity against $NaHCO_3$ indicated that the basicity of the bicarbonate and methylcarbonate anions when coupled with an onium cation are enhanced by up to two orders of magnitude with respect to inorganic bicarbonate salts.[27]

While the most popular strategy for immobilisation of an organocatalyst in an ionic liquid (or a polar solvent) is undoubtedly the attachment of an ionic tag, a non-immobilised siloxy serine organocatalyst has been developed that gives high yields and good ee's (up to 92%) in the room temperature aldol reaction between aromatic aldehydes and cyclic ketones in $[C_4mim][BF_4]$. Significant leaching of the siloxy-L-serine in the extraction step of the recycle experiment was overcome by addition of water prior to the work-up.[494]

Catalytic olefin isomerisation is a key process in the petrochemical industry and an early computational study concluded that the $[C_2mim][F]$-catalysed isomerisation of 1-butene to 2-butene involved a concerted process with a six-membered transition state and a barrier of 48 kcal mol^{-1}.[495a,b] A more recent theoretical study established that the $[C_2mim]Cl$-catalysed isomerisation occurs *via* a two-step H-migration in preference to the corresponding concerted pathway and that the two activation barriers of 35.1 and 24.7 kcal mol^{-1} required to overcome the transition states in this process are significantly lower than that of 55.1 kcal mol^{-1} for the concerted mechanism.[496]

Crown ether complex cation ionic liquids (CECIL) composed of a crown ether chelated alkali metal [18-C-6K] or [15-C-5Na] and an anion such as OAc^-, PO_4^{3-}, HPO_4^{2-}, $H_2PO_4^-$, CO_3^{2-}, NO_3^-, ClO_3^-, OH^- have been applied to a variety of C–C and C–heteroatom bond forming reactions. The Michael addition of a range of carbon, sulfur and nitrogen-based nucleophiles with electron deficient alkenes is catalysed by 10 mol% [15-C-5Na][OH] at 25 °C in MeOH as solvent to give good yields even after five recycles. Higher yields were obtained in shorter reaction times when the CECIL was used as the solvent. Good yields were also obtained for the Henry reaction between aldehydes and nitromethane as well as for the Knoevenagel condensation of aromatic aldehydes and malononitrile with 30 mol% [18-C-6K][OH] and for the oxidation of benzohydrol and benzyl alcohol with $[18\text{-}C\text{-}6K][BrO_3]$ as both solvent and oxidant.[497]

3.5.2 Metathesis

3.5.2.1 Ring Closing Metathesis

Ruthenium-catalysed ring closing metathesis (RCM) is a powerful and versatile tool for the construction of carbocyclic and heterocyclic motifs. As the catalyst is often relatively expensive and because it is also necessary to obtain product uncontaminated by metal residue various strategies have been investigated to immobilise Grubbs and Hoveyda–Grubbs-type catalysts in order to develop improved product separation and isolation protocols and to recover and reuse the catalyst. In this regard, ionic liquids appear well-suited to this task as they have a range of tuneable properties and are immiscible with organic solvents/reagents. An instructive and helpful subject specific tutorial review entitled *Olefin Metathesis in Ionic Liquids*[9] together with a supplementary article entitled *Onium-Tagged Ru Complexes as Universal Catalysts for Olefin Metathesis Reactions in Various Solvents*[498] provides a detailed coverage of this literature up to 2008.

Early studies reported that 5 mol% Grubbs I precatalyst **145** promoted the RCM of a range of dienes in [C_4mim][PF_6] at 80 °C at a substrate concentration of 30 mg mL^{-1}; the system recycled twice but conversions dropped in the last run due to significant leaching of the catalyst into the organic phase (3.9–5.3 μg mg^{-1}). Interestingly, Grubbs II precatalyst **146** recycled slightly more efficiently as evidenced by a more gradual drop off in conversion and the lower level of ruthenium contamination in the organic phase (1.3–1.6 mg mg^{-1}).[499]

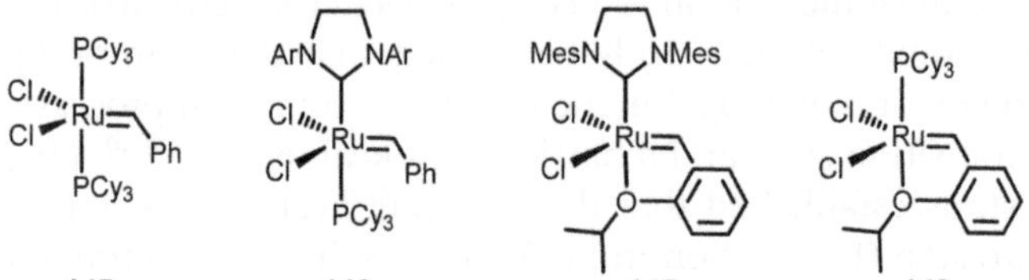

145 **146** **147** **148**

As the cationic ruthenium allenylidene complex [RuCl(=C=C=CPh_2)(P-Cy_3)(*p*-cymene)][X] forms an active catalyst for RCM its efficiency in ionic liquids has also been examined on the basis that it was expected to be highly soluble and immobilised by electrostatic interactions. Since the performance of [RuCl(=C=C=CPh_2)(PCy_3)(*p*-cymene)][X] for the RCM of non-conjugated dienes in organic solvents depends on the counteranion, $[X]^-$, it was not surprising to find that the efficiency of this catalyst for the RCM of *N,N*-diallyltosylamide in [C_4mim][X] (X = PF_6^-, BF_4^-, OTf^-) was dominated by the anion of the ionic liquid which is present in large excess and presumably exchanges with the anion of the precatalyst [Equation (3.5.5) and (3.5.6), E = NTs]. Regardless of the precursor anion, reactions conducted in [C_4mim][BF_4] and [C_4mim][PF_6] gave lower conversions and were less selective than those in [C_4mim][OTf] which reached 100% conversion with a selectivity of 97% at 80 °C using 2.5 mol% [RuCl(=C=$CCPh_2$)(PCy_3)(*p*-cymene)][OTf]. However, the catalyst only recycled twice in [C_4mim][OTf] before conversions dropped due to slow decomposition of the catalyst in the presence of water.[500] Microwave heating gave a significant acceleration in the rate of Grubbs II-catalysed RCM reactions in [C_4mim][BF_4] compared with conventional heating. A comprehensive comparison across a range of substrates gave near quantitative yields in 15–60 s under microwave heating (110 W) whereas yields varied between 0–12% in control experiments conducted under conventional thermal heating. A similar acceleration in rate was achieved for reactions conducted in dichloromethane with the same range of substrates.[501]

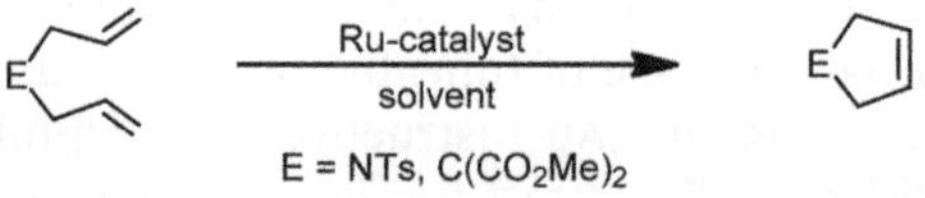

Ru-catalyst, solvent; E = NTs, $C(CO_2Me)_2$ (3.5.5)

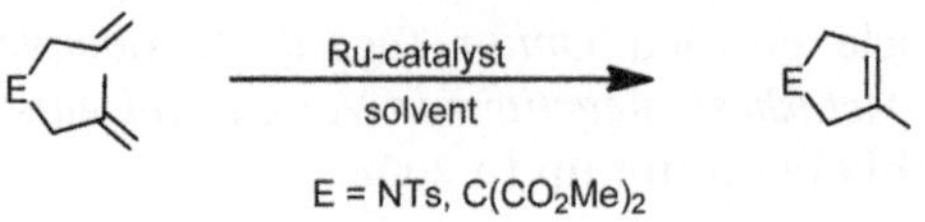

Ru-catalyst, solvent; E = NTs, $C(CO_2Me)_2$ (3.5.6)

3.5.2.2 Cross-Metathesis

The efficiency of Grubbs II catalysts for the self-cross-metathesis of styrene and its derivatives has been compared in dichloromethane, $[C_4mim][BF_4]$ and $[C_4mim][PF_6]$ and all gave good conversions after 3 h at 45 °C. Both ionic liquid–catalyst mixtures recycled in a straightforward procedure but conversions in $[C_4mim][BF_4]$ dropped from 74% to 48% over 4 runs whereas $[C_4mim][PF_6]$ recycled much more efficiently with a much smaller drop in conversions from 85% to 75%.[502] Efficient and selective cross-metathesis of methyl oleate with ethylene occurs under mild conditions in either $[C_4mim][NTf_2]$ or $[C_4C_2{}^1mim][NTf_2]$ with 5 mol% of the Hoveyda ruthenium carbene precatalyst **148** and good conversions were maintained over three consecutive runs with no loss in activity. Interestingly, catalysts tagged with an imidazolium hexafluorophosphate group recycled very poorly.[503] A marked improvement in selectivity for the ruthenium-catalysed self-metathesis of 1-octene to 7-tetradecene was obtained in ionic liquid compared with solventless conditions [Equation (3.5.7)]. Initial studies showed that 0.01 mol% of Grubbs II type catalyst **146** (Ar = Mes, 2,6-$^iPr_2C_6H_3$) gave 94–95% selectivity for 7-tetradecene at 60 °C whereas the same catalyst only reached 67–70% selectivity under solventless conditions; the poor selectivity at this temperature was attributed to secondary metathesis product formation. Even though selectivity for 7-tetradecene improved to 91% at 40 °C under solventless conditions with 0.02 mol% of Grubbs–Hoveyda catalyst **147**, reactions conducted with $[C_2C_1{}^2mim][NTf_2]$ as either solvent or additive (Ru : IL, 1 : 10) reached >98% selectivity and 95% conversion under the same conditions. The improvement in selectivity was attributed to the poor solubility of the metathesis product in the ionic liquid limiting contact between the product and the catalyst.[504]

(3.5.7)

3.5.2.3 Catalysts Modified with Ionic Tags

The overwhelming majority of studies directed towards reducing or preventing leaching of ruthenium-carbene metathesis catalysts involve modifying the catalyst with an ionic tag to improve retention in the ionic liquid, optimise recyclability of the expensive catalyst and limit/reduce contamination of the product. The first such study introduced an imidazolium hexafluorophosphate onto the aromatic ring of the chelating alkylidene-ether of **148** to afford **149**.[505] The benchmark RCM of *N,N*-diallyltosylamide [Equation (3.5.5), E = NTs] was used to assess the efficiency of **149** against its unmodified counterpart as well as Grubbs I precatalyst **145**. Comparative catalyst testing with **149** in $[C_4mim][PF_6]$ gave complete conversion after heating at 60 °C for 45 min and excellent conversions were obtained for a further nine recycles, the last of which was run after three months which attests to the stability of the ionic liquid immobilised catalyst. While **145** and **148** both gave comparable conversions in the first run activity dropped dramatically in subsequent runs due to catalyst leaching. However, a survey of different substrates revealed that **149** had some limitations as oxygen containing dienes resulted in the formation of inactive oxygen-ligated ruthenium carbene complexes and the long reaction times required for the trisubstituted diene *N*-allyl-4-methyl-*N*-(2-methylallyl)benzenesulfonamide [Equation (3.5.6), E = NTs] resulted in gradual catalyst decomposition and poor recyclability. A second generation imidazolium-tagged *N*-heterocyclic-based Hoveyda catalyst **150** (X = CH_2) was developed to improve functional group tolerance, stability and activity. Using the same trisubstituted olefin as the benchmark substrate, **150** proved to be slightly more stable than **149** under the same conditions, as catalyst activity dropped more slowly with the number of reuses. Recyclability improved when the reaction temperature was lowered to 40 °C but conversions decreased due to the higher viscosity of the ionic liquid at this temperature. A marked improvement in recyclability was achieved when the reaction was run at room temperature under biphasic conditions in [C_4mim][PF_6]–toluene (25 : 75) and under these conditions excellent conversions were obtained across eight runs in short reaction times (3 h). The authors reasoned that these imidazolium-tagged catalysts were ideally suited for olefin metathesis reactions in an ionic liquid–organic biphasic system on the basis that the release-and-return mechanism would allow the neutral active 14-electron ruthenium species to reside preferentially in the organic phase while the ionic styrenyl ether liberated from the catalyst would remain in the ionic liquid; after all the substrate had been consumed the active species would then return to the ionic liquid and re-attach to the active species. The efficiency of this immobilisation strategy was manifested by the low level of ruthenium leaching into the product which averaged 7.3 ppm across eight cycles. A select range of substrates recycled efficiently under these conditions and in all cases excellent conversions were obtained for at least six cycles, however, poor conversions were still obtained with diallyl ether-based substrates as well as highly substituted dienes. Encouraging conversions were also obtained for olefin cross-metathesis conducted in [C_4mim][PF_6]–toluene (25 : 75) using 5 mol% of the imidazolium-tagged Hoveyda–Grubbs catalyst **150**; good conversions were obtained after 3 h at room temperature for all substrate combinations examined, but in each case activity dropped quite substantially after the first run.[506a,b] Yao and co-workers incorporated a very similar *N*-methyl imidazolium tag onto the aromatic ring of the alkylidene-ether to afford **151** which was a highly efficient catalyst for the homogeneous RCM of *N*-allyl-4-methyl-*N*-(pent-4-en-1-yl)benzenesulfonamide in [C_4mim][BF_4]–CH_2Cl_2 (1 : 9 v/v). Good conversions to the corresponding seven-membered heterocycle were obtained with 5 mol% **151** (X = O) after 3 h at 50 °C and the system recycled 10 times with only a minor drop in activity; for comparison unmodified Grubbs and Hoveyda catalysts only gave good conversions in the first run.

The corresponding second generation imidazolium-tagged Hoveyda–Grubbs carbene complex **152** was subsequently shown to be a highly efficient catalyst for the RCM of a range of di- tri- and tetrasubstituted dienes and enynes; good conversions were obtained in $[C_4mim][BF_4]$–CH_2Cl_2 (1 : 9 v/v) at 45 °C with 1–3 mol% catalyst and the system recycled 17 times with only a minor loss in activity.[507a,b]

X = CH_2, n = 1; **149**
X = O, n = 3; **151**

X = CH_2, n = 1; **150**
X = O, n = 3; **152**

153

The ionophilic second generation Grubbs-type catalyst **153** bearing an imidazolium-tagged alkyldicyclohexylphosphine was developed for use in ionic liquid biphasic RCM (Figure 3.5.2). High activity was achieved for the RCM of di- and trisubstituted dienes in $[C_4mim][PF_6]$–toluene and good conversions were obtained with catalyst loadings as low as 0.25 mol%, although poor conversions were obtained with more highly substituted dienes. Recycle experiments on the cyclisation of 1,7-octadiene revealed that the $[C_4mim][PF_6]$–toluene–catalyst mixture recycled efficiently and gave good conversions over eight runs with negligible leaching of the ruthenium into the toluene phase (<1.5 % ruthenium), in contrast, the same catalyst only recycled three times in $[C_4mim][NTf_2]$ before conversions dropped and was completely inactive in $[C_4mim][FAP]$ (FAP = tris(perfluoroethyl)trifluorophosphate). Comparative recycle studies with unmodified Grubbs II

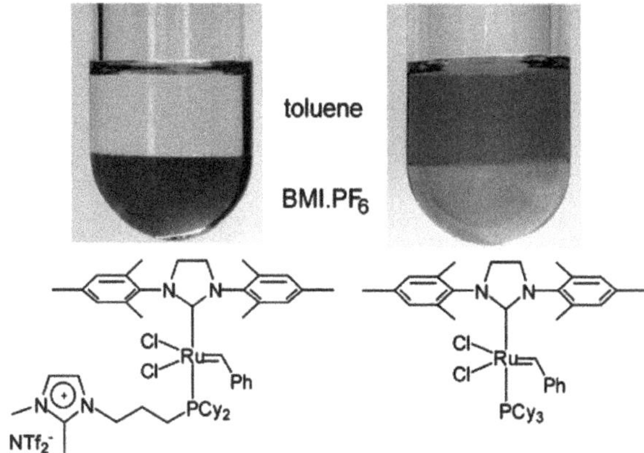

Figure 3.5.2 The affinity of ionophilic second generation Grubbs-type catalyst and unmodified Grubbs catalysts for the ionic liquid phase *versus* toluene. Reprinted with permission from reference 508. Copyright 2008 American Chemical Society.

revealed that good conversions could be obtained in short reaction times for the first run but increasingly longer reaction times were required in subsequent runs due to catalyst leaching, which underpinned the efficiency of the imidazolium-modified phosphine.[508]

Hoveyda–Grubbs catalysts have also been modified by introducing an imidazolium tag onto the oxygen atom of the coordinated ether. A series of three imidazolium-modified styrenes, varying in the substitution pattern on the tether between the oxygen atom and the imidazolium group were prepared in a straightforward three-step synthesis and the derived catalysts **154a–c** prepared by ligand substitution reactions with Grubbs-type II catalyst. In line with previous observations, $[C_4mim][PF_6]$ was shown to be the ionic liquid of choice to immobilise these ruthenium carbene complexes. Under optimised conditions, 0.0025 mol% **154a** (25 ppm) catalysed the RCM of 1,7-octadiene in $[C_4mim][PF_6]$–toluene with a TOF of 343 000 h^{-1} at 45 °C and recycled six times with no loss in activity. High conversions (>90%) were also obtained for the RCM of the sterically more challenging substrate 2-allyl-2-methylallylmalonate [Equation (3.5.6), $E = C(CO_2Me)_2$] with 0.02 mol% **154a.** Kinetic studies revealed that while **154a** retained the same rate profile over three cycles the activity of **154c** gradually decreased in each run which parallels the rate profiles for analogous catalysts under homogeneous conditions and reinforces the influence on the stability of the catalytically active species of the secondary carbon atom attached to the ruthenium-coordinated ether group. Moreover, the degree of substitution at the carbon atom of the coordinated ether appears to be responsible for the longer Ru–O bond length which also reflects the higher catalytic activity.[509]

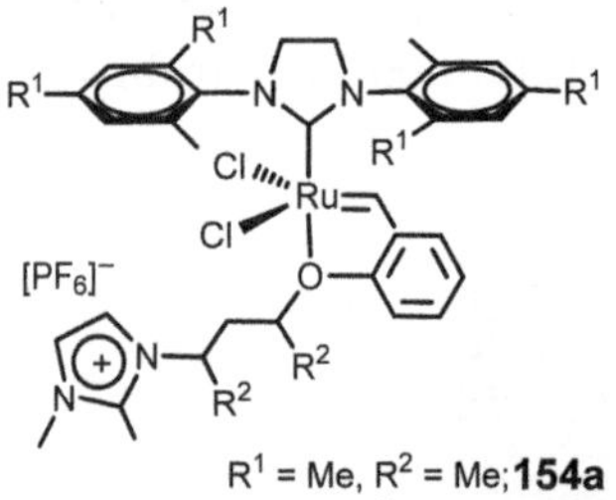

R^1 = Me, R^2 = Me; **154a**
R1 = H, R^2 = Me; **154b**
R^1 = Me, R^2 = H, **154c**

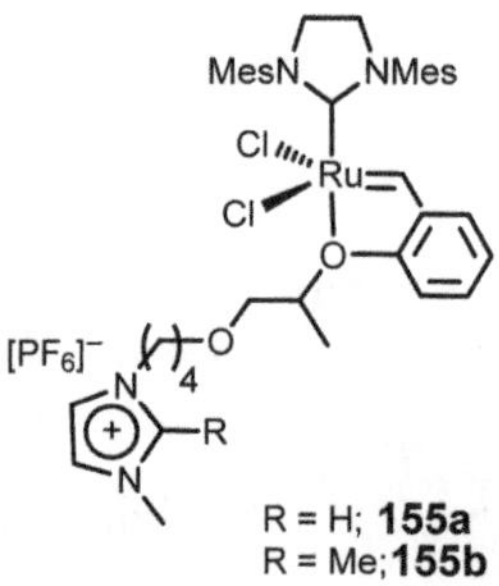

R = H; **155a**
R = Me; **155b**

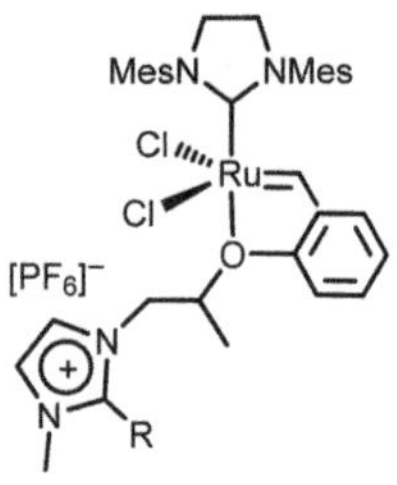

R = H; **156a**
R = Me; **156b**

The performance of ether-modified imidazolium-tagged Hoveyda–Grubbs type ruthenium metathesis catalysts **155a–b** and **156a–b** depends on the ionic liquid as well as substituent R and the length of the imidazolium tether. For example, ruthenium carbene **155a** required long reaction times (2 h) to reach moderate conversions for the RCM of *N,N*-diallyltosylamide under monophasic conditions in $[C_4mim][NTf_2]$–CH_2Cl_2 and $[C_4mim][SbF_6]$–CH_2Cl_2 whereas near quantitative conversions were obtained with each of the catalysts in $[C_4mim][PF_6]$ within 10–15 min at room temperature; however, their recyclability was very poor. Improved recyclability was obtained under biphasic conditions in a mixture of $[C_4mim][PF_6]$ and toluene

(1 : 3, v/v), although reaction rates were reduced quite significantly. A limited survey indicated that both the tether connecting the imidazolium tag to the oxygen atom and the ionic liquid had a marked influence on recyclability and **155b**/[$C_4C_1{}^2$mim][PF_6] was identified as the most efficient combination as it recycled seven times with near quantitative conversions, although runs seven and eight required longer reaction times to reach comparable conversions. The level of ruthenium leaching (21 ppm) in this combination was markedly lower than that of 87 ppm for **155a** in the same ionic liquid. Interestingly, **156a–b** were significantly more active than **155a–b** in [$C_4C_1{}^2$mim][PF_6]-toluene and gave near quantitative conversions in 30 min, however, activity decreased dramatically after the first recycle. Thus, catalyst performance (stability, activity and recyclability) is clearly a function of the location of the imidazolium anchor, its structure and proximity to the active site as well as the nature of the ionic liquid and solvent composition.[510] Highly active Hoveyda-type catalysts tagged with either carborane (*closo*-1,2-$C_2B_{10}H_{12}$) or carbollide ([*nido*-7,8-$C_2B_9H_{12}$]$^-$) have been prepared by reaction of the corresponding isopropenyl styrene with Grubbs II. Both complexes catalysed the RCM of diethyl diallylmalonate at 30 °C in dichloromethane and gave conversions that matched those obtained with unmodified Hoveyda–Grubbs catalyst. Both catalysts gave excellent conversions in the first run (>98%) for the RCM of diethyl diallylmalonate in [C_4mim][PF_6] but the carbollide-modified system recycled more efficiently than its *closo*-carborane counterpart as the former could be reused 10 times with only a minor reduction in activity whereas the latter lost the majority of its activity after the first run. Thus, [*nido*-7,8-$C_2B_9H_{12}$]$^-$ is an effective anionic tag for noncovalent immobilisation of catalysts and could find use in a wider range of transition metal-based transformations.[511] The performance of two imidazolium-tagged Hoveyda-type catalysts, **157** and **158**, have been compared on the basis of the well-documented observation that an isopropoxy substituent attached to the ruthenium-coordinated ether should lead to an active and stable catalyst and that an increase in steric hindrance *ortho* to the isoproxy group should improve catalytic activity. While the former is more active than the latter for the RCM of *N,N*-diallyltosylamide [Equation (3.5.5), E = NTs] and dimethyl diallylmalonate [Equation (3.5.5), E = C(CO_2Me)$_2$] it showed poor recyclability and lost activity gradually after the first run; for comparison the latter retained its activity over the first four runs.[512]

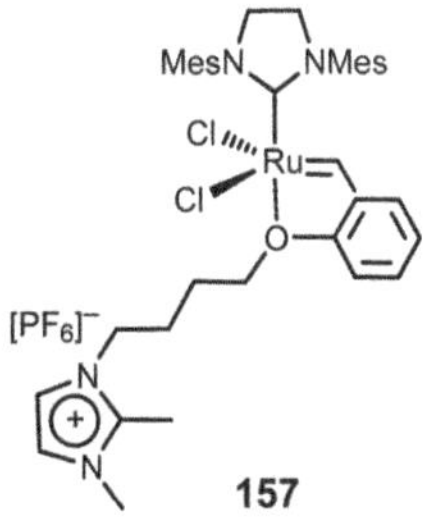

MesN NMes Cl Ru Cl O [PF6]⁻ N N **157**

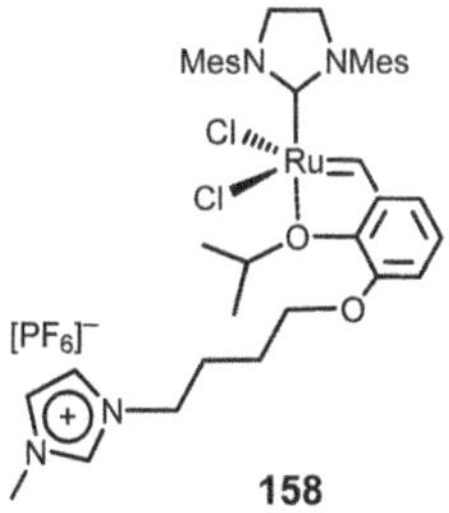

MesN NMes Cl Ru Cl O O [PF6]⁻ N N **158**

The first example of a catalytic membrane reactor for olefin metathesis was developed by immobilising an imidazolium-tagged Hoveyda catalyst in an ionic liquid supported on a solvent-resistant polyimide membrane (Starmem 228). The resulting catalytic membrane reactor was used for the RCM of *N,N*-diallyltosylamide running in a discontinuous mode in toluene; good conversions were obtained with 2.5 mol% catalyst at 35 °C in 1 h for the first two runs but conversions dropped in the third cycle. The authors considered several possible explanations for the decrease in activity including the intrinsic stability of the catalyst as well as the stability of the membrane *i.e.* leaching of the catalyst and/or ionic liquid.[513]

3.5.2.4 Continuous Flow Cross-Metathesis

Continuous flow homogeneous olefin metathesis has been achieved using a supported ionic liquid phase (SILP) catalyst comprising **152** immobilised in a thin film of [C_4mim][NTf_2] within the pores of silica with *sc*CO_2 as the transport medium (Figure 3.5.3). While the system was not stable for the metathesis of terminal olefins, good conversions and high TONs were obtained for the self-metathesis of 2-octene (TON = 3371, 5 h) and methyl oleate (TON = 4247, 10 h) at 50 °C [Scheme 3.5.7(a)]; a slight decrease in activity with time was attributed to catalyst instability. Even higher TONs (>10 000, 9 h) were obtained for the self-metathesis of methyl oleate when the temperature was reduced to 23 °C [Scheme 3.5.7(b)]. The cross-metathesis of methyl oleate with dimethyl maleate was also investigated as the resulting α,β-unsaturated terminal ester could be further carbonylated to dimethyl 1,12-dodecandioate [Scheme 3.5.7(c)]; both short chain products are potentially useful bio-derived feedstocks. Cross-metathesis occurred during the early stages of the reaction (3 h) after which homometathesis competed and eventually dominated. A marked improvement in cross-metathesis selectivity was achieved when the reaction was

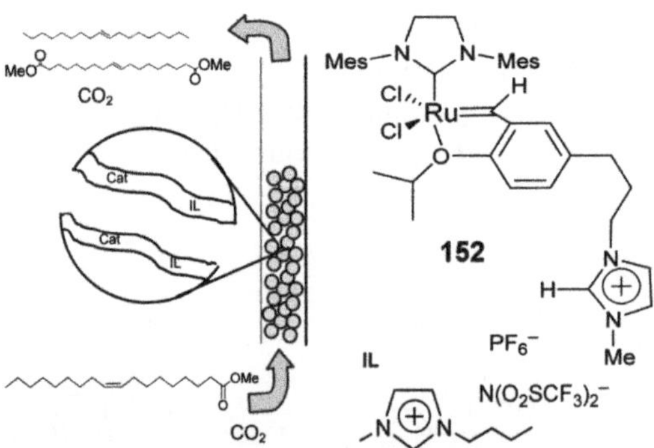

Figure 3.5.3 Diagram showing the concept of SILP process for the self-metathesis of methyl oleate.
Reproduced from reference 71.

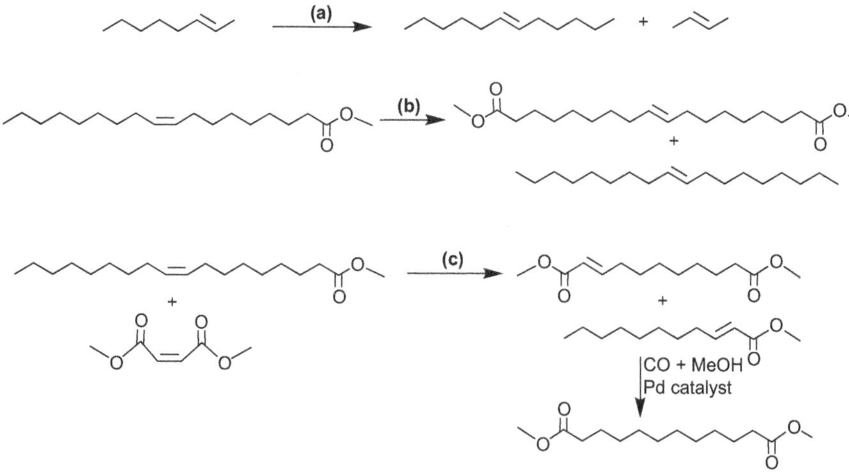

(a)

(b)

(c)

CO + MeOH
Pd catalyst

Scheme 3.5.7 (a) Self-metathesis of 2-octene; (b) self-metathesis of methyl oleate, (c) cross-metathesis of methyl oleate with dimethyl maleate, showing a possible route to dimethyl 1,12-dodecanedioate and dimethyl 1,11-undecendioate.
Adapted from reference 71.

performed under batch conditions in $[C_8mim][NTf_2]$ with 1 mol% catalyst and a four-fold excess of dimethyl maleate; under these conditions methyl oleate was completely converted into the desired cross-metathesis product in <10 min. The use of a 0.2 mol% catalyst revealed that self-metathesis of methyl oleate was much more rapid than cross-metathesis with dimethyl oleate and that the formation of cross-metathesis product was predominantly a secondary reaction of the methyl oleate self-metathesis product. Good initial selectivity for cross-metathesis in the flow system was obtained by increasing the residence time from 42 min to 130 min; this gave an initial conversion to cross-metathesis product of 70% which dropped progressively after 2 h but remained >25% even after 6 h. The products were obtained solvent free in high yield with very low levels of ruthenium contamination; typically 10–15 ppm in the self-metathesis of methyl oleate and 5–8 ppm in the cross-metathesis of methyl oleate and methyl maleate.[71]

Ring opening metathesis-derived cation-decorated monoliths have also been used as a support for a continuous biphasic process for ring closing metathesis, cross-metathesis and self-metathesis based on a SILP catalyst comprising the dicationic Grubbs–Hoveyda type catalyst $[Ru(dmf)_3(IMesH_2)(=CH\text{-}2\text{-}(2\text{-}PrO)\text{-}C_6H_4)][BF_4]_2$. The monoliths were prepared by ROMP of norbornene with tris(norborn-2-enylmethylenoxy)methylsilane as cross-linker and $[RuCl_2(PCy_3)_2(=CHPh)]$ as initiator. The inner surface was further decorated with ammonium groups to help retain the ionic liquid–catalyst mixture of $[C_4C_1{}^2mim][BF_4]$ and $[Ru(dmf)_3(IMesH_2)(=CH\text{-}2\text{-}(2\text{-}PrO)\text{-}C_6H_4)][BF_4]_2$. TONs up to 900 were obtained for the RCM which was slower than that obtained under ionic liquid–heptane biphasic conditions due to the contact time and the more efficient mixing under mechanical stirring. Continuous flow self-metathesis of methyl oleate at 80 °C and a flow rate of 0.1 mL min^{-1} reached a TON of 800 in 45 min.[514]

3.5.3 Ring Opening Polymerisation

3.5.3.1 Ring Opening Metathesis Polymerisation

Ring opening metathesis polymerisation (ROMP) is now a well-established and powerful tool that has been applied to the synthesis of technologically important polymers. While ruthenium-based Grubbs-type initiators are highly versatile in that they polymerise highly functional monomers in a living and controlled manner, the need to reduce cost by recycling, decrease metal residue contamination in the product, meet demands for cleaner chemicals synthesis and to identify solvents that are suitable for use with monomers that are 'resistant' to polymerisation in conventional organic solvents has been the major driver behind using ionic liquids as solvents for ROMP.

The cationic ruthenium allenylidene complex [(*p*-cymene)-$RuCl(PCy_3)(=C=C=CPh_2)$][OTf] was used by Dixneuf and co-workers for the first ROMP of norbornene under biphasic conditions in $[C_4C_1{}^2mim][PF_6]$–toluene. The catalyst was retained in the ionic liquid layer and recycled with reasonable efficacy over six runs; an increase in the average molecular weight in successive cycles indicated gradual loss of active catalyst due to leaching and/or deactivation. However, the allenylidene complex recycled more efficiently than either Grubbs I or Grubbs II catalysts which was presumably due to its ionic character.[515] ROMP of functional norbornene-based monomers catalysed by $[RuCl_2(py)_2(IMesH_2)(CHPh)]$ or $[RuCl_2(IMesH_2)\{=CH\text{-}(2\text{-}^iPrO)\text{-}C_6H_4)\}]$ in neat imidazolium and phosphonium ionic liquids containing non-coordinating anion such as $[PF_6]^-$ and $[BF_4]^-$ gave exceptionally high reaction rates and good yields of high molecular weight polymer within minutes whereas bromide, iodide and nitrate-based ionic liquids were not suitable as solvents. Perhaps not surprisingly ionic liquids capable of forming carbenes proved to be poor solvents and pitifully low yields of polymer were obtained in $[C_4mim]^+$-based ionic liquids; consequently, $[C_4C_1{}^2mim]^+$ was the cation of choice. A survey of ionic liquids revealed that polymer molecular weight and polydispersity depended on the properties of the ionic liquid. One of the major advantages of the use of ionic liquids for ROMP is that they enable efficient polymerisation of monomers that are hardly polymerisable in traditional organic solvents. For example 7-oxo-norborn-5-ene-2,3-dicarboxylic anhydride which forms insoluble oligomers (at degrees of polymerisation >10) in organic solvent gave high molecular weight polymer ($M_n = 290\,000$, $M_w = 568\,000$) with an acceptable polydispersity in $[C_4C_1{}^2mim][PF_6]$ with 1 mol% Grubbs–Hoveyda catalyst. Although no single ionic liquid appears to be suitable for ROMP of all functional monomers it should be possible to design and fine-tune the physicochemical properties of ionic liquids to achieve monomer specific polymerisation. Recycle experiments using 2-(2-propoxy)styrene to cleave the Ru-alkylidene resulted in a slight reduction in yield as well as M_w in the first cycle and only a poor yield was obtained after the second recycle.[516]

The reactivity of the propagating species in ROMP reactions conducted in ionic liquid and dichloromethane has been investigated by examining the microstructure of copolymer formed from two different monomers. Firstly, in the case of homopolymerisation, the propagating species formed from $[RuCl_2(PCy_3)_2(CHPh)]$ and norbornene is more reactive in $[C_4mim][X]$ ($X = BF_4^-$, PF_6^-, OTf^-) than in dichloromethane, as evidenced by the higher *cis* content of the polymer generated in ionic liquid; this enhancement in reactivity was restricted to neat ionic liquid as no so such increase was obtained under biphasic conditions. Copolymerisation studies revealed that copolymer formed from norbornene and cyclopentene in $[C_4mim][OTf]$ has a random incorporation of monomers whereas polymer formed in dichloromethane is mainly alternating in structure. In dichloromethane the propagating species (P) formed from cyclopentene has such a low reactivity that it favours reaction with the more reactive norbornene monomer, whereas in $[C_4mim][OTf]$ the relative reactivity of the two monomers towards the propagating species is unchanged; the authors proposed that the increased reactivity occurs because the viscosity of the ionic liquid reduces the rate at which the propagating species can relax into a more stable (less reactive) conformation.[517a] Previous studies have correlated the lower reactivity of $P_{cyclopentene}$ over the $P_{norbornene}$ species to its greater flexibility which allows it to relax more quickly into more stable conformations.[517b]

3.5.3.2 Ring Opening Polymerisations

A significant enhancement in the rate of the ring opening polymerisation (ROP) of ε-caprolactone catalysed by polymer-supported $Sc(OTf)_3$ was obtained in $[C_4mim][PF_6]$ compared with toluene but polydispersity broadened from 1.11 to 1.58 [Equation (3.5.8)]. In contrast, ROP was very slow in $[C_4mim][BF_4]$, $[C_4mim]Cl$ and $[C_4C_1{}^2mim][(MeO)HPO_2]$ and did not occur at all in $[C_2mim][BF_4]$ and $[C_4C_1{}^2mim][(OMe)_2PO_2]$. The authors suggested that this was due to inhibition of monomer activation as a ^{13}C NMR study of the coordination of $Sc(OTf)_3$ with ε-caprolactone in the presence of stoichiometric ionic liquid revealed that scandium triflate forms a strong interactions with $[BF_4]^-$.[518]

polystyrene resin SP-Sc(OTf)$_3$ S–O–Sc(OTf)$_2$

SP-Sc(OTf)$_3$ ionic liquid or toluene

(3.5.8)

Nomura and co-workers have also reported that the efficiency of the liquid–liquid biphasic ROP of ε-caprolactone catalysed by Lewis acid rare earth metal triflates depends on the ionic liquid. The ionic liquid–toluene biphasic system was used to immobilise the catalyst in the ionic liquid and collect the poly(ε-caprolactone) in the toluene which could then be isolated by decantation. Regardless of the rare earth metal triflate used, all polymerisations were extremely slow in $[C_4mim][BF_4]$–toluene and required up to 7 days and gave low molecular weight poly(ε-caprolactone) oligomers in 29–32% yield. In stark contrast, good yields of high molecular weight ($M_n = 1600$–4400) PCL were obtained in $[C_4mim][PF_6]$–toluene and $[C_4mim][SbF_6]$–toluene after only 29–42 h. The optimum combination of $Ce(OTf)_4$ in $[C_4mim][SbF_6]$–toluene recycled efficiently and gave quantitative conversions over three runs.[519] A markedly higher yield of atactic polyCHO was obtained in the biphasic $Sc(OTf)_3$-catalysed ROP of cyclohexene oxide (CHO) $[C_4mim][BF_4]$ compared with the bulk monomer; the former gave 77% yield after 30 min at 0 °C while the latter only reached 46% yield after 60 min at 15 °C. The number average molecular weight (M_n) of 1.3–1.6 kg mol^{-1} for the polyCHO prepared under biphasic conditions is markedly lower than that of 25.3 kg mol^{-1} obtained in bulk polymerisation which reflects the lower monomer concentration and faster chain transfer reaction in the ionic liquid phase than in bulk monomer. The authors speculated that ROP in ionic liquid occurs *via* a cationic pathway involving activation of monomer by $Sc(OTf)_3$ to generate an oxonium ion which propagates *via* CHO ring opening.[520]

The lipase-catalysed ROP of L-lactide in ionic liquid has been investigated and conversions, M_n and yields shown to vary with the anion [Equation (3.5.9)]. The M_n values of 55 000 and 50 100 obtained in $[C_4mim][BF_4]$ and $[C_4mim][NTf_2]$, respectively, are higher than that of 40 000 in the bulk but yields are lower as the high solubility of the polylactide (PLLA) in ionic liquid hampered extraction. In contrast, polymerisations conducted in $[C_4mim][PF_6]$ gave poor yields of low molecular weight polyester.[521] The lipase-catalysed ROP of lactide has been extended to hyperbranched poly-L-lactide by introducing bis(hydroxymethyl)butyric acid (BHBA) as the AB_2 co-monomer. The M_n values obtained in $[C_4mim][PF_6]$ are larger than those obtained in bulk *e*ROP and the degree of branching (DB) varies with the reaction conditions and reaches a maximum of 0.22 at an LLA to BHBA ratio of 9 : 1.[522] *e*ROP of L-lactide in $[C_6mim][PF_6]$ also generates high molecular weight (37 800 g mol^{-1}) crystalline poly-L-lactide while copolymerisation in the presence of glycolide (GA) gave poly-L-lactide-co-glycolide with up to 19% lactyl incorporation.[523]

$$\text{L-lactide} \xrightarrow[\text{[C}_4\text{mim][X]}]{\text{CALB, 90-130 °C}} \text{PLLA}_n \qquad (3.5.9)$$

Self-associating polycaprolactone-grafted cellulose brush-copolymers have been prepared by the homogeneous ROP of ε-caprolactone onto softwood dissolved pulp-derived cellulose in [C_4mim]Cl with Sn(Oct)$_2$ or (dimethylaminopyridine) DMAP as catalyst. Of the two catalysts examined DMAP was the most effective as it gave a significantly higher grafting amount of PCL, as measured by MS, DS and W_{PCL}, than those prepared with Sn(Oct)$_2$. The molecular architectures of the polymer can be modified by changing the cellulose: ε-CL feed ratio and reaction temperature with higher temperatures favouring a higher grafting ratio. The cellulose-*g*-PCL copolymers self-assemble in water to give nanoscale (average hydrodynamic radius between 20–100 nm) micelles consisting of a PCL core surrounded by a cellulose corona. The size of the micelles and the CMC can be controlled by varying the grafting ratio of PCL with a higher grafting content leading to smaller micelles.[524] High molecular weight (19.2–38.9 KDa) biodegradable co-polymer has been obtained by co-polymerisation of L-lactic acid with ε-caprolactam using imidazolium and pyridinium-based sulfonic acid functionalised Brønsted acid ionic liquids as both catalyst and solvent at 110–130 °C. The polymerisation was proposed to operate *via* an initial rapid ring opening of ε-caprolactam followed by a combination of polycondensation and simultaneous transesterification to generate polymer with a degree of randomness close to 1. Within a series of imidazolium and pyridinium-based ionic liquids the anion exerted the greatest influence on the polymerisation with [HSO4]$^-$-based systems giving polymers with the highest molecular weight (M_w) and lowest polydispersity (PD) at 130 °C. The polymer composition was shown to reflect the initial monomer ratio which could be used to modify the properties of the copolymer. Phase separation of the polymer and ionic liquid phase separated enable the system recycled four times with no loss in efficiency.[524b]

3.5.3.3 *Condensation Polymerisation*

The properties of polyphenol generated *via* acid-catalysed condensation polymerisation of phenol with paraformaldehyde under optimised conditions (100 °C, 3 h) in [C_6mim][X] (X = Cl^-, Br^-, I^-, $CF_3SO_3^-$, BF_4^-, PF_6^-) depends on the anion; *hydrophilic* ionic liquids are poor solvents and give low molecular weight polymer (M_w = 1100–3400 g mol^{-1}) whereas *hydrophobic* ionic liquids based on [BF_4]$^-$ and [CF_3SO_3]$^-$ gave high molecular weight polymer (M_w = 17 000–41 000 g mol^{-1}). The molecular weights correlate with the strength of the interaction between the phenol monomer and the imidazolium cation such that a strong interaction in a hydrophilic ionic liquid leads to low molecular weight polymer whereas the weaker interactions in a hydrophobic ionic liquid result in high molecular weight polymer (Scheme 3.5.8). A mixture of Brønsted acid ionic liquid [SO_3HC_4mim]Br and [C_4mim]Br catalysed the polymerisation of phenol with formaldehyde to give polyphenol with M_w = 7 000 g mol^{-1} while polymerisation in neat [SO_3HC_4mim]Br gave high molecular weight polymer with M_w = 120 000 g mol^{-1}, albeit with a very high polydispersity.[525]

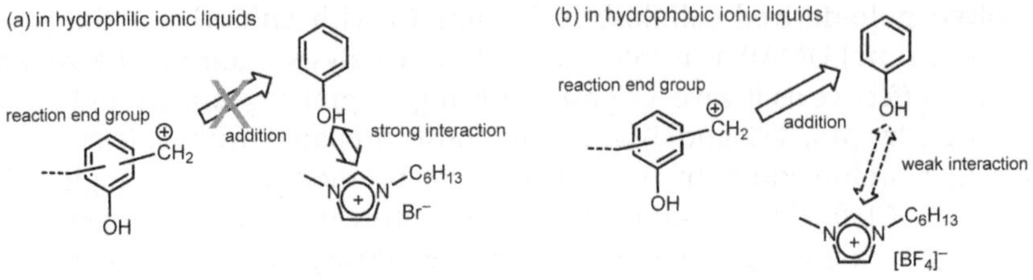

(a) in hydrophilic ionic liquids

(b) in hydrophobic ionic liquids

Scheme 3.5.8 Interaction of phenol with (a) hydrophilic ionic liquid, [C_6mim]Br, and (b) hydrophobic ionic liquid, [C_6mim][BF_4].

3.5.4 Dimerisation and Oligomerisation

3.5.4.1 Dimerisation of 1-Olefins

The dimerisation and oligomerisation of ethylene is catalysed by highly electrophilic Lewis acid metal complexes and in general the higher the electrophilicity the greater the reactivity. In this regard, ionic liquids are an ideal medium for these reactions since they are compatible with charged electrophilic metal complexes, the electrophilicity of the catalyst can be fine-tuned through a judicious choice of non-coordinating anion and the product(s) are insoluble which simplifies separation and improves selectivity (*vide infra*). The first studies on the catalytic dimerisation of propene in ionic liquid targeted 2,3-dimethylbutene (DMB) and 2-methylpentene (2MP) as their alkanes and derived ethers are suitable for use as high octane additives. Initial systems were based on a mixture of [C_4mim]Cl and $AlCl_3$ in combination with the bromide-bridged dimer [$Ni_2Br_2L_2$] (L = 2-methylallyl). When the ionic liquid was basic *i.e.* $AlCl_3$/[C_4mim]Cl ratio < 1 the mixture was unreactive whereas acidic chloroaluminates ($AlCl_3$/[C_4mim]Cl > 1) gave a mixture of dimers, trimers and oligomers; the latter originating from a cationic side reaction. Cationic side reactions were supressed by using a mixture of [C_4mim]Cl–$AlEtCl_2$ ($N = 0.7$) in combination with [Ni_2Br_2(2-methylallyl)$_2$], [Ni(acac)$_2$] or [$NiCl_2(PiPr_3)_2$] and a particularly high 2,3-dimethylbutene content was obtained with the latter. A subsequent survey of ionic liquid composition and phosphine in [$NiCl_2(PR_3)_3$] revealed that [$NiCl_2(P^iPr_3)_2$] with [C_4mim]Cl/$AlCl_3$/AlEtCl (1 : 0.82 : 0.1) gave the best combination of activity, 2,3-dimethylbutene selectivity and long term stability. Under these conditions high 2,3-dimethylbutene selectivity was obtained with nickel complexes of bulky phosphines (83%, P(*i*-Pr_3); 84% PCy_3) while less bulky phosphines gave a much lower 2,3-dimethylbutene content (P(benzyl)$_3$, 48%; PBu_3, 33%). The reduction in 2,3-dimethylbutene selectivity after long reaction times was supressed in the presence of aromatic hydrocarbons such as tetramethylbenzene.[526a,b]

The poor activity and high cost of the 1-butene dimerisation catalyst [Ni(H-COD)(hfacac)] **159** in organic solvents prompted Wasserscheid and co-workers to explore the use of biphasic ionic liquid based systems (Figure 3.5.4), however, reactions conducted in [4-MeC_4pyr]Cl/$AlCl_3$ (0.45 : 0.55, $N = 0.55$)

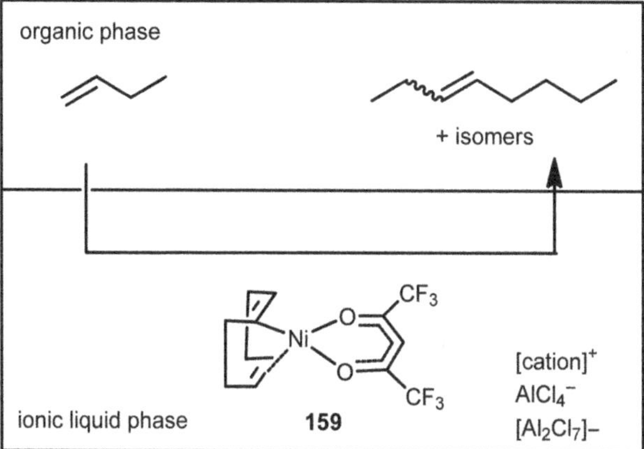

Figure 3.5.4 Biphasic dimerisation of 1-butene using chloroaluminate ionic liquids as catalyst and solvent.
Reproduced with permission from reference 527. Copyright 2001 Elsevier.

generated a large number of branched oligomers due to fast cationic oligomerisations initiated by the intrinsic acidity of the ionic liquid. These cationic side reactions can be suppressed by addition of a weak organic base which buffers the acidic ionic liquid mixture such that [4-MeC$_4$pyr]Cl/AlCl$_3$ (0.43 : 0.53) in combination with 0.04 mol% of pyrrole or *N*-methylpyrrole gave TOFs of 1350 and 2100 h^{-1}, respectively, with remarkably high selectivity for dimerisation (98%). The high dimer selectivity can be explained by the lower solubility of the C8-product in the neat ionic liquid than the C4-feedstock which results in fast and efficient extraction of product into the organic layer during reaction. The use of an ionic liquid biphasic system and a weak organic base as buffer offers a number of advantages over the homogeneous system in toluene including higher activity, markedly improved selectivity, more effective catalyst activation and facile catalyst separation with the potential to recycle.[527a,b]

159 **160** **161**

Following the discovery by Wasserscheid that weak organic bases such as quinoline, pyridine and pyrrole could be used to buffer slightly acidic chloroaluminate ionic liquids and thereby reduce their latent acidity to prevent uncontrolled cationic olefin oligomerisation reactions,[527a,b] Alt demonstrated that triphenylbismuth-buffered chloroaluminate-based ionic liquids also gave exceptionally high selectivities for Ni-catalysed propene dimerisation.[528a–c] Addition of $BiPh_3$ to buffer the acidity of chloroaluminate ionic liquids also reduces their melting point which means that it could be possible to obtain buffered room temperature ionic liquid compositions from aluminium chloride and cations that would normally be solid at ambient temperature in the absence of $BiPh_3$. The slightly acidic ionic liquid ($AlCl_3$/[C_4mim]Cl = 1.20) of the DIFASOL system gave a dimer selectivity of 89.2% with only 0.07 equivalents of $BiPh_3$ which improved to 96.0% in the presence of 0.3 equivalents of $BiPh_3$, but at the expense of productivity. A range of 100 different chloroaluminate salts of composition $AlCl_3$/[onium]Cl = 2.0 (N = 0.67) buffered with 0.3 equivalents of $BiPh_3$ were tested as solvents for the bis(imino)pyridineNi(II)-catalysed dimerisation of propene and of these only two were not liquid at ambient temperature and 95 of the 100 produced dimers, the majority of which gave >80% selectivity for dimerisation. The most efficient buffered ionic liquid based on *selectivity* and *qualitative lifetime* was identified as *N*-methylpyrrolidinium hydrochloride which gave 91.6% dimers at 25 °C. The effect of temperature on selectivity using diethylisopropylammonium and tributylammonim salts revealed that selectivity for dimerisation increased by as much as 7% by lowering the temperature to 0 °C; the same effect was reported earlier by Chauvin for $EtAlCl_2$ buffered systems.[526a,b] The acidic chloroaluminate ionic liquid [C_4mim][Al_2Cl_7] has also been buffered with the trimethylammonium-modified triphenylphosphine [$Ph_2P(4\text{-}Me_3NC_6H_4)$][X] (X = I^-, BF_4^-) to reduce the latent Lewis acidity and prevent uncontrollable cationic oligomerisation during nickel-catalysed propene dimerisation. High selectivity for propene dimerisation was obtained with bis(imino)pyridine nickel(II)-bromide **160** in buffered ionic liquid of composition [$Ph_2P(4\text{-}Me_3NC_6H_4)$][I]/[$C_4$mim][$Al_2Cl_7$] = 0.3 at 25 °C with a catalyst loading of 10^{-5} mol g^{-1} ionic liquid. Under these conditions the productivity of 6.1 g product g ionic liquid^{-1} h^{-1} and dimer selectivity of 84.1% rivalled that for $BiPh_3$-buffered ionic liquid described above. In contrast, the poor solubility of [$Ph_2P(4\text{-}Me_3NC_6H_4)$][$BF_4$] in the ionic liquid led to cationic oligomerisation and formation of higher oligomers. Moreover, addition of 0.06 equivalents of [$Ph_2P(4\text{-}Me_3NC_6H_4)$][I] to the $EtAlCl_2$-buffered [C_4mim]Cl/[$AlCl_3$] (1 : 1.2) system of the nickel-based DIFASOL process improved dimer selectivity from 79.6% to 89.1%, while maintaining high activity.[529]

A detailed kinetic study of the biphasic nickel-catalysed dimerisation of propene in acidic chloroaluminate ionic liquids ([C_4mim]Cl/$AlCl_3$ = 0.45 : 0.55) buffered with *N*-methyl pyrrole determined that 1-hexene (used as a model for the product(s) of propene dimerisation) was less soluble than propene in the ionic liquid phase by a factor of 22 at 25 °C, which explains the high selectivity (97%) obtained under biphasic conditions as the C6 products are efficiently portioned into the organic phase so that consecutive reactions to afford C9 olefins are minimised; for comparison the selectivity in toluene at the same conversion was only 60%. The C6 products obtained in ionic liquid with [Ni(H-COD)(hfacac)] as precatalyst are predominantly branched hexenes (~73%) whereas the same catalyst in toluene gave a linear selectivity of 75%; this difference was attributed to ligand degradation in the

Table 3.5.1 Propene dimerisation with different catalyst precursors in toluene and ionic liquid.[a]

Catalyst	System	Temp (° C)	Conversion	TOF (h^{-1})	$S_{hexenes}$ (%)	S_{LA} (%)	S_{MP} (%)
Ni(H-COD)(hfacac)	Ionic liquid	10	25	70 000	97	22	73
Ni(H-COD)(hfacac)	Ionic liquid	10	73	44 000	97	22	73
Ni(H-COD)(hfacac)	Toluene	75	73	770	60	75	24
$Ni(acac)_2$	Ionic liquid	30	30	530 000	97	23	71
$Ni(hfacac)_2$	Ionic liquid	15	46	480 000	94	24	71

[a]Adapted with permission from reference 530. Copyright 2009 Elsevier.

presence of the $[Al_2Cl_7]^-$ anion. The marked difference in the dimerisation activity of [Ni(H-COD)(hfacac)] in $[C_4mim]Cl/[AlCl_3]$ (70 000 h^{-1}) compared with toluene (770 h^{-1}) was attributed to activation of the nickel complex by the Lewis acid ionic liquid catalyst support to form a much more active catalyst. Moreover, considering the ionic liquid system is mass transport limited this is likely to be a minimum in the intrinsic activity difference of the nickel catalyst in ionic liquid *versus* toluene. TOFs as high as 530 000 h^{-1} were obtained by lowering the catalyst concentration and replacing [Ni(H-COD)(hfacac)] with $[Ni(acac)_2]$ or $[Ni(hfacac)_2]$ and using $AlEtCl_2$ as activator (Table 3.5.1).[530]

A Brønsted acidity scale for strong acid–ionic liquid mixtures of the type $[C_4mim][X]$–$[HNTf_2]$ based on the Hammett acidity function established that acidity decreased in the order $[SbF_6]^- > [PF_6]^- > [BF_4]^- > [NTf_2]^- > [OTf]^-$ and that acidity does not vary with the cation. This scale compares favourably with that based on conversions obtained for the acid-catalysed dimerisation of isobutene, with the exception of $[NTf_2]^-$, as isobutene conversions decreased in the order $[NTf_2]^- \sim [SbF_6]^- > [PF_6]^- > [BF_4]^- > [OTf]^- > [MeOSO_3]^-$. However, the most acidic mixtures $[C_4mim][SbF_6]$–$HNTf_2$, $[C_4mim][PF_6]$–$HNTf_2$ and $[C_4mim][BF_4]$–$HNTf_2$ were not the most efficient as this was achieved with $[C_4mim][OTf]$–$HNTf_2$ (54% conversion, 87% C8-selectivity) and was based on a compromise between conversion and selectivity because even though high acidity leads to good conversions under these conditions the carbocation is reluctant to lose a proton and parallel reactions lead to oligomers and lower selectivity; conversely, insufficient acidity leads to poor conversions.[531]

High selectivity for the biphasic dimerisation of ethylene in $[C_4mim][AlCl_3]$–toluene has been obtained using bis(imino)pyridinecobalt(II) dichloride complexes **161** activated with MAO. The optimum catalyst ($R^1 = CF_3$, $R^2 = F$, $R^3 = H$) gave a TOF of 15 300 h^{-1} and a C4-selectivity of 95.6% with 80.1% selectivity for 1-butene at 30 °C under 10 bar ethylene in the presence of 600 equivalents of MAO. The TOF increased to 17 000 h^{-1} at 50 °C but at the expense of the C4-selectivity which dropped to 89.4% with 67.1% 1-butene while at lower temperature (10 °C) the activity dropped to 6200 h^{-1} with only a marginal increase in C4-selectivity to 97.2%. Activity increased with increasing Co/MAO ratio and reached a maximum at 600 while the C4-selectivity varied only slightly. The authors claimed that this biphasic system presented the best balance of activity and selectivity for the formation of α-olefins by a Zeiger–Natta-type oligomerisation catalyst.[532]

3.5.4.2 Oligomerisation of 1-Olefins

The low nucleophilicity of ionic liquids and their suitability as solvents for ionic complexes renders them ideal for use in liquid–liquid biphasic oligomerisation of ethylene. Using [(2-methylallyl)Ni{$PPh_2CH_2PPh_2(O)$}][SbF_6] (**162**) as precatalyst, high TOFs (25 425 h^{-1}) and excellent selectivity (93.2%) for linear 1-alkenes were obtained in anhydrous chloride-free [C_4mim][PF_6] under 50 bar C_2H_4 at 25 °C; TOFs decreased and the oligomer distribution broadened from $\alpha = 0.16$ in [C_4mim][PF_6] to $\alpha = 0.24$ in [C_{10}mim][PF_6] as the length of the alkyl chain on the imidazolium cation increased. This decrease in activity was due to the high solubility of isomerisation by-products that are known to reduce catalyst activity, while the change in distribution was associated with the higher solubility of ethylene in ionic liquids with longer alkyl chains as well as the higher solubility of the growing Ni-alkyl complexes and oligomers; each of these factors favours chain growth or reinsertion of oligomers and therefore shifts the oligomer distribution. The products separated as a clear colourless layer and the catalyst mixture recycled with only a minor change in selectivity but with lower activity, probably due to deactivation by trace amounts of water. For comparison the catalyst was essentially inactive in both butane-1,4-diol, the solvent used in the biphasic SHOP process, and was much less active in dichloromethane (TOF = 1852 h^{-1}). The high activity in [C_4mim][PF_6] was explained by a combination of the weakly coordinating character of the ionic liquid and efficient extraction of the product and side-products from the catalyst layer as a high concentration of internal olefins is known to deactivate the catalyst by formation of stable Ni-olefin complexes. This biphasic system presents a number of additional advantages over the use of dichloromethane including a much shorter oligomer distribution as well as the formation of less internal alkenes. The narrower oligomer distribution was attributed to the low solubility of ethylene in the ionic liquid which limited its availability for insertion relative to β-hydride elimination while the high 1-alkene content was due to poor solubility of the oligomeric products in the ionic liquid–catalyst; this resulted in efficient phase separation of the product which prevented consecutive isomerisations at the nickel compared with the monophasic reaction in dichloromethane.[533a,b]

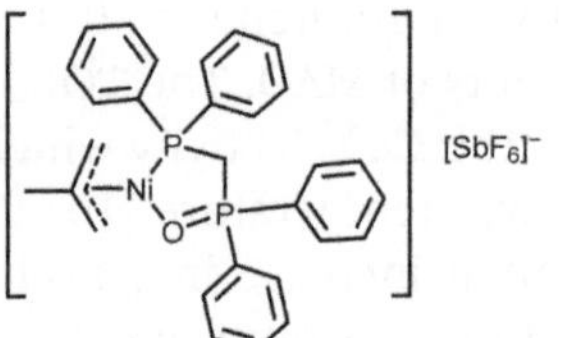

P
Ni
O=P
[SbF_6]$^-$

162

Ionic liquid-immobilised nickel complexes generated either by alkylation of the corresponding [(diimine)$NiCl_2$] or by protonation of a Ni(0) precursor are highly efficient catalysts for the biphasic oligomerisation or polymerisation of ethylene, depending on the bulkiness of the diimine.[39] Significantly higher TOFs are obtained in chloroaluminate-based ionic liquids than in toluene, which most likely reflects the higher electrophilicity of the nickel catalyst resulting from the weakly coordinating nature of the aluminate anion. High activity (78 000 g C_2H_4 g^{-1} Ni h^{-1}) and selectivity for C4 (73%) and C6 (22%) olefins was achieved in non-acidic [C_4mim]Cl/$AlCl_3$ (1 : 1) but the products were dominated by internal olefins due to facile isomerisation of the double bond. In contrast, the TOF for ethylene oligomerisation in non-chloroaluminate $[NTf_2]^-$ and $[SbF_6]^-$-based ionic liquids using the same Ni(II) precursors activated with MAO are poor compared to those in toluene. The inhibiting effect of [C_4mim][NTf_2] manifests itself even when only one equivalent of ionic liquid relative to nickel was added to a toluene solution as the activity dropped from 17 000 g C_2H_4 g^{-1} Ni h^{-1} to 12 700 g C_2H_4 g^{-1} Ni h^{-1}. A further decrease to 850 g C_2H_4 g^{-1} Ni h^{-1} in the presence 10 equivalents indicated that this inhibition/poisoning maybe associated with coordination of the $[NTf_2]^-$ anion to the nickel (section 3.1.4). The activity of ligandless nickel catalysts generated *in situ* by activation of [Ni(COD)$_2$] with [$HOEt_2$]-[BAr_f] was markedly higher in [C_4mim][X] than in toluene and for a series of $[C_4mim]^+$-based ionic liquids activity varied with the anion in the order $[SbF_6]^- > [NTf_2]^- > [PF_6]^- > [OTf]^-$, but the product distribution was largely unaffected in this series. Even in the absence of acid a mixture of [C_4mim][NTf_2] and [Ni(COD)$_2$] consumed ethylene and generated butenes and identification of the C2-ethyl substituted imidazolium cation $[C_4C_2{}^2mim]^+$ after catalysis was taken as evidence for the formation of an intermediate hydrido-carbene *via* oxidative addition of $[C_4mim]^+$ as described in Section 3.1.4; an identical pathway was proposed earlier by Clement and Cavell.[534]

The efficiency of biphasic ethylene polymerisation systems based on [$C_n$mim][$AlCl_4$] ($n = 2$, 4, 6, 8) and hexane using [Cp_2TiCl_2] as precatalyst depends on the aluminium co-catalyst and those activated by $AlEtCl_2$ gave markedly higher yields of polyethylene (30–280 kg PE mol Ti^{-1}) than those generated with $AlEt_2Cl$ (20–30 kg PE mol Ti^{-1}), regardless of the ionic liquid. A survey of ionic liquids established that an optimum activity of 289 kg PE mol Ti^{-1} was obtained in [C_8mim][$AlCl_4$] with an activator/catalyst ratio as low as 167, which was attributed to the lower density of [C_8mim][$AlCl_4$] allowing more efficient mass transfer. The polymer/hexane phase was easily separated from the ionic liquid–catalyst mixture uncontaminated by either catalyst or activator. Polymer isolated from [C_8mim][$AlCl_4$] had the highest degree of crystallinity and bulk density with a molecular weight range from 60–160 000 g mol^{-1}. Additional advantages of this biphasic polymerisation system include high catalyst stability, straightforward catalyst recycling, elimination of reactor fouling, easy separation of the product and a low activator/catalyst ratio.[535a,b]

The biphasic oligomerisation of ethylene in chloroaluminate-$[C_4mim]^+$-based ionic liquids has also been catalysed by [$Fe(MeCN)_6$][BF_4]$_2$,

$[Co(MeCN)_6][BF_4]_2$ and $[Ni(MeCN)_4][BF_4]_2$ using either MAO, $AlEt_3$, $AlEt_2Cl$ or $AlEtCl_2$ as co-catalyst and their performance as a function of activator and temperature compared; the most active catalysts are generated using $AlEtCl_2$ or $AlEt_2Cl$ as co-catalyst whereas those generated with MAO or $AlEt_3$ are much less active. The iron and cobalt systems are less active than their nickel counterpart by an order of magnitude but markedly more selective for dimerisation (C4-selectivity) and with very high α-selectivity. For example, the optimum TOFs of 19 500 h^{-1} and 33 200 h^{-1} for the iron and cobalt systems, respectively, were obtained at 50 °C whereas the nickel catalyst reached a TOF of 215 000 h^{-1} at 10 °C; the corresponding C4-selectivity for these iron and cobalt systems was 79.0% and 66.7%, respectively, whereas the nickel catalyst only reached 42.6%. Moreover, C4-selectivities as high as 100% and 98.9% could be achieved with the iron and cobalt systems, respectively, but this was at the expense of activity as reactions had to be conducted at low temperature.[536]

Imidazolium-tagged bis(salicylaldimine)Ni(II) complexes (**163**, R = H, *t*-Bu) form highly active catalysts for the biphasic oligomerisation of ethylene in either $[C_4mim][AlCl_4]$–toluene or $[C_4mim][AlCl_4]$–heptane when activated with $AlEt_2Cl$. Activities reached 41 800 h^{-1} with high C4–C8 selectivity in $[C_4mim][AlCl_4]$–solvent at 25 °C under 0.5 atm ethylene using 160 equivalents of aluminium activator; for comparison both catalysts were essentially inactive in neat *n*-heptane and toluene. The higher C8 content obtained in $[C_4mim][AlCl_4]$–heptane compared with $[C_4mim][AlCl_4]$–toluene indicated that chain transfer occurs more rapidly in toluene. Both catalysts recycled efficiently in $[C_4mim][AlCl_4]$–heptane and retained >97% of their TOF over three runs with no significant change in product distribution; for comparison the corresponding unmodified catalyst lost 58% of its activity after two runs accompanied by a significant increase in C4 and C6 products.[537] The biphasic oligomerisation of ethylene in $[C_4mim]Cl/AlCl_3$ (1 : 1.2) catalysed by $[(PPh_3)_2NiX_2]$ ($X = Cl^-, Br^-$) activated with $AlEt_2Cl$ gave slightly lower TOFs than the bis(salicylaldimine)Ni(II)-based catalyst but with high C4–C6-selectivity and a low α-olefin content. A survey of the reaction parameters identified a reaction temperature of 20 °C, an Et_2AlCl/Ni ratio of 400 and 0.5 atm ethylene as the optimum conditions which gave TOFs of 26 000 h^{-1} and 19 800 h^{-1} in $[C_4mim][AlCl_4]$–toluene and $[C_4mim][AlCl_4]$–heptane, respectively; for comparison much lower TOFs were obtained under homogeneous conditions in toluene (13 000 h^{-1}) and hexane (9000 h^{-1}). Catalyst recycle studies showed that $[(PPh_3)_2NiCl_2]/AlEt_2Cl$ immobilised in $[C_4mim][AlCl_4]$–heptane could be reused with only a minor drop in activity and no significant change in product distribution over three runs.[538]

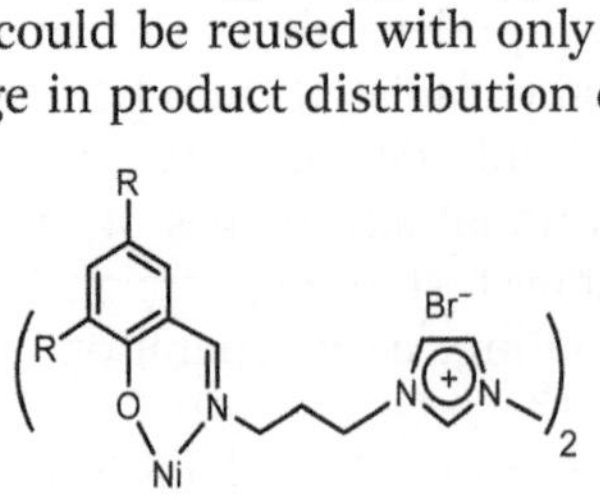

163

3.5.4.3 Dimerisation and Hydrodimerisation of Dienes

The linear selective dimerisation of 1,3-butadiene is a technologically important process as the product is used as a co-monomer in α-olefin-based polymers, plasticizers, adhesives and fragrances and as a building block for synthesis. A number of nickel and palladium complexes catalyse this transformation under homogeneous conditions but as the octatriene rapidly polymerises in the presence of air conventional separation and purification process are not practical. In this regard, palladium complexes immobilised in $[C_4mim][BF_4]$ arc highly selective for the dimerisation of 1,3-butadiene to afford 1,3,6-octatriene [Scheme 3.5.9(a)] under biphasic conditions which allowed the product to be isolated by decantation and the recovered ionic liquid phase to be reused with no significant change in activity or selectivity. The same TOFs were obtained with a range of catalyst precursors including $Pd(OAc)_2$, $PdCl_2$, $Pd(acac)_2$, $PdCl_2(MeCN)_2$ and Pd(maleic anhydride)$(PPh_3)_2$ and activities matched those obtained under homogeneous conditions.[539] In contrast, the biphasic cyclodimerisation of 1,3-butadiene in $[C_4mim][X]$ ($X = BF_4^-$, PF_6^-) catalysed by iron complexes generated by reduction of $[Fe(NO)_2Cl]_2$ with metallic zinc is exclusively selective for 4-vinyl-1-cyclohexene [Scheme 3.5.9(b)]. Reactions conducted at 30–50 °C gave 4-vinyl-1-cyclohexene as the sole product with a TOF up to 1404 h^{-1}, which is a marked improvement on that of 253 h^{-1} obtained under homogeneous conditions in toluene. The addition of phosphines to the catalyst mixture had a detrimental effect on activity but not the selectivity. The same ionic liquid–catalyst mixture also catalysed the cyclodimerisation of isoprene to afford 2,4-dimethyl-4-vinylcyclohexene and 1,4-dimethyl-4-vinylcyclohexene as the dominant adducts.[540]

The hydrodimerisation of buta-1,3-diene in ionic liquid–catalyst mixtures based on $[C_4mim][BF_4]$ and $[(\eta^3\text{-}C_4H_7)Pd(\mu\text{-}Cl)]_2$, $[(\eta^3\text{-}C_4H_7)Pd(1,5\text{-cycloocta-diene})][BF_4]$ or $Pd(OAc)_2$ in the presence of one equivalent of water at 70 °C occurs with high selectivity for telomerisation to give 90% octa-2,7-dien-1-ol and 10% 1,3,6-octatriene [Equation (3.5.10)] with a TOF of 12–15 h^{-1} which improved to 161 h^{-1} under 5 atm CO_2, however, catalyst instability led to the deposition of metallic palladium. The use of $[C_4mim]_2[PdCl_4]$ as precatalyst suppressed the formation of metallic palladium and gave high selectivity for telomerisation with a TOF of 118 h^{-1} even in the absence of CO_2; the TOF increased to 204 h^{-1} under 5 atm CO_2 but at the expense of selectivity which

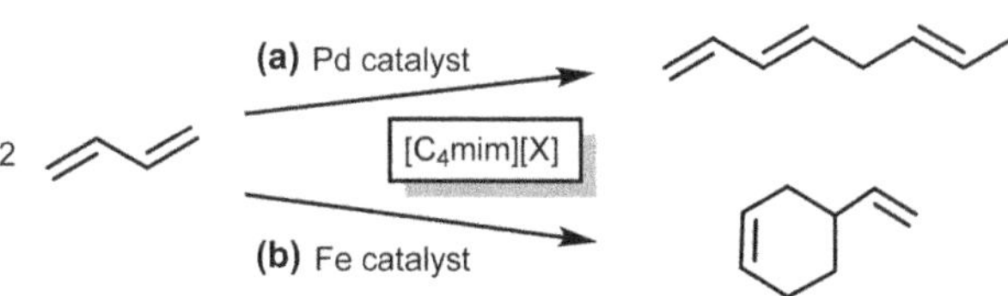

Scheme 3.5.9 Catalyst selective dependent dimerisation of 1,3-butadiene in $[C_4mim][X]$.

dropped from 94% to 84%. Interestingly, *trans*-[(*N*-methylimidazole)$_2$PdCl$_2$], isolated from the ionic liquid–catalyst mixture after reaction as well as from a solution of [C$_4$mim]$_2$[PdCl$_4$] in [C$_4$mim][BF$_4$] after treatment with water gave a TOF and selectivity similar to that obtained under the conditions of catalysis. The imidazole complex was suggested to result from successive processes involving oxidative addition of the *N*-Bu bond in the [C$_4$mim]$^+$ cation to afford a Pd(IV) species which undergoes β-elimination followed by reductive elimination.[541]

2 (butadiene) —Pd catalyst / [C$_4$mim][BF$_4$], H$_2$O→ (octadienol) OH + (octatriene)
telomerisation dimerisation

(3.5.10)

3.5.4.4 Dimerisation of Functional Monomers

A significant enhancement in the rate of the palladium-catalysed dimerisation of methylacrylate (MA) has been achieved in [C$_4$mim][BF$_4$] compared with reaction in neat methylacrylate. A catalyst mixture composed of 0.02 mol% Pd(acac)$_2$, 0.2 mol% [HPBu$_3$][BF$_4$] and [Et$_2$OH][BF$_4$] gave TONs of 2986 and 1024 after 24 h at 80 °C in [C$_4$mim][BF$_4$] and neat MA, respectively. Having dismissed the formation of a Pd-carbene on the basis that such complexes are inactive for dimerisation, the authors speculated that the enhancement in rate was due to the ionic medium lowering the activation barrier of the rate determining step by stabilising the cationic transition state. Lifetime and poisoning studies using dimethyl adipate as a model product indicated that conversions were limited by product inhibition, however, attempts to overcome this limitation by using a biphasic system based on [C$_4$mim][BF$_4$] and toluene to remove the product led to a dramatic decrease in activity upon recycle as the catalyst leached into the organic layer. More efficient recycling with no loss in activity was achieved using the ammonium-tagged phosphine [Bu$_2$PCH$_2$CH$_2$NHMe$_2$][BF$_4$] which was proposed to stabilise the catalyst by acting as a hemilabile ligand while also facilitating reoxidation of Pd(0) intermediates. The concept of a biphasic continuous flow process for the palladium-catalysed dimerisation of MA was demonstrated with Pd(OAc)$_2$/[Bu$_2$PCH$_2$CH$_2$NHMe$_2$][BF$_4$] which gave steady activity over 10 h and a TON of ~4000 after 50 h; for comparison its Bu$_3$P counterpart gave a maximum TON of 500.[542] A subsequent survey of mono- and bidentate phosphines in the palladium-catalysed dimerisation of MA found that for monodentate phosphines catalyst activity increased with increasing basicity and that bulky phosphines lowered the linear selectivity and the reaction rates while hemilabile P,N donors stabilised the catalyst. The highest activity (TOF 161 h^{-1}), linear dimer selectivity (98.0%) and long term stability was obtained with the hemilabile 1-dibutylphosphino-2-dimethylaminoethane. Although both palladium-hydride and oxidative coupling mechanisms were discussed with respect to ligand effects, no definitive evidence was provided to distinguish them but the former is perhaps

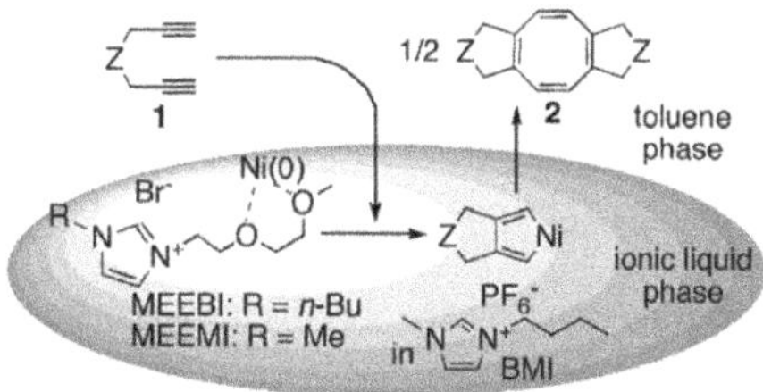

Figure 3.5.5 Dimerisation of 1,6-diynes in a biphasic IL–toluene system. Reproduced from reference 545.

intuitively the most likely.[543a,b] Interestingly, the same catalyst mixture of Pd(acac)$_2$–[HPBu$_3$][BF$_4$]–[Et$_2$OH][BF$_4$] gave a higher TOF for the tail-to-tail dimerisation of MA in protonated imidazolium-based ionic liquids demonstrating a dual use as solvent and a proton reservoir in proton- and metal-assisted catalytic reactions. The use of this mixture to catalyse the dimerisation of MA in [C$_4$mim][BF$_4$] and [HC$_4$im][BF$_4$] occurs under monophasic conditions but the latter is markedly more active with a TOF of 220 h^{-1} compared with only 100 h^{-1} for the former. In contrast, the TOF dropped to 35 h^{-1} for reactions conducted in [Hmim][BF$_4$] as the system was biphasic and under mass transfer control.[544a,b]

Highly selective dimerisation of 1,5-diynes to the corresponding 1,3,5,7-cyclooctatetraene (COT) has been achieved by conducting the reaction under biphasic conditions in [C$_4$mim][PF$_6$]–toluene with catalyst generated from NiBr$_2$.6H$_2$O/Zn and 2-(2-(2-methoxy-ethoxy)-ethyl-1-methyl-1H-imidazol-3-ium bromide as ligand. In contrast, reactions in conventional organic solvents conducted under ligand-free conditions or in the presence of 2-aryliminomethylpyridine generated large amounts of polymer and [2 + 2 + 2] cycloadducts as well as COT. The biphasic nature of this system was proposed to be responsible for the high selectivity as the catalyst is isolated and immobilised in the ionic liquid phase which promotes interaction between two intermediate nickelacycles to generate the COT product as shown in Figure 3.5.5.[545]

3.6 Biomass Transformations and Catalysis with Transition Metal Nanoparticles

3.6.1 Biomass Transformations

Diminishing supplies of fossil fuels and the rising demand for energy is placing increasing pressure on identifying alternative renewable sources of transport fuels and platform compounds to produce intermediates and speciality chemicals and within this technologically challenging arena ionic liquids are emerging as promising solvents and/or catalysts for the conversion of biomass-derived substrates. A recent review entitled *Ionic Liquids for Biofuel Production; Opportunities and Challenges*, presents an up-to-date account of the use of ionic liquids in biofuel production, including the applications and main factors affecting their use in the pre-treatment, dissolution and hydrolysis of lignocellulosic biomass to produce biofuels as well as their use in the oil transesterification process for biodiesel production.[546]

3.6.1.1 Transformations of Biomass-Derived Substrates

Carbohydrate-derived substrates are receiving considerable interest as a sustainable feedstock for the production of chemicals and transportation fuels. In this regard, transformations of hydroxymethylfurfural (HMF) have been widely investigated as a pathway to transportation fuels as it is derived from hexose by dehydration and can be further modified either through aldol chemistry to afford C9-products or ring-opened to levulinic acid (and formic acid) which can be further diversified though γ-valerolactone or pentenoic acid to afford C9-alkanes and C9–C18-alkenes. To this end, Leitner and co-workers have shown that the model biomass-derived substrate furfuralacetone (FFA) can be converted into useful intermediates such as 4-(2-tetrahydrofuryl)-2-butanol (THFA) and subsequently into platform chemicals such as 2-butyltetrahydrofuran (BTHF) and 1-octanol (Scheme 3.6.1), the former is a potential fuel additive while the latter is used for the production of detergents and surfactants, in perfumery and as a precursor to 1-octene, an important co-monomer for polyethylene. Ionic liquid-stabilised nanoparticles have proven to be particularly effective catalysts for the highly selective deep hydrogenation of FFA to afford 4-(2-tetrahydrofuryl)-2-butanol (THFA) with Ru@[C_{12}mim][NTf_2] (94.2%) and Ru@[C_2mim][NTf_2] (89.4%) at 120 °C, under 120 bar of hydrogen. Ruthenium nanoparticles stabilised by other ionic liquid anion and/or cation combinations gave vastly different selectivity patterns in terms of partial hydrogenation.[547] Further hydrogenative transformations of THFA involving dehydration–hydrogenation to BTHF and subsequent hydrogenative ring opening to 1-octanol have been catalysed by a multi-functional system based on Brønsted acid ionic liquid-stabilised RuNPs in [C_4mim][NTf_2]. The highest selectivity for hydrogenation of THFA to BTHF (74.4%) and deoxygenative ring opening hydrogenation of THFA to C8-alcohol products (76.0%) was achieved with ruthenium nanoparticles stabilised by the Brønsted acid [$SO_3HC_4N_{4,4,4}$][NTf_2] (Ru@[$SO_3HC_4N_{4,4,4}$][NTf_2]), although an

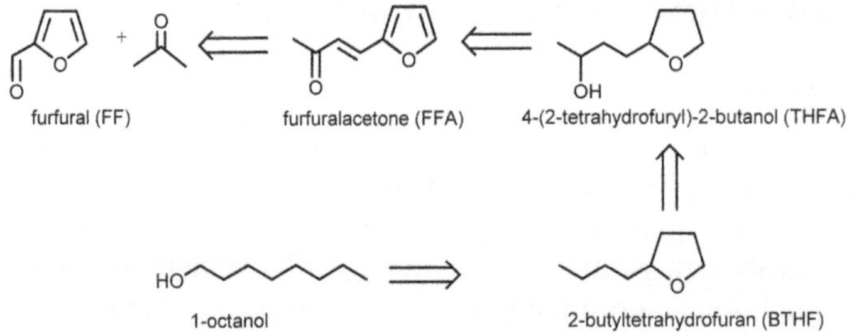

Scheme 3.6.1 Retrosynthetic analysis for a pathway to 1-octanol using platform chemicals derived from lignocellulosic feedstock.

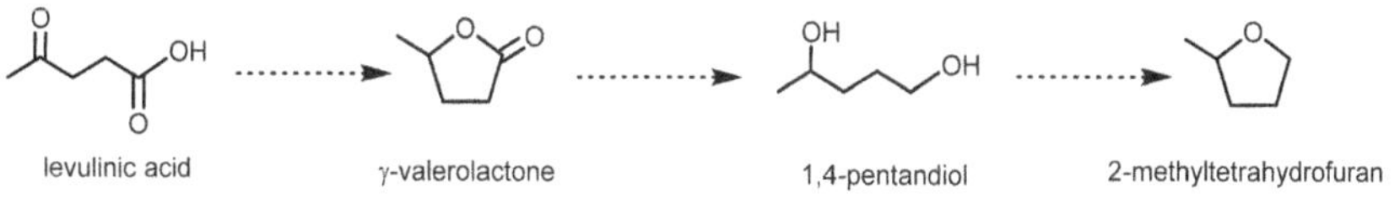

levulinic acid γ-valerolactone 1,4-pentandiol 2-methyltetrahydrofuran

Scheme 3.6.2 Reaction sequence for the selective conversion of levulinic acid (LA) into γ-valerolactone (GVL), 1,4-pentanediol (1,4-PDO), and 2-methyltetrahydrofuran (2-MTHF).

even higher selectivity of 93.0% for the latter was obtained with Ru/C using the BAIL $[SO_3HC_4N_{4,4,4}][NTf_2]$ as additive. A one-pot two-step conversion of FFA into C8-products (1-octanol and dioctyl ether) was developed by introducing $[SO_3HC_4C_4im][NTf_2]$ and $[C_2mim][NTf_2]$ into the reaction vessel after the hydrogenation of FFA to THFA.[548]

Levulinic acid (LA) is a biomass-derived platform chemical which is accessible from wood-based feedstock and can be converted into a number of potentially useful building blocks such as γ-valerolactone (GVL), 1,4-pentanediol (1,4-PDO), 2-methyltetrahydrofuran (2-MTHF), 3-methyltetrahydrofuran (3-MTHF) and 1- or 2-pentanol (PAO) (Scheme 3.6.2). To this end, the ruthenium-catalysed hydrogenation of LA affords a variety of products, the distribution of which depends on the nature of the phosphine, the acid/ionic liquid additive and the temperature. A multifunctional system composed of $Ru(acac)_3$, triphos, $[NH_4][PF_6]$, and the Brønsted acidic ionic liquid $[SO_3HC_4C_4im][p\text{-TSA}]$ gave excellent control over the hydrogenation and dehydration steps to afford 2-MTHF in 92% yield, which is a marked improvement on that of 59% obtained under the same conditions in the absence of ionic liquid. As sequential hydrogenation and dehydration are often key steps in the transformation of biomass such multifunctional catalysts could find wider applications in the utilisation of carbohydrate-based feedstocks.[549]

The dehydration and esterification of γ-valerolactone (GVL) into methyl pentenoates (M2P, M3P, M4P) is a potential route to bio-derived nylon monomers, after further elaboration by hydroformylation, hydrocyanation or olefin metathesis. Catalytic distillation of GVL in the presence of bis(imidazolium), bis(ammonium) and bis(pyridinium)-based sulfonic acid-functionalised ionic liquids such as **164–166** at 170 °C gave GVL conversions from 94–98% with high selectivity for M3P (71–88%) while their mono-imidazolium and monopyridinium counterparts gave much lower conversions (29–54%) and selectivities (2–33%). The catalytic distillation generates a dynamic equilibrium between GVL and methyl 4-methoxypentanoate (MMP) in methanol and the latter is the key intermediate that generates M3P by elimination of methanol, while M2P is generated from M3P *via* secondary isomerisation (Scheme 3.6.3).[550a] The alcoholysis of furfuryl alcohol is rapidly and efficiently catalysed by the double Brønsted acid ionic liquid $[\{C_3SO_3H)_2im][HSO_4]$ to give high yields (up to 95%) of ethyl levulinate *via* intermediates 2-ethoxymethyl furan and 4,5,5-triethoxypentan-2-one; moreover, yields of undesired dialkylether dehydration product were

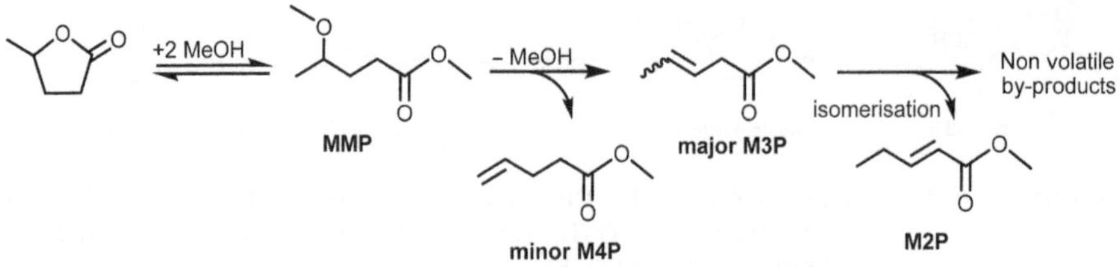

Scheme 3.6.3 Proposed reaction pathway for the acid-catalysed distillation of GVL.

negligible (<1%) and the system recycled efficiently over six runs with no loss in activity of selectivity.[550b]

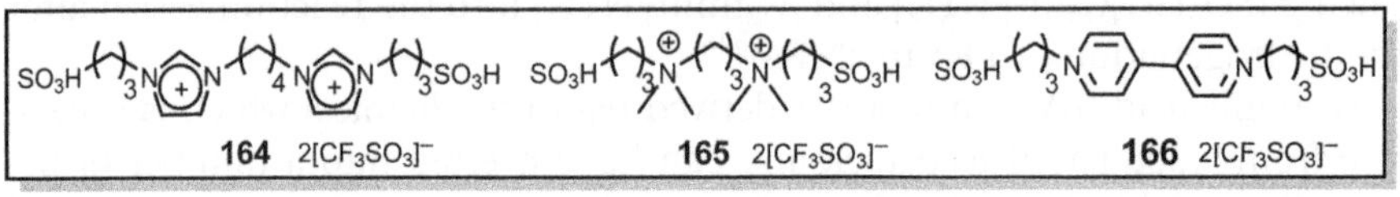

3.6.1.2 *Acid-Catalysed Hydrolysis of Lignocellulosic Biomass*

Lignocellulose is the preferred source of biomass as it is abundant and relatively inexpensive, however, if it is to become a viable renewable carbon source for conversion to biofuels and platform molecules it must first be pre-treated to increase the susceptibility of the cellulose component to depolymerisation. The complete depolymerisation of cellulose to glucose is a critical step in its use as a feedstock to bio-derived products, however, it is notoriously inert or resistant to hydrolysis and the use of enzymes or catalysis under heterogeneous conditions is either not cost effective or cannot be scaled-up. In this regard, ionic liquids provide a potential solution and have been widely studied as solvents for the acid-catalysed depolymerisation of cellulose because of their ability to dissolve up to 25 wt% cellulose.

In the first such report, mineral acids (H_2SO_4, HCl, HNO_3 and H_3PO_4) efficiently catalysed the hydrolysis of a [C_4mim]Cl solution of cellulose at 100 °C to give glucose and total reduced sugar (TRS) yields of 43% and 77%, respectively, after 540 min with an acid/cellulose mass ratio of 0.11; longer reaction times favoured glucose formation while shorter reaction times gave more TRS. A comparative study between the hydrolysis in water and [C_4mim]Cl further emphasised the beneficial influence of ionic liquid as the latter gave 59% TRS after only 3 min at an acid/cellulose mass ratio of 0.92 whereas a reaction time of 1080 min was required to reach 27% TRS in water under otherwise identical conditions (Table 3.6.1). The efficiency of hydrolysis was attributed to the accessibility of the β-glycosidic linkages when the cellulose dissolved to form a homogeneous solution compared with hydrolysis at the surface of cellulose.[551] A subsequent study reported that good yields of TRS were also obtained for the hydrolysis of corn stalk, rice straw, pine wood and bagasse in [C_4mim]Cl within 30–60 min at 100 °C using 7 wt% HCl, whereas much longer reaction times were required to

Table 3.6.1 Reaction conditions and yields of hydrolysis of cellulose in $[C_4mim]Cl$ and water at 100 °C.

Solvent	*Acid/cellulose ratio*	*Time/min*	*Yield$_{glucose}$ (%)*	*Yield$_{TRS}$ (%)*
$[C_4mim]Cl$	5	120	5	7
$[C_4mim]Cl$	0.92	3	36	59
$[C_4mim]Cl$	0.46	42	37	64
$[C_4mim]Cl$	0.11	540	43	77
Water	0.92	1080	13	27

achieve comparable yields under aqueous conditions even with extensive pre-treatment. A survey of hydrolysis as a function of the acid revealed that activity in $[C_4mim]Cl$ decreased in the order $HCl > HNO_3 > H_2SO_4 >$ maleic acid $> H_3PO_4$.[552]

The acid-catalysed hydrolysis of cellobiose in $[C_2mim]Cl$ as a model for lignocellulosic biomass hydrolysis identified a $pK_a < -2$, a water content between 5–10% (w/w), carbohydrate content less than 10% (w/w) and a temperature between 80–150 °C as optimum conditions for selective hydrolysis over sugar decomposition. Hydrolysis of cellulose and hemicellulose under the same reaction conditions as the model study demonstrated that competition between acid hydrolysis of polysaccharides and subsequent decomposition of monosaccharides in ionic liquid can be controlled through a judicious choice of acid strength and that hydrolysis of cellulose in $[C_2mim]Cl$ occurs randomly along the chain in contrast to the end group hydrolysis under aqueous conditions.[553] The trifluoroacetic acid-catalysed depolymerisation of cellulose and hemicellulose from loblolly pine wood occurs more efficiently and under milder conditions in $[C_2mim]Cl$ than required under aqueous conditions. Under optimum conditions, the vast majority of the cellulose fraction of pine wood is converted into water soluble products by heating a $[C_4mim]Cl$ solution at 120 °C for 2 h. Secondary hydrolysis experiments showed that oligosaccharides formed part of these water soluble products, together with monosaccharides and small amounts of HMF and furfural. An additional advantage associated with the use of ionic liquids is that the presence of lignin does not hinder the hydrolysis whereas under aqueous conditions lignin severely limits the accessibility of the glycosidic bonds. The efficiency of the ionic liquid-based hydrolysis depends critically on the water content as a small amount (2%) was required to facilitate hydrolysis, however, above 4% yields drop quite dramatically because the cellulose was less soluble.[554] Solid acid resins with relatively large pores also catalyse the hydrolysis of cellulose dissolved in $[C_4mim]Cl$ without formation of by-product. Amberlyst 15DRY and Amberlyst 35 in $[C_4mim]Cl$ both selectively cleave larger cellulose chains to form cellooligomers which are suitable for further processing by enzyme hydrolysis. Other solid acids including Naffion, γ-Al_2O_3 and zeolites were much less effective.[555]

Near quantitative hydrolysis of cellulosic biomass into water soluble reducing sugars has also been achieved in ionic liquid–water mixtures under relatively mild conditions (<140 °C, 1 atm) in the absence of additional acid. The efficiency of the IL–water mixture for the hydrolysis of cellulosic biomass is due to the increase in K_w by ionic liquids which can be up to three orders of magnitude higher than K_w for pure water under ambient conditions. Under optimum conditions a mixture of [C_2mim]Cl and water (1 : 4) gave 97% yield of TRS in 3 h at 140 °C; longer reaction times resulted in reduced yields due to degradation of the TRS. Cellulose can also been converted directly into HMF in [C_2mim]Cl with 10 mol% $CrCl_2$ as catalyst to give 89% yield after heating at 120 °C for 6 h; this yield implies that not only glucose, but also other water soluble reducing sugars are converted into HMF and that glucose conversion to HMF drives the hydrolysis to completion.[556]

The first Brønsted acid ionic liquids to be used as solvent and catalyst for the dissolution and hydrolysis of cellulose were [SO_3HC_nmim]Cl ($n=3$, 4) and gave TRS in moderate yield (up to 62%) after preheating for 1 h at 70 °C, then adding water and heating for a further 30 min at the same temperature.[557] A dilute aqueous solution of the same Brønsted acid ionic liquids and *p*-toluenesulfonic acid is a more efficient system for the dehydration of cellulose than aqueous sulfuric acid and affords a TRS yield of 28.5% after heating at 170 °C for 3 h[558] while hydrolysis of cellulose in [C_4mim]Cl catalysed by SO_3H-functionalised BAILs gave TRS yields up to 85% together with a minor amount of HMF after heating at 100 °C for 1 h.[559] A survey of the efficiency of BAIL catalysts for the hydrolysis of cellulose identified [SO_3HC_3-NEt_3][HSO_4] to be more efficient than the corresponding imidazolium system as a TRS yield >99% could be obtained after 1 h at 100 °C compared with only 90% for [SO_3HC_4mim][HSO_4] after 1–2.5 h. An increased reaction time or higher catalyst loading led to less TRS due to the accumulation of glucose dehydration products. Hydrolysis rates varied as a function of the ionic liquid cation/anion, acidities and viscosities and the water content as yields dropped dramatically with increasing water content due to formation of an intractable gel which restricted accessibility of the β-1,4-glycosidic bonds.[560]

The phosphotungstate-based ionic liquids [$PW_{12}O_{40}$][cation]$_3$ dissolve up to 30 wt% cellulose biomass after 2 h at 200 °C and catalyse hydrolysis to commodity monosaccharides such as glucose and xylose with a maximum yield of 4% w/w after 150 min beyond which the consumption of monosaccharides becomes faster than their formation.[561]

Lignin is a major component of lignocellulose that can either be used directly for the production of heat and electricity or transformed into aromatic compounds such as phenolic resins and bio-oils and as such is a potentially useful biorenewable feedstock/resource. Brønsted acid ionic liquids catalyse the delignification of sugarcane bagasse in ethanol–water and 100% yield of lignin was obtained with [SO_3HC_4mim][HSO_4] after heating at 200 °C for 30 min with a purity that matched that obtained by physical methods.[562a] The inexpensive and easy-to-synthesise protic ionic liquid [pyrrH][OAc] has been shown to selectively extract the lignin from lignocellulosic biomass. Commercially available model biopolymers: lignin (Kraft lignin-Indulin AT), cellulose (microcrystalline cellulose) and hemicellulose (xylan from beech wood) were first used to demonstrate the principle *i.e.* that the protic ionic liquid dissolved large quantities of lignin but little cellulose. The same ionic liquid was then extracted more than 70% of the lignin together with only a minor amount of polysaccharides and sugars from extractive-free corn stover (EF-CS) after 24 h at 90 °C. The ionic liquid was recovered by distillation under reduced pressure and the polysaccharide and sugar contaminants in the resulting lignin-rich solid residue removed by washing with water.[562b] Welton and co-workers have also conducted a thorough and systematic study of the deconstruction and fractionation of lignocellulosic biomass with 1-butylimidazolium hydrogen sulphate. A cellulose rich pulp and a lignin rich fraction were obtained. A survey of the influence of the solution acidity on the deconstruction of Miscanthus giganteus by varying the sulfuric acid to 1-butylimidazole ratio revealed that increased acidity led to shorter pre-treatment times and resulted in reduced hemicellulose content in the pulp.[562c]

3.6.1.3 Lewis Acid-Catalysed Dehydration of Sugar

In terms of biomass-derived feedstock 5-hydroxymethylfurfural (HMF) is considered to be a key renewable platform molecule. While it is readily available in high yield and selectivity from the Brønsted or Lewis acid-catalysed dehydration of fructose the more desirable feedstock glucose is much more resistant to dehydration. Interestingly, the majority of studies that report effective catalytic transformations of glucose to HMF are based on Lewis acidic chromium catalysts in an ionic liquid. High conversions of glucose to HMF were first obtained at 100 °C with 0.5 wt% $CrCl_2$ in $[C_2mim]Cl$. The efficiency was unique to $CrCl_2$ as a host of other metal halides based on La, Al, Mn, Fe, Cu, Mo, Pd, Pt, Ru, Rh and V typically gave <10% HMF under similar conditions, even though many of these metal halides catalyse the dehydration of fructose in good yield. A preliminary mechanistic model proposed that the active catalyst was $[C_2mim][CrCl_3]$ which isomerised glucose into fructose with high selectivity and then the subsequent dehydration to give HMF (Scheme 3.6.4).[563]

A subsequent *in situ* X-ray adsorption spectroscopic study coupled with DFT calculations provided convincing evidence that while the facile sugar ring opening and closure events involved a single metal centre (consistent with Scheme 3.6.4) that the rate determining H-migration of the open chain form of the carbohydrate was facilitated by transient self-organisation of the

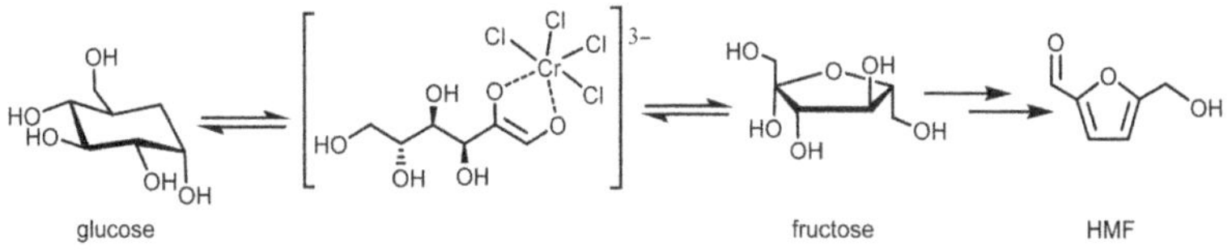

glucose fructose HMF

Scheme 3.6.4 Proposed $CrCl_2$-catalysed isomerization of glucopyranose to fructofuranose followed by dehydration to HMF.

Lewis acid Cr^{2+} centres into a binuclear complex with the open form of glucose. The efficiency of this system appears to be due the dynamic nature of the Lewis acid Cr-complex and the high density of sufficiently basic anions (chloride) in the ionic liquid as the use of strongly coordinating acetate and weakly coordinating $[BF_4]^-$ and $[PF_6]^-$ led to poor conversions. Interestingly, Cr^{III} is much more active than Cr^{II} for the glucose–fructose isomerisation as well as subsequent fructose dehydration and is therefore more selective. Qualitatively glucose isomerization follows a similar pathway for Cr(II) and Cr(III) although the latter has a substantially lower activation barrier (27 kcal mol^{-1} *cf.* 14 kcal mol^{-1}) due to more efficient stabilisation of the anionic reaction intermediate as a result of the stronger Lewis acidity of Cr^{3+}.[564a,b] The efficiency of the $CrCl_x$/[C_4mim]Cl ($x=2$, 3) system was reported to be improved in the presence of bulky *N*-heterocyclic carbenes with fructose and glucose conversions reaching 96% and 81%, respectively, after 6 h at 100 °C, which led the authors to suggest that (NHC)$CrCl_x$ complexes were the active glucose dehydration catalysts.[565] However, control experiments developed to explore the role of NHC complexes in this transformation indicated that the NHC acts as a poison to the chromium catalyst as (i) there was no enhancement in the yield of HMF on addition of additives that would generate an NHC, (ii) preformed discrete well-defined NHC-chromium complexes were poor catalysts and gave low yields of HMF and (iii) quantitative titrations with preformed NHCs conclusively showed that superstoichiometric amounts of NHC quench the catalysis.[566] The use of microwave heating for the $CrCl_3$-catalysed dehydration of glucose to HMF is more efficient and time effective than conventional heating as evidenced by the yields of 71% and 48%, respectively, achieved after 30 s at 140 °C.[567]

Since this initial disclosure various Lewis acid metal halide–ionic liquid combinations have been reported to catalyse the dehydration of glucose including lanthanide halides–[C_8mim]Cl,[568] $GeCl_4$–[C_4mim]Cl,[569] $SnCl_4$–[C_2mim][BF_4],[570] a combination of ruthenium chloride and chromium chloride–[C_2mim]Cl,[571] and LiCl-DMA–[C_2mim]Cl.[572] Tao and co-workers have shown that the use of Brønsted acid ionic liquids such as [SO_3HC_n mim][X] ($n=3$, 4; X = HSO_4^-, $H_2PO_4^-$, $CH_3CO_2^-$, $CF_3SO_3^-$, $CF_3CO_2^-$) in combination with metal salts are more efficient systems for the hydrolysis and dehydration of microcrystalline cellulose into HMF, furfural and levulinic acid than their non-acidic counterparts and that the acidity and structure of the ionic liquid influences activity. The catalytic systems were proposed to operate by promoting rapid conversion of glucose to fructose, accelerating proton transfer and facilitating mutarotation of glucose.[573a–d] The combination of $CuCl_2$ and [SO_3HC_4mim][$MeSO_3$] in [C_2mim][OAc] proved to be a particularly efficient system for the conversion of microcrystalline cellulose to HMF as conversions up to 70% were achieved after heating 160 °C for 3.5 h. Other metal salt–Brønsted acid ionic liquid combinations were less effective. The combination was proposed to form $[CuCl_2(MeSO_3)]^-$ which assists proton transfer during the glucose to fructose isomerisation in a manner similar to that proposed for

$[CrCl_3][C_2mim]$.[574] The glucose to HMF conversion is not restricted to metal halides as Lewis acid aluminium alkyls and alkoxides such as $AlEt_3$, $Al(O^iPr)_3$ and $Al(O^tBu)_3$ are also efficient catalysts for this transformation and reactions conducted in $[C_2mim]Cl$ at 120 °C typically gave yields that matched those obtained with $CrCl_2$. The aluminium salt $[AlClMe(BHT)_2][C_2mim]$ (BHT = 2,6-di-*tert*-butyl-4-methylphenol) was isolated from a solution of $AlMe(BHT)_2$ in $[C_2mim]Cl$ and proposed to be an analogue of the active species in the $CrCl_2$–$[C_2mim]Cl$ system.[575]

Heteropolyacids such as 12-molybdophosphoric (12-MPA) acid and 12-tungstosilicic acid efficiently catalyse the dehydration of glucose of HMF in ionic liquid–acetonitrile mixtures. A glucose conversion of 98% with a HMF selectivity of 99% was obtained with 12-MPA after 3 h reaction at 393 K in a mixture of $[C_2mim]Cl$ or $[C_4mim]Cl$ and acetonitrile. Hydrogenation of the HMF (after removal of the 12-MPA) with Pd/C in $[C_4mim]Cl$–MeCN gave 2,5-dimethylfuran together with a host of intermediates including 5-methylfurfural, 5-methylfurfuryl alcohol, 5-methyltetrahydrofurfuryl alcohol and 2,5-dihydroxymethylfuran that were used to map the reaction pathway.[576] Acidic ion exchange resins also catalyse the dehydration of a range of sugars including pentoses and hexoses to afford 2-furaldehydes in up to 92% with Dowex® 50 W at 100 °C after 3 h.[577]

Ionic liquids act as both solvent and catalyst for the conversion of cellulose with the optimum system of $[C_2mim]Cl$–H_2O giving monosaccharides and HMF in 24.9% and 21% yield, respectively.[578] A combinatorial survey of ionic liquids as solvent and catalyst for the conversion of fructose to HMF revealed that activity varied with the cation and anion and identified triisobutylmethylphosphonium tosylate to be highly active and selective under mild conditions (80–100 °C). However, a Brønsted or Lewis acid catalyst was required for conversion of glucose which also formed unknown by-products or intermediates that were reported but not identified.[579] Brønsted acid ionic liquids also act as solvent and catalyst for efficient and selective dehydration of fructose under biphasic conditions. Under optimum conditions $[SO_3HC_4mim][HSO_4]$ gave 100% conversion and 94.6% HMF at 120 °C with 0.3 g of ionic liquid per gram of fructose; this competes with the yields and selectivities obtained with $H_3PW_{12}O_{40}$, $AlCl_3$ and $SnCl_2$ and is a significant improvement on $[C_4mim][H_2PO_4]$, $[C_4mim]Cl$, CH_3CO_2H, H_2SO_4 and HCl.[580] Glucose dehydration has also been catalysed by the hydroxyethyl-substituted ionic liquid $[HOC_2mim][BF_4]$ to give respectable yields of HMF after 1 h heating at 180 °C.[581]

Brønsted acid ionic liquids in combination with metal salts catalyse the conversion of microcrystalline cellulose to HMF; the most efficient system of $[SO_3HC_4mim][MeSO_3]$ and $CuCl_2$ in $[C_2mim][OAc]$ gave 70% conversions.

3.6.1.4 Transesterification

Biodiesel consisting of C_{16}–C_{18}-based fatty acid methyl esters (FAME) produced from vegetable oil or animal fats by transesterification or esterification with short chain alcohols is a promising alternative diesel fuel which is

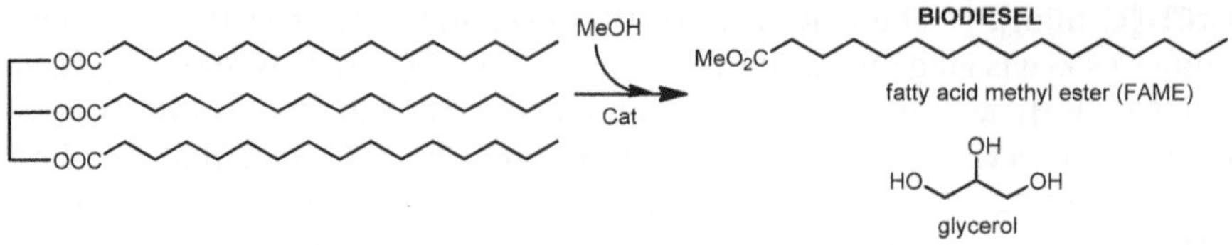

Scheme 3.6.5 Catalytic production of FAME by transesterification of oil (triglycerides) with methanol.

non-toxic, biodegradable and renewable and cleaner burning than petroleum diesel (Scheme 3.6.5). The transesterification of triglycerides is commonly catalysed by either an acid such as H_2SO_4 or a base such as an alkali metal hydroxide or alkoxide. However, both have drawbacks as the former is corrosive and requires high temperatures while the latter can lead to saponification which causes problems during separation.

3.6.1.4.1 Brønsted Acid Ionic Liquid and Lewis Acid Catalysis. Brønsted acid ionic liquids (BAILs) based on a sulfonic acid-tethered to a pyridinium, imidazolium or trialkylammonium group have proven to be potential alternatives to conventional mineral acid catalysts as they catalyse the transesterification of cottonseed oil with methanol at 170 °C. Catalyst efficiency, as measured by conversions after 5 h, decreased in the order pyridinium > *N*-methyl imidazolium > triethylmmonium and under optimum conditions 17 mol% [SO_3HC_4pyr][HSO_4] gave 92% conversion to FAME with a 12 : 1 mole ratio of methanol to oil, which compared favourably with the efficiency of concentrated sulfuric acid. However, the use of concentrated sulfuric acid presents problems as it is highly corrosive, difficult to recover and releases environmentally toxic effluents whereas Brønsted acid ionic liquids are non-corrosive, possess negligible vapour pressure, have good chemical and thermal stability, are highly active, easy to recover and can be modified in a rational manner.[582] Brønsted acid-functionalised ionic liquid-catalysed hydrolysation of soybean isoflavone glycosides such as daidzin, glycitin and genistin gives good conversions to the corresponding isoflavone. Activities varied with the ionic liquid anion the most efficient of which were [SO_3HC_nmim][HSO_4] ($n = 3$, 4) and [SO_3HC_4pyr][HSO_4] whereas much lower activities were obtained in their $[H_2PO_4]^-$ and [*p*-TSA]$^-$-based counterparts. The conversions obtained with $[HSO_4]^-$-based ionic liquids compared favourably with H_2SO_4 under the same conditions but the system recycled poorly with a significant reduction in activity after two runs due to accumulation of impurities.[583] Good yields of FAME have been obtained for the BAIL-catalysed transesterification of waste oils using sulfonic acid-functionalised ionic liquids at a reaction temperature of 170 °C and a metal : oil : catalyst ratio of 2 : 1 : 0.06. The ionic liquid can be recovered by decanting the biodiesel product and reused with no noticeable change in activity; to this end effective recyclability could ultimately be the key driver for developing an efficient biocatalytic transformation.[584] The use of ultrasound irradiation (24 kHz, 200 W) for the BAIL-catalysed transesterification of soybean oil with methanol at 60 °C gave a significant improvement in the yield of FAME (93.2%) compared with that obtained by mechanical stirring (60.8%) in 60 min.[585]

The efficiency of thermoregulated amphiphilic MPEG-based BAILs $[SO_3HC_4(PEG\text{-}N)im][HSO_4]$ (N = 350, 550, 750) as catalysts for the esterification of oleic acid with methanol decreases with increasing length of the polyether such that activity decreases in the order $H_2SO_4 \sim$ MPEG-350 > PEG-550 > PEG-750. The amphiphilicity of the MPEG-350 BAIL enabled reactions to be conducted as microemulsions (pseudo homogeneous) at 68 °C and the product to be separated by decantation as the system phase separated at room temperature.[586]

Ionic liquids grafted onto superhydrophobic mesoporous copolymers of divinylbenzene with either 1-vinylimidazolium **167** or 4-vinylpyridinium **168** are markedly more active for transesterification than homogeneous catalysts. The superhydrophobic polymers are highly active for transesterification of tripalmitin with methanol and give methyl palmitate in 93–>99.9% yield after 16 h reaction at 65 °C whereas $[C_1vim][SO_3CF_3]$ reached 89.1% yield in the same time and heterogeneous catalysts such as Amberlyst 15 and SBA-15-$[C_1vim][SO_3CF_3]$ gave yields of 24.1% and 28.6%, respectively. The efficiency of these ionic liquid grafted superhydrophobic polymers was attributed to their wettability for the reactants which increases their concentration in the catalyst and thus enhances the rate.[587]

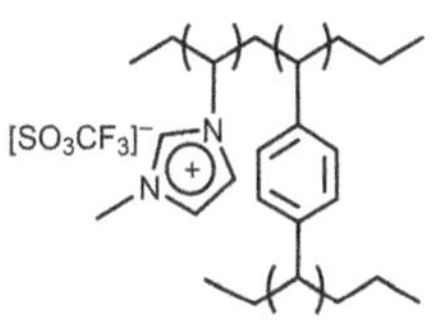

$[SO_3CF_3]^-$ N N (+)

167

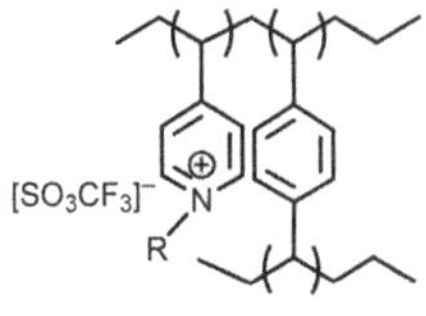

$[SO_3CF_3]^-$ N ⊕ R

168

Lewis acidic chloroaluminate-based ionic liquids also catalyse the transesterification of soybean oil with methanol under milder conditions than those required with Brønsted acid ionic liquids and gave yields that matched those obtained using concentrated H_2SO_4 and H_3PO_4. Under optimum conditions, $[NEt_3H]Cl\text{-}AlCl_3$ ($x(AlCl_3) = 0.7$) gave the methyl ester in 98.5% yield which compares favourably with that of 97.8% obtained with H_2SO_4 and 93.5% with *p*-TSA. The catalyst could be recycled by centrifugation and retained its activity over six runs.[588] Reasonable yields of FAME were obtained under mild conditions from the biphasic transesterification of soybean oil with methanol catalysed by ionic liquid immobilised $[Sn(3\text{-hydroxy-2-methyl-4-pyrone})_2(H_2O)_2]$, however, the system recycled very poorly due to leaching of the catalyst.[589a,b]

3.6.1.4.2 Enzyme Catalysis. Ionic liquids have been reported to be ideal supports for enzyme catalysis as the resulting immobilised catalysts are often highly stable, durable, more active than in traditional solvents and easy to recover.[590] The production of biodiesel *via* enzyme-catalysed transesterification overcomes several of the drawbacks associated with conventional acid and base catalysis but has several disadvantages of its own including the high cost of enzymes, low activity and inhibition of the enzyme by alcohol. The use of ionic liquids as solvent for enzyme-catalysed production of biodiesel has been explored because their miscibility or immiscibility with water or organic solvents can be tuned through a judicious choice of cation and/or anion to aid isolation of the product and recovery/reuse of the catalyst.[591] A number of studies have reported the use of hydrophobic ionic liquids for the enzymatic production of biodiesel by lipases and in all cases FAME production is markedly higher than that obtained in organic solvents and under solvent free conditions and the catalyst was often more stable.

In the first of these 2% w/w lipase (Novozym 435) immobilised in $[C_4mim][PF_6]$ catalysed the methanolysis of sunflower oil to give a 98% yield of FAME after heating a 1 : 8 oil/MeOH mole ratio at 60 °C for 10 h whereas the corresponding $[C_4mim][BF_4]$-immobilised system did not give any FAME. The hydrophobic ionic liquid appears to stabilise/protect the lipase from the methanol as rapid deactivation occurs with >3.0 equivalents of MeOH under solvent free conditions.[592] The transesterification of soybean oil has been catalysed by Novozym 435 (*Candida antarctica* type B lipase) immobilised in a host of ionic liquids and of the 20+ investigated $[C_2mim][OTf]$ proved to be the most efficient by quite some margin giving 80% conversion to FAME after 12 h at room temperature compared with 65.8% in *tert*-butanol. Good conversions were also obtained in $[C_8mim][NTf_2]$ but longer reaction times were required. In fact, FAME production in $[C_2mim][OTf]$ and $[C_8mim][NTf_2]$ was reported to be eight and three times higher, respectively, than under solvent free conditions even though the ionic liquids are highly viscous and operate under biphasic conditions. Conversions were sensitive to the methanol/soybean mole ratio and reached a maximum at 4 : 1 and decreased dramatically when the mole ratio was increased to 8 : 1, which was most likely due to inactivation of the enzyme.[593] *Pseudomona cepacia* lipase immobilised in $[C_4mim][NTf_2]$ is an exceptionally efficient system for the biphasic transesterification of soybean oil with methanol. Conversions improved with added water and the best yield of 96% was obtained after 48 h at room temperature using MeOH–water (70 : 30, v/v); the efficiency was attributed to a shift of the equilibrium by extraction of the formed glycerol into the ionic liquid–MeOH mixture. The biodiesel product was decanted and the ionic liquid–enzyme phase recycled four times with no loss in activity or selectivity.[594] Supercritical carbon dioxide can be used to selectively extract the product (butyloleate) from unreacted substrate, diolin and monoolein during the lipase-catalysed butanolysis of triolein in $[N_{8,8,8,1}][CF_3CO_2]$.[595]

Microwave heating (480 W) gives a marked enhancement in rate for the enzyme (Novozym 435)-catalysed transesterification of soybean oil with methanol in ionic liquid compared with conventional heating. Reactions

Table 3.6.2 Efficiency of enzyme-catalysed transesterification under microwave irradiation and conventional heating.

	Microwave irradiation		*Conventional heating*	
Medium	*Enzyme Activity (μmol min^{-1} g)*	*FAME (%)*	*Enzyme Activity (μmol min^{-1} g)*	*FAME (%)*
$[C_2mim][PF_6]$	441 ± 19	92 ± 3	189 ± 19	73 ± 4
$[C_4mim][PF_6]$	347 ± 15	86 ± 4	150 ± 15	62 ± 3
$[C_2mim][OTf]$	245 ± 11	79 ± 3	118 ± 11	54 ± 3
$[C_8mim][NTf_2]$	190 ± 6	53 ± 3	82 ± 6	37 ± 2
tert-BuOH	242 ± 12	70 ± 3	158 ± 12	62 ± 2
Solvent free	58 ± 4	28 ± 2	42 ± 4	25 ± 1

conducted in ionic liquid are also markedly faster than those performed in *tert*-butanol or under solvent free conditions, using either microwave or conventional heating (Table 3.6.2). For example, under microwave heating the enzyme activity of 441 μmol min^{-1} g in $[C_2mim][NTf_2]$ was about 2.3 time higher than that of 189 μmol min^{-1} g in the same solvent under conventional heating and 1.8 and 7.6 times higher than in *tert*-butanol and the solvent free system, respectively; the corresponding increase in activity under conventional heating was 1.2 and 4.6, respectively. These results suggest a possible synergistic effect between the MW–ionic liquid combination in enzyme-catalysed transesterification. The ionic liquid–enzyme system also recycled with reasonable efficacy over five runs with only an 8% drop in activity, indicating that the enzyme was not deactivated or denatured under microwave heating.[596]

Polyether-functionalised imidazolium and triethylammonium-based ionic liquids $[Me(OCH_2CH_2)_3C_2im][OAc]$, $[Me(OCH_2CH_2)_3NEt_3][OAc]$ and $[Me(OCH_2CH_2)_3NEt_3][HCO_2]$ are exceptionally good solvents for the enzyme-catalysed transesterification of Miglyol oil 812 and soybean oil with methanol. Good conversions (70–88%) were obtained with Novozym 435 immobilised in ionic liquid–methanol mixtures (70 : 30 v/v) after heating at 50 °C for 48 h. The ionic liquid stabilises the enzyme as high activity was maintained even when the methanol concentration was increased to 50 : 50.[597] A more recent survey of 15 ionic liquids identified $[C_{16}mim][NTf_2]$ to be the optimum solvent for the Novozym 435-catalysed methanolysis of triolein. Enzyme activity increased with increasing chain length of the imidazolium alkyl group up to $n = 16$ within the series $[C_nmim][NTf_2]$ and $[C_nmim][PF_6]$ ($n = 10$, 12, 14, 16, 18) which was proposed to be due to a balance between substrate solubility (ionic liquid hydrophobicity) and solvent viscosity, the latter affecting mass transfer resistance. The highest activity of 245 Units of activity per gramme of immobilised enzyme (U g^{-1} IME) obtained at 60 °C was three times higher than that obtained under solvent free conditions. Reactions were typically homogeneous but separated into a triphasic system of FAME (upper), glycerol (middle) and ionic liquid containing enzyme (lower) which facilitated product separation and catalyst recovery. For a given ionic liquid FAME production also increased as the ionic

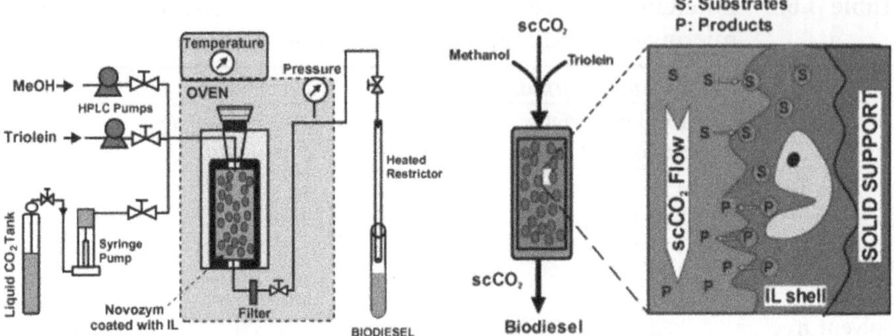

Figure 3.6.1 Schematic of the continuous packed bed reactor containing Novozym 435 coated with IL for biodiesel synthesis by methanolysis of triolein. Reproduced with permission from reference 598b. Copyright 2011 Elsevier.

liquid/methanol ratio increased from 16.6% to 46.7% to 73.7%. The authors related the improvement in efficiency in long chain alkyl containing imidazolium-based ionic liquids to the microenvironment. The reaction mixture shifted from a clear monophasic system for $[C_{18}mim][NTf]_2$ to an increasingly biphasic system as the length of the imidazolium alkyl chain decreased. Enzyme productivity/efficiency also varied dramatically with the anion and increased with increasing hydrophobicity in the order $[NTf_2]^- >> [PF_6]^- > [BF_4]^-$. A biphasic continuous process based on a biocatalytic packed-bed reactor containing immobilised lipase particles coated with $[C_{18}mim][NTf_2]$ and using *sc*CO_2 to transport the substrate to the biocatalyst microenvironment across the ionic liquid layer showed excellent long term operational stability and gave an 82% yield of biodiesel after 12 cycle of continuous operation (Figure 3.6.1).[598a,b]

Whole cell biocatalysts also show excellent stability in ionic liquid during the methanolysis of soybean oil. Good yields of FAME were obtained with the wild-type *Rhizopus oryzae* producing triacylglycerol lipase (w-ROL) immobilised in $[C_n mim][BF_4]$ ($n=2$, 4) at a reaction temperature of 30 °C and a methanol/oil ratio of 4, but yields were capped at 60% due to the 1,3-positional specificity of the lipase. However, much higher conversions were obtained by the combined use of two different whole-cell biocatalysts w-ROL and mono and diacylglycerol lipase from *Aspergillus oryzae* (r-mdlB). The yields of FAME obtained with ionic liquid-immobilised w-ROL were much higher than those under ionic liquid-free conditions and within the series of ionic liquids activity increased in the order $[C_2mim][BF_4] > [C_4mim][BF_4] > [C_2mim][OTf] >>$ ionic liquid free. As the activity of w-ROL in $[C_4mim][BF_4]$ decreased significantly over 72 h and as the robustness and stability of a biocatalyst is critical for use in an industrial setting cross-linking of w-ROL with glutaraldehyde was used to improve its stability under the conditions required for biodiesel production in ionic liquids.[599] Similarly, $[C_4mim][PF_6]$ is also a superior solvent for the *Penicillium expansum* lipase (PEL)-catalysed transesterfication of corn oil than *tert*-butanol. Both solvents were optimised for methanol/oil ratio, temperature, enzyme loading, solvent volume and water content and under these conditions PEL/$[C_4mim][PF_6]$ gave 82% yield of FAME compared with 56% in *tert*-butanol. Negligible conversions were obtained in hydrophilic ionic liquids $[C_nmim][BF_4]$ ($n=4$, 6) which is consistent with a number of earlier reports. A comparison of the efficiency and stability of PEL, Novozym 435 and Lipozyme TLIM in $[C_4mim][PF_6]$ and hexane revealed that Novozym 435 and Lipozyme TLIM respond differently to MeOH in different solvents and showed different resistance to MeOH whereas PEL is tolerant to methanol in both solvents.[600a,b] Ionic liquids have also been used for the production of microalgea-derived biodiesel by the *Penicillium expansum* lipase and Novozym 435-catalysed methanolysis of oils extracted from *Botryococus braunii*, *Chlorella vulgaris* and *Chlorella pyrenoidosa* and under optimum conditions both enzymes gave markedly higher yields in $[C_4mim][PF_6]$ (90.7% and 86.2%) than in *tert*-butanol (48.6% and 44.4%).[601]

Flavour esters can be prepared by direct biocatalytic esterification of an alkyl carboxylic acid with a flavour alcohol (*e.g.* geraniol, citronellol, nerol and isoamyl alcohol) in $[N_{16,1,1,1}][NTf_2]$ as a switchable ionic liquid/solid phase for reaction and product separation.[602] In this concept the esterification was conducted in $[N_{16,1,1,1}][NTf_2]$ at 50 °C under homogeneous conditions using Novozym 435 as the catalyst. The product was subsequently isolated by cooling the reaction mixture to room temperature and performing four consecutive centrifugations at 25 °C, 21 °C, 10 °C and 4 °C to separate the solid ionic liquid-immobilised enzyme so that the product could be decanted (Figure 3.6.2). Enzyme activity decreased in the order citronellol > geraniol > nerol > isoamyl alcohol and the highest yields were obtained in the presence of molecular sieves using a 1 : 3 mole ratio of acid to alcohol and 60% w/w ionic liquid; for comparison much lower yields were obtained under solvent free conditions, even in the presence of molecular sieves. Recycle experiments based on the above protocol showed that the hydrophobic ionic liquid provided highly effective stabilisation and immobilisation of the enzyme which retained its activity over seven runs. To this end, hydrophobic ionic liquids appear to be well-suited to lipase-catalysed esterification and transesterification reactions.[598b,603] The high stability of enzymes in long chain alkyl-containing hydrophobic ionic liquids has been attributed to a protective coating of the enzyme particles maintaining the native conformation.[604] Thus, it could ultimately be possible to tune the hydrophobicity of long chain alkyl-containing ionic liquids for use in a range of enzyme-catalysed transformations to develop green industrial processes.

3.6.1.4.3 Transition Metal Catalysis. The rhodium and ruthenium-catalysed isomerisation of linoleic acid derivatives to their conjugated counterparts (CLAs) has been investigated in ionic liquids in order to identify a system that would efficiently immobilise/retain the metal for

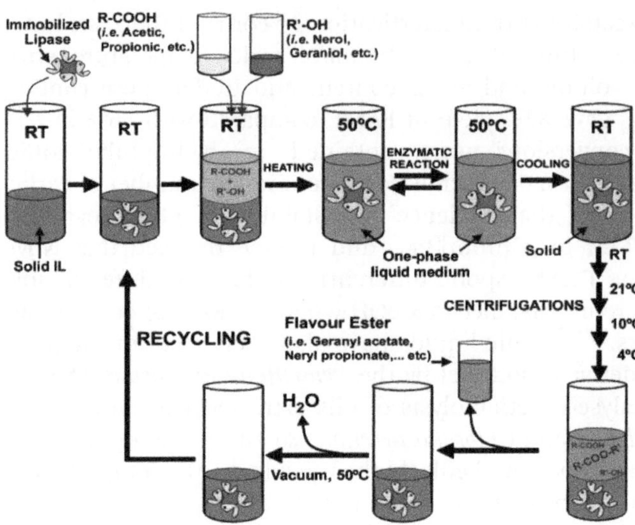

Figure 3.6.2 Schematic of the cyclic protocol for the production of flavour esters by lipase-catalysed direct esterification in switchable ionic liquid/ solid phases and reuse of the enzyme–IL system. Reproduced from reference 604.

recycling and limit contamination of the product for biomedical applications and thus be more cost effective. Biphasic isomerisation of methyl linoleate catalysed by 1 mol% $[RuH(CO)Cl(PPh_3)_3]$ immobilised in $[C_4mim][NTf_2]$ at 80 °C affords a mixture of *E,E*, *E,Z* and *Z,Z*-conjugated linoleic acid methyl esters in up to 80% yield whereas $[RhCl(PPh_3)_3]/SnCl_2$ in $[C_4C_1pyrr][NTf_2]$–EtOH gave complete conversion with a much higher *Z,E*-selectivity (85%) after 24 h at 60 °C. However, the latter system also catalysed transesterification side-reactions (up to 65%), the extent of which varied with the ionic liquid anion. Interestingly, the use of $[N_{4,4,4,4}]Br$ as additive prevented transesterification and gave good yields of conjugated product with high *Z,E*-selectivity; this was attributed to quenching of the Lewis acid SnX_2 by formation of $[SnX_3]^-$. Metal leaching into the product phase could be reduced quite dramatically by combining the rhodium or ruthenium precatalyst with an ionic liquid tagged phosphine.[605]

3.6.2 Catalysis with Transition Metal Nanoparticles

The use of noble metal NPs for catalysis is a rapidly evolving and potentially powerful technology with applications in a range of reactions including reduction and oxidation chemistry as well as carbon–carbon and carbon–heteroatom bond formation. Ionic liquids are unique media for NP catalysis and have proven to be far more than solvents and support materials as they can act as a type of template to control the size, size distribution and morphology of the particles during synthesis and thereby influence activity and selectivity; they also stabilise NPs against aggregation and facilitate separation and recycling. Functionality has been incorporated into ionic liquids (FIL) to provide additional stabilisation, anchor the NPs to a support and further tune and modify their properties while polymer grafted ionic liquids (polyionic liquids) have been shown to enhance catalyst longevity, simplify and facilitate catalyst separation and enabled advanced processes to be engineered. Since ionic liquid-stabilised NPs lie at the interface between homogeneous and heterogeneous catalysis a brief survey of this area will be presented here while a more comprehensive coverage can be found in Chapter 11. A recent review article by one of the world's leading and outstanding authorities in this area provides a perceptive and highly informative up-to-date overview on the preparation, characterisation and applications of NPs in ionic liquids.[17a]

3.6.2.1 Synthesis and Stabilisation

Ionic liquid-stabilised nanoparticles are commonly prepared by either reduction of a metal compound dissolved in an ionic liquid,[606] decomposition of a zero valent organometallic precursor dissolved in an ionic liquid,[607] transfer of preformed NPs into an ionic liquid[608] or by metal sputtering.[609] Homoleptic metal carbonyl complexes have also been used as precursors to nanoparticles because they are readily available in high purity for a wide range of transition metals but their thermal decomposition to NPs typically requires high temperatures and/or long reaction times.[610a,b] Transition metal nanoparticles have also been prepared by microwave irradiation of an ionic liquid solution of a metal carbonyl precursor ($M_x(CO)_y$, M = Cr, Mo, W, Re, Fe, Ru, Os, Co, Rh, Ir; 10 W, 3 min) and compared with those obtained using UV-photolysis (1000 W, 15 min) and conventional heating (180–250 °C, 6–12 h). The use of microwave irradiation has a number of advantages over conventional heating as it is operationally straightforward, energy saving and results in facile and rapid decomposition of the metal carbonyl precursor to afford NP–ionic liquid dispersions in extremely short reaction times.[611]

The morphology, size and size distribution of IL-stabilised NPs depends on several factors such as reaction conditions, the type of metal precursor and reducing agent as well as the nature of the ionic liquid, for example, the volume of polar and/or non-polar domains as well as the presence of functional groups (*vide infra*). Although still rather limited, it is clearly important to develop a detailed understanding of the mechanisms of solvation and stabilisation of NPs as this will be central to developing their use in catalysis and allied applications. In this regard, there is evidence for an electric double layer in which an inner solvation shell of anions surrounding the metal particle interacts with a less-ordered outer shell of cations,[612] while other studies provide compelling evidence for close interactions between imidazolium cations and NPs.[613] In fact, recent molecular dynamic simulations have shown that both cations and anions are in contact with hydrogenation active RuNPs in $[C_4mim][NTf_2]$ and that the interfacial layer is one ion thick with nonpolar groups and side chains directed away from the surface; the imidazolium ring is orientated perpendicular to the surface with the nitrogen bound methyl group close to the metal *i.e.* the H_2 and $H_{4,5}$ hydrogen atoms of the imidazolium ring are not the main sites of interaction.[614] Thus, the active sites in soluble transition metal NP catalysts can be either the entire surface, if the ionic liquid is considered as non- or weakly-coordinating, or a fraction of the surface atoms if the ionic liquid binds or contains a functional group/ligand or the NP can act as a reservoir of soluble active catalytic species. For efficient catalysis the balance between stability and activity must be optimised as strong interactions between the NP surface and the ionic liquid cation and/or anion, a ligand or tethered functional group or other species (e.g. *N*-heterocyclic carbenes, hydrides or an anionic species) will impede loss of M(0) or access of the substrate to the surface and reduce activity.

3.6.2.2 The Heck Reaction

The Heck arylation is a key C–C bond forming reaction between an aryl halide and an olefin and has been widely used for the stereoselective synthesis of substituted olefins. While PdNPs have been implicated, by virtue of either being preformed or generated *in situ*, to act as a soluble source of molecular palladium species it is still not clear whether leaching of the palladium from the NP surface occurs in the form of Pd(0) or as a Pd(II) species after oxidative addition.

3.6.2.2.1 Catalysis with NPs Stabilised by Unfunctionalised Ionic Liquids. Nanoparticles were first implicated to be involved in the phosphine-free palladium-catalysed arylation of styrene with bromobenzene using $PdCl_2(C_6H_5CN)_2$, *N,N*-dimethylglycine as additive and sodium acetate as base. Samples removed after a 1 h induction period consisted of PdNPs with an average size of 1.6 nm. In comparison, the reaction between ethyl acrylate and iodobenzene conducted under Jeffrey conditions ($Pd(OAc)_2$/*n*-Bu_4NBr, NaOAc) started immediately without induction and TEM imaging of samples confirmed the presence of 1.6 nm NPs. UV-vis and ^{13}C NMR spectroscopic studies support a process involving oxidative addition of arylbromide to a surface bound palladium to afford [ArPdI] or $[ArPdX_3]^{2-}$ with concomitant dissolution of the Pd colloid.[615] Palladium nanoparticles generated from an intermediate palladium bis(carbene) precursor have also been proposed to be responsible for the highly efficient ultrasound-promoted Heck reaction between aryl iodides and alkenes/alkynes in ionic liquids using $Pd(OAc)_2$ in $[C_4C_4im][X]$ ($X = Br^-$, BF_4^-) and sodium acetate as base; reactions typically reached near quantitative yields at ambient temperature in short reaction times. For comparison, there was no evidence for reaction under similar conditions in conventional solvents such as DMF and NMP. Ionic liquid-stabilised clusters of well-dispersed PdNPs generated from an intermediate palladium bis(carbene) precursor were identified and proposed to be responsible for the high activity.[200]

An XAFS study unequivocally identified that palladium nanoparticles between 0.8–1.6 nm are present during the Heck reaction between iodobenzene and butylacrylate at 100 °C using Pd(ethanoate)$_2$ in a range of ionic liquids; the size of these NPs depends on the ionic liquid and the presence/absence of PPh_3; those formed in the presence of phosphine are typically slightly smaller. A gradual formation of palladium observed in the XAFS explained the induction period at 50 °C and although this induction period increased in the presence of PPh_3 the resulting catalysts were more stable with respect to palladium black formation and showed better recyclability without incorporation of palladium in the product or loss of catalyst from the reaction mixture.[179]

The PdNP-catalysed Heck reaction between neutral and electron-rich aryl bromides and 1,1′-disubstituted olefins such as α-methyl styrene and *n*-butyl methacrylate in tetrabutylammonium bromide (TBAB) as solvent with tetrabutylammonium acetate (TBAA) as base affords a 3 : 1 mixture in favour of the terminal isomer at low conversions [Equation (3.6.1)]. However, at long reaction times the product distribution changes because the concentration of the minor internal alkene remains constant while the terminal isomer decreases as it converts to the double-arylated product until all the aryl bromide has been consumed, at which point the kinetically favoured terminal product slowly isomerises into the internal olefin. In this system the solvent stabilises the palladium nanoclusters while the TBAA impedes isomerisation of the terminal olefin to the thermodynamically more stable internal product by neutralising the 'active' Pd–H. In contrast, electron-deficient 4-bromoacetophenone reacts with *n*-butyl methacrylate in TBAB to afford the internal olefin as the major product even at low conversions (28 : 72) and almost exclusively after only 15 min (4 : 96). The use of TBAA also overcomes the drawback of catalyst deactivation that results from the build-up of inorganic salts generated during neutralisation of Pd–H with an inorganic base as the by-products are acetic acid and the $[n\text{-}Bu_4N]^+$ cation which sequesters bromide.[616]

R + Ar-X —Pd^0→ Ar R + Ar R + Ar Ar R

(3.6.1)

Palladium nanoparticles generated by reduction of either $Pd(OAc)_2$ or bis{(benzothiazole)carbene}PdI_2 (**169**) in TBAB are also efficient catalysts for the Heck arylation of *trans*-ethyl cinnamate and its derivatives and give good yields of the corresponding (*Z*)- or (*E*)-β,β-diaryl acrylates with excellent stereoselectivity (>99 : 1) and in short reactions times [Equation (3.6.2)]. The use of TBAB as solvent and TBAA as base was crucial to achieving high stereoselectivity and activity as other bases gave mixtures of *E*/*Z* isomers (~59 : 41) while reactions conducted in TBAA as both solvent and base were much slower but retained high stereospecificity. The authors attributed these ionic liquid effects to stabilisation of the unstable 12-electron species ArPdX by interaction with TBAB or TBAA to give the $[ArPdX_3]^{2-}$ ($X = Br^-$, AcO^-) and subsequent ammonium cation-assisted anion dissociation to render the complex more electrophilic for rapid olefin insertion. The tetra-butylammonium ions may also play a role in stabilising the NPs and extending life-time by forming a protective sheath and imposing a Coulombic barrier to collision thereby impeding cluster growth. As described above, rapid neutralisation of the Pd–H species by acetate was critical to preventing isomerisation of the reaction product to achieve the high stereoselectivity.[101] A similar effect has also been reported for PdNPs supported on chitosan which is an efficient catalyst for the Heck reaction of aryl bromides and activated aryl chlorides in TBAB–TBAA; this system exhibited remarkable long term stability and recycled over 11 times with an average conversion of 94% and a total TON of 2954 for the arylation of butyl acrylate with iodobenzene at 100 °C.[40a]

169

CO_2Et + Ar—Br → (Pd catalyst; TBAB, base, 130 °C) Ar, CO_2Et

(3.6.2)

A remarkable increase in catalyst efficacy was achieved for PdNP-catalysed Heck reactions involving aryl bromides and chlorides by increasing the proportion of TBAB in the TBAB–TBAA ionic liquid mixture; poor solvation of the bromide ions was proposed to enhance the activity of Pd(0) by providing the electron density necessary to facilitate rapid oxidative addition. Under these optimised conditions a range of electron-rich and electron-poor aryl chlorides reacted with 1,2-disubstituted olefins to afford the corresponding β,β-diaryl acrylates in high yield at 120 °C and in short reaction times. However, this study did not distinguish between oxidative addition of aryl halide to dissolved Pd(0) species or surface Pd(0).[617]

Even though Pd-clusters are good catalysts, it is extremely challenging to unequivocally establish whether catalysis occurs on the cluster surface *i.e.* catalysis by NPs or by leached palladium species serving as a reservoir of catalytically active species. An *in situ* TEM, UV-vis spectroscopic studies and ICP-AS analysis of the organic phase investigated the role of ionic liquid dispersed PdNPs in the Heck reaction and strongly indicated that the nanoparticles act as a reservoir of soluble Pd(II) by undergoing oxidative addition on the nanoparticle surface to afford an ArPdX species which detaches from the surface and enters the main catalytic cycle; the Pd(0) formed after the β-hydride elimination then either continues in the cycle or

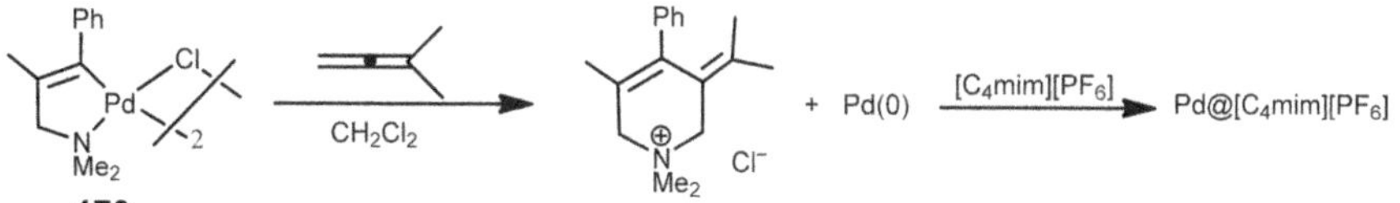

Scheme 3.6.6 Generation of [C_6mim][PF_6]-dispersed ligand-free PdNPs by reaction of N-containing palladacycle **170** with excess dimethylallene.

re-attaches to the nanoparticle reservoir. In this experiment palladium nanoparticles generated by treatment of palladacycle **170** with an excess of allene and suspended in [C_4mim][PF_6] (Scheme 3.6.6) catalysed the Heck reaction between aryl halides and *n*-butyl acrylate to give high conversions after 14 h. Poisoning tests provided strong but not definitive evidence for a homogeneous catalytically active species.[608,618]

Ionic liquids with tertiary aliphatic pendent amines tethered to the imidazolium cation are an effective solvent–base combination for the Heck reaction between iodobenzene and *n*-butyl acrylate and near quantitative conversions were obtained at 100 °C using 1 mol% Pd(OAc) as catalyst; the system also recycled efficiently across five runs with no loss in activity. In comparison, the corresponding pyridyl-based ionic liquids gave either poor or no conversions under similar conditions. The deposition of PdNPs from a refluxing solution of $Pd(OAc)_2$ and [$i$$Pr_2NC_2$mim][$PF_6$] (1 : 8) in either acetone or THF suggested that such species were responsible for the catalysis.[619]

Highly substituted arenes have been prepared in a one-pot base-specific PdNP-catalysed tandem Heck–Suzuki coupling with 4-bromochlorobenzene in molten TBAB. Good yields were obtained using TBAA as the base for the Heck reaction and Cs_2CO_3 for the Suzuki–Miyaura coupling; other combinations of base either gave poor yields or required much longer reaction times. Similar NP–base–IL systems also catalysed double Heck–Heck and Suzuki–Suzuki couplings as well as a triple Heck–Heck–Suzuki one-pot sequential coupling.[620]

3.6.2.2.2 Phase Separation Strategies and Polymer Grafted Ionic Liquids. A thermoregulated ionic liquid–organic biphasic system composed of the PEG-750-based ionic liquid system [$CH_3(OCH_2CH_2)_{10}NEt_3$][$MeSO_3$]–toluene/heptane and IL_{PEG750}-stabilised PdNPs catalyses the highly efficient Heck arylation of methyl acrylate with aryl iodides and activated bromides at 100 °C. At this temperature the IL_{PEG750} and toluene–heptane are miscible and reaction occurs under homogenous conditions but upon cooling phase separation allows the product containing organic layer to be collected while the remaining ionic liquid–catalyst phase can be recycled six times with no noticeable loss in activity. After initial leaching of 0.5 wt% Pd in the first run further leaching was below the minimum detection limit of 0.003 μg mL^{-1}.[621] Hydrophobic fluorous ionic liquid-stabilised PdNPs (Pd@[C_8mim][$CF_3CF_2CF_2CF_3SO_3$]) catalyse the Heck

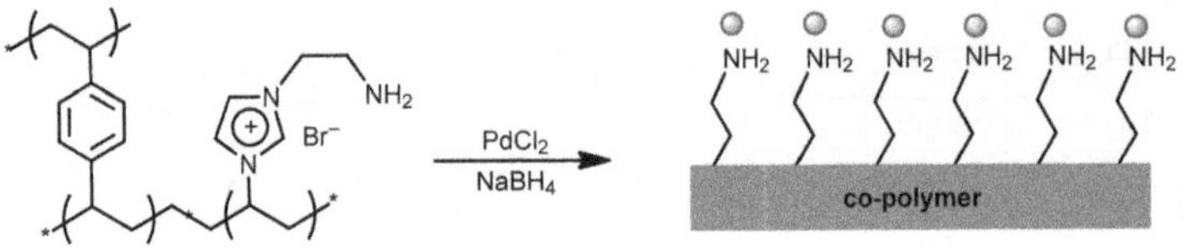

171

PdCl$_2$ / NaBH$_4$

NH$_2$ NH$_2$ NH$_2$ NH$_2$ NH$_2$ NH$_2$

co-polymer

Scheme 3.6.7 Synthesis of 1-aminoethyl-3-vinylimidazolium bromide cross-linked polystyrene immobilised palladium nanoparticles. Adapted from reference 623.

arylation of styrene and methyl acrylate in [C$_8$mim][CF$_3$CF$_2$CF$_2$CF$_3$SO$_3$] at 100 °C. Under optimum conditions good yields were obtained with a range of aryl iodides and bromides in short reaction times while aryl chlorides required much longer to achieve moderate conversions. The system recycled five times with only a minor reduction in activity.[622]

PdNPs immobilised on an ionic liquid-grafted cross-linked polymer **171** (Scheme 3.6.7) catalyse the solvent free Heck reaction between aryl iodides and acrylates to give good yields (>90%) at 120 °C with 0.20 mol% Pd and triethylamine as base; only poor conversions were obtained with inorganic bases due to their low solubility. The catalyst is stable and recycled efficiently with no loss in activity after three runs. The low solubility and high thermal stability of the cross-linked polymer together with a strong interaction between the amino group and the nanoparticles was suggested to be responsible for the high activity and recyclability. A filtration test provided evidence that catalysis occurred *via* palladium leaching and ICP-AES analysis indicated that the dissolved palladium most likely re-deposits back onto the polymer after the reactants had been completely consumed.[623]

3.6.2.2.3 Catalysis with Ligand-Stabilised Nanoparticles. Irregular shaped NPs between 5 and 10 nm in diameter and stabilised by *tris*(3-sodium sulfonatophenyl)phosphine (tppts) were identified in the Heck arylation between styrene and bromobenzene catalysed by 1 mol% PdCl$_2$/tppts in the water soluble ionic liquid [C$_n$mim][*p*-MeC$_6$H$_4$SO$_3$] ($n=1$, 2, 4, 6, 8, 10) at 110 °C; an early stage induction period and mercury poisoning tests also supported catalysis by nanoparticles. In this system catalyst activity increased with increasing length of the imidazolium alkyl chain up to $n=4$ (99.6% yield) and then decreased with a further increase in chain length ($n=10$, 15.3% yield). This trend was attributed to a balance between the solubility of the substrate in the ionic liquid (which increased with increasing n) and the solubility of the palladium-phosphine species (which decreased with increasing hydrophobicity). The optimised catalyst gave good conversions across a range of aryl bromides and recycled efficiently after extraction of the product with diethyl ether at the end of each run. Reactions conducted in conventional ionic liquids based on [PF$_6$]$^-$ and [BF$_4$]$^-$ gave much lower yields which might reflect a possible complementarity between the tppts and the anion of the ionic liquid.[196]

A tandem Heck arylation–hydrogenation sequence between aryl iodides and butanone has been catalysed by PdNPs generated from $Pd(OAc)_2$ or $Pd_2(dba)_3$ in $[C_2mim][MeHPO_3]$, both as a ligand-free system and in the presence of a phosphine stabiliser [Equation (3.6.3)]. The product distribution varied as a function of the palladium precursor and the ligand such that a mixture of **172** and **173** was obtained with preformed PdNPs generated from $Pd_2(dba)_3$/phosphine whereas the Heck product was formed exclusively in the absence of phosphine and PdNPs prepared from $Pd(OAc)_2$ both with and without added phosphine gave near equal mixtures of **172** and **173.** In stark contrast, **173** was the dominant product when PdNPs were generated *in situ* both in the presence and absence of phosphine. The PdNPs were proposed to act as a source of Pd(0) leaching from the surface after oxidative addition of aryl iodide to generate a ArPdI species that inserts alkene and undergoes β-hydride elimination to liberate **172** before regenerating the Pd(0) species by abstraction of HI with the basic ionic liquid (Scheme 3.6.8). In the case of **173**, two possible pathways were considered based on an equilibrium between the Pd-alkyl intermediate **A** and its *O*-bound tautomer **B** with $MeHPO_3$ acting as the reducing agent. The first of these involved either a concerted H-transfer process with an *O*-bound hydrogenphosphonate anion and the *O*-bound enolate or β-hydride elimination from the *O*-bound hydrogenphosphonate followed by reductive elimination from the resulting hydrido *O*-bound enolate. The second pathway involves a similar sequence of events that generates **173** *via* the *C*-bound

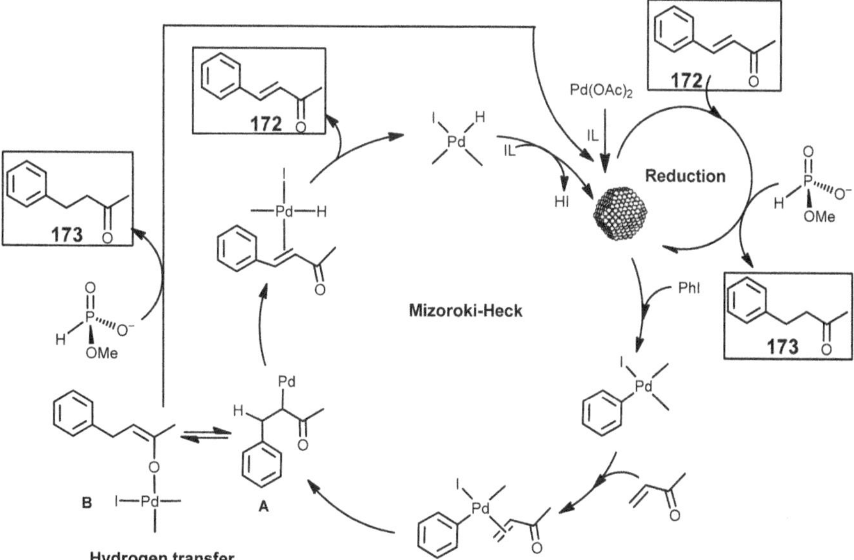

Scheme 3.6.8 Proposed mechanism for the formation of 4-phenylbut-3-en-2-one **172** and 4-phenylbutan-2-one **173** starting from iodobenzene and butenone, catalyed by palladium nanoparticles.

alkyl intermediates **C** and **D** which ultimately liberates the product *via* reductive elimiantion.[624a–c]

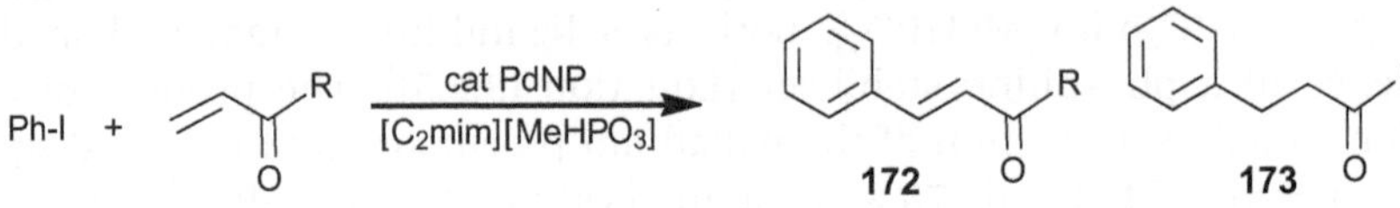

$$\text{Ph-I} + \text{CH}_2\text{=CHC(O)R} \xrightarrow[\text{[C}_2\text{mim][MeHPO}_3\text{]}]{\text{cat PdNP}} \mathbf{172} + \mathbf{173} \quad (3.6.3)$$

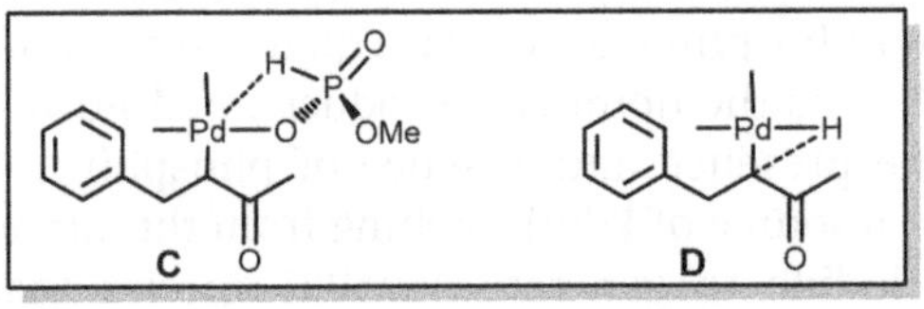

3.6.2.3 *Hydrogenation*

Hydrogenations catalysed by late transition metal NPs have been widely explored and the vast majority of studies involve olefins, arenes, dienes and α,β-unsaturated carbonyl compounds. This section will not be comprehensive in its coverage but will focus on some recent studies and emphasise developments in the area.

3.6.2.3.1 Catalysis by NPs in Unfunctionalised Ionic Liquids. While RuNPs are most often prepared by decomposition of a thermally labile Ru(0) organometallic precursor or reduction of $RuCl_3$ in the presence of a stabiliser, Dupont has shown that hydrogen reduction of hydrated RuO_2 is a simple and reproducible method for the preparation of RuNPs. The reduction of RuO_2 hydrate in $[C_4mim][X]$ ($X = PF_6^-$, BF_4^-, $SO_3CF_3^-$) generates 2.0–2.5 nm diameter monodisperse RuNPs which catalyse the solventless or biphasic hydrogenation of olefins at 75 °C and 4 atm. Although these nanoparticles are highly active under solventless conditions and give complete conversion in 12 min (TOF of 3300 h^{-1}), activity dropped rapidly after two cycles. While reaction rates were significantly lower under ionic liquid biphasic conditions (147–943 h^{-1}) due to mass transfer limitations the system was much more stable and recycled without any loss in activity. Catalyst life-time experiments conducted under biphasic conditions with a catalyst/substrate ratio of 1 : 6667 for each charge gave a total TON in excess of 110 000 after 17 recycles compared with a total TON of only 26 400 after 4 runs under solventless conditions; such an improvement in performance highlights the beneficial effect of ionic liquid on the stabilisation of the active Ru species. Poisoning experiments with mercury and CS_2 support the involvement of ruthenium nanoparticles.[625] RuNP superstructures with a mean diameter of 2.6 ± 0.4 nm, deposited by the controlled decomposition of a suspension of [Ru(cycloocta-1,5-diene)(cyclooctatetraene)] in $[C_4mim][X]$ ($X = PF_6^-$, BF_4^-, $SO_3CF_3^-$), catalyse the biphasic and solventless hydrogenation of olefins and arenes under mild conditions with markedly higher TOFs than those obtained with NPs generated in THF–alcohol mixtures. The solubility difference between benzene and cyclohexene in ionic liquids enabled the partial hydrogenation of benzene to be achieved with 39% selectivity for cyclohexene but this selectivity could only be obtained at low conversions of benzene (1%). The maximum yield of 2% is too low for technical applications and much lower than the 60% obtained with the industrial aqueous phase non-supported Ru catalyst.[626] In contrast, RuNPs generated by reduction of [Ru(cycloocta-1,5-diene)(2-methylallyl)$_2$] under 4 bar of H_2 at 50 °C in [C_4mim][X] and [C_{10}mim][X] (X = NTf_2^-, BF_4^-) catalysed the liquid–liquid biphasic hydrogenation of arenes under mild conditions (50–90 °C) and gave good conversions with a ruthenium loading of 0.25–1.0 mol%. TEM analysis showed that the mean particle size of Ru@[C_{10}mim][NTf_2] increased slightly during catalysis from 2.1 ± 0.5 nm to 3.5 ± 0.5 nm whereas the corresponding increase from 2.7 ± 0.5 nm to 3.1 ± 0.5 nm for Ru@[C_{10}-mim][BF_4] was statistically less significant. These changes are consistent with previous reports in which the smaller particles tend to agglomerate more easily than larger ones.[627] Smaller nanoparticles are formed in ionic liquids containing the less-coordinating anion [NTf_2]$^-$ than their [BF_4]$^-$ analogues and the former agglomerate more readily than the latter during catalysis as evidenced by TEM analysis before and after reactions. The activation energy E_a of 42.0 kJ mol^{-1} determined for the liquid–liquid biphasic hydrogenation of toluene with Ru@[C_4mim][NTf_2] is similar to that reported for the transition metal-catalysed hydrogenation of toluene (E_a = 40–50 kJ mol^{-1}).[628]

The selective hydrogenation of benzene to cyclohexene is of considerable interest to industry as an efficient and cost effective synthesis of cyclohexanol and subsequent conversion to adipic acid and ε-caprolactam. Addition of dicyanamide-based ionic liquids to Ru/Al_2O_3/H_2O was shown to modify benzene hydrogenation selectivity quite dramatically. In the absence of ionic liquid only cyclohexane was produced, however, the presence of even low amounts of [C_4pyr][N(CN)$_2$] resulted in a significant reduction in activity with the concomitant formation of cyclohexene, which was the major product of reaction at low conversions with an initial selectivity of up to 60%, even with a concentration of ionic liquid as low as 7.5×10^{-4} mol L^{-1} (<170 ppm). Other ionic liquids also modified the selectivity but gave varying amounts of cyclohexene. XPS studies provided convincing evidence that the decrease in activity with increasing amounts of ionic liquid was due to chemisorption on the RuNP catalyst. The use of an IL additive avoids the need for additional inorganic salts and NaOH and significantly simplifies the isolation and purification of the cyclohexene. However, further studies will be required to identify optimum conditions and reactor configuration to improve this process.[629]

The size and segregation of polar and nonpolar domains in imidazolium-based ionic liquids has a strong influence of their solvation properties and their ability to interact with different species. Santini and co-workers have reported a correlation between the size of nonpolar domains in a series of imidazolium-based ionic liquids and the size of RuNPs prepared *in situ* from [Ru(cycloocta-1,5-diene)(cyclooctatetraene)], *i.e.* size is limited by the local concentration of the organometallic precursor which is determined by the size of the domain.[630] However, as the RuNPs were not stable with respect to agglomeration under conditions of catalysis this size-control concept has been combined with the use of amines in order to prepare catalytic systems based on *stable and size-controlled* RuNPs. To this end, stable and homogeneously dispersed RuNPs were obtained by hydrogenation of [Ru(COD)(COT)] in [$C_n$mim][NTf_2] ($n=4$, 6, 8, 10) in the presence of 1-octylamine or 1-hexadecylamine. The NPs are all spherical with diameters between 1.1–1.3 nm regardless of the ionic liquid used and show no sign of agglomeration whereas those prepared in [$C_n$mim][NTf_2] in the absence of amine or in THF–amine are larger. In the absence of amine the size of the NP correlates with the alkyl chain length which appears to determine the concentration of the [Ru(COD)(COT)] precursor in the nonpolar domains *i.e.* the larger the nonpolar domains the larger the NPs but no such correlation was evident in the ionic liquid–amine mixtures. DOSY NMR studies confirm that the amine interacts with the NP surface which is most likely responsible for controlling the size while the IL prevents agglomeration. The Ru@[C_4-mim][NTf_2]-octylamine dispersion catalyses the hydrogenation of toluene with TONs that increase from 2 to 8 as the temperature increases from 30 °C to 100 °C. Minor changes in TON with the alkyl chain length of the imidazolium cation may be due to the different solubilities of toluene, methylcyclohexane and dihydrogen in the different ionic liquids and/or differences in viscosity.[631]

A comparative survey of the efficiency of RuNPs embedded in phosphonium and imidazolium-based ionic liquids as catalyst for the biphasic hydrogenation of cyclohexene revealed that stability and activity depends on the ionicity of the ionic liquid such that the most stable RuNPs are formed in ionic liquids having low ionicity in which strong interactions between the anions and cations are present. RuNPs stabilised by [$P_{4,4,4,n}$][NTf_2] ($n=1$, 8, 14) are all efficient catalysts and gave good conversions to cyclohexane in 3 h and within the series [$P_{4,4,4,14}$][X] the anion exerts a marked effect on catalyst efficacy which decreases in the order [NTf_2]$^-$>[Cl]$^-$>[OTf]$^-$>[PF_6]$^-$. The reduced activity of RuNPs in [$P_{4,4,4,14}$][X] (X = PF_6^-, OTf^-) appears to be due to aggregation of the NPs under the conditions of catalysis (TEM analysis) and/or decreased solubility of the substrates in these ionic liquids. The long terms stability of [$P_{4,4,4,n}$][NTf_2]-stabilised RuNPs increases with increasing alkyl chain length and Ru@[$P_{4,4,4,14}$][NTf_2] retained its activity over eight recycles while the activity of [$P_{4,4,4,8}$][NTf_2] and [$P_{4,4,4,1}$][NTf_2] dropped after four and two recycles, respectively, due to growth/aggregation of the NPs. While RuNPs stabilised by imidazolium ionic liquids also catalyse the hydrogenation of cyclohexene, those stabilised by [C_4mim][NTf_2] are markedly more active and stable than those protected by [$C_4C_1{}^2$mim][NTf_2], which may be associated with the C2-methyl group either preventing formation of a stabilising carbene and/or disruption of the supramolecular organisation based on an extended hydrogen-bonded network of anions and cations.[632]

RuNP dispersions generated by microwave irradiation of $Ru_3(CO)_{12}$ in $[C_4mim][BF_4]$ catalyse the biphasic hydrogenation of cyclohexene at 90 °C and 10 bar H_2 to give 95% conversion after 216 min, corresponding to a TOF of 293 mol product mol cat^{-1} h^{-1}. The ruthenium dispersions could be reused and activity increased steadily up to 522 mol product mol cat^{-1} h^{-1} in run seven; CS_2 poisoning experiments provided evidence that this increase in activity was due to surface reconstruction creating more active sites. The unused RuNPs are spherical with a diameter of 1.6 ± 1.3 nm while after seven runs TEM analysis showed that they had transformed into rods 18 ± 7 nm in length and 5 ± 2 nm in width. Microwave generated RhNPs in $[C_4mim][BF_4]$ were more efficient catalysts for the biphasic hydrogenation of cyclohexene and gave a TOF of 884 mol product mol cat^{-1} h^{-1} at 90 °C and 10 bar H_2 with no significant rhodium leaching after three runs. For comparison, IrNPs derived from $Ir_6(CO)_{16}$ were less active and gave a TOF of 222 mol product mol cat^{-1} h^{-1} under the same conditions.[611]

The first IrNP–ionic liquid dispersions for biphasic hydrogenation were prepared by reduction of $[IrCl(cycloocta\text{-}1,5\text{-}diene)]_2$ with H_2 (4 atm) in $[C_4mim][PF_6]$ at 75 °C. The nanoparticles were irregular in shape with a mean diameter of 2 nm. The biphasic hydrogenation of olefins under mild conditions reached a TOF of 6000 h^{-1} at 1200 rpm and 75 °C, which is significantly higher than that obtained under biphasic conditions with classical transition metal precursors in ionic liquid under similar conditions.[633] Comparative recycle experiments showed that the activity of Crabtree's catalyst dropped quite steeply after the first recycle whereas the IrNP–IL dispersion retained its activity over seven runs to give a total TON $>$ 8400.[606] The same IrNPs (and their Rh counterparts) also catalyse the hydrogenation of arenes in ionic liquid (liquid–liquid biphasic), acetone (homogeneous) or under solventless conditions (heterogeneous). Under similar conditions the IrNPs were much more active for the hydrogenation of benzene than their Rh counterparts with TOFs of 219 and 41 h^{-1}, respectively. Reactions under solventless or homogeneous conditions required shorter times than the biphasic system as the latter is under mass transfer control. The IrNPs were isolated essentially unchanged after catalysis and reused seven times to give a total TON of 3509 in 32 h (4 atm, 75 °C, 0.4 mol% Ir) whereas their rhodium counterparts showed significant agglomeration and loss of activity. Interestingly, RhNPs redispersed in $[C_4mim][PF_6]$ (Rh@$[C_4mim][PF_6]$) catalyzed the hydrogenation of anisole to give methoxycyclohexane as the sole product with a TOF of 82 h^{-1}, whereas the use of IrNPs resulted in competitive hydrogenolysis of the C–O bond to give a mixture of methoxycyclohexane and cyclohexane. To this end, hydrogenolysis is characteristic of a surface metal catalyst and indicative of the IrNPs behaving as a "heterogeneous catalyst" rather than a "homogeneous catalyst" in terms of active sites.[634]

High selectivity for the partial hydrogenation of 1,3-butadiene to butene has been achieved with 4.9 ± 0.8 nm PdNPs embedded in $[C_4mim][BF_4]$. PdNPs prepared by reduction of $Pd(OAc)_2$ dissolved in $[C_4mim][BF_4]$ gave 98% selectivity for butenes (72% 1-butene) at complete conversion with negligible amounts of *cis*-but-2-ene and butane. For comparison, a much lower selectivity was obtained under solvent free conditions. The selectivity profile was taken as an indication that the ionic liquid soluble PdNPs behave as surface-like (multisite) rather than a homogeneous-like (single-site) catalyst. The high selectivity obtained in ionic liquids was attributed to the differential solubility of 1,3-butadiene and butenes as the former is at least three times more soluble in $[C_4mim][BF_4]$ than the latter. The selectivity pattern for Pd@$[C_4mim][PF_6]$ shows that more butane is formed which most likely reflects the less pronounced difference in the miscibility of 1,3-butadiene and butenes in this ionic liquid. Recycle experiments gave irreproducible activity due to agglomeration of the PdNPs into larger clusters *i.e.* $[C_4mim][BF_4]$ does not provide effective stabilisation/protection under the conditions of catalysis.[635]

PtNPs 2–3 nm in diameter, prepared by decomposition of $Pt_2(dba)_3$ in $[C_4mim][PF_6]$ under 4 bar of H_2 at room temperature, and redispersed in the ionic liquid (biphasic) or in acetone (homogeneous) or used under solventless conditions (heterogeneous) catalyse the hydrogenation of hex-1-ene, cyclohexene, methylcyclohexene, 2,3-dimethyl-1-butene, 1,3-cyclohexadiene and arenes under mild conditions. Reactions conducted under solventless and homogeneous conditions required much shorter reaction times than those in $[C_4mim][PF_6]$ due to mass transfer control associated with biphasic catalysis. Notably, these PtNPs were more active than Adam's catalyst under the same conditions and no unsaturated intermediates were observed for the hydrogenation of benzene, toluene or cyclohexa-1,3-diene even at early stages of the reaction.[636]

NiNPs with diameters ranging from 4.9 ± 0.9 to 5.9 ± 1.4 nm have been prepared by decomposition of a solution of $Ni(COD)_2$ in $[C_nmim][NTf_2]$ ($n = 4$, 8, 10, 14, 16) under 4 bar of H_2 at 75 °C and the average particle size of the NPs decreased with increasing alkyl chain length of the imidazolium ring up to $n = 14$; thereafter the diameter increased to 5.5 ± 1.1 for $n = 16$. Although the variation in size is not particularly marked the authors claimed that SAXS data collected on colloidal suspensions showed that more regular and smaller diameter and size-distribution nanoparticles were obtained in the more 'pre-organised' ionic liquids. The NiNPs were proposed to be composed of a small cap layer of NiO around a core of Ni metal rather than co-existing Ni metal and NiO nanoparticles. The NiNPs dispersed in $[C_4mim][NTf_2]$ catalyse the biphasic hydrogenation of cyclohexene under 4 bar of H_2 at 100 °C to give TOFs of 91 h^{-1}.[637] The rate constant of 9.2×10^{-4} s^{-1} is two orders of magnitude greater than that reported for cyclohexene hydrogenation by classical supported nickel catalysts at 60 °C and 5 atm.[638]

Magnetically recoverable PtNPs supported on ionic liquid-modified nanomagnetite catalyse the selective reduction of alkynes to *Z*-alkene as well as α,β-unsaturated aldehydes to the corresponding allylic alcohols. Selective hydrogenation of the alkyne was attributed to a bulky ligand effect of the magnetite support which prevents access of the alkene to the active supported catalyst while selective hydrogenation of the carbonyl group in α,β-unsaturated aldehydes was attributed to polarisation of the PtNP surface by the magnetite support which favours selective adsorption and activation of the polar functional group. The catalyst was separated by applying an external magnetic field and recycled with no significant loss in activity or selectivity.[639]

High selectivity for hydrogenation of the C=C double bond in cinnamaldehyde has been achieved with PdNPs immobilised on silica gel-modified with *N*-3-(-3-triethoxysilylpropyl)-3-methylimidazolium-based ionic liquids ([[(TESP)mim][X]; X = Cl^-, NO_3^-, BF_4^-, PF_6^-) with TOFs ranging from 24 260 up to >47 000 h^{-1}, which is three orders of magnitude greater than those of 18–33 h^{-1} obtained under biphasic conditions with $Pd(OAc)_2$–[C_4mim][X]. Such an increase in TOF was attributed to the high surface area of the silica gel greatly enhancing the contact of substrate and hydrogen over the surface palladium species while the poor activities of the biphasic system were suggested to be due to formation of Pd-carbene species which are inert with respect to reduction under hydrogen. Activities varied with the ionic liquid anion in the order $[Cl]^- < [BF_4]^- < [NO_3]^- < [PF_6]^-$ with that of 270 h^{-1} for $[Cl]^-$ only marginally better than the biphasic system. The long-term stability of the Pd/SiO_2-[C_4mim][PF_6] system was demonstrated in recycle studies which gave quantitative conversions over the first nine runs and 93% conversion in runs 10 and 11, corresponding to a total TON in excess of 500 000. In stark contrast, Pd/Al_2O_3 gave a moderate conversion in the first run (78%) which dropped to 40% and 28% in the second and third runs, respectively, due to leaching of palladium.[640]

Although RuNPs stabilised by ionic liquids are effective catalysts for hydrogenation, the high cost of ruthenium has prompted studies into the use of less expensive and more environmentally benign metals such as iron, particularly since iron–ruthenium systems have been shown to catalyse the Fischer–Tropsch reaction as well as the selective hydrogenation of unsaturated aldehydes. A survey of the performance of iron and ruthenium mono- and bimetallic NPs with different metal ratios as catalysts for the biphasic hydrogenation of cyclohexenone has revealed that RuNPs stabilised by [C_4mim][PF_6] are markedly more active than their iron counterparts with TOF of 104 $h^{-1}$and 53 h^{-1}, respectively, but that FeRuNPs (1 : 1) are even more active with a TOF of 111 h^{-1}, probably due to their smaller size. However, increasing the iron content further to 3 : 1 and 9 : 1 (Fe : Ru) was detrimental and resulted in TOF of 75 h^{-1} and 25 h^{-1}, respectively. While all catalysts were selective for hydrogenation of the C=C double bond the 1 : 1 and 3 : 1 FeRuNPs gave the highest ketone/alcohol ratios (96 : 4). Addition of CO_2 (600 bar) to the reaction mixture resulted in a four-fold increase in the rate of hydrogenation by FeRuNPs (1 : 1) in [C_4mim][BF_4] due to improved mass transfer and/or hydrogen solubility.[641]

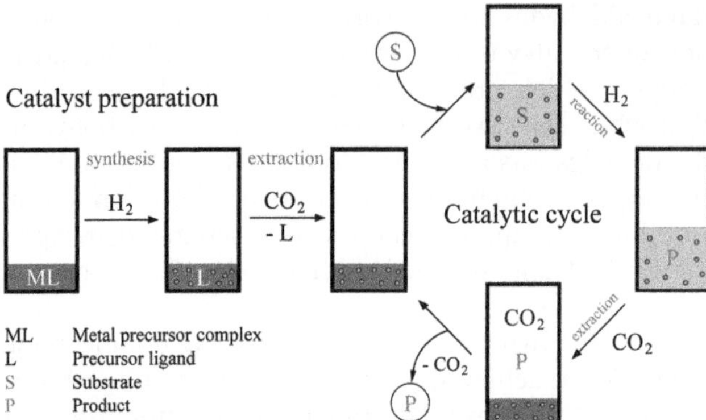

Figure 3.6.3 A schematic of the 'green' ionic liquid–$scCO_2$-based configuration. Reproduced with permission from reference 642. Copyright 2009 Elsevier.

Ionic liquids and $scCO_2$ have been combined to develop a green solventless metal-catalysed hydrogenation of acetophenone with Pd and RhNPs. In this concept, NPs are generated in an ionic liquid from a metal precursor complex and $scCO_2$ used to both extract the ligand before catalysis as well as the product after hydrogenation (Figure 3.6.3). The TOFs obtained with ionic liquid-stabilised PdNPs decreased in the order $[C_4mim][PF_6] > [C_4mim][OTf] > [N_{6,6,6,6}]Br$ while selectivity for 1-phenylethanol was generally highest in $[C_4mim][OTf]$ and reached 98% at 96.6% conversion. Selectivity towards 1-phenylethanol was lower with ionic liquid-stabilised RhNPs as was activity and although higher reaction temperatures led to higher conversions this further reduced the selectivity and under these conditions cyclohexylethanol was the major by-product. The PdNP-based catalyst was readily recycled by extraction of the 1-phenylethanol and ethylbenzene with $scCO_2$ with no loss in selectivity or activity over six runs. Long term storage of $[C_4mim][PF_6]$-stabilised PdNPs resulted in partial decomposition due to hydrolysis of the $[PF_6]^-$ anions, however, the resulting minor reduction in activity and selectivity could be restored by a single extraction with $scCO_2$ to remove the HF prior to hydrogenation. Even within this narrow series of ionic liquids there are remarkable differences in the particle size, size distribution, morphology and extent of agglomeration of the nanoparticles and interestingly the larger-sized and apparently agglomerated PdNPs are more active, selective and recyclable than the smaller ones. Thus, provided a precise and detailed understanding of how the ionic liquid influences the properties of NPs can be developed it may be possible to optimise catalyst activity and selectivity in a rational manner.[642]

The selective and efficient hydrogenation of biomass-derived and related substrates is catalysed by ionic liquid stabilised RuNPs prepared by reduction of an ionic liquid suspension of [Ru(cycloocta-1,5-diene)(2-methylallyl)$_2$]. The size, activity and selectivity of the derived NP's can be controlled by varying the ionic liquid, further details of which are discussed in section 3.6.1.1.[547]

3.6.2.3.2 Catalysis with Ligand-Stabilised or Modified Nanoparticles. Phenanthroline-stabilised PdNPs, generated by the reduction of $Pd(OAc)_2$ in $[C_4mim][PF_6]$ under 1 bar of dihydrogen, are highly active and selective catalysts for the hydrogenation of olefins and give good conversions under mild conditions in short reaction times. The ionic liquid–catalyst mixture recycles much more efficiently than that without added phenanthroline and could be reused over 10 cycles with no loss in activity and no evidence of aggregation.[643]

Bipyridine-stabilised RhNPs with a mean diameter of 2 nm and a monomodal size distribution have been prepared by sodium borohydride reduction of $RhCl_3$ in THF–IL and used for the hydrogenation of styrene at 80 °C under 4 atm of H_2. Optimisation studies identified that a bipyridine/Rh ratio of 0.5 was necessary since a lower ratio resulted in agglomeration of the colloidal suspension under the conditions of catalysis. A study of the influence of the ionic liquid revealed a marked and dramatic anion dependent selectivity as reactions conducted in $[C_4mim][BF_4]$ were highly selective in favour of ethylcyclohexane over ethylbenzene (92/8) while those in $[C_4mim][N(CN)_2]$ gave ethylbenzene as the sole product. The effect of the cation within a series of $[NTf_2]^-$-based ionic liquids was somewhat less-pronounced and ethylbenzene/ethylcyclohexane ratios varied between 85 : 15 and 70 : 30. The hydrogenation of a range of aromatic compounds was investigated and activity decreased with increasing bulk on the aromatic ring.[644] A comparison of various bipyridine isomers and 2,4,6-tris(2-pyridyl)-*s*-triazene (TPTZ) revealed that NPs with a mean diameter of 2.0–2.5 nm were obtained regardless of whether 2,2′- 3,3′- or 4,4′-bipyridine was used as stabiliser while those stabilised by TPTZ are slightly larger (3.5–4.5 nm). There was, however, a dramatic influence of the bypyridine on selectivity for the hydrogenation of styrene in $[C_4mim][PF_6]$. Under the optimised conditions described above, 4,4′-bipyridine-protected RhNPs gave complete conversion of styrene to ethylbenzene whereas those stabilised by 2,2′-bipyridine gave near quantitative conversion to a mixture of ethylbenzene (60%) and ethylcyclohexane (40%). The hydrogenation of benzene and its mono-substituted derivatives also showed a marked influence on the bipyridine stabiliser as 4,4′-bipyridine-stabilised RhNPs gave complete conversion for a range of substituted benzenes whereas the activity of 2,2′-bipyridine-protected RhNPs decreased with increasing bulk in the order cumene < propylbenzene < ethylbenzene < toluene. The difference in behaviour between the bipyridine isomers was rationalised on the basis of a coordination model in which bidentate coordination of the 2,2′-bipyridine to the NP surface stabilises the particles but impedes/blocks access of the substrate whereas 3,3′- and 4,4′-bipyridine both provide stabilisation through monodentate coordination while also allowing access of the substrate due to decreased steric hindrance. Support for this model was provided by a comparison of the efficiency of RhNPs protected by mono-alkylated 2,2′- and 4,4′-bipyridinium salts as both RhNP systems gave complete conversion of styrene to ethylbenzene in $[C_4mim][PF_6]$. As a corollary, complete conversion to ethylbenzene was also achieved with RhNPs stabilised by one equivalent of pyridine, but the nanoparticles were visually destabilised at the end of the reaction. The hydrogenation of anisole as a model lignin compound went to completion with 2,2′-bipyridine and TPTZ-stablised RhNPs and gave methoxycyclohexane and cyclohexanone as a 77 : 23 and 97 : 3 mixture, respectively. The same *N*-stabilised RhNPs also catalyse the hydrogenation of acetophenone but with vastly disparate selectivities as those capped with 2,2′-bipyridine gave a mixture of acetylcyclohexane (22%), cyclohexylethanol (40%) and phenylethanol (38%) while TPTZ-stabilised RhNPs gave acetylcyclohexane (68%) and cyclohexylethanol (68%) with no evidence for hydrogenolysis products.[645a,b] A comparative study showed that the introduction of an imidazolium tag onto 2,2′-bipyridine does not significantly modify the activity of RhNPs compared with those stabilised by unmodified 2,2′-bipyridine for the hydrogenation of a range of substituted benzenes, which suggests that it could be possible to develop task-specific ligands to optimise the activity and/or selectivity in NP catalysis.[646] However, Dyson and co-workers also explored the efficiency of imidazolium-functionalised 2,2′-bipyridine-stabilised RhNPs for the hydrogenation of toluene, ethylbenzene, propylbenzene, styrene, *o*-xylene, *m*-xylene and *p*-xylene (35 °C, 40 bar of H_2, *ca.* 0.1 mol% Rh) and showed that activity is strongly influenced by the stabiliser and decreases in the order [4,4′-bis{7-(2,3-dimethylimidazolium)heptyl}-2,2′-bipyridine][NTf_2]$_2$ **174a** ([BIHB][NTf_2]$_2$) > bipy > [4,4′-bis{(1,2-dimethylimidazolium)methyl}-2,2′-bipyridine][NTf_2]$_2$ **174b** ([BIMB][NTf_2]$_2$). The low activity of the [BIMB][NTf_2]$_2$-stabilised NPs was considered to result from the weaker interaction of the ligand with the metal surface due to the electron-withdrawing effect of the imidazolium, the greater steric bulk and the close proximity of the positive charges to the NP surface that, combined, reduce the binding affinity of the 2,2′-bipyridine fragment and result in decomposition of the NPs under the conditions of catalysis. Analysis of the ether extract after catalysis showed that *ca.* 1.4% of the 2,2′-bipyridine had leached from the ionic liquid whereas extraction of [BIHB][NTf_2]$_2$ was not detected. Thus, it may well be crucial to consider the design of the bipyridine stabiliser (and presumably those of other heteroatom donors) in terms of the location and length/type of immobilising tag in order to minimise detrimental effects on performance.[647]

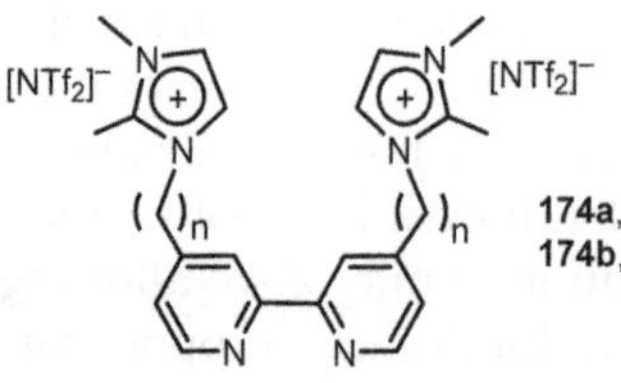

[NTf_2]$^-$ [NTf_2]$^-$

174a, n = 7, [BIHB][NTf_2]$_2$
174b, n = 1, [BIMB][NTf_2]$_2$

174a-b

The first immobilisation of a NP catalyst onto a solid surface by an ionic liquid was achieved with PdNPs using molecular sieves as the support and 1,1,3,3-tetramethylguanidinium lactate (TMGL) as the ionic liquid. The resulting catalyst was highly active for the solvent free hydrogenation of 1-hexene, cyclohexene and 1,3-cyclohexadiene and gave markedly higher TOFs (67 min^{-1}) than either biphasic ionic liquid systems (1.67 h^{-1}) or directly supported phenanthroline-protected PdNPs (0.3 h^{-1}) in $[C_4mim][PF_6]$. The efficiency and long term stability of the catalyst system was attributed to a synergism between the high surface area of the support (383 m^2 g^{-1}), small particle size (1–2 nm) and partial exposure of the ionic liquid-stabilised PdNPs.[648]

The selective hydrogenation of aromatic compounds containing a nitro group and a C=C double bond is a demanding challenge as both groups are readily hydrogenated. To this end, highly chemoselective hydrogenation of either the C=C double bond or the nitro group in 3-nitrostyrene can be obtained with ionic liquid-stabilised PtNPs by a judicious choice of reaction conditions (acidic/basic) and support. High selectivity for ethylnitrobenzene (<95%) can be obtained with PtNPs generated in 1-butyl-3-methylimidazolium tris(pentafluoroethyl)trifluorophosphate with bipyridine as co-stabiliser under acidic conditions while high selectivity for aminostyrene (89%) can be obtained with PtNPs generated in the same IL but with quinoline as co-stabiliser under basic conditions (NEt_3) and supported on carbon nanotubes; in the absence of carbon nanotubes a mixture of aminostyrene and the intermediate 3-(*N*-hydroxylamino)styrene was obtained. A comparative survey with ionic liquid free catalyst (1% Pt/CNT) under basic conditions revealed that the presence of ionic liquid was crucial to achieve high chemoselectivity.[103]

An ionic liquid functionalised with a pendent 2,2-dipyridylamide **175** (2,3-dimethyl-1-[2-*N,N*-bis(2-pyridyl)-propylamide]imidazolium hexafluorophosphate, $[BMMDAP][PF_6]$) plays an important role in stabilising and dispersing PdNPs as well as controlling selectivity in the hydrogenation of functional olefins. For example, PdNPs generated by reduction of $Pd(OAc)_2$ in a solution of acetone and $[BMMDAP][PF_6]$ catalysed highly efficient and chemoselective hydrogenation of the C=C double bond in α,β-unsaturated aldehydes/ketones in $[C_4C_1{}^2mim][PF_6]$ under mild conditions. Poor selectivity was obtained in the absence of ligand and PdNPs generated in the presence of 2,2′-dipyridylamine were much less active than those stabilised by $[BMMDAP][PF_6]$, possibly due to its limited solubility in $[C_4C_1{}^2mim][PF_6]$ and/or poisoning by strong coordination to the active Pd(0) nanoparticles.[649] The same $[BMMDAP][PF_6]$-stabilised PdNPs also catalysed the biphasic hydrogenation of cyclohexene under extremely mild conditions (35 °C, 0.1 MPa H_2) to give good yields of cyclohexane after 3 h. Catalyst activity decreased slightly as the palladium concentration was increased from 1 mmol L^{-1} to 10 mmol L^{-1} at 35 °C due to aggregation of the NPs and activity also decreased when the reaction temperature was raised above 35 °C.[650]

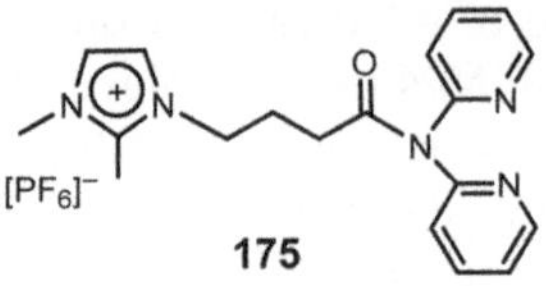

175

Nitrile-functionalised ionic liquid-stabilised RuNPs prepared by hydrogenation of [Ru(cycloocta-1,5-diene)(2-methylallyl)$_2$] in [N≡CC$_3$mim][NTf$_2$] catalyse the hydrogenative coupling of nitrile containing aromatic compounds with quite remarkable selectivity. For example, the Ru@[N≡CC$_3$-mim][NTf$_2$]-catalysed hydrogenation of benzonitrile results in reductive condensation to afford (*E*)-*N*-benzylidene-1-phenylmethanamine [Equation (3.6.4)] as the sole product, even in the presence of toluene which is typically hydrogenated by RuNPs in non-functionalised ionic liquids. In contrast, the same RuNPs catalyse the hydrogenation of aromatic ketones with high selectivity for the aromatic ring. The high selectivity for reductive coupling was proposed to be due to coordination of the nitrile group to the surface of the RuNP preventing interaction of the arene. A similar approach could be used for the selective hydrogenation of other functional groups.[651]

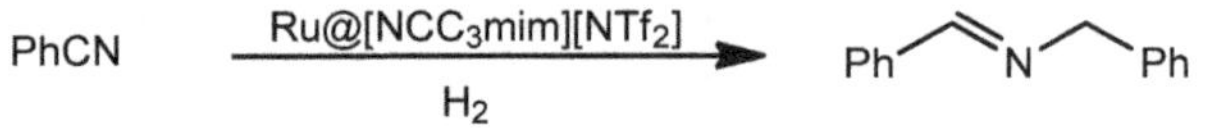

$$\text{PhCN} \xrightarrow[\text{H}_2]{\text{Ru@[NCC}_3\text{mim][NTf}_2]} \text{PhCH=NCH}_2\text{Ph} \quad (3.6.4)$$

Highly selective partial hydrogenation of internal alkynes to *Z*-alkenes has been achieved with monodisperse PdNPs generated by heating Pd(OAc)$_2$ in the nitrile-functionalised ionic liquid [C≡NC$_3$mim][NTf$_2$] under reduced pressure. Good conversions were obtained under 1 atm of H$_2$ at 25 °C and TOFs reached 1282 h^{-1} based on the metal atoms exposed on the NP surface. The catalyst recycled efficiently with no loss in activity or selectivity over four runs and, moreover, the system exhibited exceptional stability with no loss in activity for at least four months. Moreover, there was no isomerisation during the hydrogenation of hex-3-yne which is consistent with surface-like (multi-site) catalysis.[652]

The concept of employing an ionic liquid functionalised with a metal binding site such as an amine, bipyridine, carboxylate, nitrile or thiolate as solvent and/or stabiliser has been widely investigated. In a rather simplistic, elementary and straightforward extension of this approach phosphine-functionalised ionic liquids [PPh$_2$C$_n$C$_1{}^2$mim][X] **176** (PFIL; $n=3$, 11; X = NTf$_2{}^-$, OTf$^-$, PF$_6{}^-$) have also been employed as stabilising ligands for PdNPs [Equation (3.6.5)]. The dark brown suspension generated by reduction of a [C$_4$C$_1{}^2$mim][X] solution of Pd(acac)$_2$ in the presence of 1.0 equivalent of PFIL under 4 bar of H$_2$ at 80 °C consists of 3 nm PdNPs and while XPS analysis suggested that the primary stabilisation was provided by the phosphine group of the PFIL, stabilisation though interaction of OTf and/or a metal oxide layer can not be ruled out. These PFIL-stabilised PdNPs are efficient catalysts for the biphasic hydrogenation of styrene under 4 bar

H_2 at 50 °C. An increase in catalytic activity upon recycling was due to either incomplete reduction of the Pd(II) precursor or restructuring of the PdNPs under conditions of catalysis to generate more active catalytic sites. The PFIL-stabilised PdNPs are more active than those generated in $[C_4C_1{}^2mim][X]$ in the absence of phosphine stabiliser due to the larger number of catalytically active sites resulting from the larger surface-volume ratio of the smaller PFIL-stabilised NPs. The length of the alkyl chain and the anion both influence catalyst activity with $[PPh_2C_1C_1{}^2mim][NTf_2]$ giving the highest conversions and the best long-term stability with an average conversion of 85% over 10 recycles.[653]

$$\textbf{176}\ \xrightarrow[\text{[C}_4\text{C}_1{}^2\text{mim][X]}]{\text{H}_2\ \text{(4 bar)}}\ \text{PdNP-bound phosphine} \tag{3.6.5}$$

The same group also employed phosphine-functionalised ionic liquids to stabilise RhNPs for use in the biphasic hydrogenation of arenes. RhNPs generated by reduction of $Rh(allyl)_3$ with H_2 (4 atm, 50 °C) and stabilised by $[PPh_2C_nC_1{}^2mim][NTf_2]$ ($n = 3$, 11) were monodisperse with an average diameter of 2.0–2.4 nm whereas those stabilised by PPh_3 were slightly smaller with a diameter of 1.5 nm and only aggregates were formed in the absence of stabiliser. Ionic liquid-based biphasic hydrogenation of toluene in $[C_4C_1{}^2mim][NTf_2]$ showed a marked dependence of catalytic activity on the stabiliser and the phosphine-functionalised ionic liquid-stabilised RhNPs experienced an induction period over the first 3–4 cycles and then gave consistently high conversions to methylcyclohexane over the next seven runs with no loss in activity; this induction was proposed to be due to incomplete reduction of the rhodium precursor. These conversions compared favourably with those obtained using bipyridine-functionalised ionic liquids under similar conditions.[649] In contrast, the activity of RhNPs generated in the absence of stabiliser gave erratic conversions which varied between 40% and 90% after the first two cycles due to aggregation and plating of the metal on the reaction vessel. While PPh_3-stabilised RhNPs also gave good conversions after an induction period the poor solubility of PPh_3 in $[C_4mim][NTf_2]$ resulted in significant leaching of the phosphine into the organic layer and even though this did not appear to affect catalyst activity, as measured by conversions over 10 runs, leaching may ultimately cause destabilisation of the system in the long term.[654]

A study of ligand effects on the performance of RuNPs generated in $[C_4mim][NTf_2]$ by hydrogenation of [Ru(COD)(COT)] in the presence of *N*-octylamine, H_2O, PPh_2H, $PPhH_2$ revealed that for a series of RuNPs with a mean size of 2.2 nm activity increased with σ-donor ligands such as *N*-octylamine and H_2O but is lower with bulkier π-acceptors such as PPh_2H, $PPhH_2$ and CO.[655] A similar role of σ-donor π-acceptor properties has previously been formulated for the hydrogenation of methylanisole in the presence of diphosphine-stabilised RuNPs. In this study, RuNPs stabilised by PAr_3-type mono-and diphosphines showed no or very poor activity whereas those stabilised by *in situ* generated or preformed PCy_2Ar-based phosphines gave good conversions.[656]

Water soluble PdNPs stabilised by the PEG-modified dicationic imidazolium-based ionic liquid **177** are efficient catalysts for the aqueous biphasic hydrogenation of a range of olefins. Rates were higher than those obtained with Pd/C in the aqueous phase, good yields were obtained under mild conditions and recycle experiments gave consistent activities over nine runs. The efficacy and stability of this Pd@C_{12}im-PEG-IL system was attributed to formation of amphiphilic ionic liquid micelles which acted as a nanoparticle stabilising nanoreactor and emulsification-promoted transfer of the substrate to the interface of the two phases where the substrate contacted the active palladium nano-catalyst. Replacement of the C_{12}Im-PEG with unmodified C_{12}-PEG-C_{12} resulted in a dramatic reduction in activity from 12 300 h^{-1} to 4360 h^{-1} under similar conditions, which reflects the critical role of the imidazolium cations.[657]

177

While ionic liquid-supported noble metal NPs have been widely used as catalysts for the hydrogenation of simple substrates such as olefins and arenes, their use in asymmetric hydrogenation is much more recent. In this regard, ionic liquid-stabilised PtNPs generated by formic acid reduction of $H_2Pt(OH)_6$ in the *presence of cinchonidine* as co-stabiliser (Pt_{CD}@IL) catalyse the asymmetric hydrogenation of methyl benzoylformate with remarkable efficiency (Figure 3.6.4).[658] Platinum nanoparticles prepared in this manner

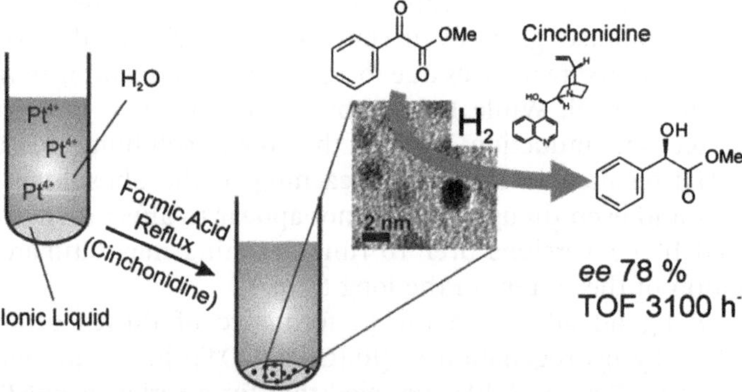

Figure 3.6.4 Schematic of cinchonidine-modified ionic liquid-supported PtNPs for enantioselective hydrogenation.
Reprinted with permission from reference 659. Copyright 2012 American Chemical Society.

in $[C_4mim][FEP]$ (FEP = tris(pentafluoroethyl)trifluorophosphate) gave *R*-mandelate in 78% ee with a TOF of 3100 h^{-1} using a catalyst loading of 0.05 mol% together with 0.25 mol% added cinchonidine under 30 bar of H_2 at 25 °C. Markedly higher TOFs were obtained with PtNPs generated in the presence of cinchonidine as co-stabiliser (Pt_{CD}@IL) compared to those prepared without (Pt@IL), which correlated with the smaller size of Pt_{CD}@IL compared with Pt@IL. The type of ionic liquid did not affect the ee and had only a limited influence on activity which increased in the order $[C_4C_1{}^2mim][PF_6] < [C_4C_1{}^2mim][BF_4] < [C_4C_1pyr][BF_4] < [C_4C_1{}^2mim][FEP]$.

Much lower ee's (45%) and TOFs (380 h^{-1}) were obtained with Pt_{CD}@IL in the absence of added cinchonidine due to agglomeration. A control reaction with 5% Pt/Al_2O_3-cinchonidine gave 61% ee with a TOF of 460 h^{-1} without pre-treatment and 87% ee with a TOF of 1600 h^{-1} after pre-treatment at 400 °C under a flow of H_2. The ee's matched those obtained with PtNPs stabilised by cinchonidine and used for the asymmetric hydrogenation of ethyl pyruvate.[659] The effectiveness of this simple concept offers immense opportunity for applications across a much wider range of noble metal NP-catalysed transformations.

3.6.2.3.3 Catalysis by Polymer-Stabilised NPs in Heteroaton Donor-Functionalised Ionic Liquids. Polyvinylpyrrolidone (PVP) has been widely used for the stabilisation of transition metal NPs synthesised in molecular solvents. However, the negligible solubility of PVP in ionic liquids severely limits its use for the stabilisation of ionic liquid-embedded NPs. Solutions to this problem include incorporation of imidazolium ionic liquid-like groups into the polymer chain or functionalisation of ionic liquids with groups to improve the solubility of PVP. To this end, hydroxyethyl-functionalised imidazolium, pyridinium and piperidinium-based ionic liquids are more compatible with PVP than $[C_2mim][BF_4]$ as the saturation concentration of PVP in $[HOC_2mim][BF_4]$ is >5% compared with less than 0.5% for $[C_2mim][BF_4]$. The combination of $[HOC_2mim][X]$ with PVP provided effective stabilisation for highly dispersed RhNPs which catalysed the biphasic hydrogenation of styrene to give good yields of ethylbenzene (>99%) in 2 h at 40 °C under 50 bar H_2 using a catalyst loading of 0.025%. For comparison the corresponding non-functionalised IL-system Rh@$[C_2mim][BF_4]$ gave a yield of only 20% under the same conditions, demonstrating the positive influence of the hydroxyethyl-functionalised ionic liquid in terms of activity, stability and leaching and consequently recyclability (Figure 3.6.5). To this end, the optimum catalyst recycled efficiently over seven runs to give a total TON of 28 000 and TEM analysis of the used catalyst confirmed that the size and distribution of the NPs remained unchanged, reflecting the high stability of PVP-protected Rh@$[HOC_2mim][BF_4]$. A slight decrease in activity after run seven was attributed to a build-up of polystyrene. For comparison, the corresponding PVP-protected Rh@$[C_2mim][BF_4]$ recycled poorly and yields varied due to the heterogeneity of the system. Anion effects also proved to be

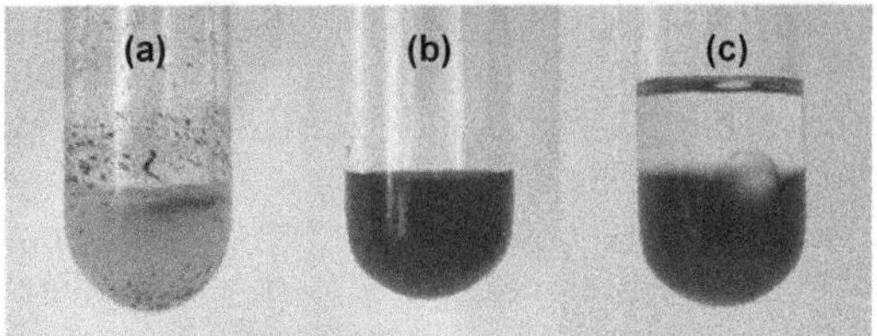

Figure 3.6.5 Photographs of RhNPs dispersed in (a) [C_2mim][BF_4], (b) [HOC_2-mim][BF_4], and (c) [HOC_2mim][BF_4] after hydrogenation of styrene. Reprinted with permission from reference 660. Copyright 2008 American Chemical Society.

important with activity increasing in the order $[NTf_2]^- > [PF_6]^- > [BF_4]^-$, which paralleled their relative coordinating power. Good conversions were also obtained for a host of other substrates including cyclopentene, 1-hexene, 1,3-cyclohexadiene, 1-octyne, 1-decene and 1-decyne.[660]

3.6.2.3.4 Catalysis with NPs Stabilised by Polymer-Grafted Ionic Liquids. The use of ionic liquid polymers to stabilise NPs is proving to be an effective strategy that could ultimately enable particle size and size distribution as well as morphology to be controlled by varying the charge density, charge distribution, regions of hydrophobicity/hydrophilicity and cross-linking while the introduction of a functional co-monomer could be used to modify the nanoparticle surface and thereby influence its activity and/or selectivity. Ionic copolymers comprising 1-vinyl-3-alkyl imidazolium salts and *N*-vinyl-2-pyrrolidone **178** have been used to prepare RhNPs in order to combine the stabilisation provided by PVP with that of ionic liquids while also overcoming the low solubility of PVP in ionic liquids. To this end 2.9 nm RhNPs stabilised by poly(*N*-vinyl-2-pyrrolidone)-co-(1-vinyl-3-alkylimidazolium chloride)] are remarkably long-lived and catalyse the hydrogenation of benzene at 75 °C and 40 bar H_2 to give a total TON in excess of 20 000 after five cycles (TOF 250 h^{-1}), which is 5.7 times higher than the previous highest value of 3509 obtained with Ir-NP dispersed in [C_4mim][PF_6].[634] The stabilisation provided by this ionic polymer results from a synergism between the ionic liquid and the polymer as hydrogenations carried out in [C_4mim][BF_4] in the absence of copolymer or in methanol in the presence of copolymer gave poor activities with formation of black precipitate. The hydrogenation of substituted arenes gave a mixture of fully hydrogenated product (cyclohexanes) and partially hydrogenated product (cyclohexene), but no diene, and TOFs decreased with increasing size and number of substituents. The formation of partially hydrogenated product was proposed to be due to a combination of the high solubility of the arene and diene relative to the monoene in the ionic liquid resulting in spontaneous separation of the product from the catalyst phase as well as the steric and electronic properties of the substitutents on the aromatic ring affecting the rate of hydrogenation.[661a,b]

The same polymers also provide efficient stabilisation for PtNPs with a mean diameter of 5.1 (Pt-I) and 1.7 nm (Pt-II). Both exhibit excellent activity and high selectivity for *o*-chloroaniline (*o*-CAN) in the hydrogenation of *o*-chloronitrobenzene in [C_4mim][BF_4] with the smaller NPs recycling more efficiently than the larger ones, reaching an unprecedented total TON >25 900 after four runs as a result of being longer lived and retaining their activity. In addition, the initial TOFs of 7697 h^{-1} and 6639 h^{-1} obtained with Pt-I and Pt-II, respectively, are well in excess of an order of magnitude higher than those of 217–701 h^{-1} obtained with conventional poly(*N*-vinyl-2-pyrrolidone)-stabilised PtNPs. IR studies and DFT calculations indicate that intermolecular interactions between the imidazolium ions of [C_4mim][BF_4] and the NO_2 weakens the N=O bond which may be responsible for the high *o*-CAN selectivity.[662]

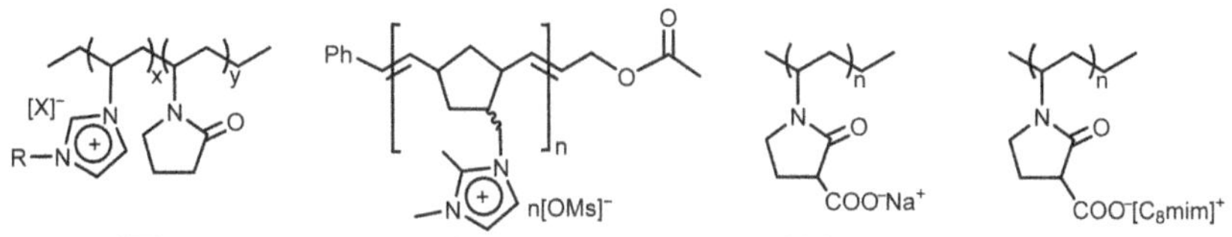

[X]⁻ R–N N x y N O **178** Ph n N –N n[OMs]⁻ O O **179** N O n COO⁻Na⁺ **180** N O n COO⁻[C₈mim]⁺ **181**

Imidazolium-decorated styrene-based homopolymers stabilise gold, platinum and palladium nanoparticles through a combination of steric/electrostatic effects and micelle formation. The NPs can be transferred from the aqueous phase to hydrophobic ionic liquids without aggregation or degradation by anion exchange of the ionic liquid polymer.[663] Similarly, small highly stable gold NPs have also been prepared by reduction of gold(III) in the presence of the ROMP-derived ionic homopolymer **179** and the size and distribution of the NPs was shown to decrease as the polymer to gold ratio increased from 3 : 1 to 50 : 1. The smallest NPs (1.8–2.9 nm) catalyse the reduction of nitrophenol with remarkable efficiency under mild conditions and are moderately selective towards reduction of the C=O bond of cinnamaldehyde with no sign of aggregation under the conditions of catalysis (Figure 3.6.6).[664] Polyethylenimine-based ionic liquid polymer-stabilised gold NPs between 1.79 ± 0.5 nm to 3.47 ± 1.3 nm have been prepared by increasing the concentration of the gold(III) precursor. The NPs are stable for at least one month at room temperature in water and exhibit good electrocatalytic activity towards NADH oxidation to NAD^+.[665]

A general strategy for the stabilisation of NPs without significantly sacrificing their performance is based on stabilisers that incorporate several weak stabilisation mechanisms such that each contributes to the overall stability without hindering the activity. Dyson and co-workers have utilised this design concept and shown that RhNPs protected by carboxylate-modified PVP such as **180** and **181** are thermally more stable than those protected by unmodified PVP (PVP-Rh). The former catalysed the hydrogenation of toluene at 60 °C and gave conversions that consistently exceeded 90% over

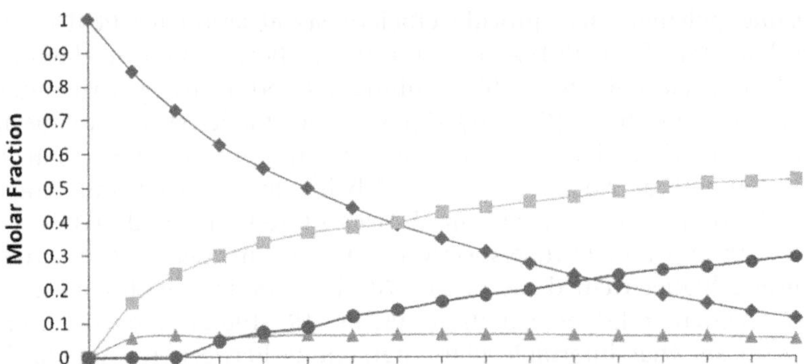

Figure 3.6.6 Kinetic curves for cinnamaldehyde hydrogenation at 60 °C and 100 bar H_2 using the AuNPs stabilised by **179** (50 : 1; polymer : Au) solution: diamonds (♦), Cinnamaldehyde; squares (■), Cinnamyl alcohol; triangles (▲), Hydrocinnamaldehyde; circles (●), Hydrocinnamyl alcohol.
Reprinted with permission from reference 664. Copyright 2011 American Chemical Society.

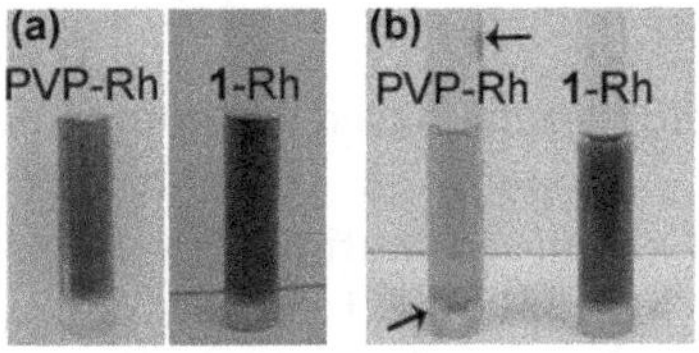

Figure 3.6.7 Thermal stability of PVP-Rh and **180**-Rh (labelled as **1**-Rh); (a) before heating; (b) after heating at 200 °C for 2 h. Arrow indicates the formation of metallic deposits.
Reproduced from reference 666.

five runs whereas the latter gave a near quantitative conversion in the first run but less than 70% conversion in run five; this decrease in activity correlated with NP agglomeration which was observed visually (Figure 3.6.7) and confirmed by TEM. XPS studies confirmed the presence of surface-absorbed carboxylate and IR spectroscopy supported an interaction between the carboxylate and RhNP surface suggesting that agglomeration was prevented by weak coordination of the carboxylate and an electrical double layer.[666]

Building on the concept of combining polymer-stabilisation with ionic liquids, sulfonic acid groups have been introduced into ionic liquid-like polymers to immobilise metal NPs by supporting them in water rather than in an IL or on a surface. Water soluble PdNPs stabilised by IL-like polymers **182** or **183** and a heteropolyacid additive catalyse the highly selective hydrogenation of phenol to cyclohexanone with near quantitative conversion and >99% selectivity after 3–10 h at 80 °C. The catalyst preparation and reaction can be conducted in a single-pot process and the system recycled with no loss in selectivity over three runs. The selectivity was suggested to be due to the intrinsic properties of the ionic liquid-like polymer and the formation of macromolecular cages decorated with hydrophilic ion pairs (imidazolium) and sulfonic acid groups which stabilise a uniform dispersion of PdNPs but also function to enrich the hydrophilic active site with phenol (through H-bonding) and efficiently disperse the cyclohexanone before it reacts further while the HPA functioned as a Brønsted superacid to generate a phenol-derived dication. The use of functional copolymers to control selectivity (and activity) is an innovative concept that could be broadly applied to a range of transformations.[79]

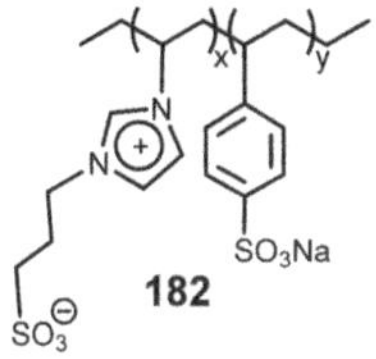

182 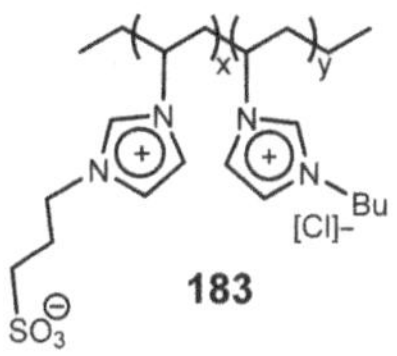 **183**

3.6.2.4 *Suzuki–Miyaura and Stille Cross-Coupling*

Arguably, the Suzuki–Miyaura cross-coupling is one of the most technologically and commercially important C–C bond forming reactions and as such the use of transition metal NP catalysts in combination with ionic liquids has been the subject of numerous investigations, selected details of which are described below.

3.6.2.4.1 Nanoparticle Catalysts Stabilised with Heteroatom Donor-Functionalised Ionic Liquids. The Suzuki–Miyaura cross-coupling in ionic liquids has been well-documented but in several cases the presence of water (or alcohol) has been found to be necessary to achieve activity. Arguing that the role of the water/alcohol was necessary for effective solvation of salt species that would otherwise be poorly solvated and steadily shut down reaction, Dyson and co-workers showed that hydroxyl-functionalised imidazolium-based ionic liquids gave a marked enhancement in activity for the ligand-free palladium-catalysed cross-coupling of aryl bromides and iodides with phenyl boronic acid compared with the same reaction in $[C_4mim][NTf_2]$ or neat isopropanol. To this end, the Suzuki–Miyaura coupling between bromobenzene and phenyl boronic acid occurs with remarkable efficiency in $[HOC_2mim][NTf_2]$ and 1 mol% $PdCl_2$ gave 96% conversion after heating at 120 °C for 4 h whereas only 26% conversion was obtained in $[C_4mim][NTf_2]$. High yields were also obtained in $[HOC_2C_1{}^2mim][NTf_2]$ indicating that *N*-heterocyclic carbenes are unlikely to play a role in the catalysis. The catalyst in $[HOC_2mim][NTf_2]$ showed remarkable resistance to poisoning by chloride with only a slight reduction in yield even in the presence of 100 equivalents of chloride; in contrast the same catalyst was inactive in [C_4mim][NTf_2] under the same conditions, probably due to poor solvation of the chloride. Although nanoparticles were identified they were viewed as reservoirs of molecular palladium rather than the actual catalyst and those in [HOC_2mim][NTf_2] were better dispersed than in [C_4mim][NTf_2]. The hydroxyl-functionalised ionic liquid was suggested to play a multifunctional role by facilitating generation and stabilisation of the catalyst, assisting in activation of the C–X bond by hydrogen bonding with the X atom and solvating salts generated during the catalytic cycle *i.e.* facilitating reaction and preventing poisoning.[667] A subsequent study of the formation of PdNPs in a range of hydroxy-tethered ionic liquids [HOC_2mim][X] (X = NTf_2^-, PF_6^-, BF_4^-, OTf^-, $CF_3CO_2^-$) as a function of the anion explored the role of these ionic liquids in generating and stabilising PdNP reservoirs. Compared to PdNPs generated in [C_4mim][NTf_2] the hydroxyl group accelerated their formation and provided protection from oxidation once formed. Within the series [HOC_2mim][X], the ease of formation of the PdNPs decreased in the order $[NTf_2]^- > [PF_6]^- > [BF_4]^- > [OTf]^- > [TFA]^-$ which relates at least in part to the nucleophilicity of the anion such that the most nucleophilic interacts with the metal precursor more strongly to reduce the nucleation process. The hydroxyl group also appears to help protect the NPs once formed with resistance to oxidation decreasing in the order $[NTf_2]^- > [PF_6]^- > [TFA]^- > [OTf]^- > [BF_4]^-$. These are important parameters to consider when using ionic liquid-stabilised PdNPs as catalyst reservoirs for cross-coupling and provide some insight into identifying an optimum anion, which so far appears to be $[NTf_2]^-$.[668] Following a lead from Alper and co-workers,[639] magnetically recoverable PdNPs immobilised on and stabilised by amino-functionalised imidazolium-based ionic liquid-modified magnetic SiO_2/Fe_3O_4 particles catalysed the Suzuki–Miyaura coupling of aryl bromides and iodides and gave good yields in reasonable reaction times at room temperature. The catalyst was separated from the reaction mixture by applying an external permanent magnet and recycled several times with only a minor reduction in activity.[669]

Task-specific nitrile-functionalised ionic liquids were introduced as an 'immobilisation strategy' on the basis that the nitrile group would interact with the metal and improve retention of the catalyst. In this regard, Dyson and co-workers demonstrated that [N≡CC_3pyr][X] (X = PF_6^-, BF_4^-, NTf_2^-) are markedly more effective immobilisation solvents for palladium-catalysed Suzuki–Miyaura and Stille couplings than their *N*-butyl pyridinium counterparts, [C_4pyr][X]. Each of the palladium complexes [$PdCl_2$(N≡CC_3pyr$)_2$][$PdCl_4$] **184** and [$PdCl_2$(N≡CC_3pyr$)_2$][X$]_2$ **185** (X = PF_6^-, BF_4^-, NTf_2^-) formed an efficient catalyst for the Suzuki–Miyaura coupling between iodobenzene and benzene boronic acid in [N≡CC_3pyr][NTf_2] or [C_4pyr][NTf_2]. While high conversions were obtained in both ionic liquids at 100 °C in the presence of 1 mol% Pd, the catalyst was efficiently immobilised in [N≡CC_3pyr][NTf_2] and recycled nine times with no loss in activity whereas catalyst immobilised in [C_4pyr][NTf_2] was completely inactive by the fifth cycle. The efficiency of the nitrile-functionalised system relative to its alkyl pyridinium counterpart was attributed to highly effective retention of the catalyst (<5 ppm palladium leached compared with loss of 28 ppm from $[C_4pyr][NTf_2]$) as well as improved catalyst stability. The same nitrile-based catalyst immobilisation strategy also extended to the palladium-catalysed Stille cross-coupling between iodobenzene and Bu_3SnPh in $[N{\equiv}CC_3pyr][NTf_2]$ and $[C_4pyr][NTf_2]$. Similar yields were obtained in both ionic liquids for the first run, however, while reactions in $[N{\equiv}CC_3pyr][NTf_2]$ gave similar conversions across nine runs, conversions in $[C_4pyr][NTf_2]$ began to drop after the first recycle and the system was completely inactive after the fourth run. In contrast to the Suzuki coupling, the Stille reaction mixture rapidly turned black and deposited PdNPs and TEM analysis revealed that the nitrile-based ionic liquid stabilised nanoparticles with respect to agglomeration more efficiently than $[C_4pyr][NTf_2]$ while ICP confirmed that leaching reduced from 46 ppm in $[C_4pyr][NTf_2]$ to 7 ppm in $[N{\equiv}CC_3pyr][NTf_2]$.[670]

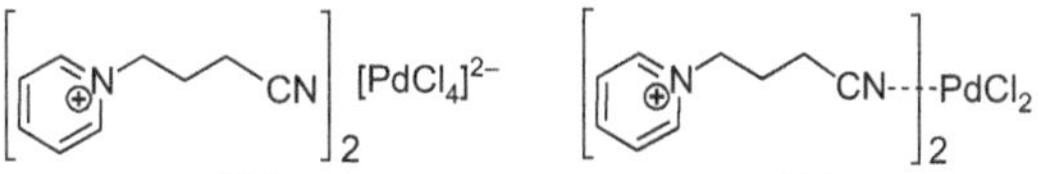

$[(N{\equiv}C(CH_2)_3C_5H_5N^+)]_2\,[PdCl_4]^{2-}$ **184** $\quad$ $[(C_5H_5N^+(CH_2)_3CN)\cdots PdCl_2]_2$ **185**

Early studies showed that Stille cross-coupling reactions were particularly sensitive to the composition of the ionic liquid and that $[NTf_2]^-$-based systems generally gave the highest activity for ligand-free coupling, although low stability led to poor recyclability.[41a] A subsequent study of anion and cation effects on the Stille reaction in a range of ionic liquids reinforced the positive influence of the nitrile functionality as consistently higher yields were obtained in $[N{\equiv}CC_3mim][X]$ ($X = NTf_2^-$, BF_4^-) compared with their $[C_4mim]$-based counterparts. IR spectroscopic studies confirmed an interaction between the nitrile group and palladium which was proposed to improve performance by stabilising the reservoir of nanoparticles with respect to aggregation. To this end, a comparison of PdNPs generated in $[C_4mim][NTf_2]$ and $[N{\equiv}CC_3mim][NTf_2]$ revealed that the former were small (2–4 nm) and highly aggregated which resulted in a low catalytically accessible surface area while NPs in $[N{\equiv}CC_3mim][NTf_2]$ were highly dispersed and open to activation. In contrast, the performance in $[C(CN)]^-$-based ionic liquids did not vary between $[C_4mim][C(CN)_2]$ and $[N{\equiv}CC_3mim][C(CN)_2]$ which was rationalised in terms of coordination of the palladium only involving the nucleophilic anion thus rendering performance insensitive to the nature of the cation.[671] Further evidence that PdNPs might serve as a reservoir of Pd(0) from which the active catalyst is generated by oxidative addition of the substrate was provided by a study of the Suzuki–Miyaura cross-coupling in a series of pyrrolidinium-based ionic liquids as the initially orange reaction mixture rapidly turned black but for reactions that gave high yields the solution subsequently turned orange whereas the solution remained black for low yielding reactions. These changes were interpreted as initial formation of PdNPs which act as a reservoir of the actual Pd(II) catalyst (orange) following oxidative addition of the aryl halide to a Pd atom on the surface of the nanoparticle. To this end, preformed PdNPs gave similar yields to those obtained with $PdCl_2$ for the same reaction. Early stage analysis of the reaction mixture revealed that the nanoparticles generated under the conditions of catalysis ranged from 3.5–10.5 nm in size and were more heterogeneous than those prepared by reduction of [$PdCl_2$(COD)] or $PdCl_2$ in [C_4mim][PF_6] and broader than those isolated from the Stille reaction conducted in nitrile-functionalised imidazolium- and pyridinium-based ionic liquids.[666] In contrast, the black solutions that formed during the Stille reaction between aryl iodides and vinyltributylstannane in pyrrolidinium-based ionic liquids persisted throughout the entire reaction. Good yields were obtained after 4 h at 80 °C with 5 mol% $PdCl_2$ as precatalyst and similar yields were also obtained with preformed nanoparticles. Interestingly, higher yields of Stille product were obtained in pyrrolidinium-based ionic liquids compared with their imidazolium and pyridinium [NTf_2]-based counterparts which was attributed to a nucleophilic assistance effect of the weakly coordinating $[NTf_2]^-$ anion in the transmetalation step due to the following order of intrinsic bond strength to the $[NTf_2]^-$ anion: [C_2mim]>[C_4mim]>[C_4pyr]>[C_4C_1pyrr]. While there is little doubt that nitrile-based ionic liquids help to solubilise precursors such as $PdCl_2$, stabilise nanoparticles by weakly interacting with their surface and can potentially stabilise the active catalyst *via* coordination, it is still not clear whether oxidative addition occurs at the surface of the NP or at leached Pd(0).[42] A report on the development of nitrile-functionalised imidazolium-based ionic liquids for C–C coupling reactions presented further evidence for the beneficial stabilising and/or immobilising role of a tethered nitrile group. For example, $PdCl_2$/[N≡CC_3mim][BF_4] is stable and shows no sign of decomposition during the Suzuki–Miyaura cross-coupling between iodobenzene and phenyl boronic acid and recycles efficiently over six runs whereas its unfunctionalised counterpart $PdCl_2$/[C_4mim][BF_4] turns black and is essentially inactive after three runs due to facile leaching of palladium (ICP analysis) *i.e.* regardless of the true nature of the catalyst, the nitrile tether improves catalyst stability and retention and therefore performance. Reaction efficiency also improved for Heck reactions involving iodobenzene and ethyl acrylate in nitrile-tethered ionic liquids such that conversions increased in the order [C_4mim][NTf_2]<[(N≡CC_3)mim][NTf_2]<[(N≡C-C_3)$_2$im][NTf_2]. Similarly, TEM analysis of Stille reactions conducted in [(N≡CC_3)mim][NTf_2] and [C_4mim][NTf_2] with $PdCl_2$ as catalyst precursors established the presence of nanoparticles with similar diameters (5 nm) but those in [(N≡CC_3)mim][NTf_2] were spatially separated whereas those in [C_4mim][NTf_2] aggregated into clusters up to 30 nm. The authors claimed that these results support the role of the nitrile preventing aggregation by forming a protective layer around the NP and stabilising/solubilising the active Pd(II) species by transient coordination.[672]

Palladium nanoparticles immobilised by a combination of a nitrile-functionalised ionic liquid and an imidazolium-decorated polymer exhibit exceptional long term stability over two years without any special precautions. This all-in-one PdNP-IP-IL system catalysed the Suzuki–Miyaura coupling of aryl iodides and bromides with phenyl boronic acid and was markedly superior to the corresponding PdNP-IL solution. The appearance of an orange colouration following reaction was consistent with the presence of molecular Pd(II) species. The same PdNP-IP-IL system also catalysed the Heck arylation between ethyl acrylate and aryl iodides and gave good conversions at 80 °C with only 0.5 mol% Pd as well as the Stille cross-coupling between aryl iodides and phenyltributylstannane with the same Pd loading; this is significantly lower than the 5 mol% required by Pd@[N≡CC_3pyr][NTf_2] to achieve comparable yields. As much lower loadings of this all-in-one PdNP-IP-IL system (0.5–1.0 mol%) are typically required compared with Pd/C or the corresponding IL-stabilised PdNPs which typically require 2–5 mol% it represents an economically viable alternative system which could find applications in a range of other C–C bond forming reactions.[673]

3.6.2.4.2 Catalysts Stabilised by Unfunctionalised Ionic Liquids. PdNPs stabilised by tetralkylammonium bromide salts catalyse the Suzuki–Miyaura coupling of activated aryl chlorides at temperatures as low as 60 °C and electron-rich chlorides at 90 °C in the presence tetrabutylammonim hydroxide, which activates the boronic acid and helps preserve the protecting shell surrounding the nanoparticles; other bases were much less effective. The same system also catalysed the Stille coupling of aryl chlorides or bromides with tributylphenylstannane at temperatures between 90–130 °C. Higher conversions and more efficient recycling was obtained in tetraheptylammonium bromide (THAB) compared with tetrabutylammonium bromide (TBAB) due to stronger hydrophobic interactions in the former which impeded nanoparticle aggregation and limited catalyst inactivation. The bromide was proposed to have a dual role as co-catalyst to form a more active anionic Pd(0) species and to accelerate the transmetalation step by expanding the coordination sphere of tin, in much the same manner proposed earlier for the PdNP-catalysed Stille reaction.[41b]

The formation of PdNPs has been strongly implicated in the Suzuki–Miyaura cross-coupling between aryl bromides and aryl boronic acids in [C_4mim][PF_6] using $Pd(OAc)_2$ as precursor in the presence of norborn-5-ene-2,3-dicarboxylic anhydride-derived amines. The highest conversions were obtained using amines functionalised with a pendent/remote hydroxyl group which appear to improve the stability of the nanoparticles. Post catalysis TEM micrographs of the reaction mixture revealed that the formation of PdNPs was required to give an active catalytic system, although it was not possible to determine whether the nanoparticles were active catalysts or acted as reservoir of active Pd(0) species.[674] Ionic liquid-stabilised palladium nanoparticles generated by reduction of [$PdCl_2$(cycloocta-1,5-diene)] with hydrogen and exhibiting star-like shaped interparticle organisation catalyse the Suzuki–Miyaura coupling between bromobenzene and phenyl boronic acid to give near quantitative conversion after 1 h at 100 °C with a palladium loading as low as 0.25 mol%. Lower conversions were obtained with NPs generated from $PdCl_2$ and $Pd_2(dba)_3$. The palladium content in the biphenyl was determined by ICP-MS and shown to be within the allowed levels for pharmaceutical production (3–5 ppm). The ionic liquid–catalyst phase recycled 10 times after work-up with only a minor reduction in yield after the eighth run and with consistently low levels of Pd-leaching.[675]

Phosphonium-stabilised palladium nanoparticles have been prepared without the use of additional reducing agent. The resulting nanoparticles are approximately 7 nm in diameter with shapes that depend on the anion of the ionic liquid and the Pd(II) precursors. A solution of $[PdCl_2(MeCN)_2]$ in trihexyl(tetradecyl)phosphonium dodecylbenzenesulfonate, $[P_{6,6,6,14}]$DBS, formed face-centred cubic nanoparticles with a diameter of 9 nm while $Pd(OAc)_2$ gave particles with a variety of crystalline forms including octahedrons, triangular plates and truncated octahedrons. Two possible pathways for reduction were considered, the first involved dealkylation to afford the corresponding phosphine that then reduced Pd(II) to Pd(0) in the presence of an oxygen containing anion (DBS) or adventitious water and the second involved H_2 reduction *via* dehydrogenation of an alkyl chain [Equation (3.6.6) and (3.6.7)]. The *in situ* generated PdNPs in $[P_{6,6,6,14}]$[DBS] catalysed the Suzuki–Miyaura reaction between 4-bromotoluene and $PhB(OH)_2$ under microwave heating (100 °C) in the presence of base to afford good yields of the biphenyl in short reaction times, however, the precise role of the NPs was not established. Since there is an increasing body of evidence that nanocrystals with different shapes can have different catalytic activities and selectivities in a chemical reaction it will be important to develop a fundamental understanding of the relationship between morphology and reactivity.[676]

Concluding Remarks

While scientific curiosity alone is a sufficiently sound justification for exploring the use of ionic liquids as alternative solvents in homogeneous catalysis, the high tuneability of their physiochemical properties provides a much sounder academic basis. This chapter has attempted to introduce some of the key basic concepts underlying the use of ionic liquids as solvents in homogeneous catalysis while also providing an overarching coverage of some of the major developments and applications across a diverse range of fundamental transformations.

Numerous benefits associated with the use of ionic liquids as solvent and/or support for catalysis have been identified including increased catalyst stability, efficient recyclability, effective catalyst retention and recovery (*i.e.* reduced catalyst leaching), facile isolation of product by extraction or distillation, remarkable and substantial enhancements in reaction rate as well as significant improvements in selectivity (*e.g.* chemoselectivity and stereoselectivity) compared with the same reaction in organic solvent. In cases where the improved catalyst performance requires an ionic liquid to be decorated either with functionality or a task specific group the additional expense and effort required for their synthesis must be justified and compensated in terms of efficiency.

The tuneable polarity and miscibility of ionic liquids with either organic solvents or compressed carbon dioxide ($scCO_2$) has also enabled biphasic systems to be developed and continuous flow processes to be engineered. To this end, biphasic systems based on $scCO_2$ as the mobile phase and an ionic liquid-immobilised catalyst have been shown to combine rapid mass transfer with quantitative phase separation which has enabled integration of catalytic reaction and product separation in a single-step process. However, for practical applications it will be necessary to address drawbacks such as any reduction in catalyst performance (*e.g.* selectivity) caused by the immobilisation matrix.

Ionic liquids have also played a central role in the development of catalysis by soluble metal nanoparticles, principally by preventing aggregation/agglomeration through surface stabilisation by protective layers of imidazolium aggregates and more recently either by coordination of a heteroatom donor-functionalised ionic liquid to the surface of the nanoparticle or by the formation of micelles in organic or aqueous solvent. The use of heteroatom donor-functionalised ionic liquids (or polyionic liquids) in nanoparticle catalysis is an emerging area/strategy which will enable catalyst longevity and recyclability to be improved and selectivity to be modified and controlled. While early developments in this area focused on expensive metals such as palladium and rhodium, future advances are likely to be focused on the use of more abundant and less expensive non-noble metals such as iron, cobalt, copper and nickel.

The concept of SILP catalysis in which a catalyst dissolved in an ionic liquid is immobilised on a high surface area porous support has presented new engineering opportunities because the resulting catalyst is a solid which behaves as a homogeneous system in which the active species is solubilised in the ionic liquid.[677] On a fundamental level, SILP combines the advantages of homogeneous and heterogeneous catalysis such as high activity and selectivity with ease of product separation, good catalyst stability and efficient recycling. However, SILP systems operate most efficiently for gas phase processes and their use in liquid phase reactions is more challenging as leaching of the catalyst and ionic liquid must be minimised; one promising solution has been to introduce a charged group/tag on to the catalyst to improve its retention in the ionic liquid. Ultimately, it will be important to develop a detailed understanding of the nature of the interaction between the support and the catalyst and how it influences the activity and selectivity if systems are to be optimised and used for industrial applications. In a closely related concept, polymers decorated with ionic groups (polyionic liquids) have been used to immobilise a range of catalysts including nanoparticles,[678] peroxotungstates,[73] heteropolyacids,[679] organocatalysts[680] Brønsted acids[681] and Lewis acids[74a] to combine the favourable properties of ionic liquids with the advantages of a support such as reduced leaching, ease of separation/recycling and continuous flow processing; the resulting polymer immobilised ionic phase catalysts complement existing systems and configurations based on supported ionic liquids such as Supported Ionic Liquid Catalysts (SILC), Solid Catalysts with Ionic Liquid Layers (SCILL), Supported Ionic Liquid Nanoparticles (SILNPs) and Periodic Mesoporous Organosilicas with Ionic Liquid frameworks (PMOs-IL).[64] Although Polymer Immobilised Ionic Liquid Phase (PIILP) catalysis is an evolving concept, the ability to modify the ionic microenvironment (charge density, type and distribution), porosity, microstructure, functionality and hydrophobicity/hydrophilicity of ionic polymers in a systematic and rational manner should enable catalyst-surface interactions, substrate accessibility and efficiency to be optimised, property-function relationships to be elucidated and new activity-selectivity relationships to be established.

While the use of ionic liquids has been shown to improve the efficiency of a reaction compared with that in molecular solvents, this alone does not justify their categorisation as green solvents as it is necessary to include the environmental impact associated with the use of these solvents in any life cycle assessment, *i.e.* their biocompatibility, toxicity, biodegradation, bioaccumulation, persistence as well as the use of volatile organic solvents during IL synthesis and product isolation must all be considered. In this regard, Gathergood has been exploring the use of environmentally friendly 1,2-dimethylimidazolium-based salts containing an amide or ester functionality in the side chain and triazolium salts as new low antimicrobial toxicity ionic liquids for use in catalysis, with some particularly encouraging results.[682]

Finally, this chapter has not been extensive in its coverage of topics due to the author's limited breadth of knowledge and expertise and one topic that has received only limited coverage in section 3.6.1.4.2 is enzymatic reactions in ionic liquids. However, a review entitled *Recent Advances of Enzymatic Reactions in Ionic Liquids* provides an up-to-date account and describes the benefits of using ionic liquids for enzymatic catalysis which include high thermal and operational stability, high conversion rates, excellent regio- and enantioselectivity, and better recoverability and recyclability than in conventional solvents.[683]

References

1. V. I. Parvulescu and C. Hardacre, *Chem. Rev.*, 2007, **107**, 2615.
2. H. Olivier-Bourbigou, K. Magna and D. Morvan, *Appl. Catal., A*, 2010, **373**, 1.

3. R. Giernoth, *Top. Curr. Chem.*, 2007, **276**, 1.
4. S. Liu and J. Xiao, *J. Mol. Catal. A: Chem.*, 2007, **270**, 1.
5. J. P. Hallett and T. Welton, *Chem. Rev.*, 2011, **111**, 3508.
6. R. Šebesta, I. Kmentová and S. Toma, *Green Chem.*, 2008, **10**, 484.
7. S.-g. Lee, *Chem. Commun.*, 2006, 1049.
8. Q. Zhang, S. Zhang and Y. Deng, *Green. Chem.*, 2011, **13**, 2619.
9. P. Śledź, M. Mauduit and K. Grela, *Chem. Soc. Rev.*, 2008, **37**, 2433.
10. D. Betz, P. Altmann, M. Cokoja, W. A. Hermann and F. E. Kühn, *Coord. Chem. Rev.*, 2011, **255**, 1518.
11. J. Muzart, *Adv. Synth. Catal.*, 2006, **348**, 275.
12. N. Isambert, D. del, M. S. Duque, J.-C. Plaquevent, Y. Génisson, J. Rodriguez and T. Constantieux, *Chem. Soc. Rev.*, 2011, **40**, 1347.
13. B. Ni and A. D. Headley, *Chem. Eur. J.*, 2001, **16**, 4426.
14. C. Van Doorslaer, J. Wahlen, P. Mertens, K. Binnemans and D. De Vos, *Dalton Trans.*, 2010, **39**, 8377.
15. Š. Toma, M. Mečiarová and R. Šebesta, *Eur. J. Org. Chem.*, 2009, 321.
16. N. V. Plechkova and K. R. Seddon, *Chem. Soc. Rev.*, 2008, **37**, 123.
17. (a) J. D. Scholten, B. C. Leal and J. Dupont, *ACS Catal.*, 2012, **2**, 184. Other relevant reviews include; (b) J. Dupont and J. D. Scholten, *Chem. Soc. Rev.*, 2010, **39**, 1780; (c) M. H. G. Prechtl, J. D. Scholten and J. Dupont, *Molecules*, 2010, **15**, 3441; (d) K. L. Luska and A. Moores, *ChemCatChem*, 2012, **4**, 1534.
18. J. Yuana, D. Mecerreyes and M. Antonietti, *Prog. Polym. Sci.*, 2013, **38**, 1007.
19. (a) N. Papaiconomou, J.-M. Lee, J. Salminen, M. von Stosch and J. M. Prausnitz, *Ind. Eng. Chem. Res.*, 2008, **47**, 5080; (b) T. V. Hoogerstraete, B. Onghena and K. Binnemanns, *J. Phys. Chem. Lett.*, 2013, **4**, 1659.
20. (a) D. E. Camper, D. L. Gin and R. D. Noble, *Acc. Chem. Res.*, 2010, **43**, 152; (b) B. E. Gurkan, J. C. de la Fuenle, E. M. Mindrup, L. E. Ficke, B. F. Goodrich, E. A. Price, W. F. Schneider and J. F. Brennecke, *J. Am Chem. Soc.*, 2010, **132**, 2116.
21. M. Hira, K. Ito-Akita and J. Ohno, *J. Mater. Chem.*, 2001, **11**, 1057.
22. M. D. Sliger, S. J. P'Pool, R. K. Traylor, J. McNeill III, S. H. Young, N. W. Hoffman, M. A. Klingshirn, R. D. Rogers and K. H. Shaugnessy, *J. Organomet. Chem.*, 2005, **690**, 3540.
23. (a) J. M. Fraile, J. I. Garcia, C. I. Herrerías, J. A. Mayoral, D. Carrie and M. Vaultier, *Tetrahedron: Asymmetry*, 2001, **12**, 1891; (b) J. M. Fraile, J. I. García, C. I. Herrerías, J. A. Mayoral, S. Gmough and M. Vaultier, *Green Chem.*, 2004, **6**, 93.
24. M. Solinas, A. Pfaltz, P. G. Cozzi and W. Leitner, *J. Am. Chem. Soc.*, 2004, **126**, 16142.
25. (a) S. Doherty, P. Goodrich, C. Hardacre, J. G. Knight, M. T. Nguyen, V. I. Parvulescu and C. Paun, *Adv. Synth. Catal.*, 2007, **349**, 951; (b) D. Evans, S. Miller and T. Lectka, *J. Am. Chem. Soc.*, 1993, **115**, 6460.
26. H. Srour, H. Rouault, C. C. Santini and Y. Chauvin, *Green Chem.*, 2013, **15**, 1341.

27. M. Fabris, V. Lucchini, M. Noé, A. Perosa and M. Selva, *Chem. Eur. J.*, 2009, **15**, 12273.
28. A. Aupoix, B. Pégot and G. Vo-Thanh, *Tetrahedron*, 2010, **66**, 1352.
29. P. J. Dyson, D. J. Ellis, W. Henderson and G. Laurenczy, *Adv. Synth. Catal.*, 2003, **345**, 216.
30. (a) C. C. Daguenet and P. J. Dyson, *Organometallics*, 2006, **25**, 5811; (b) C. Daguenet and P. J. Dyson, *Organometallics*, 2004, **23**, 6080.
31. (a) K. R. Seddon, A. Stark and M.-J. Torres, *Pure Appl. Chem.*, 2000, **72**, 2275; (b) Y. Yasaka, C. Wakai, N. Matubayasi and M. Nakahara, *Anal. Chem.*, 2009, **81**, 400.
32. R. Ge, R. W. K. Allen, L Aldous, M. R. Brown, N. Doy, C. Hardacre, J. M. MacInnes, G. McHale and M. I. Newton, *Anal. Chem.*, 2009, **81**, 1628.
33. F. Maier, J. M. Gottfried, J. Rossa, D. Gerhard, P. S. Schulz, W. Schwieger, P. Wasserscheid and H. P. Steinrück, *Angew. Chem., Int. Ed.*, 2006, **45**, 77.
34. E. Amigues, C. Hardacre, G. Keane, M. Migaud and M. O'Neil, *Chem. Commun.*, 2006, 71.
35. A. L. Johnson, M. G. Davidson, M. D. Jones and M. D. Lunn, *Inorg. Chim. Acta*, 2010, **363**, 2209.
36. D. B. William, M. E. Stoll, B. E. Scott, D. A. Costa and W. J. Oldham Junior, *Chem. Commun.*, 2005, 1438.
37. A. S. K. Hshmai, A. Loos, S. Doherty, J. G. Knight, K. J. Robson and F. Rominger, *Adv. Synth. Catal.*, 2011, **353**, 749.
38. I. Krossing and I. Raabe, *Angew. Chem., Int. Ed.*, 2004, **43**, 2066.
39. V. Lecocq and H. Olivier-Bourbigou, *Oil Gas Sci. Technol.*, 2007, **62**, 761.
40. (a) V. Caló, A. Nacci, A. Monopoli, A. Fornaro, L. Sabbatini, N. Cioffi and N. Ditaranto, *Organometallics*, 2004, **23**, 5154; (b) C. Amatore and A. Jutand, *Acc. Chem. Res.*, 2000, **33**, 314.
41. (a) C. Chiappe, G. Imperato, E. Napolitano and D. Pieraccini, *Green Chem.*, 2004, **6**, 33; (b) V. Caló, A. Nacci, A. Monopoli and F. Montingelli, *J. Org. Chem.*, 2005, **70**, 6040.
42. Y. Cui, H. Biondi, M. Chaubey, X. Yang, Z. Fei, R. Scopelliti, C. G. Hartinger, Y. Li, C. Chiappe and P. J. Dyson, *Phys. Chem. Chem. Phys.*, 2010, **12**, 1834.
43. (a) C. J. Mathews, P. J. Smith, T. Welton, A. J. P. White and D. J. Williams, *Organometallics*, 2001, **20**, 3848; (b) F. McLachlan, C. J. Mathews, P. J. Smith and T. Welton, *Organometallics*, 2003, **22**, 5350.
44. (a) N. D. Clement, K. J. Cavell, C. Jones and C. J. Elsevier, *Angew. Chem., Int. Ed.*, 2004, **43**, 1277; (b) D. S. McGuinness, K. J. Cavell and B. F. Yates, *Chem. Commun.*, 2001, 255.
45. L. Magna, Y. Chauvin, G. P. Niccolai and J.-M. Basset, *Organometallics*, 2003, **22**, 4418.
46. J. D. Scholten and J. Dupont, *Organometallics*, 2008, **27**, 4439.
47. D. Bacciu, K. J. Cavell, I. A. Fallis and L. Ooi, *Angew. Chem., Int. Ed.*, 2005, **44**, 5282.

48. F. Ye and H. Alper, *Adv. Synth. Catal.*, 2006, **348**, 1855.
49. T. Avilés, S. Jansat, M. Martínez, F. Montilla and C. Rodríguez, *Organometallics*, 2011, **30**, 3919.
50. J. M. Fraile, J. I. García, C. I. Herrerías, J. A. Mayoral, O. Reiser and M. Vaultier, *Tetrahedron Lett.*, 2004, **45**, 6765.
51. J. V. D. Broeke, F. Winter and B. J. Deelman, *Org. Lett.*, 2002, **4**, 3851.
52. P. J. Dyson, D. J. Ellis and T. Welton, *Can. J. Chem.*, 2001, **79**, 705.
53. (a) B. Tan, J. Jiang, Ya. Wang, L. Wei, Di. Chen and Z. Jin, *Appl. Organomet. Chem.*, 2008, **22**, 620; (b) Y. Zeng, Y. Wang, J. Jiang and Z. Jin, *Catal. Commun.*, 2012, **19**, 70; (c) Z. Zeng, Y. Wang, Y. Xu, Y. Song, J. Zhao, J. Jiang and Z. Jin, *Chin. J. Catal.*, 2012, **33**, 402; (d) Y. Zeng, Y. Wang, Y. Xu, Y. Song, J. Jiang and Z. Jin, *Catal. Lett.*, 2013, **143**, 200.
54. Y. Leng, J. Wang, D. Zhu, X. Ren, H. Ge and L. Shen, *Angew. Chem., Int. Ed.*, 2009, **48**, 168.
55. H. Zhi, C. Lu, Q. Zhang and J. Luo, *Chem. Commun.*, 2009, 2878.
56. F. Liu, M. B. Abrams, R. T. Baker and W. Tumas, *Chem. Commun.*, 2001, 433.
57. R. A. Brown, P. Pollet, E. McKoon, C. A. Eckert, C. L. Liotta and P. G. Jessop, *J. Am. Chem. Soc.*, 2001, **123**, 1254.
58. M. F. Sellin, P. B. Webb and D. J. Cole-Hamilton, *Chem. Commun.*, 2001, 781.
59. P. B. Webb, T. E. Kunene and D. J. Cole-Hamilton, *Green Chem.*, 2005, **7**, 373.
60. J. Theuekauf, G. Franciò and W. Leitner, *Adv. Synth. Catal.*, 2013, **355**, 209.
61. A. Riisager, R. Fehrmann, M. Haumann and P. Wasserscheid, *Eur. J. Inorg. Chem.*, 2006, 695.
62. C. P. Mehnert, *Chem. Eur. J.*, 2005, **11**, 50.
63. P. Wasserscheid, *J. Ind. Eng. Chem.*, 2007, **13**, 325.
64. T. Selvam, A. Machoke and W. Schwieger, *Appl. Catal., A*, 2012, **445-446**, 92.
65. C. van Doorslaer, J. Wahlen, P. Merlens, K. Binnemans and D. De Vos, *Dalton Trans.*, 2010, **39**, 8377.
66. M. H. Valkenberg, C. deCastro and W. F. Hölderich, *Stud. Surf. Sci. Catal.*, 2001, **135**, 179.
67. M. H. Valkenberg, C. deCastro and W. F. Hölderich, *Top. Catal.*, 2001, **14**, 139.
68. C. P. Mehnert, R. A. Cook, N. C. Dispenziere and M. Afeworki, *J. Am. Chem. Soc.*, 2002, **124**, 12932.
69. (a) C. P. Mehnert, E. J. Mozelesk and R. A. Cook, *Chem. Commun.*, 2002, 3010; (b) M. V. Chedkar, T. Sasaki and B. M. Bhange, *ACS Catalysis*, 2013, **3**, 287.
70. U. Hintermair, G. Zhao, C. C Santini, M. J. Muldoon and D. J. Cole-Hamilton, *Chem. Commun.*, 2007, 1462.
71. R. Duque, E. Öchsner, H. Clavier, F. Caijo, S. P. Nolan, M. Mauduit and D. J. Cole-Hamilton, *Green Chem.*, 2011, **13**, 1187.

72. U. Hintermair, G. Franciò and W. Leitner, *Chem. Eur. J.*, 2013, **19**, 4538.
73. S. Doherty, J. G. Knight, J. R. Ellison, D. Weekes, R. W. Harrington, C. Hardacre and H. Manyar, *Green Chem.*, 2012, **14**, 925.
74. (a) S. Doherty, J. G. Knight, J. R. Ellison, P. Goodrich, L. Hall, C. Hardacre, M. Muldoon, A. Ribeiro, C. Afonso, P. Davey and S. Park, *Green Chem.*, 2013, 10.1039/c3gc41378k; (b) R. Alleti, W. S. Oh, M. Perambuduru, C. V. Ramana and V. P. Reddy, *Tetrahedron Lett.*, 2008, **49**, 3466.
75. M. I. Burguete, E. García-Verdugo, I. García-Villar, F. Gelat, P. License, S. V. Luis and V. Sans, *J. Catal.*, 2010, **269**, 150.
76. V. San, N. Karbass, M. I. Burguete, V. Compañ, E. García-Verdugo, S. V. Luis and M. Pawlak, *Chem. Eur. J.*, 2011, **17**, 1894.
77. P. Lozano, E. Garcia-Verdugo, R. Piamtongkam, N. Larbass, T. De Diego, M. I. Burguete, S. V. Luis and J. L. Iborra, *Adv. Synth. Catal.*, 2010, **352**, 3013.
78. P. Lozano, E. Garcia-Verdugo, R. Piamtongkam, N. Karbass, T. De Diego, M. I. Burguete, S. V. Luis and J. L. Iborra, *Adv. Synth. Catal.*, 2007, **349**, 1077.
79. (a) A. Chen, G. Zhao, J. Chen, L. Chen and Y. Yu, *RSC Adv.*, 2013, **3**, 4171; (b) L. Han, H.-J. Choi, S.-J. Choi, B. Lin and D.-W. Park, *Green Chem.*, 2011, **13**, 1023; (c) Y. Zhang, S. Yin, S. Luo and C. T. Au, *Ind. Eng. Chem. Res.*, 2012, **51**, 3951; (d) W.-L. Dai, B. Jin, S.-L. Luo, X.-B. Luo, X.-M. Tu and C-T. Au, *Catal. Sci. Technol.*, 2014, DOI: 10.1039/c3cy00659j; (e) W.-L. Dai, L. Chen, S.-F. Yin, W.-H. Li, Y.-Y. Zhang, S.-L. Luo and C.-T. Au, *Catal. Lett.*, 2010, **137**, 74; (f) J. Sun, W. Cheng, W. Fan, Y. Wang, Z. Meng and S. Zhang, *Catal. Today*, 2009, **148**, 361; (g) R. A. Watile, K. M. Deshmukh, K. P. Dhake and B. M. Bhange, *Catal. Sci. Technol.*, 2012, **3**, 1051; (h) R. A. Watile, D. B. Bagal, K. M. Deshmukh, K. P. Dhake and B. M. Bhange, *J. Mol. Catal. A: Chemical*, 2011, **351**, 196.
80. J. W. Lee, J. Y. Shin, Y. S. Chun, H. B. Jang, C. E. Song and S.-G. Lee, *Acc. Chem. Res.*, 2010, **43**, 985.
81. C. E. Song, W. H. Shim, E. J. Roh, S.-g. Lee and J. H. Choi, *Chem. Commun.*, 2001, 1122.
82. B. Y. Park, K. Y. Ryu, J. H. Lee and S.-g. Lee, *Green Chem.*, 2009, **11**, 946.
83. C. E. Song, W. H Shim, E. J. Roh and J. H. Choi, *Chem. Commun.*, 2000, 1695.
84. C. E. Song, D.-u. Jung, S. Y. Choung, E. J. Roh and S.-g. Lee, *Angew. Chem., Int. Ed.*, 2004, **43**, 6183.
85. M. Y. Yoon, J. H. Kim, D. S. Choi, U. S. Shin, J. Y. Lee and C. E. Song, *Adv. Synth. Catal.*, 2007, **349**, 1725.
86. J. H. Kim, J. W. Lee, U. S. Shim, J. Y. Lee, S.-g Lee and S. E. Song, *Chem. Commun.*, 2007, 4683.
87. D. S. Choi, J. H. Kim, U. S. Shin, R. R. Deshmuhk and C. E. Song, *Chem. Commun.*, 2007, 3482.
88. R. R. Deshmukh, J. W. Lee, U. S. Shin, J. Y. Lee and C. E. Song, *Angew. Chem., Int. Ed.*, 2008, **47**, 8615.

89. I. Meracz and T. Oh, *Tetrahedron Lett.*, 2003, **44**, 6465.
90. C.-E. Yeom, H. W. Kim, Y. J. Shin and B. M. Kim, *Tetrahedron Lett.*, 2007, **48**, 9035.
91. S. Doherty, P. Goodrich, C. Hardacre, C. Paun and V. Parvulescu, *Adv. Synth. Catal.*, 2008, **350**, 295.
92. J. Y Shin, D. J. Jung and S. G. Lee, *ACS Catal.*, 2013, **3**, 525.
93. K. Bica, P. Gärtner, P. J. Gritsch, A. K. Ressmann, C. Schröder and R. Zirbs, *Chem. Commun.*, 2012, **48**, 5013.
94. A. Aggarwal, N. L. Lancaster, A. R. Sethi and T. Welton, *Green Chem.*, 2002, **4**, 517.
95. F. Liu, L. Wang, Q. Sun, L. Zhu, X. Meng and F.-S. Xiao, *J. Am. Chem. Soc.*, 2012, **134**, 16948.
96. W. Cabri, I. Candiani, A. Bedeschi and R. Santi, *J. Org. Chem.*, 1993, **58**, 7421.
97. W. Cabri, I. Candiani, A. Bedeschi and R. Santi, *J. Org. Chem.*, 1992, **57**, 3558.
98. W. Cabri, I. Candiana, A. Bedeschi and A. Penco, *J. Org. Chem.*, 1992, **57**, 1481.
99. J. Mo, L. Xu and J. Xiao, *J. Am. Chem. Soc.*, 2005, **127**, 751.
100. C. Gürtler and S. L. Buchwald, *Chem. Eur. J.*, 1999, **5**, 3107.
101. C. Caló, A. Nacci, A. Monopoli, A. Detomaso and P. Iliade, *J. Org. Chem.*, 2003, **68**, 2929.
102. S. Doherty, P. Goodrich, C. Hardacre, J.-K. Luo, D. W. Rooney, K. R. Seddon and P. Styring, *Green Chem.*, 2004, **6**, 63.
103. M. J. Beier, J.-M. Andanson and A. Baiker, *ACS Catal.*, 2012, **2**, 2587.
104. M. N. Kashid, A. Renken and L. Kiwi-Minsker, *Chem. Eng. Sci.*, 2011, **66**, 1480.
105. Md. T. Rahman, T. Fukuyama, N. Kamata, M. Sato and I. Ryu, *Chem. Commun.*, 2006, 2236.
106. S. Liu, T. Fukuyama, M. Sato and I. Ryu, *Org. Process Res. Dev.*, 2004, **8**, 477.
107. M. Ruta, I. Yuranov, P. J. Dyson, G. Laurenczy and L. Kiwi-Minsker, *J. Catal.*, 2007, **247**, 269.
108. M. Ruta, G. Laurenczy, P. J. Dyson and L. Kiwi-Minsker, *J. Phys. Chem. C*, 2008, **112**, 17814.
109. P. Virtanen, H. Karhu, K. Kordas and J.-P. Mikkola, *Chem. Eng. Sci.*, 2007, **62**, 3660.
110. P. Virtanen, J.-P. Mikkola and T. Salmi, *Ind. Eng. Chem. Res.*, 2007, **46**, 9022.
111. K. Scott, N. Basov, R. J. J. Jachuck, N. Winterton, A. Cooper and C. Davies, *Chem. Eng. Res. Des.*, 2005, **83**, 1179.
112. E. Mizushima, T. Hayashi and M. Tanaka, *Green Chem.*, 2001, **3**, 76.
113. (a) E. Mizushima, T. Hayashi and M. Tanaka, *Green Chem.*, 2001, **3**, 76; (b) E. Mizushima, T. Hayashi and M. Tanaka, *Top. Catal.*, 2004, **29**, 163.
114. X. Zhao, H. Alper and Z. Yu, *J. Org. Chem.*, 2006, **71**, 3988.

115. F. Ye and H. Alper, *J. Org. Chem.*, 2007, **72**, 3218.
116. C. S. Consorti, G. Ebeling and J. Dupont, *Tetrahedron Lett.*, 2002, **43**, 753.
117. (a) H. Cao, W.-J. Xiao and H. Alper, *J. Org. Chem.*, 2007, **72**, 8562; (b) H. Cao, W.-J. Xiao and H. Alper, *Adv. Synth. Catal.*, 2006, **348**, 1807.
118. H. Cao, L. McNamee and H. Alper, *J. Org. Chem.*, 2008, **73**, 3530.
119. Q. Yang and H. Alper, *J. Org. Chem.*, 2010, **75**, 948.
120. H. Cao, L. McNamee and H. Alper, *Org. Lett.*, 2008, **10**, 5281.
121. J. McNulty, J. J. Nair and A. Robertson, *Org. Lett.*, 2007, **9**, 4575.
122. Q. Lin, C. Yang, W. Jiang, H. Chen and X. Li, *J. Mol. Catal. A: Chem.*, 2007, **264**, 17.
123. B. Gabriele, R. Mancuso, E. Lupinacci, G. Salerno and L. Veltri, *Tetrahedron*, 2012, **66**, 6156.
124. Y. Li, Z. Yu and H. Alper, *Org. Lett.*, 2007, **9**, 1647.
125. Q. Yang, A. Robertson and H. Alper, *Org. Lett.*, 2008, **10**, 5079.
126. F. Shi, Y. He, D. Li, Y. Ma, Q. Zhang and Y. Deng, *J. Mol. Catal. A: Chem.*, 2006, **244**, 64.
127. F. Shi, J. Peng and Y. Deng, *J. Catal.*, 2003, **219**, 372.
128. (a) Y. Li, H. Alper and Z. Yu, *Org. Lett.*, 2006, **8**, 5199; (b) E. J. Garcia-Suárez, S. G. Khokarale, O. N. van Buu, R. Fehrmann and A. Riisager, *Green Chem.*, 2014, **16**, 161.
129. F. Shi, Y. Deng, T. S. Ma, J. Peng, Y. Gu and B. Qiao, *Angew. Chem., Int. Ed.*, 2003, **42**, 3257.
130. Y. N. Shim, J. K. Lee, J. K. Im, D. K. Mukherjee, D. Q. Nguyen, M. Cheong and H. S. Kim, *Phys. Chem. Chem. Phys.*, 2011, **13**, 6197.
131. H. S. Kim, Y. J. Kim, H. Lee, K. Y. Park, C. Lee and C. S. Chin, *Angew. Chem., Int. Ed.*, 2002, **41**, 4300.
132. H. S. Kim, Y. J. Kim, J. Y. Bae, S. J. Kim, M. S. Lah and C. S. Chin, *Organometallics*, 2003, **22**, 2498.
133. Y.-S. Choi, Y. N. Shim, J. Lee, J. H. Yoon, C. S. Hong, M. Cheong, H. S. Kim, H. G. Jang and J. S. Lee, *Appl. Catal., A*, 2011, **404**, 87.
134. F. Shi, Q. Zhang, D. Li and Y. Deng, *Chem. Eur. J*, 2005, **11**, 5279.
135. M. Haumann and A. Riisager, *Chem. Rev.*, 2008, **108**, 1474.
136. Y. Chauvin, L. Mussmann and H. Olivier, *Angew. Chem., Int. Ed.*, 1995, **34**, 2698.
137. A. Sharma, C. J. Lebigue, Ra. M. Deshpande, A. A. Kelkar and H. Delmas, *Ind. Eng. Chem. Res.*, 2010, **49**, 10698.
138. M. Kranenburg, Y. E. M. van der Burgt, P. C. J. Kamer, P. W. N. M. van Leeuwen, K. Goubitz and J. Fraanje, *Organometallics*, 1995, **14**, 3081.
139. L. A. van der Veen, M. D. K. Boele, F. R. Bregman, P. C. J. Kamer, P. W. N. M. van Leeuwen, K. Goubitz, J. Fraanje, H. Schenk and C. Bo, *J. Am. Chem. Soc.*, 1998, **120**, 11616.
140. P. Wasserscheid, H. Waffenschmidt, P. Machnitzki, K. W. Kottsieper and O. Stelzer, *Chem. Commun.*, 2001, 451.
141. L. A. van der Veen, P. C. J. Kamer and P. W. M. N. van Leeuwen, *Organometallics*, 1999, **18**, 4765.

142. L. A. van der Veen, P. C. J. Kamer and P. W. M. N. van Leeuwen, *Angew. Chem., Int. Ed.*, 1999, **38**, 336.
143. R. P. J. Bronger, S. M. Silva, P. C. J. Kamer and P. W. N. M. van Leeuwen, *Chem. Commun.*, 2002, 3044.
144. R. P. J. Bronger, S. M. Silva, P. C. J. Kamer and P. W. N. M. van Leeuwen, *Dalton Trans.*, 2004, 1590.
145. J. Dupont, S. M. Silva and R. F. de Souza, *Catal. Lett.*, 2001, **77**, 131.
146. R. M. Deshpande, A. A. Kelkar, A. Sharma, C. Julcour-Lebigue and H. Delmas, *Chem. Eng. Sci.*, 2011, **66**, 1631.
147. F. Favre, H. Olivier-Bourbigou, D. Commereuc and L. Saussine, *Chem. Commun.*, 2001, 1360.
148. K. W. Kottsieper, O. Stelzer and P. Wasserscheid, *J. Mol. Catal. A: Chem.*, 2001, **175**, 285.
149. A. Behr, F. Naendrup and S. Nave, *Eng. Life Sci.*, 2003, **3**, 325.
150. P. B. Webb, M. F. Sellin, T. E. Kunene, S. Williamson, A. M. Z. Slawin and D. J. Cole-Hamilton, *J. Am. Chem. Soc.*, 2003, **125**, 15577.
151. P. Wasserscheid and W. Keim, *Angew. Chem., Int. Ed.*, 2000, **39**, 3773.
152. T. E. Kunene, P. B. Webb and D. J. Cole-Hamilton, *Green Chem.*, 2011, **13**, 1476.
153. A. Riisager, R. Fehrmann, M. Haumann, B. S. K. Gorle and P. Wasserscheid, *Ind. Eng. Chem. Res.*, 2005, **44**, 9853.
154. M. Haumann, K. Dentler, J. Joni, A. Riisager and P. Wasserscheid, *Adv. Synth. Catal.*, 2007, **349**, 425.
155. M. Haumann, M. Jakuttis, S. Werner and P. Wasserscheid, *J. Catal.*, 2009, **263**, 321.
156. A. Behr, D. Obst, C. Schulte and T. Schosser, *J. Mol. Catal. A: Chem.*, 2003, **206**, 179.
157. M. Jakuttis, A. Schönweiz, S. Werner, R. Franke, K.-D. Wiese, M. Haumann and P. Wasserscheid, *Angew. Chem., Int. Ed.*, 2011, **50**, 4492.
158. U. Hintermair, Z. Gong, A. Serbanovic, M. J. Muldoon, C. C. Santini and D. J. Cole-Hamilton, *Dalton Trans.*, 2010, **39**, 8501.
159. C. Kolbeck, N. Paape, T. Cremer, P. S. Schulz, F. Maier, H-P. Steinrück and P. Wasserscheid, *Chem. Eur. J.*, 2010, **16**, 12083.
160. N. Sieffert and G. Wipff, *J. Phys. Chem.*, 2007, **111**, 4951.
161. N. Sieffert and G. Wipff, *J. Phys. Chem. C*, 2008, **112**, 6450.
162. S. Shylesh, D. Hanna, S. Werner and A. T. Bell, *ACS Catal.*, 2012, **2**, 487.
163. T. Kato, N. Mizoshita and K. Kishimoto, *Angew. Chem., Int. Ed.*, 2006, **118**, 44.
164. J. Yang, F.-F. Li, J. Zhang and W.-X. Wang, *Helv. Chim. Acta*, 2010, **93**, 1653.
165. Y. Diao, J. Li, L. Wang, P. Yang, R. Yan, H. Zhang and S. Zhang, *Catal. Today*, 2013, **200**, 54.
166. F. Belina and C. Chiappe, *Molecules*, 2010, **15**, 2211.
167. X. Wu, J. Mo, X. Li, Z. Hyder and J. Xiao, *Prog. Nat. Sci.*, 2008, **18**, 639.
168. D. Kaufmann, M. Nouroozian and H. Henze, *Synlett*, 1996, 1091.

169. W. Hermann and V. P. W. Böhm, *J. Organomet. Chem.*, 1999, **572**, 141.
170. V. P. W. Böhm and W. Hermann, *Chem. Eur. J.*, 2000, **6**, 1017.
171. A. J. Carmichael, M. J. Earle, J. D. Holbrey, P. B. McCormac and K. R. Seddon, *Org. Lett.*, 1999, **1**, 997.
172. S. Bouquillon, B. Ganchegui, B. Estrine, F. Hénin and J. Muzart, *J. Organomet. Chem.*, 2001, **634**, 153.
173. G. Battistuzzi, S. Cacchi and G. Fabrizi, *Synlett*, 2002, 439.
174. M. Reetz and M. Maase, *Adv. Mater.*, 1999, **11**, 773.
175. V. Calò, A. Nacci, L. Lopez and N. Mannarini, *Tetrahedron Lett.*, 2000, **41**, 8973.
176. V. Calò, A. Nacci, A. Monopoli, L. Lopez and A. di Cosmo, *Tetrahedron*, 2001, **57**, 6071.
177. K. S. A. Vallin, P. Emilson, M. Larhead and A. Hallberg, *J. Org. Chem.*, 2002, **67**, 6243.
178. S. B. Park and H. Alper, *Org. Lett.*, 2003, **5**, 3209.
179. N. A. Hamill, C. Hardacre and S. E. J. McMath, *Green Chem.*, 2002, **4**, 139.
180. L. Xu, W. Chen and J. Xiao, *Organometallics*, 2000, **19**, 1123.
181. C.-M. Jin, B. Twamley and J. M. Shreeve, *Organometallics*, 2005, **24**, 3020.
182. M. R. Gyton, M. L. Cole and J. B. Harper, *Chem. Commun.*, 2011, **47**, 9200.
183. C. Amatore, E. Carre and A. Jutand, *Acta Chem. Scand.*, 1998, **52**, 100.
184. L. Xu, W. Chen, J. Ross and J. Xiao, *Org. Lett.*, 2001, **3**, 295.
185. M. Portnoy, Y. Ben-David, I. Rousso and D. Milstein, *Organometallics*, 1994, **13**, 3465.
186. J. Mo, J. Ruan, L. Xu, Z. Hyder, O. Saidi, S. Liu, W. Pei and J. Xiao, *J. Mol. Catal. A: Chem.*, 2007, **261**, 267.
187. Z. Hyder, J. Mo and J. Xiao, *Adv. Synth. Catal.*, 2006, **348**, 1699.
188. J. Mo, L. Xu, J. Ruan, S. Liu and J. Xiao, *Chem. Commun.*, 2006, 3591.
189. V. Caló, A. Nacci, A. Monopoli and V. Ferola, *J. Org. Chem.*, 2007, **72**, 2596.
190. J.-C. Xiao, B. Twamley and J. M. Shreeve, *Org. Lett.*, 2004, **6**, 3845.
191. R. Wang, M. M. Piekarski and J. M. Shreeve, *Org. Biomol. Chem.*, 2006, **4**, 1878.
192. H. M. Lee, P. L. Chiu, C. H. Hu, C. L. Lai and Y. C. Chou, *J. Organomet. Chem.*, 2005, **690**, 403.
193. R. Wang, J.-C. Xiao, B. Twamley and J. M. Shreeve, *Org. Biomol. Chem.*, 2007, **5**, 671.
194. Q.-X. Wan, Y. Liu and Y.-Q. Cai, *Catal. Lett.*, 2009, **127**, 386.
195. R. Wang, B. Twamley and J. M. Shreeve, *J. Org. Chem.*, 2006, **71**, 426.
196. J. Wei, H.-Y. Fu, R.-X. Li, H. Chen and X.-J. Li, *Catal. Commun.*, 2011, **12**, 748.
197. Y. Liu, M. Li, Y. Lu, G.-H. Gao, Q. Yang and M.-Y. He, *Catal. Commun.*, 2006, 7, 985.

198. Z. D. Petrović, D. Simijonović, V. P. Petrović and S. Marković, *J. Mol. Catal. A: Chem.*, 2010, **327**, 45.
199. J. Liu, H. Liu and L. Wang, *Appl. Organomet. Chem.*, 2010, **24**, 386.
200. R. V. Deshmukh, R. Rajagopal and K. V. Srinivasan, *Chem. Commun.*, 2001, 1544.
201. J. Mo and J. Xiao, *Angew. Chem., Int. Ed.*, 2006, **45**, 4152.
202. J. C. Càrdenas, L. Fadini and C. A. Sierra, *Tetrahedron Lett.*, 2010, **51**, 6867.
203. R. Roszak, A. M. Trzeciak, J. Pernak and N. Borucka, *Appl. Catal., A*, 2011, **490**, 148.
204. R. Kumar, A. Shard, R. Bharti, Y. Thopate and A. K. Sinha, *Angew. Chem., Int. Ed.*, 2012, **51**, 2636.
205. J. C. Pastre, Y. Genisson, N. Saffon, J. Dandurand and C. R. D. Corraia, *J. Braz. Chem. Soc.*, 2010, **21**, 812.
206. R. G. Kalkhambkar and K. K. Laali, *Tetrahedron Lett.*, 2011, **52**, 1733.
207. S. Volland, M. Gruit, T. Régnier, L. Viau, O. Lavastre and A. Vioux, *New J. Chem.*, 2009, **33**, 2015.
208. C. J. Mathews, P. J Smith and T. Welton, *Chem. Commun.*, 2000, 1249.
209. C.-H. Yang, C.-C. Tai, T.-T. Huang and I.-W. Sun, *Tetrahedron*, 2005, **61**, 4857.
210. R. Rajagopal, D. V. Jarikote and K. V. Srinivasan, *Chem. Commun.*, 2002, 616.
211. C. J. Matthews, P. J. Smith and T. Welton, *J. Mol. Catal. A: Chem.*, 2004, **214**, 27.
212. J. McNulty, A. Capretta, J. Wilson, J. Dyck, G. Adjabeng and A. Robertson, *Chem. Commun.*, 2002, 1986.
213. J. D. Revell and A. Ganesan, *Org. Lett.*, 2002, **4**, 3071.
214. X. Yang, Z. Fei, T. J. Geldbach, A. D. Phillips, C. G. Hartinger, Y. Li and P. J. Dyson, *Organometallics*, 2008, **27**, 3971.
215. M. Lombardo, M. Chiarucci and C. Trombini, *Green Chem.*, 2009, **11**, 574.
216. A. Papagni, C. Trombini, M. Lombardo, S. Bergantin, A. Chams, M. Chiarucci, L. Miozzo and M. Parravicini, *Organometallics*, 2011, **30**, 4325.
217. L. Li, J. Wang, T. Wu and R. Wang, *Chem. Eur. J.*, 2012, **18**, 7842.
218. N. Iranpoor, H. Firouzabadi and Y. Ahmadi, *Eur. J. Org. Chem.*, 2012, 305.
219. S. T. Handy and X. Zhang, *Org. Lett.*, 2001, **3**, 233.
220. W. Hao, Z. Xi and M. Cai, *Synth. Commun.*, 2012, **42**, 2396.
221. J. McNulty, A. Capretta, J. Wilson, J. Dyck, G. Adjabeng and A. J. Robertson, *Chem. Commun.*, 2002, 1986.
222. J. McNulty, S. Cheekoori, J. J. Nair, V. Larichev, A. Capretta and A. J. Robertson, *Tetrahedron Lett.*, 2005, **46**, 3641.
223. J. McNulty, S. Cheekoori, T. P. Bender and J. A. Coggan, *Eur. J. Org. Chem.*, 2007, 1423.
224. V. Conte, G. Fiorani, B. Floris, P. Galloni and S. Woodward, *Appl. Catal., A*, 2010, **381**, 161.

225. P. S. Bäuerlein, I. J. S. Fairlamb, A. G. Jarvis, A. F. Lee, C. Müller, J. M. Slattery, R. J. Thatcher, D. Vogt and A. C. Whitwood, *Chem. Commun.*, 2009, 5734.
226. S. Liu and J. Xiao, *J. Mol. Catal. A: Chem.*, 2007, **107**, 1.
227. (a) W. A. Hermann, R. W. Fischer and D. W. Marz, *Angew. Chem., Int. Ed.*, 1991, **30**, 1638; (b) M. Crucianelli, R. Saladino and F. D. Angilis, *ChemSusChem*, 2010, **3**, 524.
228. G. S. Owens and M. M. Abu-Omar, *Chem. Commun.*, 2000, 1165.
229. G. S. Owens, A. Durazo and M. M. Abu-Omar, *Chem. Eur. J.*, 2002, **8**, 3053.
230. K. A. Srinivas, A. Kumar and M. S. Chauhan, *Chem. Commun.*, 2002, 2456.
231. Y. Liu, H.-J. Zhang, Y. Lu, Y.-Q. Cai and Z.-L. Liu, *Green Chem.*, 2007, **9**, 1114.
232. C. E. Song and E. J. Roh, *Chem. Commun.*, 2000, 837.
233. L. Chen, J. Wei, N. Tang and F. Cheng, *Catal. Lett.*, 2012, **142**, 486.
234. L. Chen, F. Cheng, L. Jia, A. Zhang, J. Wu and N. Tang, *Chirality*, 2011, **23**, 69.
235. R. Tang, D. Yin, N. Jin, H. Zhao and D. Yin, *J. Catal.*, 2008, **255**, 287.
236. A. A. Valente, Z. Petrovski, L. C. Branco, C. A. M. Afonso, M. Pillinger, A. D. Lopes, C. C. Romão, C. D. Nunes and I. S. Gonçalves, *J. Mol. Catal. A: Chem.*, 2004, **218**, 5.
237. D. Betz, W. A. Hermann and F. E. Kühn, *J. Organomet. Chem.*, 2009, **694**, 3320.
238. A. Günyar, D. Betz, M. Drees, E. Herdtweck and F. E. Kühn, *J. Mol. Catal. A: Chem.*, 2010, **331**, 117.
239. M. Herbert, F. Montilla, R. Moyano, A. Pastor, E. Álvarez and A. Galindo, *Polyhedron*, 2009, **28**, 3929.
240. M. Herbert, A. Galindo and F. Montilla, *Catal. Commun.*, 2007, **8**, 987.
241. S. K. Maiti, S. Dinda and R. Bhattacharyya, *Tetrahedron Lett.*, 2008, **49**, 6205.
242. J. A. Brito, S. Ladeira, E. Teuma, B. Royo and M. Gómez, *Appl. Catal., A*, 2011, **398**, 88.
243. M. Herbert, E. Álvarez, D. J. Cole-Hamilton, F. Montilla and A. Gallindo, *Chem. Commun.*, 2010, **46**, 5933.
244. M. Herbert, F. Montilla, A. Galindo, R. Moyano, A. Pastor and E. Álvarez, *Dalton Trans.*, 2011, **40**, 5210.
245. C. Bibal, J.-C. Daran, S. Deroover and R. Poli, *Polyhedron*, 2010, **29**, 639.
246. C. Conte, F. Fabbianesi, B. Flors, P. Galloni, D. Sordi, I. W. C. E. Arends, M. Bonchio, D. Rehder and D. Bogdal, *Pure Appl. Chem.*, 2009, **81**, 1265.
247. L. L. Liu, C. C. Chen, X. F. Hu, T. Mohamood, W. H. Ma, J. Lin and J. C. Zhao, *New J. Chem.*, 2008, **32**, 283.
248. A. Kumar, *Catal. Commun.*, 2007, **8**, 913.
249. L. Gharnati, O. Walter, U. Arnold and M. Döring, *Eur. J. Inorg. Chem.*, 2011, 2756.
250. H. Li, Z. Hou, Y. Qiao, B. Feng, Y. Hu, X. Wang and X. Zhao, *Catal. Commun.*, 2010, **11**, 470.

251. S.-S. Wang, W. Liu, Q.-X. Wan and Y. Liu, *Green Chem.*, 2009, **11**, 1589.
252. K. Yamaguchi, C. Yoshida, S. Uchida and N. Mizuno, *J. Am. Chem. Soc.*, 2005, **127**, 530.
253. M. Dakkach, X. Fontrodona, T. Parella, A. Atlamsani, I. Romero and M. Rodríguez, *Adv. Synth. Catal.*, 2011, **353**, 231.
254. A. J. Kotlewska, F. van Rantwijk, R. A. Sheldon and I. W. C. F. Arends, *Green Chem.*, 2011, **13**, 2154.
255. V. V. Namboodiri, R. S. Varma, E. Sahle-Demessie and U. R. Pillai, *Green Chem.*, 2002, **4**, 170.
256. I. A. Ansari, S. Joyasawal, M. K. Gupta, J. S. Yadav and R. Gree, *Tetrahedron Lett.*, 2005, **46**, 7507.
257. J. Peng, J. Li, H. Qiu, J. Jiang, K. Jiang, J. Mao and G. Lai, *J. Mol. Catal. A: Chem.*, 2006, **255**, 16.
258. C. Chiappe, A. Sanzone and P. J. Dyson, *Green Chem.*, 2011, **13**, 1437.
259. I. A. Ansari and R. Gree, *Org. Lett.*, 2002, **4**, 1507.
260. N. Jiang and A. J. Ragauskas, *Org. Lett.*, 2005, **7**, 3689.
261. L. Lin, J. Liuyan and W. Yunyang, *Catal. Commun.*, 2008, **9**, 1379.
262. N. Jiang and A. J. Ragauskas, *Tetrahedron Lett.*, 2005, **46**, 3323.
263. X. E. Wu, L. Ma, M. X. Ding and L. X. Gao, *Synlett*, 2005, 607.
264. L. Lin, M. Juanjuan, L. Liuyan and W. Yungang, *J. Mol. Catal. A: Chem.*, 2008, **291**, 1.
265. C.-X. Miao, L.-N He, J.-Q. Wang and J.-L. Wang, *Adv. Synth. Catal.*, 2009, **351**, 2209.
266. R. Hu, M. Lei, H. Wei and Y. Wang, *Chin. J. Chem.*, 2009, **27**, 587.
267. A. Chrobok, A. Baj, W. Pudło and A. Jarzębski, *Appl. Catal., A*, 2010, **389**, 179.
268. V. Farmer and T. Welton, *Green Chem.*, 2002, **4**, 97.
269. A. Wolfson, S. Wuyts, D. E. De Vos, I. F. J. Vankelcom and P. A. Jacobs, *Tetrahedron Lett.*, 2002, 8107.
270. B. S. Chhikara, R. Chandra and V. Tandon, *J. Catal.*, 2005, **230**, 436.
271. (a) K. P. Peterson and R. C. Larock, *J. Org. Chem.*, 1998, **63**, 3185; (b) K. R. Seddon and A. Stark, *Green Chem.*, 2002, **4**, 119.
272. C. Van Dooslaer, Y. Schellekens, P. Mertens, K. Binnemans and D. De Vos, *Phys. Chem. Chem. Phys.*, 2010, **12**, 1741.
273. D. Ramakrishna, B. Ramachandra and R. Karvembu, *Catal. Commun.*, 2010, **11**, 498.
274. Q. Yao, *Org. Lett.*, 2002, **4**, 2197.
275. R. Yanada and Y. Takemoto, *Tetrahedron Lett.*, 2002, **43**, 6849.
276. (a) L. C. Branco and C. A. M. Afonso, *Chem. Commun.*, 2002, 3036; (b) L. C. Branco and C. A. M. Afonso, *J. Org. Chem.*, 2004, **69**, 4381.
277. C. E. Song, D.-a. Jung, E. J. Roh, S.-g. Lee and D. Y. Chi, *Chem. Commun.*, 2002, 3038.
278. L. C. Branco, A. Serbanovic, M. N. de Ponte and C. A. M. Afonso, *Chem. Commun.*, 2005, 107.
279. L. C. Branco, A. Serbanovic, M. N. da Ponte and C. A. M. Afonso, *ACS Catal.*, 2011, **1**, 1408.

280. H. Li, L. He, J. Lu, W. Zhu, X. Jiang, Y. Wang and Y. Tan, *Energy Fuels*, 2009, **23**, 1354.
281. (a) W. Zhu, H. Li, X. Jiang, Y. Yan, J. Lu, L. He and J. Xia, *Green Chem.*, 2008, **10**, 641; (b) L. He, H. Li, W. Zhu, J. Guo, X. Jiang, J. Ldu and Y. Yan, *Ind. Eng. Chem. Res.*, 2008, **47**, 6890.
282. H. Li, X. Jiang, W. Zhu, J. Lu, H. Shu and Y. Yan, *Ind. Eng. Chem. Res.*, 2009, **48**, 9034.
283. D. Xu, W. Zhu, H. Li, J. Zhang, F. Zou, H. Shi and Y. Yan, *Energy Fuels*, 2009, **23**, 5929.
284. W. Huang, W. Zhu, H. Li, H. Shi, G. Zhu, H. Li and G. Chen, *Ind. Eng. Chem. Res.*, 2010, **49**, 8998.
285. (a) Y. Ding, W. Zhu, H. Li, W. Jeng, M. Zhang, Y. Duan and Y. Chang, *Green Chem.*, 2011, **13**, 1210; (b) W. Zhu, Y. Ding, H. Li, J. Qin, Y. Yanhong, J. Xiong, Y. Xu and H. Liu, *RSC Advances*, 2013, **3**, 3893.
286. A. Mota, N. Butenko, J. P. Hallett and I. Correia, *Catal. Today*, 2012, **196**, 119.
287. (a) W. Li, Y. Zhu, J. Wang, J. Zhang, J. Li and Y. Yan, *Green Chem.*, 2009, **11**, 810; (b) Y. Jiang, W. Zhu, H. Li, S. Yin, H. Liu and Q. Xie, *ChemSusChem*, 2011, **4**, 399.
288. (a) D. Zhao, J. Wang and E. Zhou, *Green Chem.*, 2007, **9**, 1219; (b) J. Wang, D. Zhao and K. Li, *Energy Fuels*, 2009, **23**, 3831.
289. L. Lu, S. Chen, J. Gao, G. Gao and M-y. He, *Energy Fuels*, 2007, **21**, 383.
290. H. Gao, C. Guo, J. Xing, J. Zhao and H. Liu, *Green Chem.*, 2012, **12**, 1220.
291. (a) J. Gui, D. Liu, Z. Sun, D. Liu, D. Min, B. Song and X. Peng, *J. Mol. Catal. A: Chem.*, 2010, **331**, 64; (b) C. Yansheng, L. Changping, J. Qingzhu, L. Qingshan, L. Xiumei and U. Welz-Biermann, *Green Chem.*, 2011, **13**, 1224.
292. J. Xu, S. Zhao, W. Chen, M. Wang and Y.-F. Song, *Chem. Eur. J.*, 2012, **18**, 4775.
293. J. Xu, S. Zhao, Y. Ji and Y-F. Song, *Chem. Eur. J.*, 2013, **19**, 709.
294. X. Fan, Y. Wang, Y. He, X. Zhang and J. Wang, *Tetrahedron Lett.*, 2010, **51**, 3493.
295. X. Fan, Y. He, Y. Wang, X. Zhang and J. Wang, *Tetrahedron Lett.*, 2011, **52**, 899.
296. A. A. Lindén, M. Johansson, N. Hermann and J.-E. Bäckvall, *J. Org. Chem.*, 2006, **71**, 3849.
297. S. Wang, L. Wang, M. Davovic, Z. Popovic, H. Wu and Y. Liu, *ACS Catal.*, 2012, **2**, 230.
298. X.-Y. Shi and J.-F. Wei, *J. Mol. Catal. A: Chem.*, 2008, **280**, 142.
299. X. Shi, X. Han, W. Ma, J. Wei, J. Li, Q. Zhang and Z. Chen, *J. Mol. Catal. A: Chem.*, 2011, **341**, 57.
300. Ch. V. Reddy and J. G. verkade, *J. Mol. Catal. A: Chem.*, 2007, **272**, 233.
301. L. Palomb, C. Bocchino, T. Caruso, R. Villano and A. Scettri, *Tetrahedron Lett.*, 2008, **49**, 5611.
302. R. Bernini, A. Coratti, G. Fabrizi and A. Goggiamani, *Tetrahedron Lett.*, 2003, **44**, 8991.

303. V. Conte, B. Foris, P. Galloni, V. Mirruzzo, A. Scarso, D. Sordi and G. Strukul, *Green Chem.*, 2005, **7**, 262.
304. S. Baj, A. Chrobok and R. Slupska, *Green Chem.*, 2009, **11**, 279.
305. A. Chrobok, S. Baj, W. Pudło and A. Jarzębski, *Appl. Catal., A*, 2009, **366**, 22.
306. A. Chrobok, *Tetrahedron*, 2010, **66**, 2940.
307. P. J. Dyson, G. Laurenczy, C. A. Ohlin, J. Vallance and T. Welton, *Chem. Commun.*, 2003, 2418.
308. C. J. Boxwell, P. J. Dyson, D. J. Ellis and T. Welton, *J. Am. Chem. Soc.*, 2003, **124**, 9334.
309. P. J. Dyson, D. J. Ellis, D. G. Parker and T. Welton, *Chem. Commun.*, 1999, 25.
310. H. Zhou, Z. Li, Z. Wang, T. Wang, L. Xu, Y. He, Q.-H. Fan, J. Pan, L. Gu and A. S. C Chan, *Angew. Chem., Int. Ed.*, 2008, **47**, 8464.
311. (a) P. A. Z. Suarez, J. E. L. Dullius, S. Einloft, R. F. De Souza and J. Dupont, *Polyhedron*, 1996, **15**, 1217; (b) P. A. Z. Suarez, J. E. L. Dullius, S. Einloft, R. F. De Souza and J. Dupont, *Inorg. Chim. Acta*, 1997, **255**, 207.
312. D. Zhao, P. J. Dyson, G. Laurenczy and J. S. McIndoe, *J. Mol. Catal. A: Chem.*, 2004, **214**, 19.
313. A. L. Monterio, F. K. Zinn, R. F. de Souza and J. Dupont, *Tetrahedron: Asymmetry*, 1997, **2**, 177.
314. A. Berger, R. D. de Souza, M. R. Delgado and J. Dupont, *Tetrahedron: Asymmetry*, 2001, **12**, 1825.
315. S. Guernik, A. Wolfson, M. Herskowitz, N. Greenspoon and S. Geresh, *Chem. Commun.*, 2001, 2314.
316. P. G. Jessop, R. R. Stanley, R. A. Brown, C. A. Eckert, C. L. Liotta, T. T. Ngo and P. Pollet, *Green Chem.*, 2003, **5**, 123.
317. B. Pugin, M. Studer, E. Kuesters, G. Sedelmeier and X. Feng, *Adv. Synth. Catal.*, 2004, **246**, 1481.
318. A. Wolfson, I. F. J. Vankelecom and P. A. Jacobs, *J. Organomet. Chem.*, 2005, **690**, 3558.
319. M. V. Escárcega-Bobadilla, L. Rodríguez-Pérez, E. Teuma, P. Serp, A. M. Masdeu-Bultó and M. Gómez, *Catal. Lett.*, 2011, **141**, 808.
320. S.-G. Lee, Y. J. Zhang, J. Y. Piao, H. Yoon, C. E. Song, J. H. Choi and J. Hong, *Chem. Commun.*, 2003, 2624.
321. X. Feng, B. Pugin, E. Kuster, G. Sedelmeier and H.-U. Blaser, *Adv. Synth. Catal.*, 2007, **349**, 1803.
322. K. N. Gavrilov, S. E. Lyubimov, O. G. Bondarev, M. G. Maksimova, S. V. Zheglov, P. V. Petrovskii, V. A. Davankov and M. T. Reetz, *Adv. Synth. Catal.*, 2007, **349**, 609.
323. Y. Zhao, H. Huang, J. Shao and X. Xia, *Tetrahedron: Asymmetry*, 2011, **22**, 769.
324. M. Berthod, J.-M. Joerger, G. Mignani, M. Vaultier and M. Lemaire, *Tetrahedron: Asymmetry*, 2004, **15**, 2219.
325. X. Jin, X.-F. Xu and K. Zhao, *Tetrahedron: Asymmetry*, 2012, **23**, 1058.
326. J. K. Kassube and L. H. Gade, *Adv. Synth. Catal.*, 2009, **351**, 739.

327. (a) K. Mikami, M. Terada, T. Korenaga, Y. Matsumoto, M. Ueki and R. Angleaud, *Angew. Chem., Int. Ed.*, 2000, **39**, 3532; (b) J. W. Faller, A. R. Lavoie and J. Parr, *Chem. Rev.*, 2003, **103**, 3345; (c) K. Mikami and M. Yamanaka, *Chem. Rev.*, 2003, **103**, 3369.
328. M. Schmitkamp, D. Chen, W. Leitner, J. Klankermayer and G. Franció, *Chem. Commun.*, 2007, 4012.
329. D. Chen, M. Schmitkamp, G. Franció, J. Klankermayer and W. Leitner, *Angew. Chem., Int. Ed.*, 2008, **47**, 7339.
330. C. Comyns, N. Karodia, S. Zeler and J.-A. Andersen, *Catal. Lett.*, 2000, **67**, 113.
331. T. J. Geldbach and P. J. Dyson, *J. Am. Chem. Soc.*, 2004, **126**, 8114.
332. I. Kawasaki, K. Tsunoda, T. Tsuji, T. Yamaguchi, H. Shibuta, N. Uchida, M. Yamashita and S. Ohta, *Chem. Commun.*, 2005, 2134.
333. Z. Zhou, Y. Sun and A. Zhang, *Cent. Eur. J. Chem.*, 2011, **9**, 175.
334. W. Xiong, Q. Lin, H. Ma, H. Zheng, H. Chen and X. Li, *Tetrahedron: Asymmetry*, 2005, **16**, 1959.
335. Z. Yinghuai, K. Carpenter, C. C. Bun, S. Bahnmuller, C. P. Ke, V. S. Srid, L. W. Kee and M. F. Hawthorne, *Angew. Chem., Int. Ed.*, 2003, **42**, 3792.
336. A. Hu, H. L. Ngo and W. Lin, *Angew. Chem., Int. Ed.*, 2004, **43**, 2501.
337. N. L. Ngo, A. Hu and W. Lin, *Chem. Commun.*, 2003, 1912.
338. E. Öchsner, K. Schneiders, K. Junge, M. Beller and P. Wasserscheid, *Appl. Catal., A*, 2009, **364**, 8.
339. I. Cerna, P. Kluson, M. Bendova, T. Floris, H. Pelantova and T. Pekarek, *Chem. Eng. Process*, 2011, **50**, 264.
340. E. Öchsner, M. J. Schneider, C. Meyer, M. Haumann and P. Wasserscheid, *Appl. Catal., A*, 2011, **399**, 35.
341. M. J. Schneider, M. Haumann and P. Wasserscheid, *J. Mol. Catal. A: Chem.*, 2013, **376**, 103.
342. R. Giernoth and M. S. Krumm, *Adv. Synth. Catal.*, 2004, **346**, 989.
343. S. Kanz, A. Brinkemann, W. Leitner and A. Pfaltz, *J. Am. Chem. Soc.*, 1999, **121**, 6421.
344. P. S. Campbell, A. Podgoršek, T. Gutel, C. C. Santini, A. A. H. Pádua, M. F. Costa Gomes, F. Bayard, B. Fenet and Y. Chauvin, *J. Phys. Chem. B.*, 2010, **114**, 8156.
345. A. Podgoršek, G. Salas, P. S. Campbell, C. C. Santini, A. A. H. Pádua, M. F. C. Gomes, B. Fenet and Y. Chauvin, *J. Phys. Chem. B.*, 2011, **115**, 12150.
346. S. Aubin, F. le Floch, D. Carrie, J. P. Guegan and M. Vaultier, *ACS Symp. Ser.*, 2002, **818**, 334.
347. B. Weyershausen, K. Hell and U. Hesse, *Green Chem.*, 2005, **7**, 283.
348. T. J. Geldbach, D. Zhao, N. C. Castillo, G. Laurenczy, B. Weyershausen and P. J. Dyson, *J. Am. Chem. Soc.*, 2006, **128**, 9773.
349. H. Maciejewski, K. Szubert, B. Marciniec and J. Pernak, *Green Chem.*, 2009, **11**, 1045.
350. H. Maciejewski, K. Szubert, R. Fiedorow, R. Giszter, M. Niemczak, J. Pernak and W. Klimas, *Appl. Catal., A*, 2013, **451**, 168.

351. N. Hofmann, A. Bauer, T. Frey, M. Auer, V. Stanjek, P. S. Schulz, N. Taccardi and P. Wasserscheid, *Adv. Synth. Catal.*, 2008, **350**, 2599.
352. N. Taccardi, M. Fekete, M. Berger, V. Stanjek, P. S. Schulz and P. Wasserscheid, *Appl. Catal., A*, 2011, **399**, 69.
353. M. Cai, Y. Wang and W. Hao, *Eur. J. Org. Chem.*, 2008, 2983.
354. P. Goodrich, C. Hardacre, C. Paun, V. I. Parvulescu and I. Podolean, *Adv. Synth. Catal.*, 2008, **350**, 2473.
355. Z.-M. Zhou, Z.-H. Li, X.-Y. Hao, X. Dong, X. Li, L. Dai, Y.-Q. Liu, J. Zhang, H.-f. Huang, X. Li and J.-l. Wang, *Green Chem.*, 2011, **13**, 2963.
356. S. Doherty, P. Goodrich, C. Hardacre, H.-K. Luo, M. Nieuwenhuyzen and R. K. Rath, *Organometallics*, 2005, **24**, 5945.
357. X. Liu, Z. Pan, X. Shu, X. Duan and Y. Liang, *Synlett*, 2006, **12**, 1962.
358. I. Ambrogio, A. Arcadi, S. Cacchi, G. Fabrizi and F. Marinelli, *Synlett*, 2007, 1775.
359. O. Aksun and N. Krause, *Adv. Synth. Catal.*, 2008, **350**, 1106.
360. X. Moreau, A. Hours, L. Fensterbank and J.-P. Goddard, *J. Organomet. Chem.*, 2009, **694**, 561.
361. D .-M. Cui, Y.-N. Ke, D.-W. Zhuang, Q. Wang and C. Zhang, *Tetrahedron Lett.*, 2010, **51**, 980.
362. (a) A. Dzudza and T. J. Marks, *Org. Lett.*, 2009, **11**, 1523; (b) A. Dzudza and T. J. Marks, *Chem. Eur. J.*, 2010, **16**, 3403.
363. Y. Muyanohana, H. Inoue and N. Chatini, *J. Org. Chem.*, 2004, **69**, 8541.
364. J. A. Boon, J. A. Levisky, J. L. Pflug and J. S. Wilkes, *J. Org. Chem.*, 1986, **51**, 480.
365. C. J. Adams, M. J. Earle, G. Roberts and K. R. Seddon, *Chem. Commun.*, 1998, 2097.
366. (a) Z.-K. Zhao, W.-H. Qiao, Z.-S. Li, G.-R. Wang and L.-B. Cheng, *J. Mol. Catal. A: Chem.*, 2004, **222**, 207; (b) Z. Zhao, Z. Li, G. Wang, W. Qiao and L. Cheng, *Appl. Catal., A*, 2004, **262**, 69.
367. H. Xin, Q. Wu, M. Han, D. Wang and Y. Jin, *Appl. Catal., A*, 2005, **292**, 354.
368. Z.-W. Wang and L.-S. Wang, *Appl. Catal., A*, 2004, **262**, 101.
369. M. V. Alexander, A. C. Khandekar and S. D. Samant, *J. Mol. Catal. A: Chem.*, 2004, **223**, 75.
370. J. Gao, J.-Q. Wang, Q.-W. Song and L.-N. He, *Green Chem.*, 2011, **13**, 1182.
371. K. Yoo, V. V. Namboodiri, R. S. Varma and P. G. Smirniotis, *J. Catal.*, 2004, **222**, 511.
372. (a) S. Csihony, H. Mehdi and I. T. Horváth, *Green Chem.*, 2001, **3**, 307; (b) S. Csihony, H. Mehdi, Z. Homonnay, A. Vértes, O. Farkas and I. T. Horváth, *Dalton Trans.*, 2002, 680.
373. C. Meyer and P. Wasserscheid, *Chem. Commun.*, 2010, **46**, 7625.
374. Y. Xaio and S. V. Malhotra, *J. Mol. Catal. A: Chem.*, 2005, **230**, 129.
375. Y. Xiao and S. V. Malhorta, *J. Organomet. Chem.*, 2005, **690**, 3609.
376. S. Berardi, V. Conte, G. Fiorani, B. Floris and P. Galloni, *J. Organomet. Chem.*, 2008, **693**, 3015.

377. J. Ross and J. Xiao, *Green Chem.*, 2002, **4**, 129.
378. M. J. Earle, U. Hakala, B. J. McAuley, M. Nieuwenhuyzen, A. Ramani and K. R. Seddon, *Chem. Commun.*, 2004, 1368.
379. P. Goodrich, C. Hardacre, H. Mehdi, P. Nancarrow, D. W. Rooney and J. M. Thomson, *Ind. Eng. Chem. Res.*, 2006, **45**, 6640.
380. C. Hardacre, P. Nancarrow, D. W. Rooney and J. N. Thompson, *Org. Process Res. Dev.*, 2008, **12**, 1156.
381. S. Gmouh, H. Yang and M. Vaultier, *Org. Lett.*, 2003, **5**, 2219.
382. P. H. Tran, F. Duus and T. N. Le, *Tetrahedron Lett.*, 2012, **531**, 222.
383. Y. Wang, J. Yao, H. Li, D. Su and M. Antonietti, *J. Am. Chem. Soc.*, 2011, **133**, 2362.
384. W. Shen, L. Wang, J. Tang, Z. Qian and X. Tong, *Chin. J. Chem.*, 2010, **28**, 443.
385. F. Yao, W. Hao and M.-Z. Cai, *J. Organomet. Chem.*, 2013, **723**, 137.
386. R. Kurane, J. Jadhav, S. Khanapure, R. Salunkhe and G. Rashinkar, *Green Chem.*, 2013, **15**, 1849.
387. S. L. Jain and B. Sain, *Green Chem.*, 2006, **8**, 943.
388. I. Cano, M. C. Nicasio and P. J. Pérez, *Dalton Trans.*, 2009, 730.
389. P. Rodríguez, A. Caballero, M. M. Diaz-Requejo, M. C. Nicasio and P. J. Pérez, *Org. Lett.*, 2006, **8**, 557.
390. P. Rodríguez, M. C. Nicasio and P. J. Pérez, *Organometallics*, 2007, **26**, 6661.
391. D. L. Davies, S. K. Kandola and R. K. Patel, *Tetrahedron: Asymmetry*, 2004, **15**, 77.
392. B. Y. Park, K. Y. Y. Ryu, J. H. Park and S.-g. Lee, *Green Chem.*, 2009, **11**, 946.
393. A. Marra, A. Vecchi, C. Chiappe, B. Melai and A. Dondoni, *J. Org. Chem.*, 2008, **73**, 2458.
394. C. E. Song, C. R. Oh, E. J. Roh and D. J. Choo, *Chem. Commun.*, 2000, 1743.
395. J. Peng and Y. Deng, *New J. Chem.*, 2001, **25**, 639.
396. V. Caló, A. Nacci, A. Monopoli and A. Fanizzi, *Org. Lett.*, 2002, **4**, 2561.
397. H. Kawanami, A. Sasaki, K. Matsui and Y. Ikushima, *Chem. Commun.*, 2003, 896.
398. H. S. Kim, J. J. Kim, H. Kim and H. G. Jang, *J. Catal.*, 2003, **220**, 44.
399. J. Sun, S.-I. Fujita, F. Zhao and M. Arai, *Green Chem.*, 2004, **6**, 613.
400. (a) L. Xiao, D. Lv and W. Wu, *Catal. Lett.*, 2011, **141**, 1838; (b) J. Sun, L. Han, W. Cheng, J. Wang, X. Zhang and S. Zhang, *ChemSusChem*, 2011, **4**, 502; (c) L.-F. Xiao, F.-W. Li, J.-J. Peng and C.-G. Xia, *J. Mol. Catal. A: Chemical*, 2006, **253**, 265.
401. (a) W. Cheng, X. Chen, J. Sun, J. Wang and S. Zhang, *Catal. Today*, 2013, **200**, 117; (b) W.-L. Dai, L. Chen, S.-F. Yin, W.-H. Li, Y.-Y. Zhang, S.-L. Luo and C.-T. Au, *Catal. Lett.*, 2010, **137**, 74; (c) J. Sun, W. Cheng, W. Fan, Y. Wang, Z. Meng and S. Zhang, *Catal. Today*, 2009, **148**, 361; (d) R. A. Watile, K. M. Deshmukh, K. P. Dhake and B. M. Bhange, *Catal. Sci. Technol.*, 2012, **3**, 1051; (e) R. A. Watile, D. B. Bagal,

K. M. Deshmukh, K. P. Dhake and B. M. Bhange, *J. Mol. Catal. A: Chem.*, 2011, **351**, 196.
402. Y. Du, Y. Wu, A.-H. Liu and L.-N. He, *J. Org. Chem.*, 2008, **73**, 4709.
403. Y.-N. Zhao, Z.-Z. Yang, S.-H. Luo and L.-N. He, *Catal. Today*, 2013, **200**, 2.
404. Z.-Z. Yang, Y.-N. Zhao, L.-N. He, J. Gao and Z.-S. Yin, *Green Chem.*, 2012, **14**, 519.
405. Z.-Z. Yang, Y.-N. Li, Y.-Y. Wei and L.-N. He, *Green Chem.*, 2011, **13**, 2351.
406. Z.-Z. Yang, L.-N. He, C.-X. Miao and S. Chanfreau, *Adv. Synth. Catal.*, 2010, **352**, 2233.
407. Z.-Z. Yang, L.-N. He, S.-Y. Peng and A.-H. Liu, *Green Chem.*, 2010, **12**, 1850.
408. Q. Gong, H. Luo, J. Cao, Y. Shang, H. Zhang, W. Wang and X. Zhou, *Aust. J. Chem.*, 2012, **65**, 381.
409. T. Yu and R. G. Weiss, *Green Chem.*, 2012, **14**, 209.
410. M. Alvaro, C. Baleizao, D. Das, E. Carbonell and H. Garciía, *J. Catal.*, 2004, **228**, 254.
411. (a) Y. Xie, Z. Zhang, T. Jiang, J. He, B. Han, T. Wu and K. Ding, *Angew Chem., Int. Ed.*, 2007, **46**, 7255; (b) Y. Zhang, S. Yin, S. Luo and C. T. Au, *Ind. Eng. Chem. Res.*, 2012, **51**, 3951; (c) W.-L. Dai, B. Jin, S.-L. Luo, X.-B. Luo, X.-M. Tu and C.-T. Au, *Catal. Sci. Technol.*, 2014, DOI: 10.1039/c3cy00659j; (d) L. Han, H.-J. Choi, S.-J. Choi, B. Lin and D.-W. Park, *Green Chem.*, 2011, **13**, 1023.
412. J. Li, L. Wang, F. Shi, S. Liu, Y. He, L. Lu, X. Ma and Y. Deng, *Catal. Lett.*, 2011, **141**, 339.
413. J.-Q. Wang, J. Sun, W.-G. Cheng, C.-Y. Shi, K. Dong, X.-P. Zhang and S.-J. Zhang, *Catal. Sci. Technol.*, 2012, **2**, 600.
414. (a) Y. Kishi, H. Nagura, S. Inagi and T. Fuchigami, *Chem. Commun.*, 2008, 3876; (b) Y. Kishi, S. Inagi and T. Fuchigami, *Eur. J. Org. Chem.*, 2009, 103.
415. E. Hasegawa, N. Hiroi, C. Osawa, E. Tayama and H. Iwamoto, *Tetrahedron Lett.*, 2010, **51**, 6535.
416. C. Chiappe and S. Rajamani, *Eur. J. Org. Chem.*, 2011, 5517.
417. A. C. Cole, J. L. Jensen, I. Ntai, K. L. T. Tran, K. J. Weaver, D. C. Forbes and J. H. Davis Jr., *J. Am. Chem. Soc.*, 2002, **124**, 5962.
418. K. Kondamudi, P. Elavarasan, P. J. Dyson and S. Upadhyayula, *J. Mol. Catal. A: Chem.*, 2010, **321**, 34.
419. X. Liu, M. Liu, X. Guo and J. Zhou, *Catal. Commun.*, 2008, **9**, 1.
420. X. Liu, J. Zhou, X. Guo, M. Liu, X. Ma, C. Song and C. Wang, *Ind. Eng. Chem. Res.*, 2008, **47**, 5298.
421. M. A. Harmer, C. P. Junk, V. V. Rostovtsev, W. J. Marshall, L. M. Grieco, J. Vickery, R. Miller and S. Work, *Green Chem.*, 2009, **11**, 517.
422. S. Tang, A. M. Scurto and B. Subramaniam, *J. Catal.*, 2009, **268**, 243.
423. P. Cui, G. Zhao, H. Ren, J. Huang and S. Zhang, *Catal. Today*, 2013, **200**, 30.
424. D. C. Forbes and K. J. Weaver, *J. Mol. Catal. A: Chem.*, 2004, **214**, 129.

425. J. Gui, X. Cong, D. Liu, X. Zhang, Z. Hu and Z. Sun, *Catal. Commun.*, 2005, **5**, 473.
426. H. Xing, T. Wang, Z. Zhou and Y. Dai, *Ind. Eng. Chem. Res.*, 2005, **44**, 4147.
427. D. Fang, Z.-L. Zhou, Z.-W. Ye and Z.-L. Liu, *Ind. Eng. Chem. Res.*, 2006, **45**, 7982.
428. X. Li and W. Eli, *J. Mol. Catal. A: Chem.*, 2008, **279**, 159.
429. Y. Zhao, J. Long, F. Deng, X. Liu, Z. Li, C. Xia and J. Peng, *Catal. Commun.*, 2009, **10**, 732.
430. H. Xing, T. Wang, Z. Zhou and Y. Dai, *J. Mol. Catal. A: Chem.*, 2007, **264**, 53.
431. R. Juárez, R. Martín, M. Álvero and H. García, *Appl. Catal., A*, 2009, **369**, 133.
432. C. Fehér, E. Kriván, J. Hancsók and R. Skoda-Földes, *Green Chem.*, 2012, **14**, 403.
433. (a) X. Liu, L. Xiao, H. Wu, Z. Li, J. Chen and C. Xia, *Catal. Commun.*, 2009, **10**, 424; (b) X. Liu, L. Xiao, H. Wu, J. Chen and C. Xia, *Helv. Chim. Acta*, 2009, **92**, 1014.
434. F. Han, L. Yang, Z. Li and C. Xia, *Org. Biomol. Chem.*, 2012, **10**, 346.
435. C.-J. Yu and C.-J. Liu, *Molecules*, 2009, **14**, 3222.
436. X. Chen, R. Liu, Y. Xu and G. Zou, *Tetrahedron*, 2012, **68**, 4813.
437. F. Han, L. Yang, Z. Li and C. Xia, *Adv. Synth. Catal.*, 2012, **384**, 1052.
438. H. Zhou, F. Shi, X. Tian, Q. Zhang and Y. Deng, *J. Mol. Catal. A: Chem.*, 2007, **271**, 89.
439. S. Majumdar, J. De, J. Hossain and A. Basak, *Tetrahedron Lett.*, 2013, **54**, 262.
440. H. Song, J. Chen, C. Xia and Z. Li, *Synth. Commun.*, 2012, **42**, 266.
441. M. A. Zolfigol, A. Khazaei, A. R. Moosavi-Zare, A. Zare and V. Khakyzadeh, *Appl. Catal., A*, 2011, **400**, 70.
442. J. Luo and Q. Zhang, *Monatsh. Chem.*, 2011, **142**, 923.
443. H. Song, Z. Li, J. Chen and C. Xia, *Catal. Lett.*, 2012, **142**(8), 1.
444. W. Li, Y. Wang, Z. Wang, L. Dai and Y. Wang, *Catal. Lett.*, 2011, **141**, 1651.
445. N. G. Khaligh, *J. Mol. Catal. A: Chem.*, 2011, **349**, 63.
446. J. Akbari, A. Heydari, L. Ma'mani and S. H. Hosseini, *C. R. Chim.*, 2010, **13**, 544.
447. L. Myles, R. Gore, M. Spulak, N. Gathergood and S. J. Connon, *Green Chem.*, 2010, **12**, 1157.
448. B. Procuranti and S. J. Connon, *Org. Lett.*, 2008, **10**, 4935.
449. Y. Wang, X. Gong, Z. Wang and L. Dai, *J. Mol. Catal. A: Chem.*, 2010, **322**, 7.
450. Y. Jeong, D.-Y. Kim, Y. Choi and J.-S. Ryu, *Org. Biomol. Chem.*, 2011, **9**, 374.
451. W.-L. Wong, K.-P. Ho, L. Y. S. Lee, K.-M. Lam, Z.-Y. Zhou, T. H. Chan and K.-Y. Wong, *A.C.S. Catal.*, 2011, **1**, 116.
452. (a) P. Kotrusz, I. Kmentová, B. Gotov, S. Toma and E. Solčániová, *Chem. Commun.*, 2002, 2510; (b) T.-P. Loh, L.-C. Feng, H.-Y. Yang and J.-Y. Yang, *Tetrahedron Lett.*, 2002, **43**, 8741.

453. H.-M. Guo, L.-F. Cun, L.-Z. Gong, A.-Q. Mi and Y.-Z. Jiang, *Chem. Commun.*, 2005, 1450.
454. A. Córdova, *Tetrahedron Lett.*, 2004, **45**, 3949.
455. J. Shah, H. Blumenthal, Z. Yacob and J. Liebscher, *Adv. Synth. Catal.*, 2008, **350**, 1267.
456. P. Kotrusz, S. Toma, H.-G. Schmalz and A. Adler, *Eur. J. Org. Chem.*, 2004, 1577.
457. M. S. Rasalkar, M. K. Potdar, S. S. Mohile and M. M. Salunkhe, *J. Mol. Catal. A: Chem.*, 2005, **235**, 267.
458. M. Mečiarová, S. Toma, A. Berkessel and B. Koch, *Lett. Org. Chem.*, 2006, **3**, 437.
459. L.-C. Feng, Y.-S. Sun, W.-J. Tang, L.-J. Xu, K.-L. Lam, Z. Zhou and A. S. C. Chan, *Green Chem.*, 2010, **12**, 949.
460. T. Zhang, L. Cheng, S. Hameed, L. Liu, D. Wang and Y.-J. Chen, *Chem. Commun.*, 2011, **47**, 6644.
461. A. De Niro, O. Bortolini, L. Maiuolo, A. Farofalo, B. Russo and G. Sindona, *Tetrahedron Lett.*, 2011, **52**, 1415.
462. X. Zhang, W. Zhao, C. Qu, L. Yang and Y. Cui, *Tetrahedron: Asymmetry*, 2012, **23**, 468.
463. M. Lombardo, F. Pasi, S. Easwar and C. Trombini, *Adv. Synth. Catal.*, 2007, **349**(206), 1.
464. L. Zhou and L. Wang, *Chem. Lett.*, 2007, **36**, 628.
465. D. E. Siyutkin, A. S. Kucherenko, M. I. Struchkova and S. G. Zlotin, *Tetrahedron Lett.*, 2008, **49**, 1212.
466. S. Luo, X. Mi, L. Zhang, S. Liu, H. Xu and J.-P. Cheng, *Tetrahedron*, 2007, **63**, 1923.
467. S. Luo, X. Mi, L. Zhang, S. Liu, H. Xu and J.-P. Cheng, *Angew. Chem., Int. Ed.*, 2006, **45**, 3093.
468. H. Sun, D. Zhang, C. Zhang and C. Liu, *Chirality*, 2010, **22**, 813.
469. W. Miao and T. H. Chan, *Adv. Synth. Catal.*, 2006, **348**, 1711.
470. S. Luo, L. Zhang, X. Mi, Y. Qiao and J.-P. Cheng, *J. Org. Chem.*, 2007, **72**, 9350.
471. L. Cui, X. Li, J. Li, S. Luo and J.-P. Cheng, *Chem. Eur. J.*, 2010, **16**, 2045.
472. D.-Q. Xu, B.-T. Wang, S.-P. Luo, D.-D. Yue, L.-P. Wang and Z.-Y. Xu, *Tetrahedron: Asymmetry*, 2007, **18**, 1788.
473. D. Xu, S. Luo, H. Yue, L. Wang, Y. Liu and Z. Xu, *Synlett*, 2006, 2569.
474. L.-Y. Wu, Z.-Y. Yan, Y.-X. Xie, Y.-N. Niu and Y.-M. Liang, *Tetrahedron: Asymmetry*, 2007, **18**, 2086.
475. W.-H. Wang, X.-B. Wang, K. Kodama, T. Hirose and G.-Y. Zhang, *Tetrahedron*, 2010, **66**, 4970.
476. Z.-L. Shen, K. K. K. Goh, C. H. A. Wong, W.-Y. Loo, Y.-S. Yang, J. Lu and T.-P. Loh, *Chem. Commun.*, 2012, **48**, 5856.
477. S. V. Kochetkov, A. S. Kucherenko and S. G. Zlotin, *Eur. J. Org. Chem.*, 2011, 6128.
478. S. V. Kochetkov, A. S. Kucherenko, G. V. Kryshtal, G. M. Zhdankina and S. G. Zlotin, *Eur. J. Org. Chem.*, 2012, 7129.

479. (a) B. Ni, Q. Zhang and A. D. Headley, *Green. Chem.*, 2007, **9**, 737; (b) Q. Zhang, B. Ni and A. D. Headley, *Tetrahedron*, 2008, **64**, 5091.
480. B. Ni, Q. Zhang, K. Dhungana and A. D. Headley, *Org. Lett.*, 2009, **11**, 1037.
481. Y. Qian, X. Zheng and Y. Wang, *Eur. J. Org. Chem.*, 2010, 3672.
482. V. Gauchot and A. R. Schmitzer, *J. Org. Chem.*, 2012, **77**, 4917.
483. M. Vasiloiu, D. Rainer, P. Gaertner, C. Reichel, C. Schröder and K. Bica, *Catal. Today*, 2013, **200**, 80.
484. B. C. Ranu, A. Saha and S. Banerjee, *Eur. J. Org. Chem.*, 2008, 519.
485. S. R. Roy and A. K. Chakraborti, *Org. Lett.*, 2010, **12**, 3866.
486. A. Sarkar, S. R. Roy, N. Parikh and A. K. Chakraborti, *J. Org. Chem.*, 2011, **76**, 7132.
487. A. K. Chakraborti and S. R. Roy, *J. Am. Chem. Soc.*, 2009, **131**, 6902.
488. V. Lucchini, M. Noe, M. Selva, M. Fabris and A. Perosa, *Chem. Commun.*, 2013, **48**, 5178.
489. A. Y. Arzephoni, M. R. Naimi-Jamal, A. Sharifi, M. S. Abaee and M. Mirzaei, *J. Chem. Res.*, 2013, 216.
490. A. Zhu, R. Liu, L. Li, L. Li, L. Wang and J. Wang, *Catal. Today*, 2013, **200**, 17.
491. X. Chen, X. Li, H. Song, Y. Qian and F. Wang, *Tetrahedron Lett.*, 2011, **52**, 3588.
492. Z. Kelemen, O. Hollóczki, J. Nagy and L. Nyulászi, *Org. Biomol. Chem.*, 2011, **9**, 5362.
493. M. Fabris, M. Noè, A. Perosa, M. Selva and R. Ballini, *J. Org. Chem.*, 2012, **77**, 1805.
494. Y.-C. Teo and G.-L. Chua, *Tetrahedron Lett.*, 2008, **49**, 4235.
495. (a) M. Pu, B. H. Chen and D. C. Fang, *Struct. Chem.*, 2006, **17**, 377; (b) M. Pu, B. H. Chen and H. X. Wang, *Chem. Phys. Lett.*, 2005, **410**, 441.
496. L. Xie, D. Zhang, X. Liu, X. Zhang and P. Duan, *Comput. Theor. Chem.*, 2011, **963**, 344.
497. Y. Song, H. Jing, B. Li and D. Bai, *Chem. Eur. J.*, 2011, **17**, 8731.
498. L. Gułajski, M. Mauduit and K. Grela, *Pure Appl. Chem.*, 2009, **81**, 2001.
499. R. C. Buijsman, E. van Vuuren and J. G. Sterrenburg, *Org. Lett.*, 2001, **3**, 3785.
500. D. Semeril, H. Olivier-Bourbigou, C. Bruneau and P. H. Dixneuf, *Chem. Commun.*, 2002, 146.
501. K. G. Mayo, E. H. Nearhoof and J. Kiddle, *Org. Lett.*, 2002, **4**, 1567.
502. X. Ding, X. Lv, B. Hui, Z. Chen, M. Xiao, B. Guo and W. Tang, *Tetrahedron Lett.*, 2006, **47**, 2921.
503. C. Thurier, C. Fischmeister, C. Bruneau, H. Olivier-Bourbigou and P. H. Dixneuf, *ChemSusChem*, 2008, **1**, 118.
504. D. Bradley, G. Williams, M. Ajam and A. Ranwell, *Organometallics*, 2006, **25**, 3088.
505. N. Audic, H. Clavier, M. Mauduit and J.-C. Guillemin, *J. Am. Chem. Soc.*, 2003, **125**, 9248.

506. (a) H. Clavier, N. Audic, M. Mauduit and J.-C. Guillemin, *Chem. Commun.*, 2004, 2282; (b) H. Clavier, N. Audic, J.-C. Guillemin and M. Mauduit, *J. Organomet. Chem.*, 2005, **690**, 3585.
507. (a) Q. Yao and Y. Zhang, *Angew. Chem., Int. Ed.*, 2003, **42**, 3395; (b) Q. Yao and M. Sheets, *J. Organomet. Chem.*, 2005, **690**, 3577.
508. C. S. Consorti, G. L. P. Aydos, G. Ebeling and J. Dupont, *Org. Lett.*, 2008, **10**, 237.
509. C. S. Consorti, G. L. P. Aydos, G. Ebeling and J. Dupont, *Organometallics*, 2009, **28**, 4527.
510. S.-W., J. H. Kim, K. Y. Ryu, W.-W. Lee, J. Hong and S.-g. Lee, *Tetrahedron*, 2009, **65**, 3397.
511. G. Liu, J. Zhang, B. Wu and J. Wang, *Org. Lett.*, 2007, **9**, 4263.
512. C. Thurier, C. Fischmeister, C. Bruneau, H. Olivier-Bourbigou and P. H. Dixneuf, *J. Mol. Catal. A: Chem.*, 2007, **268**, 127.
513. A. Keraani, M. Rabiller-Baudry, C. Fischmeister and C. Bruneau, *Catal. Today*, 2010, **156**, 268.
514. B. Autenrieth, W. Frey and M. B. Buchmeiser, *Chem. Eur. J.*, 2012, **18**, 14069.
515. S. Csihony, C. Fischmeister, C. Bruneau, I. T. Hovárth and P. H. Dixneuf, *New J. Chem.*, 2002, **26**, 1667.
516. Y. S. Vygodskii, A. S. Shaplov, E. I. Lozinskaya, O. A. Fillippov, E. S. Shubina, R. Bandari and M. R. Buchmeiser, *Macromolecules*, 2006, **39**, 7821.
517. (a) V. Amir-Ebrahami and J. J. Rooney, *J. Mol. Catal. A: Chem.*, 2004, **208**, 115; (b) M. M. Gallagher, A. D. Rooney and J. J. Rooney, *J. Mol. Catal. A: Chem.*, 2009, **303**, 78.
518. M. Oshimura, A. Takasu and K. Negata, *Macromolecules*, 2009, **42**, 3086.
519. N. Nomura, A. Taira, A. Nakase, T. Tomioka and M. Okada, *Tetrahedron*, 2007, **63**, 8478.
520. J. Ling, L. You, Y. Wang and Z. Shen, *J. Appl. Polym. Sci.*, 2012, **124**, 2537.
521. M. Yoshizawa-Fujita, C. Saito, Y. Takeoka and M. Rikukawa, *Polym. Adv. Technol.*, 2008, **19**, 1396.
522. M. Mena, A. López-Luna, K. Shirai, A. Tecante, M. Gimeno and E. Barzana, *Bioprocess Biosyst. Eng.*, 2013, **36**, 383.
523. S. Chanfreau, M. Mena, J. R. Porras-Domínguez, M. Ramírez-Gilly, M. Gimeno, P. Roquero, A. Tecante and E. Bárzana, *Bioprocess Biosyst. Eng.*, 2010, **33**, 629.
524. (a) Y. Guo, X. Wang, Z. Shen, X. Shu and R. Sun, *Carbohydr. Polym.*, 2013, **92**, 77; (b) Q. Peng, K. Mahmood, Y. Wu, L. Wang, Y. Liang, J. Shen and Z. Liu, *Green Chem.*, DOI: 10.1039/3cgc42477d.
525. T. Ogoshi, T. Onodera, T-a. Yamagishi and Y. Nakamoto, *Macromolecules*, 2008, **41**, 8533.
526. (a) Y. Chauvin, B. Gilbert and I. Guibard, *Chem. Commun.*, 1990, 1715; (b) Y. Chauvin, S. Einloft and H. Olivier, *Ind. Eng. Chem. Res.*, 1995, **34**, 1149.

527. (a) E. Ellis, W. Keim and P. Wasserscheid, *Chem. Commun.*, 1999, 337; (b) P. Wasserscheid and M. Eichmann, *Catal. Today*, 2001, **66**, 309.
528. (a) M. Dötterl and H. G. Alt, *Adv. Synth. Catal.*, 2012, **354**, 399; (b) M. Dötterl and H. G. Alt, *Chem. Cat. Chem.*, 2011, **3**, 1799; (c) M. Dötterl and H. G. Alt, *Chem. Cat. Chem.*, 2012, **4**, 370.
529. M. Dötterl, P. Thoma and H. G. Alt, *Adv. Synth. Catal.*, 2012, **354**, 389.
530. M. Eichmann, W. Keim, M. Haumann, B. U. Melcher and P. Wasserscheid, *J. Mol. Catal. A: Chem.*, 2009, **314**, 42.
531. L. Magna, J. Bildé, H. Olivier-Bourbigou, T. Robert and B. Gilbert, *Oil Gas Sci. Technol.*, 2009, **64**, 669.
532. D. Thiele and R. F. de Souza, *J. Mol. Catal. A: Chem.*, 2011, **340**, 83.
533. (a) P. Wasserscheid, C. M. Gordon, C. Hilgers, M. J. Muldoon and I. R. Dunkin, *Chem. Commun.*, 2001, 1186; (b) P. Wasserscheid, C. Hilgers and W. Keim, *J. Mol. Catal., A*, 2004, **214**, 83.
534. N. D. Clement and K. J. Cavell, *Angew. Chem., Int. Ed.*, 2004, **43**, 3845.
535. (a) W. Ochędzan-Siodłak, K. Dziubek and D. Siodłak, *Eur. Polym. J.*, 2008, **44**, 3608; (b) W. Ochędzan-Siodłak and B. Sacher-Majewska, *Eur. Polym. J.*, 2007, **43**, 3688.
536. D. Thiele and R. F. de Souza, *Catal. Lett.*, 2010, **138**, 50.
537. K.-M. Song, H.-Y. Gao, F.-S. Liu, J. Pan, L.-H. Guo, S.-B. Zai and Q. Wu, *Catal. Lett.*, 2009, **131**, 566.
538. L. Pei, X. Liu, H. Gao and Q. Wu, *Appl. Organomet. Chem.*, 2009, **23**, 455.
539. S. M. Silva, P. A. Z. Suarez, R. F. de Souza and J. Dupont, *Polym. Bull.*, 1998, **40**, 401.
540. R. A. Ligabue, J. Dupont and R. F. de Souza, *J. Mol. Catal. A: Chem.*, 2001, **169**, 11.
541. J. E. L. Dullius, P. A. Z. Suarez, S. Einloft, R. F. de Souza, J. Dupont, F. Fischer and A. De Cian, *Organometallics*, 1998, **17**, 815.
542. J. Zimmermann, P. Wasserscheid, I. Tkatchenko and S. Stutzmann, *Chem. Commun.*, 2002, 760.
543. (a) J. Zimmermann, P. Wasserscheid and I. Tkatchenko, *Adv. Synth. Catal.*, 2003, **345**, 402; (b) M. Picquet, S. Stutzmann, I. Tkatchenko, I. Tommasi, J. Zimmermann and P. Wasserscheid, *Green Chem.*, 2003, **5**, 153.
544. (a) M. Picquet, I. Tkatchenko, I. Tommasi, P. Wasserscheid and J. Zimmerman, *Adv. Synth. Catal.*, 2003, **345**, 959; (b) M. Picquet, D. Poinsot, S. Stutzmann, I. Tkatchenko, I. Tommasi, P. Wasserscheid and J. Zimmerman, *Top. Catal.*, 2004, **29**, 139.
545. A. Goswami, T. Ito, N. Saino, K. Kase and C. Matsumo, *Chem. Commun.*, 2009, 439.
546. (a) C.-Z. Liu, F. Wang, A. R. Stiles and C. Guo, *Appl. Energy*, 2012, **92**, 406; (b) For a highly informative overview of catalytic conversions of biomass to biofuels see: D. M. Alonso, J. Q. Bond and J. A. Dumesic, *Green Chem.*, 2010, **12**, 1493; (c) for an insightful and highly informative review dedicated to transformations of HMF see: A.-J. van Putten, J. C. van der Waal, E. de Jong, C. B. Rasrendra, H. J. Heeres and J. G. de Vrieze, *Chem. Rev.*, 2013, **113**, 1499.

547. J. Julis, M. Hölscher and W. Leitner, *Green Chem.*, 2010, **12**, 1634.
548. J. Julis and W. Leitner, *Angew. Chem., Int. Ed.*, 2012, **51**, 8615.
549. F. M. A. Geilen, B. Engendahl, A. Harwardt, W. Marquardt, J. Klankermayer and W. Leitner, *Angew. Chem., Int. Ed.*, 2010, **49**, 5510.
550. (a) F.-X. Zeng, H.-F. Liu, L. Deng, B. Liao, H. Pang and Q.-X. Guo, *ChemSusChem*, 2013, **6**, 600; (b) G. Wang, Z. Zhang and L. Song, *Green Chem.*, 2014, DOI: 10.1039/c3gc41693c.
551. C. Li and Z. K. Zhao, *Adv. Synth. Catal.*, 2007, **349**, 1847.
552. C. Li, Q. Wang and Z. K. Zhao, *Green Chem.*, 2008, **10**, 177.
553. L. Vanoye, M. Fanselow, J. D. Holbry, M. P. Atkins and K. R. Seddon, *Green Chem.*, 2009, **11**, 390.
554. C. Sievers, M. B. Valenzuela-Olarte, T. Marzialetti, I. Musin, P. K. Agrawal and C. W. Jones, *Ind. Eng. Chem. Res.*, 2009, **48**, 1277.
555. R. Rinaldi, R. Palkovits and F. Schüth, *Angew. Chem., Int. Ed.*, 2008, **47**, 8047.
556. Y. Zhang, H. Du, X. Qian and E. Y.-X. Chen, *Energy Fuels*, 2010, **24**, 2410.
557. A. S. Amarasekara and O. S. Owereh, *Ind. Eng. Chem. Res.*, 2009, **48**, 10152.
558. A. S. Amarasekara and B. Wiredu, *Ind. Eng. Chem. Res.*, 2011, **50**, 12276.
559. F. Jiang, Q. Zhu, D. Ma, X. Liu and X. Han, *J. Mol. Catal. A: Chem.*, 2011, **334**, 8.
560. Y. Liu, W. Xiao, S. Xia and P. Ma, *Carbohydr. Polym*, 2013, **92**, 218.
561. J. A. Abia and R. Ozer, *BioResources*, 2013, **8**, 2924.
562. (a) J. Long, X. Li, B. Guo, L. Wang and N. Zhang, *Catal. Today*, 2013, **200**, 99; (b) E. C. Achinivu, R. M. Howard, G. Li, H. Gracz and W. A. Henderson, *Green Chem.*, 2014, DOI: 10.1039/c3gc42306a; (c) P. Verdia, A. Brandt, J. P. Hallet, M. J. Ray and T. Welton, *Green Chem.*, 2014, DOI: 10.1039/c3gc41742e.
563. H. Zhao, J. E. Holladay, H. Brown and Z. C. Zhang, *Science*, 2007, **316**, 1597.
564. (a) E. A. Pidko, V. Degirmenci, R. A. van Santen and E. J. M. Hensen, *Angew. Chem., Int. Ed.*, 2010, **49**, 2530; (b) Y. Zhang, E. A. Pidko and E. J. M. Hensen, *Chem. Eur. J.*, 2011, **17**, 5281.
565. G. Yong, Y. Zhang and J. Y. Ying, *Angew. Chem., Int. Ed.*, 2008, **47**, 9345.
566. E. F. Dunn, D. J. Liu and E. Y.-X. Chen, *Appl. Catal., A*, 2013, **460-461**, 1.
567. X. Qi, M. Watanabe, T. M. Aida and R. L. Smith, *ChemSusChem*, 2010, **3**, 1071.
568. T. Ståhlberg, M. G. Sørensen and A. Rüsager, *Green Chem.*, 2001, **12**, 321.
569. Z. Zhang, Q. Wang, H. Xie, W. Liu and Z. Zhao, *ChemSusChem*, 2011, **4**, 131.
570. S. Q. Hu, Z. F. Zhang, J. L. Song, Y. X. Zhou and B. X. Han, *Green Chem.*, 2009, **11**, 1746.
571. B. Kim, J. Jeong, D. Lee, S. Kim, H.-J. Yoon, Y.-S. Lee and K. K. Cho, *Green Chem.*, 2011, **13**, 1503.

572. J. B. Binder and R. T. Raines, *J. Am. Chem. Soc.*, 2009, **131**, 1979.
573. (a) Tao, F. H. Song and L. Chou, *Carbohydr. Res.*, 2011, **346**, 58; (b) F. Tao, H. Song and L. Chuo, *J. Mol. Catal. A: Chem.*, 2012, **357**, 11; (c) F. Tao, H. Song and L. Chou, *Bioresour. Technol.*, 2011, **102**, 9000; (d) F. Tao, H. Song and L. Chou, *ChemSusChem.*, 2010, **3**, 1298.
574. Z.-D. Ding, J.-C. Shi, J.-J. Xiao, W.-X. Gu, C.-G. Zheng and H.-J. Wang, *Carbohydr. Polym.*, 2012, **90**, 792.
575. D. Liu and E. Y.-X. Chen, *Appl. Catal., A*, 2012, **435-436**, 78.
576. M. Chidambaram and A. T. Bell, *Green Chem.*, 2010, **12**, 1253.
577. V. Heguaburu, J. Franco, L. Reina, C. Tabarez, G. Moyna and P. Moyna, *Catal. Commun.*, 2012, **7**, 88.
578. W.-H. Hsu, Y.-Y. Lee, W.-H. Peng and K. C-W. Wu, *Catal. Today*, 2011, **174**, 65.
579. W. Liu and J. Holladay, *Catal. Today*, 2013, **200**, 106.
580. F. Tao, H. Song and L. Chou, *RSC Adv.*, 2011, **1**, 672.
581. Y. Qu, C. Huang, Y. Song, J. Zhang and B. Chen, *Bioresour. Technol.*, 2012, **121**, 462.
582. Q. Wu, H. Chen, M. Han, D. Wang and J. Wang, *Ind. Eng. Chem. Res.*, 2007, **46**, 7955.
583. Q. Yang, Z. Wei, H. Xing and Q. Ren, *Catal. Commun.*, 2008, **9**, 1307.
584. M. Han, W. Yi, Q. Wu, Y. Liu, Y. Hong and D. Wang, *Bioresour. Technol.*, 2009, **100**, 2308.
585. W. Guo, H. Li, G. Ji and G. Zhang, *Bioresour. Technol.*, 2012, **125**, 332.
586. Q. Wu, H. Wan, H. Li, H. Song and T. Chu, *Catal. Today*, 2013, **200**, 74.
587. F. Liu, L. Wang, Q. Sun, L. Zhu, X. Meng and F.-S. Xiao, *J. Am. Chem. Soc.*, 2012, **134**, 16948.
588. X. Liang, G. Gong, H. Wu and J. Yang, *Fuel*, 2009, **88**, 613.
589. (a) F. R. Abreu, M. B. Alves, C. C. S Macêdo, L. F. Zara and P. A. Z. Suarez, *J. Mol. Catal. A: Chem.*, 2005, **227**, 263; (b) B. A. DaSilveira Neto, M. B. Alves, A. A. M. Lapis, F. M. Nachtigall, M. N. Eberlin, J. Dupont and P. A. Z. Suarez, *J. Catal.*, 2007, **249**, 154.
590. F. Fischer, J. Mutschler and D. Zufferey, *J. Ind. Microbiol. Biotechnol.*, 2011, **38**, 447.
591. M. Moniruzzaman, K. Nakashima, N. Kamiya and M. Goto, *Biochem. Eng. J.*, 2010, **48**, 295.
592. S. Sunitha, S. Kanjilal, P. S. Reddy and R. B. N. Prasad, *Biotechnol. Lett.*, 2007, **29**, 1881.
593. S. H. Ha, M. N. Lan, S. H. Lee, S. M. Hwang and Y.-M. Koo, *Enzyme Microb. Technol.*, 2007, **41**, 480.
594. M. Gamba, A. A. M. Lapis and J. Dupont, *Adv. Synth. Catal.*, 2008, **350**, 160.
595. O. Miyawaki and M. Tatsumo, *J. Biosci. Bioeng.*, 2008, **105**, 61.
596. D. Yu, C. Wang, Y. Yin, A. Zhang, G. Gao and X. Fang, *Green Chem.*, 2011, **13**, 1869.
597. H. Zhao, Z. Song, O. Olubajo and J. V. Cowins, *Appl. Biochem. Biotechnol.*, 2010, **162**, 13.

598. (a) T. De Diego, A. Manjon, P. Lozano, M. Vaultier and J. L. Iborra, *Green Chem.*, 2011, **13**, 444; (b) P. Lozano, J. M. Bernal and M. Vaultier, *Fuel*, 2011, **90**, 3461.
599. S. Arai, K. Nakashima, T. Tanino, C. Ogino, A. Kondo and H. Fukuda, *Enzyme Microb. Technol.*, 2010, **46**, 51.
600. (a) K.-P. Zhang, J.-Q. Lai, Z.-L. Huang and Z. Yang, *Bioresour. Technol.*, 2011, **102**, 2767; (b) Z. Yang, K.-P. Zhang, Y. Huang and Z. Wang, *J. Mol. Catal. B: Enzym.*, 2010, **63**, 23.
601. J.-Q. Lai, Z.-L. Hu, P.-W. Wang and Z. Yang, *Fuel*, 2012, **95**, 329.
602. P. Lozano, J. M. Bernal and A. Navarro, *Green Chem.*, 2012, **14**, 3026.
603. P. Lozano, J. M. Bernal, R. Piamtongkam, D. Fetzer and M. Vaultier, *ChemSusChem.*, 2010, **3**, 1359.
604. (a) T. De Diego, P. Lozano, S. Gmouh, M. Vaultier and J. L. Iborra, *Biomacromolecules*, 2005, **6**, 1457; (b) P. Lozano, T. De Diego, S. Gmouh, M. Vaultier and J. L. Iborra, *Biocatal. Biotransform.*, 2005, **23**, 169.
605. C. S. Consorti, G. L. P. Aydos, G. Ebeling and J. Dupont, *Appl. Catal., A*, 2009, **371**, 114.
606. J. Dupont, G. S. Fonseca, A. P. Umpierre, P. F. P. Fichtner and S. R. Teixeira, *J. Am. Chem. Soc.*, 2002, **124**, 4228.
607. T. Gutel, J. Garcia-Anton, K. Pelzer, K. Philppot, C. Santini, Y. Chauvin, B. Chaudret and J. M. Basset, *J. Mater. Chem.*, 2007, **17**, 3290.
608. C. C. Cassol, A. P. Umpierrie, G. Machado, S. I. Wolke and J. Dupont, *J. Am. Chem. Soc.*, 2005, **127**, 3298.
609. T. Torimoto, T. Tsuda, K. Okazaki and S. Kuwabata, *Adv. Mater.*, 2010, **22**, 1196.
610. (a) J. Kramer, E. Redel, R. Thomann and C. Janiak, *Organometallics*, 2008, **27**, 1976; (b) E. Redel, R. Thomann and C. Janiak, *Chem. Commun.*, 2008, 1789.
611. C. Vollmer, E. Redel, K. Abu-Shandi, R. Thomann, H. Manyar, C. Hardacre and C. Janiak, *Chem. Eur. J*, 2010, **16**, 3849.
612. G. Fonseca, G. Machado, S. Teixeira, G. Fecher, J. Morais, M. C. M. Alves and J. Dupont, *J. Colloid Interface Sci.*, 2006, **301**, 193.
613. L. S. Otto, M. L. Cline, M. Deeflefs, K. R. Seddon and R. G. Finke, *J. Am. Chem. Soc.*, 2005, **127**, 5758.
614. A. S. Pensado and A. A. H. Pádua, *Angew. Chem., Int. Ed.*, 2011, **50**, 8683.
615. M. T. Reetz and W. Westerman, *Angew. Chem., Int. Ed.*, 2000, **39**, 165.
616. V. Caló, A. Nacci, A. Monopoli, A. Detomaso and P. Iliade, *Organometallics*, 2003, **21**, 4193.
617. V. Caló, A. Nacci, A. Monopoli and P. Cotugno, *Angew. Chem., Int. Ed.*, 2009, **48**, 6101.
618. C. S. Consorti, F. R. Flores and J. Dupont, *J. Am. Chem Soc.*, 2005, **127**, 12054.
619. C. Ye, J.-C. Xiao, B. Twamley, A. D. Lalonde, M. G. Norton and J. M. Shreeve, *Eur. J. Org. Chem.*, 2007, 5095.
620. P. Cotugno, A. Monopoli, F. Ciminale, N. Cioffi and A. Nacci, *Org. Biomol. Chem.*, 2012, **10**, 808.

621. Y. Zeng, Y. Wang, Y. Xu, Y. Song, J. Jiang and Z. Jin, *Catal. Lett.*, 2013, **143**, 200.
622. D. S. Gaikward, Y. Park and D. M. Pore, *Tetrahedron Lett.*, 2012, **53**, 3077.
623. G. Liu, M. Hou, J. Song, T. Jiang, H. Fan, Z. Zhang and B. Hu, *Green Chem.*, 2010, **12**, 65.
624. (a) E. Raluy, I. Favier, A. M. Lopez-Vinasco, C. Pradel, E. Martin, D. Madec, E. Teuma and M. Gomez, *Phys. Chem. Chem. Phys.*, 2011, **13**, 13579; (b) S. Jansat, J. Durand, I. Favier, F. Malbosc, C. Pradel, E. Teuma and M. Gomez, *ChemCatChem.*, 2009, **1**, 244; (c) L. Rodriguez-Perez, C. Pradel, P. Serp, M. Gomez and E. Teuma, *ChemCatChem.*, 2011, **3**, 749.
625. L. M. Ross, G. Machado, P. F. P. Fichtner, S. R. Teixeira and J. Dupont, *Catal. Lett.*, 2004, **92**, 149.
626. E. T. Silveira, A. P. Umpierre, L. M. Rossi, G. Machado, J. Morais, G. V. Soares, I. J. R. Baumvol, S. R. Teixeira, P. F. P. Fichtner and J. Dupont, *Chem. Eur. J.*, 2004, **10**, 3734.
627. D. Li and R. B. Kaner, *J. Am. Chem. Soc.*, 2006, **128**, 968.
628. M. H. G. Prechtl, M. Scariot, J. D. Scholten, G. Machado, S. R. Teixeira and J. Dupont, *Inorg. Chem.*, 2008, **47**, 8995.
629. F. Schwab, M. Lucas and P. Claus, *Angew. Chem., Int. Ed.*, 2011, **50**, 10453.
630. T. Gutel, C. C. Santini, K. Philippot, A. Padua, B. Chaudret, Y. Chauvin and J.-M. Basset, *J. Mater. Chem.*, 2009, **19**, 3624.
631. G. Salas, C. C. Santini, K. Philippot, V. Colliére, B. Chaudret, B. Fenet and P. F. Fazzini, *Dalton Trans.*, 2011, **40**, 4660.
632. K. L. Luska and A. Moores, *Green Chem.*, 2012, **14**, 1736.
633. J. E. L. Dullius, P. A. Z. Suarez, S. Einloft, R. F. de Souza, J. Dupont, J. Fischer and A. De Cian, *Organometallics*, 1998, **17**, 815.
634. G. S. Fonseca, A. P. Umpierre, P. F. P. Fichtner, S. R. Teixeira and J. Dupont, *Chem. Eur. J.*, 2003, **9**, 3263.
635. A. P. Umpierre, G. Machado, G. H. Fecher, J. Morais and J. Dupont, *Adv. Synth. Catal.*, 2005, **347**, 1404.
636. C. W. Scheeren, G. Machado, J. Dupont, P. F. P. Fichtner and S. R. Texeira, *Inorg. Chem.*, 2003, **42**, 4738.
637. P. Migowski, G. Machado, S. R. Texeira, M. C. M. Alves, J. Morais, A. Traverse and J. Dupont, *Phys. Chem. Chem. Phys.*, 2007, **9**, 4814.
638. P. C. L'Argentiere, E. A. Cagnola, M. G. Canon, D. A. Liprandi and D. V. Marconetti, *J. Chem. Technol. Biotechnol.*, 1998, **71**, 285.
639. R. Abu-Reziq, D. Wang, M. Post and H. Alper, *Adv. Synth. Catal.*, 2007, **349**, 2145.
640. Y. Kume, K. Qiao, D. Tomida and C. Yokoyama, *Catal. Commun.*, 2008, **9**, 369.
641. J.-M. Andanson, S. Marx and A. Baiker, *Catal. Sci. Technol.*, 2012, **2**, 1403.

642. F. Jutz, J.-M. Andanson and A. Baiker, *J. Catal.*, 2009, **268**, 356.
643. J. Huang, T. Jiang, B. Hu, H. Gao, Y. Chang, G. Zhao and W. Wu, *Chem. Commun.*, 2003, 1654.
644. B. Léger, A. Denicourt-Nowicki, A. Roucoux and H. Olivier-Bourbigou, *Adv. Synth. Catal.*, 2008, **350**, 153.
645. (a) B. Léger, A. Denicourt-Nowicki, H. Olivier-Bourbigou and A. Roucoux, *Inorg. Chem.*, 2008, **47**, 9090; (b) A. Denicourt-Nowicki, B. Léger and A. Roucoux, *Phys. Chem. Chem. Phys.*, 2011, **13**, 13510.
646. B. Leger, A. Denicourt-Nowicki, H. Olivier-Bourbigou and A. Roucoux, *Tetrahedron Lett.*, 2009, **50**, 6531.
647. R. R. Dykeman, N. Yan, R. Scopelliti and P. J. Dyson, *Inorg. Chem.*, 2011, **50**, 717.
648. J. Huang, T. Jiang, H. Gao, B. Hu, Z. Liu, W. Wu, Y. Chang and G. Zhao, *Angew. Chem., Int. Ed.*, 2004, **43**, 1397.
649. Y. Hu, H. Yang, Y. Zhang, Z. Hou, X. Wang, Y. Qiao, H. Liu, B. Feng and Q. Huang, *Catal. Commun.*, 2009, **10**, 1903.
650. Y. Hu, Y. Yu, Z. Hou, H. Li, X. Zhao and B. Feng, *Adv. Synth. Catal.*, 2008, **350**, 2077.
651. M. H. G. Prechtl, J. D. Scholten and J. Dupont, *J. Mol. Catal. A: Chem.*, 2009, **313**, 74.
652. R. Venkatesan, M. H. G. Prechtl, J. D. Scholten, R. P. Pezzi, G. Machado and J. Dupont, *J. Mater. Chem.*, 2011, **21**, 3030.
653. K. L. Luska and A. Moores, *Adv. Synth. Catal.*, 2011, **353**, 3167.
654. S. A. Stratton, K. L. Luska and A. Moores, *Catal. Today*, 2012, **183**, 96.
655. G. Salas, P. S. Campbell, C. C. Santini, K. Philippot, M. F. Costa Gomes and A. A. H. Padua, *Dalton Trans.*, 2012, **41**, 13919.
656. D. Gonzalez-Galvez, P. Nolis, K. Philippot, B. Chaudret and P. W. N. M. van Leeuwen, *ACS Catal.*, 2012, **2**, 317.
657. W. Zhu, H. Yang, Y. Yu, L. Hua, H. Li, B. Feng and Z. Hou, *Phys. Chem. Chem. Phys.*, 2011, **13**, 13492.
658. M. J. Beier, J.-M. Andanson, T. Mallat, F. Krumeich and A. Baiker, *ACS Catal.*, 2012, **2**, 337.
659. J. Keilitz, S. Nowag, J.-D. Marty and R. Haag, *Adv. Synth. Catal.*, 2010, **352**, 1503.
660. X. Yang, N. Yan, Z. Fei, R. M. Crespo-Quesada, G. Laurenczy, L. Kiwi-Minsker, Y. Kou, Y. Li and P. J. Dyson, *Inorg. Chem.*, 2008, **47**, 7444.
661. (a) X.-D. Mu, J.-q. Meng, Z.-C. Li and Y. Kou, *J. Am. Chem. Soc.*, 2005, **127**, 9694; (b) C. Zhao, H.-z. Wang, N. Yan, C.-x. Xiao, X.-d. Mu, P. J. Dyson and Y. Kou, *J. Catal.*, 2007, **250**, 33.
662. C.-x. Xiao, H.-z. Wang, X.-d. Mu and Y. Kou, *J. Catal.*, 2007, **250**, 25.
663. D. Zhao, Z. Fei, W. H. Ang and P. J. Dyson, *Small*, 2006, **2**, 879.
664. I. Biondi, G. Laurenczy and P. J. Dyson, *Inorg. Chem.*, 2011, **50**, 8038.
665. C. Shan, F. Li, F. Yuan, G. Yang, L. Niu and Q. Zhang, *Nanotechnology*, 2008, **18**, 285601.

666. N. Yan, Y. Yuan and P. J. Dyson, *Chem. Commun.*, 2011, **47**, 2529.
667. N. Yan, X. Yang, Z. Fei, Y. Li, Y. Kou and P. J. Dyson, *Organometallics*, 2009, **28**, 937.
668. X. Yuan, N. Yan, S. A. Katsyuba, E. E. Svereva, Y. Kou and P. J. Dyson, *Phys. Chem. Chem. Phys.*, 2012, **14**, 6026.
669. J. Wang, B. Xu, H. Sun and G. Song, *Tetrahedron Lett.*, 2013, **54**, 238.
670. D. Zhao, Z. Fei, T. J. Geldbach, R. Scopelliti and P. J. Dyson, *J. Am. Chem. Soc.*, 2004, **126**, 15876.
671. C. Chiappe, D. Pieraccini, D. Zhao, Z. Pei and P. J. Dyson, *Adv. Synth. Catal.*, 2006, **348**, 68.
672. Z. Fei, D. Zhao, D. Pieraccini, W. H. Ang, T. J. Geldbach, R. Scopelliti, C. Chiappe and P. J. Dyson, *Organometallics*, 2007, **26**, 1588.
673. X. Yang, Z. Fei, D. Zhao, W. H. Ang, Y. Li and P. J. Dyson, *Inorg. Chem.*, 2008, **47**, 3292.
674. F. Fernández, B. Cordero, J. Durand, G. Muller, F. Malbosc, Y. Kihn, E. Teuma and M. Gómez, *Dalton Trans.*, 2007, 5572.
675. J. Durand, E. Teuma, F. Malbosc, Y. Kihn and M. Gómez, *Catal. Commun.*, 2008, **9**, 273.
676. H. A. Kalviri and F. M. Kerton, *Green Chem.*, 2011, **13**, 681.
677. For a recent example see F. T. U. Kohler, K. Gartner, V. Hager, M. Haumann, M. Sternberg, X. Wang, N. Szesni, K. Meyer and P. Wasserschied, *Catal. Sci. Technol.*, 2014, DOI:10.1039/c3cy00905j.
678. C. Pavia, E. Ballerini, L. A. Bivona, F. Giacalone, C. Aprile, L. Vaccaro and M. Gruttadauria, *Adv. Synth. Catal.*, 2013, **355**, 2007.
679. J. Chen, X. Fang, X. A. Duan, L. Ye, H. Lin and Y. Yuan, *Green Chem.*, 2014, **16**, 294.
680. S. Martin, R. Porcar, E. A. Peris, M. I. Burguete, E. García-Verdugo and S. V. Luis, *Green Chem.*, 2014, **16**, DOI: 10.1039/c3gc42238k.
681. W.-I. Dai, B. Jin, S.-L. Luo, X.-B. Luo, X.-M. Tu and C.-T. Au, *Catal. Sci. Technol.*, 2014, DOI: 10.1039/c3cy00659j.
682. R. G. Gore, T.-H. Truong, M. Pour, L. Myles, S. J. Connon and N. Gathergood, *Green Chem.*, 2013, **15**, 2727 and references therein.
683. M. Moniruzzaman, K. Nakashima, N. Kamiya and M. Goto, *Biochem. Eng. J.*, 2010, **48**, 295.

CHAPTER 4

Catalysis in Ionic Liquid–Supercritical CO_2 Systems

MARK J. MULDOON

School of Chemistry and Chemical Engineering, Queen's University Belfast, David Keir Building, Stranmillis Road, Belfast, BT9 5AG, Northern Ireland, UK
Email: m.j.muldoon@qub.ac.uk

4.1 Introduction

Ionic liquids (ILs) and supercritical carbon dioxide ($scCO_2$) can be combined in a manner that allows efficient catalytic processes to be developed. These two neoteric solvent systems have very different properties from one another and until 1999 they were investigated as separate technologies for green chemistry.[1] As it turns out these solvents have complementary properties and their combination can result in systems that enable processes to be developed with integrated catalyst separation; which is often an important hurdle when employing enzymes or homogeneous organometallic catalysts.

The physiochemical properties of $scCO_2$ and the application of $scCO_2$ for a variety of different green chemical processes is well documented elsewhere[2] and will not be repeated in detail here. It is however worth introducing some of the important fundamental properties of $scCO_2$. The critical temperature of CO_2 $T_c = 31$ °C and the critical pressure $P_c = 73.8$ bar. Beyond this *critical point,* no distinct liquid or vapour phase can exist and the new supercritical phase exhibits properties that are reminiscent of both states. Liquid like densities mean that $scCO_2$ has sufficient cohesion to dissolve materials and act as a solvent. Whereas the viscosity is lower and the diffusivity is higher

RSC Catalysis Series No. 15
Catalysis in Ionic Liquids: From Catalyst Synthesis to Application
Edited by Chris Hardacre and Vasile Parvulescu

Published by the Royal Society of Chemistry, www.rsc.org

compared to liquid solvents and this means $scCO_2$ has superior mass and heat transport properties. Importantly, the solvent properties of $scCO_2$ can be tuned by changes in temperature and pressure; however, there are limitations to how much it can be altered. $ScCO_2$ is a relatively "non-polar" solvent and materials which are highly polar or with a low-volatility will exhibit poor solubility. It is also worth clarifying that the critical point that is mentioned above is for pure CO_2. Utilising $scCO_2$ for chemical processes means that you want to dissolve some organic material in $scCO_2$. In order to obtain a single phase system once you have added other materials will require higher pressures and temperatures. The point at which a single phase is obtained when other components are added is called the *mixture critical point* and this will depend on the nature of these other components and their concentrations.

$scCO_2$ has been studied for catalysis for many years and it can have a number of benefits. The combination of $scCO_2$ with heterogeneous catalysts has probably been studied the most and is arguably more straightforward; as you might expect heterogeneous catalysts allow facile catalyst separation and reuse. There are numerous examples of where $scCO_2$ is advantageous for heterogeneous catalysis.[3] For example, reaction rates and product selectivity can be improved because other gases are completely miscible in $scCO_2$; which means it is possible to eradicate gas–liquid mass transfer resistance. Utilising $scCO_2$ for homogeneous catalysis results in some challenges due to its solvent strength.[4] Many metal complexes are poorly soluble in *unmodified*[5] $scCO_2$, therefore, this resulted in researchers developing catalysts with CO_2-philic ligands (for example perfluorinated ligands) which would solubilise organometallic catalysts in $scCO_2$. The drawback of this is that such modification makes catalysts more complicated and expensive to prepare. Furthermore, if a homogeneous catalyst is made completely soluble in $scCO_2$ it can make separating and recycling the catalyst more challenging, requiring temperature and/or pressure swings in order to separate the catalyst. As we will discuss, continuous flow operation is highly desirable and that is difficult if homogeneous catalysts are used with $scCO_2$ alone.

In the case of ILs as solvents for catalysis, the properties of ILs can be beneficial for the separation and recycling of homogeneous catalysts. The use of ILs for homogeneous catalysis has been well studied and a number of strategies have been developed to separate and recycle the catalyst and the IL.[6] It is possible to develop multi-phase systems whereby the product is extracted by an immiscible solvent; often an organic solvent such as hexane or diethyl ether. In some cases, the product that is formed is immiscible with the IL and separates without the need for an additional solvent; a nice example of this is the oligomerisation of ethylene.[7] The low volatility of ILs means that removing the products by distillation is an option in some cases. Nonetheless there are occasions when product separation and catalyst recycling is challenging for ILs. The product may not separate and it might not be possible to find an immiscible solvent which will preferentially extract the product. Distillation may not be desirable due to the high boiling point of the product and the low thermal stability of the catalyst. So it is clear that both $scCO_2$ and ILs pose challenges when it comes to their application in the field of homogeneous catalysts.

The polarity and solubility limitations of $scCO_2$ have led researchers to combine it with other solvents, creating so called "gas expanded liquids".[8] There are mutual benefits to be had when CO_2 is combined with other solvents. CO_2 can be combined with volatile organic solvents, water, liquid polymers (*e.g.* polyethylene glycols) and ILs. A number of reviews have been published that discuss a range of these expanded solvent systems [8–11] and there has also been a recent review looking solely at IL–$scCO_2$ systems.[12] This chapter is focussed on the combination of $scCO_2$ with ILs and aims to highlight the possible benefits these systems offer when designing catalytic processes that utilise homogeneous catalysts.

In the last decade or so there has been significant interest in IL–$scCO_2$ systems for catalysis[12] and most people would agree that this stemmed from a report in 1999 by Brennecke and co-workers.[1] They demonstrated that ILs and $scCO_2$ could be combined in a complementary fashion. They found that while $scCO_2$ had substantial solubility in the IL (in that study it was [bmim][PF_6]) the IL did not dissolve in the upper $scCO_2$ phase. They also showed that $scCO_2$ could be used to extract organic molecules from an IL (naphthalene in this example), see Figure 4.1 for details. Due to the fact that the IL was not soluble in $scCO_2$, the organic molecule could be obtained without any contamination with the IL. This study had clear implications to those interested in separating organic products from ILs and homogeneous catalysts.

This study opened up new possibilities for the separation and recycling of homogeneous catalysts. The premise being that ILs could act as an immobile phase, trapping the catalyst in the reactor. $scCO_2$ could then be utilised to extract organic products from the IL; leaving behind the IL and catalyst for

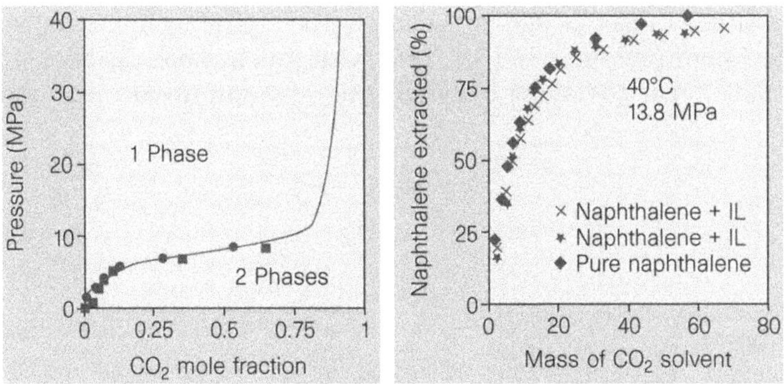

Figure 4.1 Phase behaviour of CO_2–[bmim][PF_6] at 40 °C (LHS) and extraction of naphthalene from [bmim][PF_6] using $scCO_2$ (RHS).
Reprinted with permission from reference 1 © 1999, Macmillan Publishers Ltd.

reuse. In some cases it is necessary to modify the catalyst, for example by using ligands with ionic tags to ensure that the catalyst remains anchored in the IL phase. Utilising CO_2 to extract materials in this manner had a number of clear advantages from the outset. Along with separation and recycling of the IL and catalyst, it could enable the extraction of organic products (many with relatively high boiling points) from ILs at milder temperatures. Furthermore, when CO_2 is depressurised and returns to a gas, it loses its solvent power; leaving behind the extracted material free from any solvent contamination.

Initial reports of catalysis in IL–$scCO_2$ systems utilised batch conditions, with the products being extracted using $scCO_2$ at the end of each reaction. Relatively quickly after these initial reports, Cole-Hamilton and co-workers demonstrated that these systems could also be operated under continuous flow conditions, as illustrated in Figure 4.2 [13]

This was an important development because continuous processing greatly improves the chances of such processes being implemented on an industrial scale. Continuous operation can greatly increase the efficiency and safety of a process, and this mode of operation is routinely used in the bulk chemical industry. Indeed, the benefits of continuous operation have meant that a move to flow processes has recently been highlighted as a goal by the pharmaceutical industry;[14] an industry which has relied on batch operations for the most part. Continuous flow systems avoid the need for time consuming and energy inefficient steps, such as pressurisation/depressurisation and heating/cooling of the reactor. It is possible to introduce real-time reaction monitoring and optimisation of the operating conditions. Flow systems allow much smaller reactors to be employed and these have a number of benefits, for example higher space–time–yields (STYs) are obtained and less solvent waste is produced. Smaller reactors reduce the risks associated with hazardous reactions and processes (*e.g.* those having toxic substrates and products or reactions prone to thermal runaway). In the case of an IL–$scCO_2$ process, this will have a relatively high pressure; therefore smaller reactors will be seen as a must by many industrialists. The dangers of high pressures can be dealt with much more easily on a small scale. Smaller reactors allow better mixing and improved mass and heat

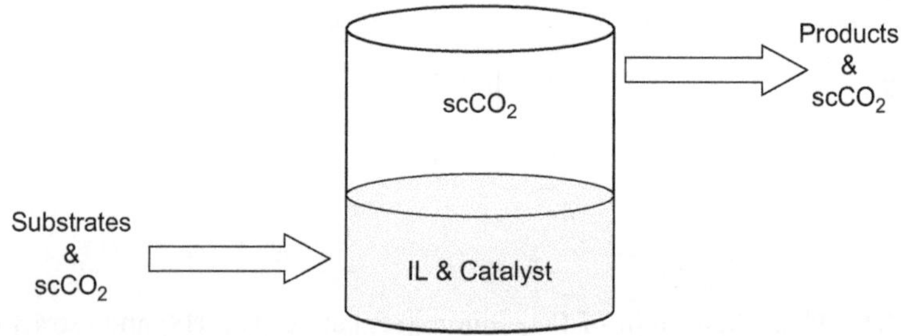

Figure 4.2 The general concept of continuous flow catalysis with IL–$scCO_2$ systems.

transfer. Indeed more efficient heat management in flow systems is an environmental and safety benefit. Since the initial development of IL–scCO_2 continuous flow systems there has been a move towards utilising supported ionic liquid phase (SILP) catalysis with scCO_2. We will discuss examples of catalytic reactions in more detail later, but first it is important to highlight some of the physicochemical properties of these IL–scCO_2 systems. In order to understand catalyst performance and optimise these multiphase systems, it is necessary to take into account a number of different factors such as solubility and phase behaviour.

4.2 Physicochemical Properties of IL–scCO_2 Systems

4.2.1 Phase Behaviour

The solubility of CO_2 in ILs depends on the structure of the IL and the temperature and pressure. In terms of temperature and pressure, increasing temperature decreases solubility and increasing pressure increases solubility. A significant number of studies have been carried out examining CO_2 solubility in a wide variety of structurally diverse ILs, this is because ILs have also been identified as potential materials for CO_2 separation and capture.[15] In the case of IL–scCO_2 systems for applications in catalysis it is unlikely that the choice of IL would be based on it having the highest CO_2 solubility. When using IL–scCO_2 systems for catalysis (particularly when used under continuous flow conditions) the pressure will generally be quite high. High pressures of scCO_2 are required if the scCO_2 is to have sufficient solvent power to extract the products. A significant quantity of CO_2 will dissolve in most ILs at the pressures typically used for extraction of products (>100 bar). Figure 4.3 shows the solubility of CO_2 in a range of ILs with increasing pressure at 25 °C. Although this is subcritical at this temperature, it exemplifies the general trend that is seen in all temperatures. It can be seen that there is not a linear increase in solubility with increasing pressure. There comes a point when the IL is essentially "full" of CO_2 and little increase in solubility will be observed with increasing pressures. As you might expect, increasing the temperature weakens the interactions between the ILs and CO_2 and, therefore, for a given pressure the solubility decreases with increasing temperature.

ILs do not dissolve CO_2 to the same extent as most organic solvents and a consequence of this is that, even at very high pressures, ILs do not swell in volume greatly; not compared to volatile organic solvents.[8] Figure 4.4 illustrates the different swelling behaviour of ILs (in this case [bmim][BF_4]) and organic solvents (exemplified with ethyl acetate and acetonitrile). Crude oil, poly(propylene glycol) (PPG) and poly(ethylene glycol) (PEG) are also shown. It is clear from Figure 4.4 that ILs do not swell a great deal. While in the case of volatile organic solvents, there is an enormous increase in volume when CO_2 pressure is applied. As the CO_2 pressure increases, more and more of the organic solvent dissolves in the CO_2 phase. If an appropriate

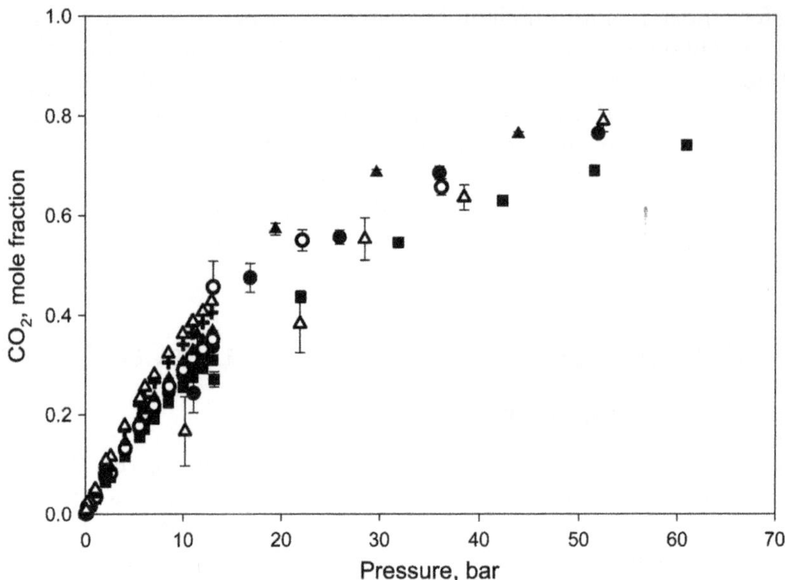

Figure 4.3 CO_2 solubility in a range of ILs at 25 °C. (■) [hmim][Tf$_2$N]; (●) [C$_6$H$_4$F$_9$mim][Tf$_2$N]; (○) [C$_8$H$_4$F$_{13}$mim][Tf$_2$N]; (▲) [hmim][eFAP]; (+) [hmim][pFAP]; (△) [p$_5$mim][bFAP].
Reproduced with permission from reference 16 © 2003 American Chemical Society.

temperature is being used, increasing pressure will eventually lead to the *mixture critical point* being reached and a single phase supercritical fluid will be formed. In the case of ILs this type of behaviour is not observed and for the purposes of designing flow processes with IL–scCO$_2$ systems, this (relative) lack of volumetric expansion and insolubility are advantages. It means that reactors do not need to be designed to incorporate very large volume changes. It also means it is possible to ensure that the IL remains in the reactor and does not get extracted by the scCO$_2$.

It has been shown that ILs do not dissolve in scCO$_2$ even at very high pressures and, therefore, two phases exist at all pressures. But of course when it comes to carrying out catalytic reactions, the system will become more complex as more components will be in the system; consequently the phase behaviour will become more complex also. As we have just discussed, organic solvents swell a great deal, even at moderate CO_2 pressures (tens of bars), consequently when CO_2 pressure is applied to reaction mixtures the organic components will often swell and begin to separate from the IL phase; increasing the number of phases. The phase behaviour of reaction mixtures will be influenced by temperature, pressure, the structure of the IL, the nature of the reactants and the concentrations of all these components. It is therefore important that phase behaviour studies are incorporated into catalytic reaction studies, as it has the potential to have a significant impact

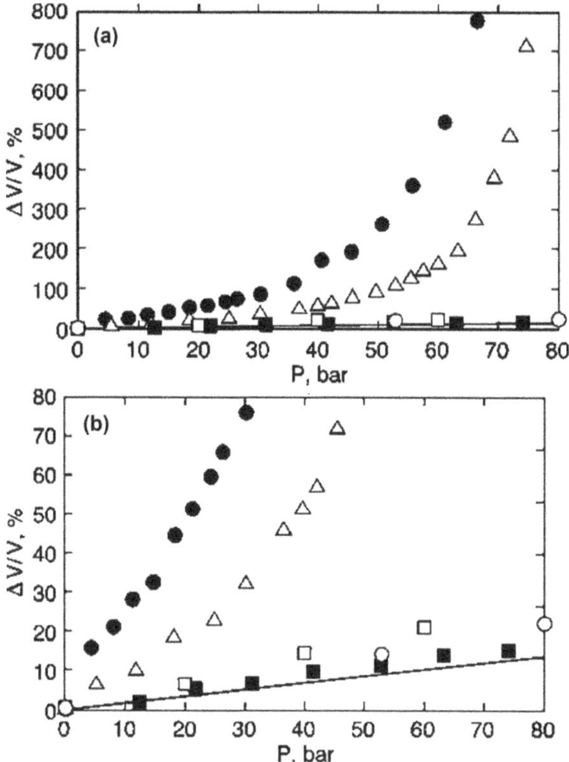

Figure 4.4 Volumetric expansion of solvents as a function of the pressure of CO_2 at 40 °C, for ethyl acetate (●), acetonitrile (△), [bmim][BF_4] (■), crude oil (line, at 43 °C), PPG (□), and PEG (○). (b) shows an expanded section of the data shown in (a).
Reproduced with permission from reference 8(*a*) © 2003 American Chemical Society.

on the reaction. Although it may be possible to model and predict the phase behaviour of multi-component mixtures, perhaps the best approach (at present) it to utilise a view cell that allows the phase behaviour under reaction conditions to be verified. Figure 4.5 gives an example of how the phase behaviour can change with increasing CO_2 pressure for a relatively simple mixture of IL + ethanol + water.[17]

It can be seen in Figure 4.5 that as CO_2 pressure is increased, turbidity is initially observed (B), then after several minutes, the turbidity disappears (C) and a third phase (L4) forms between the CO_2-rich vapour phase (V3) and the IL-rich liquid (L3) at the bottom. Increasing the pressure further (D) leads to an expansion of the middle phase and a shrinking of the bottom liquid phase. Eventually with increasing CO_2 pressure it goes through the critical point of the mixture (E) and at pressures higher than this point, the system remains biphasic in nature. Similar studies were first reported by Brennecke and co-workers for an IL + methanol + CO_2 system.[18] In this case they were

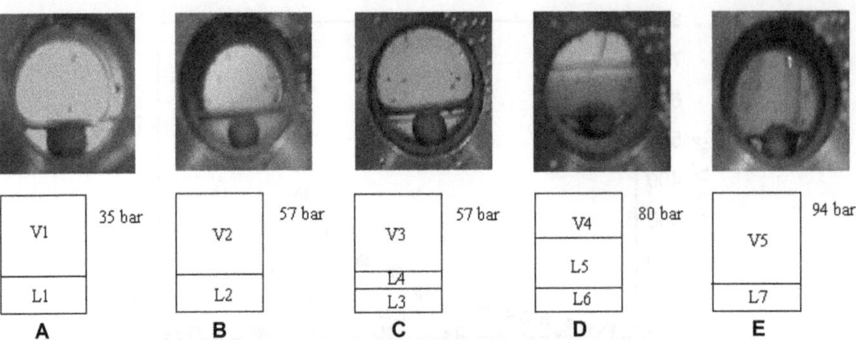

Figure 4.5 Phase changes with increasing carbon dioxide pressure for a mixture containing [bmim][PF_6], water, ethanol and CO_2 at 313.15 K. Reproduced with permission from reference 17 © 2003 Wiley-VCH Verlag GmbH & Co. KGaA, Weinheim.

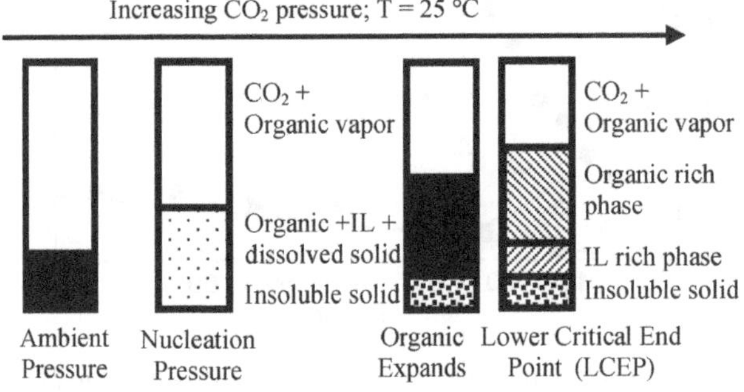

Figure 4.6 Illustration of CO_2 induced precipitation in ILs. Reproduced from reference 19.

able to sample the phases and it was found that once the critical point of the CO_2–methanol mixture had been reached then there was no IL in the upper phase. At high CO_2 pressures, the solvent strength of the organic components is greatly reduced and the supercritical fluid is unable to dissolve the IL. This is good news if $scCO_2$ is to be used for extracting organic compounds from ILs as otherwise the products would be contaminated with the IL. However, the phase changes illustrated in Figure 4.5 demonstrate that it is important to be aware of the phase behaviour under reaction conditions. Clearly such behaviour could greatly influence the reaction; for example, organic substrates may separate from the IL phase (which contains the catalyst), and this would impact reaction rates and selectivity. Brennecke and co-workers also demonstrated that inorganic salts could precipitate from IL–organic solvent mixtures.[19] They found that CO_2 pressure could precipitate ammonium bromide, ammonium chloride and zinc acetate from IL–organic solvent (acetonitrile and DMSO) mixtures, as shown in Figure 4.6.

Kroon *et al.* found that the product of an asymmetric hydrogenation reaction (*N*-acetyl-(*S*)-phenylalanine methyl ester) could not only be extracted from an IL using $scCO_2$ but could also be precipitated from the IL using $scCO_2$.[20] Such examples demonstrate that CO_2 pressure can act as an anti-solvent and could result in the precipitation of reaction components (*e.g.* catalysts), which once again has clear implications for reactions. So it is crucial that phase behaviour is taken into account when carrying out catalytic studies.

4.2.2 Solubility of Other Gases

Many important catalytic reactions involve gaseous reactants and the majority of these gases have quite poor solubility in ILs, and this can cause problems. For example, industrially important gases such oxygen and hydrogen exhibit very little solubility in ILs and this can lead to mass transfer problems when carrying out reactions. A benefit of using an IL–$scCO_2$ system is that that $scCO_2$ can help improve the solubility of other gases in the IL (catalyst) phase. Brennecke and co-workers demonstrated that CO_2 pressure improved the solubility of oxygen and methane in [hmim][Tf_2N].[21] While Leitner and co-workers demonstrated that CO_2 pressure improved the solubility of hydrogen (see later for more details of this study).[22] It is worth pointing out that use of CO_2 results in a high overall pressure. In other words when you look at the comparisons of pure gas *versus* gas + CO_2, the overall pressure needed to obtain a certain concentration might be quite similar to what would be needed if just the pure gas was used. Nonetheless, in the case of IL–$scCO_2$ systems we are usually interested in using $scCO_2$ to also extract products from the catalyst phase. Additionally, it seems sensible that it is more desirable to have such flammable gases diluted with a non-flammable gas such as CO_2. Certainly when aerobic oxidation reactions are carried out industrially the O_2 concentration has to be diluted to a lower concentration (often to less than 8%) with a non-flammable gas, to avoid explosive mixtures forming.

4.2.3 Solvent–Solute Interactions

The term "polarity" is a widely used term in chemistry, but it is usually used in a vague manner. Ultimately, reactions are influenced by the strength and nature of a number of different solvent–solute interactions. In some cases, certain types of interactions will dominate the outcome of the reaction. Measuring and understanding solvent effects is a complex endeavour.[23] One well-respected approach is to utilise spectroscopic probes which have solvatochromic behaviour. The Kamlet–Taft parameters utilise probe molecules which have solvent dependent UV–Visible spectra. This is a multi-parameter approach and uses three dyes to measure dipolarity/polarizability and hydrogen bond accepting and donating ability.[24] Fredlake *et al.* examined the Kamlet–Taft Parameters in IL–CO_2 mixtures.[25] In these multiphase catalysis systems the catalyst will reside in the IL phase and this is where the reaction will take place; but there will be significant quantities of CO_2 in the IL phase. Fredlake *et al.* found that the application of CO_2 pressure did not greatly affect the solvent strength of ILs, as measured using these probes. It seems that the presence of CO_2 does not weaken the strength of interactions between the ILs and the probe molecules. This also fits with the volumetric expansion studies discussed earlier; where CO_2 is not strong enough to overcome the IL–IL interactions.

Such studies should not be taken to mean that CO_2 will be an innocent co-solvent in reactions. As we discussed earlier, CO_2 can cause the phase separation of reaction components. Additionally, if water, alcohols or amines are present in the reaction mixture then CO_2 can react with them. Water will produce carbonic acid while alcohols can form alkylcarbonic acids. Depending on the reaction, the presence of such acids could be a problem or a benefit (*i.e.* an acid catalyst). It is, however, possible to control the pH, even when water is present in large quantities, for example, Roosen *et al.* demonstrated that buffer salts could be used to control the pH in water–CO_2 biphasic systems.[26] CO_2 can also interact and react with Lewis bases. In the case of amines, CO_2 can react to form carbamic acids or carbamate salts. Once again this could be seen as a benefit or a problem for an IL–scCO_2 system. If products form carbamic acids or carbamate salts, then it will be unlikely that scCO_2 can extract these from the IL. On the other hand this could be used as a method of immobilising catalysts in the IL phase, for example if the ligand has suitable amine groups attached. The formation of acids and carbamate salts has also been used to generate ILs from amines and the combination of amines and alcohols,[27–31] as shown in Figure 4.7.

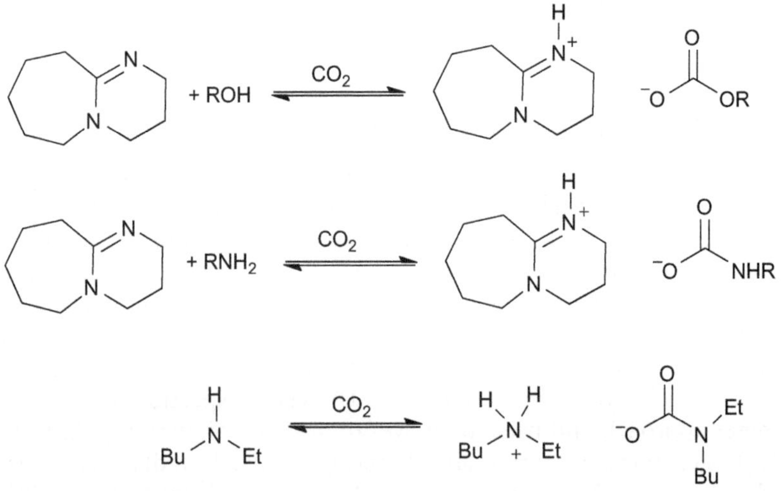

Figure 4.7 Examples of "CO_2 switchable" ionic liquids.

4.2.4 Properties that Influence Mass Transport

Properties such as melting point, viscosity, diffusion and interfacial tension are all potential challenges when it comes to utilising ILs. In some cases, the desired organic salt may not be a liquid at the reaction temperature, while it is known that ILs tend to have viscosities that are much higher than most organic solvents. A high viscosity can lead to mass transfer problems which can ultimately influence the performance of many catalytic reactions. The addition of $scCO_2$ to ILs can improve these properties.

A number of studies have shown that CO_2 pressure can be used to greatly reduce the melting point of organic salts;[32–34] thereby widening the number of ILs that could possibly be used in IL–$scCO_2$ systems. In some cases, it has been shown that the melting point of salts could be decreased by more than 100 °C. For example, Scurto and Leitner demonstrated that tetrabutylammonium tetrafluoroborate ([NBu_4][BF_4]) could be used as a solvent in an IL–$scCO_2$ system for a number of reactions; hydrogenation, hydroformylation, and hydroboration.[33] Figure 4.8 illustrates this approach. This salt has a melting point of 156 °C but it could be used as a solvent 100 °C below its melting point when CO_2 pressure was applied.

ILs are significantly more viscous than most commonly used organic solvents and such high viscosities could lead to diminished catalytic performance in some cases. It has been shown that the viscosity of ILs can be reduced with CO_2 pressure.[35] An example of the effect of CO_2 pressure on ILs is shown in Figure 4.9.

When CO_2 dissolves in ILs it leads to a reduction in the intermolecular forces between the ions (to a certain degree). The extent of the reduction is greater at lower temperatures. At higher temperatures, the viscosity of the IL has already been greatly reduced and additionally the solubility of CO_2 in the IL is less at higher temperatures. As the CO_2 pressure increases the effect becomes less, and this pattern makes sense if we look at CO_2 solubility studies. As discussed earlier, CO_2 solubility does not increase linearly with increasing CO_2 pressure; after a certain point there are diminishing returns and even going to very high pressures leads to only a small increase in the

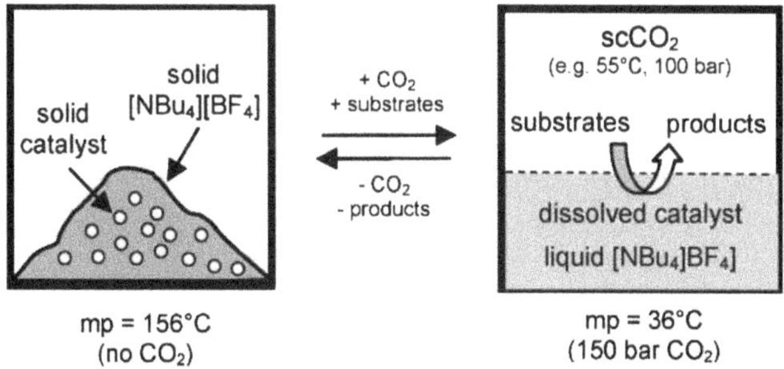

Figure 4.8 CO_2 induced melting of organic salts for catalysis. Reproduced from reference 33.

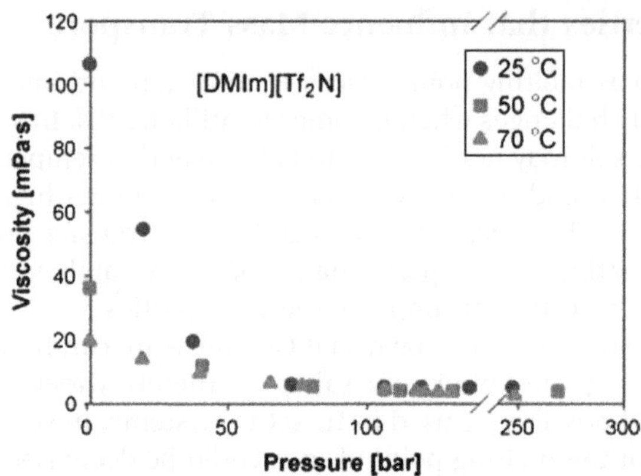

Figure 4.9 The influence of CO_2 pressure and temperature on the viscosity of 1-decyl-3-methylimidazolium bis(trifluoromethanesulfonyl)amide ([DMIm][Tf_2N]).
Reproduced with permission from reference 35 © 2009 Elsevier.

concentration of CO_2 in the IL phase. Overall, it is clear from Figure 4.9 that $scCO_2$ does reduce the viscosity of ILs and this has the potential to improve performance in reactions and processes. The diffusion coefficient for a given molecule is inversely proportional to the viscosity of the liquid; therefore lowering of the viscosity will improve diffusivity. In terms of measuring diffusivities in ILs,[36] more work needs to be carried out in this area, examining a wider range of solutes, in order to fully understand the contribution of factors such as molar volumes and charge.[37] Nonetheless, a reduction in the viscosity of the IL will undoubtedly lead to improved mass and heat transfer. Scurto and co-workers carried out detailed studies to correlate factors such as phase equilibria and mass transfer with the kinetics for Rh catalysed hydroformylation of 1-octene.[38] In those studies they showed that lowering of the viscosity and increased diffusivity allowed the reaction to proceed more easily into the kinetically controlled regime.

Interfacial tension affects the transport across the gas–liquid boundary and ILs generally have high surface tensions compared to most organic solvents.[39] Jaeger and Eggers demonstrated that CO_2 pressure decreases the interfacial tension in ILs.[40] As shown in Figure 4.10, application of CO_2 pressure can lead to significant reductions in interfacial surface tension. These results once again indicate that combining CO_2 with ILs should lead to improved mass transfer.

4.2.5 Partition Coefficients

The efficiency of these multiphase systems will be dependent on favourable partitioning of reactants and products between the IL and $scCO_2$ phases. Arguably this is even more important in continuous flow systems, where

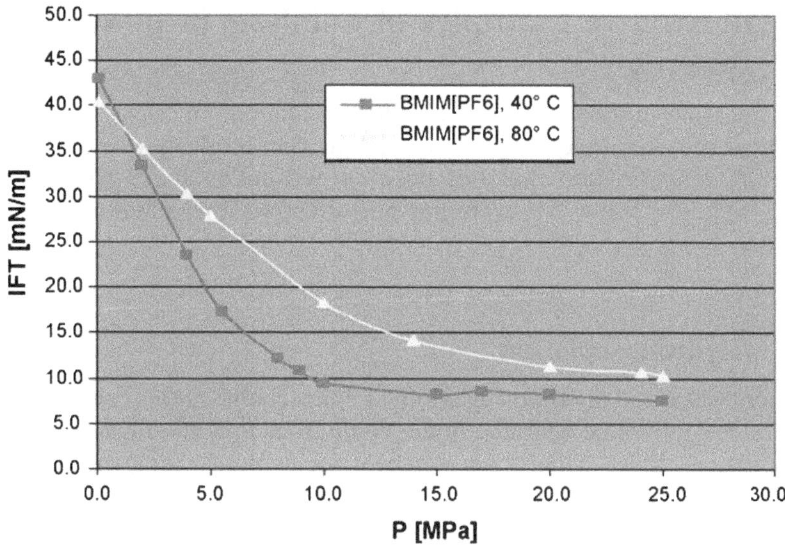

Figure 4.10 Reduction of interfacial tension (IFT) for [bmim][PF_6] with increasing CO_2 pressure.
Reproduced with permission from reference 40 © 2009 Elsevier.

materials will have a finite residence time. In IL–scCO_2 systems both phases can be tuned to some degree. The structure of the IL can be altered to increase or decrease the solubility of substrates and products. Whilst the solvent strength of scCO_2 can be varied by variation of temperature and pressure. A number of studies have measured partition coefficients between ILs and scCO_2.[41] A popular technique is to examine solute partitioning in capillary column chromatography with scCO_2 as mobile phase and the IL as stationary liquid phase. Using this method it is possible to convert retention factors into infinite-dilution solute partition coefficients. These studies have examined a range of temperatures and pressures and as might be expected the density of the scCO_2 has a significant influence on the partition coefficients.

Although a significant amount of work has been carried out in this area, it seems that it is still difficult to effectively model these systems and predict this type of data. In general, molecules with a higher polarity (*i.e.* those with hydrogen bonding capabilities) and low volatility will have a higher affinity for the IL phase. While those with lower polarity and high volatility will more readily dissolve in the scCO_2 phase. A challenge for many catalytic reactions is that there will be a significant number of variables; more than are often studied in model systems. For example, many reactions will have other reactive gases (such as O_2, CO and H_2) present and these will affect the solvent strength of scCO_2. But it is important that partitioning is taken into account, as we will show in examples later, this is key in understanding catalyst performance and optimising the operating conditions.

4.3 Illustrative Examples of Catalysis in IL–$scCO_2$ systems

A considerable number of studies exploring catalysis in IL–$scCO_2$ systems have now been reported.[12] It is not our intention to give a comprehensive review of every reaction that has been carried out in these systems, but rather highlight some key examples and illustrate the potential of these systems.

4.3.1 Hydroformylation

Rhodium catalysed hydroformylation of long chain olefins was the first catalytic reaction to be operated under continuous flow conditions in an IL–$scCO_2$ system. These reports are excellent for demonstrating the potential and indeed the complexity of these multiphase systems.

Compared to cobalt based catalysts, rhodium catalysts demonstrate high activity and good regioselectivity under mild conditions. Rhodium modified with sulfonated triphenylphosphine ligands is used industrially for the hydroformylation of short chain olefins. The Ruhrchemie/Rhône–Poulenc aqueous biphasic process is a model process for industrial green chemistry and chemical engineering.[42] Unfortunately this process is not suitable for longer chain olefins, as the water soluble ligands anchor the catalyst in the water phase and lipophilic substrates have very poor water solubility; resulting in very slow reaction rates for olefins $>C_5$. An alternative for long chain olefins would be separation of the catalyst using distillation, however the high boiling points of the aldehyde products coupled with the limited thermal stability of the Rh catalyst make this route problematic. There is, therefore, a need to develop an industrial process that will allow the separation and recycling of Rh based catalysts for the hydroformylation of long chain olefins. Consequently, academic researchers have looked at a variety of approaches to address this challenge. The Cole-Hamilton group has considerable expertise in this area and recognised the potential of IL–$scCO_2$ from an early stage.[13] Initial studies utilised a continuously stirred tank reactor (CSTR) as shown in Figure 4.11.[13,43–45] The IL remained in the reactor and $scCO_2$, syngas and olefin were all flowed into the reactor.

Figure 4.11 also demonstrates how a collection chamber is used to obtain the products. A collection chamber is situated downstream of a back pressure regulator which maintains the reactor system at the desired overall pressure. In the collection chamber the pressure is reduced or in some cases a series of collection chambers is used with multiple back pressure regulators allowing the pressure to be stepped down gradually. When the pressure drops in these chambers the $scCO_2$ returns to CO_2 gas which can no longer solvate the products and the product readily separates in the chamber. In this case, the long chain aldehydes that are being produced are liquids and these can be easily obtained by simply opening the tap at the

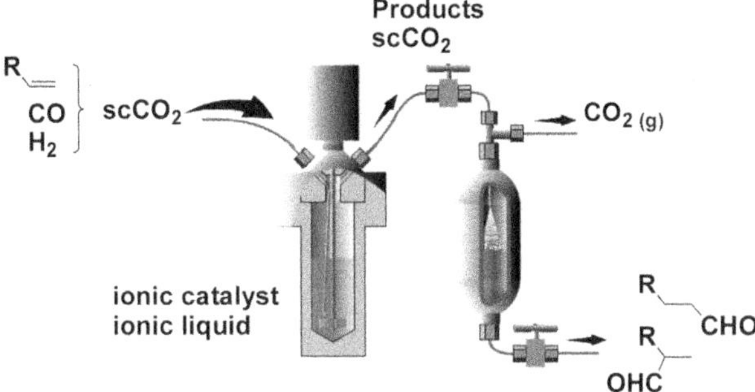

Figure 4.11 Continuous flow Rh catalysed hydroformylation of long chain olefins in an IL–$scCO_2$ system.
Reproduced with permission from reference 43 © 2003 American Chemical Society.

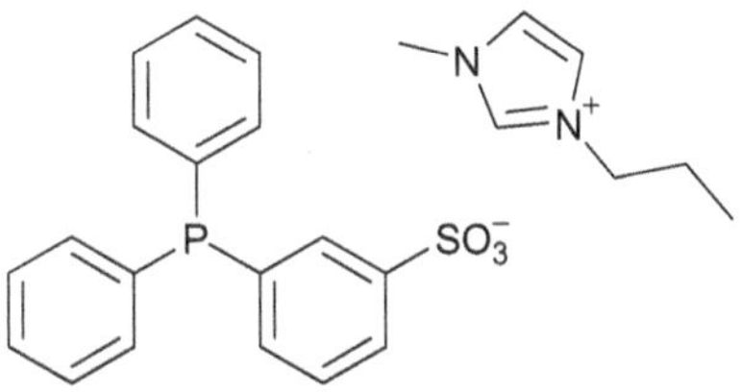

Figure 4.12 [PrMIM][TPPMS] an IL soluble and $scCO_2$ insoluble phosphine ligand.

bottom of the collection chamber. In academic studies such as these the CO_2 gas is usually vented in a fume cupboard, however, in an industrial process the CO_2 could be recompressed and reused.

To ensure that the catalyst remained anchored in the IL phase, they utilised sulfonated phosphine ligands (many of which are commercially available as these are also used for aqueous biphasic catalysis) such as sodium diphenyl(3-sulfonatophenyl)phosphine. To improve the solubility of the catalyst in the IL phase they first exchanged the sodium counterion for an imdazolium cation, resulting in a ligand such as 1-propyl-3-methylimidazolium diphenyl(3-sulfonatophenyl)phosphine ([PrMIM][TPPMS]) (Figure 4.12).

During the initial studies it was found that oxidation of the phosphine ligand was a significant problem if the reaction was carried out in a semi-batch manner. The steps involved in this approach made it difficult to exclude air from the system. Continuous flow operation was far superior in this regard, however it was noted that the grade of CO_2 did have an effect. The degree of oxidation was less when very high purity CO_2 (99.9995%) was used compared to when a lower grade (99.995%) was used.

There are a significant number of variables in this system and in order to operate efficiently under continuous flow conditions a number of factors had to be studied and optimised. As discussed earlier, partitioning of substrates and products is very important and the choice of IL had a

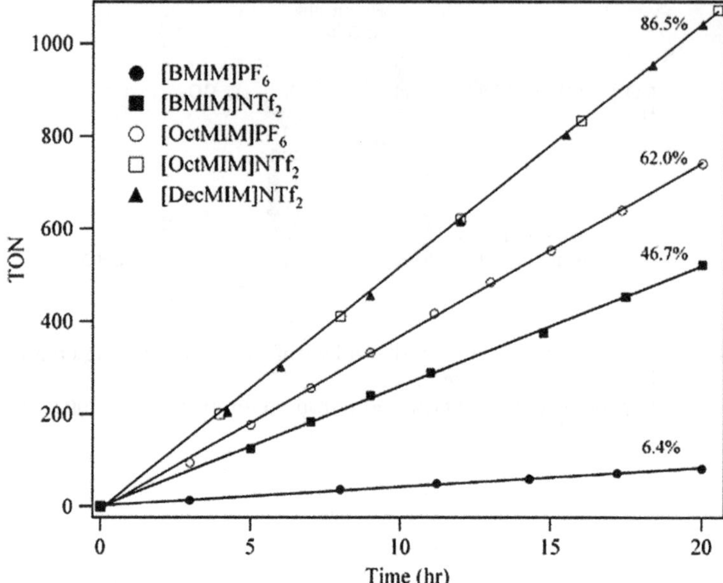

Figure 4.13 Influence of IL structure on the turnover number (TON) for the Rh catalysed hydroformylation of 1-dodecene in a continuous flow IL–$scCO_2$ system.
Reproduced with permission from reference 43 © 2003 American Chemical Society.

significant impact on the turnover frequency (TOF). It can be seen in Figure 4.13 that for a given set of conditions (*e.g.* pressure, temperature, flow rate) the choice of IL had a large impact on the amount of dodecanal produced in the Rh catalysed hydroformylation of 1 dodecene.

Both the cation and anion had an influence on the TOF. They found that the $[Tf_2N]^-$ anion gave better results than $[PF_6]^-$ based ILs and they felt that possible reasons for this were due to the fact that $[PF_6]^-$ based ILs could foam (act like a surfactant) and that they were prone to hydrolysis. There was also a clear effect from the alkyl chain length on the cations. ILs with longer alkyl chains (*i.e.* 1-octyl-3-methylimidazolium and 1-decyl-3-methylimidazolium cations) have a higher solubility for 1-dodecene resulting in a greater quantity of substrate in the catalyst phase. In the case of ILs with shorter chains ($[bmim]^+$) it seems that much of the substrate is extracted by $scCO_2$ before it has a chance to react. The partial pressure of CO and H_2 also influenced the partition coefficients in the system. It is known that in most cases, the rate of hydroformylation is negative order with respect to CO, while there is usually a zero or positive order in H_2. As can be seen in Figure 4.14 it was found that when the partial pressure of CO and H_2 was increased (ratio of gases fixed at 1 : 1) then the rate continued to increase. They believed that this is also due to the influence of these gases on the partitioning of substrates. The higher concentrations of CO and H_2 reduce

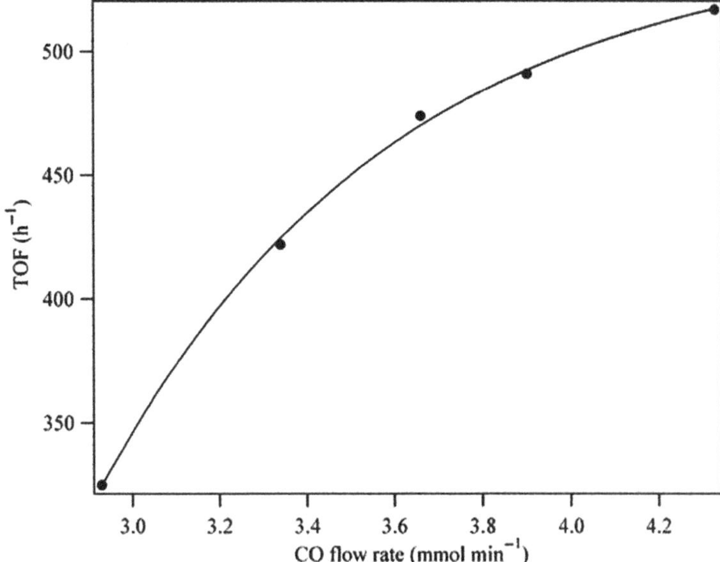

Figure 4.14 Influence of CO–H_2 (1 : 1) flow rate (equivalent to partial pressure) on the rate of hydroformylation of 1-octene (2.65 mmol min^{-1}) catalysed by Rh/[PrMIM][TPPMS] in [omim][Tf_2N]-scCO_2 system. The CO_2 flow rate is kept constant at 1.00 nL min^{-1}.
Reproduced with permission from reference 43 © 2003 American Chemical Society.

the solvent power of scCO_2, which will then increase the concentration of substrates in the IL (catalyst) phase.

There is a limit to this positive effect and with increasing CO and H_2 concentrations the ability of scCO_2 to extract substrates and products continues to weaken. There comes a point when scCO_2 is unable to extract products (and substrate) from the reactor at the same rate at which fresh substrate is being added; consequently the reactor will fill up and flood. This demonstrates the fine balance that is required when attempting to optimise such multiphase catalysis systems.

The IL–scCO_2 system was able to operate for long periods of time and deliver good rates (TOF of 517 h^{-1}). The Rh leaching at steady state operation was also low at <0.1 ppm. The liner aldehyde selectivity was 76% which is not as good as commercial processes, however this is arguably more a catalyst dependent factor and could be improved by utilising a different ligand.

The Cole-Hamilton group followed up these studies by examining the use of supported ionic liquid phase (SILP) catalysis.[46,47] There are a number of advantages to using ILs in this manner, for example SILPs enable much smaller amounts of IL to be utilised which reduces the cost and increases their industrial viability. The SILP materials appear as a "dry" powder and can be employed in reactors typically used for heterogeneous catalysts, for

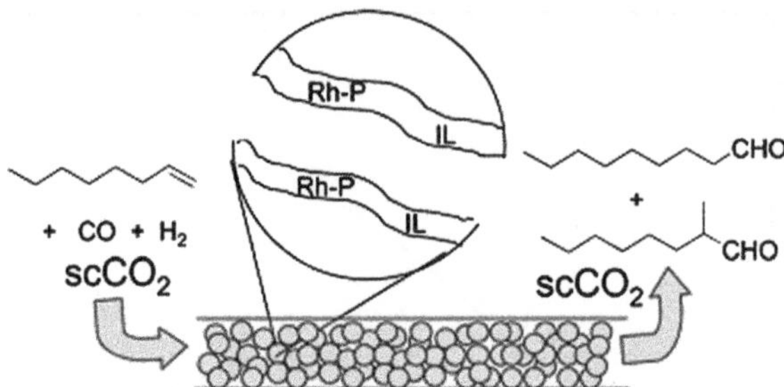

Figure 4.15 Continuous flow Rh catalysed hydroformylation long chain olefins in an SILP–$scCO_2$ system.
Reproduced from reference 46.

example simple tubular reactors, as illustrated in Figure 4.15. Because the catalyst resides in a thin layer of IL, it is very close to the interface and this can reduce mass transport limitations.

Due to the number of variables they decided to employ statistical experimental design software as part of their optimisation studies.[47] It was found that the optimal conditions were when the system was just below the mixture critical point, resulting in a CO_2 expanded organic phase. Using the optimised conditions (total pressure of 100 bar), the catalyst was found to be stable over at least 40 hrs of continuous catalysis with a steady state TOF of 500 h^{-1} and Rh leaching of 0.2 ppm.

It is worth mentioning that SILP catalysis had previously been used for the Rh catalysed hydroformylation of shorter chain alkenes (*e.g.* butene).[48] In this case volatile substrates and products allowed the reactions to be carried out under "gas phase" conditions. Unfortunately, over time the catalyst became deactivated due to the build up of non-volatile by-products (from aldol condensation of aldehydes). This was not an issue when CO_2 was utilised as a mobile phase, as such by-products could be extracted from the SILP by the flowing CO_2.

4.3.2 Hydrogenation

Hydrogenation reactions were the first to be examined in IL–$scCO_2$ systems and there have since been a number of reports of hydrogenation reactions in IL–$scCO_2$ systems. The first studies were reported in 2001. Jessop and coworkers studied the asymmetric hydrogenation of tiglic acid in [bmim][PF_6] (Scheme 4.1).[49] $Ru(O_2CMe)_2((R)\text{-TolBINAP})$ was used as the catalyst giving 2 methylbutanoic acid with high enantioselectivity (*ca.* 88%) and conversion (*ca.* 98%). In this study, $scCO_2$ was not used during the reaction but utilised to extract the product and facilitate the recycling of the catalyst–ionic liquid system. Consistent conversion and enantioselectivity over five cycles was achieved. The authors also reported that they were able to carry out the asymmetric hydrogenation of isobutylatropic acid using the same catalyst in [bmim][PF_6] with added methanol to give ibuprofen.

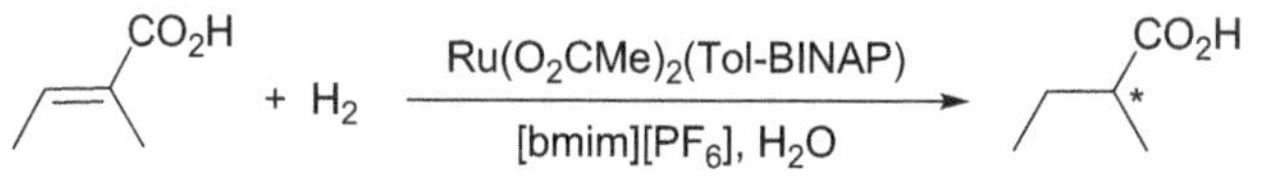

$$\text{tiglic acid} + H_2 \xrightarrow[\text{[bmim][PF}_6\text{], }H_2O]{Ru(O_2CMe)_2\text{(Tol-BINAP)}} \text{2-methylbutanoic acid (*)}$$

Scheme 4.1 Asymmetric hydrogenation of tiglic acid in [bmim][PF_6].

Catalyst

$CO_2 + H_2 \rightleftharpoons HCO_2H$

$HCO_2H + R_3N \rightleftharpoons [HNR_3]^+[HCO_2]$

$HNR_2 + CO_2 \rightleftharpoons R_2NCO_2H$

$R_2NCO_2H + NHR_2 \rightleftharpoons [NH_2R_2]^+[O_2CNR_2]^-$

$[NH_2R_2]^+[O_2CNR_2]^- + 2HCO_2H \longrightarrow 2HCON(R)_2 + 2H_2O + CO_2$

Scheme 4.2 Hydrogenation of CO_2 in the presence of dialkylamines.

Around the same time, Tumas and co-workers demonstrated that scCO_2 could be utilised with ILs in a biphasic system for the catalytic hydrogenation of alkenes and carbon dioxide.[50] Hydrogenation of dec-1-ene and cyclohexene using Wilkinson's catalyst resulted in TOFs of 410 h^{-1} and 220 h^{-1} respectively. In this case they utilised 48 bar of H_2 pressure and with the addition of scCO_2 the total pressure of the system was 207 bar. It was shown that scCO_2 allowed facile product separation and efficient recycling of the IL–catalyst system. Nonetheless, it was found that such reactions could be carried out equally well in a *n*-hexane–[bmim][PF_6] system, suggesting that there was no advantage in reactivity using scCO_2 in these particular examples. On the other hand, they also studied the hydrogenation of CO_2 in the presence of dialkylamines to produce *N*,*N*-dialkylformamides. Such reactions proceed *via* ionic carbamate intermediates as shown in Scheme 4.2. The authors suggested that the intermediates are highly soluble in ILs, as was the ruthenium catalyst used for this experiment. Previous work had shown that the selective production of DMF by such a reaction could be carried out in scCO_2 with very high rates of formation, but that the activity and selectivity (formic acid *vs.* formamide production) decreased rapidly when using longer chain amines. In contrast it was found that for the IL–scCO_2 system the activity and selectivity was higher for amines with higher molecular weights than dimethylamine. Although the formamide products were not easily extracted into the scCO_2 phase initially, after several cycles the level of formamide in the IL reached saturation level, allowing good levels of product separation.

Herein, we will highlight some more recent hydrogenation studies that exemplify the potential benefits of IL–scCO_2 systems. The chosen examples are from the Leitner Group who have been one of the pioneers of catalysis in IL–scCO_2 systems. They have contributed greatly to this area, with detailed studies which combine molecular *and* reaction engineering.

As discussed earlier in section 4.2.2, the presence of CO_2 pressure can lead to improved solubility of other gases in the IL phase. Leitner and co-workers

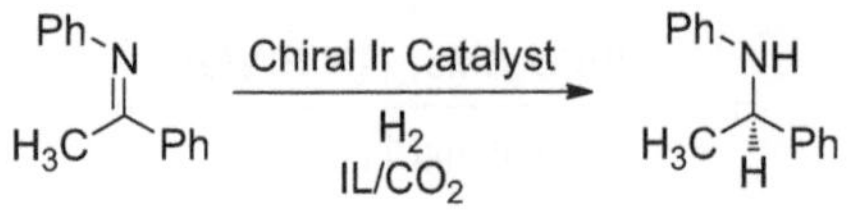

Ph–N=C(CH$_3$)Ph → (Chiral Ir Catalyst, H_2, IL/CO_2) → Ph–NH–CH(CH$_3$)Ph

Scheme 4.3 Catalytic hydrogenation of imines by Leitner and co-workers.[22]

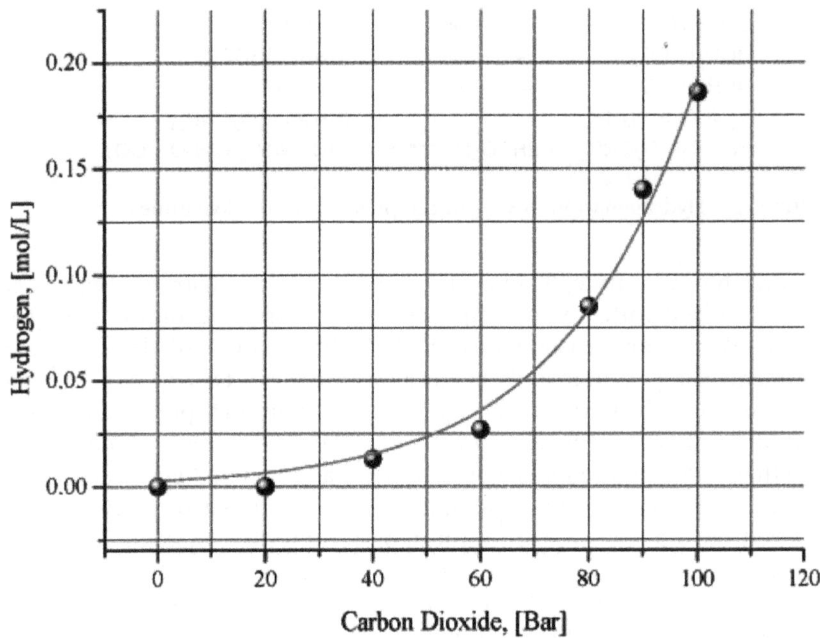

Figure 4.16 Solubility of hydrogen in [emim][Tf$_2$N] at a constant partial pressure ($p(H_2) = 30$ bar as a function of the added CO_2 pressure as determined by high-pressure ^{1}H NMR spectroscopy at $T = 297$ K). Reproduced with permission from reference 22 © 2004 American Chemical Society.

carried out an elegant study that explored the influence of CO_2 and H_2 partial pressures on the performance of catalytic imine hydrogenation as shown in Scheme 4.3.[22]

They were able to utilise ^{1}H NMR spectroscopy to measure the concentration of H_2 in the IL phase. Figure 4.16 shows the increasing solubility of H_2 in the IL phase. It was found that in the absence of CO_2 pressure, 30 bar of H_2 pressure resulted in very little reactivity (3% conversion). It required 100 bar of H_2 pressure to obtain the good catalytic activity. When CO_2 pressure was added to 30 bar of H_2 pressure the catalyst delivered essentially quantitative conversion. The products could be extracted using scCO_2 and the catalyst was recycled a number of times. As discussed earlier, scCO_2 does not greatly alter the polarity of ILs, and in this case CO_2 only influenced the reaction rate by improving H_2 solubility. It was found that CO_2 did not affect

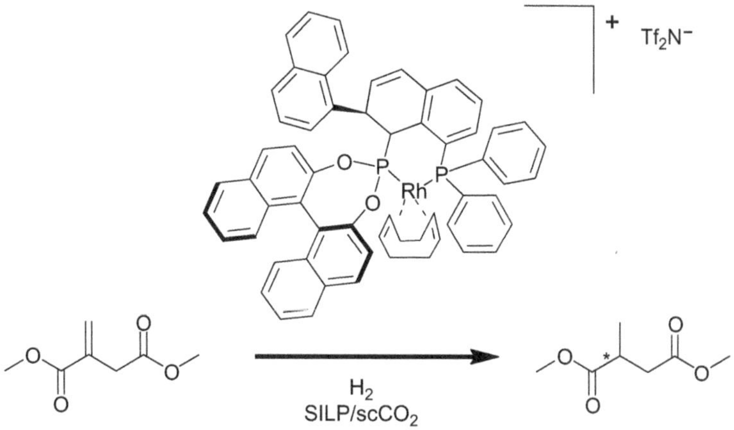

+ Tf_2N^-

O—P Rh P O

H$_2$
SILP/scCO$_2$

Scheme 4.4 Catalytic hydrogenation of dimethylitaconate in a SILP–$scCO_2$ system.

the enantioselectivity, and this was governed by the structure of the IL. For these reactions it was found that better enantioselectivity was obtained with ILs possessing weakly coordinating anions.

More recently, the Leitner Group has published a number of reports that examined the catalytic hydrogenation of dimethylitaconate (Scheme 4.4).[51–53] A chiral transition metal complex was immobilised in a SILP with $scCO_2$ as the mobile phase and operated under continuous flow conditions. The studies were all-encompassing and demonstrated the need to understand and control performance from the molecular scale (the catalyst) to the macroscale (engineering solutions). They developed a fully automated flow system that allowed conditions to be easily altered and monitored. For example, they could monitor phase behaviour using an in-line view cell and this was utilised to determine the conditions that would deliver a homogeneous supercritical phase. Supercritical chromatography was coupled to the reactor system to allow real time reaction monitoring.[52] Furthermore, high pressure ^1H NMR was used to examine the partition coefficients between the IL and $scCO_2$ phase, as shown in Figure 4.17.

The information from these studies was used to determine the size of the reactor, flow rates and residence times for the continuous flow process. As they were utilising a SILP system, they examined the nature of the support in detail. It was found that a variety of support materials could deliver good catalyst performance in the shorter term. However, it was found that water poisoned the catalyst and the longer term stability of the catalyst could be improved by using a silica with perfluoro-alkylated surface. Alternatively, dehydroxylated silica could be used as a water scavenger and placed upstream of the SILP catalyst (Figure 4.18). Once this portion of silica was saturated, catalyst deactivation would once again resume, however such a scavenger "cartridge" could easily be replaced prior to saturation and allow the catalyst to maintain its activity.

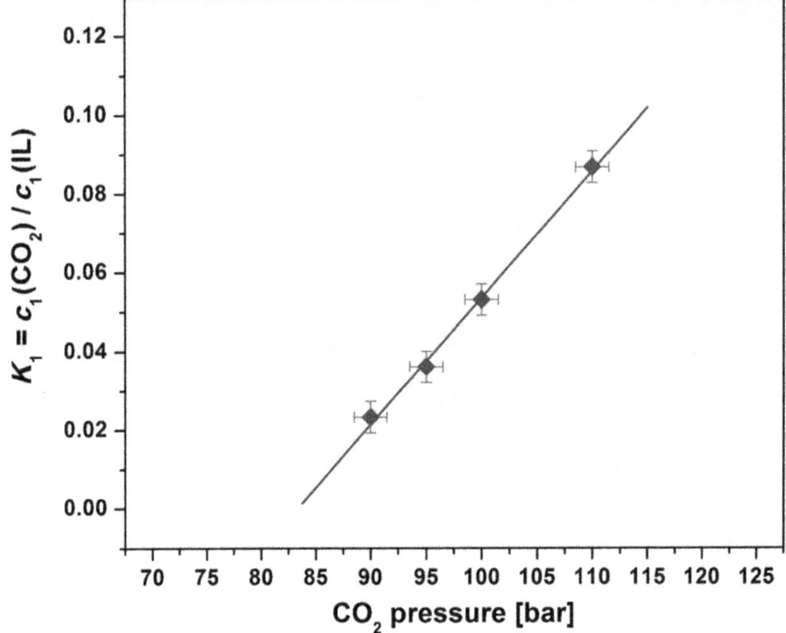

Figure 4.17 Plot of partition coefficient of dimethylitaconate between CO_2 and [EMIM][Tf_2N] at 40 °C, calculated from ^1H NMR spectroscopy measurements in the IL phase.
Reproduced with permission from reference 53 © 2013 Wiley-VCH Verlag GmbH & Co. KGaA, Weinheim.

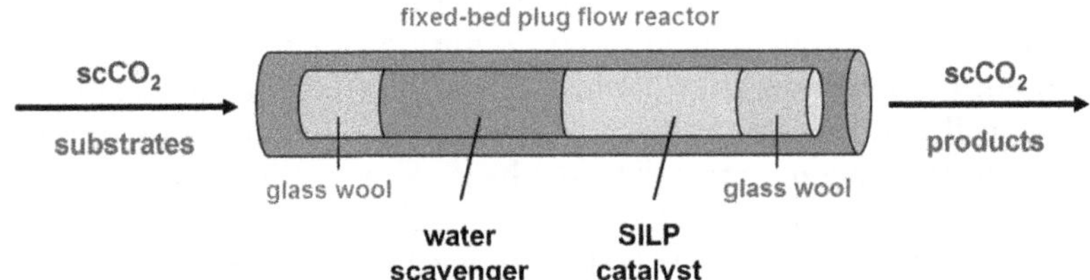

Figure 4.18 Illustration of a SILP–$scCO_2$ system utilising a water scavenger prior to the SILP catalyst in order to improve catalyst performance.
Reproduced with permission from reference 53 © 2013 Wiley-VCH Verlag GmbH & Co. KGaA, Weinheim.

In these studies they were able to obtain quantitative conversion and enantioselectivity of >99% ee utilising a homogeneous catalyst (Scheme 4.4). The products were obtained analytically pure and free from catalyst or solvent impurities. Employing a single operating unit they could obtain space–time yields of up to 0.7 kg L^{-1} h^{-1} and productivities of more than 100 kg of product per gram of rhodium or 14 kg per gram of ligand. We believe that performance such as this demonstrates the industrial potential of these types of systems.

4.3.3 Hydrovinylation

Bösmann *et al.* reported one of the first examples of a continuous flow IL–CO_2 system, with their studies on the hydrovinylation of styrene using Wilke's Ni catalyst (Scheme 4.5).[54] In this case the reaction was carried out at 25 °C and therefore the CO_2 was *subcritical*. Nonetheless, this paper is a particularly nice example because it demonstrates that continuous flow operation is a significant advantage in this case. It was found that when the reaction was carried out batch-wise (with the products extracted using $scCO_2$) the catalyst quickly deactivated. This was thought to be due to the catalyst having decreased stability in the absence of substrate. Therefore, when the system was operated under continuous flow conditions it was found that the catalyst was stable.

4.3.4 Oxidation Reactions

There have been relatively few catalytic oxidation reactions carried out in IL–$scCO_2$ systems. Arguably this is not a surprise as most IL–$scCO_2$ studies are often aimed at enabling the separation and continual use of homogeneous catalysts. Compared to other areas of catalysis such as hydrogenations and carbonylations, selective oxidation catalysis using homogeneous catalysts is a less mature area. There are very few aerobic homogeneous oxidation catalysts that are stable and can deliver high TONs; consequently there is a less of a driver to separate and recycle such catalysts. This area is worth mentioning though as IL–$scCO_2$ systems have properties that are well suited to oxidation reactions that involve O_2. As mentioned earlier in section 4.2.2, CO_2 pressure can help increase the concentration of other gases, such as O_2, in the IL phase. The use of non-flammable CO_2 to increase O_2 concentrations is a far safer option than increasing the pressure of pure O_2. Indeed, industrial oxidations that utilise O_2 generally require that the O_2 is diluted to a percentage that ensures the gas phase is below *limiting oxygen conditions*; in order to prevent any potential explosions. So although there are relatively few reports on oxidation reactions in IL–$scCO_2$ systems, they are well suited

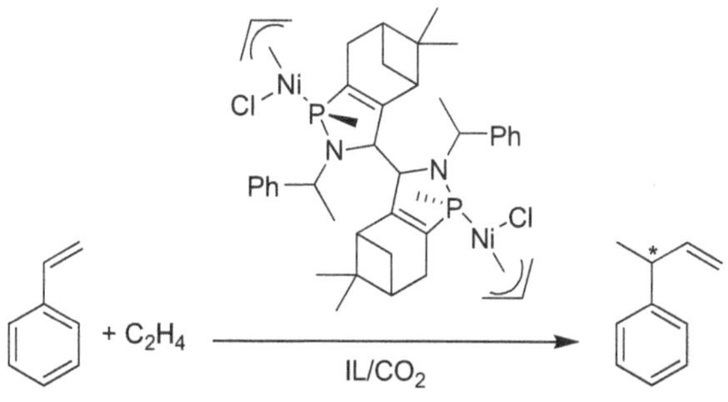

+ C_2H_4

IL/CO_2

Scheme 4.5 Hydrovinylation of styrene using Wilke's Ni catalyst.

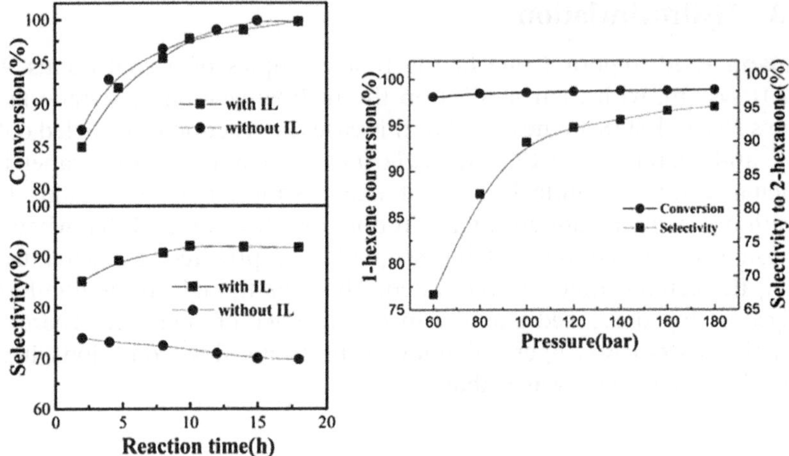

Figure 4.19 LHS: Conversion and selectivity against reaction time at 333.2 K and 125 bar where the IL is [bmim][PF_6]. RHS: Dependence of conversion and selectivity on reaction pressure in [bmim][PF_6]/scCO_2 at 333.2 K with a reaction time of 17 h.
Reproduced from reference 55.

to these reactions and could be a viable approach once better catalysts have been developed.

Han and co-workers reported the Wacker oxidation of 1-hexene in [bmim][PF_6]–scCO_2 using a $PdCl_2$, $CuCl_2$ catalyst system along with methanol as a nucleophilic reagent.[55] It was found that the catalyst was more stable and selective (to 2-hexanone) in the IL–scCO_2 system, compared to when scCO_2 was used alone (Figure 4.19). The products could be extracted using scCO_2 and the catalyst was reused six times. Over time the catalyst performance decreased and this was believed to be due to catalyst degradation rather than the catalyst being extracted by the scCO_2, as Pd was not detected in the products. The catalyst was found to be more stable in [bmim][PF_6]–scCO_2 compared to using scCO_2 without the IL.

Aerobic alcohol oxidation reactions utilising immobilised perruthenate catalysts have been studied by two different groups, both employing IL type functionalised solid supports. Ciriminna *et al.* utilised a catalyst that consisted of a silica xerogel prepared by copolycondensation of silane monomers with an ionic imidazolium-functionalized organosilane.[56] The anion was then exchanged for $[RuO_4]^-$ resulting in a catalyst that had no detectable leaching when used with scCO_2. An illustration of the system is shown in Figure 4.20. They demonstrated catalysis for both activated (benzyl alcohol and 1-phenylethanol) and unactivated alcohols (1- and 2-octanol), with the catalyst exhibiting good selectivity to the desired aldehyde or ketone.

The Han group studied the oxidation of benzyl alcohol to benzaldehyde in scCO_2 where the perruthenate catalyst was immobilised on an IL type polymer.[57] In this case, the study utilises an ionic polymer prepared from

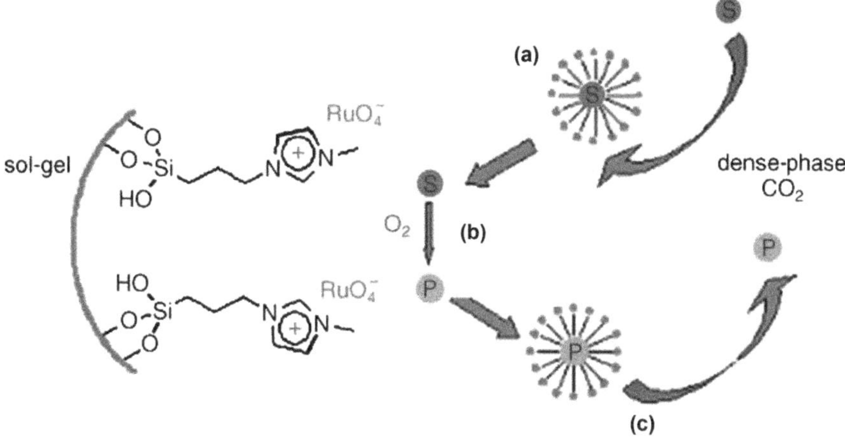

Figure 4.20 Outline of how CO_2 facilitates the transport, reaction, and adsorption–desorption steps of the process. Dense phase CO_2 transports the substrate (S) into the catalyst tethered to the ionic liquid moiety where the catalytic process takes place. CO_2 then carries the product (P) back into solution.
Reproduced with permission from reference 56 © 2006 Wiley-VCH Verlag GmbH & Co. KGaA, Weinheim.

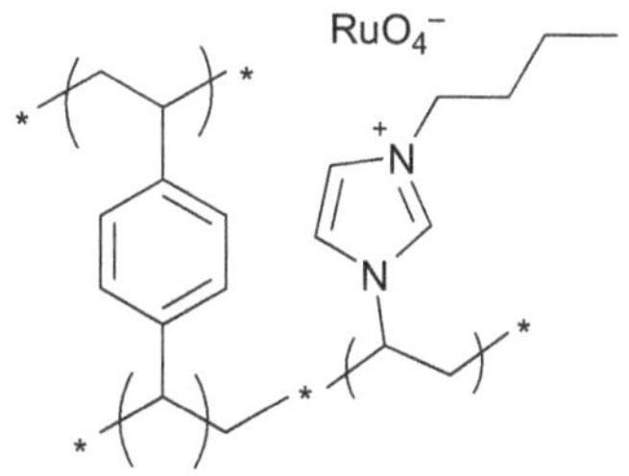

Figure 4.21 Perruthenate catalyst immobilised on an IL based polymer.[57]

1-vinyl-3-butyl imidazolium chloride ([VBIM]Cl). The Cl^- anions were then exchanged for $[RuO_4]^-$ as shown in Figure 4.21. It was found that $scCO_2$ delivered superior performance compared to when toluene or methylene chloride was used as the solvent, with quantitative yield possible in $scCO_2$. Unfortunately, the recycling experiments indicated that the catalyst was not stable and yields on subsequent runs were significantly diminished.

It is also worth highlighting a few examples of oxidation reactions that are not aerobic in nature. The Han group examined the electro-oxidation of benzyl alcohol in an IL–$scCO_2$ system.[58] ILs can obviously act as electrolytes as well as solvents and this paper showed that it is even possible to carry out electrochemical synthesis in IL–$scCO_2$ systems. They demonstrated that products could be extracted from the IL using $scCO_2$ and the IL reused.

Another example of a non aerobic oxidation reaction is the epoxidation of olefins facilitated by H_2O_2 and base (NaOH) which was reported by Bortolini *et al.*[59] They showed that the epoxide products could be quantitatively extracted from the IL by $scCO_2$. It was found that after $scCO_2$ extraction the conversion was severely decreased on subsequent runs. This was believed to be due to NaOH being neutralised and forming Na_2CO_3. If the IL was washed with water prior to another reaction and additional base was added, high conversions in the IL could once again be obtained.

4.3.5 Enzymatic Catalysis

IL–$scCO_2$ systems have been explored for enzyme catalysis from around the same time they were initially exploited for organometallic homogeneous catalysis.[60–62] The benefits of separating products and continually using the catalyst apply equally as much to enzymes as they do to homogeneous transition metal catalysts. Furthermore, studies have shown that ILs can improve the activity and stability of enzymes.[63] In the case of IL–$scCO_2$ systems, ILs also offer some protection against the negative effects of $scCO_2$. $scCO_2$ can be detrimental to enzymes,[64] as it can strip water molecules and also alter the pH of water around the enzymes. CO_2 can also react with amino groups on the surface of the enzyme and form carbamates. As discussed in recent reviews of biocatalysis in IL–$scCO_2$ systems, enzymes can be used continuously for long periods of time with high temperatures and CO_2 pressures.[65,66] Although a number of different biocatalysts have been tested in ILs the majority of studies have utilised lipases.[63] Lipases are robust enzymes that are known to be able operate in non-aqueous systems and furthermore they can catalyse a range of different reactions. Here we will highlight two recent examples that illustrate the types of continuous processes that can be developed using the combination of ILs and $scCO_2$.

Lozano *et al.* reported continuous kinetic resolution (KR) and dynamic kinetic resolution (DKR) using ILs and $scCO_2$.[67] In this study they immobilised *Candida antarctica* lipase B (CALB) on to polymeric "Supported Ionic Liquid-Like Phases" (SILLPs). These SILLPs essentially have IL type structures covalently tethered on to a polymer support. These materials are able to provide the enzyme with the same sort of local environment that using bulk/liquid ILs do, enabling stabilisation and activation of the enzyme. The advantage of these tethered systems is that it is not possible for the IL to leach and less IL is required compared to other methods. Initially, the SILLPs were optimised by studying the KR of *rac*-1-phenylethanol with vinyl propionate (Scheme 4.6) in hexanes at 50 °C. They examined the influence of different types of polymer, anion and loadings. These studies led them to utilise the following combination for further studies; a macroporous polymer (*i.e.* a rigid polymer with permanent pore structure as opposed to a swellable "gel type" polymer), the IL had a chloride anion and CALB was loaded on the SILLP at *ca.* 10 mg per g of support.

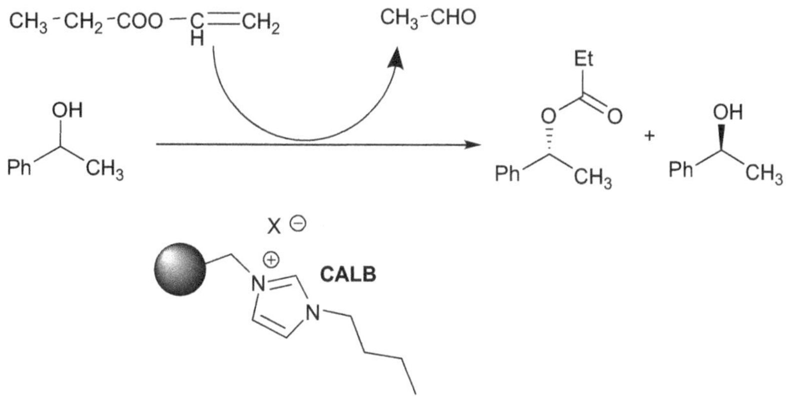

Scheme 4.6 Kinetic resolution of *rac*-1-phenylethanol.

They then went on to study this CALB–SILLP for KR in a continuous flow process using $scCO_2$, demonstrating that the system could operate well even in the presence of $scCO_2$. The issue with KR is that the maximum yield that can be obtained is 50%, therefore, they wanted to develop a DKR process. DKR involves the addition of a chemical catalyst that can racemise the unwanted enantiomer, and therefore potentially increase the yield up to 100%. In this case, they decided to explore the use of heterogeneous acid catalysts (zeolites) to carry out the racemisation. The challenge here is that generally the enzyme and zeolite cannot be in the same reactor as one step requires dry conditions (enzymatic KR) and the other requires water. They addressed this by initially testing a flow system that had the different catalysts in sequence, as shown in Figure 4.22. Using this approach would require a number of sequences to achieve 100% yield and using the sequence that is shown in Figure 4.22, a maximum yield of 75% was possible. The results were impressive, demonstrating that >99.9% ee could be obtained and the CALB–SILLP could operate continuously for 19 days. Nevertheless, they wanted to develop a "one-pot" approach with a single reactor.

Initial attempts at carrying out the DKR in a "one-pot" system led to an unselective process with both enantiomers of the ester being formed. This was due to the zeolite catalysing the esterification of the racemic substrate. To address this they coated the zeolite with a layer of IL, which reduced the rate of both the racemisation reaction and the undesired acylation reaction. The setup for this "one-pot" approach is shown in Figure 4.23. Using this system it was possible to obtain up to 92% of the desired *R*-ester with >99.9% ee.

This study is a good example of how both ILs and $scCO_2$ can be engineered to develop efficient flow processes for high value products.

Another nice demonstration of reaction engineering applied to enzyme catalysis was reported by Greiner and co-workers.[68] In this study they developed a continuous flow system for the biocatalytic synthesis of (*R*)-2-octanol which employed a *Lactobacillus brevis* alcohol dehydrogenase

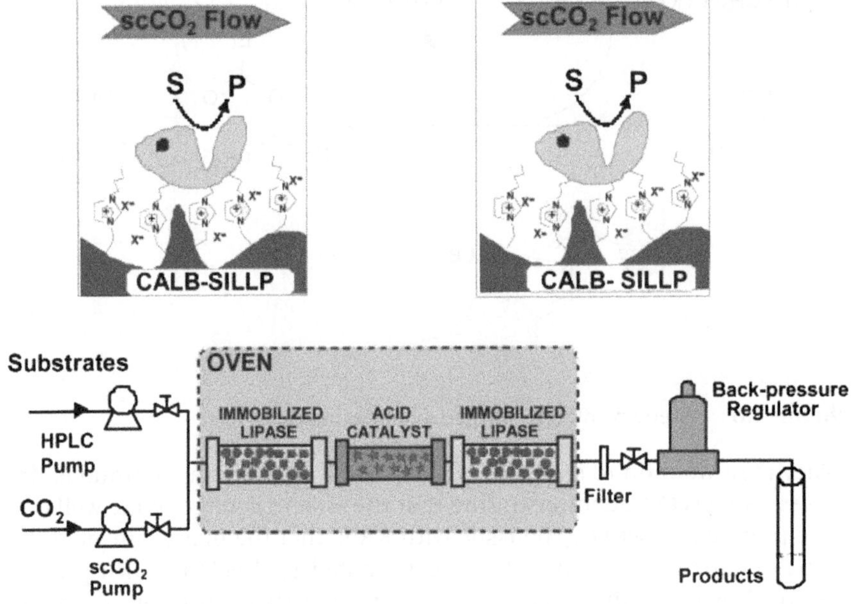

Figure 4.22 Consecutive biocatalytic–chemocatalytic–biocatalytic reactors in a $scCO_2$ flow system.
Reproduced from reference 67.

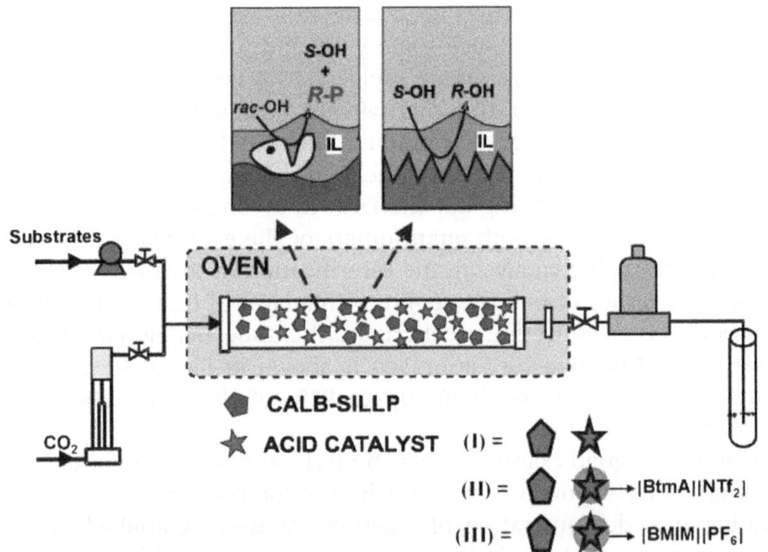

Figure 4.23 Continuous "one-pot" chemo-enzymatic DKR of *rac*-1-phenylethanol.
Reproduced from reference 67.

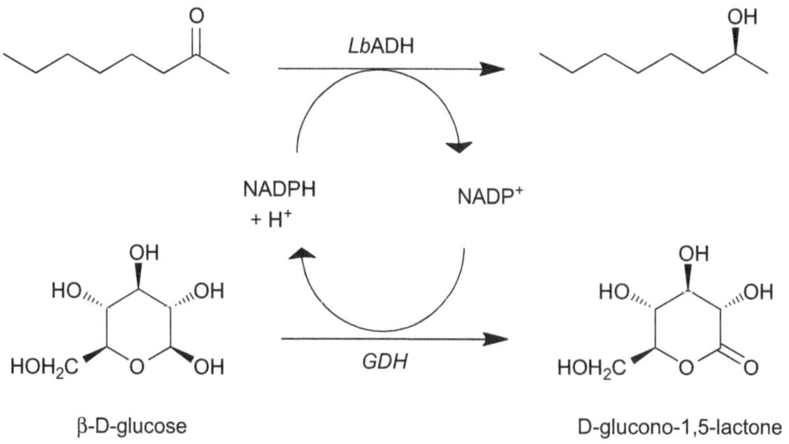

Scheme 4.7 *Lb*ADH reduction of 2-octanone to (*R*)-2-octanol with GDH catalysed cofactor regeneration.

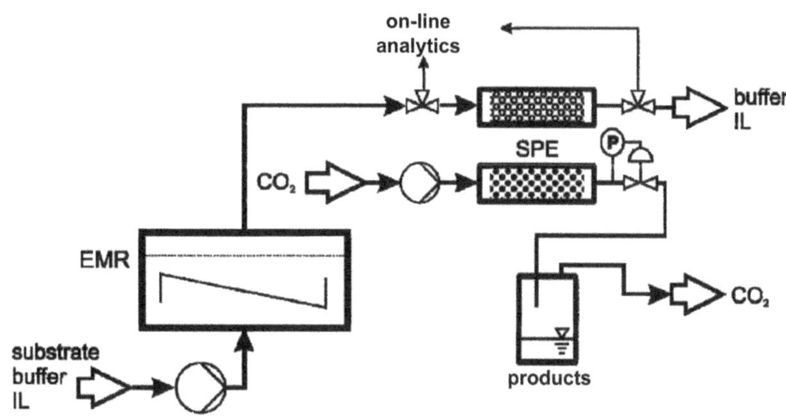

Figure 4.24 Continuous synthesis using an enzyme membrane reactor coupled with solid phase extraction (SPE) and subsequent downstream product extraction from the SPE column using $scCO_2$.
Adapted from reference 68.

(*Lb*ADH) enzyme. As shown in Scheme 4.7, for this enzyme, the necessary cofactor regeneration was achieved using the glucose dehydrogenase (GDH)-catalysed oxidation of β-D-glucose.

In this case, the IL AMMOENG™ 101 was used as an additive in an aqueous solvent system. The addition of this IL increased the substrate concentration and improved space time yields and turnover numbers. The system also demonstrated excellent selectivity with >99.9% ee obtained. However, they did not wish to add $scCO_2$ to this system in the same way as the previous example as this would have been detrimental to the catalyst. As shown in Figure 4.24 a continuous flow system was developed using a membrane reactor. This allows the enzymes to remain in the reactor and be continually used, with the products then separated using a solid phase extraction system. Once the solid phase was full, it was replaced by another cartridge and the products were removed from the solid phase with flowing $scCO_2$. It was shown that these solid phase cartridges could be reused more than 30 times without any detrimental impact on their capacity. This innovative process not only delivered excellent catalytic performance (improved by the use of an IL), but it also produced very little waste as the use of organic solvents was completely avoided.

4.3.6 Miscellaneous

As mentioned earlier, it is not our intention to list every type of reaction that has been carried out in an IL–$scCO_2$ system. It is possible to utilise these systems for a wide range of reactions and many have now been reported. Quite simply for a reaction to be suitable, the catalyst has to be stable under the applied conditions and the products need to be soluble in $scCO_2$. Herein, we will mention a few other types of reactions that have been reported in these systems.

A significant number of papers have reported studies that have not only utilised CO_2 as the co-solvent but also as a reactant. For example, a number of papers have studied the synthesis of cyclic organic carbonates from epoxides and CO_2 (Scheme 4.8), a reaction that can be catalysed by ILs themselves in some cases. These reactions (and a number of others utilising CO_2) have recently been discussed in detail in the review by Baiker and co-workers,[12] therefore we will not cover this area in any detail.

A recent addition to this area was reported by Leitner and co-workers who demonstrated a continuous flow system for the hydrogenation of CO_2 to formic acid.[69] The synthesis of pure formic acid from CO_2 and H_2 is strongly disfavoured by entropy shifting the equilibrium far to the left. In this case, the continuous removal of product from the reactor adjusts this equilibrium. Figure 4.25 outlines this IL–$scCO_2$ system.

They utilised a homogeneous ruthenium catalyst and a non-volatile base (*e.g.* an amine functionalised IL) which helps stabilise the product. The results obtained were extremely promising with performance superior to previous reports in the literature. Nonetheless they identified extraction of formic acid as the limiting factor. As we have already discussed in this chapter partition coefficients are very important for these systems. In this

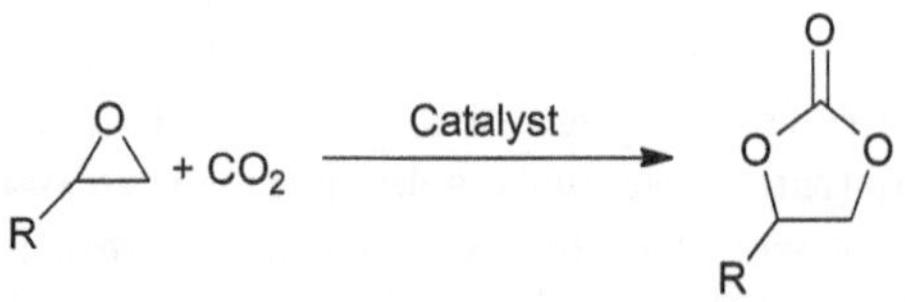

Scheme 4.8 Synthesis of cyclic organic carbonates from epoxides and CO_2.

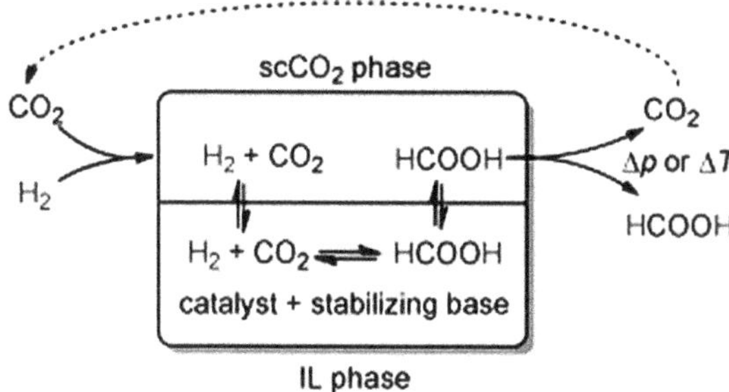

Figure 4.25 Continuous flow hydrogenation of carbon dioxide to formic acid in IL–$scCO_2$ system.
Reproduced with permission from reference 69 © 2012 Wiley-VCH Verlag GmbH & Co. KGaA, Weinheim.

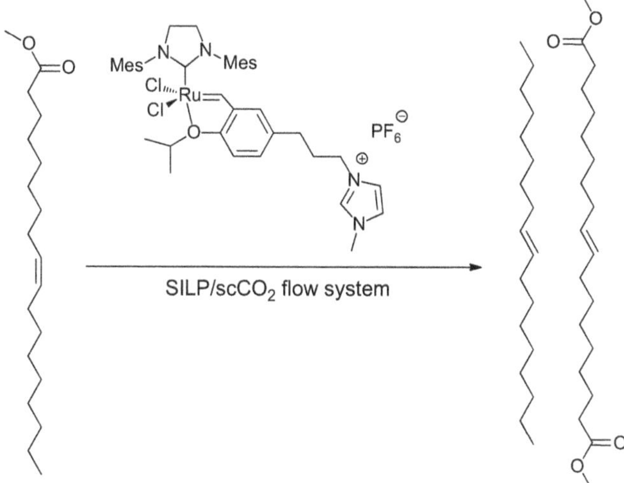

Scheme 4.9 Continuous flow self-metathesis of methyl oleate in IL–$scCO_2$.

case, formic acid with its strong hydrogen bonding capabilities will not have a high solubility in $scCO_2$ and will have a much greater solubility in ILs. Consequently, for this approach to be viable methods need to be developed that allow the acid to be extracted more efficiently.

Duque *et al.* reported alkene metathesis under continuous flow operation using an SILP–$scCO_2$ system, with a homogeneous catalyst.[70] This report is the first time this reaction has been carried out under flow using a homogeneous catalyst. As shown in Scheme 4.9 they utilised a ruthenium complex with an imidazolium tag to anchor it in the SILP. Continuous operation was possible for the self-metathesis of methyl oleate with only a slight loss of activity after 10 h. It was found that cross-metathesis of dimethyl maleate with methyl oleate was more challenging under flow conditions but could be carried out under batch conditions. The poorer performance of the cross-metathesis reaction was thought to be partly due to a low residence time within the reactor, as improved flow performance could be achieved by lowering the flow rate of the substrates.

4.4 Conclusions and Perspective

IL–scCO_2 systems have the potential to allow very sustainable catalytic processes to be developed that could be utilised on an industrial scale. As far as we are aware there are no IL–scCO_2 industrial processes at the present time, however, we believe that they will eventually find commercial applications. We hope that the examples given in this chapter illustrate the promise of these systems. It is possible to obtain high space–time yields with homogeneous catalysts and this makes these systems very attractive. The integration of catalyst separation into the process enables very efficient utilisation of enzymes and organometallic catalysts. Catalysts can be "anchored" in an IL phase and a continuous process operated whereby the scCO_2 delivers substrates and extracts products. Upon depressurisation of the scCO_2, the products can be obtained free from catalyst and solvent waste. The ability to operate these systems in a continuous flow manner further improves the efficiency and indeed the practicality of these systems.

Undoubtedly these systems will not be the answer to every problem, but there are potentially a wide number of reactions that could utilise this approach. To be viable, the product(s) should be readily soluble in scCO_2 so that they can be easily extracted from the IL (catalyst) phase. It is also important that the catalyst has good stability and performance, in other words, this is only feasible if the catalyst can deliver high turnover numbers. The catalyst performance has to be such that it is worthwhile addressing the separation problems to enable the continual use of the catalyst. In the case of hydrogenation reactions, many homogeneous catalysts have the potential to deliver high turnover numbers and their cost (due to expensive metals and in some cases even more expensive ligands) means there are significant drivers to develop new processes that allow better use of the catalyst. Conversely, (at the current time) most homogeneous oxidation catalysts do not have sufficiently good performance to warrant the use of an IL–scCO_2 system. If the catalyst can only deliver a few hundred turnovers then there is no point trapping it in a relatively expensive IL. But it should be remembered that IL–scCO_2 systems are well suited for aerobic oxidation reactions. As we discussed, scCO_2 can improve mass transfer and the solubility of other gases in the IL phase, and in the case of O_2 this has safety advantages. Therefore when more stable catalysts are eventually developed, IL–scCO_2 systems could find practical applications in this area.

A potential barrier to IL–scCO_2 systems being studied more widely is the requirement for specialist equipment and there is also a fair degree of complexity associated with these multiphase systems. It is evident from some of the studies that we have highlighted that it is necessary to take into account a number of factors; ranging from catalyst design to phase behaviour. Successful implementation of these systems requires both molecular and reaction engineering. Nonetheless, with the recent boom in "flow chemistry" and "process intensification" in general, an appreciation for such challenges is becoming more widespread and it is becoming commonplace to find chemists and chemical engineers working closely on such challenges. We are, therefore, optimistic that IL–scCO_2 catalytic processes will earn their place in the green engineering tool box alongside other emerging technologies.

References

1. L. A. Blanchard, D. Hâncu, E. J. Beckman and J. F. Brennecke, *Nature*, 1999, **399**, 28.
2. (*a*) E. J. Beckman, *J. Supercrit. Fluids*, 2004, **28**, 121; (*b*) P. G. Jessop and W. Leitner, ed. *Chemical Synthesis Using Supercritical Fluids*, Wiley-VCH, Weinheim, 1999.
3. A. Baiker, *Chem. Rev.*, 1999, **99**, 453.
4. P. G. Jessop, T. Ikariya and R. Noyori, *Chem. Rev.*, 1999, **99**, 475.
5. For some applications, for example extraction of active ingredients from plants or super critical chromatography, the solvent power of scCO_2 is typically "modified" by adding organic solvents. The addition of small percentages of solvents such as ethanol can greatly increase the solvent strength of scCO_2.
6. (*a*) T. Welton, *Chem. Rev.*, 1999, **99**, 2071; (*b*) V. I. Pârvulescu and C. Hardacre, *Chem. Rev.*, 2007, **107**, 2615; (*c*) J. P. Hallett and T. Welton, *Chem. Rev.*, 2011, **111**, 3508.
7. P. Wasserscheid, C. M. Gordon, C. Hilgers, M. J. Muldoon and I. R. Dunkin, *Chem. Commun.*, 2001, 1186.
8. (*a*) P. G. Jessop and B. Subramaniam, *Chem. Rev.*, 2007, **107**, 2666; (*b*) G. R. Akien and M. Poliakoff, *Green Chem.*, 2009, **11**, 1083.
9. W. Keim, *Green Chem.*, 2003, **5**, 105.
10. M. J. Muldoon, *Dalton Trans.*, 2010, **39**, 337.
11. U. Hintermair, G. Francio and W. Leitner, *Chem. Commun.*, 2011, **47**, 3691.
12. F. Jutz, J.-M. Andanson and A. Baiker, *Chem. Rev.*, 2011, **111**, 322.
13. M. F. Sellin, P. B. Webb and D. J. Cole-Hamilton, *Chem. Commun.*, 2001, 781.
14. C. Jiménez-González, P. Poechlauer, Q. B. Broxterman, B.-S. Yang, D. am Ende, J. Baird, C. Bertsch, R. E. Hannah, P. Dell'Orco, H. Noorman, S. Yee, R. Reintjens, A. Wells, V. Massonneau and J. Manley, *Org. Process Res. Dev.*, 2011, **15**, 900.
15. M. Ramdin, T. W. de Loos and T. J. H. Vlugt, Review on CO_2 capture using ionic liquids which includes detailed discussions on CO_2

solubility in ILs and the potential of ILs for CO_2 separations and capture, *Ind. Eng. Chem. Res.*, 2012, **51**, 8149.
16. M. J. Muldoon, S. N. V. K. Aki, J. L. Anderson, J. K. Dixon and J. F. Brennecke, *J. Phys. Chem. B*, 2007, **111**, 9001.
17. V. Najdanovic-Visak, A. Serbanovic, J. M. S. S. Esperança, H. J. R. Guedes, L. P. N. Rebelo and M. Nunes da Ponte, *Chem Phys Chem*, 2003, **4**, 520.
18. A. M. Scurto, S. N. V. K. Aki and J. F. Brennecke, *J. Am. Chem. Soc.*, 2002, **124**, 10276.
19. E. M. Saurer, S. N. V. K. Aki and J. F. Brennecke, *Green Chem.*, 2006, **8**, 141.
20. M. C. Kroon, J. van Spronsen, C. J. Peters, R. A. Sheldon and G. Witkamp, *Green Chem.*, 2006, **8**, 246.
21. D. G. Hert, J. L. Anderson, S. N. V. K. Aki and J. F. Brennecke, *Chem. Commun.*, 2005, 2603.
22. M. Solinas, A. Pfaltz, P. G. Cozzi and W. Leitner, *J. Am. Chem. Soc.*, 2004, **126**, 16142.
23. C. Reichardt, T. Welton, *Solvents and Solvent Effects in Organic Chemistry*, Wiley-VCH, Weinheim, 4th edn, 2010.
24. (*a*) M. J. Kamlet and R. W. Taft, *J. Am. Chem. Soc.*, 1976, **98**, 377; (*b*) R. W. Taft and M. J. Kamlet, *J. Am. Chem. Soc.*, 1976, **98**, 2886; (*c*) M. J. Kamlet, J. L. Abboud and R. W. Taft, *J. Am. Chem. Soc.*, 1977, **99**, 6027.
25. C. P. Fredlake, M. J. Muldoon, S. N. V. K. Aki, T. Welton and J. F. Brennecke, *Phys. Chem. Chem. Phys.*, 2004, **6**, 3280.
26. C. Roosen, M. Ansorge-Schumacher, T. Mang, W. Leitner and L. Greiner, *Green Chem.*, 2007, **9**, 455.
27. P. G. Jessop, D. J. Heldebrant, X. Li, C. A. Eckert and C. L. Liotta, *Nature*, 2005, **436**, 1102.
28. L. Phan, J. R. Andreatta, L. K. Horvey, C. F. Edie, A.-L. Luco, A. Mirchandani, D. J. Darensbourg and P. G. Jessop, *J. Org. Chem.*, 2008, **73**, 127.
29. T. Yamada, P. J. Lukac, M. George and R. G. Weiss, *Chem. Mater.*, 2007, **19**, 967.
30. L. Phan, D. Chiu, D. J. Heldebrant, H. Huttenhower, E. John, X. Li, P. Pollet, R. Wang, C. A. Eckert, C. L. Liotta and P. G. Jessop, *Ind. Eng. Chem. Res.*, 2008, **47**, 539.
31. Review on CO_2 switchable materials: P. G. Jessop, S. M. Mercer and D. J. Heldebrant, *Energy Environ. Sci.*, 2012, **5**, 7240.
32. S. G. Kazarian, N. Sakellarios and C. M. Gordon, *Chem. Commun.*, 2002, 1314.
33. A. M. Scurto and W. Leitner, *Chem. Commun.*, 2006, 3681.
34. A. M. Scurto, E. Newton, R. R. Weikel, L. Draucker, J. Hallett, C. L. Liotta, W. Leitner and C. A. Eckert, *Ind. Eng. Chem. Res.*, 2008, **47**, 493.
35. A. Ahosseini, E. Ortega, B. Sensenich and A. M. Scurto, *Fluid Phase Equilib.*, 2009, **286**, 72.

36. (*a*) D. Camper, C. Becker, C. Koval and R. Noble, *Ind. Eng. Chem. Res.*, 2006, **45**, 445; (*b*) L. Ferguson and P. Scovazzo, *Ind. Eng. Chem. Res.*, 2007, **46**, 1369.
37. H. Weingärtner, *Angew. Chem., Int. Ed.*, 2008, **47**, 654.
38. A. Ahosseini, W. Ren and A. M. Scurto, *Ind. Eng. Chem. Res.*, 2009, **48**, 4254.
39. M. Tariq, M. G. Freire, B. Saramago, J. A. P. Coutinho, J. N. C. Lopes and L. P. N. Rebelo, *Chem. Soc. Rev.*, 2012, **41**, 829.
40. P. Jaeger and R. Eggers, *Chem. Eng. Process*, 2009, **48**, 1173.
41. (*a*) Reviews: M. Roth, *J. Chromatogr., A*, 2009, **1216**, 1861; (*b*) H. Machidaa, T. Kawasumib, W. Endob, Y. Satoa and R. L. Smith Jr., *Fluid Phase Equilib.*, 2010, **294**, 114.
42. (*a*) B. Cornils, *Org. Process Res. Dev.*, 1998, **2**, 121; (*b*) B. Cornils, *J. Mol. Catal. A: Chem.*, 1999, **143**, 1; (*c*) F. Joó, É. Papp and Á. Kathó, *Top. Catal.*, 1998, **5**, 113.
43. P. B. Webb, M. F. Sellin, T. E. Kunene, S. Williamson, A. M. Z. Slawin and D. J. Cole-Hamilton, *J. Am. Chem. Soc.*, 2003, **125**, 15577.
44. P. B. Webb, T. E. Kunene and D. J. Cole-Hamilton, *Green Chem.*, 2005, **7**, 373.
45. T. E. Kunene, P. B. Webb and D. J. Cole-Hamilton, *Green Chem.*, 2011, **13**, 1476.
46. U. Hintermair, G. Zhao, C. C. Santini, M. J. Muldoon and D. J. Cole-Hamilton, *Chem. Commun.*, 2007, 1462.
47. U. Hintermair, Z. Gong, A. Serbanovic, M. J. Muldoon, C. C. Santini and D. J. Cole-Hamilton, *Dalton Trans.*, 2010, **39**, 8501.
48. A. Riisager, R. Fehrmann, M. Haumann and P. Wasserscheid, *Eur. J. Inorg. Chem.*, 2006, **4**, 695.
49. R. A. Brown, P. Pollet, E. McKoon, C. A. Eckert, C. L. Liotta and P. G. Jessop, *J. Am. Chem. Soc.*, 2001, **123**, 1254.
50. F. Lui, M. B. Abrams, R. T. Baker and W. Tumas, *Chem. Commun.*, 2001, 433.
51. U. Hintermair, T. Höfener, T. Pullmann, G. Franciò and W. Leitner, *Chem Cat Chem*, 2010, **2**, 150.
52. U. Hintermair, C. Roosen, M. Kaever, H. Kronenberg, R. Thelen, S. Aey, W. Leitner and L. Greiner, *Org. Process Res. Dev.*, 2011, **15**, 1275.
53. U. Hintermair, G. Franciò and W. Leitner, *Chem. Eur. J.*, 2013, **19**, 4538.
54. A. Bösmann, G. Franciò, E. Janssen, M. Solinas, W. Leitner and P. Wasserscheid, *Angew. Chem., Int. Ed.*, 2001, **40**, 2697.
55. Z. Hou, B. Han, L. Gao, T. Jiang, Z. Liu, Y. Chang, X. Zhang and J. He, *New J. Chem.*, 2002, **26**, 1246.
56. R. Ciriminna, P. Hesemann, J. J. E. Moreau, M. Carraro, S. Campestrini and M. Pagliaro, *Chem. Eur. J.*, 2006, **12**, 5220.
57. Y. Xie, Z. Zhang, S. Hu, J. Song, W. Li and B. Han, *Green Chem.*, 2008, **10**, 278.

58. G. Zhao, T. Jiang, W. Wu, B. Han, Z. Liu and H. Gao, *J. Phys. Chem. B*, 2004, **108**, 13052.
59. O. Bortolini, S. Campestrini, V. Conte, G. Fantin, M. Fogagnolo and S. Maietti, *Eur. J. Org. Chem.*, 2003, 4804.
60. J. A. Lazlo and D. L. Compton, *Biotechnol. Bioeng.*, 2002, **75**, 181.
61. P. Lozano, T. de Diego, D. Carrie, M. Vaultier and J. L. Iborra, *Chem. Commun.*, 2002, 692.
62. M. T. Reetz, W. Wiesenhöfer, G. Franciò and W. Leitner, *Chem. Commun.*, 2002, 992.
63. Reviews of biocatalysis in ILs: (*a*) R. A. Sheldon, R. M. Lau, M. J. Sorgedrager, F. van Rantwijk and K. R. Seddon, *Green Chem.*, 2002, **4**, 147; (*b*) F. van Rantwijk and R. A. Sheldon, *Chem. Rev.*, 2007, **107**, 2757; (*c*) R. P. Müller and L. Greiner, *Appl. Microbiol. Biotechnol.*, 2008, **81**, 607; (*d*) M. Moniruzzaman, N. Kamiya and M. Goto, *Org. Biomol. Chem.*, 2010, **8**, 2887; (*e*) H. Zhao, *J. Chem. Technol. Biotechnol.*, 2010, **85**, 891.
64. A. J. Mesiano, E. J. Beckman and A. J. Russell, *Chem. Rev.*, 1999, **99**, 623.
65. P. Lozano, *Green Chem.*, 2010, **12**, 555.
66. Y. Fan and J. Qian, *J. Mol. Catal. B: Enzym.*, 2010, **66**, 1.
67. P. Lozano, E. García-Verdugo, N. Karbass, K. Montague, T. De Diego, M. I. Burguete and S. V. Luis, *Green Chem.*, 2010, **12**, 1803.
68. C. Kohlmann, S. Leuchs, L. Greiner and W. Leitner, *Green Chem.*, 2011, **13**, 1430.
69. S. Wesselbaum, U. Hintermair and W. Leitner, *Angew. Chem., Int. Ed.*, 2012, **51**, 8585.
70. R. Duque, E. Öchsner, H. Clavier, F. Caijo, S. P. Nolan, M. Mauduit and D. J. Cole-Hamilton, *Green Chem.*, 2011, **13**, 1187.

CHAPTER 5

Heterogeneous Catalysis in Ionic Liquids

PETER GOODRICH,*[a] CHRISTOPHER HARDACRE[a] AND VASILE I. PARVULESCU[b]

[a] School of Chemistry and Chemical Engineering/QUILL, Queen's University, Stranmillis Road, Belfast, Northern Ireland BT9 5AG, UK;
[b] University of Bucharest, Department of Organic Chemistry, Biochemistry and Catalysis, B-dul Regina Elisabeta 12, Bucharest 030016, Romania
*Email: p.goodrich@qub.ac.uk

5.1 Introduction

The use of heterogeneous catalysts in connection with ionic liquids (ILs) has attracted special attention and several review papers have been devoted to the subject.[1] In many cases this combination has resulted in higher rates and more selective reactions. Moreover, the use of ILs in many catalytic processes, especially when using transition metal catalysts, has enabled simpler work up procedures so as to circumvent time-consuming and laborious purification steps.[2]

Although heterogeneous catalysis employing ILs is possible *via* a number of pathways[3] three basic strategies exist. One such method involves the dispersion of bulk or nanoparticle solid materials together with substrates in the ILs, thus employing the IL as a catalyst immobiliser and reaction solvent. The second strategy involves the use of a dispersed thin supported ionic liquid film (SILF) as a support-modifying functional layer.[4] Another strategy is the synthesis of composite support systems where the ILs act as

RSC Catalysis Series No. 15
Catalysis in Ionic Liquids: From Catalyst Synthesis to Application
Edited by Chris Hardacre and Vasile Parvulescu

Published by the Royal Society of Chemistry, www.rsc.org

templates.[5] In particular, the use of SILFs, taking advantage of the tunability of the physico-chemical properties of the IL and their potential to chemically interact with supported catalysts, has attracted detailed attention for a range of supports for which many acronyms exist.[3] Two main concepts have emerged from SILFs, immobilisation of a catalyst into an IL followed by impregnation onto a support called supported ionic liquid phase (SILP) catalysis and traditional heterogeneous catalysts impregnated with an IL layer, called supported catalyst with ionic liquid layer (SCILL). SILP technology is a useful alternative to traditional heterogenisation methods for homogeneous catalysis. The main drawbacks of this concept are catalyst leaching during the reaction and product extraction from the reaction mixture before recycling. The use of specifically designed or modified catalysts, with physical–chemical properties similar to that of ILs, has emerged as a possible solution for the problem of catalyst leaching. For example, incorporation of an ionic moiety into the catalyst structure is one such approach. However, an important premise in the introduction of an ionic tag to a catalyst is that the tag is catalytically silent and inert in reaction conditions.[6]

The field of heterogeneous catalysis utilising ILs is vast and, therefore, this chapter will mainly focus on the use of metal-catalysed transformations. Where possible, examples will be selected to highlight the potential for process optimisation of systems originally based on the use of the IL as a bulk solvent through to biphasic liquid–liquid systems, and yet further on to SILP systems. Within this chapter, 1-alkyl-3-methyl imidazolium cations are denoted as $[C_nC_1im]^+$, 1-alkyl-3-alkylimidazolium cations are denoted as $[C_nC_nim]^+$ and tetraalkylammonium cations are denoted as $[N_{w,x,y,z}]^+$ where the letters represent the alkyl chain length attached to the N centre. $[C_2pic]^+$ refers to the *N*-ethyl-3-methylpiccolinium cation. The anion bis(trifluoromethylsulfonyl)imide, *i.e.* $[(CF_3\text{-}SO_2)_2N]^-$, is denoted as $[NTf_2]^-$, hexafluorophosphate is denoted $[PF_6]^-$, tetrafluoroborate is denoted $[BF_4]^-$ and triflate is denoted as $[OTf]^-$. BINAP 2,2′-bis(diphenylphosphino)-1,10-binaphthyl, QUINAPHOS 2-(1-naphthyl)-8-diphenylphosphino-1-[3,5-dioxa-4-phospha-cyclohepta[2,1-a;3,4-a0]dinaphthalen-4-yl]-1,2-dihydroquinoline, DPEN 1,2-diphenylethylenediamine, biphephos 6,6′-[(3,3′-di-t-butyl-5,5′-dimethoxy-1,1′-biphenyl-2,2′-diyl)bis(oxy)] bis(dibenzo[d,f] [1,3,2] dioxaphosphepin)hemiethylacetateadduct, TRIPHOS trisodium phosphate, Xantphos 4,5-bis(diphenylphosphino)-9,9-dimethylxanthene.

5.2 Hydrogenation Reactions

Hydrogenation is one of the most studied catalytic transformations in the presence of ILs.[7] The vast majority of studies involving homogeneous systems showed lower reaction rates compared with molecular solvents due to the low hydrogen solubility,[8] in particular, when ILs are used as bulk solvents. The solubility of the substrate and reaction products in an IL can also govern the hydrogenation rate of that particular substrate.[9] For example,[8] essentially no difference in the rate of benzene reduction was found using $[H_4Ru_4(\eta^6\text{-}C_6H_6)_4][BF_4]_2$ as the catalyst in a biphasic system using ILs and molecular solvents. In this case, the higher solubility of hydrogen in benzene compared with the IL mitigated the effect of the low hydrogen solubility in the pure IL. The low gas solubility in ILs has opened the way for the application of SCILL and SILP based catalysts, where the thin film of IL employed would dramatically reduce the mass transfer limitations.

In order to study the various mass transfer limitations and understand the lower rates of reaction in the IL compared with molecular solvents, the hydrogenation of phenylacetylene to styrene in heptane and $[C_4C_1im][NTf_2]$ was examined in detail by Hardacre *et al.*[10] By comparing the kinetics of the Pd–$CaCO_3$-catalysed reaction in a stirred tank reactor and a rotating disc reactor, the reaction rate in the IL was found to be limited by the mass transfer of dissolved hydrogen. Recently, the mechanism of hydrogenation in ILs has been evaluated. Therein, parahydrogen-induced polarisation has been observed with Pd nanoparticles under both homogeneous[11] and SILP[12] conditions.

5.2.1 Solubility of Hydrogen in ILs and $scCO_2$ Systems

Combining ILs with supercritical fluids is also a method employed to increase the hydrogen solubility in the bulk IL. Typically an increase in pressure is accompanied by an increase of the solubility of CO_2 in the IL.[13] In IL-rich phases the solubility of CO_2 decreases with temperature, and depends on the nature of both the anion and cation. CO_2 forms weak Lewis acid–base complexes with the anions in ionic liquids and the strength of the interaction depends on the Lewis basicity of the anion.[14] These flexible systems offer the ability to select the optimum solvent for a specific reaction process.[15]

5.2.2 Heterogeneous Hydrogenation Reactions

5.2.2.1 Hydrogenations using Metal Complexes

Initial hydrogenation studies using conventional catalysts indicated little difference in selectivity between ILs and molecular solvents. For example, Dupont and co-workers were the first to report that $[C_4C_1im][BF_4]$ could be used as a stabiliser and immobiliser for metal complexes based on ruthenium[16,17] in the biphasic hydrogenation of olefins. For a $RuCl_2(PPh_3)_3$ catalyst, similar behavior was observed in IL compared with the classical system whereby the structure of the molecule and not the IL played a role in reactivity, namely the sterically hindered cyclohexene reacted more slowly than hex-1-ene. In the hydrogenation of 1,3-butadiene moderate activities and selectivities to but-1-ene were observed in the IL. The major product was *trans*-but-2-ene again indicating little difference in catalytic behavior between ILs and molecular solvents.

The advantage of using an IL, albeit as an IL–organic solvent biphasic system over an aqueous–organic system was demonstrated using a [Ru(η^6-*p*-cymene)(η^2-TRIPHOS)Cl][PF_6] catalyst for the hydrogenation of arenes in both CH_2Cl_2 and IL–organic systems.[18] Although the latter IL system is biphasic, significantly higher turnover numbers (TONs) were reported. Catalyst decomposition was observed in CH_2Cl_2, whereas in [C_4C_1im][BF_4] no loss in activity was observed after five runs. IL-biphasic systems were also found to be superior compared with other biphasic systems for the Ru complex catalysed hydrogenation of sorbic acid proceeding with enhanced activity and up to 90% selectivity.[19]

The choice of solvent is important for any catalysed chemical process due to catalyst stability and solvent–solute interactions. This is of particular importance when dealing with ILs. The polarity of most ILs is thought to be similar to that of short chain alcohols[20] or acetonitrile.[21] The ability of the ILs to be able to act as both hydrogen bond donors and hydrogen bond acceptors[22] can result in the selective solubility of substrates or products. The complexes [Pd(NCC_3mim)$_2Cl_2$][BF_4]$_2$ and [$Ru_6C(CO)_{16}$]$^{2-}$ were independently used for hydrogenation of 1,3-cyclohexadiene under biphasic conditions.[23,24] With both catalysts, cyclohexene was formed with high selectivity (>97%), whereas in molecular solvents a significantly lower selectivity was achieved. The use of an IL immiscible co-solvent system was crucial in determining the selectivity of the reaction. The higher solubility of the diene compared with the monoene in the IL was thought to be responsible for the preferential hydrogenation. Moreover, due to the biphasic nature of the reactions facile separation of the products from the IL–catalyst phase was possible. Dyson *et al.*[25] used an IL – water catalyst system that undergoes a temperature-controlled and reversible two phase – single phase transition. At room temperature [C_8C_1im][BF_4] is immiscible with water; however, on heating to 80 °C, a single phase is formed. Using this system and [Rh(η-C_7H_8)(PPh_3)$_2$][BF_4] as the catalyst, butyne-1,4 diol was hydrogenated producing a mixture of but-2-ene-1,4-diol and butane-1,4-diol. Although poor selectivity was observed this methodology enables the simple separation of the catalyst/products while increasing the mass transfer characteristics of the system.

Following the ionic tag concept, Dyson and co-workers[26] also reported a ruthenium complex bearing imidazolium functionalised η^6-arene ligands for biphasic aqueous and IL hydrogenation of styrene, Figure 5.1. In water, the complexes showed moderate activity, but leached, while [C_2pic][OTf] afforded markedly lower reaction rates but the leaching was below the detectable limit.

A [Rh(NBD)(PPh_3)$_2$]PF_6 catalyst immobilised onto silica gel using [C_4mim][PF_6] was first employed in the Rh-catalysed hydrogenation of a range of alkenes; hex-1-ene, cyclohexene and 2,3-dimethylbut-2-ene.[27] The enhanced activity of the Rh complex under SILP conditions over biphasic or homogeneous processes was due to a larger catalyst concentration effect in the supported IL phase. The SILP catalyst also showed good long-term

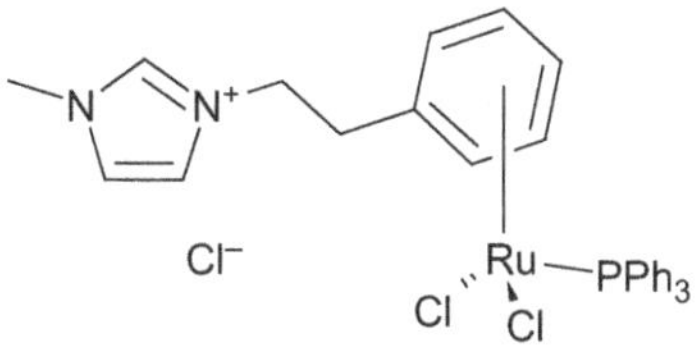

Figure 5.1 Ionic tagged ruthenium complex employed in the biphasic hydrogenation of styrene.
Adapted with permission from the American Chemical Society. Taken from reference 26.

stability enabling 18 batch runs without any significant loss of activity. A similar metal complex $Rh(H)_2Cl(PPh_3)_3$ has also been confined to the surface of a structured support (*i.e.* sintered metal fibres SMFs) in the IL layer.[28] This catalyst was used in a fixed bed reactor for the gas phase hydrogenation of 1,3-cyclohexadiene to cyclohexene, where due to the regular structure of the support, turnover frequencies (TOFs) of 150–250 h^{-1} and a high selectivity (>96%) were achieved. The practicability of a SILP concept and the use of an IL film immobilised on the carbon nanotubes (CNT)–SMF, provided an efficient use of the IL without any mass transfer limitations while maintaining isothermal conditions during the hydrogenation reaction.

5.2.2.2 Hydrogenations using Nanoparticles

Nanoparticles (NPs) prepared using many different methodologies have found many applications as heterogeneous catalysts.[29] Solid supported NP catalysts are typically prepared from a metal salt, a reducing agent and a stabiliser that are supported on an oxide, charcoal or zeolite.[30] More recently polymers, dendrimers and CNT supports have also been utilised. This field, sometimes named "semi-heterogeneous catalysis", is at the frontier between homogeneous and heterogeneous catalysis and significant progress has been made in the efficiency and selectivity of reactions, recovery and recycle of the catalytic materials.[30] Recently, NP immobilisation using liquids including alternative solvents such as ILs and their use as precursors to heterogeneous nanocluster catalysts or as catalysts has been intensively investigated.[31]

5.2.2.2.1 Nanoparticles Suspended in ILs. NP catalysts have attracted large interest for hydrogenations as well as other reactions because they exhibit a high surface area to bulk metal ratio allowing more efficient use. This aspect is frequently accompanied by large enhancements in the activity and selectivity where they are used as catalysts.[32] However, metal nanocatalysts are prone to aggregation and, therefore, require stabilisation. The low surface tension of many ILs can result in high nucleation rates and formation of smaller particles.[33,34] The stabilisation of NPs in ILs can be due to the presence of hydrides at the NP surface and the confinement of NPs in the non-polar domains of the structured IL.[35]

The most extensively studied are Pd-NPs as hydrogenation catalysts in the selective hydrogenation of terminal and internal double bonds. Therein, dispersion of Pd-NPs in room temperature ILs prevented the phenomenon of agglomeration and retained the catalytic activity.[36] For example, Pd-NPs in $[C_4C_1im][PF_6]$ have been shown to be very active and selective for hydrogenation of C_6-olefins (hex-1-ene, cyclohexene, cyclohexadiene) at low temperature under low pressures of hydrogen.[37] The 100% selectivity to cyclohexene in the cyclohexadiene reduction was attributed to the higher adsorption strength of the diene compared with the monoene on the Pd metal. These NPs were also reusable under the mild reaction conditions. In this example, phenanthroline was added to the reaction mixture to prevent the aggregation of the Pd-NP and facilitate recycle. Polynitrogen ligands such as 2,4,6-tris(2-pyridyl)-s-triazine have been used to stabilise colloidal suspensions of Rh NP when prepared in $[C_4C_1im][PF_6]$.[38] The resulting suspensions of ligand-stabilised NP were also successful catalysts in the hydrogenation of substituted arenes.

Microwave irradiation has also been used for the synthesis of transition metal NPs in $[C_4C_1im][BF_4]$ from metal carbonyl precursors forming NP with <5nm dispersions.[39] The long-term stability of the metal-NP–IL dispersions were characterised by tunneling electron microscope (TEM), transmission electron diffraction (TED), and dynamic light scattering (DLS). In particular, Ru, Rh and Ir-NP–IL dispersions were highly active and easily recyclable for the biphasic liquid–liquid hydrogenation of cyclohexene to cyclohexane with activities of up to 522 (mol product)(mol Ru)$^{-1}$ h^{-1} and 884 (mol product)(mol Rh)$^{-1}$ h^{-1}. Catalyst poisoning experiments with CS_2 (0.05 equiv. per Ru) suggested that heterogeneous catalysis was occurring with the Ru-NPs. The same NPs have also been moderately successful for the selective hydrogenation of benzene in the same IL.[40] The lower solubility of cyclohexene compared with benzene enabled the extraction of cyclohexene during hydrogenation. However, a maximum selectivity of 39% was only obtainable at very low benzene conversion. Although, this selectivity/yield is too low for industrial applications, it represents a rare example of partial hydrogenation of benzene by soluble transition metal NPs. Recently, it has been reported that even small changes in Ru-NP size can have a profound effect on reaction and selectivity.[41] For example in the hydrogenation of 1,3-cyclohexadiene the selectivity to cyclohexene decreased from 97% to 80% when the Ru-NP size increased from 1.1 to 2.9 nm.

A highly effective Pt-NP catalyst immobilised in $[C_4C_1im][BF_4]$ has been developed for the selective synthesis of aromatic chloroamines from aromatic chloronitro compounds.[42] Employing NPs or heterogeneous catalysts in organic solvents resulted in inferior selectivities with 66% and 89% found, respectively, compared with the Pt-NP–IL system which exhibited >99% selectivity. Infra-red studies confirmed that the higher selectivity was due to weak non-covalent interactions between the nitro-group and the IL. In the case of $[C_2OHC_1im][BF_4]$ an IL with a hydroxyl-functionalised system, improved ligand stability of the NPs allowed extensive recycling with high

TONs. The improved stability was attributed to the lack of aggregation of Pt-NP (3.7–3.8nm) in [C_2OHC_1im][BF_4]. In contrast Pt-NPs immobilised in [C_4C_1mim][BF_4] which performed poorly during recycle showed an increase to 7.9 nm.

Nickel-NPs dispersed in an aqueous phase have been conveniently prepared by reducing a nickel(II) salt with hydrazine in the presence of the functionalised ionic liquid 1-(3-aminopropyl)-2,3-dimethylimidazolium bromide.[43] These catalysts were able to hydrogenate 4-phenyl-3-buten-2-one with high conversions and selectivities to 4-phenyl-2-butanone in water, but were almost inactive in ILs, Scheme 5.1. Similar behavior was also observed for the hydrogenation of C=C double bonds of various functionalised alkenes; cyclohexene, 2-propen-1-ol, 3-methyl-2-buten-1-ol, styrene, ethyl acrylate and cyclohex-2-enone.

Imidazolium ILs have been used to control the formation and stabilisation of Ir^0, Pt^0 and Rh^0 nanoparticles. Ir^0 was indicated as an efficient catalyst for the hydrogenation of arenes[44] and Pt^0 for alkenes.[45] However, the hydrogenation of arenes containing functional groups, such as acetophenone or anisole, by these NPs occurred with concomitant hydrogenolysis of the C–O bond, suggesting that they behave as heterogeneous catalysts rather than homogeneous catalysts, Scheme 5.2.

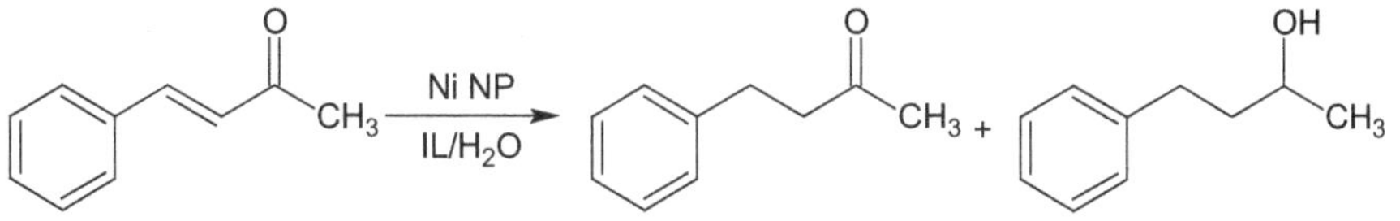

O CH3 Ni NP IL/H2O O CH3 + OH CH3

Scheme 5.1 Hydrogenation products from the Ni catalysed hydrogenation of 4-phenyl-3-buten-2-one.
Reproduced with permission from John Wiley and Sons. Taken from reference 43.

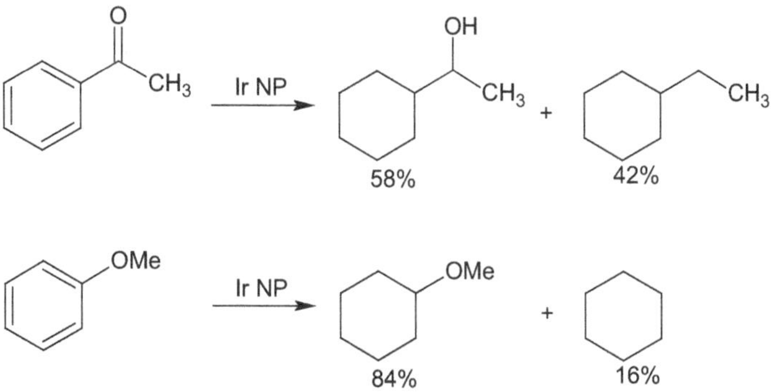

O CH3 Ir NP OH CH3 58% + CH3 42%

OMe Ir NP OMe 84% + 16%

Scheme 5.2 Hydrogenolysis of aryl ketones and ethers.
Adapted with permission from John Wiley and Sons. Taken from reference 44.

5.2.2.2.2 Supported Nanoparticles. Although, in many cases, ILs can be classified as involatile, separation of the IL and products still requires liquid phase processing especially if high distillation temperatures are required. In this regard, heterogeneous catalyst systems offer a significantly simpler work up *via* filtration/decantation of the solid material and ILs supported on solids facilitate this possibility. Xu *et al.*[46] showed that a series of $[C_nC_n\text{im}][PF_6]$ and $[C_nC_n\text{im}][BF_4]$ ILs as bulk solvents are excellent media for the hydrogenation of halonitrobenzenes. Using Raney Ni or 5wt% Pt or Pd supported on carbon, excellent selectivities (>98.7%) to the corresponding haloanilines were achieved, albeit at lower rates compared with the reactions in methanol. In contrast, a wide range of selectivities from 27.4–98.3% were observed in methanol. The extent of unwanted dehalogenation was lower in ILs due to mass transfer limitations and a decrease in chemisorption of the haloaniline in the IL.

The selective hydrogenation of an α,β-unsaturated aldehyde, citral to citronellal (Scheme 5.3) was performed using 10% Pd/C catalyst in a wide range of ammonium, pyridinium and imidazolium ILs.[47] Although lower reaction rates were generally observed in the more viscous ILs due to mass transfer limitations, >99% selectivity was still possible at complete conversion. In contrast, 77% selectivity at 69% conversion was the highest achieved in a range of molecular solvents. Although some deactivation due to pore blocking was found on the first recycle, thereafter, the activity remained constant maintaining high selectivity. The mechanism by which the IL enhanced the selectivity was thought to be due to an increased IL – carbonyl interaction which adjusts the adsorption geometry of the aldehyde by weakening the interaction of the carbonyl with the surface of the catalyst. Similar results were achieved with the same catalyst–IL system for the reduction of cinnamaldehyde to hydrocinnamaldehyde, Scheme 5.4.

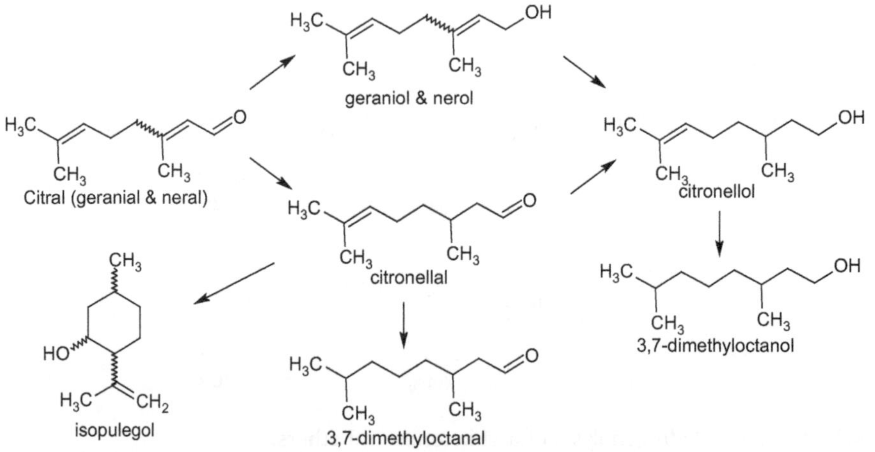

Scheme 5.3 Main Pd catalysed citral hydrogenation pathway.

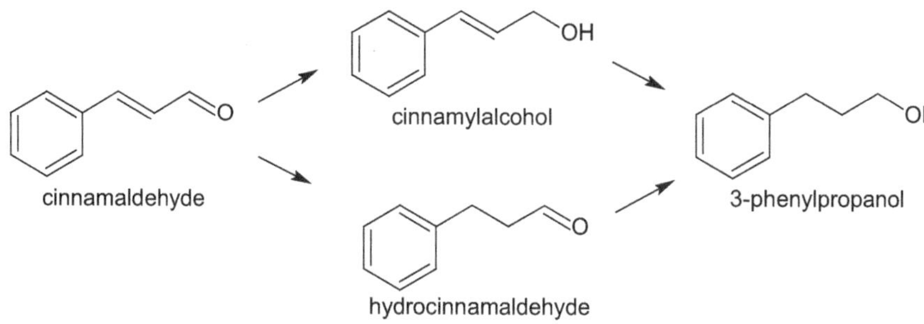

cinnamaldehyde → cinnamylalcohol → 3-phenylpropanol
cinnamaldehyde → hydrocinnamaldehyde → 3-phenylpropanol

Scheme 5.4 Cinnamaldehyde hydrogenation pathway.

Pd-NPs have also been immobilised on an active carbon cloth (ACC) using a thin film of $[C_nC_n\text{im}]^+$ and $[N_{w,x,y,z}]^+$ based ILs for the hydrogenation of citral in *n*-hexane.[48] Therein, the thin film of IL was sufficient to stabilise the active Pd species; however, low selectivity to citronellal was observed in all cases. Catalyst deactivation from batch-to-batch experiments was also observed resulting in a change in product distribution with the over hydrogenated compound 3,7-dimethyloctanal being formed as the main product. Further studies showed that by using an alkylpyridinium $[BF_4]^-$ based IL, the rate was found to increase by between 3–5 times together with >95% selectivity to citronellal using the same Pd/ACC catalyst indicating significant changes in selectivity with different ILs.[49] The hydrogenation of cinnamaldehyde has also been conducted under similar conditions but the moderate selectivity to hydrocinnamaldehyde (~80%) was due to over-hydrogenation forming 3-phenylpropanol.[50] The selective hydrogenation of cinnamaldehyde to hydrocinnamaldehyde has also been obtained with Pd-NPs using a $[C_4C_1\text{im}][PF_6]$–SiO_2 system.[51] The high activity of the system, which showed complete conversion after 20 min, was explained by the high surface area of the silica gel which enhanced the contact between substrate and H_2 gas over the Pd surface.

Support effects have also been reported for the selective hydrogenation of α,β-unsaturated alkynes and aldehydes to the corresponding *cis*-alkenes and alcohols, respectively.[52] In the case of the alkyne substrate, the magnetite support employed was reported to act as a bulky ligand preventing an excess of alkene product accumulating on the Pt surface. For aldehyde reduction, the support was also capable of inducing a slight positive charge on the surface of Pt and activating the more polar carbonyl group leading to the selective formation of alcohols.

2,2′-Bipyridine and an imidazolium-functionalised bipyridine ligand were used in the synthesis of Pd-NPs in water and in IL that were subsequently deposited onto carbon nanofibre (CNF) supports.[53] The hydrogenation of 1-hexyne occurred selectively to *trans*-hexene (up to 98.5%). The high selectivity to the alkene was thought to be due to the bipyrididine based ligands blocking sites which led to over hydrogenation. In these catalysts, the IL-functionalised ligand enhanced the binding of the Pd with the CNF support and reduced leaching. Electronic effects on coordinating to the cationic palladium species also contributed to the reported catalytic performances.

The design of nanocomposites consisting of functional metals and different matrices was recently reported as a promising research area for the fabrication of catalysts.[54] Pd nanoparticles immobilised on the molecular sieve SBA-15 by 1,1,3,3-tetramethyl-guanidinium lactate have been found to exhibit high catalytic activity for hydrogenation of cyclohexene, 1-hexene, and 1,3-cyclohexadiene. Ru-NPs deposited on montmorillonite using 1,1,3,3-tetramethylguanidinium trifluoroacetate as an immobiliser exhibited much higher activities than the conventional Ru catalysts in the hydrogenation of aromatic compounds. The immobilisation of the metal-NPs is a combination of strong electrostatic forces and coordination resulting in a very stable catalyst.

The nature of the ILs anion is not negligible in the behavior of the metal-NPs.[12] Pd-NPs $[C_4C_1im][PF_6]$ and $[C_4C_1imOH][NTf_2]$ in the hydrogenation of propyne from a mixture consisting of H_2 (86%) and propyne (14%) led to propane : propylene ratios of *ca.* 3.3 : 1. In these experiments, normal H_2 was used to evaluate conversion and selectivity of hydrogenation, while para-hydrogen-enriched H_2 (*ortho*-H_2/*para*-H_2 ratio *ca.* 1 : 1) was used to observe the parahydrogen-induced polarisation effects. Arras *et al.*[55] used dicyanamide $[N(CN)_2]^-$ and $[NTf_2]^-$ anion containing ILs as selective modifiers for the heterogeneously catalysed citral hydrogenation (Scheme 5.3) on Pd-supported catalysts in *n*-hexane. Excellent selectivities (>99%) towards citronellal were observed by using the $[N(CN)_2]^-$ IL as a catalyst additive or coating on 10 wt% Pd/C or Pd/SiO_2 in comparison to 37–41% selectivity in the absence of IL. XPS analysis showed the presence of Pd(II) species where the basic $[N(CN)_2]^-$ IL is able to modify the active Pd catalyst sites similar to a bimetallic system thus promoting large changes in selectivity. An increase in Pd–O bond distance also suggested partial removal of Pd-NP from the surface of the support forming stable Pd colloidal IL solution which has also been reported for other IL coated Pt/SiO_2 catalysts.[56] Interestingly, the use of an $[NTf_2]^-$ based IL in hexane, which resulted in only 62% selectivity, showed a much weaker Pd–$[NTf_2]$ interaction under XPS analysis. Time-resolved IR reflection absorption spectroscopy during preparation of ultra-thin IL films (typically 0.5–50 Å) by physical vapor deposition showed a chemical interaction between the Pd and $[NTf_2]^-$ through the $-SO_2-$ groups on the anion.[57] Moreover, it was found that the IL was most likely to interact with the most active Pd-NPs deposited on the support at edge and defect sites.

A $[C_4C_1im][n\text{-}C_8H_{17}OSO_3]$ IL film has also been used to improve the selectivity in the sequential liquid phase hydrogenation of cyclooctadiene to cyclooctene and cyclooctane using a commercial catalyst (37 wt% Ni on SiO_2).[58] Although the IL coating had a negative influence on the rate of cyclooctadiene conversion, an increase in selectivity to *cis*-cyclooctene from 40% to 70% was observed. This strong influence of selectivity with IL coating was not due to the different concentrations of substrate and intermediate in the IL monolayer. In this case it was proposed that the IL acted as a co-catalyst inhibiting adsorption of the cyclooctadiene on the Ni sites.

Under a continuous flow operation the hydrogenation of acetylene to ethane employing Pd-NP supported on a CNF or SMF combined with ILs showed high selectivity and excellent long-term stability. Hydroxyl functionalisation of the cation and the use of a weakly coordinating anion in the IL $[C_4OHC_1im][NTf_2]$ resulted in a very stable Pd catalyst which suppressed the formation of oligomers and, therefore, catalyst deactivation. The low solubility of ethane compared with acetylene was also responsible for the high selectivity. Moreover, the presence of a SILP system was responsible for the prevention of oligmerisation which resulted in catalyst poisoning.[59]

5.2.2.3 Asymmetric Hydrogenation

While the use of ILs as catalyst immobilisers has been shown to be effective for transition metals and their complexes, for asymmetric metal catalysts, retention, stability and recycle is a necessity due to the added expense of chiral ligands. Wolfson *et al.*[60] developed a SILP, based on a Ru-BINAP catalyst in $[C_4C_1im][PF_6]$ and a poly(diallyldimethylammonium chloride) polymeric support for the liquid phase hydrogenation of methyl acetoacetate in IPA. The SILP reaction showed higher activity than the biphasic reaction but was lower than the homogeneous reaction. The main advantage of the heterogeneous system was the catalyst reusability without significant loss in activity, selectivity and enantioselectivity (ee).

Catalyst reusability can also be achieved by introduction of an ionic tag, Figure 5.2. Josiphos ligands modified with an imidazolium tag were successfully employed in Rh-catalysed asymmetric hydrogenation of methyl acetamidoacrylate and dimethyl itaconate.[61]

Rh and Ru BINAP complexes have been immobilised onto SILPs using phosphonium based ILs allowing them to be employed as heterogeneous

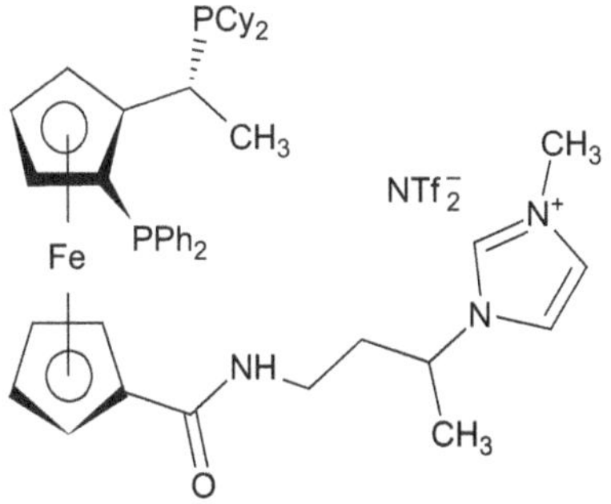

Figure 5.2 Imidazoilum tagged Josiphos ligand employed for asymmetric hydrogenations.
Adapted with permission from John Wiley and Sons. Taken from reference 61.

catalysts for the hydrogenation of acetophenone. Compared with the analogous homogeneous reactions (4% ee), 1-cyclohexylethanol was formed with ees up to 74%. Nitrogen adsorption analysis showed that most of the IL was residing in the pores of the silica. Sievers *et al.*[62] speculated that the IL in the pores gave rise to so called "*solvent cages*", which consisted of supramolecular aggregates of one complex molecule and 16–33 ion pairs of the IL. A binding model was proposed where acetophenone was binding preferentially *via* both the carbonyl group and phenyl ring to the metal centre. The resultant η^2/η^2-coordinated acetophenone fitted tightly into the pocket formed by the phenyl rings of the chiral BINAP ligand, resulting in high stereoselection.

Lou *et al.*[63] immobilised a chiral ruthenium complex $RuCl_2(PPh_3)_2$(*S*,*S*-dpen) in an imidazolium based IL phase confined on the surface of a series of mesoporous materials (MCM-48, MCM-41, SBA-15) as well as amorphous SiO_2 onto which was covalently grafted an IL, Figure 5.3. The corresponding heterogeneous catalysts were evaluated in the asymmetric hydrogenation of acetophenone in IPA. All of the catalysts exhibited increased activity compared with the homogeneous equivalent systems. In the absence of either the chemisorbed or physisorbed IL poor conversions were noted upon recycle indicating a synergistic effect of both ILs. The modified SILP MCM-48 Ru-IL complex which showed the highest ee was attributed to the confinement effect of the 3-D channel of MCM-48, which is better than that of the one 1-D channel of MCM-41 and SBA-15. However, the better stability of the SiO_2-supported catalysts that showed comparable conversion and ee value after the fifth run may suggest that the external surface is more effective in this process.[64]

Enantioselective hydrogenation of the C=N double bonds has also been investigated using chiral supported ionic liquid phase catalysts (Scheme 5.5).[65]

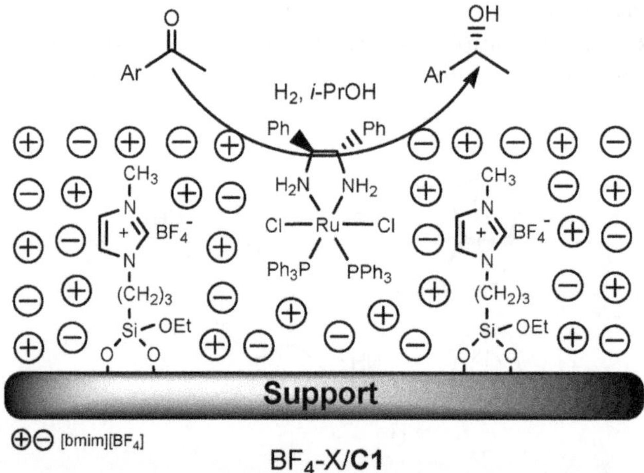

Figure 5.3 Supported ionic liquid phase catalyst BF4-X/C1 for asymmetric hydrogenation of aromatic ketone.
Reproduced with permission from Elsevier. Taken from reference 64.

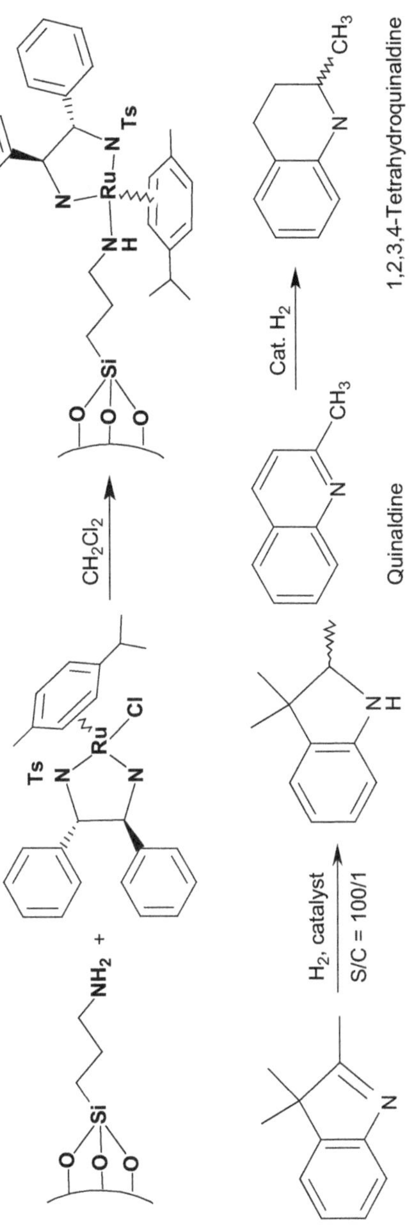

Scheme 5.5 Covalent grafting of RuCl(*p*-cymene)[(*S*,*S*)-Ts-DPEN] complex on aminopropyl MCM-41 carrier for the asymmetric hydrogenation of C=N double bond using SILP catalysts.
Reproduced with permission from Elsevier. Taken from reference 65.

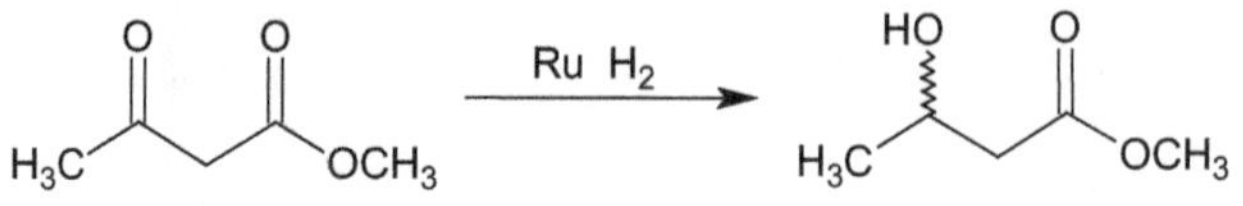

Ru H_2

Scheme 5.6 Hydrogenation of methylacetoacetate to 3-hydroxyl methylbutyrate. Adapted from reference 66.

These were prepared either by physical adsorption (within highly porous carbons or mesoporous silica) of Ir, Ru and Rh complexes or by covalent grafting of these complexes on aminopropyl modified carriers, Scheme 5.5. Several ligands: (*S*,*S*)-BDPP, (*S*)-BINAP, (*S*,*S*)-Ts-DPEN,(*S*,*S*)-DIPAMP or (*R*,*R*)-Me-DuPHOS, and the ILs $[C_2C_1im][NTf_2]$, $[C_4C_1im][BF_4]$ and $[C_4C_1im][PF_6]$ were used for this purpose. The conversion and enantioselectivity were found to depend on the nature of the complex (metal and ligand), the immobilisation method used, nature of the IL, nature of the support and the experimental conditions.

Asymmetric hydrogenation using a SILP catalyst has also been reported under continuous gas phase conditions.[66] The hydrogenation of methyl acetoacetate (Scheme 5.6) over silica immobilised dibromo[3-(2,5-(2*R*,5*R*)-dimethylphospholanyl-1)-4-di-*o*-tolylphosphino-2,5-dimethylthio-phene]-ruthenium catalyst occurred with enantiomeric excesses in the range 65–82% for more than 100 h time-on-stream continuous operation. The relatively low volatility of the substrate was overcome by using helium as carrier gas in the reactor.

Combining ILs with supercritical CO_2 has also been reported to bring some advantages.[67] A highly efficient continuous-flow asymmetric catalysis has been achieved by combining a SILP catalyst with $scCO_2$ as the mobile phase, for the hydrogenation of dimethyl itaconate in the presence of a Rh-QUINAPHOS complex.[68] The immobilisation of the catalyst in the SILP system did not affect the high selectivity of 99% ee. No reaction was observed when the Rh catalyst was supported on SiO_2 without the IL, confirming the homogeneous nature of the active catalytic species formed in the SILP matrix under $scCO_2$ flow conditions. Under flow conditions, partial decomposition of the active catalyst resulted in a decrease in ee due to the unselective Rh hydrogenation catalyst. However, in direct comparison to other immobilisation techniques[69] such as the concept of self-supported catalysts, roughly one order of magnitude higher space–time yields are achieved with the SILP–$scCO_2$ system.

It has also been reported that ILs as coated catalysts or additives tremendously alter the selectivity pattern of a heterogeneous solid catalyst in the selective hydrogenation.[70] This effect was exploited using a conventional monometallic Ru/Al_2O_3 catalyst combined with an IL in supercritical carbon dioxide. It was thus possible to conduct the one-pot synthesis of the *p*-menthene intermediate with selectivities higher than 99% through limonene hydrogenation using $[C_{10}C_1im][NTf_2]$ as an additive, Scheme 5.7. Further hydrogenation of *p*-menthene was strongly inhibited by employing the

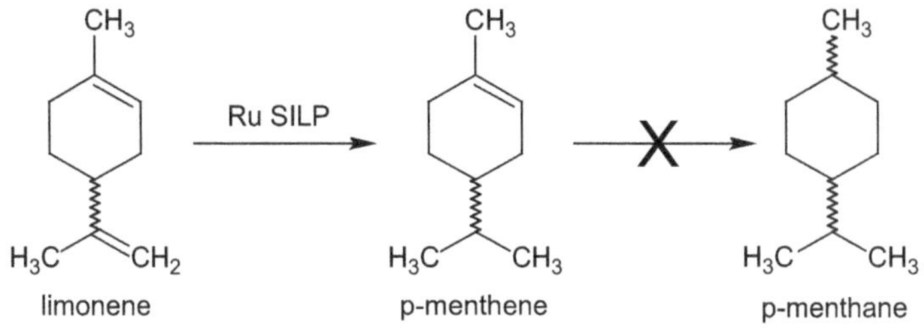

CH_3 CH_3 CH_3

Ru SILP

H_3C CH_2 H_3C CH_3 H_3C CH_3

limonene p-menthene p-menthane

Scheme 5.7 Selective hydrogenation of limonene. Adapted from reference 70.

ionic liquid. Catalyst recycling indicated that there was no depletion of catalyst reactivity even after four successive cycles.

5.2.2.4 *Hydroprocessing Treatments*

Hydroprocessing treatments typically refer to the reductive cleavage of a C–X bond by highly reactive atomic hydrogen. Hydrodechlorination is the specific case where X is a chlorine atom and is used for the treatment of toxic waste streams. Hydrodechlorination of dieldrin and DDT was carried out using a multiphase system composed of an organic phase and aqueous KOH, a quaternary ammonium IL promoter (Aliquat 336), and a metal catalyst, *e.g.* 5% Pd/C, 5% Pt/C, or Raney Ni with the latter showing the highest activity.[71] For example, at 50 °C under an atmospheric pressure of hydrogen a quantitative hydrodechlorination of DDT was achieved in 40 min.

5.2.2.5 *Catalytic Heterofunctionalisation*

Catalytic heterofunctionalisation represents an elegant route of functionalisation of alkenes and alkenes allowing the formation of a wide variety of bonds between carbon and other elements by adding compounds or mixtures containing CO and H_2 (hydroformylation), N–H (hydroamination) and Si–H (hydrosilylation).

5.2.2.6 *Hydroformylation*

The entrapment of $[Rh(cod)Cl]_2$ using two different sol–gel routes led to active catalysts for the hydroformylation of vinylarenes in aqueous media (Scheme 5.8).[72] In the first route, $[Rh(cod)Cl]_2$ was confined with 1-butyl-3-[(3-trimethoxysilyl)propyl]imidazolium chloride together with $Na[Ph_2P(C_6H_4\text{-}3\text{-}SO_3)]$ within silica sol–gel, while in the second method, the same rhodium compound was encaged within an IL-free hydrophobicized sol–gel. Working with H_2/CO ratios of between 1.0 and 1.1, both methods allowed very good yields to the 2-propanal isomer. While the regioselectivity

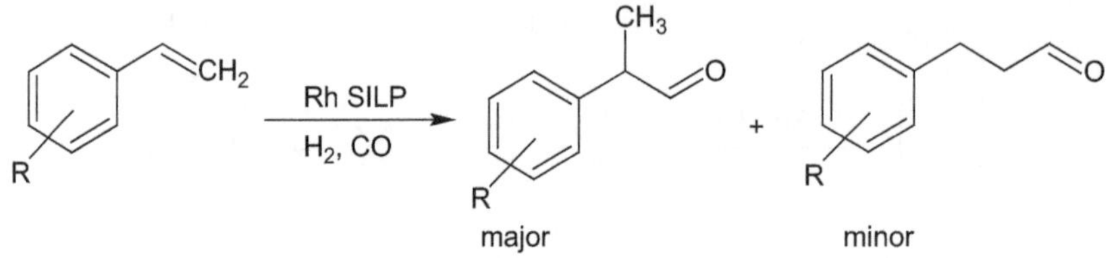

Scheme 5.8 Hydroformylation of vinylarenes in aqueous media. Adapted from reference 72.

and the yield were unaffected by the electronic nature of the substrates, they were significantly dependent on the reaction temperature, the surfactant employed and the hydrophobicity of the support of the catalyst. Under these conditions despite the use of H_2 in the reactions, no transformation of the organometallic catalyst into metallic-NPs was detected.

Rh complexed with a bisphosphine ligand sulfoxantphos and dissolved in both halogen containing and halogen-free IL ($[C_4C_1im][X]$ ($X = PF_6$ or n-$C_8H_{17}OSO_3$)) has been shown to catalyse the gas phase hydroformylation of propene in a fixed bed reactor when it is physisorbed on the unmodified silica support showing good activity.[73] However, such a system deactivated after prolonged use (reaction time >24 h) regardless of the type of IL or loading. Using the SILP concept allowed the application of this reaction in continuous catalytic gas phase processes and eliminated these drawbacks.[74] Under these conditions the same Rh–bisphosphine ligand sulfoxantphos catalyst remained active, highly selective and stable over extended periods in a continuous gas phase process. Therein, the degree of hydroxylation of silica is very important. Thus the partly dehydroxylated silica support exhibited the better performances ($TOF = 44\ h^{-1}$, $TON \approx 2600$, $n/iso > 20$). The catalytic performance and catalyst stability also depended on the catalyst composition. Catalysts having low ligand contents were initially very active compared with the catalyst with high ligand contents but less selective. In addition, the catalysts with low ligand content were deactivated relatively quickly within the first 24 h.

Immobilisation of a Rh–biphephos catalyst in a silica SILP system made it possible to extend the SILP hydroformylation catalysis to reactions with a highly diluted, technical C_4 feedstock containing 1.5% 1-butene, 28.5% 2-butenes, and 70% of inert *n*-butane.[75] Such a system allowed consecutive isomerization/hydroformylation activity resulting in a conversion up to 81% of the reactive butenes, and a selectivity in the *n*-pentanal greater than 92%. No significant loss of the phosphate ligand through ligand oxidation during the reaction has been detected.

5.2.2.7 *Hydrosilylation*

Polysiloxanes with a long alkyl chain (>C_8) as a pendant group represent one of the most important classes of modified polysiloxanes due to a much higher molecular weight and viscosity in the liquid state than do paraffin waxes. The synthesis of silicone waxes is based on catalytic hydrosilylation of alkenes with poly(hydromethyl)siloxanes. The immobilisation of a siloxide rhodium complex $[\{Rh(\mu\text{-}OSiMe_3)(cod)\}_2]$ in ILs resulted in a highly selective and very active catalytic biphasic system that enabled easy product separation and repeated use of the catalytic system, Scheme 5.9.

Excellent selectivities were also observed in the case of hydrosilylation using a typical heterogeneous system in the presence of supported metallic catalysts. Monometallic catalysts with 1 wt% Pt or Rh loading and bimetallic catalysts with Pt and Cu loadings of 1 and 0.1 wt%, respectively, were prepared by the incipient wetness method using different polymers, copolymers and carbon supports.[76] Within this group of catalysts, Pt supported on a highly hydrophobic styrene–divinylbenzene resin was particularly active which enabled product yields of about 90% in reactions with 1-octene and over 90% in reactions with 1-hexadecene. For the bimetallic Pt–Cu system with supports such as activated carbon and carbon black, product yields were higher than those obtained on monometallic catalysts. In contrast, supported Rh was either inactive or poorly active under the reactions conditions studied.

5.2.2.8 *Hydroamination*

Hydroamination is an atom efficient reaction involving the addition of an N–H across an olefinic or acetylinic bond, Scheme 5.10. Despite substantial efforts catalyst recyclability and in some cases reaction selectivity remain poor.

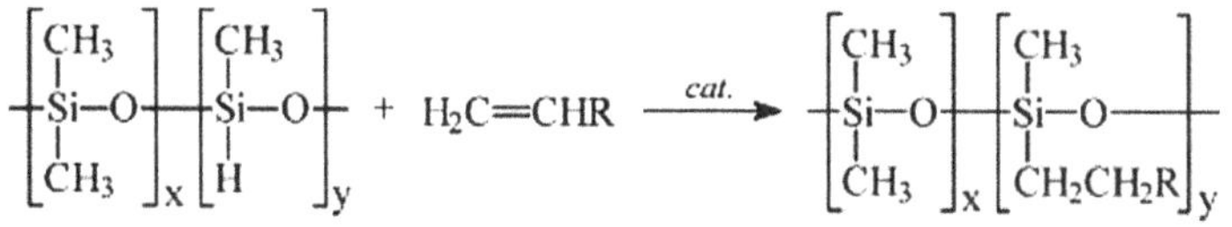

Scheme 5.9 Synthesis of polysiloxanes.
Reproduced with permission from Elsevier. Taken from reference 76.

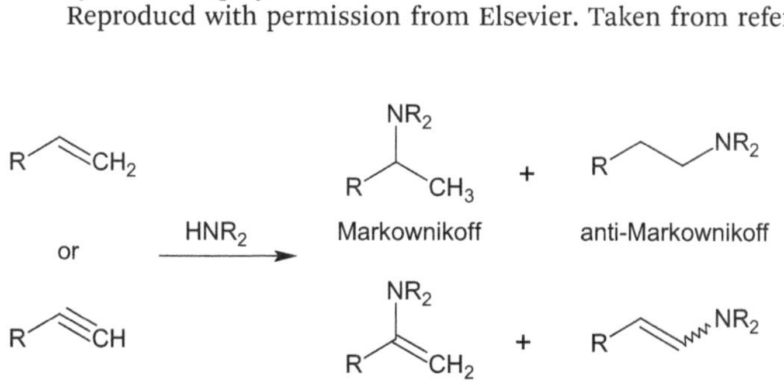

Scheme 5.10 Hydroamination of terminal alkenes and alkynes.

Employing ILs and working under biphasic conditions could help improve these reactions due to a simple and complete separation of the product from the catalyst. In this regard, a system comprised of a polar catalyst phase containing Rh(I), Pd(II), Cu(II) and Zn(II) complexes in $[C_2C_1im][OTf]$ and a substrate mixture in heptanes has been used to exemplify this concept.[77] Bifunctional catalysts combining soft Lewis acidic function (activation of the alkene) and strong Brønsted acidic function (acceleration of the rate determining step) were reported to provide high catalytic activities in this reaction.[78] Electron rich anilines react more readily and the presence of a Brønsted acid also accelerated the reaction. The role of the acidic promoter was explained through protonolysis of the precatalyst to give cationic complexes, which are the active species involved in the mechanism. Thus, further immobilisation of organometallic complexes like Pd-1,1′-bis(diphenylphosphino)-ferrocene combined with TfOH in a thin film of supported IL on a silica support showed good catalytic activity for the cyclisation of 3-aminopropyl-vinylether and the addition of aniline to styrene, providing the Markovnikov product under kinetically controlled conditions and mainly the *anti*-Markovnikov product in the thermodynamic regime.[79] The simple addition of an acid in the IL was also observed to promote intermolecular hydroamination which is more easily achieved compared with intramolecular hydroamination.[80] In contrast to these studies it was shown the presence of n-Bu_4PBr has no beneficial effect for the Pt-catalysed hydroamination of terminal alkynes. Pristine $PtBr_2$ catalyses the hydroamination of terminal alkynes with aniline affording the branched imine Markovnikov product in high selectivity.[81] Efficient and recyclable hydroamination of alkenes with amines, amides, sulfonamides, and carbamates was reported using an Amberlyst-15 resin immobilised in $[C_4C_1im][BF_4]$.[82] Therein, the solvent effect was believed to be responsible for the high product selectivities observed.

5.3 Dehydrogenation

Tetralin dehydrogenation was investigated on a Pt-supported activated carbon catalyst in a microwave-induced system.[83] In comparison to the reaction carried out in a conventional reactor, tetralin conversion increased from 31% to 56% under 190 W microwave irradiation. Therein, the addition of $[C_4C_1im][PF_6]$ to the system to function as a heating promoter remarkably enhanced the heating rate of tetralin. However, the presence of the IL resulted in deactivation of catalyst.

5.4 Oxidation

Oxidation represents a direct route to prepare many valuable compounds; however, the selectivity remains the key issue in this reaction. Working under mild, green conditions can improve the selectivity and the recent reports presenting results on oxidations carried out in ILs confirmed the efficiency of these solvents. Selective oxidation can be carried out using both homogeneous and heterogeneous catalysts. There are also many examples in which the homogeneous catalyst has been heterogenised by immobilisation leading to active and even stereoselective heterogeneous catalysts, for example immobilised Salen complexes have been reported.[84]

5.4.1 Oxidation of Hydrocarbons

5.4.1.1 Oxidation of Alkanes

Fe-ZSM-5 was reported for cyclohexane oxidation with *tert*-butyl-hydroperoxide in $[C_2C_1im][BF_4]$ under mild conditions.[85] Good yields and high selectivity of products were found in the IL towards the mixture of cyclohexanone, cyclohexanol and cyclohexyl hydroperoxide compared with molecular solvent. Higher catalyst activity for the Fe-ZSM-5 over other metal-ZSM-5 or the parent H-ZSM-5 in the IL was reported. Heterogeneous poly(4-vinylpyridine)–methyltrioxorhenium and microencapsulated polystyrene–methyltrioxorhenium systems in $[C_4C_1im][PF_6]$ have been investigated in the selective oxidation of triphenylmethane.[86] Benzophenone was recovered as the only product and no significant differences were observed between the two catalytic systems. The efficiency of oxygen atom insertion by these catalysts in ILs was further demonstrated through their ability to oxidise saturated hydrocarbons, such as *cis*-1,2-dimethylcyclohexane and adamantine. The reaction of *cis*-1,2-dimethylcyclohexane with catalysts performed in $[C_4C_1im][PF_6]$ at 45 °C, gave the corresponding *cis*-1,2-dimethylcyclohexan-1-ol both in high yield and conversion of substrate showing a selective oxidation at the tertiary C–H bond. In the oxidation of adamantane in $[C_2C_1im][NTf_2]$ 1-adamantanol was selectively obtained in acceptable yield and conversion of substrate.

5.4.1.2 Oxidation of Alkenes

Oxidation of cyclohexene was carried out with a mesoporous silica with incorporated titanium dioxide prepared using $[C_{16}C_1im]Cl$ as a template.[87] In the presence of hydrogen peroxide as oxidant, the major products at 25 °C were the mono and diketone, while at 50 °C the diketone and adipic acid (Scheme 5.11). The formation of epoxide occurred only to a small extent.

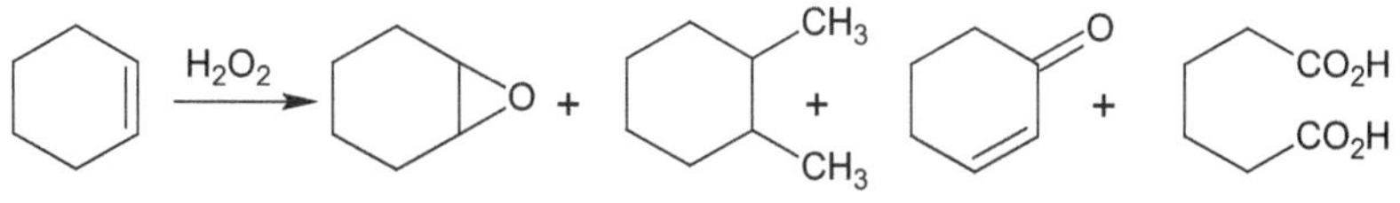

H_2O_2 → O + CH₃ CH₃ + O + CO₂H CO₂H

Scheme 5.11 Products of cyclohexene oxidation.

5.4.1.3 *Oxidation of Aromatic Compounds*

A heteropolyanion based ionic hybrid was prepared by combining a divalent IL cation of 1,1′-(butane-1,4-diyl)-bis(3-methylimidazolium) with the Keggin-structured V-containing heteropolyanion.[88] This compound exhibits a semi-amorphous structure and presented good activity in the hydroxylation of benzene with aqueous H_2O_2, Scheme 5.12 The reaction occurs in a liquid–solid biphasic system leading to good yields of phenol (26%) allowing a convenient recovery and steady reuse. However, the system is sensitive to the nature of the solvent and the best results were obtained in a solution of acetonitrile–acetic acid.

These features were associated with a strong ionic interaction between the divalent IL cation and the heteropolyanion with the V^{5+}/V^{4+} redox pairs thought to be the active centres. Oxidation of styrene with H_2O_2 to produce benzaldehyde was reported using a Ni^{2+}-containing ionic liquid catalyst immobilised on silica.[89] The catalyst was prepared *via* hydrothermal synthesis using a mixture of Ni^{2+}-1-methyl-3-[(triethoxysilyl)propyl] imidazolium chloride, tetraethoxysilane, cetyltrimethylammoniun bromide, and ammonia. The reaction proceeded with good selectivities and moderate conversions. The advantage of using this catalyst was that the reaction could be carried out under solvent-free conditions as both of the reactants, styrene and H_2O_2, are miscible with the IL. The Ni^{2+} catalytic centre was coordinated by the immobilised IL that allowed both reactants access to active sites of the catalyst effectively.

5.4.2 Oxidation of Alcohols

Over the last few decades the search for efficient and environmentally friendly oxidation procedures has intensified. The replacement of toxic of oxochromium(VI) based stoichiometric oxidants with transition metal catalysts and of chlorinated solvents with green solvents like ILs in the oxidation of alcohols has been extensively investigated.

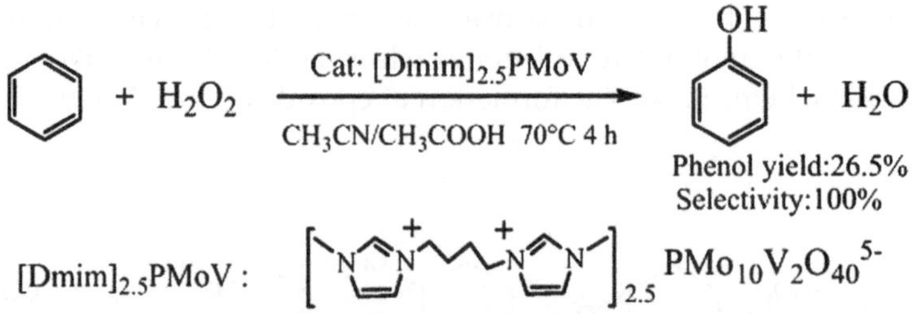

Cat: $[Dmim]_{2.5}PMoV$
$+ H_2O_2$ → $+ H_2O$
CH_3CN/CH_3COOH 70°C 4 h
OH
Phenol yield:26.5%
Selectivity:100%
$[Dmim]_{2.5}PMoV$: $[\ldots]_{2.5}$ $PMo_{10}V_2O_{40}^{5-}$

Scheme 5.12 Oxidation of benzene using polyoxometallate–IL catalyst. Reproduced with permission from Elsevier. Taken from reference 88.

5.4.2.1 Aliphatic Alcohols

The oxidation of 1-octanol and 2-octanol to the corresponding carbonyl compounds was achieved with ordered mesoporous hexagonal or lamellar silicas. These catalysts were prepared by tethering various imidazolium IL moieties onto the ordered mesoporous silica.[90] Thus, the integrated approach combines sol–gel entrapped perruthenate as an aerobic catalyst, an encapsulated IL as solubility promoter and $scCO_2$ as the reaction solvent, Figure 5.4.[91]

Another approach to aerobic oxidation of these alcohols using supported ILs consisted of coating a polystyrene–nitroxy radical 2,2,6,6-tetramethylpiperidyl-1-oxy (TEMPO) resin with $[C_4C_1im]PF_6$ and $CuCl_2$.[92] The oxidation of primary or secondary afforded the corresponding aldehydes or ketones with good yield. The supported ionic liquid layer substantially enhanced catalytic activity and facilitated catalyst recycle. IL-modified polystyrene resin beads were also demonstrated to be an appropriate support for polyoxometalate catalysts.[93] In this heterogeneous catalytic system, alcohols can be efficiently oxidised to corresponding carbonyl groups with H_2O_2 in CH_3CN. The catalyst can be easily recovered by filtration and recycled without apparent loss of catalytic performance.

Three phase oxidation of cinnamyl alcohol was carried out in toluene and $[C_4C_1im][NTf_2]$ using a 5 wt% Pd/Al_2O_3 catalyst and a rotating disc reactor (RDR) and a stirred tank reactor (STR).[94] Among the various products (Scheme 5.13) cinnamaldehyde is the product of interest of this reaction.

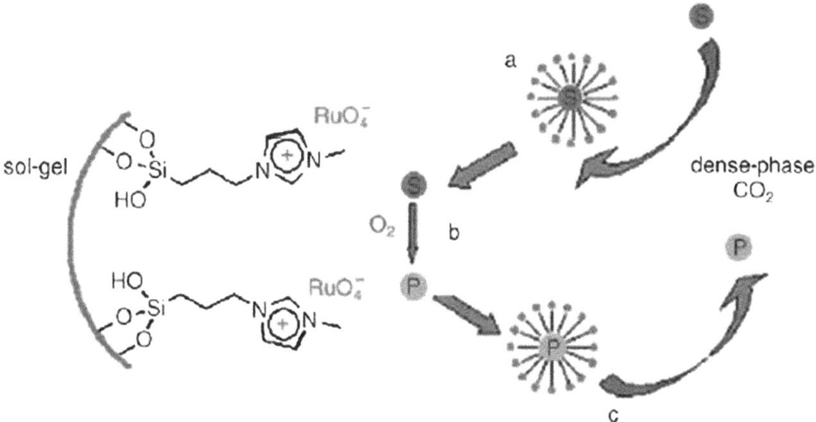

Figure 5.4 Aerobic catalytic oxidation of alcohols in dense phase CO_2: (a) transfer of the substrate to the catalyst tethered to the ionic liquid moiety; (b) catalytic reaction; (c) transfer of the reaction product back into solution.
Reproduced with permission from John Wiley and Sons. Taken from reference 91.

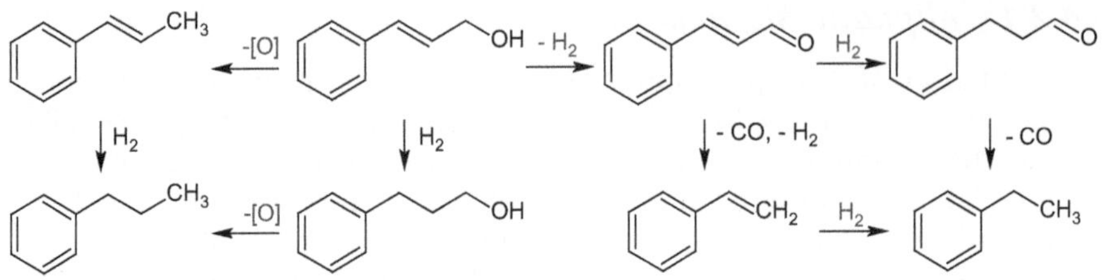

Scheme 5.13 Products from the oxidation of cinnamyl alcohol under STR and RDR conditions.

In the STR higher reaction rates were found for both solvent systems compared to the analogous reactions conducted in the RDR primarily due to internal diffusion limitations in the latter. Despite the lower reaction rates mass transfer resistance resulted in higher cinnamaldehyde selectivity.

5.4.3 Selective Oxidation

The oxidation of benzyl alcohol with aqueous H_2O_2 has been examined using an amino-functionalised bipyridine–Keggin-structured heteropolyacid ionic hybrid prepared by protonating and anion-exchanging the amino-attached bipyridine ionic liquid with phosphotungstic acid.[95] The hybrid catalyst proved to be a highly efficient solid catalyst for solvent-free oxidation of benzyl alcohol with H_2O_2, leading to TOFs around 350 for selectivities higher than 93%.

Aromatic alcohols (benzyl alcohol and 1-phenylethanol) have also been aerobically oxidised to the corresponding carbonyl compounds by ionically tethering RuO_4^- to the imidazolium moiety of the ordered mesoporous supported ionic liquid catalysts described above.[96] Aerobic oxidation of 4-methoxybenzyl alcohol, benzyl alcohol, 4-nitrobenzyl alcohol, 4-chlorobenzyl alcohol, 4-methylbenzyl alcohol and 2-methylbenzyl alcohol using the polystyrene–TEMPO resin with the ionic liquid $[C_4C_1im][PF_6]$ and $CuCl_2$ system also afforded high conversions and high selectivities.[94] The selective oxidation of various alcohols into their corresponding aldehydes and ketones was also achieved by ruthenium species stabilised on the nanocrystalline magnesium oxide by the incorporation of a basic ionic liquid, choline hydroxide.[97] The catalyst was shown to be recyclable with little loss in activity.

Recently it was reported that a new Pd containing IL based periodic mesoporous organosilica (PMO) catalyst was very active in the aerobic oxidation of a large number of primary and secondary alcohols including non-activated alcoholic substrates using molecular oxygen and air.[98] The catalyst system could be successfully recovered and reused several times without any significant decrease in activity and selectivity and without any Pd leaching in reaction solution. The reactivity and recyclability of the catalyst was thought to be due to the PMO-IL mesostructure which immobilises and prevents the aggregation of the Pd-NPs.

In addition to the effects on the catalytic reaction, the use of a suitable IL allows the separation of the polar reaction products without the need for organic solvent extraction or distillation. Experiments carried out in the oxidation of non-activated aliphatic alcohols with molecular oxygen in the presence of palladium(II) acetate demonstrated this effect.[99] In $[C_4C_1im][BF_4]$, yields of ketones of 79 and 86% were obtained for 2-octanol and 2-decanol oxidation, respectively. After cooling to room temperature, the ionic liquid enabled the ketone product phase to phase separate allowing it to be isolated by simple decantation. As in previous examples, the ionic liquid acts as an immobilisation medium for the palladium catalyst, allowing efficient catalyst recycling.

5.4.4 Stereoselective Oxidation

The SIL strategy has also been applied for stereoselective synthesis of alcohols. In this scope the immobilisation of a chiral Mn(III) salen complex onto an ionic liquid modified siliceous material (SBA-15, MCM-41, MCM-48, amorphous silica) allowed the heterogeneous oxidative kinetic resolution of secondary alcohols.[100] The immobilisation occurred through a thin film of covalently anchored imidazolium ionic liquid. The immobilised catalyst provided very high enantioselectivity in this reaction. Changing the support made practically no difference in this oxidative kinetic resolution of secondary alcohols. However, the nature of the oxidant was very important. While diacetoxyiodobenzene was found to be the best oxidant in this oxidative kinetic resolution, no oxidation was observed using H_2O_2 and *tert*-butylhydrogenperoxide as oxidants.

5.4.5 Heterocyclic Compounds

Immobilisation of ILs, such as 1-methyl-3-(triethoxysilylpropyl)imidazolium hydrogensulfate, onto a silica solid support led to an acid solid catalyst active for the Baeyer–Villiger reaction.[101] Cyclic ketones were readily oxidised on these catalysts with 68% hydrogen peroxide in dry dichloromethane to their corresponding lactones in high yields (60–91%) at 50 °C within a short time (5–20 h). The heterogeneous catalyst was easily recovered by simple filtration and reused in another oxidation process.

5.4.6 Sulfoxidation

5.4.6.1 *Selective Oxidation to Sulfoxides*

Heterogeneous catalytic oxidation of a series of thioethers (2-thiomethylpyrimidine, 2-thio-methyl-4,6-dimethylpyrimidine, 2-thiobenzylpyrimidine, 2-thiobenzyl-4,6-dimethylpyrimidine, thioanisole, and *n*-heptyl methyl

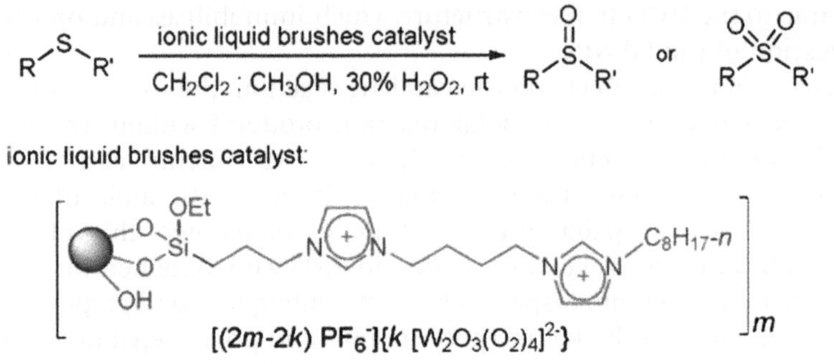

Figure 5.5 Silica immobilised peroxotungstate ionic liquid brush used for sulfoxidation.
Reproduced with permission from Elsevier. Taken from reference 103.

sulfide) was performed in ILs by using MCM-41 and UVM-type mesoporous catalysts containing Ti, or Ti and Ge in a range of $[OTf]^-$, $[BF_4]^-$, $[CF_3CO_2]^-$, $[NTf_2]^-$ and lactate based ILs.[102] The oxidations were carried out by using anhydrous hydrogen peroxide or the urea–hydrogen peroxide adduct and showed that ILs are very effective solvents, achieving greater reactivity and selectivity than reactions performed in dioxane. The addition of Ge to Ti was found to increase the rate of oxidation, but reduced the selectivity towards the sulfoxide.

Peroxotungstates immobilised in a supported ionic liquid brush catalyst have also been shown to have high catalytic activity towards the selective oxidation of sulfides to sulfoxides with aqueous hydrogen peroxide under mild conditions (Figure 5.5).[103] The catalyst's activities were found to be determined by the number of supported ionic liquid layers and N-end-capped alkyl group on the imidazolium ring. The catalysts were able to oxidise both aliphatic (also with unsaturated double bonds) and aromatic sulfides resulting in high yields towards sulfur groups (over 83%) even though the experiments were carried out in excess of hydrogen peroxide in order to generate high catalytic activity. There was no apparent loss of catalytic efficiency until the 8th cycle.

5.4.6.2 *Stereoselective Oxidation to Sulfoxides*

The immobilisation of a chiral Ti–binol complex onto an IL modified SBA-15 led to a catalyst that was found to be highly enantioselective in the heterogeneous asymmetric oxidation of prochiral sulfides to sulfoxides and subsequent oxidative kinetic resolution of the sulfoxides using aqueous *tert*-butylhydroperoxide as the oxidant.[104] Using this supported catalyst, a positive non-linear effect was reported in the oxidation-kinetic resolution of thioanisole.

Active titanium (IV)-salan asymmetric complexes for asymmetric sulfoxidation of thioanisole were also immobilised on polymer matrices.[105] Using the homogeneous reaction, the oxidation of thioanisole with H_2O_2 as oxidant in organic solvents led to high H_2O_2 conversion and moderate enantioselectivities (up to 51% ee). In selected ionic liquids with [Ti{5-MeO-sal(*R*,*R*-chan)}] covalently bound to a polystyrene matrix and *tert*-butyl hydroperoxide as oxidant the Ti–salan complexes showed better activity but with reduced enantioselection of 18% ee.

5.4.7 Epoxidation

The use of the organometallic methyltrioxorhenium ($MeReO_3$) complex in oxidation reactions, particularly for the epoxidation of olefins, was extensively reported due to the high yields and selectivities obtained with this homogeneous catalyst.[106] This interest is based on its capability to activate H_2O_2 and the urea hydrogen peroxide adduct. Upon heterogenisation as poly(4-vinylpyridine)-methyltrioxorhenium or polystyrene–methyltrioxorhenium microcapsules, active catalytic systems in ILs were reported to eliminate the disadvantages of the homogeneous catalysis leading to sensitive terpenic epoxides in excellent yields (Scheme 5.14).[107] The combination with the ionic liquid provided more efficient conditions compared with molecular solvents.

The same catalysts proved efficient yielding domino epoxidation–methanolysis of glycals by oxidation with urea hydrogen peroxide adduct and H_2O_2 in ionic liquids.[108] The facial diastereoselectivity of the oxidations were dependent on the substrate with the reactions performed with urea hydrogen peroxide adduct proceeded with a higher degree of diastereoselectivity than those performed with H_2O_2.

Although monomeric cyclopentadiene molybdenum oxides were among the first synthesized high oxidation state organometallics their applications were less investigated compared with their rhenium(VII) congeners.[109] However, there are several reports indicating good catalytic performance, especially in olefin epoxidation with *tert*-butylhydroperoxide,[110] which could be achieved with a variety of organomolybdenum complexes.[111] Investigations carried out in a series of imidazolium based ionic liquids immobilised (cyclopentadienyl)tricarbonyl molybdenum complexes showed good performances in epoxidation of cyclooctene. Complete recycling of the catalyst $CpMo(CO)_3Me$ was achieved in a mixture of $[C_4C_1im][NTf_2]$ and $[C_4C_1im][PF_6]$ in a volume ratio of 4:1. Increasing the amount of $[C_4C_1im][PF_6]$ increased the formation of diols. The supported catalysts combine properties of homogeneous catalysts (reactivity, control and selectivity) and heterogeneous catalysts (enhanced stability and reusability). Homogeneous catalysts have a clearly defined composition, and mechanistic examinations are usually less problematic than with bulk (heterogeneous) catalysts, where the active sites are less clearly defined. Peroxotungstate immobilised on dihydroimidazolium based ionic liquid-modified SiO_2 also behaved as an efficient heterogeneous epoxidation catalyst with

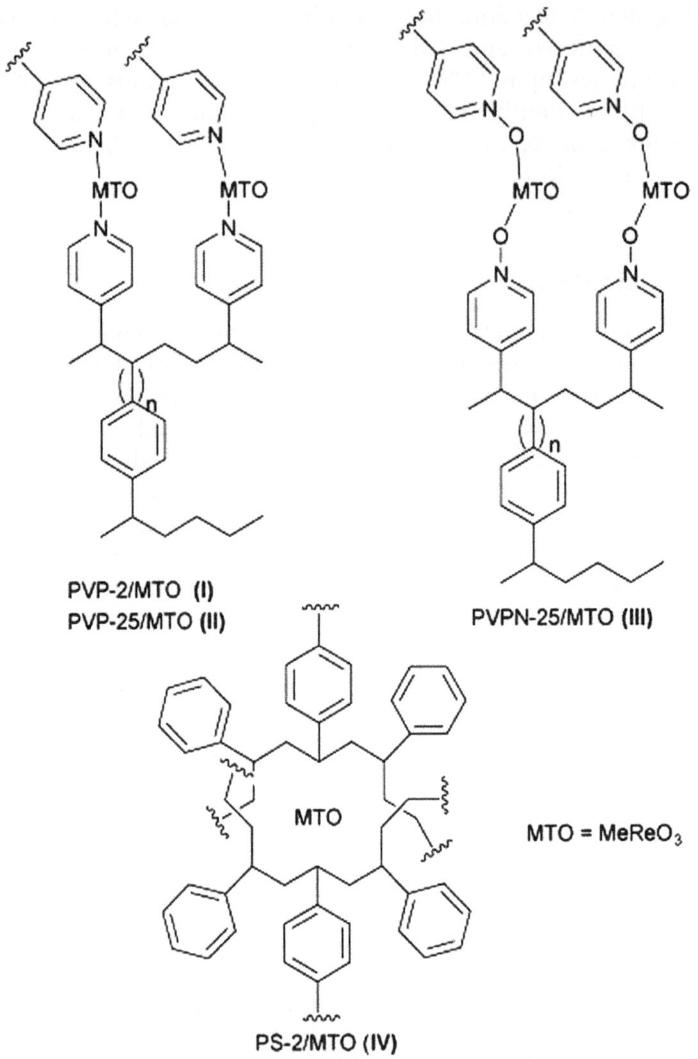

PVP-2/MTO (I)
PVP-25/MTO (II)

PVPN-25/MTO (III)

MTO = $MeReO_3$

PS-2/MTO (IV)

Scheme 5.14 Heterogeneous poly(4-vinylpyridine)–methyltrioxorhenium and polystyrene–ethyltrioxorhenium microcapsules employed in epoxidation.
Reproduced with permission from Elsevier. Taken from reference 108.

H_2O_2 for a large series of alkenes.[112] These were prepared by treating 1-octyl-3-(3-triethoxysililpropyl)-4,5-dihydroimidazolium chloride with sodium hexafluorophosphate in acetonitrile resulting in the desired ionic liquid that was further immobilised on SiO_2. Manganese(III) porphyrins immobilised in ionic liquids have also been successfully applied in olefin epoxidation.[113]

5.4.7.1 Asymmetric Epoxidation

Asymmetric epoxidation of unfunctionalised olefins was efficiently carried out with a supported ionic liquid system in which chiral Mn(III) salen complexes were immobilised.[114] These heterogeneous catalysts exhibited excellent activity and enantioselectivity, especially in the epoxidation of α-methylstyrene where both the conversion and ee exceeded 99%. The immobilised catalysts also exhibited very good stability.

5.5 Heterogeneous Carbon–Carbon Bond Formation Employing Ionic Liquids

5.5.1 Supported Heck and Similar Carbon–Carbon Bond Forming Reactions

The first heterogeneous catalysed Heck reaction was reported by Hagiwara *et al.* using a SCILL based Pd/C in $[C_4C_1im][PF_6]$.[115] Moreover, similar to homogeneous reactions previously conducted in ionic liquids the presence of a phosphine or similar ligand was not required. Utilising this SCILL catalyst high yields and excellent selectivities to *trans* cinnamates were obtained. Analysis of the IL phase post reaction revealed that only 0.31% of the Pd leached from the surface of the support indicating that the reaction was a true heterogeneous catalyst system. This enabled the SCILL catalyst to be recycled five times with yields averaging 88%, provided the $NHBu_3I$ by-product was removed *via* washing with water. This group also demonstrated that SILP-type catalysts using $Pd(OAc)_2$, $Pd(PPh_3)_4$ or Pd black supported in $[C_4C_1mim][PF_6]$ within silica pores were active.[116] SEM analysis and leaching experiments suggested that the IL was residing inside the pores of the silica *via* hydrogen bonding between the C_2-imidazolium protons and the surface silanol groups. Similar to previous results with the SCILL system, high conversions for a range of electron withdrawing and electron donating substrates were observed. TONs of 90 000 were observed for these Pd–SILP systems which showed up to 0.28% catalyst leaching from the support into the organic phase. Again water washing was required to remove the buildup of the alkylammonium iodide from the IL layer.

Uniform Pd-NPs (~5 nm) have been immobilised onto clays such as sepiolite using a guanidinium based IL.[117] The coupling of iodobenzene with methylacrylate in the presence of NEt_3 achieved quantitative conversions under solvent-free conditions. The linear biopolymer chitosan has also been reported to be an efficient support for the Pd-catalysed Heck reaction of aryl bromides and activated aryl chlorides in $[N_{4,4,4,4}]Br$ as solvent employing the analogous acetate as the base.[118] Rapid reaction times were observed due to the stabilisation of Pd colloids by the bromide and an efficient PdH neutralisation by the acetate base. In contrast, no reaction occurs when utilising imidazolium based ILs.

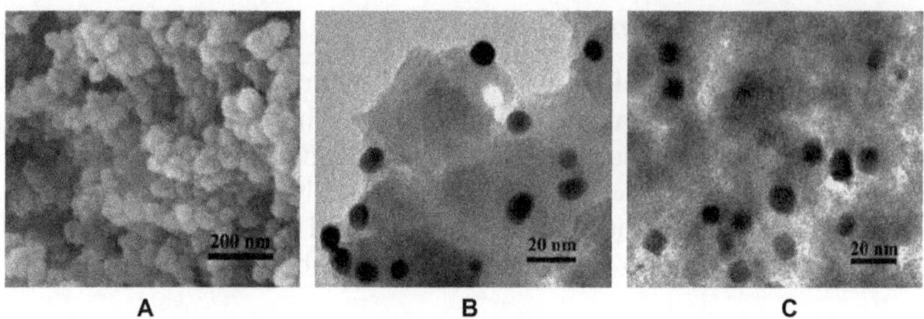

Figure 5.6 SEM image (left) and TEM image (centre) of PDVB–IL–Pd. PDVB–IL–Pd after being recycled three times (right).
Reproduced with permission from the Royal Society of Chemistry. Taken from reference 119.

A DVB cross-linked copolymer with chemically grafted amino tethered imidazolium IL has also been used as a support for Pd-NPs.[119] The NPs immobilised on the amorphous copolymers showed high catalytic activity for Heck coupling reactions between iodobenzenes and olefins. Coordination between the amine group and the Pd-NPs resulted in a stable system that could be easily separated from the products. Good recycle was achieved as the Pd-NP size was in the range 7–8 nm with a very narrow distribution which maintained uniformity post reaction, see Figure 5.6.

Heck coupling between iodobenzene and methylacrylate was also effectively catalysed by various moisture and air stable Pd–encapsulated IL polymers.[120] Mechanistic studies revealed the formation of a precatalyst reservoir which released Pd into the reacting medium. Post reaction the polymer was able to recapture the Pd to regenerate the reservoir needed for successive recycles. Moreover, the IL modified surface limited aggregation into large, non-active particles which also aided recycle. These catalysts have also been shown to efficiently catalyse Sonogashira and Suzuki reactions. IL–ionogels encapsulating $Pd(OAc)_2$ were successfully used as catalysts in Heck reactions with reaction rates comparable to homogeneous systems.[121] Leaching tests showed that catalysis actually took place in the IL phase confined within the silica matrix.

In order to produce a more practical Pd–SILP catalyst for recycle use in C–C bond forming reactions, the SILP catalyst was coated with polymer to improve chemical as well as mechanical stability. This was achieved by immersion of a pellet of Pd–SILP into the saturated solution of polystyrene (PS) in *n*-hexane or polyethylene terephthalate (PET) in hexafluoroisopropanol.[122] For the polymer-coated Pd–SILP or PET–Pd–SILP catalysts, aminopropylated silica gel was the most stable towards a variety of substrates and products. The thickness of the PET coat was evaluated to be *ca.* 20–60 μm by means of the SEM image. For the Suzuki–Mizuri reaction involving arylbromides and aryltriflates longer reactions times were required due to the polymer coating. Despite the longer reaction times PET–Pd–SILP could be reused up to ten times for a reaction between 4-bromoacetophenone and 2-methylphenylboronic acid. Leaching of palladium after the reaction was not observed at all by ICP-AES analysis. In the case of PS–Pd–SILP, the TON and TOF were 170 000 and 2400 h^{-1}, respectively.

Palladium acetate has been immobilised with $[C_4C_1im][PF_6]$ onto nanosilica gel cores modified with either polyamidoamine (PAMAM) and polyamidodimethylamine (PAMDMAM) forming Pd–nano–PAMAM–SILP; and Pd–nano–PAMDMAM–SILP, respectively. Both these materials formed free-flowing fine powders which showed excellent dispersion in a 50% aqueous ethanol without forming gels, Figure 5.7.[123] These were found to efficiently catalyse the Suzuki–Miyaura reaction at room temperature without additional liquids being necessary under aerobic conditions in the presence of potassium carbonate. A range of substituted aryl bromides and arylboronic acids which could not be catalysed by the Pd–SILP on amorphous silica gel pellets or showed low conversions with Pd–nano–PAMDMAM were found to be efficiently catalysed by Pd–nano–PAMDMAM–SILP. The corresponding Pd–nano–PAMAM–SILP; system also showed low catalytic

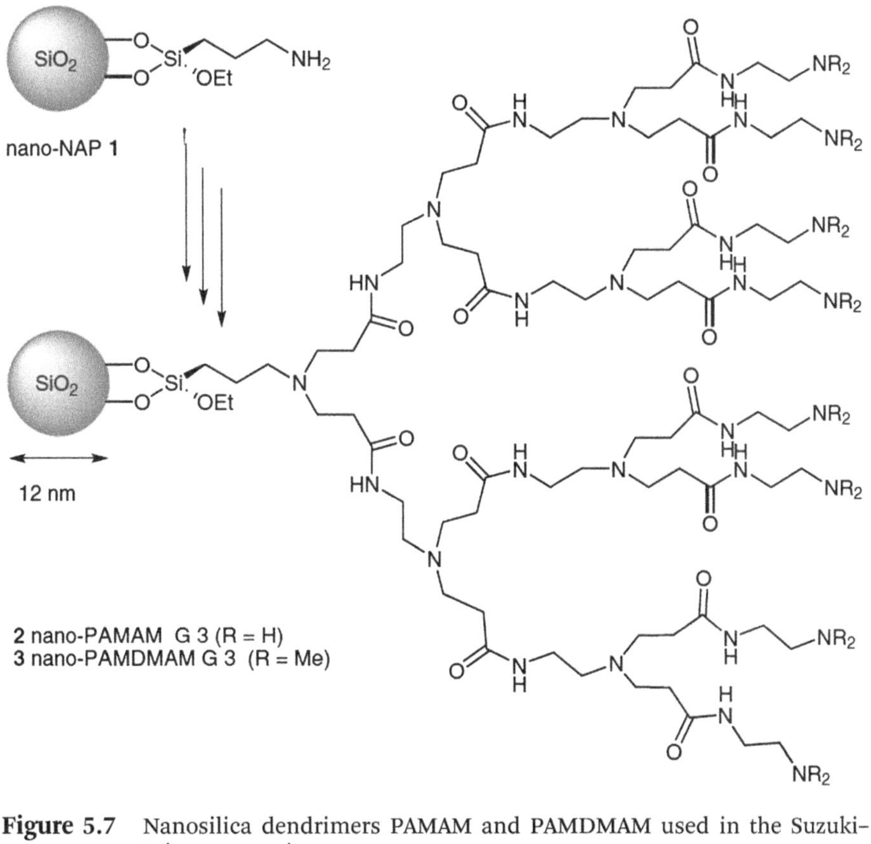

Figure 5.7 Nanosilica dendrimers PAMAM and PAMDMAM used in the Suzuki–Miyaura reaction.
Reproduced with permission from Thieme. Taken from reference 123.

activity. Transmission electron microscopy (TEM) showed that the Pd–nano-PAMDMAM–SILP forms a three-dimensional wormhole spherical cluster structure, in which exists palladium nanoparticles of 1–2 nm size. This structure was not observed with the PAMAM based material suggesting that the cluster structure might be responsible for the higher catalytic activity.

5.5.2 Heterogeneous Friedel–Crafts Reactions in Ionic Liquids

Following on from the early Friedel–Crafts reactions utilising $AlCl_3$ as the Lewis acid, dialkyl imidazolium or alkylpyridinium chloride–$AlCl_3$ melts or $AlCl_3$ dissolved in non-halide containing ILs have been extensively studied.[124] The ionic liquid chloride–$AlCl_3$ melts offered better control of the Lewis acidity than previously found in conventional systems. This is achievable through the formation of the $[Al_2Cl_7]^-$ or $[Al_3Cl_{10}]^-$ species which is dependent on the mole fraction of metal halide to chloride from the ionic liquid.[125] One major problem with these IL melts was the air and moisture sensitivity of the $AlCl_3$ and the problems associated with work up and recycle. In an attempt to overcome this issue, Holderich and co-workers supported $[C_nC_1im]Cl$–$AlCl_3$ melts onto a range of metal oxides.[126] The corresponding catalysts showed higher initial catalytic activity compared with the conventional H-β-zeolite under similar reaction conditions for the liquid phase alkylation of arenes with dodecene.[127]

The stability of the anion-immobilised chloroaluminates depended greatly on the nature of the surface. Low surface area supports such as TiO_2 and ZrO_2 with reduced hydroxyl groups compared to SiO_2 resulted in lower anion chemisorptions. Moreover, these supports showed rapid leaching of the IL melt during reaction resulting in lower catalytic activity. For other supports such as zeolites or MCM-41, XRD showed that partial destruction of the surface was observed due to the formation of HCl. The support structure was found to be less problematic if the IL-cation was chemisorbed onto the surface of the support prior to the addition of the metal halide.[128] Elimination of the surface silanol groups resulted in an ionic liquid supported chlorolauminate material which was successfully employed in the gas phase isopropylation of cumene and toluene.[129] Moreover, reactions with dry feedstocks for toluene alkylation resulted in over-equilibrium formation of the *meta*-cymene. This shift in thermodynamic equilibrium was explained by the preferential evaporation of *meta*-product from the support.

Chloroferrate ionic liquids have also been covalently immobilised *via* cation-tethering onto different silica and charcoal supports for the acylation of *m*-xylene.[130] Both batch and continuous liquid phase reactions resulted in lower conversions but a higher selectivity than homogeneously catalysed reactions. The lower conversion was believed to be due to strong leaching of the IL from the support. Gas phase reactions were also tested in an attempt to reduce leaching rates; however, low conversions of up to 5% were obtained. This was attributed to the formation of polyaromatic products which do not easily desorb from the catalyst support. Chloroferrate ILs have also been immobilised *via* the covalent grafting of a methylimidazolium chloride–$FeCl_3$ melt onto an MCM-41 silica.[131] Higher efficiency in the synthesis of diphenylmethane from benzylchloride was observed for the MCM-41–IL–Fe catalyst compared to the homogeneous $[C_4C_1im]Cl–FeCl_3$ system. A similar technique has been used to covalently incorporate $InCl_3$ into a mesoporous SBA-15 silica.[132] In comparison with SBA-15–physisorbed $InCl_3$, the covalently supported $InCl_3$ catalyst material showed better reusability for the reaction between benzene and benzyl chloride.

Zeolites in the presence of ILs have also been shown to catalyse Friedel–Crafts acylation.[133] Further insights into the reaction mechanism showed that, while initial activity in the ILs was higher than for molecular solvents, strong deactivation under continuous flow was observed.[134] Therein, the role of the IL anion was instrumental in determining the true catalyst species. The zeolite was simply acting as a catalyst precursor and undergoing cation exchange with the IL to form a homogeneous acid in solution. This acid was responsible for the main catalytic species. The activity of the zeolite could be recovered by calcining the solid, removing the exchanged organic cation and regenerating the proton sites.

5.5.3 Asymmetric C–C Bond Forming Reactions

5.5.3.1 Diels–Alder Reactions

Doherty *et al.* have reported that ILs can increase the ee and yield for the reaction between oxazolidinones and cyclopentadiene using platinum complexes of BINAP as well as conformationally flexible NUPHOS-type diphosphines, Figure 5.8.[135]

Using a $[C_2C_1im][NTf_2]$–Et_2O biphasic system, significant enhancements in the enantioselectivity as well as reaction rate were achieved compared with the organic media. This was attributed to the ILs limiting the extent of

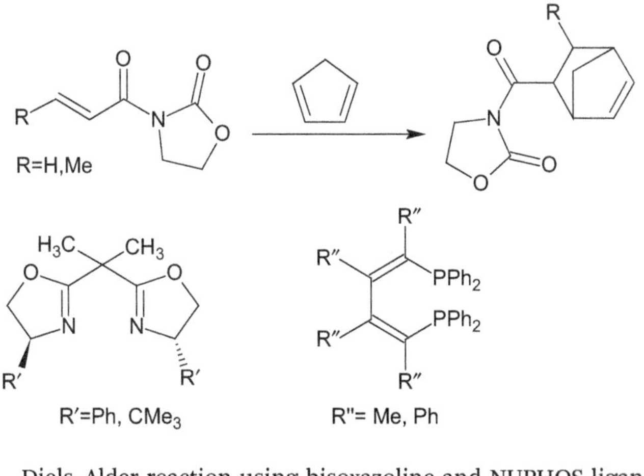

Figure 5.8 Diels–Alder reaction using bisoxazoline and NUPHOS ligands.

NUPHOS atropinversion observed in dichloromethane. In addition, the IL allowed the catalyst to be recycled in air without hydrolysis or oxidation of the phosphine ligand. Goodrich *et al.*[136,137] reported the optimisation and the comparison of homogeneous and SILP (*S*)-bis(oxazoline) complexes of Mg(II), Cu(II) and Zn(II) for the asymmetric Diels–Alder reaction between acryloyloxazolidinone (R=H) and cyclopentadiene. Under homogeneous conditions in $[C_2C_1im][NTf_2]$ all metal complexes showed accelerated rates of reaction with ees higher than reactions conducted in molecular solvents. The use of the corresponding SILP catalysts was also employed in the DA reaction using various silica (SiO_2, MCM-41, SBA-15) and carbon supports ((Activated Carbon (AC), Graphite (G), Carbon NanoTube (CNT)). While Cu and Mg catalysts showed slight enhancements in ee between SILP and homogeneous IL reactions the use of SILP Zn catalyst showed significant ee enhancement. For example, ee's up to 25–30% higher were observed for Zn catalyst supported on high surface area MCM-41 and low surface area G highlighting the complex support–catalyst interactions even in the presence of an IL.

5.5.3.2 Mukaiyama Aldol

Supported ionic liquid systems in the presence of a copper Lewis acid have also been successful for the asymmetric Mukaiyama aldol reaction between 1-phenyl-1-trimethylsiloxyethene and methyl pyruvate (Scheme 5.15).[138] As found with the Diels–Alder reactions, it was observed that complete conversions were achieved within 2 min at room temperature in all $[NTf_2]^-$ based ILs examined, whereas in CH_2Cl_2 only moderate to good conversions (55–90%) were achieved after 1 h. Lower chemoselectivity was observed due to the formation of a by-product resulting from the Mukaiyama aldol reaction of 1-phenyl-1-trimethylsiloxyethene and acetophenone. Formation of this by-product was suppressed, without any reduction in the enantioselectivity, by supporting the catalyst on SILP-type systems.

The heterogenised systems were less active than the homogeneous in ILs, but more active than the analogue in dichloromethane. Importantly, grafting the ionic liquid onto silica was not necessary for high conversions and ees, provided that a supporting ionic liquid film on the silica was present. Poor recycle of this catalyst was observed which was explained by leaching of the chiral ligand.

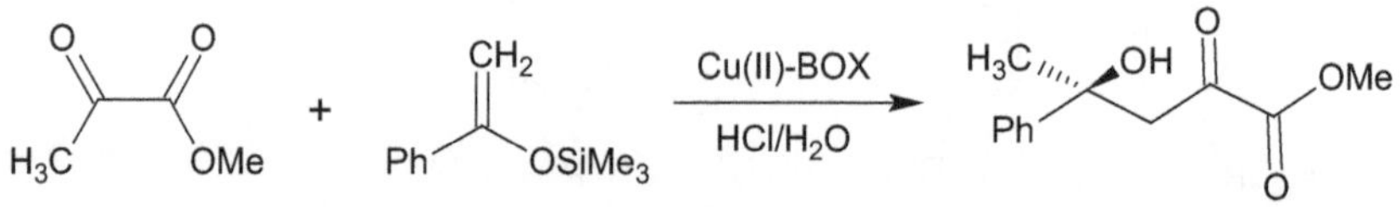

O O
H_3C OMe
+
CH_2
Ph $OSiMe_3$
Cu(II)-BOX
HCl/H_2O
H_3C OH O
Ph OMe
O

Scheme 5.15 The asymmetric Mukaiyama aldol condensation between methyl pyruvate and 1-phenyl-1-trimethylsiloxyethene.

Scheme 5.16 Enantioselective cyanosilylation of benzaldehyde. Adapted from reference 139.

5.5.3.3 *Cyanosilylation of Benzaldehyde*

As well as finding extensive deployment in oxidation reactions, immobilised chiral, metal–salen have also been used for the asymmetric cyanosilylation of benzaldehyde, Scheme 5.16.[139] The catalytic activity and the asymmetric induction ability of vanadyl–salen complexes were tested under liquid–liquid and solid–liquid systems. Unsurprisingly, it was found that the liquid–liquid system employing the IL-tagged salen complex dissolved in $[C_4C_1im][PF_6]$ was 5–6 times more active than the non-tagged salen complex dissolved in IL or those of the supported silica, single-wall carbon nanotube (SWNT) and activated carbon (AC) systems.

However, the enantioselectivity of the IL-tagged catalyst, 57% ee, was lower than the silica based system, 89% ee. A possible explanation for this reduced asymmetric induction ability of the catalyst is the interaction of the vanadyl group of the complex with the associated chloride anion present in the ionic liquid. However, this was never verified by exchanging chloride to a non-coordinating anion. Both the carbon supports also showed a significant decrease in ee (48–66%) compared with the silica system due to the ill defined structure and/or the presence of heteroatoms on the support.

5.5.4 Cyclopropanations

Chiral bis(oxazoline)–copper complexes have also been easily immobilised onto anionic solids or in ILs and used as catalysts in cyclopropanation reactions, Scheme 5.17.[140]

The use of 2,2′-isopropylidenebis[(4*S*)-4-*tert*-butyloxazoline] ($R = CMe_3$, Figure 5.8) as the chiral ligand resulted in significant differences between the homogeneous reaction in dichloromethane, the use of heterogeneous laponite support and the reaction in the IL. While little change in the *trans* : *cis* ratio was reported, the (1*R*) ee, efficiency and recovery of the catalysts are strongly dependent on the reaction system employed. For reactions where the chiral Cu(II)–bis(oxazolline) catalyst was supported on laponite enantiomeric excesses (69%) were not comparable to those obtained in the homogeneous reaction which showed an ee of 94%. The reason for this behaviour is the existence of an equilibrium between free and complexed copper leading to the formation of non-chiral catalytic centres (Figure 5.9). This was further evidenced by a further reduction in ee upon recycle.

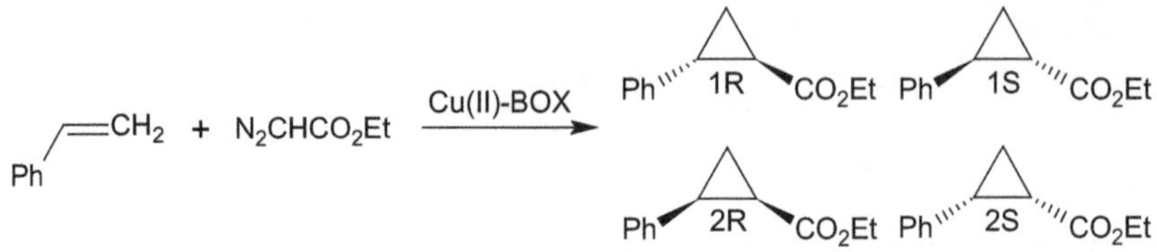

Cu(II)-BOX
=CH2 + N2CHCO2Et
Ph
Ph 1R CO2Et
Ph 1S CO2Et
Ph 2R CO2Et
Ph 2S CO2Et

Scheme 5.17 Cylopropanation reaction between styrene and ethyl diazoacetate. Reproduced with permission from The Royal Society of Chemistry. Taken from reference 140.

N N
Cu2+
anionic support
N N
Cu2+
anionic support
N N
BOX Ligand

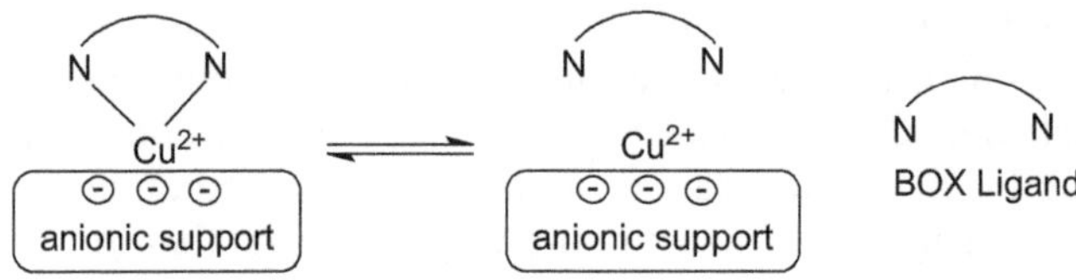

Figure 5.9 Equilibrium between the free and ligand complexed copper on an anionic support.

In contrast, for experiments conducted in a [C_2C_1im][OTf] no such support interactions were possible providing a more strongly complexed catalyst. This catalyst could be recycled up to five times with a slight reduction in ee (85% to 63%) due to ligand leaching into the organic extraction phase. Addition of a fresh portion of ligand on the sixth recycle resulted in initial catalytic activity and performance. The purity of the ionic liquids was found to have a marked effect on the catalysis. Therefore, particular care should be taken in the purification of hydrophilic ionic liquids prepared by metathetical anion exchange to ensure that all the dialkyl imidazolium halide is removed.[141]

5.6 Conversion of CO_2 into Valuable Compounds

5.6.1 Synthesis of Cyclic Carbonates

The use of heterogeneous catalysts for the synthesis of cyclic and dimethyl carbonates from CO_2 through various routes has recently elicited high interest[142] with the synthesis of cyclic carbonates *via* cycloaddition of CO_2 to epoxides being one of the few processes that has been commercialised. Among many effective catalysts, ILs and SILs have attracted attention. Covalent immobilisation of reactive-group-containing ILs on organic and inorganic supports with functional surfaces was achieved, based on the fact that the reactive group can actively react with different nucleophilic, electrophilic, neutral or free-radical species.[143] The typical supports included polymers with amino- and/or carboxyl group-functionalised surfaces and silica functionalised with amino groups attached as inorganic supports. In

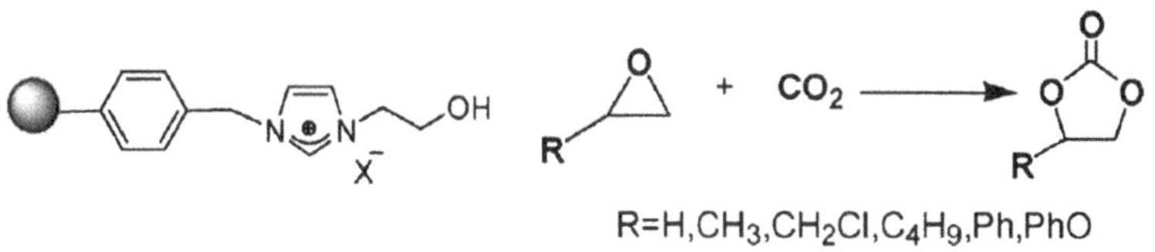

R=H,CH_3,CH_2Cl,C_4H_9,Ph,PhO

Scheme 5.18 Synthesis of cyclic carbonate from CO_2 and epoxide with ionic liquids grafted on highly cross-linked porous polymers.
Reproduced with permission from Elsevier. Taken from reference 146.

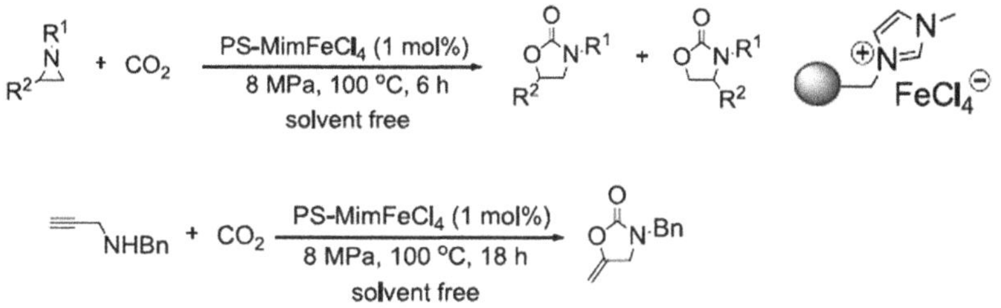

PS-Mim$FeCl_4$ (1 mol%), 8 MPa, 100 °C, 6 h, solvent free

PS-Mim$FeCl_4$ (1 mol%), 8 MPa, 100 °C, 18 h, solvent free

Scheme 5.19 PS–Mim$FeCl_4$-catalysed carboxylation of aziridines and propargyl amines using CO_2.
Reproduced with permission from Elsevier. Taken from reference 147.

particular, the polymer supports generated synergistic effects with the ionic liquid in the coupling reaction of CO_2 with epoxides.[144] Synthesis of cyclic carbonate from CO_2 and epoxide was achieved using both non-porous[145] and highly cross-linked porous poly(*N*-vinylimidazole-co-divinylbenzene) beads of various pore sizes modified with various alkyl halides to obtain various ionic liquid-grafted porous polymer beads, Scheme 5.18.[146]

Catalytic reactions carried out under mild conditions in the absence of organic solvents showed the effect of texture properties of the polymer matrix, molecular compositions of the ionic liquid moieties and catalytic reaction parameters (reaction temperature, pressure, and time). Grafted ILs with more nucleophilic anions, bulkier alkyl chains, or hydroxyl groups led to an enhanced reactivity. Additionally, the incorporation of a small amount of ethanol or water into the reaction system also significantly increased the reactivity.

Recently it has been found that a Lewis acidic $[FeCl_4]^-$ based IL supported on polystyrene was effective in converting CO_2 and aziridines or propargyl amines into cyclic carbonates under solvent free conditions at low temperature (Scheme 5.19).[147]

IL functionalised porous or mesoporous silicas were also reported as heterogeneous catalysts for the solventless synthesis of cyclic carbonate from epoxides or allyl glycidyl ether and carbon dioxide.[148] High temperature, high carbon dioxide pressure, and the presence of $ZnBr_2$ cocatalyst were favorable for the conversion of allyl glycidyl ether.

Multilayered, covalent SILP materials synthesized by the grafting of bis-vinylimidazolium salts on different types of silica or polymeric supports were tested as well as catalysts in the reaction of supercritical carbon dioxide with various epoxides to produce cyclic carbonates, Figure 5.10.[149] Among the different materials, supported bromide bis-imidazolium salt on the ordered mesoporous silica SBA-15 led to the most efficient catalyst. However, working under supercritical carbon dioxide conditions required rather energetic conditions.

Further improvements of the catalytic performances of these ionic liquid catalysts were achieved by incorporating various metal chlorides ($CoCl_2$, $NiCl_2$, $CuCl_2$, $ZnCl_2$, and $MnCl_2$) into silica-grafted 1-methyl-3-[(triethoxysilyl)propyl]imidazolium chloride.[150]

Importantly, the surface area and pore dimensions of the silica supports has been found to have a significant effect on the catalyst activity. These parameters effect the diffusion of the reactant and the amount of the IL immobilised on the support. For a series of imidazolium based ionic liquids

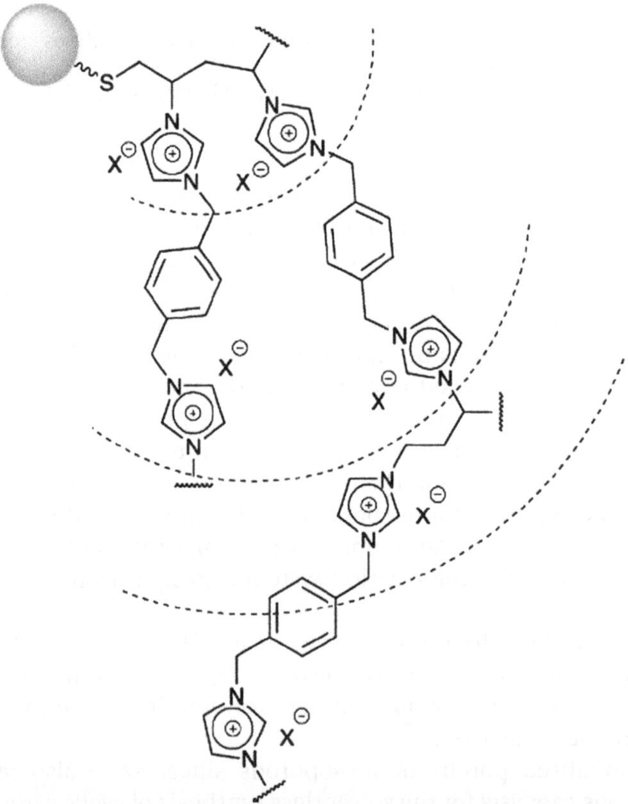

Figure 5.10 Multilayered, covalently supported ionic liquid catalysts. Reproduced with permission from John Wiley and Sons. Taken from reference 149.

with different alkyl chain lengths bearing different anions (Cl^-, Br^- and I^-), increasing the chain length and the nucleophilicity of the anion resulted in higher activity.[151]

5.6.2 Hydrogenation of CO_2 to Formic Acid

Hydrogenation of CO_2 to formic acid used ILs as solvents for heterogeneous catalysts. Coupling the basic ionic liquids 1-(*N*,*N*-dimethylaminoethyl)-2,3-dimethylimidazolium trifluoromethanesulfonate or 1,3-di(*N*,*N*-dimethylaminoethyl)-2-methylimidazolium trifluoromethanesulfonate and the heterogeneous catalyst "Si"$(CH_2)_3$ $(CSCH_3)$-{$RuCl_3(PPh_3)$} led to satisfactory activity and selectivity in this reaction. The addition of water and an increase in pressure and temperature led to an increase in the rate of reaction.[152]

5.7 Ionic Liquids in Valorization of Biomass

The valorisation of biomass for the synthesis of valuable chemicals and biofuels has become of great interest because of limitation in fossils resources and new environmental legislative regulations. The concept of biorefineries for the production of chemicals as well as materials and energy products is a key to ensuring a sustainable future for the chemical and allied industries. It integrates green chemistry principles and the use of low environmental impact technologies. The first step in these biorefineries is the benign extraction of surface chemicals and ionic liquids are expected to exhibit a high importance.[153] While hot water (244 °C) was claimed to convert cellulose into various alcohols, even without added acids,[154] chloride containing ILs were reported to also be promising reaction media where the ability of the IL to break the hydrogen bonds resulted in increased dissolution of the cellulose/lignin.[155]

Chemocatalytic depolymerisation of cellulose, hemicelluloses and lignins to simple building blocks is another very important step in this approach; however it is important to first dissolve the cellulose/lignin. In this regard the depolymerisation of these natural polymers can be carried out in an IL using a heterogeneous catalyst and hydrogen. For example, experiments carried out in $[C_4C_1im]Cl$ in the presence of hydrogen gas and a combination of heterogeneous (Pt/C or Rh/C) and homogeneous ($HRuCl(CO)(PPh_3)_3$) catalysts[156] led to complete conversion of cellulose where sorbitol was the dominant product in 51–74% yields. The role of the ruthenium compound has been suggested to enhance the transfer of hydrogen to the metallic surface.

Furthermore, the transformation of the molecular entities that have resulted from the depolymerisation of cellulose/lignin into valuable chemicals can also be achieved in ILs. The selective glucose dehydration to 5-hydroxymethylfurfural was reported using $CrCl_2$ catalysis in a thin immobilised ionic liquid layer (1-(triethoxysilyl)-propyl-3-methyl-imidazolium chloride on SBA-15.[157] The reactivity of $CrCl_2$ in $[C_2C_1im]Cl$ was assigned to the dynamic nature of this IL–catalytic system.[158] As a result these reactions can be achieved even in an aqueous medium.

Enzymes can also be used in this approach. Bioreactors with covalently tethered SILP systems were prepared as polymeric monoliths based on styrene–divinylbenzene or 2-hydroxyethyl methacrylate–ethylene dimethacrylate and with imidazolium units loadings. They were able to absorb *Candida antarctica* lipase B resulting in the formation of active and robust heterogeneous biocatalysts for the continuous flow synthesis[159] of citronellyl propionate in supercritical carbon dioxide by transesterification, Scheme 5.20.

Depolymerisation of cellulosic substrates dissolved in ILs was also reported using macroreticulated styrene–divinylbenzene resins functionalised with sulfonic groups (-SO_3H) like Amberlyst.[160] However, it is still not clear if in this case the catalyst is the leached sulfuric acid or the heterogeneous one.

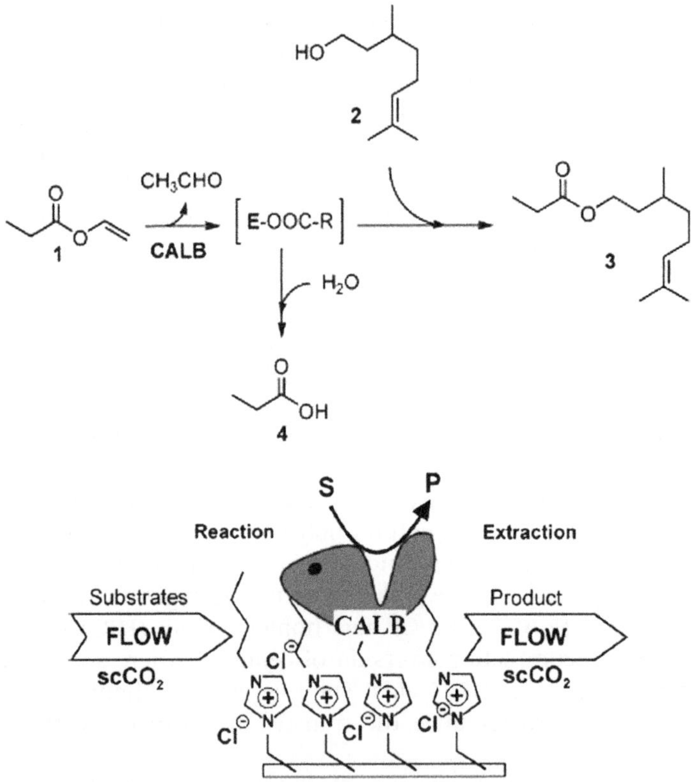

Scheme 5.20 CALB-catalysed citronellyl propionate (3) synthesis from vinyl propionate (1) and citronellol (2) by transesterification and the reactor with immobilised CALB on a monolith-supported ionic liquid phase for continuous operation under flow conditions in $scCO_2$.
Reproduced with permission from John Wiley and Sons. Taken from reference 159.

Conclusions

The use of ILs in the very diverse area of heterogeneous catalysis has shown, in general, that significant improvements may be obtained in reaction selectivity and recyclability over a wide range of reactions resulting in some systems being run under continuous flow operations. Overall, the deployment of an IL in the field of heterogeneous catalysis, regardless of its nature, has also resulted in a positive change in reaction activity. There are only limited examples where detrimental effects are observed when employing IL, due to mass transport limitations. However, these can be overcome with a better understanding of the IL, catalyst and support interaction, and the reactivity, stability and reaction pathways can be influenced to further improve chemical yield.

References

1. (*a*) D. Zhao, M. Wu, Y. Kou and E. Min, *Catal. Today*, 2002, **74**, 157; (*b*) P. Migowski and J. Dupont, *Chem. Eur. J.*, 2007, **13**, 32; (*c*) Y. Gu and G. Li, *Adv. Synth. Catal.*, 2009, **351**, 817.
2. C. Ch. Tzschucke, C. Markert, W. Bannwarth, S. Roller, A. Hebel and R. Haag, *Angew. Chem., Int. Ed.*, 2002, **41**, 3964.
3. T. Selvam, A. Machoke and W. Schwieger, *Appl. Catal., A*, 2012, **445–446**, 92–101.
4. (*a*) A. Rissager, R. Fehrmann, M. Haumann and P. Wasserscheid, *Top. Catal.*, 2006, **40**, 91; (*b*) H.-P. Steinrück, J. Libuda, P. Wasserscheid, T. Cremer, C. Kolbeck, M. Laurin, F. Maier, M. Sobota, P. S. Schulz and M. Stark, *Adv. Mater.*, 2011, **23**, 2571.
5. S. Miao, Z. Liu, B. Han, J. Huang, Z. Sun, J. Zhang and T. Jiang, *Angew. Chem., Int. Ed.*, 2006, **45**, 266.
6. R. Sebesta, I. Kmentov and S. Toma, *Green Chem.*, 2008, **10**, 484.
7. P. J. Dyson, T. Geldbach, F. Moro, C. Taeschler and D. Zhao, Hydrogenation Reactions in Ionic Liquids: Finding Solutions for Tomorrow's World, in Ionic Liquids IIIB: Fundamentals, Progress, Challenges and Opportunities: Transformations and Processes, *ACS Symp. Ser.*, 2005, **902**, 322.
8. P. J. Dyson, G. Laurenczy, C. A. Ohlin, J. Vallance and T. Welton, *Chem. Commun.*, 2003, 2418.
9. E. Bogel-Łukasik, S. Santos, R. Bogel-Łukasik and M. Nunes da Ponte, *J. Supercrit. Fluids*, 2010, **54**, 210.
10. C. Hardacre, E. A. Mullan, D. W. Rooney, J. M. Thompson and G. S. Yablonsky, *Chem. Eng. Sci.*, 2006, **61**, 6995.
11. T. Gutmann, M. Sellin, H. Breitzke, A. Stark and G. Buntkowsky, *Phys. Chem. Chem. Phys.*, 2009, **11**, 9170.
12. K. V Kovtunov, V. V. Zhivonitko, L. Kiwi-Minsker and I. V. Koptyug, *Chem. Commun.*, 2010, **46**, 5764.

13. L. A. Blanchard, Z. Gu and J. F. Brennecke, *J. Phys. Chem. B*, 2001, **105**, 2437.
14. S. G. Kazarian, B. J. Briscoe and T. Welton, *Chem. Commun.*, 2000, 2047.
15. P. G. Jessop, R. R. Stanley, R. A. Brown, C. A. Eckert, C. L. Liotta, T. T. Ngo and P. Pollet, *Green Chem.*, 2003, **5**, 123.
16. P. A. Z Suarez, J. E. L. Dullius, S. Einloft, R. F. de Souza and J. Dupont, *Inorg. Chim. Acta*, 1997, **255**, 207.
17. P. A. Z. Suarez, J. E. L. Dullis, S. Einloft, R. F. de Souza and J. Dupont, *Polyhedron*, 1996, **15**, 1217.
18. C. J. Boxwell, P. J. Dyson, D. J. Ellis and T. Welton, *J. Am. Chem. Soc.*, 2002, **124**, 9334.
19. S. Steines, P. Wasserscheid and B. Driessen-Hölscher, *J. Prakt. Chem.*, 2000, **342**, 348.
20. A. J. Carmichael and K. R. Seddon, *J. Phys. Org. Chem.*, 2000, **13**, 591.
21. S. N. V. K. Aki, J. F. Brennecke and A. Samanta, *Chem. Commun.*, 2001, 413.
22. L. Crowhurst, P. R. Mawdsley, J. M. Perez-Arlandis, P. A. Salter and T. Welton, *Phys. Chem. Chem. Phys.*, 2003, **5**, 2790.
23. D. Zhao, Z. Fei, R. Scopelliti and P. J. Dyson, *Inorg. Chem.*, 2004, **43**, 2197.
24. D. Zhao, P. J. Dyson, G. Laurenczy and J. S. McIndoe, *J. Mol. Catal., A*, 2004, **214**, 19.
25. P. J. Dyson, D. J. Ellis and T. Welton, *Can. J. Chem.*, 2001, **79**, 705.
26. (*a*) P. J. Dyson, *Dalton Trans.*, 2003, 2964; (*b*) T. J. Geldbach, G. Laurenczy, R. Scopelliti and P. J. Dyson, *Organometallics*, 2006, **25**, 733.
27. C. P. Mehnert, E. J. Mozeleski and R. A. Cook, *Chem. Commun.*, 2002, 3010.
28. M. Ruta, I. Yuranov, P. J. Dyson, G. Laurenczy and L. K-. Minsker, *J. Catal.*, 2007, **247**, 269.
29. M. Boutonnet, S. Lögdberg and E. Elm Svensson, *Curr. Opin. Colloid Interface Sci.*, 2008, **13**, 270.
30. D. Astruc, F. Lu and J. R. Aranzaes, *Angew. Chem., Int. Ed.*, 2005, **44**, 7852.
31. (*a*) P. J. Dyson, *Coord. Chem. Rev.*, 2004, **248**, 2443; (*b*) P. Migowski and J. Dupont, *Chem. Eur. J.*, 2007, **13**, 32; (*c*) Y. Zhu, C. Nong Lee, R. A. Kemp, N. S. Hosmane and J. A. Maguire, *Chem.-Asian J.*, 2008, **3**, 650.
32. A. Roucoux, J. Schulz and H. Patin, *Chem. Rev.*, 2002, **102**, 3757.
33. A.-V. Mudring, T. Alammar, T. Bäcker and K. Richter in *Ionic Liquids: From Knowledge to Application*, ed. N. V. Plechkova, R. D. Rogers and K. R. Seddon, American Chemical Society, Washington, 2009, pp. 177–188.
34. L. D. Pachón and G. Rothenberg, *Appl. Organomet. Chem.*, 2008, **22**, 6.
35. P. S. Campbell, C. C. Santini, D. Bouchu, B. Fenet, K. Philippot, B. Chaudret, A. A. H. Pádua and Y. Chauvin, *Phys. Chem. Chem. Phys.*, 2010, **12**, 4217.

36. D. Mukherjee, *J. Nanopart. Res.*, 2008, **10**, 429.
37. J. Huang, T. Jiang, B. X. Han, H. X. Gao, Y. H. Chang, G. Y. Zhao and W. Z. Wu, *Chem. Commun.*, 2003, 1654.
38. B. Léger, A. Denicourt-Nowicki, A. Roucoux and H. Olivier-Bourbigou, *Adv. Synth. Catal.*, 2008, **350**, 1.
39. C. Vollmer, R. Engelbert, K. Abu-Shandi, R. Thomann, H. Manyar, C. Hardacre and C. Janiak, *Chem.-Eur. J.*, 2010, **16**, 3849.
40. E. T. Silveira, A. P. Umpierre, L. M. Rossi, G. Machado, J. Morais, G. V. Soares, I. J. R. Baumvol, S. R. Teixeira, P. F. P. Fichtner and J. Dupont, *Chem.-Eur. J.*, 2004, **10**, 3734.
41. P. S. Campbell, C. C. Santini, F. Bayard, Y. Chauvin, V. Collière, A. Podgoršek, M. F. Costa Gomes and J. Sá, *J. Catal.*, 2010, **275**, 99.
42. X. Yuan, N. Yan, C. X. Xiao, C. N. Li, Z. F. Fei, Z. P. Cai, Y. Kou and P. J. Dyson, *Green Chem.*, 2010, **12**, 228.
43. Y. Hu, Y. Yu, Z. Hou, H. Yang, B. Feng, H. Li, Y. Qiao, X. Wang, L. Hua, Z. Pan and X. Zhao, *Chem.-Asian J.*, 2010, **5**, 1178.
44. G. S. Fonseca, A. P. Umpierre, P. F. P. Fichtner, S. R. Teixeira and J. Dupont, *Chem.-Eur. J.*, 2003, **9**, 3263.
45. C. W. Scheeren, G. Machado, J. Dupont, P. F. P. Fichtner and S. Ribeiro Texeira, *Inorg. Chem.*, 2003, **42**, 4738.
46. D.-Q. Xu, Z.-Y. Hu, W.-W. Li, S.-P. Luo and Z. Y. Xu, *J. Mol. Catal. A: Chem.*, 2005, **235**, 137.
47. K. Anderson, P. Goodrich, C. Hardacre and D. Rooney, *Green Chem.*, 2004, **5**, 448.
48. J.-P. Mikkola, P. Virtanen, H. Karhu, T. Salmi and D. Y. Murzin, *Green Chem.*, 2006, **8**, 197.
49. P. Virtanen, J.-P. Mikkola and T. Salmi, *Ind. Eng. Chem. Res.*, 2007, **46**, 9022.
50. J.-P. T. Mikkola, P. P. Virtanen, K. Kordas, H. Karhu and T. O. Salmi, *Appl. Catal., A*, 2007, **328**, 68.
51. Y. Kume, K. Qiao, D. Tomida and C. Yokoyama, *Catal. Commun.*, 2008, **9**, 369.
52. R. Abu-Reziq, D. Wang, M. Post and H. Alper, *Adv. Synth. Catal.*, 2007, **349**, 2145.
53. M. Crespo-Quesada, R. R. Dykeman, G. Laurenczy, P. J. Dyson and L. Kiwi-Minsker, *J. Catal.*, 2011, **279**, 66.
54. (*a*) H. Huang, T. Jiang, H. Gao, B. Han, Z. Liu, W. Wu, Y. Chang and G. Zhao, *Angew. Chem., Int. Ed.*, 2004, **43**, 1397; (*b*) S. Miao, Z. Liu, B. Han, J. Huang, Z. Sun, J. Zhang and T. Jiang, *Angew. Chem., Int. Ed.*, 2006, **45**, 266.
55. J. Arras, M. Steffan, Y. Shayeghi and P. Claus, *Chem. Commun.*, 2008, 4058.
56. R. Knapp, A. Jentys and J. A. Lercher, *Green Chem.*, 2009, **11**, 656.
57. M. Sobota, M. Schmid, M. Happel, M. Amende, F. Maier, H.-P. Steinruck, N. Paape, P. Wasserscheid, M. Laurin, J. M. Gottfried and J. Libuda, *Phys. Chem. Chem. Phys.*, 2010, **12**, 10610.

58. U. Kernchen, B. Etzold, W. Korth and A. Jess, *Chem. Eng. Technol.*, 2007, **30**, 985.
59. M. Ruta, G. Laurnczy, P. J. Dyson and L. Kiwi-Minsker, *J. Phys. Chem. C*, 2008, **112**, 17814.
60. A. Wolfson, J. F. J. Vankelecom and P. A. Jacobs, *Tetrahedron Lett.*, 2003, **44**, 1195.
61. X. Feng, B. Pugin, E. Kusters, G. Sedelmeier and H.-U. Blaser, *Adv. Synth. Catal.*, 2007, **349**, 1803.
62. C. Sievers, O. Jimenez, T. E. Müller, S. Steuernagel and J. A. Lercher, *J. Am. Chem. Soc.*, 2006, **128**, 13990.
63. L.-L. Lou, X. Peng, K. Yu and S. Liu, *Catal. Commun.*, 2008, **9**, 1891.
64. L.-L. Lou, Y. Dong, K. Yu, S. Jiang, Y. Song, S. Cao and S. Lui, *J. Mol. Catal. A: Chem.*, 2010, **333**, 20.
65. I. Podolean, C. Hardacre, P. Goodrich, N. Brun, R. Backov, S. M. Coman and V. I. Parvulescu, *Catal. Today*, 2013, **200**, 63.
66. E. Öchsner, M. J. Schneider, C. Meyer, M. Haumann and P. Wasserscheid, *Appl. Catal., A*, 2011, **399**, 35.
67. S. Keskin, D. Kayrak-Talay, U. Akman and O. Hortacsu, *J. Supercrit. Fluids*, 2007, **43**, 150.
68. U. Hintermair, T. Höfener, T. Pullmann, G. Franciò and W. Leitner, *ChemCatChem*, 2010, **2**, 150.
69. S. Shi, X. Wang, C. A. Sandoval, Z. Wang, H. Li, J. Wu, L. Yu and K. Ding, *Chem. Eur. J.*, 2009, **15**, 9855.
70. E. Bogel-Łukasik, S. Santos, R. Bogel-Łukasik and M. Nunes da Ponte, *J. Supercrit. Fluids*, 2010, **54**, 210.
71. S. S. Zinovyev, N. A. Shinkova, A. Perosa and P. Tundo, *Appl. Catal., B*, 2005, **55**, 39.
72. Z. Nairoukh and J. Blum, *J. Mol. Catal. A: Chem.*, 2012, **358**, 129.
73. A. Riisager, P. Wasserscheid, R. van Hal and R. Fehrmann, *J. Catal.*, 2003, **219**, 252.
74. A. Riisager, R. Fehrmann, S. Flicker, R. van Hal, M. Haumann and P. Wasserscheid, *Angew. Chem., Int. Ed.*, 2005, **44**, 815.
75. M. Haumann, M. Jakuttis, R. Franke, A. Schçnweiz and P. Wasserscheid, *ChemCatChem*, 2011, **3**, 1822.
76. H. Maciejewski, A. Wawrzynczak, M. Dutkiewicz and R. Fiedorow, *J. Mol. Catal. A: Chem.*, 2006, **257**, 141.
77. (*a*) V. Neff, T. E. Muller and J. A. Lercher, *Chem. Commun.*, 2002, 906; (*b*) J. Bódis, T. E. Müller and J. A. Lercher, *Green Chem.*, 2003, **5**, 227; (*c*) S. Breitenlechner, M. Fleck, T. E. Müller and A. Suppan, *J. Mol. Catal. A: Chem.*, 2004, **214**, 175.
78. J. Penzien, C. Haeßner, A. Jentys, K. Kohler, T. E. Muller and J. A. Lercher, *J. Catal.*, 2004, **221**, 302.
79. (*a*) O. Jimenez, T. E. Muller, C. Sievers, A. Spirkl and J. A. Lercher, *Chem. Commun.*, 2006, 2974; (*b*) C. Sievers, O. Jimenez, R. Knapp, X. Lin, T. E. Muller, A. Turler, B. Wierczinski and J. A. Lercher, *J. Mol. Catal. A: Chem.*, 2008, **279**, 187.

80. A. A. M. Lapis, B. A. Da Silveira Neto, J. D. Scholten, F. M. Nachtigall, M. N. Eberlin and J. Dupont, *Tetrahedron Lett.*, 2006, **47**, 6775.
81. J.-J. Brunet, N. C. Chu, O. Diallo and S. Vincendeau, *J. Mol. Catal. A: Chem.*, 2005, **240**, 245.
82. Z. S. Qureshi, K. M. Deshmukh, P. J. Tambade, K. P. Dhake and B. M. Bhanage, *Eur. J. Org. Chem.*, 2010, 6233.
83. Y. Suttisawat, S. Horikoshi, H. Sakai, P. Rangsunvigit and M. Abe, *Fuel Process Technol.*, 2012, **95**, 27.
84. (*a*) A. Corma and H. Garcia, *Top. Catal.*, 2008, **48**, 8; (*b*) L. Protesescu, M. Tudorache, S. Neatu, M. Nicoleta, E. Kemnitz, P. Filip, V. I. Parvulescu and S. M. Coman, *J. Phys. Chem. C*, 2011, **115**, 1112.
85. J.-Y. Wang, F.-Y. Zhao, R.-J. Liu and Y.-Q. Hu, *J. Mol. Catal. A: Chem.*, 2008, **279**, 153.
86. G. Bianchini, M. Crucianelli, F. De Angelis, V. Neri and R. Saladino, *Tetrahedron Lett.*, 2005, **46**, 2427.
87. D. S. Gopala, R. R. Bhattacharjee, R. Haerr, B. Yeginoglu, O. D. Pavel, B. Cojocaru, V. I. Parvulescu and R. M. Richards, *ChemCatChem*, 2011, **3**, 408.
88. Y. Leng, J. Wang, D. Zhu, L. Shen, P. Zhao and M. Zhang, *Chem. Eng. J*, 2011, **173**, 620.
89. G. Liu, M. Hou, J. Song, Z. Zhang, T. Wu and B. Han, *J. Mol. Catal. A: Chem.*, 2010, **316**, 90.
90. B. Gadenne, P. Hesemann and J. J. E. Moreau, *Chem. Commun.*, 2004, 1768.
91. R. Ciriminna, P. Hesemann, J. J. E. Moreau, M. Carraro, S. Campestrini and M. Pagliaro, *Chem. Eur. J.*, 2006, **12**, 5220.
92. L. Liu, D. Liu, Z. Xia, J. Gao, T. Zhang, J. Ma, D. Zhang and Z. Tong, *Monatsh. Chem.*, 2013, **144**, 251.
93. X. Lang, Z. Li and C. Xia, *Synth. Commun.*, 2008, **38**, 1610.
94. C. Hardacre, E. A. Mullan, D. W. Rooney and J. M. Thompson, *J. Catal.*, 2005, **232**, 355.
95. Y. Leng, P. Zhao, M. Zhang and J. Wang, *J. Mol. Catal. A: Chem.*, 2012, **358**, 67.
96. R. Ciriminna, P. Hesemann, J. J. E. Moreau, M. Carraro, S. Campestrini and M. Pagliaro, *Chem. Eur. J.*, 2006, **12**, 5220.
97. M. Lakshmi Kantam, U. Pal, B. Sreedhar, S. Bhargava, Y. Iwasawa, M. Tada and B. M. Choudary, *Adv. Synth. Catal.*, 2008, **350**, 1225.
98. B. Karimi, D. Elhamifar, J. H. Clark and A. J. Hunt, *Org. Biomol. Chem.*, 2011, **9**, 7420.
99. C. Van Doorslaer, Y. Schellekens, P. Mertens, K. Binnemans and D. De Vos, *Phys. Chem. Chem. Phys.*, 2010, **12**, 1741.
100. S. Sahoo, P. Kumar, F. Lefebvre and S. B. Halligudi, *Appl. Catal., A*, 2009, **354**, 17.
101. A. Chrobok, S. Baja, W. Pudło and A. Jarzebski, *Appl. Catal., A*, 2009, **366**, 22.

102. V. Cimpeanu, V. I. Parvulescu, P. Amoros, D. Beltran, J. M. Thompson and C. Hardacre, *Chem. Eur. J.*, 2004, **10**, 4640.
103. X. Shia, X. Hana, W. Ma, J. Wei, J. Li, Q. Zhang and Z. Chen, *J. Mol. Catal. A: Chem.*, 2011, **341**, 57.
104. S. Sahoo, P. Kumar, F. Lefebvre and S. B. Halligudi, *J. Catal.*, 2009, **262**, 111.
105. P. Adão, F. Avecilla, M. Bonchio, M. Carraro, J. C. Pessoa and I. Correia, *Eur. J. Inorg. Chem.*, 2010, 5568.
106. (*a*) I. R. Beattie and P. Jones, *J. Inorg. Chem.*, 1979, **18**, 2318; (*b*) W. A. Herrmann, R. W. Fischer and D. W. Marz, *Angew. Chem., Int. Ed. Engl.*, 1991, **30**, 1638; (*c*) P. Huston, J. H. Espenson and A. Bakac, *Inorg. Chem.*, 1993, **32**, 4517; (*d*) K. N. Brown and J. H. Espenson, *Inorg. Chem.*, 1996, **35**, 7211; (*e*) W. Adam and C. Mitchell, *Angew. Chem., Int. Ed. Engl.*, 1996, **35**, 533; (*f*) C. C. Romao, F. E. Kuhn and W. A. Herrmann, *Chem. Rev.*, 1997, **97**, 3197; (*g*) W. Adam, C. M. Mitchell, C. R. Saha-Moller and O. Werchold, *J. Am. Chem. Soc.*, 1991, **121**, 2097; (*h*) C. S. Owens, J. Arias and M. M. Abu-Omar, *Catal. Today*, 2000, **55**, 317.
107. R. Saladino, R. Bernini, V. Neri and C. Crestini, *Appl. Catal., A*, 2009, **360**, 171.
108. R. Saladino, C. Crestini, M. Crucianelli, G. Soldaini, F. Cardona and A. Goti, *J. Mol. Catal. A: Chem.*, 2008, **284**, 108.
109. Ch. Freund, M. Abrantes and F. E. Kuhn, *J. Organomet. Chem.*, 2006, **691**, 3718.
110. F. E. Kuhn, J. Zhao, M. Abrantes, W. Sun, C. A. M. Afonso, L. C. Branco, I. S. Goncalves, M. Pillinger and C. C. Romao, *Tetrahedron Lett.*, 2005, **46**, 47.
111. C. Freund, W. Herrmann and F. E. Kühn, *Top. Organomet. Chem.*, 2007, **22**, 39.
112. K. Yamaguchi, C. Yoshida, S. Uchida and N. Mizuno, *J. Am. Chem. Soc.*, 2005, **127**, 530.
113. (*a*) Z. Li and C.-G. Xia, *Tetrahedron Lett.*, 2003, **44**, 2069; (*b*) Z. Li, C.-G. Xia and M. Ji, *Appl. Catal., A*, 2003, **252**, 17.
114. L. L. Lou, K. Yu, F. Ding, W. Zhou, X. Peng and S. Li, *Tetrahedron Lett.*, 2006, **47**, 6513.
115. H. Hagiwara, Y. Shimizu, T. Hoshi, T. Suzuki and M. Ando, *Tetrahedron Lett.*, 2001, **42**, 4349.
116. H. Hagiwara, Y. Sugawara, K. Isobe, T. Hoshi and T. Suzuki, *Org. Lett.*, 2004, **6**, 2325.
117. R. T. Tao, S. D. Miao, Z. M. Liu, Y. Xie, B. X. Han, G. M. An and K. L. Ding, *Green Chem.*, 2009, **11**, 96.
118. D. B. Zhao, Z. F. Fei, T. J. Geldbach, R. Scopelliti and P. J. Dyson, *J. Am. Chem. Soc.*, 2004, **126**, 15876.
119. G. Liu, M. Hou, J. Song, T. Jiang, H. Fan, Z. Zhang and B. Han, *Green Chem.*, 2010, **12**, 65.
120. M. I. Burguete, E. García-Verdugo, I. Garcia-Villar, F. Gelat, P. Licence, S. V. Luis and V. Sans, *J. Catal.*, 2010, **269**, 150.

121. S. Volland, M. Gruit, T. Régnier, L. Viau, O. Lavastre and A. Vioux, *New J. Chem.*, 2009, **33**, 2015.
122. H. Hagiwara, K. Sato, T. Hoshi and T. Suzuki, *Synlett*, 2011, **17**, 2545.
123. H. Hagiwara, H. Sasaki, N. Tsubokawa, T. Hoshi, T. Suzuki, T. Tsuda and S. Kuwabata, *Synlett*, 2010, **13**, 1990.
124. J. A. Boon, J. A. Levinsky, J. I. Pflug and J. S. Wilkes, *J. Org. Chem.*, 1986, **51**, 480.
125. L. Hussey, T. B. Scheffler, J. S. Wilkes and A. A. Fannin, Jr., *J. Electrochem. Soc.*, 1986, **133**, 1389.
126. C. de Castro, E. Sauvage, M. H. Valkenberg and W. F. Holderich, *J. Catal.*, 2000, **196**, 86.
127. M. H. Valkenberg, C. de Castro and W. F. Holderich, *Top. Catal.*, 2001, **14**, 1.
128. M. H. Valkenberg, C. de Castro and W. F. Holderich, *Green Chem.*, 2002, **4**, 88.
129. J. Joni, M. Haumann and P. Wasserscheid, *Appl. Catal., A*, 2010, **372**, 8.
130. M. H. Valkenberg, C. de Castro and W. F Hölderich, *Appl. Catal., A*, 2001, **215**, 185.
131. G. Wang, N. Yu, L. Peng, R. Tan, H. Zhao, D. Yin, H. Qiu, Z. Fu and D. Yin, *Catal. Lett.*, 2008, **123**, 252.
132. H. Zhao, N. Yu, J. Wang, D. Zhuang, Y. Ding, R. Tan and D. Yin, *Microporous Mesoporous Mater.*, 2009, **122**, 240.
133. S. Katdare, J. M. Thompson, C. Hardacre and D. Rooney, 2002, WO 03028882.
134. C. Hardacre, S. P. Katdare, D. Milroy, P. Nancarrow, D. W. Rooney and J. M. Thompson, *J. Catal.*, 2004, **227**, 44.
135. S. Doherty, P. Goodrich, C. Hardacre, H.-K. Luo, D. W. Rooney, K. R. Seddon and P. Styring, *Green Chem.*, 2004, **6**, 63.
136. P. Goodrich, C. Hardacre, C. Paun, V. I. Parvulescu and I. Podolean, *Adv. Synth. Catal.*, 2008, **350**, 2473.
137. P. Goodrich, C. Hardacre, C. Paun, A. Ribeiro, S. Kennedy, M. J. V. Lourenço, H. Manyar, C. A. Nieto de Castro, M. Besnea and V. I. Pårvulescu, *Adv. Synth. Catal.*, 2011, **353**, 995.
138. S. Doherty, P. Goodrich, C. Hardacre, V. I. Parvulescu and C. Paun, *Adv. Synth. Catal.*, 2008, **350**, 302.
139. C. Baleizão, B. Gigante, H. Garcia and A. Corma, *Tetrahedron*, 2004, **60**, 10461.
140. J. M. Fraile, J. I. García, C. I. Herrerías, J. A. Mayoral, S. Gmough and M. Vaultier, *Green Chem.*, 2004, **6**, 93.
141. D. L. Davies, S. K. Kandola and R. K. Patel, *Tetrahedron: Asymmetry*, 2004, **15**, 77.
142. W. L. Dai, S. L. Luo, S. F. Yin and C. T. Au, *Appl. Catal., A*, 2009, **366**, 2.
143. Y. Xie, K. Ding, Z. Liu, J. Li, G. An, R. Tao, Z. Sun and Z. Yang, *Chem. Eur. J.*, 2010, **16**, 6687.

144. (*a*) Y. Xie, Z. Zhang, T. Jiang, J. He, B. Han, T. Wu and K. Ding, *Angew. Chem., Int. Ed.*, 2007, **46**, 7255; (*b*) Y. Xie, K. Ding, Z. Liu, J. Li, G. An, R. Tao, Z. Sun and Z. Yang, *Chem. Eur. J.*, 2010, **16**, 6687.
145. J. Sun, W. Cheng, W. Fan, Y. Wang, Z. Meng and S. Zhang, *Catal. Today*, 2009, **148**, 361.
146. L. Han, H.-J. Choi, D.-K. Kim, S.-W. Park, B. Liu and D.-W. Park, *J. Mol. Catal. A: Chem.*, 2011, **338**, 58.
147. J. Gao, Q. W. Song, L. N. He, C. Liu, Z.-Z. Yang, X. Han, X.-D. Li and Q.-C. Song, *Tetrahedron*, 2012, **68**, 3835.
148. (*a*) H.-L. Shim, S. Udayakumar, J.-I. Yu, I. Kim and D.-W. Park, *Catal. Today*, 2009, **148**, 350; (*b*) X. Zhang, D. Wang, N. Zhao, A. S. N. Al-Arifi, T. Aouak, Z. A. Al-Othman, W. Wei and Y. Sun, *Catal. Commun.*, 2009, **11**, 43.
149. C. Aprile, F. Giacalone, P. Agrigento, L. F. Liotta, J. A. Martens, P. P. Pescarmona and M. Gruttadauria, *ChemSusChem*, 2011, **4**, 1830.
150. L. Han, M.-S. Park, S.-J. Choi, Y.-J. Kim, S.-M. Lee and D. W. Park, *Catal. Lett.*, 2012, **142**, 259.
151. L. Han, S.-W. Park and D. W. Park, *Energy Environ. Sci.*, 2009, **2**, 1286.
152. (*a*) Z. Zhang, Y. Xie, W. Li, S. Hu, J. Song, T. Jiang and B. Han, *Angew. Chem., Int. Ed.*, 2008, **47**, 1127; (*b*) Z. Zhang, S. Hu, J. Song, W. Li, G. Yang and B. Han, *Chem. Sus. Chem.*, 2009, **2**, 234.
153. J. H. Clark, F. E. I. Deswarte and T. J. Farmer, *Biofuels, Bioprod. Biorefin.*, 2009, **3**, 72.
154. C. Luo, S. Wang and H. Liu, *Angew. Chem., Int. Ed.*, 2007, **46**, 7636.
155. (*a*) R. P. Swatloski, R. Rogers and J. D. Holbrey, WO03029329, 2003; (*b*) R. P. Swatloski, S. K. Spear, J. D. Holbrey and R. D. Rogers, *J. Am. Chem. Soc.*, 2002, **124**, 4974; (*c*) D. A. Fort, R. C. Remsing, R. P. Swatloski, P. Moyna, G. Moyna and R. D. Rogers, *Green Chem.*, 2007, **9**, 63; (*d*) R. Rinaldi and F. Schuth, *Energy Environ. Sci.*, 2009, **2**, 610.
156. I. A. Ignatyev, C. Van Doorslaer, P. G. N. Mertens, K. Binnemans and D. E. De Vos, *ChemSusChem*, 2010, **3**, 91.
157. V. Degirmenci, E. A. Pidko, P. C. M. M. Magusin and E. J. M. Hensen, *ChemCatChem*, 2011, **3**, 969.
158. E. A. Pidko, V. Degirmenci, R. A. van Santen and E. J. M. Hensen, *Angew. Chem., Int. Ed.*, 2010, **49**, 2530.
159. P. Lozano, E. Garcoa-Verdugo, R. Piamtongkam, N. Karbass, T. De Diego, M. I. Burguete, S. V. Luis and J. L. Iborra, *Adv. Synth. Catal.*, 2007, **349**, 1077.
160. R. Rinaldi, R. Palkovits and F. Schuth, *Angew. Chem., Int. Ed.*, 2008, **47**, 8047.

CHAPTER 6

Modification of Supports and Heterogeneous Catalysts by Ionic Liquids: SILP and SCILL Systems

PETER CLAUS* AND FREDERICK SCHWAB

Technische Chemie II, Ernst-Berl-Institut/Fachbereich Chemie,
TU Darmstadt. Alarich-Weiss-Str. 8, D-64287 Darmstadt
*Email: claus@ct.chemie.tu-darmstadt.de

6.1 Introduction

The improvement of catalysts is one of the most important challenges to create environmentally benign processes and to develop a greener chemistry.[1] Despite their benefits heterogeneous as well as homogeneous catalysts have some drawbacks. Homogeneous catalysts exhibit excellent selectivities and yields but the separation of the products and the recovery of the catalyst often prevent the use of these in industrial scale applications. The disadvantages of heterogeneous catalysis are mass and heat transfer limitations and often a loss of activity. Many efforts have been made to combine the advantages of both homogeneous and heterogeneous catalysis.[2]

One possible way is the usage of ionic liquids (ILs) as additive or solvent. Liquid–liquid biphasic catalysis with ionic liquids as solvent is an approach to combine the benefits of both types of catalysis. The advantages of this

RSC Catalysis Series No. 15
Catalysis in Ionic Liquids: From Catalyst Synthesis to Application
Edited by Chris Hardacre and Vasile Parvulescu

Published by the Royal Society of Chemistry, www.rsc.org

biphasic catalysis are high stability of metal complex, high selectivities and good recovery of the catalyst. The drawbacks of reactions with ionic liquids as bulk solvents are the costs of the ILs and mass transfer limitations caused by the viscosity of the IL.[3]

The modification of catalysts with ionic liquids is a further development of combining both homogeneous and heterogeneous catalysis. There are three different concepts for modification of catalysts with ionic liquids: SILP catalysis, the SCILL concept and nanoparticles in ILs.

Based on homogeneous catalysis the Supported Ionic Liquid Phase (SILP) catalysis was introduced by Mehnert and co-workers[4] and Wasserscheid and co-workers[5] in 2002 and 2003, respectively. A metal complex is dissolved in an ionic liquid which is immobilised on an inert support material. Here, the benefits of homogeneous catalysts (high activity and selectivity) are combined with long-time stability and easy separation of product and catalyst. By reducing the amount of IL to a minimum the costs are decreased dramatically and no mass transfer limitation can be observed. Because of the adjustable properties of the ionic liquids like polarity and solubility, by combining cations and anions SILP catalysis has been used in various reactions, *e.g.* organic syntheses like Heck and Suzuki reactions,[6] hydrogenation and hydroamination of alkenes[7] and hydroformylation of alkenes.[8]

Another concept is the Solid Catalyst with an Ionic Liquid Layer (SCILL). The catalyst system consists of a heterogeneous catalyst which is coated by a thin layer of an ionic liquid. The presence of the IL changes both the chemical properties of the active metal and the diffusion of the reactants. Kobayashi and co-workers[9] and Jess and co-workers[10] investigated this catalyst system first in organic reactions and in the selective hydrogenation of 1,5-cyclooctadiene, respectively. The main benefits of the SCILL concept are an increased activity for organic reactions like the Mukaiyama aldol reaction and an increased selectivity of the intermediate in selective hydrogenations.

Metal nanoparticles can be considered as highly dispersed unsupported catalysts with excellent activity in many reactions.[11] The catalytic behaviour of the nanoparticles depends on size and shape of the metal. Therefore, the stabilisation of these particles is the most important step towards an active catalyst. Dupont *et al.*[12] first described the ability of ILs to stabilise transition metal nanoparticles which have been investigated in several reactions like hydrogenation of olefins and typical organic reactions.[13]

In the following sections the SILP and SCILL concepts are explained in detail with several applications. Especially the SILP catalysed hydroformylation of propene and 1-butene and the SCILL catalysed selective hydrogenation of citral and alkynes will be used to show the development of the catalyst towards industrial application. Furthermore the influence of the IL on the performance of the catalyst will be explained by means of a spectroscopy investigating model and real catalysts.

The most important step to a successful utilisation of these two concepts is the determination of a suitable ionic liquid. Characteristic properties of almost all ILs are low vapour pressure and reasonable thermal stability. Other properties, like solubility of reactants and gases, immiscibility with polar or non-polar solvents, coordinating or non-coordinating character, can be adjusted by the choice of cation and anion.[14] Because of the large diversity of ionic liquids the reaction system must be well known to create an adequate catalyst system. Further premises must be conformed to use an ionic liquid in SILP or SCILL catalysis. Stability of the ionic liquid is required especially under reaction conditions. However, not only reaction temperature and pressure are relevant, the presence of nucleophiles, bases and water could decompose the IL as well. For example, basic and nucleophilic ions could lead to deprotonation and dealkylation and, as well, the presence of water could be responsible for hydrolysis. Another important criterion for a suitable ionic liquid is the solubility of reactants as well as the immiscibility with products. This property has a huge impact on the performance of the catalyst system. Only if these parameters are successfully confirmed could SILP or SCILL systems serve as catalysts with long-time stability and excellent performance.

6.2 Applications of SILP Systems

Supported Ionic Liquid Phase (SILP) catalysts are immobilised homogeneous catalysts. A transition metal complex is dissolved in an ionic liquid which coats an inert support material. The preparation of this type of catalysts is really easy. Under inert conditions, the metal complex, the ligands and the ionic liquid are dissolved in an organic solvent and the support is added. The catalyst is obtained by evaporation of the solvent (Scheme 6.1).[15]

Since the introduction by Mehnert and co-workers[4] and Wasserscheid and co-workers[5] SILP catalysts have become a commonly used catalyst system in various reactions. Due to the enhanced stability and reusability of the SILP catalysts these are used in Pd catalysed C–C coupling reactions like Mizoroki–Heck and Suzuki–Miyaura,[6] as well as in olefin metathesis, methanol carbonylation and hydroamination of alkenes.[7] The increased activity of the catalysts is explained by the larger interface between the active sites and the reactants which means there is a higher local concentration of the metal complex because of the dispersion of the IL on the support material. Another reason for the enhanced activity is the non-coordinating character of the IL which leads to a better availability of the active sites. The main drawbacks are leaching problems in liquid phase reactions and a requirement for fine-tuning of support materials to avoid deactivation.

Mehnert and co-workers investigated the hydrogenation[4] and the hydroformylation[4,17] of 1-hexene in liquid phase catalysed by Rh–SILP (Scheme 6.2). In comparison with the biphasic ionic liquid and homogeneous system the hydrogenation activity is increased by 1 or 2 orders of magnitude to a TOF of 26 820 h^{-1} when using a SILP catalyst because of the larger interface between IL layer and organic phase or the absence of

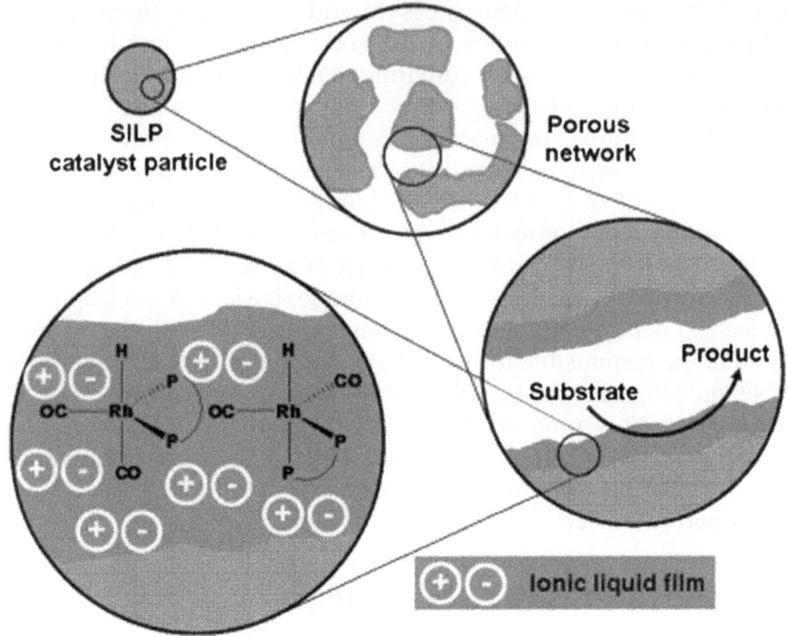

Scheme 6.1 SILP catalyst (from reference 16).

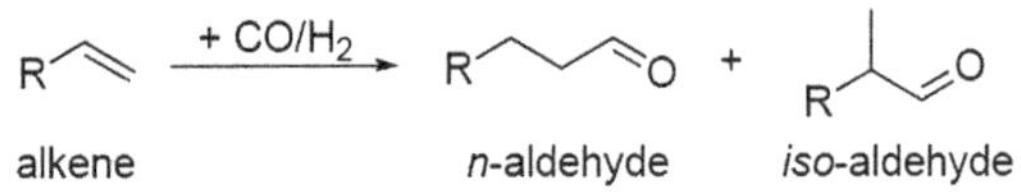

R–CH=CH₂ (alkene) $\xrightarrow{+ CO/H_2}$ *n*-aldehyde + *iso*-aldehyde

alkene *n*-aldehyde *iso*-aldehyde

Scheme 6.2 Hydroformylation of *n*-alkenes.

coordinating solvents, respectively. Other benefits of this catalysis are long-time stability with a reusability of the catalyst for 18 runs and no leaching of rhodium. The hydroformylation of 1-hexene was performed with a Rh–tppti–SILP catalyst and reached a TOF of 3600 to 3900 h^{-1} for [BMIM][PF_6] and [BMIM][BF_4], respectively, at an *n*/*i*-heptanal ratio of 2.4. The biphasic system exhibits a lower TOF of only 1320 to 1380 h^{-1} at comparable selectivities. Drawbacks of this catalyst system are deactivation and leaching of the catalyst.

Wasserscheid and co-workers investigated the continuous gas phase hydroformylation of propene and 1-butene and developed a catalyst system which outperformed most of the industrial homogeneous catalysts. In 2003, a new catalyst system was introduced which reached excellent activity and selectivity in the hydroformylation of propene.[5] The so-called SILP catalyst contained a Rh complex with different ligands which was dissolved in the ionic liquids [BMIM][PF_6] or [BMIM][$OctSO_4$] supported by silica. Using the bidentate ligand sulfoxantphos (Scheme 6.3) an excellent TOF of 37 h^{-1} and an *n*/*i*-butanal ratio of 23.3 was reached. The reaction was strongly

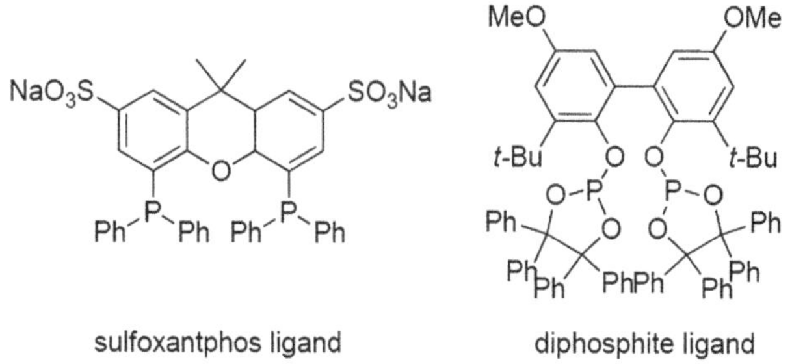

sulfoxantphos ligand

diphosphite ligand

Scheme 6.3 Ligands for Rh–SILP catalysts.

influenced by L/Rh ratio and pore filling degree of IL ($\alpha = V(\text{IL})/V(\text{pore})$). At an L/Rh ratio of 2.5 the selectivity as well as the activity is drastically decreased. The explanation for the decrease in activity was that the active sites were "surface-immobilised ligand-free complexes" under these conditions and mass transfer was limited because of low solubility of hydrogen and carbon monoxide. By increasing the ratio to 10 the active rhodium complex $[HRh(L)(CO)_2]$ was formed and both activity (TOF = 37 h^{-1}) and selectivity (96%) were tremendously enhanced. The pore filling degree slightly influenced the activity of the catalyst from 37 h^{-1} to 25.4 h^{-1} for $\alpha = 0.08$ and 0.52 because the mass transfer limitation increased. The choice of ionic liquid had a slight impact on the activity of the catalyst at comparable selectivities. In comparison with monodentate ligands the Rh sulfoxantphos complex was less active but more selective. A maximum selectivity of 73.8% to *n*-butanal was reached with a Rh monophosphine SILP catalyst. This indicated that a multidentate ligand was required to achieve excellent selectivities. The only disadvantage was the deactivation of the catalyst due to an interaction of ligand and support.

The deactivation of the SILP catalyst was prevented by calcination of the support material prior to the catalyst preparation.[18] This pre-treatment led to a partial dehydroxylation of surface silanol groups which formerly reacted with the ligands. The long-time stability of Rh sulfoxantphos–[BMIM][$OctSO_4$]–SiO_2 ($\alpha = 0.1$, L/Rh = 10) was verified within a time-on-stream (TOS) of 60 h without loss in both activity (TOF = 44 h^{-1}) and selectivity (*n*/*iso* = 20). In contrast to these excellent results, the stability of the catalyst system rapidly decreased when the L/Rh ratio or the pore filling degree was lowered. The catalyst deactivated within 15 h time-on-stream. These results indicated a sufficient amount of ligand and ionic liquid was required to obtain a stable and active catalyst, otherwise diffusion of the ligands from the rhodium complex to the silanol groups of the support occurred. In other words, the deactivation of the catalyst was due to a change of an active monomeric rhodium complex to an inactive dimeric complex which was verified *via* FT-IR spectroscopy. The other important information from FT-IR studies was that the SILP catalyst was still a homogeneous catalyst which was shown by CO absorption bands compared to those of a homogeneous catalyst without an IL layer.

The stability of catalyst was extended to 200 h TOS because the deactivation mechanism was analysed and overcome.[19] The loss in activity within 180 h was 17% without a decrease of selectivity. This indicated there was no decomposition of the catalyst itself, but a formation of aldol condensation products like 2-ethyl-hexanal which was further hydrogenated to 2-ethyl-hexanol. These high boiling side products were soluble in the ionic liquid phase and decreased the concentration of active sites. By evaporation the catalyst could be reactivated and a long-time stability of 700 h could be achieved. Kinetic studies with this Rh–SILP catalyst showed first order kinetics in propene and negative order in CO which were, together with the activation energy of 63 kJ mol^{-1}, in good agreement with the literature for a homogeneous catalyst. This strongly indicated that the SILP system is a homogeneous complex dissolved in an ionic liquid film immobilised on an inert oxide.

The variation of support material of a Rh–sulfoxantphos–SILP catalyst showed the highest initial activity (TOF = 21 h^{-1}) and selectivity (95% *n*-butanal) for silica support but also a complete deactivation in 24 h.[20] The catalysts with other support materials like Al_2O_3, TiO_2 and ZrO_2 reached lower activities (5–10 h^{-1}) and selectivities of 90% to 95% but these catalysts hardly deactivated during a time-on-stream of 55 h. The pore volume of support seemed to influence the performance more than the specific BET surface area. Because of the high initial activity and selectivity as well as the known deactivation mechanism mentioned above, silica was chosen to be the standard support material.

In 2007, SILP catalysis was extended to the gas phase hydroformylation of 1-butene in a continuous fixed-bed reactor.[21] The SILP catalyst showed both excellent activity (TOF up to 563.5 h^{-1}) and selectivity (97.1–99.9% *n*-pentanal) in the hydroformylation of 1-butene. These results showed characteristics like a truly homogeneous complex as previously described for propene hydroformylation. In comparison with propene as substrate the SILP catalyst exhibited a 2.5-fold higher activity at 80 °C and an increased selectivity (95% to 98% *n*-pentanal) in 1-butene hydroformylation. The solubility of propene and 1-butene in the ionic liquid [BMIM][$OctSO_4$] were determined *via* microbalance and a solubility difference of 2.4 was found which expressed perfectly the measured difference in activity. Kinetic studies confirmed the general observations found for propene. A first order in 1-butene and a negative order in CO were obtained in excellent accordance with the aforementioned values of propene hydroformylation and the literature. The variation of rhodium content between 0.1 wt% and 0.3 wt% showed a first order dependency on reaction rate as well. Due to the same TOF in this concentration range the Rh complex acted like a typical homogeneous catalyst due to the complete availability of the active sites. Moreover, these results indicated that there was no mass transfer limitation possibly caused by the viscosity of the IL. This was indicated by an activation energy of 63 kJ mol^{-1} which was in good agreement with values for both homogeneous and biphasic systems. A rhodium content of 0.9 wt% in the

SILP catalyst lowered the TOF from 252 h^{-1} (0.3 wt%) to 159 h^{-1} and an activation energy of 55 kJ mol^{-1} was determined. This decrease in activity was explained by a higher ligand concentration in the ionic liquid film increasing the viscosity of the latter and thereby the mass transfer limitation.

To verify the kinetic data already described, a gradient-free gas phase Berty type reactor was used in the hydroformylation of 1-butene.[22] The catalyst system was the aforementioned Rh–sulfoxantphos–SILP system with $\alpha = 0.1$. This reactor type guaranteed the determination of true microkinetic parameters without any influence of concentration or temperature gradients. The activation energy was determined to be 64.1 kJ mol^{-1} at low conversions of about 10%. This was in good agreement with the results from fixed bed reactor studies under similar conditions. These values strongly indicated the absence of mass transfer limitation of gas into the ionic liquid phase. To exclude pore and film diffusion limitation, the Weisz–Prater modulus (1.11×10^{-2}) and Mears criterion (4.75×10^{-3}) were calculated and the results showed that the SILP system was neither influenced by pore nor film diffusion. The stability of the sulfoxantphos ligand was tested *via* NMR spectroscopy by analysing a fresh and used SILP catalyst. After a time-on-stream of 120 h only small amounts of oxidised phosphines were found which hardly influenced the performance of the catalyst. The SILP catalyst was really stable and no mass transfer limitation occurred during the reaction.

The only drawback for an industrial application of this catalyst system is the feed. Pure 1-butene is not the desired feed for the hydroformylation of C_4 at industrial scale because of its costs of purification. The industrial feedstock is a mixture of C_4 components called raffinate II (45% 1-butene, 30% 2-butene and 25% butane).[23] A similar mixture of C_4 compounds (raffinate I, consisting of butane, butenes and butadiene) was successfully hydroformylated in 2011.[15] Due to the use of the new diphosphite ligand (Scheme 6.3), the SILP catalyst was able to isomerise 2-butenes prior to the hydroformylation so an exceptional high selectivity to *n*-pentanal up to 99.5% was reached. Additional to a good activity of 20% no side reactions of butane or isobutene occurred. The decomposition of the ligand, and thereby a quick deactivation of the catalyst due to hydrolysis of the diphosphite ligand by traces of water in the feed, was the only drawback. The introduction of an acid scavenger to catch the formed phosphoric acid and a dried feed led to an excellent long-time stability of more than 800 h time-on-stream with no selectivity loss and a TOF of 410 h^{-1}. A TOF of 3600 h^{-1} was reached by increasing reaction temperature and pressure to 120 °C and 25 bar, respectively. At these conditions an *n*-pentanal space–time yield of 850 kg m^{-3} h^{-1} was achieved which depicted a higher value than industrial catalysts. This SILP catalyst combined excellent activity and selectivity with long-time stability reaching the required industrial performance.

In 2012, kinetic studies[24] of propene hydroformylation catalysed by Rh sulfoxantphos–[BMIM][$OctSO_4$]–SiO_2 revealed a significant impact of reaction temperature on selectivity and the rate-determining step (RDS). At low temperature (90 °C), the RDS was the alkene insertion into the Rh–H bond and at high temperature (140 °C), the oxidative addition of hydrogen became rate-determining. At both temperatures, the order with respect to propene hardly changed from 0.9 to 0.7 in contrast to the reaction orders in hydrogen and CO which were dependent on temperature. At 90 °C, a zero order in hydrogen and a slightly negative order in CO were determined which were in good agreement with the results of Wasserscheid and co-workers at 100 °C mentioned above. At high temperature, the partial dependence of hydrogen and carbon monoxide was first and zero order, respectively.

In recent studies, the SILP catalysis was applied in two further reactions. The industrially important water–gas shift reaction (WGSR) was performed in a continuously operating fixed-bed reactor.[25] The SILP catalysts contained a series of known or novel homogeneous complexes of different metals which were known for application in the water–gas shift reaction. In agreement with literature, ruthenium based catalysts showed the highest activity and were the only stable catalysts over a TOS of 24 h. Despite a long activation period $RuCl_3$ precursor exhibited the highest activity among the Ru complexes. From FT-IR analysis, it was suggested that a Ru carbonyl species was the active site because several Ru carbonyl signals were detected for the used SILP catalysts explaining the long activation period of $RuCl_3$ by a formation of a carbonyl complex. However, the applied Ru carbonyl SILP catalysts were less active than $RuCl_3$. An anionic Ru complex with imidazolium cations showed the highest activity and best stability resulting in a TOF of 2 h^{-1} which was still too low for an industrial application, but the reaction temperature could be decreased from 200–350 °C to only 120 °C.

Another promising approach for SILP catalysis was the alkylation of aromatics.[26] Friedel–Crafts alkylation is usually catalysed by $AlCl_3$ which produces large amounts of wastewater as by-product and the drastic reaction conditions lead to coke formation. A pre-treatment of the silica support with the acidic ionic liquid was essential for a stable catalyst because of the cleavage of the basic surface groups. The highly acidic [EMIM]Cl–$AlCl_3$–SiO_2–SILP catalyst which was used in the continuous, gas phase isopropylation of cumene still deactivated during a time-on-stream of 20 h. The deactivation was explained by formation of higher alkylation products (triisopropylbenzene and tetraisopropylbenzene) which were dissolved in the IL layer and reduced the availability of the active anions. The other reason for deactivation was traces of water in the feed which reacted with the strongly Lewis acidic chloroaluminate to form less acidic species. The first drawback could be circumvented by a smaller pore filling degree of IL. By changing the feed to toluene with less than 40 ppm water the catalyst system exhibited long-time stability. The selectivities to mono-alkylated products were tremendously increased by using a high aromatic-to-propylene ratio (7 : 1 molar ratio). As well, a higher acidity of the ionic liquid film on the support had a positive effect on selectivity to mono-alkylated products. The SILP catalyst for the isopropylation of toluene exhibited both high activity (up to 99%) and excellent selectivity (up to 95%) with a long-time stability of more than 210 h time-on-stream with no deactivation.

6.3 Applications of SCILL Systems

Solid Catalysts with an Ionic Liquid Layer (SCILL) are heterogeneous catalysts which are covered by a thin IL layer. In contrast to SILP catalysts the active metal is deposited on the support material and not dissolved in the ionic liquid. Usually the preparation of these catalysts is a two step incipient-wetness impregnation. First, the inert support material is impregnated with a metal precursor and then, the reduced catalyst is impregnated with the desired amount of ionic liquid (Scheme 6.4).

In 2007, Jess and co-workers[10] investigated the selective hydrogenation of cyclooctadiene (COD) to cyclooctene (COE) catalysed by a SCILL containing a commercial Ni catalyst coated with the ionic liquid 1-butyl,3-methyl imidazolium octylsulfate [BMIM][$OctSO_4$] (Scheme 6.5). A pore filling degree of about 10% was required to achieve the complete SCILL effect. This led to a decrease in pore volume and surface area. The IL coating of the Ni catalyst increased the yield of COE by a factor of almost 2, from 36% for the neat catalyst to 70% for the SCILL. The reaction rate was decreased dramatically because of the inferior diffusion of COD to the active sites. The large improvement in yield was explained in two possible ways: a "physical solvent effect" or a "co-catalytic effect". The physical solvent effect changed the concentrations of the reactants at the active sites because of the different solubilities of the reactants in ionic liquids. The co-catalytic effect changed the properties of the active metal in some way, maybe as a result of a ligand or an ensemble effect known from bimetallic catalysts as postulated later by Claus and co-workers.[28] However, if the SCILL effect was only a physical solvent effect, yield would only be increased from 36% to 39% as the authors determined. So to understand the catalyst performance, a co-catalytic effect

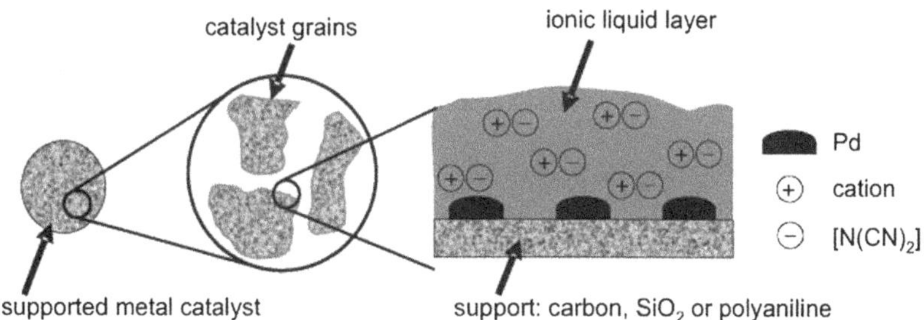

Scheme 6.4 SCILL system (from reference 27).

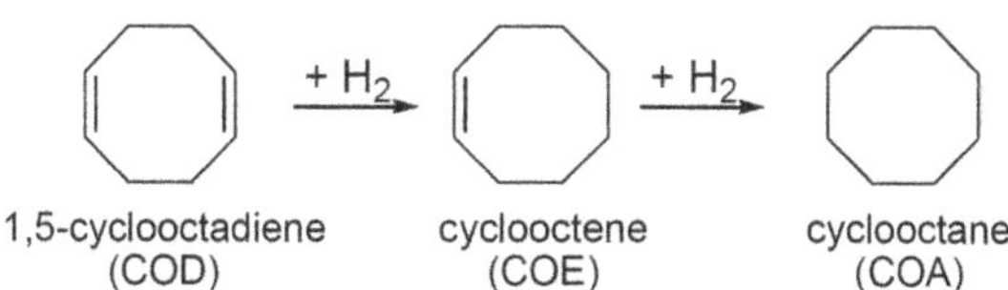

Scheme 6.5 Hydrogenation of cyclooctadiene.

which changed the hydrogenation properties of the active sites has to be considered as well (*vide infra*).

For further examinations of the impact of SCILL on the selectivity of hydrogenation reactions Ni catalysed hydrogenation of octyne and cinnamaldehyde and Ru catalysed hydrogenation of naphthalene were investigated.[29] A pore filling degree of 15% [BMIM][$OctSO_4$] was used. The trends found for the selective hydrogenation of cyclooctadiene could be verified for these reactions. In comparison with the uncoated catalyst, the yield was increased if the intermediate had a lower solubility in the IL layer than the educt. The physical solvent effect can not explain the improvement in yield. The activity of the catalyst was decreased if the substrate was more soluble in organic solvent than in IL. After these experiments in liquid phase hydrogenation, the gas phase hydrogenation of 1,3-butadiene in a fixed-bed reactor was tested. In contrast to the hydrogenation with the uncoated catalyst, the selectivity to the intermediates (1-butene and 2-butene) was about 99% over the whole conversion range indicating the positive impact of IL on the performance of the catalyst. In combination with these excellent selectivities the reaction rate was dramatically lowered by a factor of 100.

Araujo and co-workers[30] further increased the yield of COE in the selective hydrogenation of COD. In comparison with the neat Pd/C catalyst the yield was improved from 10% to more than 80% when using Pd/C-2.5[HMIM][Br].[†] The consecutive hydrogenation to cyclooctane was completely inhibited. It was explained by a "membrane-like effect" caused by the thin ionic liquid layer which hindered the re-adsorption of COE. This was explained by different toluene–IL phase partition coefficients for 1,5-cyclooctadiene and cyclooctene.

In 2008, Claus and co-workers[31] applied the SCILL concept in the selective hydrogenation of citral to citronellal (Scheme 6.6). The hydrogenation of citral is an important reaction towards the production of perfumes. Starting with usage of the IL [EMIM][NTf_2] as bulk solvent in combination with the heterogeneous catalyst Pd/C an excellent yield of citronellal was obtained. However, the main drawback of this catalyst system was the necessary product extraction from the ionic liquid phase.[32] The SCILL system Pd/C-50[BMIM][DCA] exhibited a quantitative yield of citronellal by only using a small fraction of the IL amount used in the bulk experiment (Scheme 6.7).

Due to the properties of the SCILL system, no extraction problems and no increased reaction time by diffusion limitation occurred. The IL layer strongly inhibited the consecutive hydrogenation to dihydrocitronellal which was proven by a hydrogenation experiment in which the intermediate was used as starting material and a conversion below 1% was obtained. The amount of butyl-methylimidazolium dicyanamide had a beneficial effect on selectivity of the reaction. The selectivity to citronellal was increased from almost 50% for the neat Pd catalyst to 100% for the SCILL system with an IL

[†]Nomenclature for SCILL catalyst: Pd/C-2.5[HMIM][Br] = active metal/support-wt%[IL].

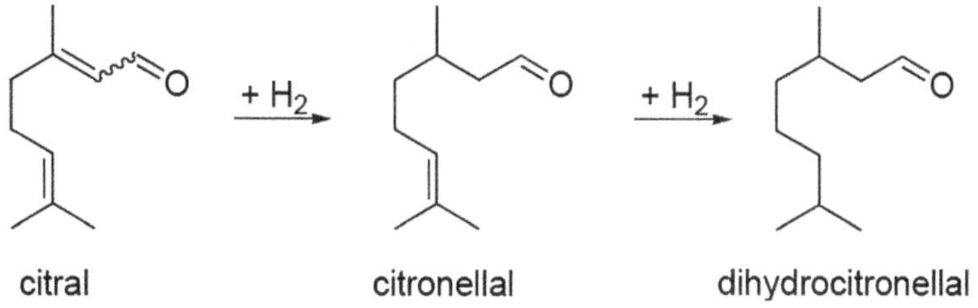

+ H_2 → + H_2 →

citral citronellal dihydrocitronellal

Scheme 6.6 Hydrogenation of citral.

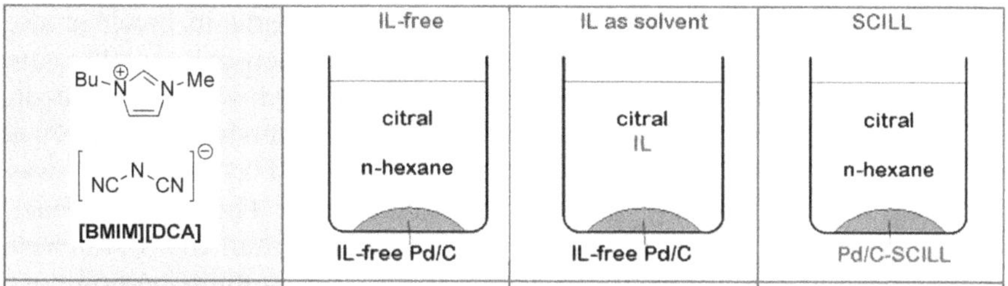

Bu-N⊕ N-Me [NC-N-CN]⊖ [BMIM][DCA]	IL-free citral n-hexane IL-free Pd/C	IL as solvent citral IL IL-free Pd/C	SCILL citral n-hexane Pd/C-SCILL
Product extraction	–	*necessary*	*no*
X_{citral} [%]	100	100	100
$S_{citronellal}$ [%]	41	97	>99
$S_{dihydrocitronellal}$ [%]	49	1	<1
$S_{other\ products}$ [%]	10	2	<1

conditions: 200 mg Pd (10 wt%)/C, 50 °C, 10 bar H_2, 1.1 mol/L citral, 1200 rpm, 360 min.

Scheme 6.7 Hydrogenation of citral with Pd/C without IL (solvent: *n*-hexane), with IL as bulk solvent and modified by an ionic liquid layer (SCILL system), from left to right, IL = [BMIM][DCA].[31]

loading of 50 wt%. However, an IL loading of more than 50 wt% led to a lower conversion of only 40%.

The influence of several ionic liquids ([BMIM][NTf_2], [BMIM][PF_6], [BMPL][NTf_2], [BMIM][DCA], [BMPL][DCA], [B3MPYR][DCA]) was investigated in the SCILL catalysed liquid phase hydrogenation of citral.[27] The impregnation of the catalyst with an ionic liquid led to an enhanced selectivity to the desired intermediate in all cases. The anion of the IL had the most important effect on the catalyst's performance. With $[PF_6]^-$ and $[NTf_2]^-$ based ILs a maximum selectivity of 60% at X = 70% was reached. The DCA based ionic liquids all exhibited far higher selectivities even up to a quantitative yield. A "quantitative one-pot synthesis of citronellal" was achieved under optimised conditions with the SCILL system Pd/SiO_2-29[B3MPYR][DCA]. The choice of metal precursor, support material and preparation technique only slightly influenced the performance of the catalyst. Even the hydrogenation of pure citral catalysed by Pd/SiO_2-50[BMIM][DCA] could be performed with high selectivities of 86% at moderate conversion of 52%. *Via* characterisation, a decrease of surface area and pore volume was detected for the SCILL system in comparison with the neat Pd catalyst. A blue shift of 15 cm^{-1} of the nitrile vibrations in the IR spectra strongly indicated an interaction of IL and Pd active sites (*vide infra*).

To enable this technology to be assessed for industrial application, the catalyst system was successfully applied in the continuous hydrogenation of citral in a trickle-bed reactor.[33] The selectivity to citronellal was enhanced from about 40% for the neat Pd catalyst to almost 100% for the SCILL system in combination with a decrease in activity. The yield of citronellal was about four times higher for the catalyst Pd/SiO_2-32[BMIM][DCA]. These findings were in good agreement with the above mentioned result in batch mode. Again, no dihydrocitronellal could be detected which meant that the consecutive hydrogenation was completely inhibited. An increase in temperature from 70 °C to 100 °C led to enhanced conversion of about 70% with only a small loss in selectivity. A smaller feed rate of 0.75 mL min^{-1} was associated with the highest yield of citronellal of 83%. The only drawback was desorption of ionic liquid at a temperature of more than 100 °C. At lower temperature, the catalyst exhibited excellent long-time stability with no loss in activity for 78 h time-on-stream.

To achieve full conversion at lower temperature, a SCILL system with higher Pd loading was applied. Unfortunately, the selectivities to citronellal were lower (36%) but in comparison with the neat Pd catalyst the yield of citronellal could be doubled. In all experiments with the SCILL catalyst selectivity could be enhanced in combination with a loss in activity leading to a higher yield of citronellal.

The SCILL system used for the selective hydrogenation of citral was applied in the selective hydrogenation of acetylene to ethylene.[34] Two different ionic liquids ([MMIM][MeHPO_3] and [BMIM][DCA]) were used for the catalyst system because of better solubility of acetylene than ethylene in the first and good performance of the latter in citral hydrogenation. In both cases selectivity to ethylene was enhanced in combination with a lowered formation of ethane (product of the consecutive hydrogenation) and C_4 hydrocarbons (unwanted side products, source of "green oil"). However, the increased selectivity was associated with a loss of activity from 89% for the neat Pd catalyst to 79% and 45% for Pd–[BMIM][DCA] and Pd–[MMIM][MeHPO_3], respectively. Decreased activity was explained by a change in rate determining step or transport phenomena indicated by a doubled value of activation energy of 41 kJ mol^{-1} and 47 kJ mol^{-1} for the SCILL systems compared to 19 kJ mol^{-1} for the neat Pd catalyst.

Lercher and co-workers[35] investigated the SCILL catalysed water–gas shift reaction. For the neat Cu/Al_2O_3 catalyst, an activity maximum was identified for a specific surface area of support material of 150 m^2 g^{-1}. For catalysts with a lower support area, larger copper particles decreased the activity while the catalysts with BET surface areas of 257 and 360 m^2 g^{-1} had very small particles which were partially oxidised leading to lower activity. This meant that the activity of the Cu catalyst was strongly influenced by the fraction of oxidised Cu and its particle size. At low temperature (180 °C), the Cu/Al_2O_3-[BMMIM][F_3MeSO_4] catalyst was more active than the neat catalyst and even five times more active than a commercial water–gas shift catalyst which had a ten times higher metal loading. The ionic liquid layer increased the concentration of reacting species by lowering the interactions with the surface. The dissociation of water was facilitated by a higher amount of oxygen leading to a decreased CO concentration which was necessary for a maximum in activity.

In a recent study, Wasserscheid and co-workers[36] applied bifunctional SCILL catalysis in the isomerisation of *n*-octane to *iso*-octane. A catalyst system consisting of Pd/SiO_2-[BMIM]Cl-$AlCl_3$ exhibited both good activity ($X = 73\%$) and selectivity to *iso*-octane ($S = 33\%$). The performance of the catalyst was strongly influenced by hydrogen pressure and reaction temperature. Activity and selectivity were considerably enhanced compared to a liquid–liquid biphasic experiment catalysed only by the acidic chloroaluminate ionic liquid and an experiment with a monofunctional SILP catalyst without platinum. This was evidence for platinum and hydrogen being necessary to realise an active and selective catalyst as the uncoated catalyst did not show any activity. The role of platinum was characterised as minimising side products by hydrogenating olefinic intermediates which were normally alkylated. The enhancement in activity was explained by an increasing acidity of the SCILL system because activity in isomerisation of alkanes was a function of acidity and platinum itself was not active. The SCILL system with platinum, dissolved hydrogen and acidic IL led to a heterolytic dissociation of hydrogen molecules forming surface Pt-hydride and protons in the ionic liquid increasing the acidity of the catalyst.

6.4 Applications of Other IL Systems

Transition metal nanoparticles in ionic liquids are excellent catalysts in various reactions. Dupont and co-workers[13] and Gu and Li[37] reviewed the improvements in the last decade recently. The role of the ILs was characterised as solvent, stabiliser or ligand for the nanoparticles depending on choice of reaction and ionic liquid. These catalyst systems exhibited great catalytic performances in selective hydrogenation of arenes under very mild reaction conditions. Another area of application was organic C–C coupling reactions like Heck, Sonogashira and Suzuki reactions. Non-functionalised ionic liquids acted as non-coordinating solvents improving the availability of active sites. The usage of functionalised ionic liquids led to a stabilisation of nanoparticles *via* coordination to the metal surface comparable to organometallic catalysis. The stability of the transition metal nanoparticles was still a problem because of the weak coordination of ILs to metal which could be replaced by reactants leading to agglomeration of nanoparticles. In combination with other stabilisers long-time stability could be achieved and transfer limitation could be reduced as well because of the small amounts of ionic liquids used.

The so-called Supported Ionic Liquid Catalysis (SILCA) was introduced by Mikkola and co-workers[38] and consisted of transition metal nanoparticles in ionic liquid which were dispersed on active carbon cloth (ACC). The SILCA system was applied in the selective hydrogenation of citral with different ionic liquids. All catalysts coated with an ionic liquid layer were more active than the neat Pd/ACC catalyst. The catalyst containing [B4MPyr][BF_4] exhibited an increase in activity by a factor of two. The enhanced activity was explained by the non-coordinating character of the ionic liquids which increased the availability of the active sites. This was a similar behaviour of the catalyst as was determined for nanoparticles in IL indicating a similar chemical nature. Dihydrocitronellal was the main product in all experiments except for the one with [BMIM][PF_6] where a selectivity of 49% to citronellal was reached. The differences in selectivity of the SILCA systems were explained by different solubilities of hydrogen and reactants in the ionic liquids (physical solvent effect as cited before). Further modification of Pd–[B4MPYR][BF_4]/ACC with Lewis acid $ZnCl_2$ led to a good selectivity to menthols of 41%. The catalyst Pd–[(C3OH)PYR][NTf_2]/ACC in combination with KOH exhibited an enhanced selectivity to citronellal of 74% compared to 16% without the alkaline modifier. Deactivation of catalyst was explained by agglomeration of Pd nanoparticles and an accumulation of hydrogenation products in the ionic liquid film.

The utilisation of ionic liquids as additive was investigated by Claus and co-workers[39] in the selective hydrogenation of benzene to cyclohexene. The main drawbacks of catalyst systems used in this reaction were the large amounts of ruthenium as active metal and zinc sulfate as additive creating high costs and corrosion problems. In the case of usage of ionic liquid, the catalyst system could be facilitated consisting only of supported ruthenium in water and [B3MPyr][DCA] in the ppm-range. It is notable to say that the used amounts of Ru and additive are about 2 orders of magnitude smaller than that of the commonly used catalyst systems. Moreover, the usage of sodium hydroxide and zinc sulfate, usually necessary to achieve high selectivities to cyclohexene, was completely avoided. This simple catalyst system exhibited good selectivities to cyclohexene of 60% at low conversions under moderate reaction conditions ($T = 100$ °C, $p(H_2) = 20$ bar). The $[DCA]^-$ anions adsorbed on the catalyst surface and selectively poisoned the most active ruthenium sites thus inhibiting the consecutive hydrogenation to cyclohexane.

6.5 Understanding the Role of IL

SILP and SCILL systems exhibited excellent performances in various reactions. To understand the interaction of ionic liquid and active metal several studies have been made. The active complex in SILP catalysts was described by Wasserscheid and co-workers[18] as [$HRh(CO)_2(L)$] which was confirmed by FT-IR spectroscopy. It was concluded that the metal complex in SILP systems remained a homogeneous catalyst because of identical CO bands for Rh–SILP and homogeneous Rh complexes. ^{31}P NMR studies showed identical signals for the supported and unsupported Rh complex in IL so the interaction with the support material did not change the metal complex and the ligand.[15] The role of the ionic liquid was the immobilisation and the uniform dispersion of the metal complex on the support and controlling the diffusion of reactants. In other words, the IL acted like a solvent for a homogeneous catalyst. Different solubilities of reactants changed the activity of the catalyst dramatically. For example, the solubility of 1-butene in [BMIM][$OctSO_4$] was 2.4 times higher than that of propene in the IL which was in excellent accordance with a 2.5 fold higher activity in hydroformylation catalysed by Rh–SILP.[21] In contrast to the interpretation made by Wasserscheid and co-workers, Bell and co-workers proposed a different interaction of IL and active metal.[40] The active complex [$HRh(CO)_2(L)$] was stabilised by the support through binding of the sulfonate groups of the ligand with silanol groups of the support. The dehydroxylation of the support at temperatures above 100 °C led to a smaller amount of surface silanol groups enhancing their acidity. In this case, an increasing fraction of phosphine groups of the ligand interacted with the support, thereby reducing the availability of these groups to coordinate to the active metal. At high IL loadings, a competitive interaction of IL anions and sulfonate groups of the ligand with the silanol groups occurred leading to a reduced fraction of coordinated ligand. The inactive dimeric complex $[Rh(CO)(\mu\text{-}CO)(L)]_2$ was formed due to a larger amount of free ligands available. A ratio of L/Rh of 10 was required to enable an enhanced fraction of sulfoxantphos bound to the surface *via* sulfonate groups stabilising the active metal. In other words, the role of the ionic liquid was the spatial distribution of the active complexes on the support surfaces.

In SCILL catalysis, Jess and co-workers[10] investigated the effect of ionic liquids on the physical properties of the Ni catalyst which was used in the selective hydrogenation of 1,5-cyclooctadiene. Impregnating the catalyst with an ionic liquid led to decreased surface areas and pore diameters. Depending on the pore filling degree of the catalyst with ionic liquid, both the surface area and the pore diameter further decreased which was in good accordance with the calculated decrease. Due to a simple model it was possible to determine the IL layer thickness depending on the pore filling degree, at least for filling below 0.15.

Lercher and co-workers[41] characterised a Pt/SiO_2-[BMMIM][OTf] SCILL catalyst used for ethylene hydrogenation. The interaction of the ionic liquid with silica increased the viscosity of the IL leading to a change in diffusion of reactants. In the presence of platinum the imidazolium ring coordinated to the active metal enhancing the basicity of the IL. In comparison with the uncoated catalyst the SCILL system exhibited a similar activity in ethylene hydrogenation indicating a weaker interaction of Pt and IL than that of the active metal with the reactants. So transfer limitation due to lower solubility of reactants did not occur.

In contrast to Lercher and co-workers' conclusion, Claus and co-workers[28] determined a dramatic decrease in hydrogen uptake for Pd/SiO_2-IL catalyst compared to that of neat Pd/SiO_2. The hydrogen uptake strongly depended on the IL, the IL loading as well as temperature. For $[DCA]^-$ based ionic liquids, which were the most effective ones in selective hydrogenation of citral, the amount of hydrogen was decreased by a factor of 9. XPS analysis revealed partially oxidised palladium in the presence of ionic liquid in contrast to the neat Pd catalyst. Furthermore the heats of adsorption were lowered significantly when analysing the SCILL system which was a strong indication for a new ligand effect similar to those known from bimetallic catalysts. Here, the ionic liquid influenced the catalyst system in two ways. By reducing the available amount of hydrogen on Pd surface and by changing the chemical structure of the active metal the formation of the intermediate is favoured.

In good agreement with the interpretation of Claus and co-workers, Libuda and co-workers[42] proposed a ligand effect for SCILL systems as well. A model SCILL system containing Pt and Pd nanoparticles on a well-defined alumina film and [BMIM][NTf_2] was investigated. A strong interaction of the ionic liquid with the supported nanoparticles was observed and a partial replacement of adsorbed CO occurred. The proposed ligand effect was specific to the nature of metal and selective to particular sites: the ionic liquid replaced CO preferentially from the facets and CO remained adsorbed onto the edges for Pt nanoparticles. In contrast to Pt, desorption of CO started at the edges of Pd clusters followed by the depopulation of the terraces. Modification of the Pt–CO bond by electronic effects strongly indicated a ligand-type interaction of the adsorbed IL with the active metal.

6.6 Conclusions

Modification of homogeneous and heterogeneous catalysts with ionic liquids improves the performance of the catalysts. The most promising concepts are the Supported Ionic Liquid Phase (SILP) catalysis and the Solid Catalyst with an Ionic Liquid Layer (SCILL) concept. These two concepts were intensively investigated in various reactions during the last decade. Both concepts reduce the required amount of ionic liquid to a minimum leading to cost efficient catalyst systems which is the most important drawback of ionic liquid biphasic catalysis. The advantages of SILP catalysts compared to homogeneous catalysts are long-time stability and good recovery of the catalyst without separation problems. These systems exhibited excellent activities and selectivities in the hydroformylation of 1-butene and reached industrial application level. Furthermore, SILP catalysed hydroformylation of C_4 feed surpasses the industrial process. Based on a heterogeneous catalyst the ionic liquid in SCILL systems changes the chemical properties of the active metal leading to enhanced selectivities to intermediates in selective hydrogenation for example. In acetylene hydrogenation selectivity to ethylene was increased in combination with a lowered production of "green oils" using a Pd based SCILL system. The SCILL catalysed selective hydrogenation of citral to citronellal reached quantitative yield of citronellal. In other words a one-pot synthesis of citronellal was realised with this new catalyst system. These two catalyst systems show huge potential for developing environmentally benign processes leading to a greener chemistry. Even in complex reactions like the selective hydrogenation of benzene, the modification of the catalyst system with ionic liquids led to an enhanced selectivity of the intermediate cyclohexene. The development of this new catalyst system is the first step towards a simple and green catalyst for this complicated hydrogenation.

References

1. P. T. Anastas and J. C. Warner, *Green Chemistry: Theory and Practice*, Oxford University Press, Oxford, 1998; S. L. Y. Tang, R. L. Smith and M. Poliakoff, *Green Chem.*, 2005, **7**, 761.
2. C. Coperet, M. Chabanas, R. P. Saint-Arroman and J.-M. Basset, *Angew. Chem., Int. Ed.*, 2003, **42**, 156; S. M. Thomas, *ChemCatChem*, 2010, **2**, 127.
3. P. S. Campbell, A. Podgoršek, T. Gutel, C. C. Santini, A. A. H. Pádua, M. F. Costa Gomes, F. Bayard, B. Fenet and Y. Chauvin, *J. Phys. Chem. B*, 2010, **114**, 8156.
4. C. P. Mehnert, E. J. Mozeleski and R. A. Cook, *Chem. Commun.*, 2002, 3010; C. P. Mehnert, R. A. Cook, N. C. Dispenziere and M. Afeworki, *J. Am. Chem. Soc.*, 2002, **124**, 12932.
5. A. Riisager, K. M. Eriksen, P. Wasserscheid and R. Fehrmann, *Catal. Lett.*, 2003, **90**, 149; A. Riisager, P. Wasserscheid, R. van Hal and R. Fehrmann, *J. Catal.*, 2003, **219**, 452.
6. H. Hagiwara, *Synlett*, 2012, **23**, 837.
7. C. van Doorslaer, J. Wahlen, P. Mertens, K. Binnemans and D. de Vos, *Dalton Trans.*, 2010, **39**, 8377.
8. M. Haumann and A. Riisager, *Chem. Rev.*, 2008, **108**, 1474.
9. Y. Gu, C. Ogawa, J. Kobayashi, Y. Mori and S. Kobayashi, *Angew. Chem., Int. Ed.*, 2006, **45**, 7217.
10. U. Kernchen, B. Etzold, W. Korth and A. Jess, *Chem. Eng. Technol.*, 2007, **30**, 985.
11. D. Astruc, F. Lu and J. R. Aranzaes, *Angew. Chem., Int. Ed.*, 2005, **44**, 7852.
12. J. Dupont, G. S. Fonseca, A. P. Umpierre, P. F. P. Fichtner and S. R. Teixeira, *J. Am. Chem. Soc.*, 2002, **124**, 4228.
13. J. D. Scholten, B. C. Leal and J. Dupont, *ACS Catal.*, 2012, **2**, 184.
14. R. Sheldon, *Chem Commun.*, 2001, 2399.
15. M. Jakuttis, A. Schönweiz, S. Werner, R. Franke, K.-D. Wiese, M. Haumann and P. Wasserscheid, *Angew. Chem., Int. Ed.*, 2011, **50**, 4492.
16. P. Wasserscheid, *J. Ind. Eng. Chem.*, 2007, **13**, 325.
17. C. P. Mehnert, *Chem.-Eur. J.*, 2005, **11**, 50.

18. A. Riisager, R. Fehrmann, S. Flicker, R. van Hal, M. Haumann and P. Wasserscheid, *Angew. Chem., Int. Ed.*, 2005, **44**, 815.
19. A. Riisager, R. Fehrmann, M. Haumann, B. S. K. Gorle and P. Wasserscheid, *Ind. Eng. Chem. Res.*, 2005, **44**, 9853.
20. A. Riisager, R. Fehrmann, M. Haumann and P. Wasserscheid, *Eur. J. Inorg. Chem.*, 2006, 695.
21. M. Haumann, K. Dentler, J. Joni, A. Riisager and P. Wasserscheid, *Adv. Synth. Catal.*, 2007, **349**, 425.
22. M. Haumann, M. Jakuttis, S. Werner and P. Wasserscheid, *J. Catal.*, 2009, **263**, 321.
23. F. Obenaus, W. Droste and J. Neumeister, *"Butenes," Ullmann's Encyclopedia of Industrial Chemistry*, Wiley-VCH, Weinheim, 2002.
24. D. G. Hanna, S. Shylesh, S. Werner and A. T. Bell, *J. Catal.*, 2012, **292**, 166.
25. S. Werner, N. Szesni, R. W. Fischer, M. Haumann and P. Wasserscheid, *Phys. Chem. Chem. Phys.*, 2009, **11**, 10817; S. Werner, N. Szesni, A. Bittermann, M. J. Schneider, P. Härter, M. Haumann and P. Wasserscheid, *Appl. Catal., A*, 2010, **377**, 70.
26. J. Joni, M. Haumann and P. Wasserscheid, *Adv. Synth. Catal.*, 2009, **351**, 423; J. Joni, M. Haumann and P. Wasserscheid, *Appl. Catal., A*, 2010, **372**, 8.
27. J. Arras, M. Steffan, Y. Shayeghi, D. Ruppert and P. Claus, *Green Chem.*, 2009, **11**, 716.
28. J. Arras, E. Paki, C. Roth, J. Radnik, M. Lucas and P. Claus, *J. Phys. Chem. C*, 2010, **114**, 10520.
29. A. Jess, C. Kern and W. Korth, *Oil Gas Eur. Mag.*, 2012, **38**, 38.
30. E. C. O. Nassor, J. C. Tristao, E. N. dos Santos, F. C. C. Moura, R. M. Lago and M. H. Araujo, *J. Mol. Catal.*, 2012, **363–364**, 74.
31. J. Arras, M. Steffan, Y. Shayeghi and P. Claus, *Chem. Commun.*, 2008, 4058.
32. K. Anderson, P. Goodrich, C. Hardacre and D. W. Rooney, *Green Chem.*, 2003, **5**, 448.
33. N. Wörz, J. Arras and P. Claus, *Appl. Catal., A*, 2011, **391**, 319.
34. T. Herrmann, L. Rößmann, M. Lucas and P. Claus, *Chem. Commun.*, 2011, **47**, 12310.
35. R. Knapp, S. A. Wyrzgol, A. Jentys and J. A. Lercher, *J. Catal.*, 2010, **276**, 280.
36. C. Meyer, V. Hager, W. Schwieger and P. Wasserscheid, *J. Catal.*, 2012, **292**, 157.
37. Y. Gu and G. Li, *Adv. Synth. Catal.*, 2009, **351**, 817.
38. P. Virtanen, J.-P. Mikkola, E. Toukoniitty, H. Karhu, K. Kordas, K. Eränen, J. Wärna and T. Salmi, *Catal. Today*, 2009, **147S**, S144; P. Virtanen, H. Karhu, G. Toth, K. Kordas and J.-P. Mikkola, *J. Catal.*, 2009, **263**, 209; P. Virtanen, T. O. Salmi and J.-P. Mikkola, *Top. Catal.*, 2010, **53**, 1096; E. Salminen, P. Virtanen, K. Kordas and J.-P. Mikkola, *Catal. Today*, 2012, **196**, 126.

39. F. Schwab, M. Lucas and P. Claus, *Angew. Chem., Int. Ed.*, 2011, **50**, 10453.
40. S. Shylesh, D. Hanna, S. Werner and A. T. Bell, *ACS Catal.*, 2012, **2**, 487.
41. R. Knapp, A. Jentys and J. A. Lercher, *Green Chem.*, 2009, **11**, 656.
42. M. Sobota, M. Happel, M. Amende, N. Paape, P. Wasserscheid, M. Laurin and J. Libuda, *Adv. Mater.*, 2011, **23**, 2617.

CHAPTER 7

SILP and SCILL Catalysis

MARCO HAUMANN* AND PETER WASSERSCHEID*

Institute of Chemical Reaction Engineering, University Erlangen-Nuremberg, Egerlandstrasse 3, D-91058 Erlangen, Germany
*Email: marco.haumann@fau.de; wasserscheid@crt.cbi.uni-erlangen.de

7.1 Introduction

Natural and synthesized solid materials that are commonly used as heterogeneous catalysts in industry are generally characterized by a non-uniform and largely undefined surface. Often, only certain sites (*e.g.* step sites, defects, corners or ad-atoms) of a nanoparticulate metal on support are advantageous with regard to the desired catalytic functionality (Figure 7.1). Consequently, the design of solid materials with greatly enhanced surface uniformity and specificity is a task of great relevance for the engineering of advanced catalytic materials.[1] In addition, technologies to access completely new surface reactivity on solid surfaces are desired.

One elegant way to achieve a uniform solid surface is the coating of such solid material with a thin film of a uniform liquid. This approach makes use of the fact that the synthesis and purification of uniform liquids is typically much easier than the technical synthesis of a uniform surface. Thus by coating a surface with defined organic or ionic liquids the solid surface adopts some of the chemical properties of the liquid coating.

Three major classes of liquids have been applied for this type of catalytic hybrid materials:[2–4]

- High-boiling molecular solvents, leading to so-called Supported Liquid Phase (SLP) catalysts;

RSC Catalysis Series No. 15
Catalysis in Ionic Liquids: From Catalyst Synthesis to Application
Edited by Chris Hardacre and Vasile Parvulescu

Published by the Royal Society of Chemistry, www.rsc.org

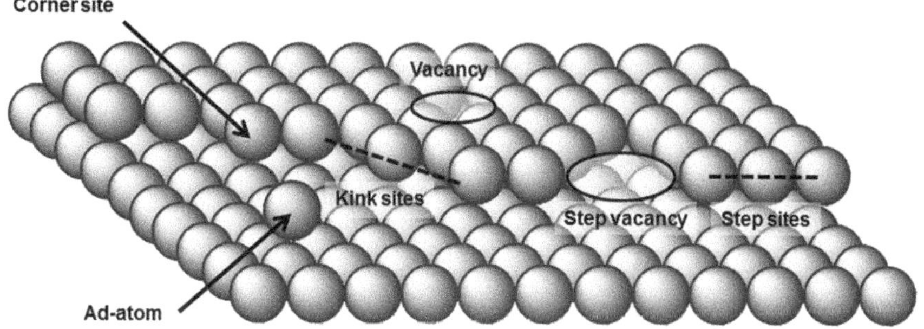

Figure 7.1 Schematic representation of the inhomogeneous nature of solid materials.

- Water, resulting in so-called Supported Aqueous Phase (SAP) catalysts, and
- Ionic liquids (ILs), forming either Supported Ionic Liquid Phase (SILP) or Solid Catalyst with Ionic Liquid Layer (SCILL) systems depending on the nature of the active species being either homogeneously dissolved or part of the solid surface.

Due to the extremely low vapor pressure of ionic liquids under the conditions of commercial catalytic gas-phase reactions, SILP and SCILL materials represent a fundamentally new approach to realize long-term stable, surface coated catalytic materials. Both technologies involve surface modification of a porous solid by dispersing a thin film of ionic liquid in close analogy to the SLP and SAP concept. While the SILP technology aims for the heterogenization of an ionic catalyst solution with the microscopic nature of the catalysis remaining homogeneous, SCILL systems aim for a modified surface reactivity by ionic liquid co-adsorption effects and represent a kind of fluid-side, ligand-type modification of a classical heterogeneous catalyst by the liquid salt. Both fundamentally different types of ionic liquid thin film catalysis are illustrated in Figure 7.2.[5]

Ionic liquids are salts consisting typically of organic cations and inorganic or organic anions. The narrow definition of "ionic liquids" relates to salt systems with a melting point below 100 °C.[6] Such low melting points are realized by applying salts composed of mono-charged ions with low charge densities. Typically the charges are well distributed in large, often unsymmetrical ions to prevent dense ion packing. Due to their salty nature, ionic liquids are characterized by extremely low vapor pressures. This property enables the high stability of IL coatings on surfaces in contact with a continuous gas phase. This is a significant difference between SLP/SAP systems on the one hand and SILP/SCILL systems on the other hand.[7] Another very important success factor of SILP/SCILL materials is the chemical and physico-chemical variability of ionic liquids. With respect to material and surface design, ionic liquids are characterized by representing

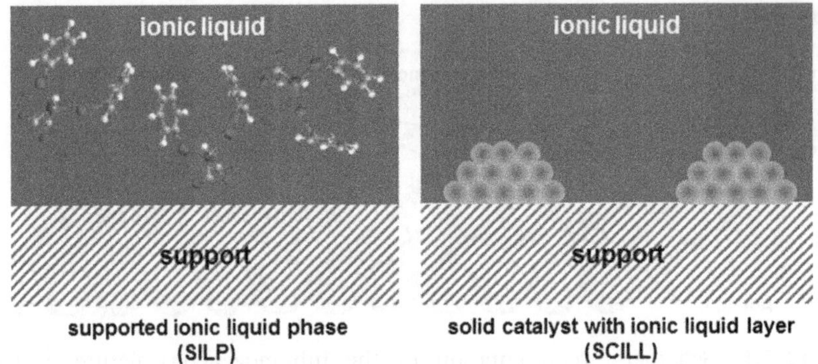

Figure 7.2 Schematic representation of two fundamentally different types of ionic liquid thin film catalysis. Left: Supported Ionic Liquid Phase (SILP). Right: Solid Catalyst with Ionic Liquid Layer (SCILL).

an often pre-organized, homogeneous liquid structure with distinctive physicochemical characteristics and these—often unique—characteristics are exclusively governed by the combination of ions forming the IL material.[8] Hence, by an appropriate choice of the ions (and eventually additives) contained in the ionic liquid material, it is possible to define all relevant liquid properties for the specific needs of the SILP and SCILL materials.

For SILP materials, important selection criteria for the applied IL are catalyst complex, substrate and product solubility as well as an appropriate solvent nucleophilicity to stabilize the dissolved catalyst complex without deactivation. In the SCILL concept the selection criteria for the IL selection are surface wettability, specific IL–catalyst interactions and product solubility. Both methodologies, SILP and SCILL constitute attractive ways to circumvent the lack of uniformity of solids in traditional heterogeneous catalysis. Additionally, both approaches provide great potential to create materials with new surface properties, as the transfer of specific ionic liquid properties to the solid surface may result in "designer surfaces" having properties which are impossible to realize with any other synthetic approach.

In principle, all ionic liquids can be contacted with a solid surface and therefore, looking at the tremendous numbers of publications in the field of "ionic liquids" (exceeding 4800 in the year 2012), it is anticipated that the concept of "supported ionic liquids" will benefit from this scientific input.[9]

A common way to contact ionic liquids with surfaces is the covalent anchoring of ionic liquid fragments onto a—usually pre-treated—support.[10–20] The first ionic liquid monolayer becomes part of the support material, thereby losing certain bulk phase properties like solvation strength, conductivity, viscosity. Such a monolayer can be enough to modify the catalytic properties of a solid catalyst in the sense of SCILL catalysis. Multilayers of ionic liquid are required, however, to provide the uniform solvent environment required to properly immobilize a homogeneous complex in the SILP approach. Alternatively, the coated IL may itself be an active catalyst (*e.g.* in the case of supported Lewis- or Brønsted-acidic ILs) or may contain the active catalyst in the form of dispersed nanoparticles.[21] According to a more detailed differentiation of different IL functions in ionic liquid thin film catalysis, the following abbreviations can be found in the literature:

- Supported ionic liquid catalysis/catalysts (SILC/SILCA);[22]
- Solid catalysts with ionic liquid layer (SCILL);[23,24]
- Supported ionic liquid nanoparticles (SILnPs);[25]
- Supported ionic liquid phase (SILP);[26]
- Supported ionic liquid phase catalyst (SILPC);[27,28]
- Ionic liquid crystalline-SILP (ILC-SILP);[29]
- Structured SILP (SSILP);[30]
- Supported ionic liquid-like phase (SILLP);[31]
- Polymer supported ionic liquid (PSIL),[32] and
- Supported ionic liquid membrane (SILM).[33]

The synthesis of all types of SILP and SCILL materials is usually straightforward. The thin film of ionic liquid is fixed in most cases on the surface by physisorption, only in few cases by chemisorption.[34] The ionic liquid is mixed with the support and the catalyst complex (if applied) in a low boiling solvent. The solvent is then removed by evaporation yielding a dry, free-flowing powder as the SILP or SCILL catalyst. Depending on the amount of ionic liquid and the pore structure of the support material film thicknesses between 3 and 30 nm are typically obtained.

Solid-state NMR studies of different amounts of ionic liquid on silica support indicated that below a critical value of 10 vol% ionic liquid loading (with respect to the pore volume of the uncoated support) small islands of ionic liquids exist on the support.[35,36] At values higher than 10 vol%, a complete surface coverage with ionic liquid was observed (see below for more details). Complete surface coverage is an important prerequisite for the efficient immobilization of homogeneous catalyst complexes in SILP systems. Most catalytically active transition metal complexes would lose activity and, more important, selectivity upon interaction with the support surface or in the constrained environment of a chemically imperfect solid surface.

From an engineering point of view SILP and SCILL materials offer many advantages compared to classical gas–liquid or liquid–liquid systems, especially:

- A high specific fluid–fluid and fluid–surface contact area provided by the porous support;
- A thin film of ionic liquid that minimizes mass transport problems at the IL-side of the fluid–fluid phase boundary;
- Adjustable solvent properties of the IL, *e.g.* solubility of formed products;

- Thermal stability of most ionic liquids up to 200 °C with minimum vapor pressures;
- Application of fixed-bed or fluidized-bed reactor technology;
- Efficient immobilization of transition metal complexes without the need for sophisticated ligand modification (as the only requirement to prevent ligand leaching is a low vapor pressure of the ligand in the IL solvent).

The use of SILP materials has been highlighted in several actual reviews, including both gas-phase and liquid phase, slurry applications.[21,37] Also the actual achievements in SCILL catalysis have been summarized recently with a strong focus on explaining the observed selectivity effects.[20b] With respect to the application of SILP and SCILL materials in liquid-phase slurry reactions, the leaching of the ionic liquid from the support by either cross-solubility or convective forces is the most crucial issue. While many SCILL systems are based on some kind of chemical IL-site interaction that prevents rapid leaching, purely physisorbed ILs in SILP materials tend to lose their catalytic ionic liquid phase in prolonged contact with liquids unless the liquid is of extremely unipolar nature (*e.g.* in the case of liquid alkane isomerization[38]).

In contrast, the SILP and SCILL catalysts have demonstrated in many applications very high stability and lifetime when contacted with purely gaseous reactants and products.[21,37] As this approach builds on the volatility of the reaction products it is clearly limited to feedstock and products with considerable vapor pressure. Note, that every molecule that can be analyzed by gas chromatography is in principal eligible for SILP or SCILL gas-phase reactions. The removal of high boiling reactants from the SILP or SCILL catalyst requires however a high amount of inert gas stripping which is economically less attractive at least for the production of bulk chemicals. A suitable alternative for performing continuous reactions with high boiling substrates is the combination of SILP catalysis with a supercritical fluid as mobile extraction phase, in particular $scCO_2$.[39,40] To the best of our knowledge this approach has not yet been tested for SCILL catalysis.

In this contribution, we'd like to focus in particular on the influence of the support material and ionic liquid loading on the performance of SILP and SCILL materials using specific examples to illustrate the most relevant effects. Moreover, recent developments in the larger scale production of these materials are highlighted. Most examples presented in more detail will deal with SILP catalysis which historically has been developed first. Therefore more detailed work is available on the optimization of these materials for specific catalytic purposes.

In the first studies on SILP catalysis, work was mainly focusing on the immobilization of homogeneous transition metal complexes within the thin ionic liquid film. Homogeneous catalysts, in contrast to their heterogeneous counterparts, have a uniform molecular structure and can be easily modified by the use of dedicated ligands in terms of reactivity, selectivity and stability.[41] The main drawback of homogeneous catalysts is, however, the elaborate recycling of the dissolved catalyst from the reaction mixture, usually accomplished by distillation or extraction.[42] This issue, which currently limits more applications of homogeneous catalysts in continuous processes, can be circumvented elegantly by the SILP technology if the reaction is performed in SILP gas-phase contact. Since the ionic liquid does not have any technically relevant vapor pressure, it is not removed *via* gas-phase leaching and catalyst stabilities have been found to be very high. Moreover, the gas-phase has no solution power for the catalyst which means that catalyst immobilization in SILP–gas-phase systems does not require any dedicated ligand modification.

7.2 Support Texture Influence Exemplified for SILP Hydroformylation Catalysis

A variety of different support materials was already tested by Riisager *et al.* in the first study of SILP catalysis for the gas-phase hydroformylation of propene using a Rh-SX (SX = sulfoxantphos) complex dissolved in a thin film of [BMIM][n-$C_8H_{17}OSO_3$].[43] Table 7.1 summarizes the textural properties and the performance of the SILP materials. The ionic liquid loading differed slightly for the different support materials used, but from the results it becomes obvious that the pore volume and pore diameter are crucial parameters for efficient SILP catalysis, probably more important than high surface area. Silica, having the highest pore volume and the largest mean pore diameter, showed the highest turn-over frequency of 19.9 h^{-1}, more than two times higher than all other maximum activities. The three other support materials all differed in their surface area and possessed significantly smaller pore volumes and pore diameters. All three SILP catalysts based on these materials exhibited lower but constant activity and selectivity. The silica based SILP catalyst deactivated rapidly within the first

Table 7.1 Textural properties and catalytic performance of Rh-SX–SILP catalysts for gas-phase propene hydroformylation.[a]

	Silica	*Alumina*	*Zirconia*	*Titania*
Chemical formula	SiO_2	Al_2O_3	ZrO_2	TiO_2
BET surface area/$m^2\ g^{-1}$	298	103	100	309
Pore volume/mL g^{-1}	1.02	0.25	0.13	0.37
Average pore diameter/nm	13.7	9.8	5.0	4.7
Particle size/μm	63–200	63–200	15	n.a.
Max. activity TOF_{max}/h^{-1}	19.9	8.6	7.4	9.9
Final activity TOF_{60h}/h^{-1}	2.1[b]	7.0	3.0	5.4
Max. selectivity/$\%_{n\text{-butanal}}$	95.2	91.9	95.0	94.0
Final selectivity/$\%_{n\text{-butanal}}$	0[c]	91.6	93.9	90.0

[a]Reaction conditions: $T = 100$ °C, $p = 10$ bar, $C_3H_6 : H_2 : CO = 1 : 1 : 1$, $w_{Rh} = 0.2$ wt%, L/Rh = 10, α_{IL} approx. 10 vol%.
[b]Activity after 20 h.
[c]Selectivity after 20 h.

20 h of reaction, the reason being an interaction between the excess ligand and the surface silanol groups.[4b,44]

The small pore volume and pore diameter of alumina, titania and zirconia allowed only a minor amount of ionic liquid to be dispersed onto the support. Higher loading with ionic liquid would immediately result in flooding of the pore system and this blockage would hamper efficient mass transport inside the catalyst particle, hence low overall activity.

Ideal supports for SILP catalysis, therefore, allow a large amount of ionic liquid to be deposited and still leave enough pore space for efficient mass transport by diffusion. Especially hierarchical materials with small meso-pores (2 nm $< d_{pore} <$ 5 nm) for reaction and larger meso-pores (5 nm $< d_{pore} <$ 20 nm) for transport would constitute an ideal SILP support material.[45]

7.3 Surface Acidity Exemplified for SILP Hydroformylation Catalysis

Different forms of pre-treated silica were tested for propene hydroformylation, namely dried silica (110 °C, 0.1 mbar, 24 h), partially dehydroxylated silica (500 °C, 15 h) and silylated silica (silylating agent = $(CH_3)_2Si(OCH_3)_2$, 90 °C).[46] High-temperature treatment will remove the surface bound silanol groups by release of water while the silylation coats the silica surface with a monolayer of dimethyl-silane upon release of methanol as shown in Scheme 7.1.

The relevance of reducing the number of surface silanol groups is linked to the detrimental interaction of basic phosphine ligand in an acid–base type fashion with the support material as depicted in Scheme 7.2.

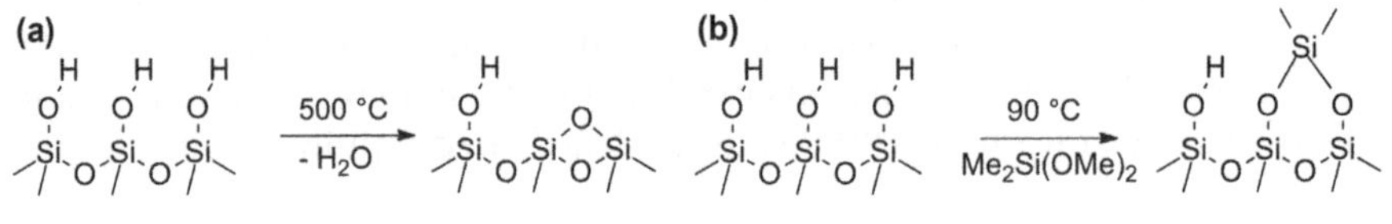

(a) 500 °C, - H_2O (b) 90 °C, $Me_2Si(OMe)_2$

Scheme 7.1 Surface treatment of silica: (a) thermal dehydroxylation, (b) silylation.

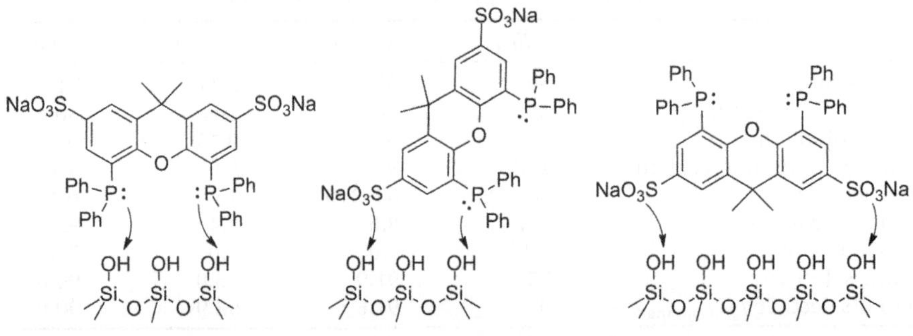

Scheme 7.2 Detrimental interaction between surface silanol groups Si–OH and the basic diphosphine ligand sulfoxantphos at the solid surface according to Shylesh *et al.*[44]

Since the SX ligand excess is determining the catalyst stability and selectivity, reduction of the excess *via* this surface interaction will deactivate the catalyst.[4b] Inactive Rh clusters or metallic Rh can be formed if not enough ligand is present in the ionic liquid film to stabilize the active species $HRh(CO)_2(SX)_2$. The reason for the low but constant activity and selectivity of alumina, titania and zirconia in the previous experiments stems from the fact that these materials have less acidic groups at the surface.

Interestingly, the silylation resulted in a rather poor catalyst performance as shown in Figure 7.3. The poor activity can be related to the now poor wettability of the hydrophilic ionic liquid on the hydrophobic support. It can be assumed that this poor wettability results in formation of rather large droplets of ionic liquid in the pores of the support, thereby reducing the reactive surface and clogging the transport pores. The selectivity, however, should be not affected by this behavior, but also decreases over time. The highest activity and best stability was achieved with the partially dehydroxylated silica material. Here, the number of silanol groups that can interact with the excess ligand was reduced to such an extent that enough SX ligand was present in the ionic liquid film to maintain the active species intact. Furthermore, the wettability of the surface was only slightly changed, since some silanol groups remained on the surface. The latter have been found to be important to interact with the ionic liquid *via* hydrogen bonding to allow an even distribution of the ionic liquid over the support surface.

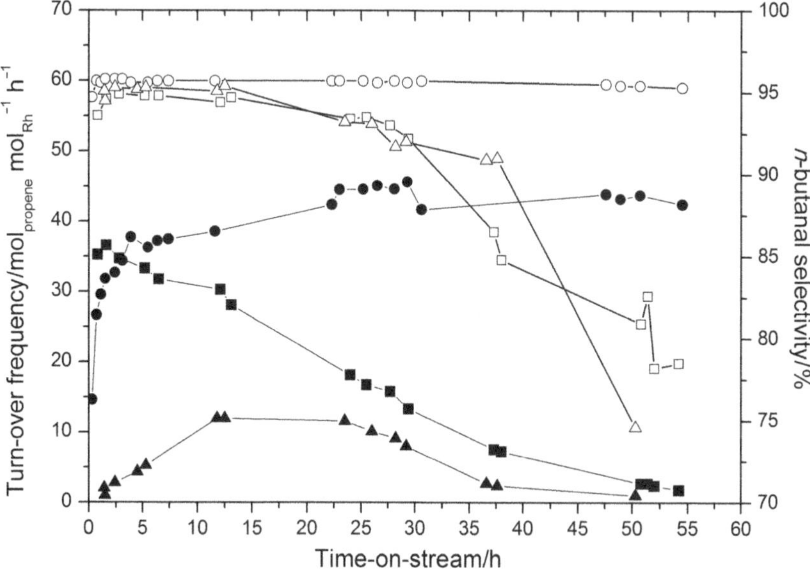

Figure 7.3 Continuous propene hydroformylation activity (closed symbols) and selectivity (open symbols) of SILP Rh-SX–[BMIM][n-$C_8H_{17}OSO_3$] catalysts (SX/Rh = 10) over time. Dried support (α_{IL} = 0.5, ■, □), partially dehydroxylated support (α_{IL} = 0.1, ●, ○) and silylated support (α_{IL} = 0.5, ▲, △).

Table 7.2 Results from 1-butene hydroformylation using Rh-BzP–SILP catalysts based on different silica support materials.[a]

	Silica	*Aerolyst N*[b]	*Aerolyst T*[c]	*Aerolyst Cs*[d]
Chemical formula	SiO_2	SiO_2	SiO_2	SiO_2
BET surface area/m^2 g^{-1}	298	228	187	302
Pore volume/mL g^{-1}	1.02	0.89	0.55	0.94
Average pore diameter/nm	13.7	15.5	11.7	12.4
Particle size/μm	63–200	2000–5000	2000–5000	2000–5000
Conversion/%	39.7	30.0	32.4	26.0
Selectivity/%$_{n\text{-pentanal}}$	99.4	99.7	99.6	100.0

[a]Reaction conditions: $T=80$ °C, $p=10$ bar, $C_4H_8:H_2:CO=1:2.5:2.5$, $w_{Rh}=0.2$ wt%, L/Rh = 10, $\alpha_{IL}=10$ vol%, [EMIM][NTf_2].
[b]Native support.
[c]Thermally treated support.
[d]Thermally treated and Cs doped support.

Pore blocking is avoided by using a low ionic liquid loading; hence both activity and stability were excellent.

Industrial support materials have been tested in 1-butene hydroformylation using Rh-BzP–SILP (BzP = 1,1′-((3,3′-di-*tert*-butyl-5,5′-dimethoxy-[1,1′-biphenyl]-2,2′-diyl)bis(oxy))bis(3,3,4,4-tetraphenylphos-pholane) "benzopinacol") catalysts.[47] Aerolyst 3038, a silica material without micro-pores, has been used for catalyst synthesis as native (Aerolyst A), partially dehydroxylated (Aerolyst T) and Cs doped (Aerolyst Cs) support material, respectively. Note that for the hydrolytically labile diphosphite ligand BzP the hydrophobic ionic liquid [EMIM][NTf_2] has been used instead of [BMIM][n-$C_8H_{17}OSO_3$], which was the standard IL for Rh-SX systems (Table 7.2).

Similar to the results obtained with silylated silica, the complete blocking of surface acidic silanol groups is not favorable for the SILP catalyst activity. While all catalysts showed excellent selectivity higher than 99% *n*-pentanal, the activity for the Cs doped support was lowest. Since the non-coated support exhibited high surface area of 303 m^2 g^{-1} as well as large pore volume and pore diameter close to the standard silica material, the textural properties cannot be the reason for the lower activity. Again, it can be assumed that the ionic liquid did not coat the complete surface but rather formed islands inside the particle, thereby blocking the pore system.[35] The native support as well as the partially dehydroxylated support showed better activity since here a small amount of acidic groups was left on the surface, enhancing the dispersion of the ionic liquid film *via* hydrogen bonding.

7.4 Influence of Surface Reactivity in SILP Catalysis

Undesired surface reactivity can be suppressed by coating the surface sites with appropriate inert compounds as schematically summarized in Scheme 7.3. Coating the silica surface with a monolayer of ionic liquid fragments has been used by Mehnert *et al.* in their pioneering work on SILP catalysts for hydroformylation in 2002.[4a]

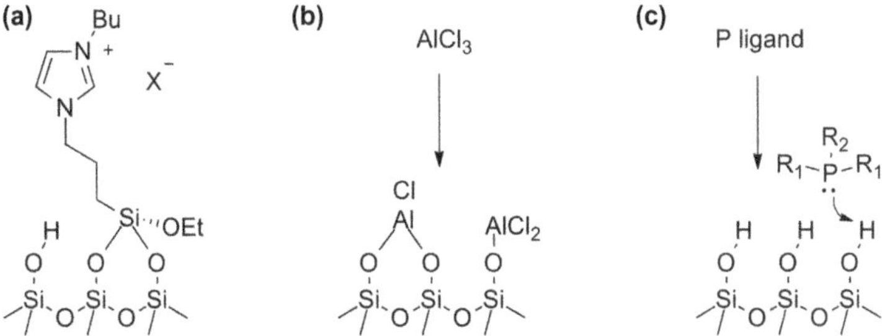

Scheme 7.3 Different possibilities for the chemical modification of a silica support surface: (a) covalent anchoring of IL fragment, (b) reaction with $AlCl_3$, (c) sacrificing phosphine ligand.

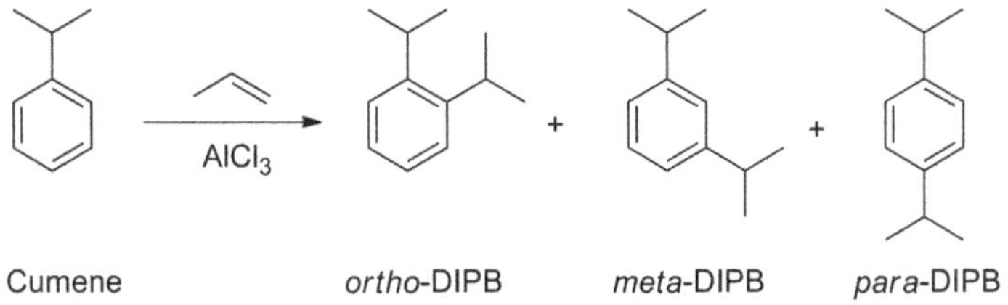

$AlCl_3$

Cumene *ortho*-DIPB *meta*-DIPB *para*-DIPB

Scheme 7.4 Friedel–Crafts alkylation of cumene with propene catalyzed by Lewis acid $AlCl_3$.

Additional ionic liquid can be deposited onto this monolayer, if required. Although this surface modification can result in uniform dispersion of the ionic liquid film on the surface, the elaborate multi-step synthesis of the ionic liquid fragment limits the practicability of this approach for industrial applications with larger catalyst inventories.

Chemical modification of the support surface has been found to be crucial also in SILP-type Friedel–Crafts alkylation reactions where an acidic ionic liquid film on silica served as the catalyst phase for the isopropylation of cumene (Scheme 7.4).[48] It was found that surface Si–OH groups had to be removed by reaction with $AlCl_3$ to obtain a catalyst of sufficient acidity after the subsequent surface coating with acidic [EMIM][$AlCl_4$].

When using non-modified silica support in the SILP catalyst preparation, the Friedel–Crafts alkylation of cumene with propene produced the kinetically favored diisopropylbenzene (DIPB) isomers indicating that the obtained catalyst was not acidic enough to produce the thermodynamic mixture of DIPBs. In contrast, in the biphasic systems in the absence of silica the thermodynamically more stable *meta*-DIPB product is formed predominantly as summarized in Table 7.3. The isomerization of the formed products to yield *meta*-DIPB predominantly is catalyzed by the most acidic IL ions, namely $[Al_2Cl_7]^-$ and $[Al_3Cl_{10}]^-$. During the IL coating of non-modified silica support a certain share of these acidic anions react with the surface Si–OH groups reducing the acidity of the remaining ionic liquid film.

Table 7.3 Product composition in the Friedel–Crafts isopropylation of cumene: relative energy of different DIPB isomers and comparison of the DIPB selectivity for the biphasic and the SILP catalyzed reaction.[a]

	Time/min[b]	*meta-DIPB*	*ortho-DIPB*	*para-DIPB*
Rel. energy/kJ mol^{-1}[c]	—	0	18.05	0.25
Biphasic	23	60	~0	40
SILP, native[d]	4	44	5	51
SILP, coated[e]	4	57	2	41

[a]Reaction conditions: $T = 150$ °C, $p = 6.9$ bar, cyclohexane, $c_{cumene} = 3.8$ mmol mL^{-1}, cumene : catalyst = 36 : 1.
[b]Reaction time required for approx. 80% cumene conversion.
[c]Relative internal electronic energy, estimated using DFT-B3LYP calculations.
[d]Native SiO_2 support.
[e]$AlCl_3$ pre-treated SiO_2 support.

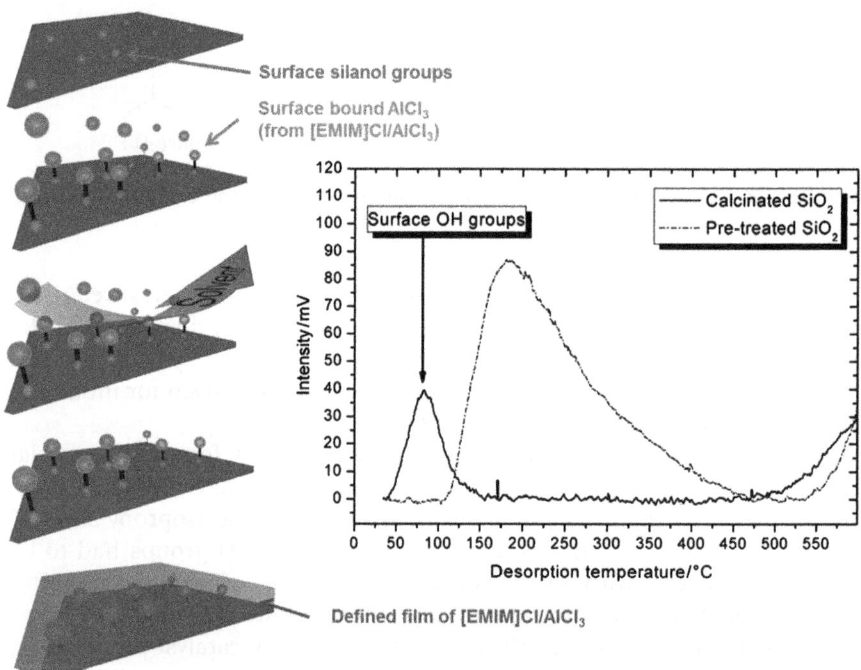

Figure 7.4 Pre-treatment of a silica support with $AlCl_3$ for use in acidic SILP catalysis (left). NH_3 TPD of both native and pre-treated support materials (right).

In a similar approach as reported by the group of Hölderich,[49] Joni *et al.* covered the silica surface by a monolayer of chloroaluminate.[48] Excess $AlCl_3$ was removed by washing with dichloromethane. All steps of the support modification prior to SILP catalyst synthesis are schematically shown in Figure 7.4.

The ammonia TPD of the native support showed a clear desorption peak between 50 and 120 °C for surface silanol groups, while the pre-treated support only gave a broad signal at significantly higher temperatures. This indicates that the support became more acidic during the surface modification, since higher thermal energy was required to desorb the ammonia. However, the resulting pre-treated SILP material showed no sign of isomerization activity when exposed to a non-equilibrated mixture of DIPB isomers, so the surface bound chloroaluminate species behave inertly towards isomerization. A SILP catalyst with 20 vol% [EMIM][$AlCl_4$] on non-coated silica isomerized a non-equilibrated DIPB mixture within 60 min due to the ionic liquid's acidity. A similar SILP catalyst containing 10 vol% [EMIM][$AlCl_4$] on pre-treated silica showed much higher isomerization activity, indicating that the acidity solely stems from the acidic ionic liquid film. When this coated SILP catalyst was used in Friedel–Crafts alkylation of cumene, the expected thermodynamically most stable isomer (*m*-DIPB) was formed with a product composition closely resembling the one from the liquid–liquid biphasic reaction without support present. From these results it can be concluded that the surface reactivity can play a crucial role in determining the SILP catalyst's performance. Clever surface modification can significantly enhance the activity and selectivity of the catalyst. Also the stability of the acidic SILP catalyst on pre-treated silica is very remarkable. The SILP catalyst has been recycled three times without significant changes in performance in slurry applications.[48a] In the continuous gas-phase alkylation the same SILP materials exhibited stable catalytic performance for more than 200 h time-on-stream if pre-dried feed was applied.[48b]

Another approach to surface modification is the use of phosphines that can act as sacrificial base to block surface acidity. For Rh-SX–SILP hydroformylation catalysts two cheap monodentate phosphines, namely TPP and its sulfonated analogue TPPTS, have been used to modify silica supports prior to their use in SILP catalyst preparation. Silica was immersed in a methanolic solution of TPP or TPPTS for several hours, followed by removal of the methanol *in vacuo*. The as pre-treated support was then used to prepare the Rh-SX–SILP catalyst using [BMIM][n-$C_8H_{17}OSO_3$]. To test the stability of the ionic liquid film on these supports, the SILP materials were heated at 80 °C in toluene for 30 h, after which the organic phase was analyzed with respect to their S and N content (with high S and N contents indicating leaching of IL and SX ligand from the support).[50]

From Figure 7.5 it becomes obvious that the best fixation of both ionic liquid (N-content and S-content) and sulfoxantphos (only S-content) is achieved with the native silica support. Here, as previously discussed, the wettability of the ionic liquid on the surface is best due to the interaction between the surface silanol groups and the ionic liquid *via* hydrogen bonding. If the silanol groups are blocked completely by TPP, high IL and ligand leaching is observed, since the polar ionic liquid cannot wet the nonpolar surface any more. The use of sulfonated TPPTS improves this

S and N content in solvent/ppm
0
5
10
15
20
25
S
N
native
TPP
TPP
TPPTS
TPPTS
Silica pre-treatment

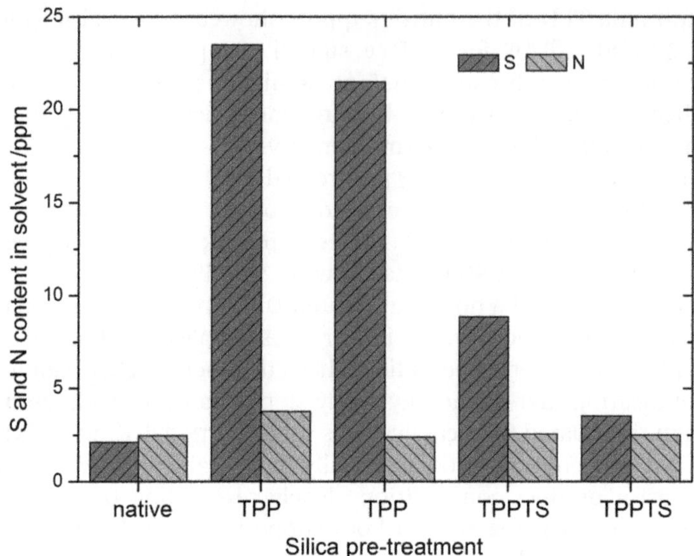

Figure 7.5 Leaching experiments of SILP materials in slurry-mode after support pre-treating with TPP and TPPTS: S and N content of the toluene phase after heating different silica based SILP materials at 90 °C for 30 h.

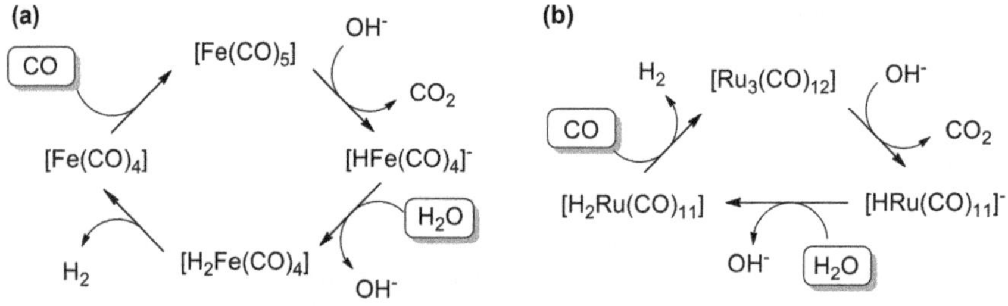

(a)
CO
[Fe(CO)5]
OH-
CO2
[Fe(CO)4]
[HFe(CO)4]-
H2O
H2
[H2Fe(CO)4]
OH-
(b)
H2
[Ru3(CO)12]
OH-
CO
CO2
[H2Ru(CO)11]
[HRu(CO)11]-
OH-
H2O

Scheme 7.5 Proposed mechanisms for the (a) iron carbonyl and (b) ruthenium carbonyl catalyzed water–gas shift reaction.

interaction; however, leaching is still slightly higher compared to the native support.

So far, we have only discussed cases of undesired surface activity. There are, however, examples, where the reactivity of the support surface is beneficial not only for a uniform distribution of the ionic liquid on the support, but also for improving the catalyst activity. One important example is the ultra-low temperature water–gas shift (ULT WGS) reaction using Ru–SILP catalysts.[51] The exact mechanism for homogeneously catalyzed WGS using iron or ruthenium complexes is still under debate, but a significant influence of the system basicity on the low temperature reactivity of homogeneous Fe and Ru-complexes has been reported by several authors (Scheme 7.5).[52]

Table 7.4 Textural properties of different support materials and catalytic results in SILP WGS.[a]

	Silica gel 100	*γ-Alumina*	*Boehmite*
Supplier	Merck KgaA	Süd-Chemie AG	Sasol Germany GmbH
Chemical formula	SiO_2	γ-Al_2O_3	AlO(OH)
Surface area/m^2 g^{-1}	334	101	160
Pore volume/mL g^{-1}	0.97	0.51	1.10
Mean pore diameter/nm	12.1	17.2	23.7
Initial activity TOF_0/h^{-1}	0.1	1.7	6.3
Induction time/h[b]	64	95	230
Max. activity TOF_{max}/h^{-1}	0.9	4.6	8.5
Productivity TON_{100}[c]	44	376	662

[a]Reaction conditions: $T = 120$ °C, $p = 1$ bar, $CO:H_2O = 1:2$, $w_{Ru} = 2$ wt%, $\alpha_{IL} = 10$ vol%, [BMMIM][OTf].
[b]Time to reach maximum activity.
[c]Turn-over number TON after 100 h.

In both the iron and ruthenium based mechanism the concentration of basic OH^- is important, since it involves activation of the most stable complex $Fe(CO)_5$ and $Ru_3(CO)_{12}$, respectively. Since these catalysts were less active at lower pHs, the rate-limiting step was presumed to be the activation of coordinated CO by a nucleophilic attack of OH^-. In order to increase the hydroxyl concentration in the ionic liquid film of a promising WGS SILP catalyst, the support was changed from standard silica (surface acidic silanol groups) to more basic alumina from Süd-Chemie AG (now Clariant SE). Indeed, the activity of the WGS reaction was increased by almost an order of magnitude when changing from silica to boehmite as summarized in Table 7.4.[51]

The boehmite structure supplied sufficient surface hydroxyl groups to also increase the concentration of the latter inside the ionic liquid film. Additionally, the [BMMIM][OTf] wettability on all supports was sufficiently high to ensure long-term stability, as indicated by the turn-over number after 100 h time-on-stream.

7.5 Surface Wettability Exemplified for SILP Hydroformylation Catalysis

To test the importance of wettability compared to other structural parameters, non-acidic carbon supports were tested in 1-butene hydroformylation using Rh-SX–SILP catalysts containing [BMIM][n-$C_8H_{17}OSO_3$]. Two different carbon supports, active charcoal (ACC) and carbon nanofibres (CNF), were applied.[47] Similar to Aerolyst, these supports did not contain any micro-pores but exhibited large surface areas and, in the case of ACC, an extremely large pore volume of 1.72 mL g^{-1} as summarized in Table 7.5. However, the carbon surface was non-functionalized and could be regarded as hydrophobic. For all carbon based supports the activity was significantly lower compared to silica, regardless of the higher surface area and larger

Table 7.5 Results from 1-butene hydroformylation using Rh-SX–SILP catalysts based on different carbon support materials.[a]

	Active charcoal			*Carbon nanofibres*
Supplier	Fluka			Univ. Bayreuth, CVT
BET surface area/m^2 g^{-1}	2055			118
Pore volume/mL g^{-1}	1.72			n.a.
Average pore diameter/nm	3.4			6.1[b]
Particle size/μm	<40			500–5000
L/Rh	4	10	4	4
Conversion/%	11.0	12.7[c]	19.3[c]	0.9
Activity TOF/h^{-1}	133	164	234	56
Selectivity/$\%_{n\text{-pentanal}}$	97.2	97.2	96.8	96.5

[a]Reaction conditions: $T = 120$ °C, $p = 10$ bar, $\tau = 28$ s, $C_4H_8 : H_2 : CO = 1 : 1.3 : 1.3$, $w_{Rh} = 0.2$ wt%, L/Rh = 4, 10, $\alpha_{IL} = 10$ vol%, [BMIM][n-$C_8H_{17}OSO_3$].
[b]Open pore network.
[c]$\tau = 100$ s, $C_4H_8 : H_2 : CO = 1 : 1 : 1$.

pore volume (in the case of ACC). Longer contact times and lowering the L/Rh ratio increased the conversion to approx. 20%. However, due to the poor wettability of the polar ionic liquid on the carbon surface no complete coverage was achieved and the ionic liquid was removed from the support over time. A guard bed, placed below the catalyst bed inside the reactor, turned yellow, resembling the color of the ionic catalyst solution, after a few hours of reaction in all cases of carbon based SILP catalysts.

In agreement with previous observations it can be concluded that wettability is an important parameter for creating a uniform and thin film of ionic liquid that remains attached to the surface during operation. Removing the surface acidity in *e.g.* silica completely by silylation or Cs doping lowers the wettability to such extents that the ionic liquid forms islands inside the pore system and can easily be removed by convective gas flow. The same effect seems to happen when using support materials that do not possess any acidic functionality, such as *e.g.* carbon.

The criteria for successful and stable ionic liquid film formation can be roughly listed in the following order:

- **Wettability:** poor wettability leads to the formation of islands inside the pore system or the ionic liquid does not penetrate the pore network completely. This results in easy removal of the ionic liquid from the support by convective flow. Catalyst activity and stability in such SILP systems tend to be poor.
- **Pore diameter:** the ideal support should have large transport pores for efficient reactant diffusion into the particle. Small pore diameters such as *e.g.* in zeolites are not ideal, since the ionic liquid film of 1 nm thickness already leads to pore flooding.
- **Surface area:** the higher the surface area, the larger the specific exchange area and thus the higher the transfer rate of substrate into the ionic catalyst solution.

- **Pore volume:** while a high pore volume is mandatory in gas cleaning applications with SILP absorber/adsorber materials (higher loading leads to higher capacities) lower film thicknesses and thus lower pore volumes can be accepted in SILP catalysis.
- **Surface reactivity:** interaction between the surface and the ionic liquid should only occur *via* hydrogen bonding to facilitate the uniform distribution of the ionic liquid. No chemical interaction between the ionic liquid and the support or the support and the dissolved catalyst is usually desired. Blocking of undesired surface reactivity can be achieved by thermal or by chemical surface modifications. If the catalytic reaction profits from an acid or basic environment, this acidity/basicity can be introduced into the system *via* an acidic/basic IL, *via* an acidic/basic support or by a combination of both.

7.6 Ionic Liquid Loading in SILP and SCILL Catalysis

SILP and SCILL materials are characterized by the amount of ionic liquid loading on the support that is typically expressed as mass or volume ratio. The mass ratio does not give any information on the material properties. Note that in a SILP or SCILL material with a mass ratio of 100 wt% the pore system of the initial contact is not necessarily filled with ionic liquid completely since most ionic liquids have densities higher than 1.

$$w_{IL} = \frac{m_{IL}}{m_{\text{support}}} \times 100\% \qquad (7.1)$$

In contrast, indication of the volumetric IL loading—the so-called "alpha-value" of the SILP or SCILL catalyst—requires knowledge of the neat support (in the case of SILP materials) or catalyst (in the case of SCILL materials) pore volume but gives a better impression of the structure of the resulting SILP or SCILL material. Small α-values indicate a thin layer or islands of ionic liquid in the support, while at $\alpha = 100\%$ the complete pore system is filled with ionic liquid.

$$\alpha_{IL} = \frac{V_{IL}}{V_{pore} \times m_{\text{support}}} \times 100\% \qquad (7.2)$$

For the system [EMIM][NTF_2] on a silica 100 support the formation of the IL film has been followed by means of solid-state NMR (MAS-NMR) measurements.[35] The native support material was analyzed with respect to its ^{29}Si and ^{1}H signals. The ^{29}Si spectrum contained the typical Q2, Q3 and Q4 group signals and remained practically unaffected by the different ionic liquid loadings in all experiments. The ^{1}H spectrum gave signals at 2.7 ppm and 2.2 ppm, which were assigned to hydrogen bonded silanols (see type D and E in Figure 7.6), and narrow ones at 1.4 and 1.6 ppm, indicative of isolated silanol groups (see type C in Figure 7.6).

However, the pattern and the intensity of the ^{1}H-signals changed upon addition of ionic liquid. At low ionic liquid loading of 5 vol% the surface is

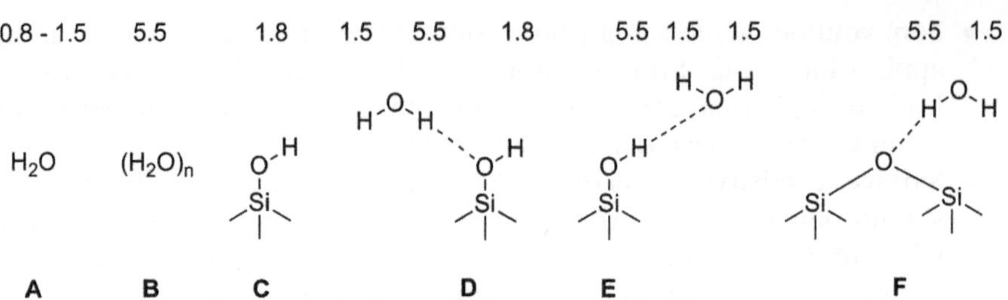

Figure 7.6 Overview of possible ¹H signals for possible interactions between water and silica.

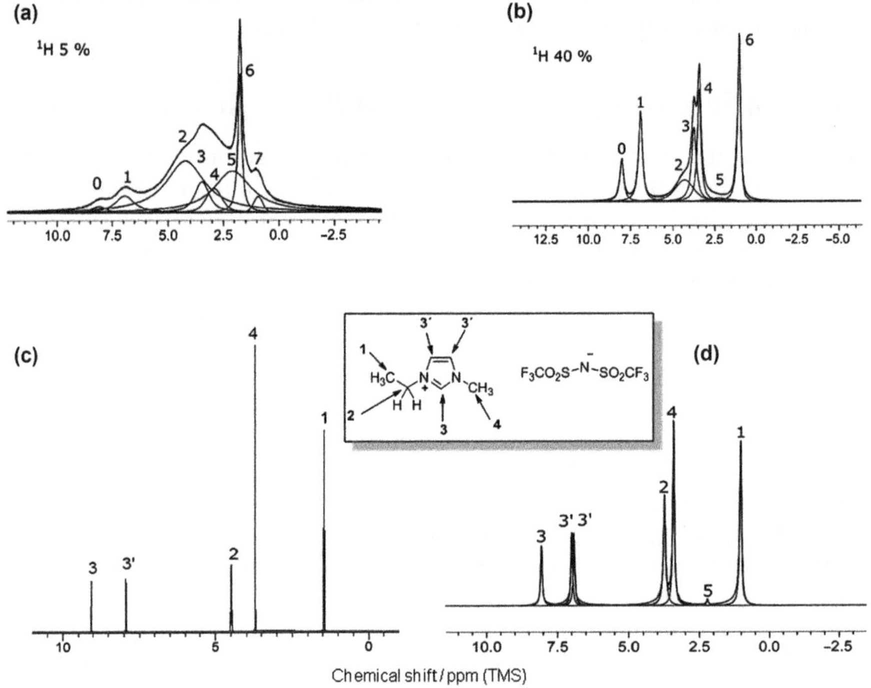

Figure 7.7 ¹H NMR spectrum of SILP material loaded with 5 vol% (a) and 40 vol% (b) of the ionic liquid [EMIM][NTf_2]. Protons are labeled according to their carbon position at the imidazolium cation as indicated in the inlet. Simulated (c) and experimental (d) ¹H NMR spectra of [EMIM][NTf_2] are also shown.

not completely covered by the ionic liquid in the NMR spectrum, the Si–OH peak at 1.76 ppm (type C) still represents the strongest peak. At 10 vol%, the surface signals are dampened completely and the signals for the pure ionic liquid become prominent. At higher loadings of 20 and 40 vol% the ¹H signals of the SILP material resemble those of the pure bulk ionic liquid, as depicted in Figure 7.7

Table 7.6 Continuous gas-phase water–gas shift reaction of a technical syngas feed using a Ru–SILP catalyst with different IL-loading ([BMIM][NTf_2]).[a]

Ionic liquid loading/vol%	0	7.2	14.3	28.6	57.2
CO conversion/%	0	0.9	9.8	5.8	4.0

[a]200 mg SILP catalyst (powder); $T=413$ K; $p=0.3$ MPa; syngas composition: H_2 75%, CO 8%, CO_2 13%, N_2 4%; steam to gas 1 : 3; flow rate (dry gas) 40 mL min^{-1}; catalyst composition: support silica-100; ionic liquid [BMIM][NTf_2]; catalyst precursor $Bz(Et)_3NCl$ [$Ru(CO)_3Cl_2$]; w_{Ru} 2 wt%.

In the continuous gas-phase water–gas shift reaction catalyzed by Ru–SILP materials the ionic liquid loading on silica gel 100 has been varied between 0 (no ionic liquid) and 57 vol% ionic liquid ([BMIM][NTf_2]) and the conversion of CO was monitored using these different catalyst materials at 393 K (Table 7.6).[35]

Clearly, when no ionic liquid was used and the Ru-precursor was purely physisorbed onto the silica support, no WGS activity could be observed under the conditions applied. Increasing the amount of ionic liquid to 7 vol% resulted in minor CO conversion of 0.9% and the maximum conversion of 9.8% was obtained at an ionic liquid loading of 14 vol%. Remarkably, a further increase of the ionic liquid content resulted again in a steady decrease of the CO conversion down to 4% for an alpha-value of 57.2%. Obviously, the homogeneous Ru-complex requires the ionic liquid to become active in the WGS. However, if the ionic liquid loading is too high, blocking of transport pores takes place thereby lowering the overall activity of the SILP material. Hence, it is an important aspect of optimizing SILP and SCILL catalysts to identify the ionic liquid loading for optimum performance of the catalyst. For many SILP systems, IL loadings of 10–20 vol% have been found to be particularly suitable.[21a]

7.7 Larger Scale Synthesis of SILP and SCILL Materials

Very recently, Petronas, a Malayan petrochemical company, disclosed at the EUCHEM Molten Salt and Ionic Liquids Conference 2012 in Celtic Manor, Wales, the commercial operation of a SILP material for mercury removal from natural gas on a technical refinery scale (60 tons of SILP-type material).[53] To the best of our knowledge this marks the first publication of a commercial SILP application but many more applications of SILP and SCILL catalysis are actually in advanced stages of industrial development.

It is obvious that the application of SILP and SCILL materials in industrial-scale quantities attracts attention to the aspect of a cost-efficient and reproducible larger scale synthesis of these materials. The state-of-the-art preparation method at the laboratory scale uses an incipient wetness-type impregnation of the support material (SILP) or heterogeneous catalyst

(SCILL) by a solution of the ionic liquid in auxiliary solvent followed by subsequent removal of that auxiliary solvent under vacuum. Typically this solvent removal is carried out in an ordinary rotary evaporator limiting this kind of preparation to batches of a few hundred grams of SILP or SCILL materials.

Recently, a new and scalable preparation method has been published.[54] This method applies fluidized-bed spray coating to disperse the ionic liquid film onto the support's surface. It involves the fluidization of the uncoated support or catalyst material by means of a temperature controlled gas flow, such as air, nitrogen or argon. Once the support material is fluidized, a solution of the ionic liquid in an auxiliary solvent is sprayed onto the material through a nozzle. The helper solvent is rapidly evaporated due to the gas-flow, resulting in a good dispersion of the catalyst containing ionic liquid onto the support. The fluidized bed with its inherent mixing pattern ensures that the support material is well mixed and evenly coated by the ionic liquid film (Figure 7.8).

The described process has been shown to be applicable to different types of support materials including powders, spheres, agglomerates and extrudates. Moreover, the process offers high flexibility with respect to the specific IL distribution in the solid material. Depending on the temperature and pressure inside the fluidized-bed the auxiliary solvent evaporates faster or slower, giving rise to either shell-type or complete coating of the solid. Post-treatment of the primary SILP or SCILL material with a volatile solvent can lead to a specific re-distribution of the ionic liquid in the solid.

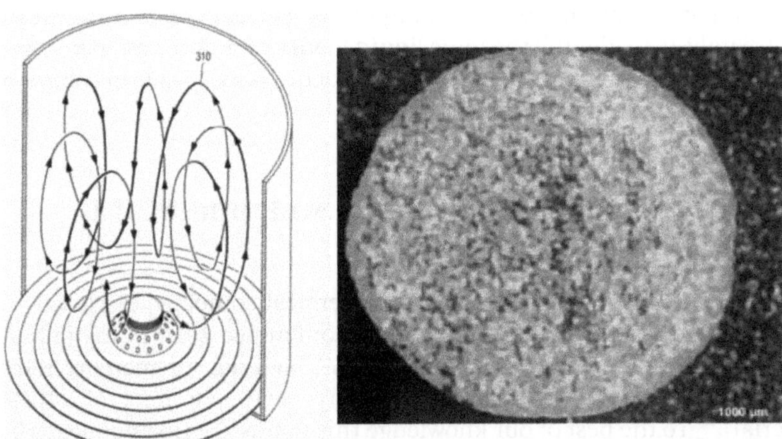

Figure 7.8 Production and quality assessment of SILP catalyst materials. Left: schematic drawing of the toroidal movement of the support particles in an aircoater during impregnation with the solvent-diluted ionic catalyst solution. Right: SEM-EDX picture of the cross section of a spherical γ-alumina coated with $[Ru(CO)_3Cl_2]_2$ in [BMMIM][OTf] to check the elemental distribution [ruthenium (dark grey); sulfur (light grey)]. Adapted from reference 54.

In general the production of SILP and SCILL materials by fluidized bed coating is easily scalable from grams to tons and allows catalyst production with high reproducibility and at low costs. It is anticipated that the recent advances in SILP and SCILL material production will significantly support the successful further implementation of such materials into more and more industrial catalytic processes.

References

1. U. S. Ozkan, ed. *Design of Heterogeneous Catalysts*, Wiley-VCH, Weinheim, 2009.
2. J. J. F. Scholten and R. van Hardeveld, *Chem. Eng. Commun.*, 1987, **52**, 75.
3. J. P. Arhancet, M. E. Davis, J. S. Merola and B. E. Hanson, *Nature*, 1989, **339**, 454.
4. (*a*) C. P. Mehnert, R. A. Cook, N. C. Dispenziere and M. Afeworki, *J. Am. Chem. Soc.*, 2002, **124**, 12932; (*b*) A. Riisager, R. Fehrmann, S. Flicker, R. van Hal, M. Haumann and P. Wasserscheid, *Angew. Chem., Int. Ed.*, 2005, **44**, 815.
5. R. Fehrmann, A. Rissager and M. Haumann, ed. *Supported Ionic Liquids – Fundamentals and Applications*, Wiley-VCH, Weinheim, 2013.
6. P. Wasserscheid and T. Welton, ed. *Ionic Liquids in Synthesis*, Wiley-VCH, Weinheim, 2007.
7. A. Riisager, R. Fehrmann, M. Haumann, B. S. K. Gorle and P. Wasserscheid, *Ind. Eng. Chem. Res.*, 2005, **44**, 9853.
8. P. Wasserscheid and W. Keim, *Angew. Chem., Int. Ed.*, 2000, **39**, 3772.
9. SciFinder search for the term "ionic liquid", January 2013
10. (*a*) K. Yamaguchi, C. Yoshida, S. Uchida and N. Mizuno, *J. Am. Chem. Soc.*, 2005, **127**, 530; (*b*) J. Kasai, Y. Nakagawa, S. Uchida, K. Yamaguchi and N. Mizuno, *Chem.–Eur. J*, 2006, **12**, 4176; (*c*) R. Ciriminna, P. Hesemann, J. J. Moreau, M. Carraro, S. Campestrini and M. Pagliaro, *Chem.–Eur. J*, 2006, **12**, 5220; (*d*) T. Sasaki, C. Zhong, M. Tada and Y. Iwasawa, *Chem. Commun.*, 2005, 2506; (*e*) C. Zhong, T. Sasaki, M. Tada and Y. Iwasawa, *J. Catal.*, 2006, **242**, 357; (*f*) W. Chen, Y. Zhang, L. Zhu, J. Lan, R. Xie and J. You, *J. Am. Chem. Soc.*, 2007, **129**, 13879.
11. B. Karimi and D. Enders, *Org. Lett.*, 2006, **8**, 1237.
12. M. H. Valkenberg, C. de Castro and W. F. Hölderich, *Green Chem.*, 2002, **4**, 88.
13. C. P. Mehnert, R. A. Cook, N. C. Dispenziere and M. Afeworki, *J. Am. Chem. Soc.*, 2002, **124**, 12932.
14. (*a*) Y. Jin, P. Wang, D. Yin, J. Liu, H. Qiu and N. Yu, *Microporous Mesoporous Mater.*, 2008, **111**, 569; (*b*) Y. Jin, D. Zhuang, N. Yu, H. Zhao, Y. Ding, L. Qin, J. Liu, D. Yin, H. Qiu, Z. Fu and D. Yin, *Microporous Mesoporous Mater.*, 2009, **126**, 159.
15. A. Monge-Marcet, R. Pleixats, X. Cattoën and M. W. C. Man, *Catal. Sci. Technol.*, 2011, **1**, 1544.

16. H. Q. Yang, X. J. Han, G. Li and Y. W. Wang, *Green Chem.*, 2009, **11**, 1184.
17. J. Y. Shin, B. S. Lee, Y. Jung, S. J. Kim and S. G. Lee, *Chem. Commun.*, 2007, 5238.
18. H. J. Yoon, J. W. Choi, H. Kang, T. Kang, S. M. Lee, B. H. Jun and Y. S. Lee, *Synlett*, 2010, 2518.
19. M. Gruttadauria, L. F. Liotta, A. M. P. Salvo, F. Giacalone, V. La Parola, C. Aprile and R. Notoa, *Adv. Synth. Catal.*, 2011, **353**, 2119.
20. (*a*) T. Cremer, M. Stark, A. Deyko, H.-P. Steinrück and F. Maier, *Langmuir*, 2011, **27**, 3662; (*b*) H.-P. Steinrück, J. Libuda, P. Wasserscheid, T. Cremer, C. Kolbeck, M. Laurin, F. Maier, M. Sobota, P. S. Schulz and M. Stark, *Adv. Mater.*, 2011, **23**, 2571 and references therein.
21. (*a*) Y. Gu and G. Li, *Adv. Synth. Catal.*, 2009, **351**, 817; (*b*) C. van Doorslaer, J. Wahlen, P. Mertens, K. Binnemans and D. de Vos, *Dalton Trans.*, 2010, **39**, 8377; (*c*) T. Selvam, A. Machoke and W. Schwieger, *Appl. Catal., A*, 2012, **445–446**, 92.
22. (*a*) P. Virtanen, H. Karhu, K. Kordas and J.-P. Mikkola, *Chem. Eng. Sci.*, 2007, **62**, 3660; (*b*) P. Virtanen, J.-P. Mikkola, E. Toukoniitty, H. Karhu, K. Kordas, J. Wärnå and T. Salmi, *Catal. Today*, 2009, **147S**, S144; (*c*) P. Virtanen, H. Karhu, G. Toth, K. Kordas and J.-P. Mikkola, *J. Catal.*, 2009, **263**, 209.
23. (*a*) U. Kernchen, B. Etzold, W. Korth and A. Jess, *Chem. Eng. Technol.*, 2007, **30**, 985; (*b*) A. Jess, C. Kern and W. Korth, *Oil Gas*, 2012, **38**, OG38–OG44.
24. (*a*) J. Arras, M. Steffan, Y. Shayeghi, D. Ruppert and P. Claus, *Green Chem.*, 2009, **11**, 716; (*b*) J. Arras, E Paki, C. Roth, J. Radnik, M. Lucas and P. Claus, *J. Phys. Chem. C*, 2010, **114**, 10520.
25. K. B. Sidhpuria, A. L. Daniel-da-Silva, T. Trindade and J. A. P. Coutinho, *Green Chem.*, 2011, **13**, 340.
26. A. Riisager, R. Fehrmann, M. Haumann and P. Wasserscheid, *Eur. J. Inorg. Chem.*, 2006, 695.
27. Y. Yang, C. Deng and Y. Yuan, *J. Catal.*, 2005, **232**, 108.
28. H. Hagiwara, *Synlett*, 2012, **23**, 837.
29. F. T. U. Kohler, B. Morain, A. Weiß, M. Laurin, J. Libuda, V. Wagner, B. U. Melcher, X. Wang, K. Meyer and P. Wasserscheid, *ChemPhysChem*, 2011, **12**, 3539.
30. M. Ruta, I. Yuranov, P. J. Dyson, G. Laurenczy and L. Kiwi-Minsker, *J. Catal.*, 2007, **247**, 269.
31. M. S. Journshari, M. Mamaghani, K. Tabatabaeian and F. Shirini, *J. Iran. Chem. Soc.*, 2012, **9**, 75.
32. (*a*) D. W. Kim, D. J. Hong, K. S. Jang and D. Y. Chi, *Adv. Synth. Catal.*, 2006, **348**, 1719; (*b*) D. W. Kim, H.-J. Jeong, S. T. Lim, M.-H. Sohn and D. Y. Chi, *Tetrahedron*, 2008, **64**, 4209.
33. L. J. Lozano, C. Godínez, A. P. De los Ríos, F. J. Hernández-Fernández, S. Sánchez-Segado and F. J. Alguacil, *J. Membr. Sci.*, 2011, **376**, 1 and references therein.

34. R. Meijboom, M. Haumann, T. E. Müller and N. Szesni, Synthetic methodologies for supported ionic liquid materials, in *Supported Ionic Liquids – Fundamentals and Applications*, ed. R. Fehrmann, A. Rissager and M. Haumann, Wiley-VCH, Weinheim, 2013.
35. M. Haumann, A. Schönweiz, H. Breitzke, G. Buntkowsky, S. Werner and N. Szesni, *Chem. Eng. Technol.*, 2012, **35**, 1421.
36. (*a*) S. Breitenlechner, M. Fleck, T. E. Mueller and A. Suppan, *J. Mol. Catal. A: Chem.*, 2004, **214**, 175; (*b*) C. Sievers, O. Jimenez, T. E. Mueller, S. Steuernagel and J. A. Lercher, *J. Am. Chem. Soc.*, 2006, **128**, 13990; (*c*) C. Sievers, O. Jimenez, R. Knapp, X. Lin, T. E. Mueller, A. Tuerler, B. Wierczinski and J. A. Lercher, *J. Mol. Catal. A: Chem.*, 2008, **279**, 187; (*d*) K. L. Fow, S. Jaenicke, T. E. Mueller and C. Sievers, *J. Mol. Catal. A: Chem.*, 2008, **279**, 239.
37. S. Werner, M. Haumann and P. Wasserscheid, *Annu. Rev. Chem. Biomol. Eng.*, 2010, **1**, 203.
38. C. Meyer, V. Hager, W. Schwieger and P. Wasserscheid, *J. Catal.*, 2012, **292**, 157.
39. U. Hintermair, G. Zhao, C. S. Santini, M. J. Muldoon and D. J. Cole-Hamilton, *Chem. Commun.*, 2007, 1462.
40. (*a*) U. Hintermair, T. Höfener, T. Pullmann, G. Francio and W. Leitner, *ChemCatChem*, 2010, **2**, 150; (*b*) U. Hintermair, G. Francio and W. Leitner, *Chem. Commun.*, 2011, **47**, 3691.
41. P. W. N. M. van Leeuwen, *Homogeneous Catalysis: Understanding the Art*, Springer, Dordrecht, 2008.
42. D. Cole-Hamilton and R. Tooze, ed. *Catalyst, Separation, Recovery and Recycling*, Springer, Dordrecht, 2006.
43. A. Riisager, P. Wasserscheid, R. van Hal and R. Fehrmann, *J. Catal.*, 2003, **219**, 452.
44. S. Shylesh, D. Hanna, S. Werner and A. Bell, *ACS Catal.*, 2012, **2**, 487.
45. Z.-Y. Yuan and B.-L. Su, *J. Mater. Chem.*, 2006, **16**, 663.
46. A. Riisager, S. Flicker, M. Haumann, P. Wasserscheid and R. Fehrmann, *Proc. – Electrochem. Soc. (Molten Salts XIV)*, 2006, 630.
47. M. Jakuttis, Gasphasenhydroformylierung von technischen Substraten mit neuartigen SILP Katalysatoren, Dissertation, University Erlangen-Nuremberg, 2013.
48. (*a*) J. Joni, M. Haumann and P. Wasserscheid, *Adv. Synth. Catal.*, 2009, **351**, 433; (*b*) J. Joni, M. Haumann and P. Wasserscheid, *Appl. Catal., A*, 2010, **372**, 8.
49. P. Kumar, W. Vermeiren, J. Dath and W. F. Hölderich, *Appl. Catal., A*, 2006, **304**, 131.
50. K. Dentler, M. Haumann and P. Wasserscheid, unpublished results.
51. (*a*) S. Werner, N. Szesni, R. W. Fischer and P. Wasserscheid, *Phys. Chem. Chem. Phys.*, 2009, **11**, 10817; (*b*) S. Werner, N. Szesni, A. Normen, M. J. Bittermann, P. Schneider, M. Härter, Haumann and P. Wasserscheid, *Appl. Catal., A*, 2010, **377**, 70; (*c*) S. Werner, N. Szesni,

M. Kaiser, R. W. Fischer, M. Haumann and P. Wasserscheid, *ChemCatChem*, 2010, **2**, 1399.

52. For reviews on homogeneous WGS see: (*a*) P. C. Ford, *Acc. Chem. Res.*, 1981, **14**, 31; (*b*) R. M. Laine and E. J. Crawford, *J. Mol. Catal.*, 1988, **44**, 357; (*c*) G. Jacobs and B. H. Davis, *Catalysis*, 2007, **20**, 122.
53. M. P. Atkins, PETRONAS TMD, Malaysia, keynote lecture "*Ionic Liquids for Hg Removal*" presented at Euchem 2012, Celtic Manor, Wales, 2012.
54. S. Werner, N. Szesni, M. Kaiser, M. Haumann and P. Wasserscheid, *Chem. Eng. Technol.*, 2012, **35**, 1962.

CHAPTER 8

Electrocatalysis in Ionic Liquids

LEIGH ALDOUS,* ASIM KHAN, MD. MOKARROM HOSSAIN AND CHUAN ZHAO

School of Chemistry, The University of New South Wales, Sydney, NSW 2052, Australia
*Email: l.aldous@unsw.edu.au

8.1 Introduction

Electrochemistry is a vast subject. It covers a significant range of biological functions, is a primary tool to obtain physical data, is used in the synthesis, isolation and destruction of a wide range of compounds, is responsible for corrosion, and underpins modern quality of life *via* our reliance on 'mobile energy' such as batteries and increasingly fuel cells.[1]

Essentially every electrochemical process involves an electrode, and frequently an electrolyte. Whereas the definition of a catalyst is relatively clear cut,[1] the electrocatalytic ability of an electrode is a purely relative measure, with every electrode system therefore possessing certain 'electrocatalytic' ability. This situation has been further complicated by the widespread application of electrochemistry into the fields of fuel cells, electrosynthesis and electroanalysis, each bringing their own criteria for an 'electrocatalytic' system.

Ionic liquids (ILs) are essentially pure electrolyte materials, and lend themselves extremely well to electrochemical studies.[2] Many ILs have moderately high ionic conductivity and relatively good to excellent

RSC Catalysis Series No. 15
Catalysis in Ionic Liquids: From Catalyst Synthesis to Application
Edited by Chris Hardacre and Vasile Parvulescu

Published by the Royal Society of Chemistry, www.rsc.org

electrochemical stability.[3] The unique solvating and stabilising effect of ILs has facilitated the direct electrochemical study of many compounds, as well as the application of ILs to a wide variety of disciplines and electrochemical applications.[4]

The goal of this chapter is to provide a basic introduction to the concept of electrocatalysis, as well as summarise a range of electrochemical topics in which ILs have been recently investigated, and in which electrocatalysis is both important and has been actively changed by the introduction of ILs to the system.

8.1.1 What is Electrocatalysis

Electrodes are employed as sites for electrochemical reactions to occur, with such reactions typically involving the loss or gain of electrons. This can occur as a galvanic cell (*i.e.* occurs spontaneously, such as rust, or energy sources such as batteries and fuel cells), or alternatively as a Faradaic cell (whereby external work is applied to drive a non-spontaneous reaction to occur, such as the electrochemical synthesis of adiponitrile or sodium hydroxide, and the electroanalytical quantification of blood glucose levels).[1]

Formal definitions of electrocatalysis are typically strictly related to the rate of an electrochemical process. For example, Appleby[5] stated that "Electrocatalysis may be defined as the relative ability of different substances, when used as electrode surfaces under the same conditions, to accelerate the rate of a given electrochemical process." The Electrochemical Dictionary[1] states that "Electrocatalysis is the catalysis of an electrode reaction. The effect of electrocatalysis is an increase of the standard rate constant of the electrode reaction, which results in an increase of the Faradaic current. As the current increase can be masked by other non-electrochemical rate-limiting steps, the most straightforward indication for the electrocatalytic effect is the shift of the electrode reaction to a lower overpotential at a given current density". In the above definitions, catalysts are not consumed and (in Appleby's[5] definition) do not influence the overall Gibbs energy change of the process. The electrocatalysts can be heterogeneous (*i.e.* a solid electrode surface) or homogeneous (*i.e.* a dissolved metal complex which reacts but is regenerated electrochemically, thus acting as a catalytic mediator).[1]

The above definitions are of particular relevance to fuel cells, and such definitions largely became formalised by early fuel cell research. In conventional fuel cells, O_2 and H_2 can be delivered rapidly and effectively to electrode surfaces. The relative performance of the device under reasonable loadings relies almost entirely upon the relative rate of electron transfer at the electrode surface, hence the concept of relative electrocatalytic ability.[5]

More recently, application of the term 'electrocatalysis' has spread to more diverse areas of electrochemistry, and has also been applied more casually. Instead, any modification made to the system which results in a favourable change in the system's observed electrochemical response (including, but not limited to increases in the rate of electron transfer) has, upon numerous occasions, been referred to as electrocatalysis. This 'evolution' of the term electrocatalysis is particularly apparent in the electroanalytical literature, and especially those featuring ionic liquids (ILs). In these cases the introduction of ILs might lead to increases in current density, and/or shifts in the overpotential required for appreciable current to flow (and therefore, technically, an electrocatalytic effect), through any number of possible mechanisms unrelated to the rate of electron transfer. These other situations can involve changed thermodynamic properties, changed reaction pathways, altered mass transport and electrode surface accessibility, pre-accumulation of reactants at electrode surfaces, *etc.*

The purpose of this chapter is not to define electrocatalysis as a term, nor deeply explore the underlying nature of the physical process. This chapter will instead cover basic electrochemistry, summarise how ILs might influence electrocatalysis, and highlight some examples in the literature where some form of 'electrocatalysis' has been noted. More detailed discussions regarding the underlying nature of electrocatalysis (purely in terms of electrochemical rate constants) can be found elsewhere.[1,5–7] This section will introduce on a very basic level 'electrocatalysis' in relation to electrochemical rates. Some examples of stated 'electrocatalysis' which is unrelated to rate will be given in the 'Electroanalysis' section (Section 8.2.8).

The ratio between the concentration of an oxidised, [Ox], and reduced species, [Rd], can theoretically be controlled at the surface of an electrode by controlling the potential of the electrode. This is expressed elegantly by the Nernst equation (8.1), where at the formal potential of the redox couple, $E^{\circ\prime}$, concentrations of the two should be equivalent. In the equation, n = electrons transferred per molecule, F = Faraday's constant, E = potential of the electrode, and all others have standard meanings.

$$E = E^{\circ\prime} + \left(\frac{\mathrm{RT}}{n\mathrm{F}}\right) \ln\left(\frac{[\mathrm{Ox}]}{[\mathrm{Rd}]}\right) \tag{8.1}$$

When the applied potential, E, differs from $E^{\circ\prime}$ by 0.1 V, the concentration at the surface of the electrode will become *ca.* 98% Ox or Rd (depending upon whether it is +0.1 or −0.1 V, for $n = 1$, $T = 298$ K); at 0.3 V this becomes *ca.* 99.999%. An applied potential can then be used to covert one compound to another (*i.e.* $2H_2O$ to H_2 and $2OH^-$, requiring $2e^-$), and typically a suitable opposite reaction occurs at a counter electrode in order to form a cell and balance the current flow (*i.e.* $2H_2O$ to O_2 and $4H^+$, losing $4e^-$). Sufficient mass transport can then result in the conversion of large quantities of material at stationary electrode surfaces, known as electrosynthesis. Alternatively, work can be performed using suitable spontaneous reactions at different electrodes (*i.e.* fuel cells utilising $2H_2$ to $4H^+$ at one electrode, O_2 and $4H^+$ to $2H_2O$ at the other). However, such relationships neglect the rate of electron transfer, and therefore time. Observed trends in current–electrode potentials can be expressed by the Butler–Volmer equation (8.2), which was first derived by studying the hydrogen evolution reaction at different electrocatalysts.[1]

$$j = k^{\circ}F\left[-[\mathrm{Ox}]_{(x=0)}\exp\left(-\frac{\alpha_c F(E-E^{\circ\prime})}{RT}\right) + [\mathrm{Rd}]_{(x=0)}\exp\left(\frac{\alpha_a F(E-E^{\circ\prime})}{RT^4}\right)\right] \quad (8.2)$$

In this equation, the current density (j, current per unit area) is a function of the applied potential, E, relative to the formal potential, $E^{\circ\prime}$ (sometimes represented as overpotential, $\eta = E - E^{\circ\prime}$). Concentrations are the concentration at the surface of the electrode ($x=0$). Both α_c and α_a are charge transfer coefficients, related to the degree of symmetry in the potential dependence of the cathodic and anodic processes, respectively (often $\alpha_c + \alpha_a = 1$). As the overpotential increases the current will increase exponentially, the direction of the current flow depends upon whether the overpotential is negative or positive. However, the overall measured current is also a function of k°, the electron transfer rate constant. The electron transfer rate constant depends upon a number of factors, including the type of ligand or solvation sphere changes involved, and the nature of any adsorbed intermediates. For example, if the transition from Rd to Ox involves the extensive reorganisation of bulky ligands, a relatively low k° might be expected, and therefore a relatively high overpotential would have to be applied in order to observe a significant current density; an order of magnitude increase in k° should theoretically lead to an order of magnitude higher current at the same potential. Favourable adsorption of intermediates at an electrode surface can lower the activation energy (*cf.* a traditional heterogeneous catalyst) and thus dramatically increase k°; k° is therefore solvent and electrode specific, and one direct means of quantifying and comparing electrocatalysis.

When the Butler–Volmer equation is expressed as overpotential, and the concentration represented by the ratio of surface concentrations to bulk concentrations, the exchange current density (j_0) is the limiting factor,[1] and is a measure of the rate of interchange between the species at the surface of the electrode when the net current density flow is zero. Therefore, either k° or j_0 can be utilised as quantitative measures of electrocatalytic ability, although k° and j_0 values are not necessarily interchangeable.[7] Alternatively, for a more qualitative analysis of electrocatalysis trends a fixed overpotential can be applied, and the measured current density values (j) compared. Finally, a fixed current density can be passed through the system, and the potential required to maintain this current measured.

Current or current density is very frequently measured by cyclic voltammetry. This technique involves increasing potential in tiny increments and measuring the resulting current, essentially sweeping from one potential to another potential, and then back again. This allows the observation of both reduction and oxidation processes. As a key example, Figure 8.1 displays a simulation of some cyclic voltammograms, to represent the current which would be recorded for a single Pt hemispherical particle (radius 1 μm) on a

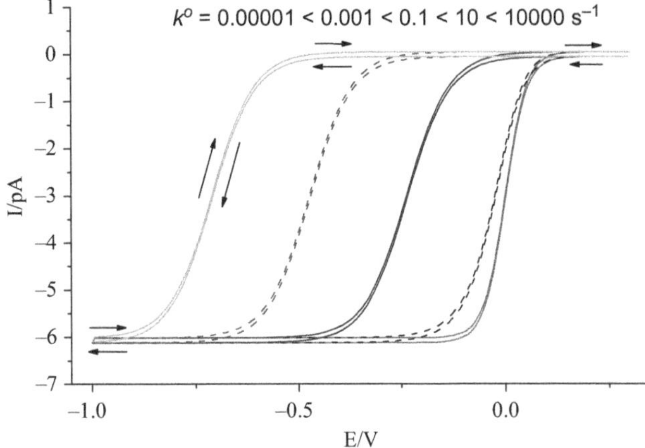

Figure 8.1 Simulated cyclic voltammograms for the one-electron reduction of a species, Ox, to Rd at a single isolated Pt hemisphere (radius 1 μm). Simulated by DigiElch 8.1, using D for both species $= 1 \times 10^{-5}$ cm^2 s^{-1}, $[\text{Ox}] = 0.001$ M, $[\text{Rd}] = 0$ M, $E^{\circ} = 0$ V, $\alpha = 0.5$, $C_{\text{DL}} = 5 \times 10^{-14}$ F, $T = 298.2$ K. Five different electron transfer rate constants (k°) were used, as indicated in the figure. The direction of the scan is indicated by the arrows.

non-electroactive substrate, as a function of applied potential when immersed in an electrolyte solution containing dissolved Ox ($E^{\circ\prime} = 0$ V). At potentials $\gg 0$ V no current flows, while at potentials < 0 V reduction is observed and the current flow is proportional to the electron transfer kinetics. When $k^{\circ} = 10\,000$ s^{-1}, the current increases exponentially before becoming limited by the rate of mass transport to the particle's surface. As the rate constant progressively becomes smaller, a larger overpotential is required before an equivalent current flow can be observed. Therefore, an electrosynthetic process using an electrode with poor electrocatalytic properties (*i.e.* low k°) will either be significantly slower, or in order to reach mass-transport limitation it will require a greater energy input and therefore be significantly more expensive in terms of electricity consumption than a system using a better electrocatalyst. Alternatively, if this represented one half of a spontaneous cell (*i.e.* a fuel cell), the power and energy of the device will strongly correlate with the electrocatalytic ability of the system.

Generally, such measurements are not made at individual, isolated particles but instead made at larger, planar electrode surfaces or dense, random arrays of particle, where diffusion boundaries overlap and mass transport is less efficient. Figure 8.2 displays a simulated cyclic voltammogram recorded for the same scenario as in Figure 8.1, but measured at a 1 cm^2 planar, circular electrode surface. In this case clear reduction peaks are observed due to the limited amount of material close the electrode on the timescale of the measurement; oxidation peaks are also observed on the reverse sweep due to the reduced material, Rd, remaining close to the electrode surface

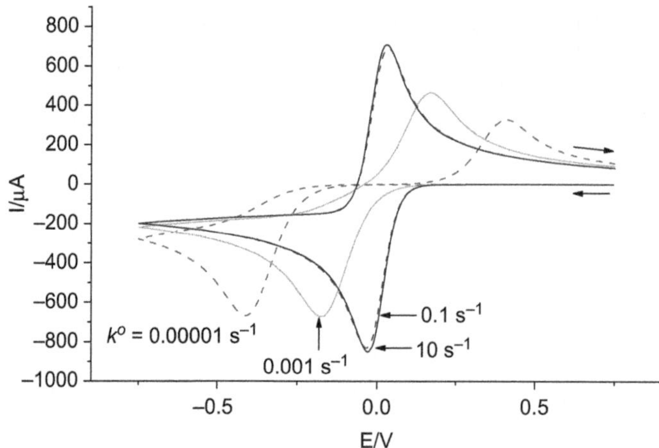

Figure 8.2 Simulated cyclic voltammograms for the one-electron reduction of a species, Ox, to Rd at a bulk Pt planar, circular electrode (area = 1 cm^2). Four different electron transfer rate constants (k°) were used, as indicated in the figure. All other parameters as described for Figure 8.1.

and being oxidised when the potential difference is reversed (*i.e.* scan is reversed). For high k° values characteristic peak shapes and 'reversibility' is observed; as the rate constant decreases the peaks become less pronounced and more separated. Cyclic voltammetry is therefore a relatively quick method for comparing relative rate constants, *e.g.* for different electrode materials, or the same system in different ionic liquids. Full simulation using commercial software such as DigiSim or DigiElch can yield quantitative data such as $E^{\circ\prime}$ and k°.

Figure 8.2 also highlights the principle of homogeneous electrocatalysis, as all systems in the figure are actually thermodynamically equivalent (with $E^{\circ\prime} = 0$). The reduction of a material with sluggish electron transfer kinetics (*i.e.* $k^\circ < 10$ s^{-1}, denoted here as A) would require a significant overpotential. Homogeneous reaction with a suitable compound, denoted as B, could have a much higher rate constant. Therefore a compound with a higher k° (*i.e.* B) can be introduced to the system and reduced at lower overpotentials, followed by thermodynamically and kinetically favourable homogeneous electron transfer to A. Compound B is thus catalytically regenerated, and an apparent improvement in the rate constant and/or decrease in potential is thus observed for the electrochemical reduction of A.

8.1.2 Ionic Liquids and their Properties

Ionic liquids (ILs) are salts with melting points less than 100 °C, and are frequently liquid at room temperature.[8] The latter are defined as room temperature ionic liquids (RTILs). ILs are entirely composed cations and anions, which can be organic or inorganic in nature. The term "ionic liquid" has only been widely adopted since the late 1970's, prior to which they were largely recognised as molten or fused salts, or as organic liquid salts.[2]

The most commonly used cations are asymmetric, quaternary nitrogen or phosphorus compounds with linear alkyl chains; frequently incorporated into aromatic or saturated heterocycles. A small number of possible structures are highlighted in Figure 8.3. The anions tend to be more diverse, covering organic anions, inorganic anions, metal salts, *etc.*, and range from strongly coordinating to weakly coordinating. The diverse range of synthesis routes have been summarised elsewhere.[9] Protic ILs are a distinct subcategory of ILs, with subtly different electrochemical properties to aprotic ILs.[10] They are prepared by proton transfer from an acid to a base, and are thus composed of ionic conjugate acid and ionic conjugate base.

A phenomenal number of different possible cation–anion combinations, combined with the ability to make minor structural changes and even introduce functionality directly into the anions and cations make ILs potential 'designer solvents'. They can be designed to be non-flammable liquids which possess excellent thermal and chemical stability, low toxicity, unique solvating abilities, negligible vapour pressure, high ionic conductivity, high polarity, moderate viscosity and large potential or electrochemical windows. Conversely, if so desired they can be designed and synthesised to possess opposite properties,[11] such as explosive, toxic, viscous ILs. Some major issues are the relative cost of most of these ILs, as well as the relatively high viscosity of ILs, which can typically be expected to be *at least* an order of magnitude higher than most aqueous and conventional non-aqueous solvents.[2]

Much of the early development of ILs focussed upon developing them as pure ionic electrolytes for electrochemical processes.[9] As ionic media, ILs possess high concentrations of ions. This would typically indicate high ionic conductivity, but low ion mobility resulting from high viscosity means ILs tend to possess ionic conductivities on par with 0.1 M electrolyte solutions in conventional non-aqueous solvents, and conductivities inferior to aqueous electrolytes.[12] Protic ILs are notable exceptions, which can have aqueous-like ionic conductivity values.[13]

Due to their viscous nature, ILs are, therefore, sometimes mixed with a co-solvent in order to balance IL-derived functionality and characteristics with favourable mass transport conditions.[14,15] The application of ILs as supporting electrolyte (dissolved in a conventional neutral solvent to introduce ionic conductivity) also has widespread and growing applications, including in influencing electrocatalysis (*cf.* section 8.2.2).

The potential window (the useable voltages which can be applied in the system without oxidation and reduction of solvent or electrolyte itself) is a key criterion for electrochemistry. Several ILs are known to possess extremely large potential windows (*ca.* 4.5–6 V),[16] making available numerous possible electrochemical and electrosynthetic processes.[17] ILs are now being actively researched for electrochemical applications including electrosynthesis, electrodeposition, solar cells, batteries, supercapacitors, fuel cells,

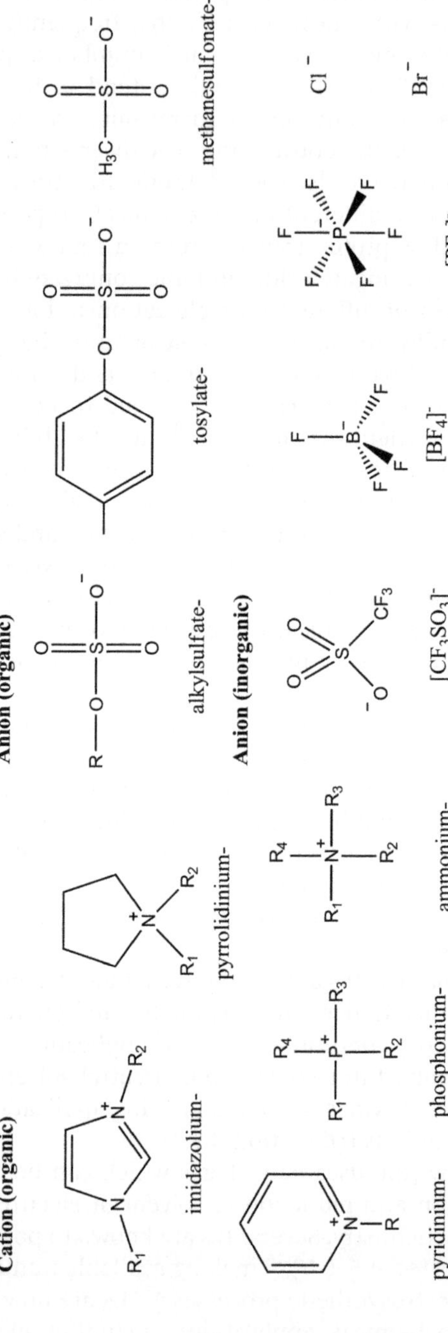

Figure 8.3 Examples of common cations and anions used to prepare ionic liquids (where R typically, but not exclusively, indicates an alkyl group).

electrochemical sensors and biosensors.[12] Many of these rely upon favourable 'electrocatalysis' in order to make the processes as effective as possible.

8.1.3 Why might Ionic Liquids influence Electrocatalysts

Figure 8.4 displays a stylised energy diagram for a reaction, highlighting the thermodynamic (Gibbs energy change, ΔG) and kinetic (activation energy, E_a) aspects. In electrochemistry, k° is proportional to the activation overpotential required to drive a reaction at a certain rate, and much as a catalyst reduces E_a so a good eletrocatalyst will increase k°. This k° value can be highly sensitive to the solvation sphere, especially in the case of outer sphere electron transfer processes, as well as electrode surface interactions, especially for inner sphere electron transfers passing through surface-adsorbed intermediates.[1]

When moving from a conventional electrolyte to an IL, or from one IL to another IL, there are many aspects which might result in changed electrocatalytic activity. Firstly, the IL will almost inevitably solvate or interact with the molecule and will therefore be influential upon the rearrangement of the solvation sphere, thus potentially changing k°. For example, the electrochemical rate constant for the 'outer sphere' oxidation of a range of phenylenediamines in a range of ILs displays a slight tendency to decrease with increasing IL viscosity,[18] although similar trends were not observed for a range of bulkier and more extensively delocalised arenes and substituted anthracenes.[19] ILs are already well established to have significant effects upon both the rate and mechanism of reactions involving dissolved reactants.[20,21] The IL solvation sphere, as well as the charged double layer formed at the electrode surface, can also potentially influence how the molecule adsorbs at the electrode surface. The double layer structure of ILs has already been established to differ from most conventional electrolytes,[22] with the formation of 'potential dependant compact layers' which might affect

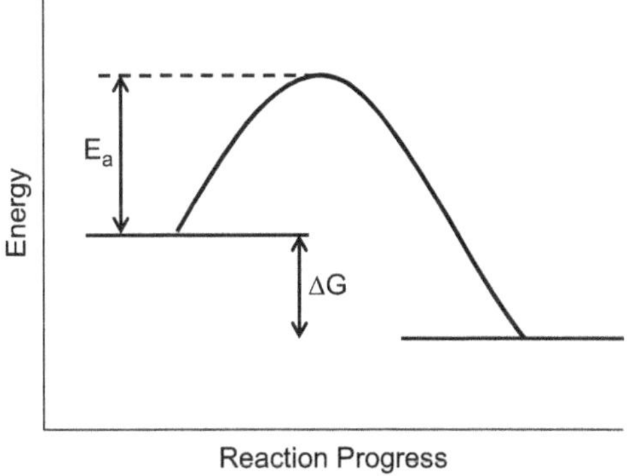

Figure 8.4 Stylised energy diagram.

the mass transport of electroactive species close to the electrode surface,[23] as well as inhibit or encourage adsorption of electroactive species.[24]

In addition to multiple ways in which the electron transfer kinetics might be influenced, many investigations simply relate current and potential to electrocatalysis, using these values as relative indicators. As seen in Figure 8.2, a higher peak current or a shift in potential *can* be related to a change in electron transfer kinetics; however, such shifts can also be related to changes in thermodynamics rather than electron transfer kinetics.

Finally, numerous non-ideal effects can be influential, especially in qualitative comparisons. Migliorini *et al.* have highlighted that alkyl-ether-functionalised methylimidazolium ILs with methylsulfonate anions will act as electrocatalyst poisons in the presence of trace water and a Pt surface,[25] blocking the available surface and thus decreasing current density. Conversely, Meng *et al.* have highlighted that different gases dissolved in ILs can result in dramatic changes in viscosity and diffusion coefficients, therefore an inert gas could lead to an increase in current density;[26] a definition of electrocatalysis in a very general sense. Shiddiky *et al.* have highlighted that in ILs, ferrocene oxidation currents and cobaltocenium reduction currents both increase by 25–35% if dissolved together in the IL, relative to the current in ferrocene-only and cobaltocenium-only solutions in ILs.[27] This observation did not extend to conventional non-aqueous solvents,[27] and highlights the interplay between observed electrochemical parameters, IL physical structure, and the significant effect solutes in ILs can have.

Yu *et al.*[28] have highlighted that molecular films of water-miscible ILs will form at the surface of a glassy carbon electrode from water–IL mixtures, resulting in 'striking electrochemical properties such as electrocatalysis toward ascorbic acid', as substantiated by a shift in the peak potential for ascorbic acid oxidation when an IL layer was pre-formed. In reality the latter could be due to changed thermodynamics of the process in an immobilised IL layer, pre-concentration of ascorbic acid in the IL layer, changed properties of the glassy carbon such as increased hydrophilicity, electrocatalysis in the conventional sense (rate of electron transfer genuinely accelerated), or any combination of the above.

When comparing IL systems against non-IL systems, or comparing between different ILs, significant changes can be expected in thermodynamics, kinetics, mass transport, electrode surface availability, *etc.* This can in turn lead to observations of electrocatalysis or electrocatalytic trends which fit a loose definition of electrocatalysis, but do not always correspond strictly to activation energies and rate constants.

8.2 Examples of Electrocatalysis in Ionic Liquids

The terms 'electrocatalysis' and 'electrocatalytic' have appeared in conjunction with 'ionic liquids' in relation to a number of areas over the past decade. The rate at which such studies are appearing is also increasing. However, in many cases the term 'electrocatalysis' has been used relatively casually and qualitatively, as noted above. Currently, claims of electrocatalysis in association with ILs infrequently correspond to quantitative investigations and even more rarely to direct comparison with other more established (non-IL) systems. Below we have summarised the research to date in relation to a number of different electrochemical systems investigated in ILs, focussing where possible on quantitative investigations and highlighting qualitative examples where appropriate. It is not intended to be a comprehensive overview of every system mentioned in relation to 'ionic liquids' and 'electrocatalysis', areas such as electrocatalytic hydrogenation[29] and fluorination[30–32] having been excluded. The subsections below are largely, but not entirely, comprehensive summaries of the literature to date in relation to noted electrocatalysis in relation to ILs.

8.2.1 Hydrogen Evolution and Oxidation in Ionic Liquids

The hydrogen evolution reaction and hydrogen oxidation reaction are both of extensive practical and fundamental interest. As such these processes have been the subject of a vast number of studies, primarily in aqueous media.[33–35] In strongly acidic solution the process is commonly simplified as $2H^+ + 2e^- \leftrightarrow H_2$, and this redox process forms the basis of fundamental reference scales such as the standard hydrogen electrode (SHE)[36] and the reversible hydrogen electrode (RHE),[37] as well as dihydrogen gas sensing apparatus, pH meters, *etc.* This process is also highly relevant to hydrogen-based fuel cells and the so-called 'hydrogen economy'.

In reality, two possible reaction pathways are frequently observed at the surface of electrodes. The Volmer reaction constitutes the first step [Equation (8.3)] to form adsorbed hydrogen atoms on the electrode surface. This is then followed by either the Tafel reaction [dimerisation of adsorbed hydrogen atoms, Equation (8.4)] or the Heyrovsky reaction [electrochemical reduction of proton in conjunction with an adsorbed hydrogen atom, Equation (8.5)].[33,36,38] As all steps pass through an adsorbed intermediate (H_{ads}), trends in electrocatalysis are dominated by the electrode material, and to a lesser degree electrolyte–electrode interactions. In strongly acidic solutions the hydronium ion is a universal feature, whereas proton solvation environments are considerably less well established in ILs, and more variable across ranges of different ILs.[37,39]

$$H^+ + e^- \rightarrow H_{ads} \tag{8.3}$$

$$2H_{ads} \rightarrow H_2 \tag{8.4}$$

$$H_{ads} + H^+ + e^- \rightarrow H_2 \tag{8.5}$$

Trends in the rate of the hydrogen evolution reaction (HER) in acidic aqueous media as a function of electrode material are widely employed as the prototypical example of electrocatalyst trends.[1,6] The exchange current density (j_0)[36] of the HER covers several orders of magnitude from 'poor' electrocatalysts such as Hg and Pb to 'good' electrocatalysts such as Pt.[33]

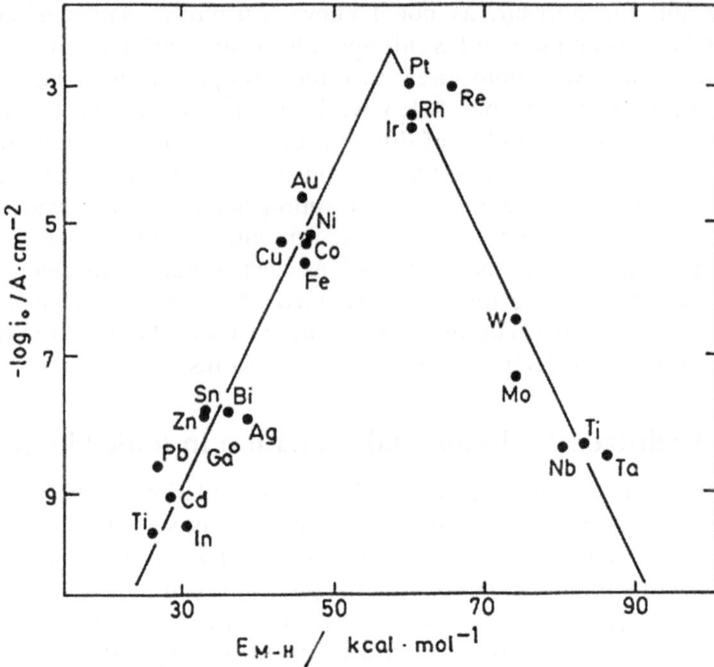

Figure 8.5 Volcano plot highlighting trends in hydrogen production electrocatalysis, represented by the logarithm of exchange current density (log i_o), as a function of the bonding adsorption strength of the intermediate metal–hydrogen (M–H) bond formed during electrolysis. Reproduced from reference 40 with permission from the *Journal of New Materials for Electrochemical Systems*.

When plotted against the metal-hydride (M–H_{ads}) bond strength of the metal in question, a 'Volcano Plot' is generated, with Pt typically found at the apex of the plot in aqueous solution (Figure 8.5).[33–35] The precise nature of this trend is still the topic of discussion long after its original discovery. The original assertion was that this trend comes from a Goldilocks-like scenario; Pt forms a strong enough M–H_{ads} bond to encourage proton reduction, without being too strong and thus inhibiting H–H bond formation, while other metals suffer from too strong or too weak M–H_{ads} affinities. However, this is being increasingly challenged by other arguments, such as higher M–H bond strengths should not necessarily lead to slower H_2 evolution, and the observed decrease in the rate constant correlates with dramatically increased propensities to form inhibiting surface metal oxide layers.[41–43] The fact that the H^+/H_2 system in aqueous media is still a topic of contention today, despite being (theoretically) the most simple electron-transfer mechanism possible and likely the most investigated system in terms of electrocatalyst trends in aqueous media, hints at the challenge facing researchers attempting to quantitatively probe electrocatalytic systems in ILs.

The direct application of the aprotic $[C_4mim][BF_4]$ in a commercial H_2–O_2 fuel cell assembly has been reported, with a good overall cell efficiency of 67%,[44] demonstrating such systems are feasible. Proton reduction to form dihydrogen has been semi-quantitatively investigated in 1-alkyl-3-methylimidazolium bis(trifluoromethylsulfonyl)imide ILs, $[C_nmim][NTf_2]$ ($n=2, 4$), at Au,[45] Pt,[26,45,46] Pd[47,48] and GC[45] electrodes, using $H[NTf_2]$ as the dissolved proton source. The oxidation of dissolved dihydrogen, presumably to form solvated protons, has also been investigated.[37,46,49,50] These studies have generally highlighted a strong dependence upon the precise nature of the electrode material, for example hydrogen evolution being observed at lower potentials on Pt than on Au and GC,[45] and the necessity of 'cleaning' the Pt surface prior to effective H_2 oxidation by extended platinum oxide formation and reduction.[15]

Quantitative studies regarding electrocatalysis of the H^+–H_2 redox system in ILs has been complicated by a number of IL characteristics. When ILs melt or fuse, there is a significant volume expansion, as is commonly observed for molten salts.[51] This results in interstices or voids between anion and cation, in which dissolved species such as H_2 are also believed to primarily reside. Meanwhile, the ionic, solvated protonic species possess a significantly larger solvodynamic radii.[26,46] Therefore, the diffusion coefficient of H_2 in ILs is typically an order of magnitude greater than that of the proton (*i.e.* 5.9×10^{-10} m^2 s^{-1} for H_2 in $[C_2mim][NTf_2]$ at 298.1 K, compared to 2.5×10^{-11} m^2 s^{-1} for $H[NTf_2]$).[26] As a direct result, equivalent concentrations of H^+ and H_2 would result in orders of magnitude different current densities. Furthermore, this results in the rapid accumulation of one species over the other at the electrode surface whenever the equilibrium is perturbed, leading to significant deviation relative to the mathematical models and Nerstian trends which have been developed in aqueous media and which assume equivalent mass transport properties.[38] This issue is still further complicated by the fact that the H_2 molecules residing between the ions change the relative permittivity and thus electrostatic interactions experienced within the IL bulk, acting as a 'lubricant' such that the viscosity of the IL decreases as H_2 concentration increases, and the diffusion coefficient of all other solutes varies accordingly.[26] It is, therefore, only recently that quantitative studies have been able to be performed in ILs (by Meng *et al.*[26,39,52]), looking at both the electrocatalytic properties of different electrode materials as well as the role that IL structure plays.

Meng *et al.*[52] were able to successfully monitor proton and H_2 concentration throughout a series of experiments, as well as quantify the diffusion coefficient of each, by simulating chronoamperometric transients. With all physical parameters known, the simulation of cyclic voltammetric data using conventional techniques could reliably yield the standard electrochemical rate constant, k^0, a measure of the rate of electron transfer at the formal potential of the redox couple and thus relative to the electrocatalytic ability of the materials in question. Figure 8.6 displays the cyclic voltammetric data, as well as simulation data, for the reduction of $H[NTf_2]$ to form H_2 in

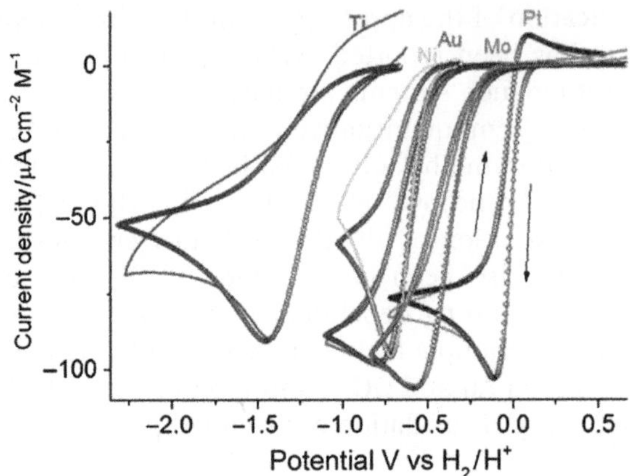

Figure 8.6 Normalised cyclic voltammetry displaying the reduction of $H[NTf_2]$ to form H_2 at different (Pt, Mo, Au, Ni and Ti) metallic electrocatalysts in $[C_2mim][NTf_2]$ at 400 mV s^{-1} at 298 K. Displays both experiment data (-) and simulation data (○).
Reproduced from reference 52 with permission from the Royal Society of Chemistry.

Table 8.1 Electrochemical rate constant for the hydrogen evolution reaction from $H[NTf_2]/[C_2mim][NTf_2]$ when performed at a range of different metals. Taken from reference 52.

k^0 (cm s^{-1}) at electrode material at 298 K	
Pt	1.3×10^{-3}
Mo	5×10^{-6}
Au	1.7×10^{-7}
Ni	1.3×10^{-7}
Ti	2.6×10^{-8}

$[C_2mim][NTf_2]$ at electrodes constructed from Pt, Mo, Au, Ni and Ti, with the more positive potential correlating to better electrocatalysis. Table 8.1 displays the k^0 values for the hydrogen evolution reaction obtained from the cyclic voltammetric data, highlighting that the relative electrocatalytic ability of the metals span more than 5 orders of magnitude.[52] Interestingly, trends in the observed electrocatalytic ability of the different metals were slightly different from those observed in aqueous acidic media. Figure 8.7 displays a log–log plot of the representative values, with Pt clearly superior as an electrocatalyst for H_2 production in both systems, and Ti a poor electrocatalyst. However, the trend Ni > Au > Mo in acidic aqueous solution was inverted in $[C_2mim][NTf_2]$ to become Mo > Au ~ Ni. These trends represent order of magnitude shifts in relative rates. The different metals are also known to evolve H_2 from acidic aqueous solution by different mechanisms

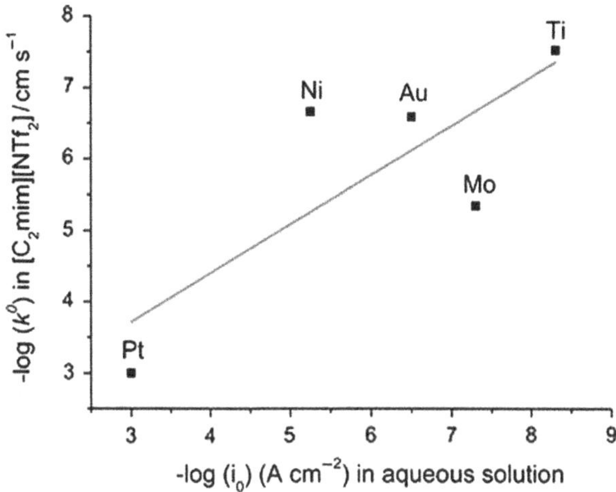

Figure 8.7 Plot comparing the relative electrocatalysis of five different metals for the hydrogen evolution reaction, comparing k^0 values obtained in [C_2mim][NTf_2] in comparison with the j_0 values for aqueous solution at the same metals.
Reproduced from reference 52 with permission from the Royal Society of Chemistry.

based upon M–H bond strength, but in [C_2mim][NTf_2] all metals had the first step [the Volmer reaction, H_{ads} formation, Equation (8.3)] as the rate limiting step.[52]

Navarro-Suárez *et al.*[53] investigated single-crystalline Pt electrodes for the electrocatalytic oxidation of H_2 in a range of 1-alkyl-3-methylimidazolium ILs. In ILs with [OTf]$^-$, [BF_4]$^-$ and [$EtSO_4$]$^-$ anions, electrocatalytic ability followed the trend Pt(111) > Pt(110) > Pt(100), while the [NTf_2]$^-$ resulted in the trend Pt(100) > Pt(111) > Pt(110). However, no obvious shift in oxidation potential was observed between the different facets, and electrocatalytic activity was assessed by current density.[53] Interestingly, Ciaccafava *et al.*[54] have reported the bioelectrooxidation of H_2 using electrodes modified with bacterial hydrogenases in contact with ILs. However, oxidation of H_2 was not observed in pure IL media, and could only be observed for IL concentrations of up to 80 : 20 IL : aqueous buffer before the bioelectrocatalytic process was deactivated.[54]

Johnson *et al.*[55] have investigated the oxidation of H_2 and reduction of O_2 at Pt electrodes in the protic IL, diethylmethylammonium trifluoromethanesulfonate, [$N_{2,2,1}$][OTf]. Interestingly, oxidation of trace H_2O in the IL resulted in passivating oxide films, which significantly reduced the electrocatalytic activity of Pt towards both processes in the IL. The authors noted the extreme difficulty in removing all traces of water from ILs in general and protic ILs in particular, and highlighted that instead future electrocatalysts for application in ILs should have a lower propensity to form surface oxide layers than Pt.[55] Decreases in the activation energy for the

Table 8.2 Electrochemical rate constants for hydrogen evolution (k_c^0) and hydrogen oxidation (k_a^0) at a Pt electrode at 298 K in a range of ILs containing both H[NTf$_2$] and H_2. Also shown in the formal potential ($E°_f$) for the H^+/H_2 redox couple, as measured in that IL. Taken from reference 39.

Ionic Liquid	$E°_f$ *(V vs. Ag/Ag$^+$)*	k_c^0 *(10^3 cm s^{-1})*	k_a^0 *(10^3 cm s^{-1})*
[C$_2$mim][NTf$_2$]	−0.338	0.45	6.5
[C$_4$mim][NTf$_2$]	−0.346	0.55	6
[C$_4$mpyrr][NTf$_2$]	−0.350	0.4	6
[C$_4$dmim][NTf$_2$]	−0.352	0.275	7.55
[C$_4$mim][OTf]	−0.457	0.3	7.5
[C$_4$mim][BF$_4$]	−0.587	0.008	0.9

electrochemical oxidation of H_2 at Pt were also noted by Rollins and Conboy[56] in both protic and aprotic hydrophobic ILs upon water-saturation.

The nature of the IL anion also plays a key role upon the observed electrocatalytic formation of H_2 at Pt.[39] In four bis(trifluoromethylsulfonyl)imide ([NTf$_2$]$^-$) based ILs with different cations, the electrochemical rate constant and formal potential for the proton reduction and hydrogen oxidation reactions were essentially unchanged (Table 8.2). However, upon moving from 1-butyl-3-methylimidazolium bis(trifluoromethylsulfonyl)imide ([C$_4$mim][NTf$_2$]) to the trifluoromethanesulfonate analogue ([OTf]$^-$), the electrocatalytic rate constant for proton reduction at Pt decreased slightly and hydrogen oxidation increased slightly. The formal potential also shifted more negative by *ca.* 100 mV, implying slightly stronger proton–anion interactions in the [OTf]$^-$-based IL. For the tetrafluoroborate analogue ([BF$_4$]$^-$), the rate constants decreased by one to two orders of magnitude and the formal potential shifted more negative by *ca.* 250 mV. The significant change, as well as a relative asymmetry in the oxidation and reduction features, implied much stronger proton–anion interactions in the [BF$_4$]$^-$-based IL,[39] potentially even corresponding to some form of proton-induced dissociation of the [BF$_4$]$^-$, analogous to its rapid and facile hydrolysis in acidic aqueous solutions.[57]

In protic ILs, the thermodynamics of the reduction of protons and oxidation of H_2 has been investigated at Pt and Pd electrodes, by assuming that electrocatalytic ability was essentially constant throughout.[37] The overpotential gap between the onset of these two processes was found to directly correlate with the energy gap between the proton free energy for the acid and base used to prepare the IL (*i.e.* pK_a difference between the pair).[37]

8.2.2 Water Electrolysis in Association with Ionic Liquids

Water is relatively abundant, and when an appropriate potential difference is applied across appropriate electrodes immersed in an electrolyte solution (such as aqueous H_2SO_4 or KOH), H_2 and O_2 are evolved at the cathode and anode, respectively. This process is commonly referred to as 'water splitting'. The H_2 can be collected and used in H_2–O_2 fuel cells, producing only H_2O as the product; the H_2, therefore, acts as an 'energy vector', allowing energy in the form of electricity (obtained by nuclear power, solar or wind energy, hydrothermal sources, *etc.*) to be readily distributed as an on-demand fuel. However, significant energy is lost during this process in the form of over-potential (slow kinetics, resistive losses, *etc.*) and electrode longevity will likely be an issue on the industrial scale that a 'hydrogen economy' would demand. Therefore, extensive research is being devoted to both the practical aspects of electrolytic cells (design, orientation, *etc.*) as well as developing suitably electrocatalytic and robust electrode materials.

Electrolytes play an equally important role as electrode in a water electrolysis cell. Electrolytes with superior charge transport could have substantially improved water splitting efficiency. Moreover, the electrolyte can have a major impact on the properties of the electrode–electrolyte interface, where the intermediate processes of water oxidation such as corrosion and the adsorption of water molecules and hydroxide ions occurs. Therefore the nature of an IL employed as an electrolyte should be influential, and the vast range of possible ILs introduces significant versatility in tuning the system's properties.

Water is rapidly taken up from the ambient atmosphere by ILs, and has been demonstrated to strongly associate with the anions of the IL *via* hydrogen bonding to form 1:2 type anion–H_2O–anion complexes.[58] Distinctly different molecular structure and reactivity have been observed when water is dissolved in ILs relative to the situation in pure liquid water, as a result of the dramatic difference in intermolecular interactions derived from van der Waals bonds, hydrogen bonds, and electrostatic interactions.[58] The disruption or absence of H_2O–H_2O hydrogen bonding, and creation of new IL–H_2O networks, correspondingly modifies the molecular structure of water and results in changed thermodynamics and kinetics of reactions featuring water,[4] including water electrolysis.[59] Significant photochemical activity of water in ILs has also been noted, including water splitting mechanisms which are only possible in the presence of IL.[60–62]

The presence of water dramatically shrinks the observed electrochemical windows of ILs, primarily due to the more facile electrolysis of water rather than the ionic constituents of the IL itself.[16] Water electrolysis has been demonstrated as an effective means to 'dry' an IL, converting the IL-bound H_2O to volatile H_2 and O_2.[63] The water content in the IL *N,N,N*-trimethyl-*N*-propylammonium bis(trifluoromethanesulfonyl)imide ($[N_{1,1,1,3}][NTf_2]$) could be reduced from 2.2% w/w ($\sim$1.8 M) to 0.060% w/w by this technique, without any observed degradation of the IL itself.[63]

Platinum is widely regarded as an excellent electrocatalyst for both the anodic oxygen evolution reaction and the cathodic hydrogen evolution reaction during water electrolysis. De Souza *et al.*[64] investigated water reduction to form H_2 at Pt electrodes in two ILs. When mixed with water in a 99:1 %v/v IL:H_2O ratio, the current density for water reduction was more than four times higher in $[C_4mim][BF_4]$ than $[C_4mim][PF_6]$, and water electrolysis began 0.6 V more positive in the former, highlighting the key role of the anion. The precise ratio of IL : H_2O was crucial, more water resulting in more negative reduction potentials but conversely higher current densities, with an optimum Faradaic efficiency with respect to H_2 evolution of 94.5% at 70 : 30 %v/v IL : H_2O.[64]

De Souza *et al.*[65] also investigated a range of electrocatalytic metals for H_2 evolution from $[C_4mim][BF_4]$–H_2O mixtures. Investigating current density at a fixed negative potential value in 10 : 90 %v/v $[C_4mim][BF_4]$: H_2O, the trend was found to be Mo > Pt > FeNi alloys > FeMn alloys > FeCr alloys > Ni. Further studies highlighted the poor electrocatalytic ability of Mo in KOH solutions, the literature precedent as a poor electrocatalyst in acidic solutions, and comparatively outstanding electrocatalytic ability for H_2 evolution from IL–H_2O solutions.[65,66] In the presence of $[C_4mim][BF_4]$, the activation energy for hydrogen evolution was 9.22 kJ mol^{-1} at Mo and 23.40 kJ mol^{-1} at Pt.[65] Further studies at the Mo–$[C_4mim][BF_4]$ interface using Electrochemical Impedance Spectroscopy (EIS) highlighted specific adsorption of the imidazolium cation and a much more compact double layer when compared to Pt and Ni electrodes (see Figure 8.8).[24] It was therefore proposed that the strongly adsorbed cations acted as a mediator, being reduced to form an adsorbed hydrogen atom and a neutral imidazole (potentially a carbene, Equation 8.6), and the latter subsequently reacted with water to be regenerated (Equation 8.7).[24] The net result is the loss of water and formation of H_{ads} on the Mo surface,[24] and ultimate formation of H_2 *via* the Tafel or Heyrovsky reaction.[52] It should be noted that this trend was not observed in anhydrous $[C_2mim][NTf_2]$ containing $H[NTf_2]$, where Pt was found to be a significantly better electrocatalyst for H_2 formation than Mo, indicating the key role of either water or pH in the IL-induced inversion in electrocatalytic ability upon moving from IL to IL–H_2O mixtures.[52]

$$[\text{R-CH}]^+ + e^- \rightarrow [\text{R-C}] + H_{ads} \tag{8.6}$$

$$[\text{R-C}] + H_2O \rightarrow [\text{R-CH}]^+ + OH \tag{8.7}$$

The long-term stability of a range of materials during H_2 evolution from $[C_4mim][BF_4]$–H_2O mixtures was also investigated, and Mo, FeMn and NiFe electrodes were found to exhibit both high Faradaic efficiency and corrosion resistance.[66] Stainless steel was found to be a poor electrocatalyst under similar conditions, while low carbon steel was found to be an excellent and economical electrocatalyst, but the latter suffered from severe corrosion if the proportion of IL in the IL–H_2O mixture was not maintained; at high enough IL concentrations the imidazolium cation was speculated to act as a corrosion inhibitor at the electrocatalyst's surface.[67]

If ILs are applied for industrial scale hydrogen generation from water electrolysis, the high cost typically associated with ILs may represent an issue.[11] Protic ILs (PILs) may represent a more promising solution than aprotic ILs, by virtue of their more straightforward synthesis by acid–base neutralisation, and the ubiquity of starting materials such as amines and

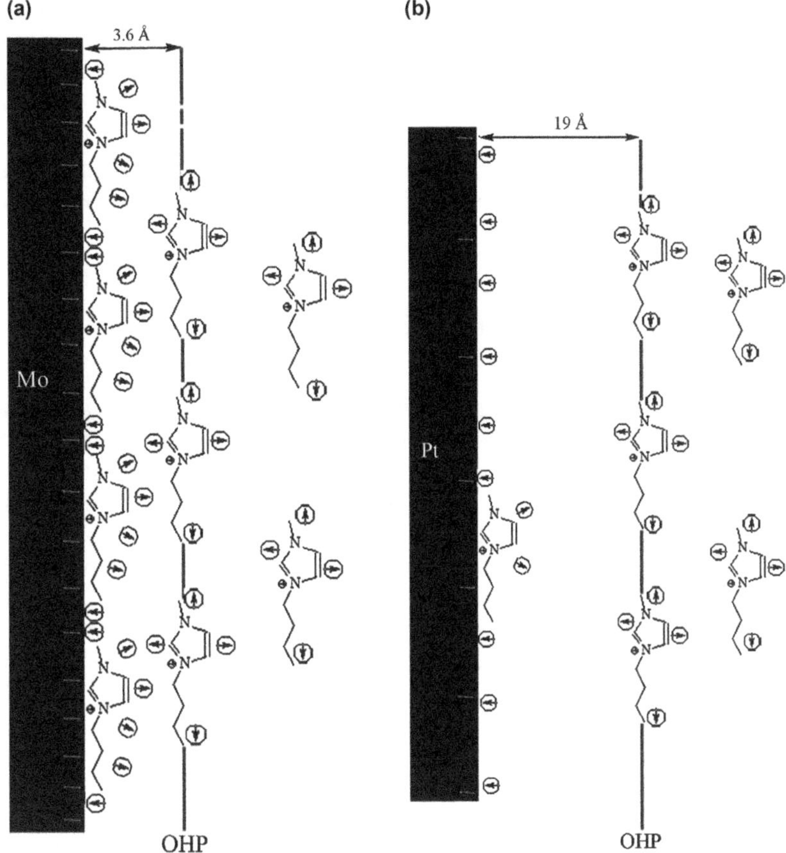

Figure 8.8 Representation of the electric double layer on the surface and Outer Helmholtz Planes for (a) Mo and (b) Pt electrodes immersed in 10 : 90 v/v% [C_4mim][BF_4] : H_2O. Anions not shown, and arrows represent water molecules of solvation in association with the IL cations. Reproduced from reference 24 with permission from Elsevier.

acids. The latter are typically well known in industry and fully characterized in terms of toxicity, biodegradability, cost, and environmental effects. An increasingly large number of PILs have emerged over the past few years with systematic accounts of their electrochemical properties becoming available.[68] Lu *et al.*[10] have recently reported the application of a protic ionic liquid, triethylammonium methylsulfonate, as an effective electrolyte (1 M in water) in water electrolysis.

Pool *et al.*[69] have recently focussed upon using a dissolved nickel complex as a homogeneous electrocatalyst, and using an IL as proton source, medium and electrolyte. In the presence of a highly acidic protic IL, dibutylformamidium bis(trifluoromethanesulfonyl)imide, electrochemical reduction of [NiL_2][BF_4]$_2$ (where L = 1,5-di(4-*n*-hexylphenyl)-3,7-diphenyl-1,5-diaza-3,7-diphosphacyclooctane) resulted in formation of H_2 and regeneration of the Ni electrocatalyst, although at very low current densities. Dilution of the system with acetonitrile resulted in a considerable increase in current density and a turn over frequency (TOF) of 7.4×10^2 s^{-1}; dilution with water (up to a maximum of 78 %mol/mol water before phase separation) resulted in an almost two-orders of magnitude increase in current density and a TOF of *ca.* 5×10^4 s^{-1}. Increasing water content did not lead to significant increases in overpotential, unlike prior studies involving heterogeneous electrocatalysts.[69]

8.2.3 Oxygen Reduction Reaction in Ionic Liquids

The electrochemical reduction of oxygen (O_2) is an important reaction in energy conversion and storage devices,[70–73] oxidation of organic molecules,[74,75] biological processes,[76,77] and oxygen sensors.[78,79] As such it has been widely studied in aqueous systems,[80,81] non-aqueous solvents[82,83] and in ionic liquids.[84–86] The protonic nature of aqueous media means that oxygen reduction typically proceeds *via* a series of electron transfer and proton transfer reactions, to yield either H_2O_2 or H_2O in two or four electron processes [Equations (8.9) and (8.10)], respectively, dependent upon the electrocatalyst employed. Various studies in anhydrous, aprotic ILs have observed the one electron reduction to superoxide [Equation (8.8)],[85] while wet ILs or protic ILs tend to result in the formation of H_2O_2 [Equation (8.9)].[87] In anhydrous ILs, the stability of the superoxide is also dependent upon the absence of protonic impurities[84] and the nature of the cation, both phosphonium[88] and imidazolium[89] cations undergoing proton abstraction reactions. The peroxide dianion, $[O_2]^{2-}$, can be formed electrochemically under aprotic conditions but is extremely reactive, and can react with many IL cations, as well as traditionally unreactive anions such as the bis(trifluoromethylsulfonyl)imide anion.[90]

$$O_2 + e^- \rightarrow O_2^{\bullet -} \tag{8.8}$$

$$O_2 + 2H^+ + 2e^- \rightarrow H_2O_2 \tag{8.9}$$

$$O_2 + 4H^+ + 4e^- \rightarrow 2H_2O \tag{8.10}$$

The cathodic reduction of O_2 is a relatively sluggish reaction, and is the primary limiting factor in H_2–O_2 fuel cells, with optimal rate constants for the oxygen reduction reaction (ORR) typically being orders of magnitude slower than the hydrogen oxidation reaction.[5] In ILs, work is focussing upon both the identification of favourable IL structures for high O_2 solubility and rapid electron transfer kinetics, as well as the development of electrocatalytic electrode materials.

As described previously for H_2,[26] O_2 is a small molecule and can reside in the interstices of the IL structure. Therefore mass transport is higher than observed for many other compounds dissolved in ILs, and diffusion poorly correlates with viscosity (does not obey the Stokes–Einstein relationship).[91]

Table 8.3 Examples of rate constants obtained by voltammetric investigations of the oxygen–superoxide redox couple in a range of ILs.

	k^0 at electrode material (10^3 cm s^{-1})		
Ionic Liquid	*Pt*	*Au*	*GC*
$[N_{6,2,2,2}][NTf_2]$[88]	3	5	—
$[C_4mpyrr][NTf_2]$[88]	0.8	3.5	—
$[P_{14,6,6,6}][NTf_2]$[88]	0.14	1.77	—
$[P_{14,6,6,6}][FAP]$[88]	0.05	0.11	—
$[C_2mim][BF_4]$[99]	0.94	2.2	6.4
$[C_3mim][BF_4]$[99]	1.5	3.3	7.3
$[C_4mim][BF_4]$[99]	0.83	1.2	7.5

However, the reduced anionic compounds interact strongly with the surrounding IL, and therefore IL-effects on the electrocatalytic reduction of O_2 are widely apparent.[92]

Both Compton[87–88,90–96] and Ohsaka[86,89,97–99] have contributed extensively to understanding of the oxygen–superoxide redox couple in ILs, including reporting kinetic data. Table 8.3 highlights a sample of k^0 values for the oxygen–superoxide redox couple, highlighting variations in the rate constant as a function of both electrode material and IL structure. Although these data only cover a very small range of materials and IL structures, it is clear from Table 8.3 that only minor changes in the IL structure can alter the rate constant by two orders of magnitude. A vast range of electrode material and IL combinations remain to be investigated, and further work is required to understand the precise roles of the IL structure on the reduction of O_2.

Significant interest has been generated by the ability to employ aprotic ILs as electrochemical O_2 sensors. Thin layers of IL can be employed as solvent and electrolyte, potentially simplifying sensor design, removing the need for an additional membrane to retain the solvent, increasing longevity by removing evaporation, and expanding the range of O_2 sensors to nonstandard conditions beyond the reach of aqueous electrolytes.[95,100] Many studies have investigated the robustness and range of applicable conditions using simple electrodes (*e.g.* Pt) in order to focus upon the role of the IL in the electroanalytical signal.[95,100] However, Toniolo *et al.*[101] have reported that the dissolution of ionic quinone moieties into an IL allows the homogeneous electrocatalytic reduction of O_2, reducing the required potential in IL-based O_2 electrochemical sensors by *ca.* 1 V. Li *et al.*[102] have highlighted that a gel comprised of potassium-doped multi-walled carbon nanotubes and $[C_4mim][PF_6]$ is effective for the *in vitro* determination of superoxide anions released from cancer cells, the potassium-doping and incorporation of the IL both significantly enhancing measured current densities.

The high O_2 solubility and high local proton concentration offered by protic ILs (encouraging the two or four electron formation of H_2O_2 or H_2O, respectively, over the one electron superoxide anion) make these promising electrolytes for fuel cells.[55] An anhydrous H_2–O_2 fuel cell has been prepared from the protic IL $[N_{2,2,1}]$[OTf] with good characteristics demonstrated at 120 °C.[103] However, a more detailed investigation of the electrocatalytic reduction of O_2 at Pt in this IL revealed the detrimental effect of water present in this system at ambient temperatures (due to platinum oxide formation), and the necessity for high temperatures to overcome this electrocatalytic-inhibiting process (remove water or overcome the activation energy barrier).[55] Therefore, at ambient temperatures protic ILs have been most successful when applied as electrocatalyst modifiers rather than as pure electrolytic media. For example, nanoporous Ni–Pt is an effective electrocatalyst for the ORR in aqueous perchloric acid solutions, but the prior impregnation of the nanopores with the protic IL 7-methyl-1,5,7-triazabicyclo[4.4.0]dec-5-ene (bis(perfluoroethylsulfonyl)imide) resulted in a significant increase in current density.[104] This was attributed to the confined IL contributing to high local O_2 solubility, efficient proton shuttling and the hydrophobic repulsion of the product (H_2O) from the electrocatalytic nanopores.[104] The same IL was combined with Pt-nanoparticle modified graphene to form a composite which was tested for the ORR in the presence of methanol.[105] Methanol is known to result in CO-poisoning of Pt electrocatalysts, and this represents a significant bottleneck in the development of efficient, long-lasting direct methanol fuel cells. In the presence of the IL modifier, lower ORR onset potentials and dramatically reduced methanol poisoning were observed. These were attributed to the IL's higher oxygenphilicity and lower methanol-philicity in comparison to the bulk aqueous solution, thus pre-concentrating O_2 and repelling methanol from the electrocatalyst surface.[105]

The efficient four electron reduction of O_2 to form water [Equation (8.10)] is imperative in order to ensure maximum power output and efficiency from fuel cell devices, as well as metal–air batteries. However, sophisticated materials or precious metals are typically required to achieve this, and most monometallic species tend to demonstrate only two electron reduction [Equation (8.9)]. Mao and co-workers[106] reported the preparation of a $[C_4mim][PF_6]$–carbon nanotube 'bucky gel', into which two electrocatalysts were *in situ* synthesised, immobilised and accommodated in the IL phase. The electrocatalysts comprised cobalt porphyrin and Prussian blue nanoparticles. When the resulting composite was immobilised at an electrode surface and immersed in an oxygenated sodium acetate buffer solution, the two electrocatalysts were effective for the reduction of O_2 to H_2O_2 and then H_2O_2 to H_2O, respectively (Figure 8.9), yielding an overall apparent electrocatalytic four electron reduction of O_2.[106]

The nature of the IL can be influential on the electrocatalytic ability of metals and non-metallic substrates. The ORR at Pt in protic ILs with the anions $[N(SO_2F)_2]^-$, $[N(SO_2CF_3)_2]^-$ and $[N(SO_2CF_2CF_3)_2]^-$ was investigated, with increasing current and decreasing potential noted as the degree of fluorination increased.[107] *In situ* FT-IR measurements of the Pt surface

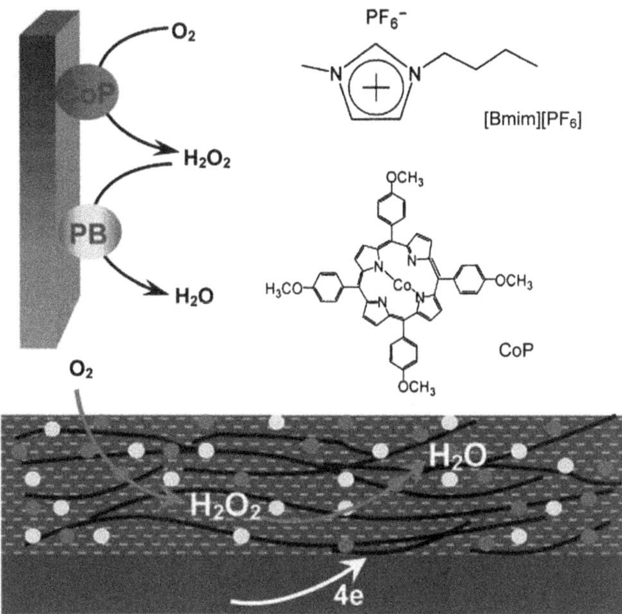

Figure 8.9 Scheme highlighting (top) the two-step electrocatalytic process required to ensure an apparent four electron reduction of O_2, using CoP (structure shown) and PB-NPs (Prussian blue nanoparticles); (bottom) schematic illustration of the two electrocatalysts immobilised in a multiwalled carbon nanotube–IL gel.
Reproduced from reference 106 with permission from the American Chemical Society, copyright © 2008.

revealed dramatically reduced anion adsorption at the Pt surface as fluorination increased, potentially correlating to the significantly enhanced electrocatalysis of the ORR at Pt as fluorination increased.[107] However, thermodynamic, concentration and mass transport effects were also acknowledged as non-quantified possible contributors. Shin *et al.*[108] investigated composites comprising Pd nanoparticles and carbon nanotubes functionalised with IL-derived polymerised imidazolium salts. The anion of the imidazolium salt was extremely influential upon the ORR at the Pd nanoparticles in aqueous 0.1 M $HClO_4$ solution, but not 0.1 M NaOH, highlighting the influential nature of the anion in controlling the local proton concentration near the catalyst surface, and thus its observed electrocatalytic ability.[108] Ding *et al.*[86] have highlighted that in a range of [BF_4]-based ILs, multiwalled carbon nanotubes have two to three times higher standard rate constants for O_2 reduction to the superoxide anion than edge-plane pyrolytic graphite electrodes, and the rate constant at both surfaces are moderately influenced by the size of the IL cation. Ernst *et al.*[87] investigated carbon materials less susceptible to adsorption, and observed that the rate constant for O_2 reduction at boron-doped diamond (BDD) is much slower than glassy carbon in ILs, with the electrocatalytic trends essentially unchanged from what has been previously observed in aqueous media. The difference was primarily ascribed to the fewer electronic density of states in BDD close to O_2 reduction values; the density of states was a bulk electrode feature which did not change significantly when moving between aqueous and IL-based systems. BDD is therefore an effective electrode in both systems when electrocatalytic O_2 reduction is *not* desired, such as sensors designed to operate in oxygenated solutions.[87]

8.2.4 Carbon Dioxide Reduction in Ionic Liquids

Carbon dioxide is rapidly accumulating in the atmosphere due to human activity.[109] In the atmosphere it acts as a Greenhouse gas and contributes significantly to global warming.[109] Electrochemical fixation is not only a potential route towards removing excess CO_2 (*i.e.* from a gas flue); it can also be a valuable step in organic synthetic processes, and represent a source of sustainable fuels and feedstock chemicals.[109–111]

The electrochemical reduction of CO_2 can result in CO, hydrocarbons, oxalic acid and formic acid.[112–115] However, CO_2 reduction typically takes place at very negative potentials, which results in high energy consumption and can frequently result in poor Faradaic efficiency due to competing processes at equivalent or lower potentials.[116] Due to the fact that ILs display wide potential windows, can exhibit high CO_2 solubility, and are typically non-volatile enough for application in harsh conditions such as those found in gas flue streams, they are promising electrolytes for CO_2 reduction.[94,117–119]

A key step in the electrocatalytic conversion of CO_2 is having enough dissolved to make the process viable. The solubility of CO_2 was simulated in 408 ILs using COSMO-RS, and the effect of anions and cations on the solubility explored. It was found that the solubility of CO_2 was highest in ILs based upon the tris(pentafluoroethyl)trifluorophosphate anion[120] (under high pressures of CO_2). Beyond solubility, studies have focussed upon both the role of the electrode surface and the role of IL–CO_2 interactions in the observed electrocatalytic reduction reaction.

The reduction of CO_2 has been investigated at Pt in the acetate-based IL, $[C_4mim][OAc]$.[121] A very high solubility of CO_2 (1520 mM at 1 bar), and relatively slow diffusion coefficient indicated coordination of CO_2 to the IL. The CO_2 was reduced *via* a single electron transfer to form $[CO_2]^{\bullet -}$, with follow-up chemistry speculated to form oxalate, CO or carbonate.[121] A novel IL, $[C_4mim][BF_3Cl]$, was synthesized and tested for the reduction of CO_2 at Pt.[122] It was speculated that the $[BF_3Cl]^-$ reacted with CO_2 to form the Lewis acid–base BF_3–CO_2 adduct, resulting in 0.52 ± 0.25 wt% CO_2 at 1 atm, and a high reduction current density of $\sim$5.7 mA cm^{-2}. The electrocatalytic reduction of CO_2 in this IL was also claimed, with comparisons in $[C_4mim][BF_4]$ and $[C_4mim][NTf_2]$ not generating appreciable currents under similar conditions.[122]

Rosen *et al.*[123] investigated the reduction of CO_2 in the presence and absence of $[C_2mim][BF_4]$ at an Ag electrode. They speculated that the formation

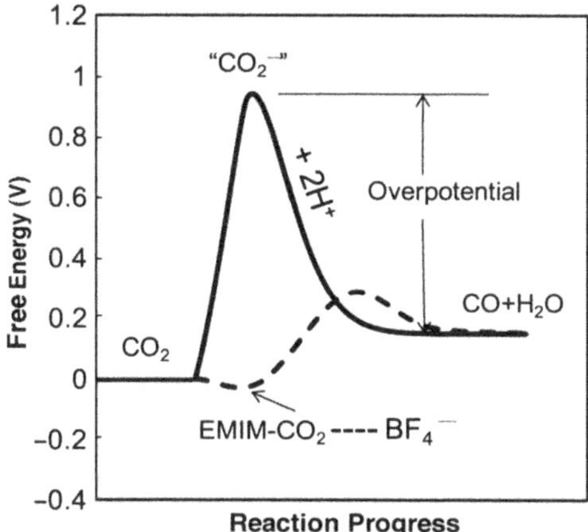

Figure 8.10 A schematic predicting how the free energy of the system might change during the reaction $CO_2 + 2H^+ + 2e^- \rightleftharpoons CO + H_2O$ in water in the presence (---) and absence (–) of [C_2mim][BF_4].
Reproduced from reference 123 with permission from The American Association for the Advancement of Science.

of [CO_2]$^-$ would be facilitated in proximity to ILs due to complex formation between this anion and an IL cation (see energy diagram in Figure 8.10). Measurable CO was evolved from a CO_2-saturated aqueous solution containing 18 mol% [C_4mim][BF_4] at an applied potential of 1.5 V, which was calculated to be an overpotential of 0.17 V. Conversion always had Faradaic efficiencies of over 96%. Conversely, in the absence of the IL (replaced with 0.5 M KCl) CO evolution was not detected until an applied potential of 2.1 V, implying a significant electrocatalytic effect of IL on the reduction of CO_2 at Ag in water–IL mixtures.[123]

Several studies have investigated CO_2 reduction in [C_4mim][BF_4].[112,118–121] Feroci *et al.* observed that the reduction of CO_2 in [C_4mim][BF_4] containing an amine could be used to synthesise carbamates, with electrocatalytic activity (based upon product yield isolated under similar conditions) following the trend $Pt > Cu \sim Ni$.[124] Similar electrocatalytic trends were also observed for the reduction of CO_2 in the presence of aromatic ketones to electrosynthesise α-hydroxycarboxylic esters in the same IL.[125] Feng *et al.*[126] investigated the reduction of CO_2 in [C_4mim][BF_4] to carboxylate 2-amino-5-bromopyridine. Again, electrocatalytic ability was assigned according to product yield, with Ag superior to Cu, Ni and stainless steel, the latter three resulting in similar yields.[126] Feng *et al.*[127] also synthesised dimethylcarbamate from CO_2 in [C_4mim][BF_4], and compared a Pt electrode with a copper skeleton-platinum shell nanostructured composite. The latter displayed higher electrocatalytic activity by virtue of a lower reduction potential for CO_2, which was attributed to its larger surface area and nanoporous structure.[127]

Chu *et al.*[128] reported the electropolymerisation of CO_2 (to low-density polyethylene) by its electrochemical reduction on a nanostructured titanium dioxide electrode immersed in a water–[C_4mim][BF_4] mixture. The nanostructured titanium dioxide (TiO_2) electrode was reported to be an excellent electrocatalyst, based upon reduction potential and selectivity. [C_4mim][BF_4] was chosen in order to have a high CO_2 solubility, but addition of water resulted in a significant increase in current density, so a 1 : 1 v/v mixture was used as a compromise between IL-improved CO_2 solubility and H_2O-reduced viscosity.[128]

Cyclic carbonates can be electrosynthesised in ILs by CO_2 reduction at Cu. Under identical conditions, the best charge efficiencies and isolated yields were obtained in [C_2mim][BF_4] and [C_4mim][BF_4], and significantly poorer results obtained in [*N*-butylpyridinium][BF_4]. Intermediate results were observed in [C_4mim][PF_6], as well as different trends in yield *vs.* the ring-forming substrate employed, highlighting the different role of both cation and anion in influencing these reactions.[129] However, it is currently not clear if these trends represent electrocatalytic effects, or solution-phase reaction directing effects.

8.2.5 Methanol, Formic Acid and Carbon Monoxide Oxidation in Ionic Liquids

Formic acid, methanol and ethanol are all key potential fuels for application in non-hydrogen based fuel cells. All three are oxidised *via* a series of complicated and competing reaction pathways, typically yielding only CO_2 and $H^+/H_2O/[OH]^-$ (depending upon pH) as products and four, six or eight electrons per molecule, respectively.[130] Their application in viable fuel cells relies upon effective electrocatalysis of the entire reaction pathway. Pt is generally the most effective electrocatalyst for the initial oxidation steps, which for all three involves sequential dehydrogenation reactions at the electrode surface.[130] However, this results in the final step consisting of strongly adsorbed carbon monoxide, CO, at the Pt surface, which is more resistant to oxidation.[130] Accumulation of CO on the electrocatalyst surface results in CO oxidation becoming the limiting step, decreasing efficiency and adding several hundred millivolts to the overpotential required for appreciable current flow through direct fuel cells.[131] One route is to raise the temperature of the fuel cell to facilitate CO oxidation; a promising route here is to replace water (operating limit of *ca.* 90 °C) with an IL (theoretical operating limit $\gg$100 °C).

The electrochemical oxidation of a solution of formic acid, HCOOH, in a pure IL has only appeared in one report to date; Martindale and Compton observed that a mixture of 50 mM HCOOH and 50 mM H[NTf_2] in [C_2mim][NTf_2] resulted in an oxidation feature at a Pt microelectrode, but this was not described in detail.[132] Instead ILs have primarily been used as synthesis media to form HCOOH electrocatalysts which are subsequently electrochemically tested in aqueous media,[133] or ILs are employed as thin-layer electrode modifiers to facilitate the electroanalytical quantification of

HCOOH dissolved in water.[134] The closest system investigated to date is the oxidation of HCOOH in 1.25 : 1.0 HCOOH : $[NH_4][HCOO]$, an ionic eutectic melt.[135] This system could be effectively electrolysed to yield H_2 gas, as a potential H_2-storage media, using Pt as electrocatalyst.[135] However, galvanostatic transients revealed that oxidation of the adsorbed CO intermediate was the limiting electrochemical step.[135]

Ejigu *et al.* have recently investigated the oxidation of methanol and CO at Pt in a protic IL, $[N_{2,2,1}]$[OTf].[131] Trace water in the IL resulted in clear Pt oxide (Pt-OH_{ads}) formation features. The oxidation potentials of CO and methanol were both found to correlate with the Pt oxide formation features, therefore indicating that the availability and oxidation of trace water in the IL effectively controls the electrocatalytic oxidation of methanol *via* oxidation of its intermediate, CO. The latter is subsequently oxidised in the presence of platinum oxide to form Pt and CO_2. Increasing the water content and increasing the temperature both had the effect of lowering the potential of oxide formation (and therefore CO and methanol oxidation),[131] indicating that the presence of water is crucial for these oxidation processes to occur in an IL. The oxidation of water is therefore primarily the limiting factor in the observed electrocatalytic ability of Pt with respect to R_yH_xCO-based fuels in ILs. This indicates that ILs are unlikely to exceed aqueous-based electrolytes in such fuel cells, as long as standard approaches and electrocatalysts (*i.e.* direct oxidation at Pt) are employed. Most of the current research into electrocatalysts for use in fuel cells operating with CO-forming fuels has focussed upon the more facile production of oxide (such as *via* the introduction of metals such as Ru[131]) to assist in CO electro-oxidation. Such electrocatalysts will likely be of little utility in high-temperature, essentially anhydrous IL systems, and instead novel electrocatalytic materials or processes dedicated solely to IL-based systems are, therefore, required.

8.2.6 Alkyl and Aryl Halides

The electrochemical dimerisation of alkyl and aryl halides is of extensive synthetic interest and widespread utility.[136] The application of an 'activating' species such as a cobalt Schiff base complex can facilitate such reactions by cleaving organohalide bonds, and electrochemistry be used to constantly regenerate such species such that the reactive complex is employed as a homogeneous (electro)catalyst.[137] Alternatively, an electrocatalytic electrode material such as silver can be employed, such that direct electrochemical activation of the organohalide bond occurs.[138] Both routes typically employ organic solvents due to the poor solubility of the starting materials and products in water, and both routes feature ionic intermediates and products. There is, therefore, scope to replace such organic solvents with a more 'green' IL based system,[136,138] and potentially even accelerate such processes by ionic intermediate–IL interactions, which are known to enhance other processes such as S_N2 reactions between charged species.[20]

Mellah *et al.*[139] used a range of inorganic complexes to catalyse the homocoupling of aryl halide compounds in a range of ILs, but found that fast ligand exchange between some catalysts and ILs was occurring and thus removing catalytically active species. Only $Ni(BF_4)_2(bipy)_3$ and $NiCl_2(bipy)$ (where bipy = bipyridine) displayed electrocatalytic activity in $[C_4mim][NTf_2]$ with respect to the dimerisation of bromobenzene and benzylbromide. The catalysts were stable in the IL and were not removed during product extraction,[139] but comparison with conventional systems was not made.

Gaillon and Bedioui.[137] subsequently performed a more detailed investigation, looking at the electrochemical mediating activity of the electrogenerated Co(I)–salen complex with respect to the electroreductive activation of various organohalogenated derivatives to yield dimers in $[C_4mim][PF_6]$. In this study they found that the electrochemical behaviour of the complex and the mechanisms involved in the activation of the organohalogenated derivatives were essentially unchanged from those found in conventional organic solvents. Therefore, the relative 'immobilisation' of the homogeneous electrocatalyst in the IL phase during product extraction was the primary benefit of ILs in such reactions.[137]

Another cobalt complex, vitamin B_{12}, has been successfully employed by Lagunas *et al.*[140] as a homogeneous electrocatalyst for the reduction of cyclic and acyclic vicinal dibromoalkanes to form their analogue alkene species. Such transformations could only be performed in ILs containing 5–20% v/v water, as vitamin B_{12} was found to be essentially insoluble in anhydrous ILs.[140]

Niu *et al.*[138] and Feng *et al.*[136] investigated the direct reductive, electrocatalytic dimerisation of organohalide species at electrode surfaces in the IL $[C_4mim][BF_4]$. In both cases, Ag electrodes were shown to be the best electrocatalytic material (Ag > Cu > Ni > Ti[138] and Ag ~ Cu > Ni ~ stainless steel,[136] based upon reduction potential and product yields), with maximum dimer yields of *ca.* 50%. Comparison was not made with non-IL systems, although the electrocatalytic trends and yields are similar to those reported in tetrahydrofuran (THF).[141] Therefore, such dimerisations can be effectively performed in ILs using either homogeneous or heterogeneous electrocatalysts. Relative *electrocatalytic trends* and *reaction mechanisms* appear to be unchanged when moving from conventional non-aqueous solvents to ILs, although it should be noted that *relative rates* have not been compared, and therefore an 'ionic liquid effect'[20] on the electrocatalysis of these processes can neither be confirmed nor denied at this stage.

8.2.7 Application of Ionic Liquids in the Synthesis of Electrocatalysts

A significant number of publications have reported the synthesis of catalysts, including electrocatalysts, using IL as a reaction solvent. The motivation behind utilising ILs in this manner includes (i) their large electrochemical windows and good chemical stability,[16] which for example allows the electrodeposition of reactive metals,[142] electrodissolution of unreactive metals[143] or facile application of reactive reagents.[4] A wide range of available anions, high metal solubility, and ILs which feature metal complexes as the anion of the IL are all characteristics which facilitate the electrochemical fabrication of high surface area, nanoporous metals for application as electrocatalytic electrode materials,[144] Other features include (ii) low interfacial tension, which makes them easy to adapt to other phases and results in high nucleation rates and generally smaller nanoparticles and larger surface area,[145] (iii) high thermal stability, which allows syntheses to be conducted at temperatures above 100 °C without using high-pressure vessels, (iv) functionalised ILs, which can modify material surfaces and control the size and improve the stability of nanoparticles,[146,147] and (v) highly structured hydrogen-bonding networks, which can provide templates for the formation of well-defined, extended nanostructures.[148]

There are too many individual applications to be comprehensively covered here. As an example, a gold-based IL (gold(III)-aminoethyl imidazolium aurate salt, $[Cl_3AuNH_2(CH_2)_2mim][AuCl_4]$) has been successfully employed for the electrosynthesis of gold nanoparticles.[149] When applied as an electrocatalyst for methanol oxidation, the IL-derived gold nanoparticles were superior to $KAuCl_4$-derived gold nanoparticles, in terms of surface coverage, electroactive surface area and catalytic efficiency. This difference was attributed to a templating effect of the amino-functionalised imidazolium moiety.[149]

Masud *et al.*[150] reported the preparation of a novel electrocatalyst for formic acid oxidation from an IL. It was prepared by the electrodeposition of Ta from an IL (1-butyl-1-methyl-pyrrolidinium bis(trifluoromethylsulfonyl)imide), $[C_4mPyrr][NTf_2]$ at a Pt electrode, followed by subsequent calcination to form a Ta_2O_5-modified Pt.[150] This electrocatalyst was superior to the bare Pt electrode for the electrocatalytic oxidation of formic acid in aqueous solution,[150] and could also be employed for the electrocatalytic oxidation of formaldehyde.[151]

Suryanto *et al.* reported the controlled electrodeposition of Ag onto various substrates from three protic ILs. Both Ag microparticles and nanoparticles could be obtained by varying parameters such as IL viscosity, and which exhibited excellent electrocatalytic activity towards the oxygen reduction reaction.[152] Gunawan *et al.* also reported the electrodeposition of Cu from protic ILs *via* reduction of Cu(II) to unstable Cu(I).[153] The slow disproportionation of Cu(I) in the PILs was utilised to generate shiny thin films of uniformly distributed Cu nanoparticles, which are known to be efficient electrocatalysts for carbon dioxide reduction.[154]

Conducting polymers can be employed as electrocatalysts, and ILs can be used as media for the electrosynthesis of a wide range of conducting polymers; a topic which has been reviewed by Dong *et al.*[155] As an example, polyaniline is an effective electrocatalyst for the oxidation of formaldehyde, and displays significantly more favourable morphology and physical properties when electrosynthesised from the protic IL 1-ethylimidazolium trifluoroacetate in comparison with the conventional H_2SO_4-based electrosynthesis route.[156]

Pt nanocatalysts supported on carbon have been prepared by impregnation then reduction on carbon black powder, from solutions of Pt salts in dichloromethane, [C_4mim][BF_4] or isopropanol, and compared with commercial samples.[157] When the IL was utilised as solvent, the resulting Pt/C nanocatalysts displayed a large surface area and good electrocatalytic activity with respect to methanol oxidation,[157] although close inspection of the data reveals that dichloromethane-based Pt/C nanocatalysts were slightly superior indicating the role of the IL was minimal. However, Zhao *et al.* have recently investigated the role of the IL 1,1-dibutyl-2,2,3,3-tetramethyl guanidinium bromide in the synthesis of Pd nanocatalysts supported on carbon (Pd/C).[133] It was observed that complex formation occurred between the Pd and IL. This strong interaction was believed to play an influential role in the subsequent nucleation and growth process of the Pd nanoparticles, and corresponding favourable morphology of the nanoparticles (Figure 8.11). In comparison to the Pd salt, the resulting Pd(0) nanoparticles only had a weak interaction with the IL such that the IL could be easily removed to result in a 'clean' Pd surface. When employed as an electrocatalyst for formic acid oxidation, the IL-derived Pd/C exhibited superior performance compared to commercial Pd/C and Pd/C prepared in the absence of the IL.[133]

ILs have been shown to facilitate the formation of nanosized metal catalysts, result in the formation of extremely stable nanoparticle colloids, and even increase the observed catalytic ability of the nanoparticles; this subject has been reviewed comprehensively by Pårvulescu and Hardacre.[158] One interesting aspect of this is that sputtering methods can be used to fabricate the metal nanoparticles in ionic liquids.[159–162] Sputtering results in the introduction of essentially pure metal atoms, while the IL plays a templating role to form small, monodisperse nanoparticles without by-products (*i.e.* surfactants or stochiometric reducing agents). The Pt nanoparticles (average size 2.3–2.4 nm) produced by sputtering from a Pt target onto trimethyl-*n*-propylammonium bis(trifluoromethylsulfonyl)imide demonstrated good electrocatalytic activity for oxygen reduction,[162] indicating a clean Pt surface.

8.2.8 Electrocatalysis in Electroanalytical Studies Employing Ionic Liquids

The most frequent mention of electrocatalysis in relation to the application of ILs arguably occurs in electroanalytical studies. This is primarily due to the nature of electroanalytical studies, whereby a calibration is obtained with respect to an analyte or a range of analytes, by measuring potential, current, resistance or some other attribute related to electrochemistry (such as electrochemiluminscence).[163] A measured response can, therefore, be

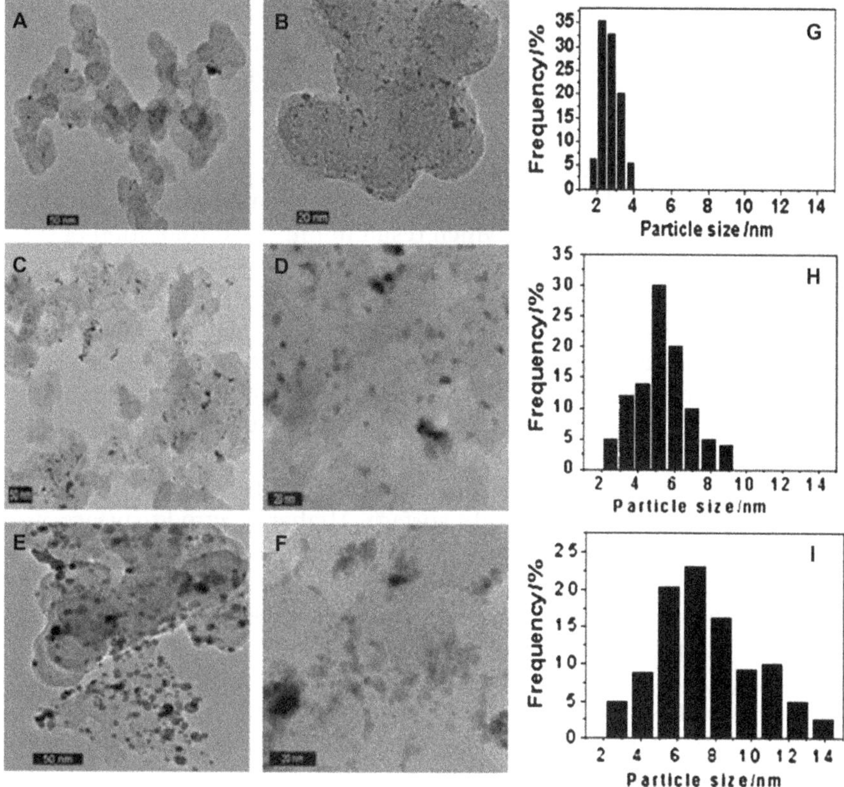

Figure 8.11 TEM images and size-histograms of Pd nanoparticles produced on carbon black by reduction of H_2PdCl_4 with $NaBH_4$ in the presence (A, B, G) and absence (E, F, I) of the IL 1,1-dibutyl-2,2,3,3-tetramethyl guanidinium bromide. A commercial 10 wt% Pd/C catalyst (C, D, H) is also shown as comparison.
Reproduced from reference 133 with permission from the International Association for Hydrogen Energy.

related to a property of the analyte, most frequently to concentration. Any change to the system which results in a decrease in potential or resistance, or increase in current is frequently labelled as improved electrocatalysis.

Having an electrocatalytic process is typically desirable in electroanalytical studies. For example, cyclic voltammetry is used widely in electroanalytical studies. Generally, the highest current response per unit change in the analyte is desired, in order to give the greatest signal-to-noise response. The Faradaic current will flow at a particular potential, and the lower the potential the better, as it means any resulting analytical device will have lower power consumption and it reduces the possibility that an interference compound will undergo a Faradaic process at or below this applied potential. Finally, the resolution of the peak is important; as displayed in Figure 8.2, slower electron transfer kinetics lead to broader peaks. Figure 8.2

also highlights ideal scenarios, where the potential dependence on the energy barriers (α) are equal, whereas many systems have asymmetric values, lower values resulting in significant peak broadening. Higher electron transfer constants typically lead to sharper, more resolved peaks, which typically result in more facile data processing, lower potentials, higher current, and ultimately improved signal-to-noise ratios.

Many studies therefore report new 'electrocatalysts' aimed at particular analytes, or report new systems which result in improved resolution of cyclic voltammograms. Generally, any change which results in a lower potential, higher current density and/or improved peak resolution is classed as an improvement in electrocatalysis. In some cases it is due to a genuine belief that the rate constant has been improved (but is rarely demonstrated), and in some cases it is due to the more general application of 'electrocatalysis' to mean 'improved characteristic(s)'. These improved electrocatalytic characteristics can result from preaccumulation of the analyte, altered thermodyanamics or other parameters such as pH at an IL-modified surface, or improved wetting of the electrode surface. Higher electrode surface area, particularly *via* the introduction of nanoparticles and nanocarbon materials, are also frequently employed. Therefore, given the vast area of ILs in electroanalysis, but the tentative manner in which noted electrocatalytic processes and trends actually relate to more conventional definitions of electrocatalysis, only a brief overview will be given of some broad areas. The application of ILs in electroanalysis has been reviewed extensively.[164–169]

One popular electrode widely employed in electroanalysis is the 'carbon paste electrode'. It is typically constructed of fine graphite powder mixed to a paste with mineral oil.[1] The surface is extremely easily refreshed, ensuring a clean and consistent surface, and additional materials can be introduced to provide electrocatalysis (*i.e.* nanoparticles) or functionality (*i.e.* chelating ligands). When ILs are employed as binders, the resulting electrode is known as a Carbon Ionic Liquid Electrode (CILE).[165] Improved characteristics have been observed for CILE, including enhanced electrocatalytic response for a wide range of species, due to the inherent conductivity of the IL, enhanced accessibility to the carbon surface (covered in a thin layer of IL as electrolyte, rather than inert mineral oil), and the ability of the IL to anion exchange to accumulate material at the surface of the electrode.[170] Electrocatalytic species can be directly incorporated into the IL–carbon powder mixture during the electrode fabrication step, such as glucose oxidase.[171] Alternatively, they can be electrodeposited onto the surface of the CILE, such as metallic nanoparticles.[172] Interestingly, it has also been noted that ILs can be too effective at making carbon surfaces 'available'; carbon nanotube–IL mixed electrodes possess detrimentally high background charging currents when applied for electroanalysis aimed at low concentrations of analytes due to the very high surface area.[173] This had to be countered by increasing mass transport, using a rotating disc electrode.[173]

ILs are also frequently employed as surface modifiers. Small quantities of ILs have been proven to be highly effective for the pre-concentration of analytes, either from solution or from the gas phase.[174] Hydrophobic ILs can be employed to form a thin layer on the surface of an electrode, which when immersed in an aqueous solution can selectively pre-concentrate certain compounds at the electrode surface,[105] the larger current (from a higher local concentration of analyte) resulting in an apparent electrocatalytic effect. Similar films have also been reported for hydrophilic ILs, which can spontaneously assemble at electrode surfaces from dilute aqueous solutions of ILs.[28] Such thin layers are known to alter the observed electrochemical response, resulting in changes in current, potential and improving resolution between Faradaic peaks which overlap in the absence of the IL layer.[28] The improved, selective determination of uric acid, dopamine and ascorbic acid in complex mixtures is one key area where thin IL layers have been highly effective.[175–180] The combination of thin IL layers on rough carbon surfaces, such as screen printed carbon electrode surfaces[181] or carbon nanotubes,[178,182–184] can result in a significant increase in the surface area and thus the kinetics of analyte uptake, as well as potential improvements in electron transfer kinetics.[86] This application of ILs as electrocatalytic electrode modifiers is frequently dramatic and effective, but can stem from altered mass transport, concentration, thermodynamics as well as kinetics. Therefore, quantitative comparisons of electrocatalytic ability (*i.e.* changes in rate constants) are highly challenging and typically cannot be performed.

Biological systems offer a vast range of elegant compounds such as proteins and enzymes, which can boast fantastic selectivity combined with unrivalled catalytic ability. Most operate on the principle of electron transfer, and if suitably utilised with respect to an electrode (such that direct electron transfer can occur, or a suitable redox mediator shuttle between them) then these biological molecules can be employed as electrocatalysts. This technique is particularly effective in the development of electrochemical biosensors. The combination of ILs with enzymes and other biomaterials can introduce unique microenvironments which can alter the electrocatalytic nature of the processes, pre-concentrate the analyte, protect the sensitive biological material from hostile analytical samples, *etc.* Primarily such systems have been employed in electroanalytical studies with significant success and have been extensively reviewed[164,165,167–169] although electrocatalytic aspects suffer from the same multifaceted role as the IL as noted above.

8.3 Conclusions

Electrocatalysis is a vital attribute to a vast range of systems. Effective electrocatalysis promises cheaper and more energy efficient electrosynthetic processes, more powerful and efficient fuel cells, electroanalytical procedures with unrivalled sensitivity and selectivity, and viable routes towards the remediation of pollutants such as carbon dioxide. Ionic Liquids (ILs) possess vast potential applications due to the sheer diversity of their available properties.

A major highlight of ILs is their ability to dramatically enhance the electrocatalysis of certain processes, by virtue of their unique physical interactions with solutes and electrode surfaces, either as pure electrolyte materials or when introduced as dilute electrolytes in more conventional solvents such as water. Although the introduction of ILs into a range of systems undeniably improves observed electrochemical responses, the precise role of the IL is often multifaceted and difficult to quantitatively analyse. This in turn inhibits the informed development of IL-electrocatalyst combinations. Additionally, many conventional electrocatalysts often rely upon the presence of water to operate effectively (*e.g.* oxide-based electrocatalysts for the oxidation of carbon monoxide-forming fuels). New materials are therefore required if ILs are to be fully utilised and ultimately exceed aqueous-based electrocatalytic systems.

All of this ultimately indicates that a significant amount of work remains to be done in order to probe the physical nature of ILs and the concept of electrocatalysis in ILs. This also makes it an extremely fascinating area, where a vast scope of improvements, discoveries and unique applications await.

References

1. A. J. Bard, G. Inzelt and F. Scholz, ed. *Electrochemical Dictionary*, Springer, Berlin, London, 2008.
2. M. C. Buzzeo, R. G. Evans and R. G. Compton, *ChemPhysChem*, 2004, **5**, 1106.
3. M. C. Buzzeo, C. Hardacre and R. G. Compton, *ChemPhysChem*, 2006, **7**, 176.
4. D. S. Silvester, L. Aldous, M. C. Lagunas, C. Hardacre and R. G. Compton, *J. Phys. Chem. B*, 2006, **110**, 22035.
5. A. J. Appleby, *Catal. Rev.*, 1970, **4**, 221.
6. J. O. M. Bockris, A. K. N. Reddy and M. Gamboa-Aldeco, *Modern Electrochemistry. Vol. 2A, Fundamentals of Electrodics*, Kluwer Academic/ Plenum, New York, London, 2000.
7. C. F. Zinola, *Electrocatalysis: Computational, Experimental, and Industrial Aspects*, CRC Press, London, Taylor & Francis Group, Boca Raton, FL, 2010.
8. T. Welton, *Chem. Rev.*, 1999, **99**, 2071.
9. P. Wasserscheid and T. Welton, *Ionic Liquids in Synthesis*, Wiley-VCH, Weinheim, 2008.
10. X. Y. Lu, G. Burrell, F. Separovic and C. Zhao, *J. Phys. Chem. B*, 2012, **116**, 9160.
11. M. Deetlefs and K. R. Seddon, *Chim. Oggi*, 2006, **24**, 16.
12. D. R. Macfarlane, M. Forsyth, P. C. Howlett, J. M. Pringle, J. Sun, G. Annat, W. Neil and E. I. Izgorodina, *Acc. Chem. Res.*, 2007, **40**, 1165.
13. W. Xu and C. A. Angell, *Science*, 2003, **302**, 422.
14. A. E. Visser, R. P. Swatloski and R. D. Rogers, *Green Chem.*, 2000, **2**, 1.

15. J. A. Widegren, A. Laesecke and J. W. Magee, *Chem. Commun.*, 2005, 1610.
16. A. M. O'Mahony, D. S. Silvester, L. Aldous, C. Hardacre and R. G. Compton, *J. Chem. Eng. Data*, 2008, **53**, 2884.
17. E. I. Rogers, B. Sljukic, C. Hardacre and R. G. Compton, *J. Chem. Eng. Data*, 2009, **4**, 2049.
18. J. S. Long, D. S. Silvester, A. S. Barnes, N. V. Rees, L. Aldous, C. Hardacre and R. G. Compton, *J. Phys. Chem. C*, 2008, **112**, 6993.
19. S. R. Belding, N. V. Rees, L. Aldous, C. Hardacre and R. G. Compton, *J. Phys. Chem. C*, 2008, **112**, 1650.
20. J. P. Hallett, C. L. Liotta, G. Ranieri and T. Welton, *J. Org. Chem.*, 2009, **74**, 1864.
21. C. A. Zhao, D. R. MacFarlane and A. M. Bond, *J. Am. Chem. Soc.*, 2009, **131**, 16195.
22. R. Hayes, N. Borisenko, M. K. Tam, P. C. Howlett, F. Endres and R. Atkin, *J. Phys. Chem. C*, 2011, **115**, 6855.
23. T. J. Simons, A. A. J. Torriero, P. C. Howlett, D. R. MacFarlane and M. Forsyth, *Electrochem. Commun.*, 2012, **18**, 119.
24. J. C. Padilha, E. M. A. Martini, C. Brum, M. O. de Souza and R. F. de Souza, *J. Power Sources*, 2009, **194**, 482.
25. M. V. Migliorini, R. K. Donato, M. A. Benvegnu, J. Dupont, R. S. Goncalves and H. S. Schrekker, *Catal. Commun.*, 2008, **9**, 971.
26. Y. Meng, L. Aldous and R. G. Compton, *J. Phys. Chem. C*, 2011, **115**, 14334.
27. M. J. A. Shiddiky, A. A. J. Torriero, C. Zhao, I. Burgar, G. Kennedy and A. M. Bond, *J. Am. Chem. Soc.*, 2009, **131**, 7976.
28. P. Yu, Y. Q. Lin, L. Xiang, L. Su, J. Zhang and L. Q. Mao, *Langmuir*, 2005, **21**, 9000.
29. A. Tsyganok, C. M. Holt, S. Murphy, D. Mitlin and M. R. Gray, *Fuel*, 2012, **93**, 415.
30. T. Fuchigami and T. Tajima, *Electrochemistry*, 2006, **74**, 585.
31. T. Sawamura, S. Kuribayashi, S. Inagi and T. Fuchigami, *Adv. Synth. Catal.*, 2010, **352**, 2757.
32. T. Sawamura, S. Kuribayashi, S. Inagi and T. Fuchigami, *Org. Lett.*, 2010, **12**, 644.
33. P. H. Rieger, *Electrochemistry*, Prentice-Hall, Englewood Cliffs, London, 1987.
34. B. E. Conway and J. O. Bockris, *J. Chem. Phys.*, 1957, **26**, 532.
35. S. Trasatti, *J. Electroanal. Chem.*, 1972, **39**, 163.
36. L. A. Kibler, *ChemPhysChem*, 2006, **7**, 985.
37. J. A. Bautista-Martinez, L. Tang, J. P. Belieres, R. Zeller, C. A. Angell and C. Friesen, *J. Phys. Chem. C*, 2009, **113**, 12586.
38. R. G. Compton and C. E. Banks, *Understanding Voltammetry*, World Scientific, Singapore, London, 2nd edn, 2011.
39. Y. Meng, L. Aldous, S. R. Belding and R. G. Compton, *Chem. Commun.*, 2012, **48**, 5572.

40. M. M. Jaksic, *J. New Mater. Electrochem. Syst.*, 2000, **3**, 153.
41. E. Santos, P. Quaino and W. Schmickler, *Phys. Chem. Chem. Phys.*, 2012, **14**, 11224.
42. E. Santos, P. Hindelang, P. Quaino, E. N. Schulz, G. Soldano and W. Schmickler, *ChemPhysChem*, 2011, **12**, 2274.
43. W. Li and A. M. Lane, *Electrochem. Commun.*, 2011, **13**, 913.
44. R. F. de Souza, J. C. Padilha, R. S. Goncalves and J. Dupont, *Electrochem. Commun.*, 2003, **5**, 728.
45. L. Aldous, D. S. Silvester, W. R. Pitner, R. G. Compton, M. C. Lagunas and C. Hardacre, *J. Phys. Chem. C*, 2007, **111**, 8496.
46. D. S. Silvester, L. Aldous, C. Hardacre and R. G. Compton, *J. Phys. Chem. B*, 2007, **111**, 5000.
47. Y. Meng, L. Aldous and R. G. Compton, *Green Chem.*, 2010, **12**, 1926.
48. Y. Meng, L. Aldous, B. S. Pilgrim, T. J. Donohoe and R. G. Compton, *New J. Chem.*, 2011, **35**, 1369.
49. D. S. Silvester, K. R. Ward, L. Aldous, C. Hardacre and R. G. Compton, *J. Electroanal. Chem.*, 2008, **618**, 53.
50. R. Fukuta, Y. Katayama and T. Miura, *ECS Trans.*, 2007, **3**, 567.
51. J. O. M. Bockris and A. K. N. Reddy, *Modern Electrochemistry. Vol. 1, Ionics*, Plenum Press, New York, London, 1998.
52. Y. Meng, L. Aldous, S. R. Belding and R. G. Compton, *Phys. Chem. Chem. Phys.*, 2012, **14**, 5222.
53. A. M. Navarro-Suárez, J. C. Hidalgo-Acosta, L. Fadini, J. M. Feliu and M. F. Suárez-Herrera, *J. Phys. Chem. C*, 2011, **115**, 11147.
54. A. Ciaccafava, M. Alberola, S. Hameury, P. Infossi, M. T. Giudici-Orticoni and E. Lojou, *Electrochim. Acta*, 2011, **56**, 3359.
55. L. Johnson, A. Ejigu, P. Licence and D. A. Walsh, *J. Phys. Chem. C*, 2012, **116**, 18048.
56. J. B. Rollins and J. C. Conboy, *J. Electrochem. Soc.*, 2009, **156**, B943.
57. M. G. Freire, C. M. S. S. Neves, I. M. Marrucho, J. A. P. Coutinho and A. M. Fernandes, *J. Phys. Chem. A*, 2010, **114**, 3744.
58. L. Cammarata, S. G. Kazarian, P. A. Salter and T. Welton, *Phys. Chem. Chem. Phys.*, 2001, **3**, 5192.
59. Z. G. Han, A. M. Bond and C. Zhao, *Sci. China: Chem.*, 2011, **54**, 1877.
60. G. Bernardini, A. G. Wedd, C. Zhao and A. M. Bond, *Proc. Natl. Acad. Sci. U. S. A*, 2012, **109**, 11552.
61. G. Bernardini, A. G. Wedd, C. Zhao and A. M. Bond, *Dalton Trans.*, 2012, **41**, 9944.
62. G. Bernardini, C. Zhao, A. G. Wedd and A. M. Bond, *Inorg. Chem.*, 2011, **50**, 5899.
63. M. M. Islam, T. Okajima, S. Kojima and T. Ohsaka, *Chem. Commun.*, 2008, 5330.
64. R. F. de Souza, J. C. Padilha, R. S. Goncalves and J. Rault-Berthelot, *Electrochem. Commun.*, 2006, **8**, 211.

65. R. F. de Souza, G. Loget, J. C. Padilha, E. M. A. Martini and M. O. de Souza, *Electrochem. Commun.*, 2008, **10**, 1673.
66. G. Loget, J. C. Padilha, E. A. Martini, M. O. de Souza and R. F. de Souza, *Int. J. Hydrogen Energy*, 2009, **34**, 84.
67. R. F. de Souza, J. C. Padilha, R. S. Goncalves, M. O. de Souza and J. Rault-Berthelot, *J. Power Sources*, 2007, **164**, 792.
68. C. Zhao, G. Burrell, A. A. J. Torriero, F. Separovic, N. F. Dunlop, D. R. MacFarlane and A. M. Bond, *J. Phys. Chem. B*, 2008, **112**, 6923.
69. D. H. Pool, M. P. Stewart, M. O'Hagan, W. J. Shaw, J. A. S. Roberts, R. M. Bullock and D. L. DuBois, *Proc. Natl. Acad. Sci. U. S. A.*, 2012, **109**, 15634.
70. X. Ren, S. S. Zhang, D. T. Tran and J. Read, *J. Mater. Chem.*, 2011, **21**, 10118.
71. L. Qu, Y. Liu, J.-B. Baek and L. Dai, *ACS Nano*, 2010, **4**, 1321.
72. C. O. Laoire, S. Mukerjee, K. M. Abraham, E. J. Plichta and M. A. Hendrickson, *J. Phys. Chem. C*, 2010, **114**, 9178.
73. L. Cecchetto, M. Salomon, B. Scrosati and F. Croce, *J. Power Sources*, 2012, **213**, 233.
74. S. F. Wang, H. S. Lu, S. B. Zhang and Y. J. Lin, *J. Appl. Electrochem.*, 1993, **23**, 387.
75. R. Poupko and I. Rosenthal, *J. Phys. Chem.*, 1973, **77**, 1722.
76. T. M. Buetler, A. Krauskopf and U. T. Ruegg, *News Physiol. Sci.*, 2004, **19**, 120.
77. J. Z. Byczkowski and T. Gessner, *Int. J. Biochem.*, 1988, **20**, 569.
78. R. Toniolo, N. Dossi, A. Pizzariello, A. P. Doherty, S. Susmel and G. Bontempelli, *J. Electroanal. Chem.*, 2012, **670**, 23.
79. J. Gebicki, A. Kloskowski and W. Chrzanowski, *Electrochim. Acta*, 2011, **56**, 9910.
80. E. Yeager, *Electrochim. Acta*, 1984, **29**, 1527.
81. Z. Guo, Y. Qiao, H. Liu, C. Ding, Y. Zhu, M. Wan and L. Jiang, *J. Mater. Chem.*, 2012, **22**, 17153.
82. D. T. Sawyer, G. Chiericato, Jr., C. T. Angelis, E. J. Nanni, Jr. and T. Tsuchiya, *Anal. Chem.*, 1982, **54**, 1720.
83. R. Henriquez, P. Grez, E. Munoz, E. A. Dalchiele, R. Marotti and H. Gomez, *Electrochem. Solid-State Lett.*, 2009, **12**, H288.
84. M. T. Carter, C. L. Hussey, S. K. D. Strubinger and R. A. Osteryoung, *Inorg. Chem.*, 1991, **30**, 1149.
85. I. M. AlNashef, M. L. Leonard, M. C. Kittle, M. A. Matthews and J. W. Weidner, *Electrochem. Solid-State Lett.*, 2001, **4**, D16.
86. K. Ding, T. Okajima and T. Ohsaka, *Electrochemistry (Tokyo, Jpn.)*, 2007, **75**, 35.
87. S. Ernst, L. Aldous and R. G. Compton, *J. Electroanal. Chem.*, 2011, **663**, 108.
88. R. G. Evans, O. V. Klymenko, S. A. Saddoughi, C. Hardacre and R. G. Compton, *J. Phys. Chem. B*, 2004, **108**, 7878.

89. M. M. Islam, T. Imase, T. Okajima, M. Takahashi, Y. Niikura, N. Kawashima, Y. Nakamura and T. Ohsaka, *J. Phys. Chem. A*, 2009, **113**, 912.
90. C. Villagran, L. Aldous, M. C. Lagunas, R. G. Compton and C. Hardacre, *J. Electroanal. Chem.*, 2006, **588**, 27.
91. X.-J. Huang, E. I. Rogers, C. Hardacre and R. G. Compton, *J. Phys. Chem. B*, 2009, **113**, 8953.
92. A. S. Barnes, E. I. Rogers, I. Streeter, L. Aldous, C. Hardacre, G. G. Wildgoose and R. G. Compton, *J. Phys. Chem. C*, 2008, **112**, 13709.
93. M. C. Buzzeo, O. V. Klymenko, J. D. Wadhawan, C. Hardacre, K. R. Seddon and R. G. Compton, *J. Phys. Chem. A*, 2003, **107**, 8872.
94. M. C. Buzzeo, O. V. Klymenko, J. D. Wadhawan, C. Hardacre, K. R. Seddon and R. G. Compton, *J. Phys. Chem. B*, 2004, **108**, 3947.
95. X. J. Huang, L. Aldous, A. M. O'Mahony, F. J. del Campo and R. G. Compton, *Anal. Chem.*, 2010, **82**, 5238.
96. E. I. Rogers, X.-J. Huang, E. J. F. Dickinson, C. Hardacre and R. G. Compton, *J. Phys. Chem. C*, 2009, **113**, 17811.
97. M. M. Islam, B. N. Ferdousi, T. Okajima and T. Ohsaka, *Electrochem. Commun.*, 2005, **7**, 789.
98. M. M. Islam and T. Ohsaka, *J. Phys. Chem. C*, 2008, **112**, 1269.
99. D. Zhang, T. Okajima, F. Matsumoto and T. Ohsaka, *J. Electrochem. Soc.*, 2004, **151**, D31.
100. Z. Wang, P. L. Lin, G. A. Baker, J. Stetter and X. Q. Zeng, *Anal. Chem.*, 2011, **83**, 7066.
101. R. Toniolo, N. Dossi, A. Pizzariello, A. P. Doherty, S. Susmel and G. Bontempelli, *J. Electroanal. Chem.*, 2012, **670**, 23.
102. X. R. Li, B. Wang, J. J. Xu and H. Y. Chen, *Nanoscale*, 2011, **3**, 5026.
103. S. Y. Lee, A. Ogawa, M. Kanno, H. Nakamoto, T. Yasuda and M. Watanabe, *J. Am. Chem. Soc.*, 2010, **132**, 9764.
104. J. Snyder, T. Fujita, M. W. Chen and J. Erlebacher, *Nat. Mater.*, 2010, **9**, 904.
105. Y. Tan, C. Xu, G. Chen, N. Zheng and Q. Xie, *Energy Environ. Sci.*, 2012, **5**, 6923.
106. P. Yu, J. Yan, H. Zhao, L. Su, J. Zhang and L. Mao, *J. Phys. Chem. C*, 2008, **112**, 2177.
107. H. Munakata, T. Tashita, M. Haibara and K. Kanamura, *ECS Trans*, 2010, **33**, 463.
108. J. Y. Shin, Y. S. Kim, Y. Lee, J. H. Shim, C. Lee and S.-g. Lee, *Chem.–Asian J*, 2011, **6**, 2016.
109. M. Wise, K. Calvin, A. Thomson, L. Clarke, B. Bond-Lamberty, R. Sands, S. J. Smith, A. Janetos and J. Edmonds, *Science (Washington, DC, U. S.)*, 2009, **324**, 1183.
110. J. Bian, M. Xiao, S.-J. Wang, Y.-X. Lu and Y.-Z. Meng, *Appl. Surf. Sci.*, 2009, **255**, 7188.
111. G. A. Olah, A. Goeppert and G. K. S. Prakash, *J. Org. Chem.*, 2009, **74**, 487.

112. S. Kaneco, H. Katsumata, T. Suzuki and K. Ohta, *Electrochim. Acta*, 2006, **51**, 3316.
113. S. Ikeda, T. Takagi and K. Ito, *Bull. Chem. Soc. Jpn.*, 1987, **60**, 2517.
114. A. J. Morris, R. T. McGibbon and A. B. Bocarsly, *ChemSusChem*, 2011, **4**, 191.
115. S. Kaneco, N.-h. Hiei, Y. Xing, H. Katsumata, H. Ohnishi, T. Suzuki and K. Ohta, *Electrochim. Acta*, 2002, **48**, 51.
116. T. Mizuno, K. Ohta and M. Kawamoto, *Energy Sources*, 1997, **19**, 249.
117. M. B. Shiflett and A. Yokozeki, *Ind. Eng. Chem. Res.*, 2005, **44**, 4453.
118. J. L. Anthony, E. J. Maginn and J. F. Brennecke, *J. Phys. Chem. B*, 2002, **106**, 7315.
119. R. E. Baltus, B. H. Culbertson, S. Dai, H. Luo and D. W. DePaoli, *J. Phys. Chem. B*, 2004, **108**, 721.
120. X. Zhang, Z. Liu and W. Wang, *AIChE J*, 2008, **54**, 2717.
121. L. E. Barrosse-Antle and R. G. Compton, *Chem. Commun.*, 2009, 3744.
122. L. L. Snuffin, L. W. Whaley and L. Yu, *J. Electrochem. Soc.*, 2011, **158**, F155.
123. B. A. Rosen, A. Salehi-Khojin, M. R. Thorson, W. Zhu, D. T. Whipple, P. J. A. Kenis and R. I. Masel, *Science*, 2011, **334**, 643.
124. M. Feroci, M. Orsini, L. Rossi, G. Sotgiu and A. Inesi, *J. Org. Chem.*, 2007, **72**, 200.
125. Q. J. Feng, K. L. Huang, S. Q. Liu, J. G. Yu and F. F. Liu, *Electrochim. Acta*, 2011, **56**, 5137.
126. Q. J. Feng, K. L. Huang, S. Q. Liu and X. Y. Wang, *Electrochim. Acta*, 2010, **55**, 5741.
127. Q. J. Feng, S. Q. Liu, X. Y. Wang and G. H. Jin, *Appl. Surf. Sci.*, 2012, **258**, 5005.
128. D. Chu, G. Qin, X. Yuan, M. Xu, P. Zheng and J. Lu, *ChemSusChem*, 2008, **1**, 205.
129. H. Z. Yang, Y. L. Gu, Y. Q. Deng and F. Shi, *Chem. Commun.*, 2002, 274.
130. L. Aldous and R. G. Compton, *ChemPhysChem*, 2011, **12**, 1280.
131. A. Ejigu, L. Johnson, P. Licence and D. A. Walsh, *Electrochem. Commun.*, 2012, **23**, 122.
132. B. C. M. Martindale and R. G. Compton, *Chem. Commun.*, 2012, **48**, 6487.
133. X. Zhao, Y. Hu, L. Liang, C. P. Liu, J. H. Liao and W. Xing, *Int. J. Hydrogen Energy*, 2012, **37**, 51.
134. J. Chai, F. H. Li, Y. W. Hu, Q. X. Zhang, D. X. Han and L. Niu, *J. Mater. Chem*, 2011, **21**, 17922.
135. L. Aldous and R. G. Compton, *Energy Environ. Sci.*, 2010, **3**, 1587.
136. Q. J. Feng, K. L. Huang, S. Q. Liu, H. M. Wang and W. B. Yan, *J. Phys. Org. Chem.*, 2012, **25**, 506.
137. L. Gaillon and F. Bedioui, *J. Mol. Catal. A: Chem.*, 2004, **214**, 91.
138. D. F. Niu, A. J. Zhang, T. Xue, J. B. Zhang, S. F. Zhao and J. X. Lu, *Electrochem. Commun.*, 2008, **10**, 1498.

139. M. Mellah, S. Gmouh, M. Vaultier and V. Jouikov, *Electrochem. Commun.*, 2003, **5**, 591.
140. M. C. Lagunas, D. S. Silvester, L. Aldous and R. G. Compton, *Electroanalysis*, 2006, **18**, 2263.
141. C. A. Paddon, F. L. Bhatti, T. J. Donohoe and R. G. Compton, *J. Phys. Org. Chem.*, 2007, **20**, 115.
142. R. Wibowo, L. Aldous, S. E. W. Jones and R. G. Compton, *Chem. Phys. Lett.*, 2010, **492**, 276.
143. J. F. Huang and H. Y. Chen, *Angew. Chem., Int. Ed.*, 2012, **51**, 1684.
144. I. W. Sun and P.-Y. Chen in *Electrodeposition from Ionic Liquids*, ed. Frank Endres, Douglas MacFarlane and Andrew Abbott, Wiley-VCH Verlag GmbH & Co. KGaA, Weinheim, 2008, 125.
145. M. Antonietti, D. B. Kuang, B. Smarsly and Z. Yong, *Angew. Chem., Int. Ed.*, 2004, **43**, 4988.
146. Z. F. Fei, T. J. Geldbach, D. B. Zhao and P. J. Dyson, *Chem.–Eur. J*, 2006, **12**, 2123.
147. H. Itoh, K. Naka and Y. Chujo, *J. Am. Chem. Soc.*, 2004, **126**, 3026.
148. A. Mele, C. D. Tran and S. H. D. Lacerda, *Angew. Chem., Int. Ed.*, 2003, **42**, 4364.
149. B. Ballarin, M. C. Cassani, M. Gazzano and G. Solinas, *Electrochim. Acta*, 2010, **56**, 676.
150. J. Masud, M. T. Alam, M. R. Miah, T. Okajima and T. Ohsaka, *Electrochem. Commun.*, 2011, **13**, 86.
151. J. Masud, M. T. Alam, T. Okajima and T. Ohsaka, *Chem. Lett.*, 2011, **40**, 252.
152. B. H. R. Suryanto, C. A. Gunawan, X. Y. Lu and C. Zhao, *Electrochim. Acta*, 2012, **81**, 98.
153. C. A. Gunawan, B. H. R. Suryanto and C. Zhao, *J. Electrochem. Soc.*, 2012, **159**, D611.
154. A. A. Peterson, F. Abild-Pedersen, F. Studt, J. Rossmeisl and J. K. Norskov, *Energy Environ. Sci.*, 2010, **3**, 1311.
155. B. Dong, J. K. Xu and L. Q. Zheng, *Prog. Chem. (Beijing, China)*, 2009, **21**, 1792.
156. C. A. Ma, M. C. Li, Y. F. Zheng and B. Y. Liu, *Electrochem. Solid-State Lett*, 2005, **8**, G122.
157. X. Xue, C. Liu, T. Lu and W. Xing, *Fuel Cells*, 2006, **6**, 347.
158. V. I. Pårvulescu and C. Hardacre, *Chem. Rev.*, 2007, **107**, 2615.
159. K.-i. Okazaki, T. Kiyama, T. Suzuki, S. Kuwabata and T. Torimoto, *Chem. Lett.*, 2009, **38**, 330.
160. T. Torimoto, K.-i. Okazaki, T. Kiyama, K. Hirahara, N. Tanaka and S. Kuwabata, *Appl. Phys. Lett.*, 2006, **89**, 243117/243111–243117/243113.
161. K.-i. Okazaki, T. Kiyama, K. Hirahara, N. Tanaka, S. Kuwabata and T. Torimoto, *Chem. Commun.*, 2008, 691.
162. T. Tsuda, K. Yoshii, T. Torimoto and S. Kuwabata, *J. Power Sources*, 2010, **195**, 5980.

163. L. C. Chen, D. J. Huang, Y. J. Zhang, T. Q. Dong, C. Zhou, S. Y. Ren, Y. W. Chi and G. N. Chen, *Analyst*, 2012, **137**, 3514.
164. M. Pumera, A. Ambrosi, A. Bonanni, E. L. K. Chng and H. L. Poh, *TrAC, Trends Anal. Chem*, 2010, **29**, 954.
165. M. J. A. Shiddiky and A. A. J. Torriero, *Biosens. Bioelectron.*, 2011, **26**, 1775.
166. A. C. Franzoi, D. Brondani, E. Zapp, S. K. Moccelini, S. C. Fernandes, I. C. Vieira and J. Dupont, *Quim. Nova*, 2011, **34**, 1042.
167. H. T. Liu, Y. Liu and J. H. Li, *Phys. Chem. Chem. Phys.*, 2010, **12**, 1685.
168. D. S. Silvester, *Analyst*, 2011, **136**, 4871.
169. Y. R. Wang, P. Hu, Q. L. Liang, G. A. Luo and Y. M. Wang, *Chin. J. Anal. Chem.*, 2008, **36**, 1011.
170. N. Maleki, A. Safavi and F. Tajabadi, *Electroanalysis*, 2007, **19**, 2247.
171. M. M. Musameh, R. T. Kachoosangi, L. Xiao, A. Russell and R. G. Compton, *Biosens. Bioelectron.*, 2008, **24**, 87.
172. A. Safavi, N. Maleki, F. Tajabadi and E. Farjami, *Electrochem. Commun.*, 2007, **9**, 1963.
173. R. T. Kachoosangi, G. G. Wildgoose and R. G. Compton, *Electroanalysis*, 2007, **19**, 1483.
174. L. Vidal, E. Psillakis, C. E. Domini, N. Grane, F. Marken and A. Canals, *Anal. Chim. Acta*, 2007, **584**, 189.
175. M. Ammam and E. B. Easton, *Electrochim. Acta*, 2011, **56**, 2847.
176. J. P. Dong, Y. Y. Hu, S. M. Zhu, J. Q. Xu and Y. J. Xu, *Anal. Bioanal. Chem.*, 2010, **396**, 1755.
177. Y. H. Li, X. S. Liu and W. Z. Wei, *Electroanalysis*, 2011, **23**, 2832.
178. Y. Liu, X. Q. Zou and S. J. Dong, *Electrochem. Commun.*, 2006, **8**, 1429.
179. W. Sun, M. X. Yang and K. Jiao, *Anal. Bioanal. Chem.*, 2007, **389**, 1283.
180. M. Pandurangachar, B. E. K. Swamy, B. N. Chandrashekar, O. Gilbert and B. S. Sherigara, *J. Mol. Liq.*, 2011, **158**, 13.
181. J. F. Ping, J. A. Wu and Y. B. Ying, *Electrochem. Commun.*, 2010, **12**, 1738.
182. Y. H. Li, X. S. Liu, X. Y. Liu, N. N. Mai, Y. D. Li, W. Z. Wei and Q. Y. Cai, *Colloids Surf., B*, 2011, **88**, 402.
183. L. N. Qu, J. Wu, X. Y. Sun, M. Y. Xi and W. Sun, *J. Chin. Chem. Soc.*, 2010, **57**, 701.
184. X. Z. Zhang, S. F. Liu, K. Jiao and Y. W. Hu, *Electroanalysis*, 2008, **20**, 1909.

CHAPTER 9

Photochemistry in Ionic Liquids

HERMENEGILDO GARCIA* AND SERGIO NAVALON*

Instituto de Tecnología Química CSIC-UPV and Departamento de Química, Universidad Politécnica de Valencia, 46022 Valencia, Spain Av. De los Naranjos s/n, 46022 Valencia, Spain, sernaol@doctor.upv.es
*Email: hgarcia@qim.upv.es; sernaol@doctor.upv.es

9.1 Definition

Ionic liquids (ILs) are ionic compounds with an organic cation that are in the liquid state near the ambient temperature. ILs are considered green alternatives to conventional volatile organic solvents since, due to the ionic bond, their vapor pressure is almost negligible, thus, contrasting with the behaviour of the volatile organic compounds (VOCs) commonly used as solvents. Other important features related to their greenness are their lack of flammability and their high thermal and chemical stability. ILs find application as solvents in organic synthesis and particularly for the development of recyclable catalysts.[1–3] Figure 9.1 shows the chemical structure of the most common cations and anions of ILs. Among the various possible structures, those containing an imidazolium core have been the most widely used due their ease and quantitative synthesis from available intermediates.

9.2 Polarity in ILs

One property that is relevant for the use of ILs as solvent is polarity. ILs are generally considered to be highly polar solvents based on structural

RSC Catalysis Series No. 15
Catalysis in Ionic Liquids: From Catalyst Synthesis to Application
Edited by Chris Hardacre and Vasile Parvulescu

Published by the Royal Society of Chemistry, www.rsc.org

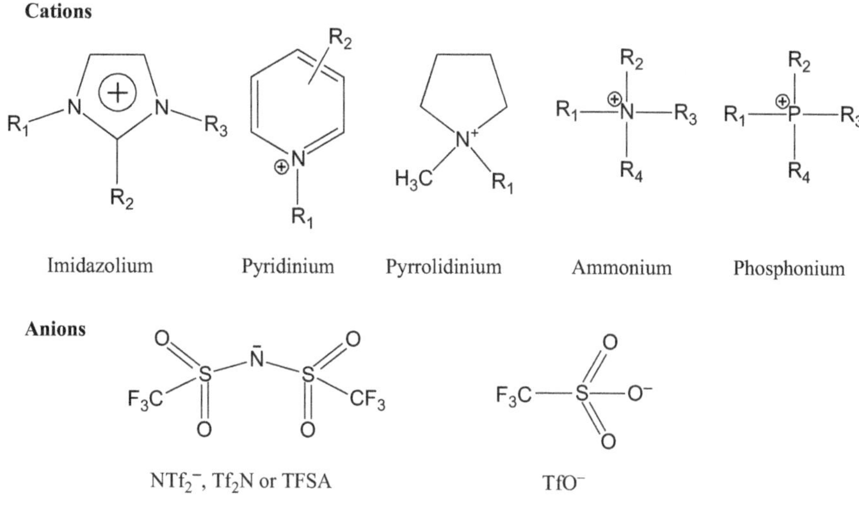

Figure 9.1 Structure of some of the common organic cations and accompanying anions that constitute ILs.

considerations and the typical behaviour of inorganic ionic compounds. However, experimental measurements have frequently established that the polarity of these media is similar to that of alcohols and other conventional organic solvents of intermediate polarity. Certainly, the macroscopic polarity of ILs is much lower than expected in view of the composition.

A simple way to determine polarity of solvents is the use of solvatochromic dyes, *i.e.* dyes that change their absorption maximum in the visible region depending on the solvent, this leading to a visual coloration change. Among solvatochromic dyes, the 2,6-diphenyl-4-(2,4,6-triphenylpyridinium-1-yl)phenolate betaine, known as the Reichardt's dye, is probably the preferred one since it exhibits the largest shift as a function of the nature of the medium.[4] The variation in the position of the absorption band is related to the so-called $E_T(30)$ that gives a quantitative value for the polarity of the medium.[5] Besides dyes and absorption spectra determination, neutral fluorescence probes are also widely applied to estimate solvent polarity, including the polarity of ILs, based on measurement of the position of the emission maxima and they also give coincident conclusions to those reached with the Reichardt's dye.

For example, the polarity of 1-butyl-3-methylimidazolium hexafluorophosphate, [bmim][PF_6], $E_T(30)$ 52.39 kcal mol^{-1}, is between that corresponding to acetonitrile (45.3 kcal mol^{-1}) and methanol (55 kcal mol^{-1}) with negligible effect on replacing the BF_6^- with NO_3^-. In the case of 1-octyl-3-methylimidazolium hexafluorophosphate, [omim][PF_6], and *N*-butylpyridinium tetrafluoroborate, [*N*-bpy][BF_4], the $E_T(30)$ values are 46.84 and 44.91 kcal mol^{-1}, respectively, corresponding to even lower polarity than [bmim][PF_6] as could be inferred from the higher C/N ratio of the organic moiety.[6] In the case of *N*-butyl-*N*-methylpyrrolidinium bis(trifluoromethanesulfonyl)imide, [bmpy][Tf_2N], an $E_T(30)$ value of 47.7 kcal mol^{-1} has been determined. These $E_T(30)$ values for ILs are in the polarity range of some alcohols such as ethanol, propan-2-ol and *tert*-butyl alcohol whose $E_T(30)$ values are 51.9, 48.5 and 43.5 kcal mol^{-1}, respectively.

The outcome of these polarity measurements is that the mean polarity experienced by a solute is similar or even lower than that experienced in alcohols and other organic solvents, probably because ILs are not isotropic media and the organic probes are located in a less polar environment, shielded from the ions present.

9.3 Apolar Domains in ILs

In fact, one of the intriguing disimilarities of ILs with respect to common organic solvents is that the liquid phase should have some kind of pre-organization and structuring, changing dynamically. Generally, gas and liquids are considered isotropic media, meaning that the space has identical properties at any point of the phase. However, a fact that is increasingly recognized is that some liquids can have some dynamic organization, the best example of this being liquid crystals.[7,8] In liquid crystals, weak intermolecular forces, typically van der Waals or dipole–dipole interactions, cause the organization of the phase. Evidence has been obtained showing that also in the case of ILs an analogous type of dynamic organization can take place, the driving force being maximization of the ion–ion interactions, hiding hydrocarbon chains in apolar pockets.

This pre-organization and ordering induces spatial directionality and anisotropy in the medium, in the sense that there will be domains with different properties and boundaries between them. The term "*IL effect*" has been coined to denote that in a certain way ILs should be considered as supramolecular solvents with some fluctuating and flexible ordering.[9]

To illustrate the kind of experimental evidence supporting the existence of nanodomains in ILs we will comment on a report on fluorescence dynamics. A systematic study using fluorescence correlation spectroscopy to study the diffusion dynamics of rhodamine 6G in a series of *N*-alkyl-*N*-methyl-pyrrolidinium bis(trifluoromethylsulfonyl)imide, [C_nMPy][Tf_2N] ($n = 3$, 4, 6, 8 and 10) as solvent has clearly revealed the existence of two regimes, one faster and the other slower.[10] As the alkyl chain of the pyrrolidinium IL increases in length the diffusion coefficients of the slow and fast dynamics decrease, while the relative contribution to the overall fluorescence correlation of the slow diffusion increases. These experimental results were interpreted as evidence of the existence of two domains in the pyrrolidinium IL, the extent of the apolar domain increasing with the alkyl chain length. It was proposed that these two domains arise from the self-assembly of the apolar alkyl chains of the [CMPy]$^+$ cation. It is clear that more information about the structuring and pre-organization of ILs is still needed in order to confirm their existence, including the microphase partitioning as well as the energetics and dynamics of the structuring in each IL. The influence of this heterogeneity in the liquid phase on the macroscopic properties of the solutes also requires further research.

9.4 Viscosity in ILs

One property that has a strong influence on the photochemistry and dynamics of the photogenerated transients is solvent viscosity.[11] Bimolecular reactions and transient quenching are processes whose dynamics are strongly dependent on the viscosity of the medium. In general, ILs exhibit high viscosity that can be orders of magnitude higher than conventional organic solvents. This high viscosity limits molecular diffusion and can be alleviated by selecting large counter anions. Table 9.1 provides some representative values of the viscosity coefficient for imidazolium ILs.[12,13] For comparison, it should be noted that water viscosity is 0.89 cP at 25 °C.

One property that is relevant in several types of general photochemical reactions is oxygen solubility. Typically the solubility of oxygen in ILs is low at atmospheric pressure (<0.2 mM) and this fact, together with high viscosity, is reflected in long lifetimes of triplet excited states, carbon-centred radicals and organic radical ions in imidazolium ILs with respect to traditional organic solvents. On the other hand, the oxygen solubility is, however, enough to generate a steady-state concentration of singlet oxygen in ILs that can produce photooxygenation of solutes.

Table 9.1 Dynamic viscosity at 20 °C (cP 0.01 g cm^{-1} s^{-1}) of some imidazolium ILs. Estimated error ± 5%.

Im^+	TfO^-	NfO^{-a}	Tf_2N^-	TA^{-b}	HB^{-c}	AcO^{-d}
3-Me						
1-Me			44			
1-Et	45		34	35	105	162
1-Bu	90	373	52	73	182	
1-*i*-Bu			83			
1-MeOEt	74		54			
1-CF_3CH_2			248			
3-Et						
1-Et	53		35	43		
1-Bu		323	48			
1-Et-2-Me						
3-Me			88			
1-Et-5-Me						
3-Me	51		37			
3-Et			36			

[a]NfO^-: nonaflate.
[b]TA^-: trifluoroacetate.
[c]HB^-: heptafluorobutanoate.
[d]AcO^-: acetate.

These unique properties of ILs are responsible for the fact that photochemistry in ILs can be notably different than in other solvents, opening a wide range of possibilities and new applications, as will be discussed in the next sections.

9.5 Photochemical Stability of ILs

One of the prerequisites of a medium to be employed as an inert solvent in photochemical reactions is photochemical stability.[14,15] As we have commented earlier, ILs exhibit high thermal and chemical stability allowing their use as inert solvents in chemical reactions. The photochemical stability of those ILs that do not have absorption bands in the UV–visible range, such as quaternary ammonium salts and pyrrolidinium ions, is generally taken for granted, because light absorption is the first necessary event in a photochemical process. However, one should be aware that these ILs could be reactive against radicals and other photogenerated transients appearing during the course of the photoreaction, thus, they can indirectly intervene in photochemical reactions.

In addition, those ILs having absorption in the UV spectral region act as strong cut-off filters avoiding excitation of the solutes and limiting the range of excitation wavelengths. Generally, irradiations of the solutes can be performed using long wavelengths where ILs are transparent and do not interfere with the photoexcitation of the probe. In the case of imidazolium ILs, this implies performing irradiations with wavelengths longer than 350 nm that should not be absorbed by this type of heterocyclic compound.

The photostability of imidazolium ILs has been a matter of study. Thus, the photodegradation of [bmim][Tf_2N] under pulsed 220 nm laser irradiation has been studied.[16] The absorption spectrum of [bmim][Tf_2N] shows a maximum band at 211 nm tailing up to 300 nm. It was observed that laser irradiation of [bmim][Tf_2N] produces an increase of the absorbance at wavelengths as high as 500 nm, attributed to the formation of unidentified polymeric species [see Equation (9.1)]. The estimated quantum efficiency of the conversion was 0.1 which is notably high for a solvent used in photochemical reactions. It can be assumed that this process will be general for imidazolium based ILs, although it would be important to have detailed information on the photoproducts and photodegradation mechanism.

$$[\text{bmim}][\text{Tf}_2\text{N}] \rightarrow [\text{bmim}][\text{Tf}_2\text{N}] \rightarrow \text{oligo-/polymer} \qquad (9.1)$$

Concerned by the possibility of the presence of ILs in water derived from their use and the need to develop efficient degradation processes, a study has focused on the efficiency of some photochemical waste water treatments to degrade ILs. In this sense, three common advanced oxidation processes (AOPs) such as UV, UV–H_2O_2 and UV–TiO_2 were studied for the degradation of aqueous solutions containing $[\text{bmim}]^+$, $[\text{hmim}]^+$, $[\text{omim}]^+$ and $[\text{dmim}]^+$ chlorides or tetrafluoroborates and [eeim][BF_4] as well as methyl imidazole [mim] as reference compound.[17] UV irradiations for 6 h using a 1000 W xenon lamp working at 16 °C resulted in degradation percentages ranging from 8 to 55% for $[bmim]^+$ and $[omim]^+$ based ILs, respectively. The presence of H_2O_2 enhances the degradation rates of all the ILs from 30 to 60% for $[hmim]^+$ and $[eeim]^+$ based ILs, probably due to the generation of highly aggressive hydroxyl radicals. Finally, the use of a UV–TiO_2 system led to the almost complete degradation of [mim] and $[bmim]^+$ after 6 h, although the degradation of the other ILs varied between 15 and 25%. From these results it was concluded that the system UV–H_2O_2 was the most efficient for degradation of ILs and that $[eeim]^+$ is one of the more resistant ILs to this treatment as well as to the UV–TiO_2 irradiation.

The presence of a solute can influence the absorption spectrum of a given IL and, therefore, its photochemical reactivity. For example, the presence of iodide can lead to the formation of charge transfer complexes with heterocyclic ILs that exhibit a characteristic absorption at wavelengths longer than those of the ILs. Thus, $[bmim]^+$ and I^- forms a charge transfer complex in $[bmim][Tf_2N]$ as a solvent that can be photolyzed at 280 nm, whereby the generation of a neutral radical [bmim] that is stable under the reaction conditions was observed [see Equation (9.2)].[16] The consequence of the presence of iodide and the formation of this stable radical is that degradation of $[bmim][Tf_2N]$ is not observed in the presence of iodide.

$$[bmim^+]I^- \rightarrow [bmim]I \rightarrow [bmim]I \rightarrow [bmim]{+}I[bmim^+] \qquad (9.2)$$

Some studies have focused on determining the stability of ILs in the presence of highly reactive intermediates. In general, these studies reveal that the possibility of a lack of stability of ILs in the presence of highly reactive species has to be considered and evaluated. For instance, hydrogen abstraction from imidazolium IL by the photogenerated triplet excited state of benzophenone has been reported.[18] Arrhenius data indicate, however, higher activation energies for hydrogen abstraction from imidazolium ILs (21.6 to 27.3 kJ mol^{-1}) compared to conventional organic solvents such as toluene, cyclohexane or 1-butanol with E_a values of 14.5, 13.7 and 13.3 kJ mol^{-1}, respectively. This process was found to be independent of the counter anion ($[PF_6]^-$ or $[Tf_2N]^-$) and of the availability of the hydrogen atom at the C2-position of the imidazolium ring. The high rate constant value observed for hydrogen abstraction in $[omim][PF_6]$ (3.4×10^5 s^{-1}) with respect to $[bmim][PF_6]$ (2.0×10^5 s^{-1}) was interpreted as evidence that the hydrogen abstraction takes place from the alkyl chain.

In contrast, some studies have shown the compatibility of ILs with other highly reactive intermediates generated photochemically, suggesting that ILs can be convenient media to avoid undesirable side reactions.[19] Thus, the photochemistry of 4-chloro-*N*,*N*-dimethylaniline (**1**) in imidazolium ILs with or without the presence of different nucleophiles such as iodide, olefins (allyltrimethylsilane and 4-penten-1-ol), aromatic (benzene) and heteroaromatic (thiophene) compounds has been evaluated and compared with that of a polar nonprotic solvent such as CH_3CN.[19] In an initial step, photolysis of

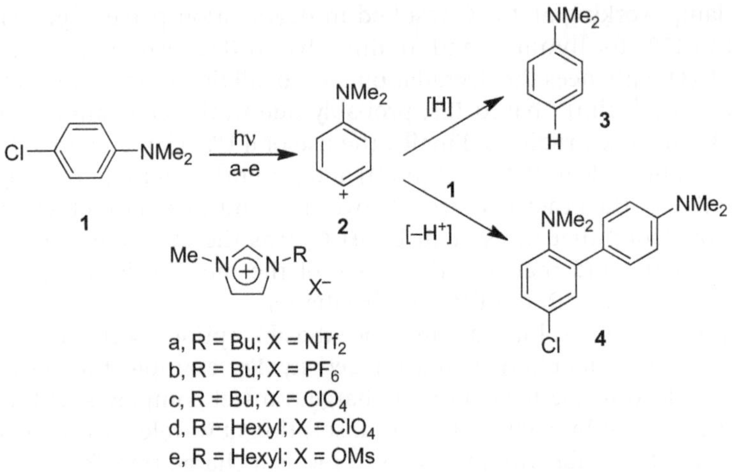

Scheme 9.1 Photochemical reactivity of 4-chloro-*N,N*-dimethylaniline in imidazolium ILs.

aniline **1** led to the formation of the *N,N*-dimethylaminophenyl cation (**2**) in its triplet excited state (Scheme 9.1). Then, in acetonitrile as reaction medium almost equimolar amounts of products **3** (reduction) and **4** (self-attack) were observed. In contrast, using ILs both either reduction (in ILs **d**, **e**) or self-attack (in ILs **a–c**) predominate. Therefore, hydrogen transfer from the IL to the aminophenyl cation leading to the formation of **3** is a phenomenon much more efficient for *N*-hexyl-*N*-methylimidazolium salts compared to *N*-butyl-*N*-methyl **a–c** analogues. This fact was related to the supramolecular aggregation of alkyl chains longer than four carbons in nonpolar domains for ILs **d** and **e** making the migration of the intermediate more difficult and facilitating the hydrogen transfer. On the other hand, the formation of **4** was found to exhibit an inverse dependence on the viscosity of the medium and, therefore, was strongly influenced by the nature of the IL anion.

Similarly, nucleophilic trapping of arenium cation was also influenced by the solvent. Thus, reaction of this photogenerated *N,N*-dimethylaminebenzenium cation with triethylamine (TEA) resulted in better yields (70%) in imidazolium **a** compared to CH_3CN (42%) as solvent, where the phenylenediamine was accompanied by a significant percentage of by-products **3** and **4**.

In spite of the above comments and yields, ESI-MS analysis of the irradiated neat IL with or without the presence of nucleophiles revealed, in the photochemical reaction of dimethylaniline **1**, the formation of by-products arising from insertion into a C–H bond of the aminophenyl moiety in both $[bmim]^+$ and $[hmim]^+$ (258 or 286 *m/z*) (**5** in Scheme 9.2). Furthermore, the observation of a small peak at 202 *m/z* was attributed to the insertion of the aminophenyl cation into the *N*-methylimidazolium ion or to the *N*-dealkylation of the *N*-butyl or *N*-hexyl imidazoliums (**6** in Scheme 9.2). Interestingly, the formation of the *N*-(4-dimethyl-aminophenyl)bis(trifluoromethanesulfonimide) arising from the trapping of benzenium cation by the imidazolium counteranion was

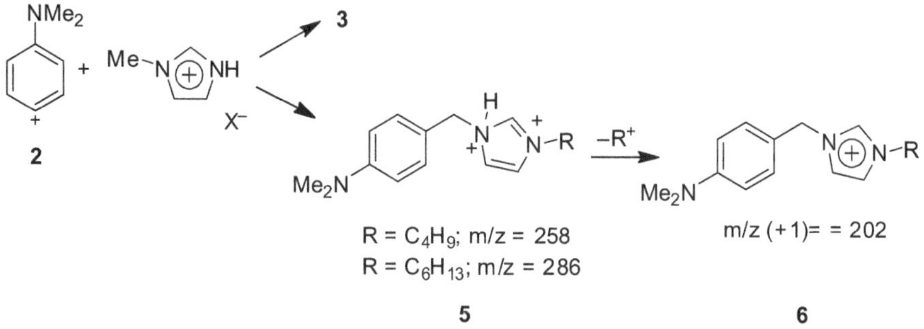

$R = C_4H_9$; m/z = 258
$R = C_6H_{13}$; m/z = 286

m/z (+1)= = 202

Scheme 9.2 Products arising from the reaction of imidazolium IL with *N,N*-dimethylaminobenzenium ion **2** in its triplet excited state.

not detected by ESI-MS, probably due to the triplet excited state nature of this arenium ion. In fact, there are precedents reporting this trapping of the anion $[Tf_2N]^-$ by a benzenium cation in the singlet state generated thermally from diazonium salts,[20] and its lack of formation in the photolysis of dimethylaniline **1** is compatible with the different reactivity of the triplets with respect to the singlet state.

9.6 Primary Photochemical Events Upon Direct Excitation of Imidazolium ILs

Transient spectroscopy in which a relatively high concentration of photochemical species is generated after a laser pulse is a convenient technique to study the photochemical behaviour of ILs.[21] All the current evidence obtained by transient spectroscopy indicates that ILs are not photochemically inert and upon light absorption exhibit a rich photochemistry involving the singlet and triplet excited states of the imidazolium heterocycle and resulting generally in photoinduced electron ejection to the medium for a long time after the laser excitation. Trapping of these electrons can finally be equivalent to photoinduced electron transfer between two IL molecules, although mediated by solvated electrons (electrons in the medium).

Thus, photoejection of electrons upon excitation of imidazolium ILs has been detected using femtosecond and nanosecond laser pulses.[22,23] These solvated electrons can be formed directly from the singlet excited state since its generation occurs during the nanosecond laser pulse. Alternatively, those singlets that do not eject electrons immediately can undergo intersystem crossing (ISC) to the triplet excited state, which also has enough energy to photoionize giving electrons but at much longer times scales than those formed from the singlet. Ejected electrons can be characterized by their transient spectra, which exhibit an intense absorption band. Solvated electrons can finally be trapped in a different imidazolium heterocycle giving rise to dipositive (the imidazolium expelling the electron) and neutral radical (the imidazolium

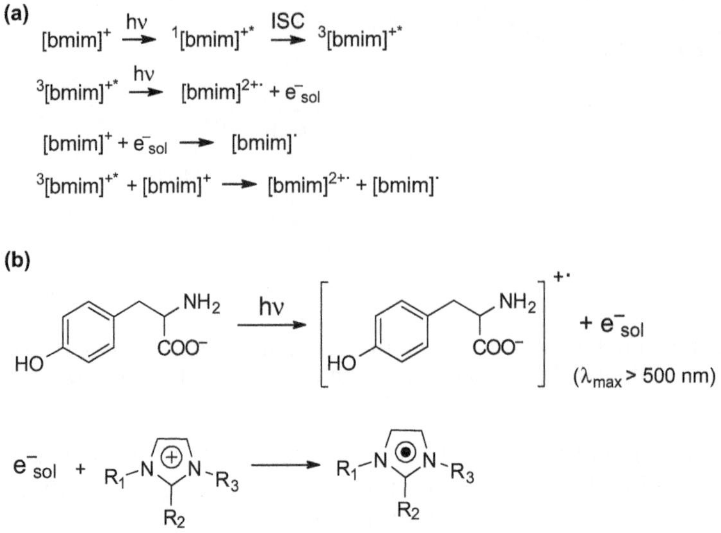

(a)

$$[bmim]^+ \xrightarrow{h\nu} {}^1[bmim]^{+*} \xrightarrow{ISC} {}^3[bmim]^{+*}$$

$${}^3[bmim]^{+*} \xrightarrow{h\nu} [bmim]^{2+\cdot} + e^-_{sol}$$

$$[bmim]^+ + e^-_{sol} \rightarrow [bmim]^\cdot$$

$${}^3[bmim]^{+*} + [bmim]^+ \rightarrow [bmim]^{2+\cdot} + [bmim]^\cdot$$

(b)

HO, NH_2, COO^- $\xrightarrow{h\nu}$ [HO, NH_2, COO^-]$^{+\cdot}$ + e^-_{sol} (λ_{max} > 500 nm)

e^-_{sol} + R_1–N⊕N–R_3 (R_2) → R_1–N•N–R_3 (R_2)

Figure 9.2 (a) Proposed photochemical pathways for photoinduced electron transfer in ILs. (b) Ability of ILs to act as electron acceptors of solvated electrons in water. The electrons are obtained by photolysis of tyrosine using 266 nm laser excitation.

accepting the electron) species, all of them detectable by transient spectroscopy. The proposed photochemical pathways from these studies are presented in Figure 9.2. Part (b) of Figure 9.2 shows how tyrosine is used to generate solvated electrons in water that are characterized by an intense absorption band starting from 500 nm and reaching the far red end of the spectra. Under these conditions, addition of increasing amounts of [bmim][BF_4] causes the quenching of these electrons following a linear relationship between the concentration of [bmim][BF_4] and the decay rate constants of solvated electrons. These spectroscopic and kinetic data constitute evidence of the third equation of part (a) in Figure 9.2 where electrons are captured by $[bmim]^+$.

It may happen that the accompanying counter anion of the imidazolium IL is not an innocent spectator in the photochemical process and could intervene giving or trapping electrons. This is the case for halide containing ILs that are exposed to pulse radiolysis. Ultrafast dynamics and femtosecond pulse excitation have provided evidence that halides can give electrons forming neutral atoms that then will continue reacting to form trihalide anions [see Equations (9.3)–(9.6)].[23–25]

$$X^- \rightarrow X + e^-_{sol} \quad (X = Cl \text{ or } I) \tag{9.3}$$

$$X + e^-_{sol} \rightarrow X^- \tag{9.4}$$

$$X + X^- \rightarrow X_2^- \tag{9.5}$$

$$2X_2^- \rightarrow X_3^- + X^- \tag{9.6}$$

These elementary steps do not necessarily require excitation of the halides, since the first equation can also take place if an electron acceptor species generated photochemically is formed.

The processes discussed above clearly show that when using ILs as medium for photochemical reactions it is necessary to show that the conditions employed do not produce changes in the imidazolium ILs. Blank controls are always advisable to ensure the photostability of the IL. This condition can frequently be met since as commented earlier typical absorption spectra of ILs show that the absorption band onset occurs for wavelengths shorter than 300 nm and there are convenient cut-off filters and light sources that guarantee that the light used in the photoirradiation does not interact with the IL.

9.7 Photoluminescence in ILs

Photoluminescence is a property that is extremely sensitive to the medium. Parameters like the emission wavelength, photoluminescence quantum yield, light decay and luminescence quenching vary depending on the solvent and can be used to understand the characteristics of the medium.[26] There have been several studies focused on the use of steady-state fluorescence spectra and time-resolved measurements to assess the polarity of various ILs.[27–30] As discussed these studies have consistently shown that the polarity of ILs is similar to that of alcohols and medium polarity organic solvents.

Other types of photoluminescence measurements have tried to take advantage of the use of ILs to enhance the photoluminescence of dissolved fluorophores. One example of these studies is the photostability enhancement of a highly luminescent europium-β-diketonate complex, namely [hmim][Eu(tta)$_4$], (tta: 2-thenoyltrifluoroacetonate), in [hmim][Tf$_2$N] with respect to acetonitrile solutions (Figure 9.3).[31] Although this europium

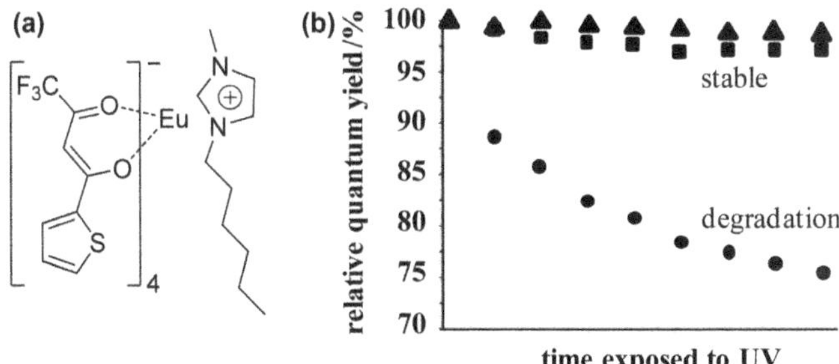

Figure 9.3 (a) 1-Hexyl-3-methylimidazolium tetrakis(2-thenoyltrifluoroacetonato) europate(III) complex. (b) Relative decrease of quantum yield under UV exposure of an acetonitrile solution compared to a solution in a [hmim][Tf$_2$N] ionic liquid and to a sample not exposed to UV radiation. Legend: (●) 1 h in acetonitrile with UV irradiation; (■) 1 h in [hmim][Tf$_2$N] with UV irradiation; (▲) 1 h in acetonitrile in the dark.

complex exhibited in IL an absolute quantum yield lower (54%) than in acetonitrile (61%), the emission remained constant under daylight irradiation for 10 days. In contrast, in acetonitrile a decrease of the emission quantum yield to 51% was observed under the same conditions. Furthermore, under UV irradiation the [hmim][Eu(tta)$_4$] complex also exhibited much higher stability in the IL compared to acetonitrile (Figure 9.3). This instability in acetonitrile was attributed to the photodegradation of the β-diketone ligand. The remaining, undegraded europium complex does not vary the lifetime of the excited state. In particular, the lifetimes of the electronic excited states of the complex [hmim][Eu(tta)$_4$] in acetonitrile and in [hmim][Tf$_2$N] are 615 and 572 μs, respectively. Strong interactions between the imidazolium cation and the emitting europium complex through hydrogen bonding between cations and ligands have been proposed to be responsible for the observed stabilization of the phosphor in the IL.

One of the possible uses of these organic complexes requires their inclusion within polymeric matrixes that can serve for constructing organic light-emitting diode (OLED) cells. ILs can also be a co-additive (plastifier) in these formulations. Thus, several flexible luminescent polymer films of europium (III) complexes embedded in poly(methyl methacrylate) and containing ILs as plasticizers have been prepared as the active emitting layer in OLED cells.[32] The list of the various europium (III) complex–IL combinations that have been tested in polymeric matrices includes [hmim][Eu(nta)$_4$] (nta: 2-naphthoyltrifluoroacetonate), [hmim][Eu(tta)$_4$] [Eu(tta)$_3$(phen)] (phen: 1,10-phenanthroline) and [choline]$_3$[Eu(dpa)$_3$] (choline: 2-hydroxyethyltrimethyl ammonium cation; dpa: 2,6-pyridinedicarboxylate or dipicolinate). As is characteristic of europium complexes, UV irradiation of these polymers incorporating the europium complex–IL assembly resulted in the emission of bright red photoluminescence for all the films as exemplified in Figure 9.4. Interestingly, the absence of an IL as plasticizer resulted in shorter lifetimes and therefore higher quenching of the luminescence showing again the benefits of the presence of an IL binding the europium complex. Comparison of the emission efficiency of the different combinations showed that [choline]$_3$[Eu(dpa)$_3$] exhibits a quantum yield and radiative lifetime of 0.47 and 3.85 ms, respectively, which exceeds those measured for the analogous europium(III)-β-diketonate complexes.

Similarly to the case of polymeric films, several luminescent europium-β-diketonates have been incorporated inside mesoporous SBA-15 containing an IL. These hybrid solids exhibited good photoluminescence properties (Scheme 9.3).[33] In particular, SBA-15–im$^+$[Eu(tta)$_4$]$^-$ and SBA-15–im$^+$[Eu(bta)$_4$]$^-$ (bta: benzoyltrifluoroacetylacetonato) present high quantum efficiencies of 55.9 and 62.3% together with long luminescent lifetimes of 685.2 and 729.2 ms, respectively. These quantum efficiencies compare favorably with respect to those achieved in previous europium based mesoporous materials and show again the advantages of IL stabilizing emissive Eu^{3+} complexes.[34]

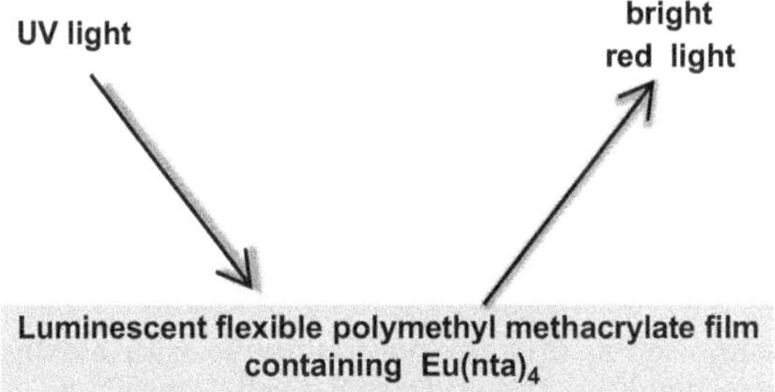

Figure 9.4 Luminescent flexible polymethyl methacrylate film comprising [hmim][Tf_2N] as plastifier and stabilizer of the emitting europium (III) complex [hmim][$Eu(nta)_4$]. Simulated image showing exposure of the film to 365 nm UV irradiation.

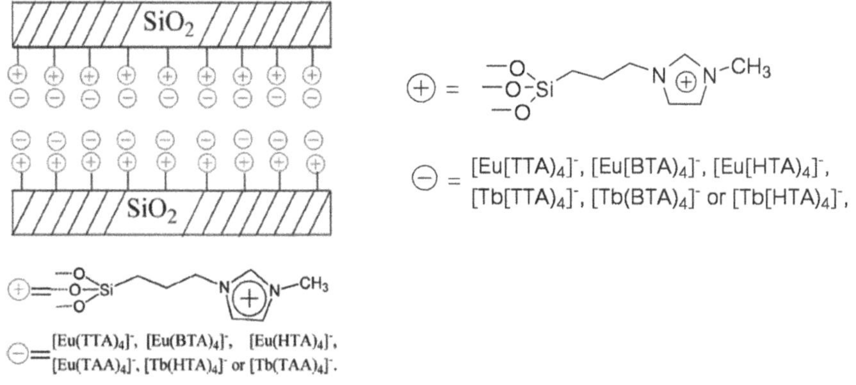

Scheme 9.3 Idealized structure of hybrid IL–SBA-15 solid comprising negatively-charged Eu^{3+} complexes.

9.8 Photo-Friedel–Crafts Acylation in ILs

Friedel–Crafts reactions promoted by strong Lewis acid catalysts are the most general process to form C–C in aromatic compounds.[35] During the work-up after the reaction, the adduct of the Lewis acid and the reaction product are decomposed by hydrolysis, whereby a large amount of waste is generated. The photochemical version of these reactions has also been reported and has the advantage of not requiring the presence of Lewis acids.

In this context the photo-Friedel–Crafts acylation of 1,4-naphthoquinone (**7**) with various aldehydes (**8**) has been studied in ILs (Scheme 9.4).[36] High conversions and selectivities were achieved using [emim][NTf_2] as solvent. Reuse experiments for three consecutive cycles did not show a decrease in the efficiency of the photoreaction. Furthermore, ^{1}H-NMR spectroscopy does not detect any change before and after the photoreaction, suggesting stability of the IL structure. Interestingly, ILs behave better as solvent compared

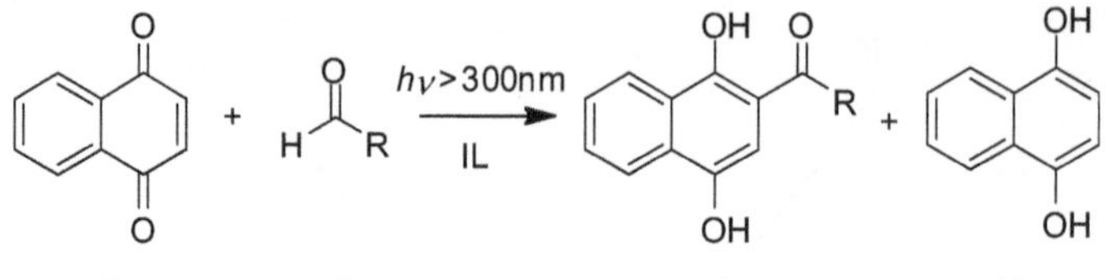

Scheme 9.4 Photoreaction of 1,4-naphthoquinone with aldehydes to afford the corresponding photo-Friedel–Crafts product accompanied by dihydronaphthoquinone.

with the use of toxic benzene, probably due to the enhanced lifetime of the triplet excited state of the quinone experienced in ILs.

It is worth commenting that increasing the alkyl chain of the aldehyde increases the percentage of undesired dihydronaphthoquinone (**10**) that appears as a side reaction product. Formation of this by-product probably arises from hydrogen abstraction by the triplet excited state of the quinone, a process that requires the presence of hydrogen donors and that appears to depend on the structure of the aldehyde.

9.9 Photochromism in ILs

Photochromism is the property of changing colour upon irradiation.[37–39] Typically this change in the colour is reversible and after irradiation the system reverts in the dark to the initial state (Scheme 9.5). The simplest photochromic systems involve the photoinduced rearrangement of a stable molecule into a coloured isomer that is thermally unstable and after its formation tends to transform again into the initial compound.[39] This conversion of the unstable form into the more stable isomer can occur spontaneously or can be promoted thermally or photochemically.[39] In the latter case the most convenient photochromic systems are those in which the wavelength promoting the interconversion of the photogenerated isomer to the original form is different from those wavelengths used to cause the colour change of the stable form. Otherwise, a stationary mixture of the two forms would be reached.

One of the most important properties of a photochromic system is the kinetics of the interconversion between the photochemically generated coloured form and the stable isomer. This interconversion has to be sufficiently slow to be visually detectable by naked eye (seconds), but not long enough to make the reversibility of the photochromic system so slow that the cyclic response between the two states could only operate at very low frequencies. Photochromism finds many commercial applications in coatings of glasses and other materials and there is much interest in developing novel photochromic systems with enhanced reversibility and durability.[40]

Most of the photochromic compounds require large molecular rearrangements with the generation (or disappearance) of conjugated double

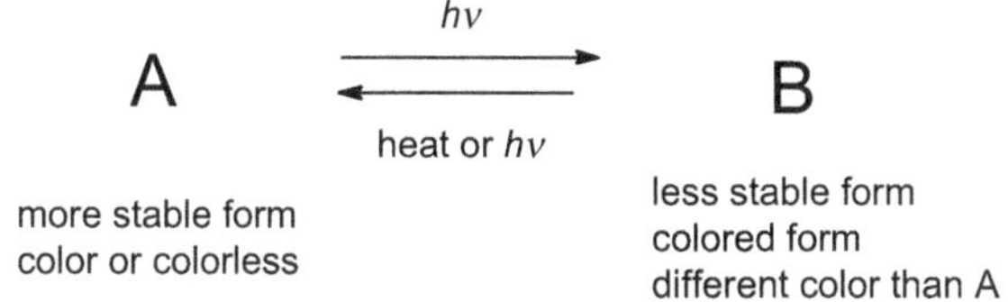

Scheme 9.5 Illustration of a simple photochromic system.

C–C bonds. Similar to molecular diffusion, solvent viscosity strongly influences the rate of these large molecular changes, varying the lifetimes of the unstable state. To gain control of the photoresponse rates, photochromism has been studied in a variety of media including porous aluminosilicates like zeolites or mesoporous MCM-41,[41–43] which can be considered solid polyelectrolytes and, therefore, the rigid analogues of ILs.

Not surprisingly, then, ILs can find application in photochromic systems, extending the lifetime and increasing the persistence of some coloured forms that will last much shorter times in other less viscous media.[44] For this reason the type of photochromic systems that can be used in ILs can be different from those found in other media, including those whose coloured form have lives too short to be observed in other solvents like water.[37]

In this context, Pina and co-workers have extensively studied a new complicated photochromic system based on flavylium compounds that may involve several equilibria among many different forms.[37,45–48] In general, substituents in the core flavylium structure, reaction medium and temperature are the main parameters determining the behaviour of these systems.[49] In this sense, Scheme 9.6 shows an example of the chemical processes taking place for the 7-(*N*,*N*-diethylamino)-4′-hydroxyflavylium in water as a function of the medium.[46] In moderately acidic aqueous solutions the main species are the flavylium cation (**11**, AH^+), the quinoidal neutral base (**12**, A), the hemiketal (**13**, B2), the *cis*-chalcone (**14**, Cc) and the *trans*-chalcone (**15**, Ct). In addition each of these structures can exhibit at least three acid–base forms depending on the protonation of the diethylamino or deprotonation of the phenolic hydroxyl group, making possible in total up to twelve different structures depending on the pH, all of them interconnected by equilibria. Some of these equilibria require light to interconvert the species, in particular the *cis* to *trans* and the *trans* to *cis* chalcone isomerization. In water it has been found that there is almost no *cis*–*trans* isomerization barrier and, therefore, the less stable *cis*-chalcone should not be considered in this solvent, because its lifetime is much too short.

Using the biphasic system water–[bmim][PF_6] the flavylium compound is partitioned as a function of the water pH leading to stable colours in water or IL such as pink (**11**, AH^+ species) or unstable blue colour (**12**, A species) in the IL phase.[46] Preferentially, neutral and monocharged species reside in the IL, while double charged species reside in water.

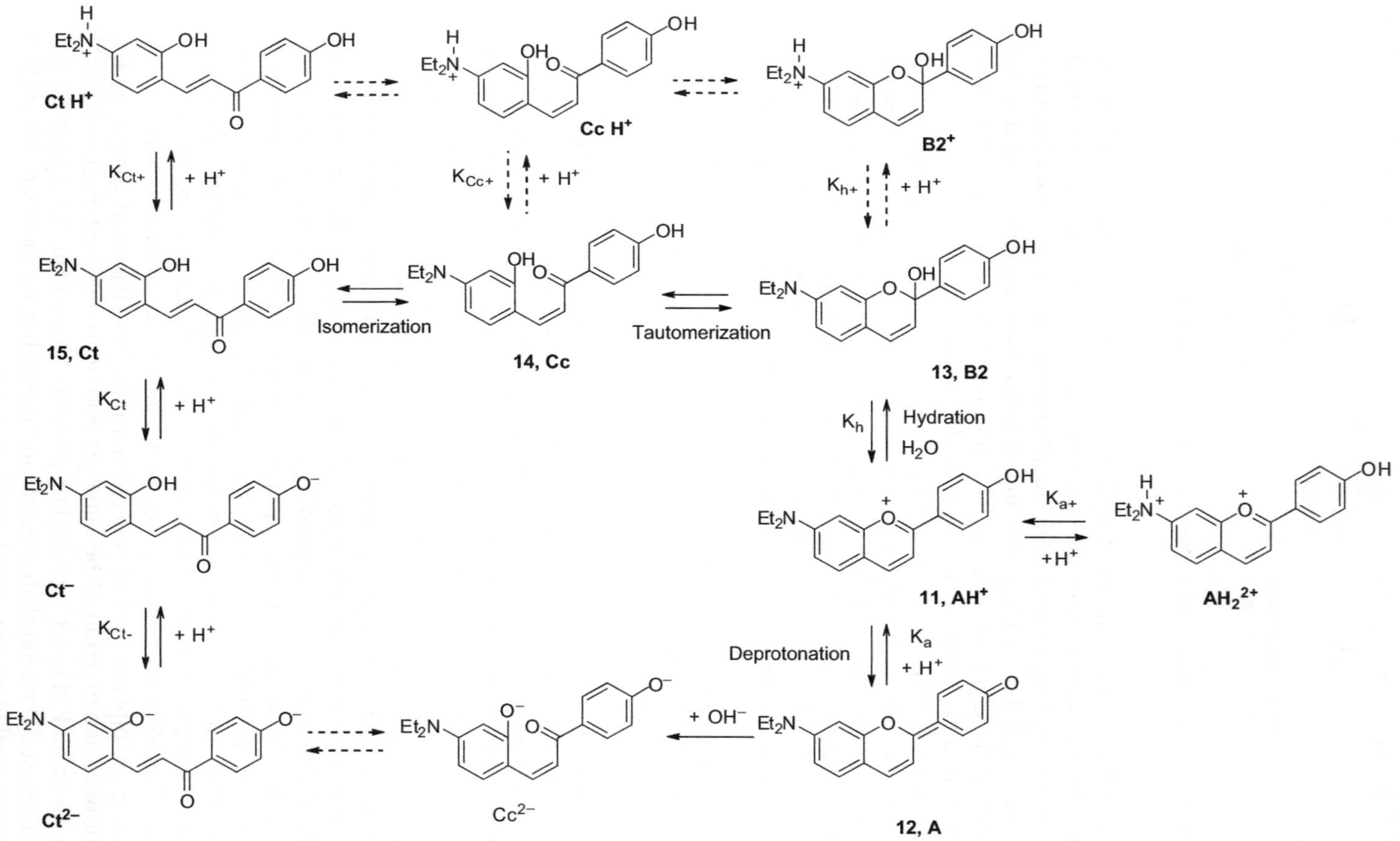

Scheme 9.6 Photochromic system derived from 7-(*N*,*N*-diethylamino)-4′-hydroxyflavylium showing some of the equilibria among the twelve possible forms.

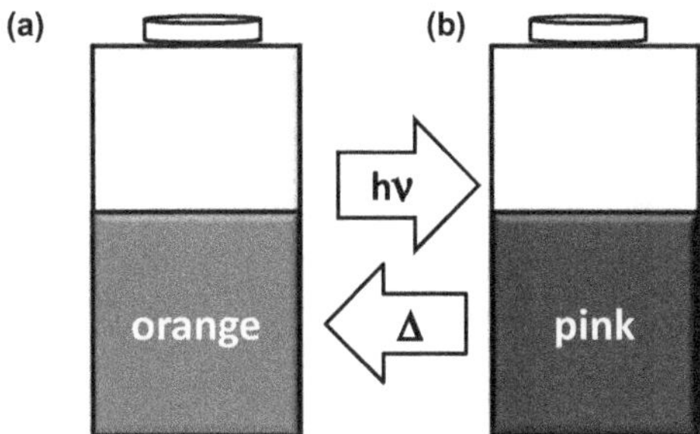

Figure 9.5 Photochromic behaviour of 7-(*N*,*N*-diethylamino)-4′-hydroxyflavylium in a water–[bmim][PF_6] biphasic system. Irradiation of IL containing Ct species (**15**, bottom) and aqueous phase at pH 6.8 (a) and IL containing the pink photoproduct (**11**, AH^+) (b).
Data taken from reference 46.

Interestingly, this 7-(*N*,*N*-diethylamino)-4′-hydroxyflavylium system does not exhibit photochromism in pure water due to the high instability of the *cis*-chalcone discussed. In contrast using this biphasic water–IL mixture a well-defined photochromic behaviour could be developed.[46] Figure 9.5 shows light irradiation induced photoisomerization from *trans* (**15**, Ct) to *cis* (**14**, Cc) leading to the subsequent formation of **11**, AH^+. Importantly, the system reverts back almost completely to **15**, Ct after 11 h in the dark at 22 °C. The key point for the observation of photochromism in this system is that in the highly viscous nanodomains of the IL, operation of intermolecular interactions is responsible for an increase of the *trans*–*cis* isomerization barrier. The influence of the pH in the system locking the isomerization by forming the flavylium ring has led to the proposal of a write–read–erase optical memory consisting of: (i) *writing* with light forming the *cis*-chalcone (**14**) that is locked into the flavylium, (ii) *reading* by detecting the optical spectrum and concentration of flavylium ion in the IL, and (iii) *erasing* by changing the pH and re-forming the *cis*-chalcone (**14**). It should be stressed that this optical memory does not work in water.

9.10 Reversible Photoresponsive Films based on ILs

The use of ILs as polymer additives opens a new possibility in designing photoresponsive materials for a large number of applications.[8,50] One example is the design of remote-controlled ion conductive devices through the preparation of photoresponsive ion conductive gels using ILs.[51,52] In this sense, an oligo(amide acid) containing azobenzene groups (**16**) was cross-linked by 1,3,5-tri(4-aminophenyl)benzene (**17**) (Figure 9.6). Further impregnation in 1,3-dibutylimidazolium bromide [bbim][Br] resulted in the formation of a self-standing film (P-Gel–IL).[52] This IL was selected due to its high polarity in order to solubilise the starting materials and the possibility of working under air atmosphere. The presence of the photoactive azobenzene moiety is responsible for the shrinking or swelling of

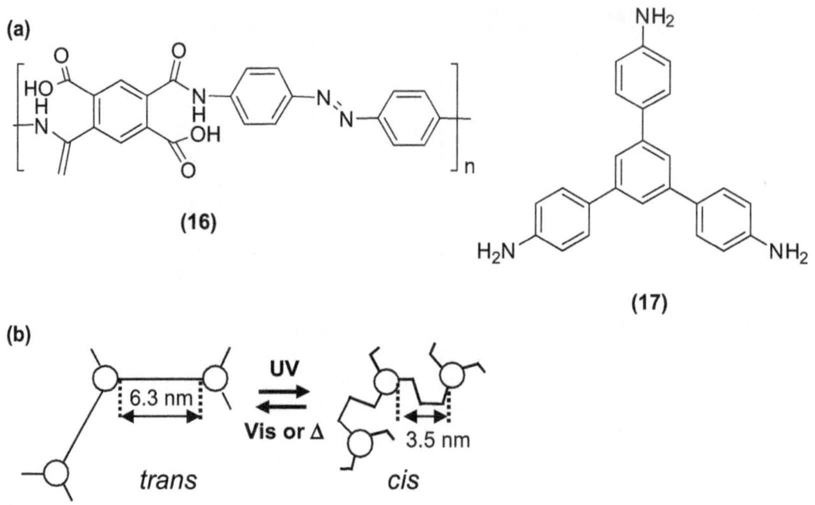

(a)

(16)

(17)

(b)

Figure 9.6 Structure of oligo(amide acid) containing azobenzene units and tri-aminophenylbenzene cross-linker (a). Photoresponsive poly-(amide acid) gel in swollen state and shrink state (b).
Adapted from reference 52.

the polymeric film upon irradiation with UV and visible, respectively [Figure 9.6(b)]. Besides the mechanical effect on the film, irradiation also controls the ionic conductivity measured at 20 °C that decreases or increases under UV or visible light irradiation from around 10^{-5} to 4×10^{-5} S cm^{-1}, respectively. This change in ionic conductivity was consistently measured for several UV–visible irradiation cycles without fatigue in the changes. These facts were explained by considering an increase of the local viscosity and then suppression of ion diffusion in the gels when *cis*-azobenzene is the predominant stereoisomer in the film and this state is reached upon UV irradiation of the *trans*-azobenzene. The opposite effect is observed when the *cis*-azobenzene is irradiated with visible light and interconverts into the *trans*-azobenzene. Kinetic measurements indicated that the recovery is faster compared to the shrink process and this faster response time was attributed to the higher thermodynamic stability of the *trans*-azobenzene form and to the fact that during the shrinkage process the ILs should be released from the network. It should be commented that this photoresponsive film requires an IL and cannot be prepared employing volatile organic solvents since their use leads to a rapid fatigue in the ionic conductivity swing due to solvent vaporization during the irradiation.

9.11 Use of ILs as Solvents in Photochirogenesis

One topic that has attracted considerable interest in photochemistry is the possibility of developing highly enantioselective photochemical reactions.[53] The most powerful methodology to obtain chiral compounds from pro-chiral

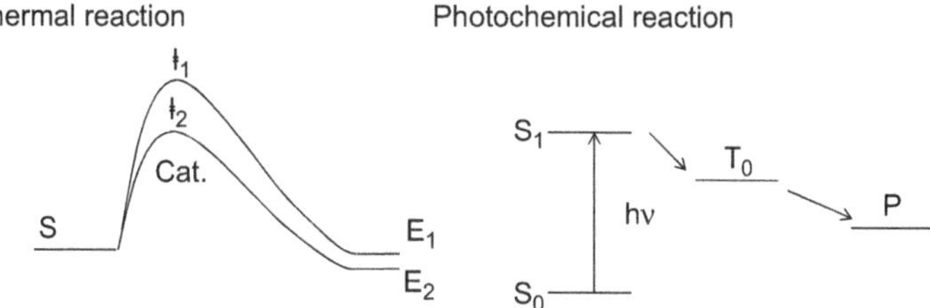

Scheme 9.7 Illustration summarizing the differences between thermal and photochemical reactions for chiral induction. In the thermal reaction (even if it is carried out at low temperature) there is an activation barrier going from the substrate (S) to the two enantiomers (E_i) that is different depending on the enantiomer because the reaction requires a chiral catalyst (Cat.) that discriminates between the two transition states. In the case of the photochemical reactions, light absorption leads to electronically excited states that can be either singlet (S_1) or triplet (T_0) and will give the product (P) with negligible E_a. In addition, the lifetime of the excited state is generally nanoseconds for the singlets and microseconds for the triplets, giving short contact time for chiral induction to take place.

substrates is the use of chiral catalysts that by asymmetric induction can promote the enantioselective formation of a single enantiomer.[54,55] This strategy, known as "*asymmetric catalysis*", is based on the fact that the homochiral catalyst and the pro-chiral substrate associate and there are two diastereomeric transition states with, similar, but different, activation energy (E_a) each of them leading to one enantiomer. (see Scheme 9.7). Asymmetric induction is the result of the preferential formation of the enantiomer with lowest E_a that is formed faster than the other with higher energy barrier. The difference between the E_a of the two enantiomers is, though, very small (in the range of 1 kcal mol^{-1} or less) compared to their absolute E_a values (varying roughly between 20 to 40 kcal mol^{-1}) and as result the reaction rate for the two enantiomers is frequently very similar.

Asymmetric catalysis is based on "thermal reactions" even though they are generally carried out at low temperatures in order to achieve the highest possible enantiomeric excess (ee). The contradiction arises from the fact that, on one hand, higher temperatures generally result in low ee values. But, on the other hand, if low temperatures are employed the reaction is very slow and the yields are low. Thus, there is a narrow temperature window in which the reaction can proceed at a reasonable rate and the enantioselectivity is not negligible.

In contrast to these methodologies based on ground state thermal reactions, photochemical processes occur in an electronically excited state and can proceed virtually without activation energy. In other words, by virtue of light irradiation chemical transformations can be performed at very low temperatures, since once the excited state is reached by light absorption, all the subsequent steps occur without any further activation. For this reason the chiral compound that is going to transfer its chirality to the products is not denoted in the case of photochemical reactions as catalyst but as "*inductor*". In other words, the reaction rate of the photochemical reaction could be the same no matter if a chiral inductor is present or not. The main problem is that due to the short lifetimes and the spontaneous decay of excited states, asymmetric induction of the photogenerated transients is very difficult. Various strategies to achieve high ee excesses in photochemical reactions have met frequently with failure, because pre-association of the chiral inductor and the substrate in the ground state strong enough to alter the decay of the excited state transient is required prior to photoexcitation. In this context, ILs offer the advantage of promoting pre-association and close interaction between the substrate that is going to undergo the photochemical reaction and the chiral inductor that has to transfer the chirality due to their high viscosity and the existence of microdomains in which self-assembly can be favored. The process of photoinduced enantioselectivity can be termed "*photochirogenesis*" and in some examples ILs have been employed as medium or even as medium and chiral inductor.

One obvious possibility is to use chiral ILs (CILs) as solvent to promote photochirogenesis by solvating the substrate providing an asymmetric environment. Few reports have been published in this area. The main advantage of CIL as solvents compared to conventional chiral solvents derives from their inherent physicochemical properties such as the presence of nanodomains with polar and non-polar character. In this sense, one of the obvious possibilities from a mechanistic point of view to induce chirality photochemically in CILs is to promote strong ion pairing or hydrogen-bond interactions between the substrate and the CIL. Although the proof of concept has been already set, the current development of photochirogenesis in CIL is clearly unsatisfactory and there is still considerable room for improvement. Irradiation of the substrate can be performed at wavelengths where the CILs are transparent, which in the case of imidazolium ILs would be above 350 nm.[56,57]

The first example of chiral induction based on the use of a CIL for an irreversible, unimolecular photoisomerization consists of the di-π-methane rearrangement of dibenzobicyclo[2.2.2]octratrienes (**18**), shown in Scheme 9.8.[58] Independently of the CIL used as solvent, irradiation of the carboxylic compound **18b** led to the formation of the corresponding photoproduct **19b** with some ee from 3 to 12%. In contrast, photolysis of the ester form (**18a**) resulted in the formation of photoproduct **19a** as a racemic mixture. It should be noted that in the case of CIL **f**, **h** and **i** the presence of a base, either chiral or nonchiral, was necessary in order to achieve some enantioselectivity. In contrast, in the case of CIL **g** the addition of a base had a minor influence varying the ee values from 5.8 to 6.8 and from −6.4 to −6.5 in the presence of (+)-**g** or (−)-**g**, respectively. Based on these data it was proposed that the photochirogenesis induction derives from ion pairing of the deprotonated acids of bicyclo[2.2.2]octatriene with the CIL cation. Importantly, control experiments using acetone, benzene and achiral

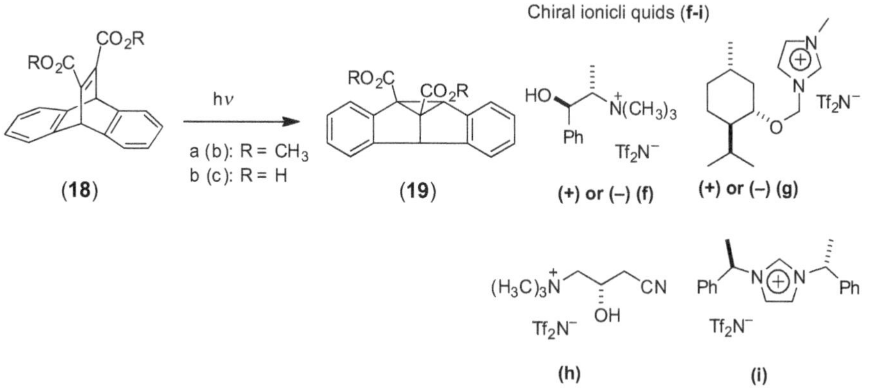

Chiral ionicli quids (**f-i**)

$h\nu$

a (b): R = CH_3
b (c): R = H

(**18**) (**19**) (+) or (–) (**f**) (+) or (–) (**g**) (**h**) (**i**)

Scheme 9.8 Photoinduced di-π-methane rearrangement of dibenzobicyclo-[2.2.2]octatrienes using CILs as solvent and chiral inductors.

[bmim]Cl in the presence of base for the photolysis of **18b** did not lead to any detectable enantioselectivity.

Although the above ee values are only modest and unsatisfactory, they prove the possibility of applying photochirogenesis in ILs and give some hints on the type of interactions that should be maximized to achieve optimal results. The superior performance of CILs with respect to conventional organic solvents was clearly demonstrated.

Another enantioselective photochemical reaction that has been frequently used as a probe reaction to compare the performance of various solvents and conditions is the photochemical dimerization of anthracene. When using monosubstituted anthracene, this photodimerization can lead to four different diastereomers depending on the orientation of the anthracene rings (head or tail) and the relative position of the substituent (*syn* or *anti*). These four diastereomeric dimers are split into the corresponding pair of enantiomers when the photoproduct does not have symmetry plane. The parameters to be considered in this model reaction are the diastereomeric distribution (head-to-tail and *syn*-to-*anti*) as well as the ee values for each diastereomer. In particular, 2-anthracenecarboxylic acid is a good substrate since the carboxylic group can establish strong hydrogen-bond interactions and can be easily transformed into derivatives in which the hydrogen bond can be replaced by other types of intermolecular force. Enantiodifferentiating photocyclodimerization of 2-anthracenecarboxylic acid (**20**, AC-H) and its alkali metal salts (AC-Li, -K and -Cs) in (*R*)-1-(2,3-dihydroxypropyl)-3-methylimidazolium bistriflimide (**j**) resulted in the formation of *syn*-HT (**22***) in 41% ee for AC-H but in 4–10% ee for AC-Li, -K or -Cs at -50 °C (Scheme 9.9). At this temperature, the reaction conversions for AC-H and AC-Li were 48 and 33%, respectively, while the HT/HH ratios were 0.7 and 1.5.[56]

Therein, the different behaviour of the carboxylic acid group *vs.* its corresponding alkali metal salts was considered as evidence that the important

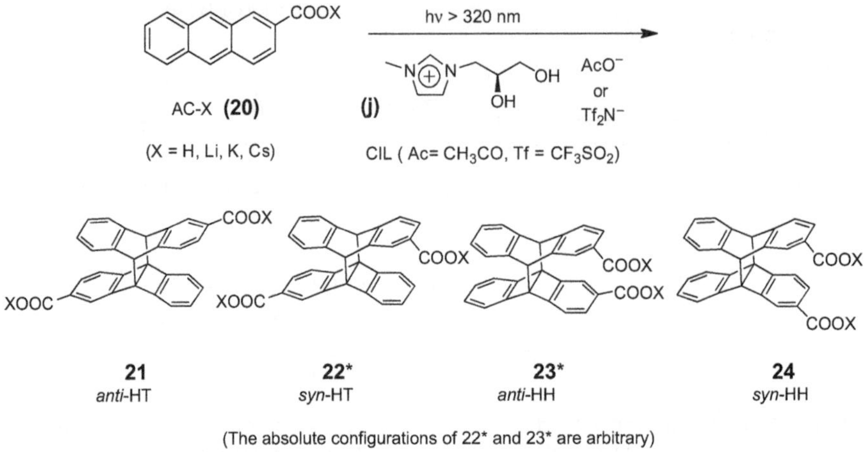

Scheme 9.9 Photochemical dimerization of 2-anthracenecarboxylic acid and its alkali metal salts in an imidazolium CIL. H and T correspond to head and tail, respectively.

intermolecular force responsible for the chiral induction in this system is the hydrogen bond between AC-H and the chiral diol moiety of the CIL (**j**). This hydrogen bridge interaction will be disrupted in the AC anions, where ion pairing should prevail. Similar conclusions about the importance of hydrogen bridges were obtained when changing the counter anion of the CIL from $[Tf_2N]^-$ to the more basic $[AcO]^-$.[57] $[AcO]^-$ should be able to abstract the proton from the AC-H giving the carboxylate anion which is more reluctant to undergo enantiodifferentiation. For example, using the CIL$[AcO]^-$ a product conversion lower than 1% with a nearly racemic mixture of the chiral dimers **22*** and **23*** (0–3% ee) was obtained at −50 °C. These results contrast with the values obtained when using CIL$[Tf_2N]$, whereby a yield of 48% and 41% ee were obtained for dimer **22***. Thus, the importance of hydrogen-bonding interaction of the substrate with the CIL rather than electrostatic interaction was confirmed. Although no reasons why hydrogen bonding should be a more suitable intermolecular force to induce chirality were given, it should be noted that the main characteristic of hydrogen bonding is directionality in the interaction, while ion–ion forces are equally strong regardless of the orientation of substrate and inductor.

In a more elaborate system, the influence of the presence of a supramolecular host, such as a cucurbituril (CB), able to incorporate AC-H within its hollow internal space in the photochirogenesis was studied by adding it to the CIL. CBs are hollow, pumkin-shaped molecular containers constituted by glycoluril units connected by methylene bridges.[59,60] The portals of CBs are flanked by polar carbonyl groups, while the internal void space of the organic capsule is apolar.[61] The size of CBs depends on the number of glycoluril units and organic molecules that can access CB[7] and CB[8]. In the latter case, the dimension of the internal empty space is about 1 nm and can even incorporate two guest molecules. In this case, the presence of CB[8] during the photocyclodimerization of AC-H in the CIL in water at 25 °C was investigated. It was observed that the presence of CB[8] in CIL[OAc]$^-$ or CIL[Tf_2N]$^-$ changes the HT : HH ratio from 77 : 23 to 9 : 91 or 1 : 99, respectively, thus rendering the dimerization highly stereoselective. However, the formation of **23*** as major product was totally racemic in both CILs. It was proposed that two AC-H molecules are aligned in an HH manner inside a single CB[8] capsule, although the interaction with the chiral dihydroxypropylimidazolium should be weak according to the experimental formation of racemic **23*** irrespective of the nature of the counter anion. Probably in the present case the dihydroxypropylimidazolium chiral inductor is preferentially ligated to the CB[8] portal.

Also it was found that the use of CIL[AcO]$^-$ as a chiral template in dichloromethane as solvent results in the exclusive formation of HH dimers and in particular *anti*-HH dimer **23*** with 14% ee at 79% yield. This result is interpreted as arising from both hydrogen bonding and electrostatic interactions between two ACs and CIL dissolved in dichloromethane. Considering the availability of conventional organic solvents compared to ILs that are prepared on a much smaller scale, the combination of ILs and conventional solvents is a simple way to increase the appeal of the process, reducing the amount of IL required.

9.12 Photochemical Synthesis Using Singlet Oxygen in ILs

Singlet oxygen ($^1O_2$) is a highly reactive species that can be conveniently generated by energy transfer from triplet excited states of photosensitizers to ambient molecular oxygen that is a rare case of a triplet ground state.[62,63] $^1O_2$ can react with allylic positions (*ene* reaction), undergo cycloadditions with conjugated π systems and produce the room temperature oxidation of heteroatoms [64–67] among other reactions.

Photochemical generation of $^1O_2$ is carried out typically in solution in which the photosensitizer and the substrates are dissolved, the efficiency of the process depending on several factors including oxygen concentration, nature of the solvent, substrate and photosensitizer.

Although the low solubility of O_2 in ILs compared to other traditional organic solvents can be a drawback in this reaction, ILs have been successfully used as reaction media to effect oxygenation promoted by $^1O_2$. One type of substrates that exhibit high reactivity towards $^1O_2$ is substituted furans and some synthetic routes to several natural products have as the key step of the total synthesis the photooxidation of furans (**24**) to butenolides (**25**) (Scheme 9.10).[68]

Alternatively, this photooxygenation has been carried out using ILs in the presence of O_2 and Rose Bengal as photosensitizer at room temperature (Scheme 9.11). Among the ILs tested [hmim]Cl gave the best results with 76% yield in 2 h in the absence of the Hünig's base. This photooxygenation

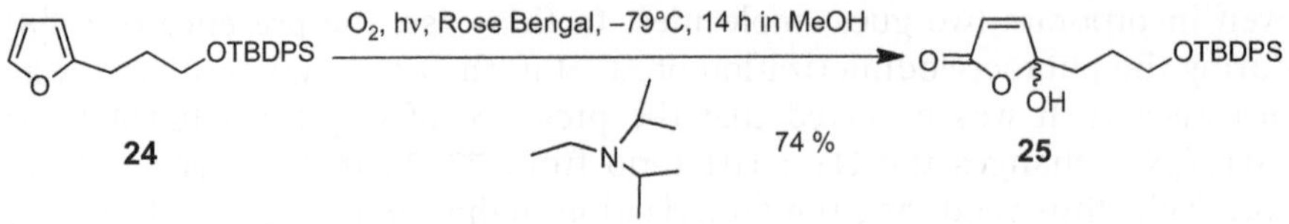

Scheme 9.10 $^1O_2$ photooxygenation of 2-(3-hydroxypropyl) substituted furan in methanol using Rose Bengal as photosensitizer in the presence of ethyldi(isopropyl)amine (Hünig's base). TBDPS corresponds to the protecting group *t*-butyldiphenylsilyl.

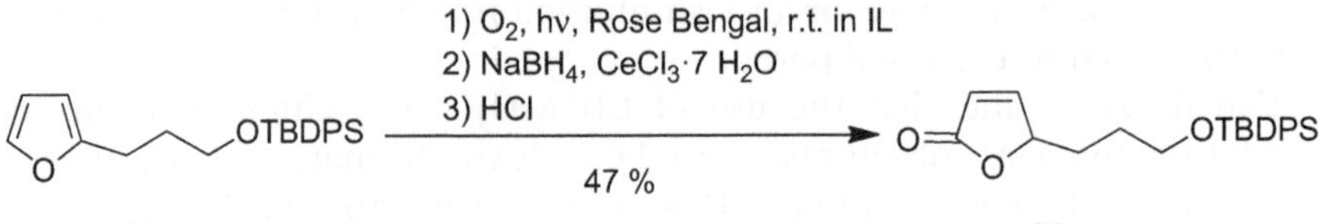

Scheme 9.11 One pot synthesis of 2-furanone by consecutive $^1O_2$ photooxygenation in IL followed by reduction of the resulting 5-hydroxy-2-furanone intermediate.

can be coupled with the reduction of the formed hydroxyl intermediate leading in one pot to butenolide **26** in 47% yield (Scheme 9.11). Although the synthesis of biologically active compounds derived from **26** as (+)-9a-epi-stemeoamide requires the enantioselective photooxygenation of 2-(3-hydroxypropyl)furan or the subsequent resolution of the resulting enantiomers, the efficient preparation of compound **26** from the corresponding furan exemplifies the potential of photooxygenation in ILs in organic synthesis of natural products.

$^1O_2$ can also effect selective heteroatom oxygenation. In particular the oxidation of sulfur containing compounds can lead to a variety of compounds including sulfides, sulfoxides, sulfones or sulfonic acid depending on the oxidizing reagent and reaction conditions. In this context, $^1O_2$ has been used to promote the selective oxidation of thioanisole (**27**) to the corresponding sulfoxide (**28**) which results more efficiently in ILs compared to conventional aprotic solvents. It has been proposed that the selective product formation derives from the ability of ILs to stabilize the persulfoxide that is the intermediate, leading to the sulfoxide.[69]

Further studies have shown that the combination of CH_3CN and IL is even more efficient at effecting the oxidation of thioanisole by $^1O_2$ than the individual solvents.[70] The maximum sulfoxide yields were obtained using imidazolium-[bmim][[Tf_2N] mixtures (Figure 9.7).

While irradiation in pure methanol outperforms the results using CH_3CN as solvent, the most efficient media result was the combination of CH_3CN and imidazolium ILs at a certain mole ratio. The good performance of the CH_3CN–imidazolium ILs was explained in terms of both electrostatic and hydrogen bonding stabilization of the persulfoxide intermediate promoted

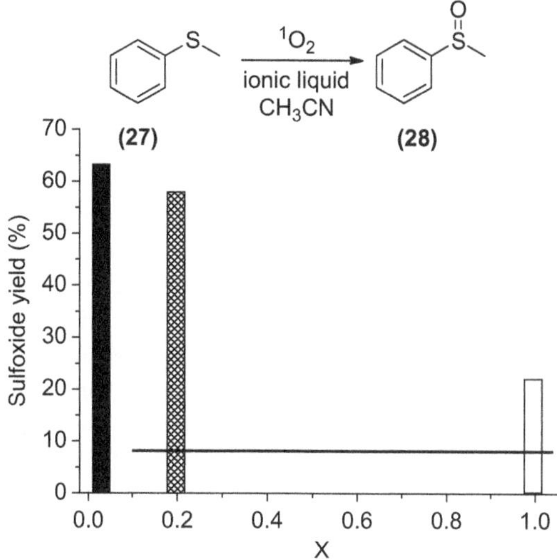

Figure 9.7 Maximum selectivity to sulfoxide in the oxidation of thioanisole depending on the ILs–CH_3CN molar fraction (X). The line corresponds to the selectivity to sulfoxide achieved with MeOH–CH_3CN mixtures in the indicated X interval. (■) [bmim][Tf_2N], (⊠) [emim][Tf_2N], (□) [bmpy][Tf_2N]. Reaction conditions: thioanisole (5×10^{-2} M), methylene blue (5×10^{-4} M), 2 mL of an oxygen-saturated solution, irradiation at 400–600 nm, 15 min reaction time, 25 °C. Mass balance>97% in all cases.

by ILs and the lower viscosity and aggregation imparted by CH_3CN with respect to neat ILs. Other reasons, however, also have to be taken into account to understand how the nature of the IL influences the photooxygenation efficiency. Thus, in contrast to imidazolium based ILs, the use of [bmpy][Tf_2N] always results in poor sulfoxide yields.

9.13 Pollutant Degradation in ILs

An area in which ILs are finding application is their use as solvent to extract organic pollutants in solid matrices to effect their subsequent photochemical degradation.[71] In this use ILs are replacing volatile solvents that have a negative impact on atmospheric pollution. Complete photochemical degradation/mineralization of the pollutant in the IL should regenerate the IL such that it would be ready for a consecutive reuse. In this sense, the photodegradation of chlorinated phenols in [bmim][PF_6] has been reported under UV irradiation (253.7 nm).[72] The key issues in this application are the selective degradation of the pollutant and IL stability under UV irradiation, particularly considering the short wavelengths needed in the photolysis. ^{1}H-NMR and UV spectroscopic studies indicate some changes in the IL as a function of photolysis time and, therefore, the long-term use of the IL for this purpose could be limited.

In a similar way, the photodegradation of pentachlorophenol (PCP) previously extracted from real soil samples by [bmim][PF_6] has been carried out using different advanced oxidation protocols, namely UV light (250–258 nm) with or without H_2O_2 and UV–Fe(II)–oxone.[73] The most efficient degradation procedure for PCP in [bmim][PF_6] was the photo-Fenton UV–Fe(II)–oxone process probably due to the massive production of hydroxyl radicals, a proposal that was supported by quenching experiments using *tert*-butyl alcohol as inhibitor of $^{\bullet}OH$ radicals. Considering the high reactivity of $^{\bullet}OH$ radicals, one important point that needs to be carefully addressed is the long term reusability of [bmim][PF_6], an issue that determines the feasibility of the whole procedure. As commented previously, different ILs have been degraded to a different extent by UV, UV–H_2O_2 or UV–TiO_2 advanced oxidations and it is likely also that the UV–Fe(II)–oxone process partially degrades the [bmim][PF_6] solvent in addition to PCP.[17] For this reason, sufficient characterization of the by-products formed upon prolonged exposure of ILs to deep advanced oxidation processes and convincing studies of the lack of toxicity of the resulting used media are necessary before large scale application of these protocols for soil recovery.[74]

9.14 Photochemical Generation of Noble and Transition Metal Nanoparticles in ILs

As we have commented earlier high viscosity is one of the main characteristics of ILs. Viscosity resulting in the reduced diffusion of solid particles through a liquid could be advantageous in the case of nanoparticle preparation because the process of particle growth could be reduced. For many purposes, such as catalysis and chemical sensors, development of reliable procedures for the preparation of colloidal noble metal nanoparticles with small particle size, in the nanometric range (<10 nm), is highly important.[75–77] It has been found that parameters like activity and efficiency decrease considerably as the size of the metal nanoparticle increases in the scale of nanometres.

It is generally encountered that preparation of noble metal nanoparticles in the liquid phase gives broad size distribution due to the easy agglomeration of unprotected nanoparticles and Oswald type growth of the initial nanoparticles. These types of phenomena require diffusion and, therefore, they could be minimized using high viscosity solvents. In order to overcome aggregation and gain control of the size and shape of the nanoparticles it is common to use water soluble polymers such as poly(*N*-vinyl-2-pyrrolidone) (PVP) or polyvinylalcohol (PVA), quaternary ammonium salts, surfactants or polyoxoanions.[78] These compounds bind at the surface of the nanoparticle acting as a ligand and inhibiting its growth.

On the other hand, preparation of metal nanoparticles in non-aqueous medium (solvato synthesis) is becoming increasing employed for the preparation of metal nanoparticles. Not surprisingly, ILs—especially those based on imidazolium salts—can also be used for the preparation and stabilization of small size nanoparticles.[9,79] It is generally accepted that anionic aggregates of IL stabilize small metal nanoparticles (<10 nm), while cationic aggregates mainly interact with larger ones.[9]

Formation of nanoparticles in pure IL using photoreduction methods has been applied for different metals. For example, photoinduced formation of gold nanoparticles has been reported using $HAuCl_4 \cdot 4H_2O$ and acetone as photoinitiator in [bmim][PF_4].[80] X-ray diffraction (XRD) measurements reveal the formation of crystalline gold nanoparticles and their interaction with the IL through surface adsorption and even covalent bonding has been proposed. Depending on the reaction conditions, different morphologies of Au nanoparticles were observed. For example, using $HAuCl_4 \cdot 4H_2O$ as gold precursor (0.1 g mL^{-1}) resulted in the formation of gold nanosheets with triangular and hexagonal shapes with maximal dimensions of 4 μm and thickness of 60 nm. Prolongation of the reaction times leads to an increase in the dimensions of the nanoparticles up to hundreds of microns. These large nanoparticles hexagonal shapes preferentially exhibited and sometimes terraces and steps. It was proposed that the ordered IL structure acting as template can promote the preferential formation of gold sheets. On the other hand, lowering the concentration of $HAuCl_4 \cdot 4H_2O$ to 0.05 or 0.025 g mL^{-1} results in the preferential formation of polyhedral gold particles of 1–5 μm and 1–3 μm, respectively. This study demonstrates that the IL [bmim][PF_4] can act as reaction medium, template and capping agent to fabricate gold particles with different shapes and sizes.

Imidazolium based ILs have also been used as solvent for the photolytic decomposition of metal carbonyl precursors with UV-light under argon resulting in the formation of metal nanoparticles. In particular, Cr, Mo and W nanoparticles have been synthesized in different ILs namely [bmim][BF_4], [bmim][OTf] and *n*-butyltrimethylammonium, abbreviated as [btma][Tf_2N].[81] Interestingly, in all cases, using [bmim][BF_4] as the solvent, small and uniform nanoparticles (1 to 1.5 nm) were obtained, while the largest nanoparticles (around 100 nm) were produced by using [btma][Tf_2N]. By using [bmim][OTf] or [btma][Tf_2N], it was demonstrated that the molecular volume of the ionic anion dramatically increases the size of the Cr, Mo and W nanoparticles. Aerobic conditions lead to larger metal oxide nanoparticles with respect to the sizes that can be obtained under argon. It was proposed that the first outer shell stabilizing the nanoparticles must be anionic, this proposal correlating well with other IL parameters such as density, viscosity, conductivity and surface tension. In any case the supramolecular pre-organization of the imidazolium based IL and the existence of different nanodomains should be considered in the generation of metal nanoparticles.

The same group has also reported the synthesis of iron, ruthenium and osmium nanoparticles with a diameter of about 1.5–2.5 nm under similar conditions as above in [bmim][BF_4] as solvent.[82] Surprisingly, the particle size of the metal nanoparticles is not affected by the concentration of precursor in the range from 0.2 to 1 wt%. Normally the concentration of the metal precursor plays a strong influence on the size of the nanoparticles.

Another alternative for the photochemical synthesis of metal nanoparticles in ILs has been the laser ablation of metallic foils[83] or plates,[84] as well as pre-formed metallic nanoparticles.[85] It is known that irradiation of metal nanoparticles with intense laser pulses focused on a small volume can produce strong mechanical effects due to the high energy concentration in short times. Dissipation of this concentrated energy can lead among other effects to remarkably high local temperatures and powerful shock waves producing melting of the nanoparticles or their fragmentation, increasing or decreasing their particle size, respectively.[86] In this context, laser ablation in [bmim][PF_6] has been reported as a convenient technique to induce fragmentation of pre-formed metallic Pd and Ru nanoparticles.[85] Thus, for instance it was observed that laser irradiation at 532 nm of preformed metallic Pd (10 nm) and Ru (15 nm) nanoparticles resulted in the formation of metallic nanoparticles with sizes of 4.2 ($\pm$0.8) and 7.2 ($\pm$1.3) nm, respectively. The resulting smaller colloidal metal nanoparticles formed after fragmentation were stable in ILs for weeks, *i.e.* 8 weeks in the case of Ru(0), as determined by transmission electron microscopy (TEM) measurements. Besides the positive influence of the high viscosity in avoiding particle growth, it was proposed that the IL also acts as a ligand of the laser-ablated nanoparticles, forming a protective and stabilizing layer on the electron deficient metal nanoparticle surface.

In addition to using pure ILs for the formation of metal nanoparticles or modification of their size, another strategy for the synthesis of metal nanoparticles in ILs is the use of two immiscible liquid phases in the form of microemulsions. Microemulsions have the advantage of overcoming problems of metal precursor solubility since the corresponding metal source can be initially dissolved in the second phase of the emulsion rather than in the IL. One of the combinations that has been studied in this context is a microemulsion based on hydrophilic ionic liquids, water and a surfactant. Depending on the nature of the surfactant, *i.e.* hydrophilic or hydrophobic, nanodomains different from the micellar structure typically found in aqueous solutions can be formed, opening new possibilities in the control of the size and morphology of the resulting metal nanoparticles.[87,88] For example, three different micro-regions of the microemulsion prepared with [bmim][BF_6], the non-ionic surfactant Tween 20 (see structure **29** in Figure 9.8) and water have been proposed to explain the electrochemical data obtained by cyclic voltammetry in this emulsion.[88] One nanodomain will be droplets of water inside a large region of [bmim][BF_6], while a second nanodomain should be the reverse situation, *i.e.* droplets of [bmim][BF_6] in a large zone of water. The third domain will constitute a true solution of [bmim][BF_6] and water. The presence of these multiple nanodomains changing dynamically in the space and dimensions makes the system very complex and provides various possibilities depending on the preferential location of the metal precursor.

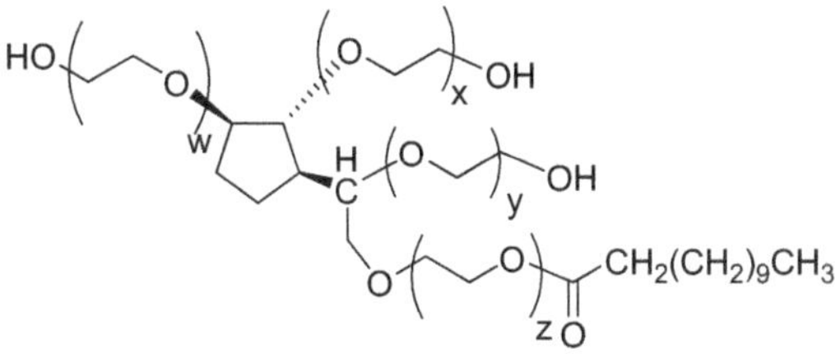

Figure 9.8 Chemical structure of surfactant Tween 20 used for the formation of microemulsions in imidazolium ILs.

In this context, metallic silver nanoparticles have been synthesized by photoreduction of silver perchlorate, using benzoin as photoiniciator, in water-in-IL microemulsions.[89] In particular, the surfactant Tween 20 (**29**), water and the ILs [bmim][BF_4] or [omim][BF_4] were used. Metallic silver nanoparticles with average diameters of 8.9 and 4.9 nm depending on the [bmim][BF_4] or [omim][BF_4] IL, respectively, were obtained as determined by TEM measurements. These nanoparticle sizes are remarkably small for silver which tends to form much larger particles.

Furthermore, the water droplet size in the water-in-IL system increases during the photoreduction from around 20 to 40 nm. This variation has consequences on the final silver particle size distribution. Interestingly, when the size of the silver nanoparticles exceeds that of the water droplet in which they are formed, precipitation of the nanoparticles from the liquid phase starts to be observed. Considering the wide range of possible combinations of surfactants, ILs and co-solvents, it is clear that microemulsions still offer many possibilities for the formation of silver and other metal nanoparticles with varied morphologies and size distributions.

Microemulsions can also be influenced by the presence of solutes or gases. Thus, photoreduction of silver nitrate using the biphasic microemulsion systems water-in-[bmim][PF_4] or water-in-[omim][PF_4] under high-pressure CO_2 (25 MPa) resulted in the formation of metallic silver nanoparticles with notably small average diameters of 3.2 and 3.7 nm, respectively.[90] It was proposed that high-pressure CO_2 might make the continuous phase more hydrophobic increasing the preference of silver nanoparticle formation in the water droplet nanodomain. The system certainly has a high level of complexity since it was also observed that, with respect to ambient conditions, the high-pressure CO_2 also controls the average size of the water droplets that are in the 30–40 nm range, preventing silver nanoparticle aggregation or precipitation.

9.15 Photochemistry of Gold Nanoparticles in ILs

Besides formation of metal nanoparticles, ILs can also be useful to modify the photochemical properties of metal nanoparticles. As commented earlier,

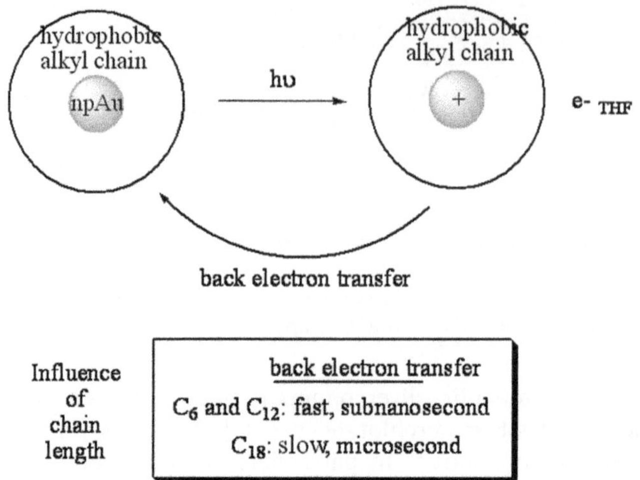

Scheme 9.12 Photoinduced electron ejection in gold nanoparticles (npAu) having alkylthiolate ligands as a function of the alkyl chain length.

laser irradiation can produce alteration of nanoparticle average size due to temperature increase and ablation. Besides this, laser excitation can also have electronic effects. Thus, Garcia, Martin and co-workers reported that irradiation of alkylthiol stabilized gold nanoparticles (5 nm) in [bmim][PF_6] allows the detection of photoejected electrons in the microsecond time scale (Scheme 2.12).[91] This transient state of charge separation with electrons in the medium and transient positive gold metal nanoparticles decays by charge recombination that in [bmim][BF_4] takes place in hundreds of microseconds. In addition, quenching experiments using methyl viologen allowed the estimation of the quantum yield of charge separation state in the IL to be 0.12.

These results contrast with the use of a conventional organic solvent, such as tetrahydrofuran (THF), where this phenomenon of electron photoionization and subsequent charge recombination occurs much faster and cannot be observed in the nanosecond time scale. However, when the apolar alkyl chain of the alkylthiol capping agent is sufficiently long, then, charge recombination can also become slow in THF and photoejected electrons are again observed during microseconds after the laser pulse. Thus, using octadecanethiol as capping agent of gold nanoparticles and upon 355 nm laser excitation, a weak signal corresponding to THF solvated electrons was observed in the submicrosecond time scale. In contrast thiol capping agents with shorter alkyl chain such as hexanethiol and dodecanethiol should make electron-hole recombination faster and it was not possible to detect any transient in THF in the microsecond scale indicating the importance of the medium (ILs) or the ligands (long alkyl chains) to achieve nanoparticle photoresponse (Scheme 9.12). These results illustrate how the unique properties of ILs can serve to control the photochemistry of the solute or nanoparticle, generally making slow those processes that require diffusion to occur and that can be very fast in other solvents.

9.16 Solar Fuels Using ILs

In the context of renewable energy alternatives to conventional fossil fuels, the production of chemical compounds using solar light can in the long term serve to develop a new generation of transportation fuels ("*solar fuels*"). These solar fuels include H_2 generation from water and CH_3OH, CH_4, CO and HCHO from CO_2 reduction among other possible chemicals complementing biomass.

In the particular case of photocatalytic water splitting into H_2 and O_2, the slowest and bottle neck process is water oxidation to O_2. This reaction can be studied separately from the overall water splitting if suitable sacrificial electron acceptors are present. In this context, Bond and co-workers have proposed that the modified structure of water present in IL facilitates its oxidation under UV irradiation with respect to water and common organic solvents such as CH_3CN or CH_3CH-CH_2Cl_2 containing water.[92,93] In addition, it was found that the formal reversible redox potentials $E^0{}_F$ for $[P_2W_{18}O_{62}]^{6-/7-/8-/9-/10-}$ couples are much more positive in IL compared to conventional solvents.[92] Thus, UV photolysis of the classic Dawson polyoxometalate salt $K_6[P_2W_{18}O_{62}]$ using the aprotic [bmim][BF_4] or the protic diethanolamine hydrogen sulfate ILs in the presence of water produces the reduction of $[P_2W_{18}O_{62}]^{6-}$ anion to $[P_2W_{18}O_{62}]^{7-}$ as well as the oxidation of water to molecular oxygen and protons.[92] While these preliminary studies can be considered a proof of principle, more comprehensive studies on the overall water splitting in ILs are still needed to assess their efficiency and additional advantages with respect to other media.

In the context of solar fuels derived from CO_2, it is interesting to note that some ILs offer some promise because they can dissolve large CO_2 amounts,[94] although they do not mix with supercritical CO_2.[95] This high CO_2 solubility in ILs should be advantageous to perform its photocatalytic reduction to CH_3OH or CH_4. In fact one of the main problems of performing this photocatalytic reaction in water is low CO_2 solubility at the acid pH values that are the most adequate for the photoreduction. It can be anticipated that ILs will also be used for CO_2 photoreduction to valuable products.

9.17 Conclusions and Future Prospects

The above sections have discussed some of the physical properties of ILs that can be relevant for their use as solvent in photochemical reactions including viscosity, polarity and photochemical stability. In the main body of this chapter we have illustrated how these properties can be advantageous to observe a photochemical pattern that does not take place in conventional solvents. Emphasis has been placed on showing the unique behaviour of ILs and how the outcome of the photochemical process can be rationalized based on the physical properties of ILs. Our aim has not been to cover comprehensively all the photochemical reactions carried out in ILs, but to exemplify a range of processes in which the use of ILs leads to a unique outcome in the process.

It can be anticipated that the use of ILs as solvent in photoreactions will continue to grow in the future in most of the reaction types that we have discussed. Photochirogenesis in ILs has given better ee values than in conventional solvents, but the present situation still needs CILs that impart much higher ee in order to have some potential applicability. One area that is particularly promising is the combination of ILs and metal nanoparticles including aspects such as the preparation of nanoparticles with well defined size and morphologies and in photocatalysis.

Finally, considering the compatibility of ILs and some polymers and the wide use of polymers as engineering materials, there is no doubt that the number of systems in which a reversible photoresponse in polymers is achieved by adding ILs in the formulation will increase and could possibly reach commercial use.

References

1. V. I. Parvulescu and C. Hardacre, *Chem. Rev.*, 2007, **107**, 2615.
2. T. Welton, *Chem. Rev.*, 1999, **99**, 2071.
3. J. P. Hallett and T. Welton, *Chem. Rev.*, 2011, **111**, 3508.
4. C. Chiappe, C. S. Pomelli and S. Rajamani, *J. Phys. Chem. B*, 2011, **115**, 9653.
5. S. N. V. K. Aki, J. F. Brennecke and A. Samanta, *Chem. Commun.*, 2001, 413.
6. P. K. Mandal and A. Samanta, *J. Phys. Chem. B*, 2005, **109**, 15172.
7. K. Binnemans, *Chem. Rev.*, 2005, **105**, 4148.
8. S. Xiao, X. Lu and Q. Lu, *Macromolecules*, 2007, **40**, 7944.
9. J. Dupont and J. D. Scholten, *Chem. Soc. Rev.*, 2010, **39**, 1780.
10. J. Guo, G. A. Baker, P. C. Hillesheim, S. Dai, R. W. Shaw and S. M. Mahurin, *Phys. Chem. Chem. Phys.*, 2011, **13**, 12395.
11. M. Alvaro, B. Ferrer, H. Garcia and M. Narayana, *Chem. Phys. Lett.*, 2002, **362**, 435.
12. P. Bonhote, A.-P. Dias, N. Papageorgiou, K. Kalyanasundaram and M. Grätzel, *Inorg. Chem.*, 1996, **35**, 1168.
13. M. Galinski, A. Lewandowski and I. Stepniak, *Electrochim. Acta*, 2006, **51**, 5567.
14. I. A. Shkrob, T. W. Marin, S. D. Chemerisov and J. F. Wishart, *J. Phys. Chem. B*, 2011, **115**, 3872.
15. I. A. Shkrob, T. W. Marin, S. D. Chemerisov, J. L. Hatcher and J. F. Wishart, *J. Phys. Chem. B*, 2011, **115**, 3889.
16. R. Katoh and K. Takahashi, *Radiat. Phys. Chem.*, 2009, **78**, 1126.
17. P. Stepnowski and A. Zaleska, *J. Photochem. Photobiol., A*, 2005, **170**, 45.
18. M. J. Muldoon, A. J. McLean, C. M. Gordon and I. R. Dunkin, *Chem. Commun.*, 2001, 2364.

19. V. Dichiarante, C. Betti, M. Fagnoni, A. Maia, D. Landini and A. Albini, *Chem.-Eur. J.*, 2007, **13**, 1834.
20. R. Bini, C. Chiappe, E. Marmugi and D. Pieraccini, *Chem. Commun.*, 2006, 897.
21. A. Gilbert and J. Baggott, *Essentials of Organic Photochemistry*, Blackwell, Oxford, 1990.
22. G. Zhu, G. Wu, X. Xu and X. Ji, *Spectrochim. Acta, Part A*, 2011, **82**, 74.
23. N. Chandrasekhar, O. Schalk and A.-N. Unterreiner, *J. Phys. Chem. B*, 2008, **112**, 15718.
24. H. Brands, N. Chandrasekhar and A.-N. Unterreiner, *J. Phys. Chem. B*, 2007, **111**, 4830.
25. C. Nese and A.-N. Unterreiner, *Phys. Chem. Chem. Phys.*, 2010, **12**, 1698.
26. J. R. Lakowicz, *Principles of Fluorescence Spectroscopy*, Springer, Berlin, 3rd edn, 2006.
27. S. Pandey, S. N. Baker, S. Pandey and G. A. Baker, *J. Fluoresc.*, 2012, **22**, 1313.
28. A. Samanta, *J. Phys. Chem. B*, 2006, **110**, 13704.
29. A. Samanta, *J. Phys. Chem. Lett.*, 2010, **1**, 1557.
30. E. Binetti, A. Panniello, L. Triggiani, R. Tommasi, A. Agostiano, M. L. Curri and M. Striccoli, *J. Phys. Chem. B*, 2012, **116**, 3512.
31. P. Nockemann, E. Beurer, K. Driesen, R. V. Deun, K. V. Hecke, L. V. Meervelt and K. Binnemans, *Chem. Commun.*, 2005, 4354.
32. K. Lunstroot, K. Driesen, P. Nockemann, L. Viau, P. H. Mutin, A. Vioux and K. Binnemans, *Phys. Chem. Chem. Phys.*, 2010, **12**, 1879.
33. Q.-P. Li and B. Yan, *Dalton Trans.*, 2012, **41**, 8567.
34. Y. Y. Li, B. Yan, L. Guo and Y. J. Li, *Microporous Mesoporous Mater.*, 2012, **148**, 73.
35. G. Sartori and R. Maggi, *Chem. Rev.*, 2006, **106**, 1077.
36. B. Murphy, P. Goodrich, C. Hardacre and M. Oelgemöller, *Green Chem.*, 2009, **11**, 1867.
37. F. Pina, M. J. Melo, C. A. T. Lai, A. J. Parola and J. C. Lima, *Chem. Soc. Rev.*, 2012, **41**, 869.
38. K. Ichimura, *Nat. Mater.*, 2005, **4**, 193.
39. M. Natali and S. Giordani, *Chem. Soc. Rev.*, 2012, **41**, 4010.
40. R. Byrne, K. J. Fraser, E. Izgorodina, D. R. MacFarlane, M. Forsyth and D. Diamond, *Phys. Chem. Chem. Phys.*, 2008, **10**, 5919.
41. C. Schomburg, M. Wark, Y. Rohlfing, G. Schulz-Ekloff and D. Wöhrle, *J. Mater. Chem.*, 2001, **11**, 2014.
42. I. Casades, M. Álvaro, H. García and M. N. Pillai, *Photochem. Photobiol. Sci.*, 2002, **1**, 219.
43. H. Okada, N. Nakajima, T. Tanaka and M. Iwamoto, *Angew. Chem., Int. Ed.*, 2005, **44**, 7233.
44. F. Pina and L. Branco in *Ionic Liquids: Theory, Properties, New Approaches*, ed. A. Kokorin, Intech open access publishers, Croatia, 2011, ch. 6, 1.
45. F. Pina, V. Petrov and C. A. T. Laia, *Dyes Pigm.*, 2012, **92**, 877.

46. F. Pina, A. J. Parola, M. J. Melo, C. A. T. Laia and C. A. M. Afonso, *Chem. Commun.*, 2007, 1608.
47. F. Pina, J. C. Lima, A. J. Parola and C. A. M. Afonso, *Angew. Chem., Int. Ed.*, 2004, **43**, 1525.
48. D. Fernandez, A. J. Parola, L. C. Branco, C. A. M. Afonso and F. J. Pina, *J. Photochem. Photobiol. A*, 2004, **168**, 185.
49. M. C. Moncada, D. Fernández, J. C. Lima, J. A. Parola, C. Lodeiro, F. Folgosa, M. J. Melo and F. Pina, *Org. Biomol. Chem.*, 2004, **2**, 2802.
50. D. Mecerreyes, *Prog. Polym. Sci.*, 2011, **36**, 1629.
51. S. Zhang, S. Liu, Q. Zhang and Y. Deng, *Chem. Commun.*, 2011, **47**, 6641.
52. M. Tamada, T. Watanabe, K. T. Horie and H. Ohno, *Chem. Commun.*, 2007, **39**, 4050.
53. Y. Inoue, *Chem. Rev.*, 1992, **92**, 741.
54. U. M. Lindstrom, *Chem. Rev.*, 2002, **102**, 2751.
55. R. Noyori and T. Ohkuma, *Angew. Chem., Int. Ed.*, 2001, **40**, 40.
56. G. Fukuhara, C. Chiappe, A. Mele, B. Melai, F. Bellina and Y. Inoue, *Chem. Commun.*, 2010, **46**, 3472.
57. T. Fukuhara, M. Okazaki, M. Lessi, C. Nishijima, T. Yang, A. Mori, F. Mele, C. Bellina, Chiappe and Y. Inoue, *Org. Biomol. Chem.*, 2011, **9**, 7105.
58. J. Ding, V. Desikan, X. Han, T. L. Xiao, F. Ding, W. S. Jenks and D. W. Armstrong, *Org. Lett.*, 2005, **2**, 335.
59. J. Lagona, P. Mukhopadhyay, S. Chakrabarti and L. Isaacs, *Angew. Chem., Int. Ed.*, 2005, **44**, 4844.
60. L. Isaacs, *Chem. Commun.*, 2009, **6**, 619.
61. E. Masson, X. Ling, R. Joseph, L. Kyeremeh-Mensah and X. Lu, *RSC Adv.*, 2012, **2**, 1213.
62. D. R. Kearns, *Chem. Rev.*, 1971, **71**, 395.
63. C. Schweitzer, *Chem. Rev.*, 2003, **103**, 1685.
64. M. N. Alberti and M. Orfanopoulos, *Chem.–Eur. J*, 2010, **16**, 9414.
65. K. Nahm, Y. Li, J. D. Evanseck, K. N. Houk and C. S. Foote, *J. Am. Chem. Soc.*, 1993, **115**, 4879.
66. A. Toutchkine and E. L. Clennan, *J. Am. Chem. Soc.*, 2000, **122**, 1834.
67. M. N. Alberti and M. Orfanopoulos, *Synlett*, 2010, 999.
68. A. Fall, M. Sène, E. Tojo, G. Gómez and Y. Fall, *Synthesis*, 2011, 3415.
69. E. Baciocchi, C. Chiappe, T. Del Giacco, C. Fasciani, O. Lanzalunga, A. Lapi and B. Melai, *Org. Lett.*, 2009, **11**, 1413.
70. E. Baciocchi, C. Chiappe, C. Fasciani, O. Lanzalunga and A. Lapi, *Org. Lett.*, 2010, **12**, 5116.
71. J. Ma and X. J. Hong, *Environ. Manage.*, 2012, **99**, 104.
72. Q. Yang and D. D. Dionysiou, *J. Photochem. Photobiol., A*, 2004, **165**, 229.
73. B. Subramanian, Q. Yang, Q. Yang, A. P. Khodadoust and D. D. Dionysiou, *J. Photochem. Photobiol., A*, 2007, **192**, 114.
74. R. P. Swatloski, J. D. Holbrey and R. D. Rogers, *Green Chem.*, 2003, **5**, 361.
75. E. C. Dreaden, A. M. Alkilany, X. Huang, C. J Murphy and M. A. El-Sayed, *Chem. Soc. Rev.*, 2012, **41**, 2740.

76. M. Stratakis and H. Garcia, *Chem. Rev.*, 2012, **112**, 4469.
77. T. K. Sau, A. L. Rogach, F. Jäckel, T. A. Klar and J. Feldmann, *Adv. Mater.*, 2010, **22**, 1805.
78. A. Roucoux, J. Schulz and H. Patin, *Chem. Rev.*, 2002, **102**, 3757.
79. M.-A. Neouze, *J. Mater. Chem.*, 2010, **20**, 9593.
80. J. Zhu, Y. Shen, A. Xie, L. Qiu, Q. Zhang and S. Zhang, *J. Phys. Chem. C*, 2007, **111**, 7629.
81. E. Redel, R. Thomann and C. Janiak, *Chem. Commun.*, 2008, 1789.
82. J. Krämer, E. Redel, R. Thomann and C. Janiak, *Organometallics*, 2008, **27**, 1976.
83. H. Wender, M. L. Andreazza, R. R. B. Correia, S. R. Teixeira and J. Dupont, *Nanoscale*, 2011, **3**, 1240.
84. Y. Kimura, H. Takata, M. Terazima, T. Ogawa and S. Isoda, *Chem. Lett.*, 2007, **36**, 1130.
85. M. A. Gelesky, A. P. Umpierre, G. Machado, R. R. B. Correia, W. C. Magno, J. Morais, G. Ebeling and J. Dupont, *J. Am. Chem. Soc.*, 2005, **127**, 4588.
86. P. V. J. Kamat, *J. Phys. Chem. B*, 2002, **106**, 7729.
87. Y. Gao, S. Han, B. Han, G. Li, D. Shen, Z. Li, J. Du, W. Hou and G. Zhang, *Langmuir*, 2005, **21**, 5681.
88. Y. Gao, N. Li, L. Zheng, X. Zhao, S. Zhang, B. Han, W. Hou and G. Li, *Green Chem.*, 2006, **8**, 43.
89. M. Harada, Y. Kimura, K. Saijo, T. Ogawa and S. Isoda, *J. Colloid Interface Sci.*, 2009, **339**, 373.
90. M. Harada, C. Kawasaki, K. Saijo, M. Demizu and Y. Kimura, *J. Colloid Interface Sci.*, 2010, **343**, 537.
91. C. Aprile, M. A. Herranz, E. Carbonell, H. Garcia and N. Martín, *Dalton Trans.*, 2008, 134.
92. G. Bernardini, C. Zhao, A. Wedd and A. M. Bond, *Inorg. Chem.*, 2011, **50**, 5899.
93. C. Z. Zhao and A. M. Bond, *J. Am. Chem. Soc.*, 2009, **131**, 4279.
94. Q. Li, Y. Zhao and L. Wang, *Adv. Mater. Res.*, 2012, **347-353**, 116.
95. J. Liu, S. Cheng, J. Zhang, X. Feng, X. Fu and B. Han, *Angew. Chem., Int. Ed.*, 2007, **46**, 3313.

CHAPTER 10

Ionothermal Synthesis

F. H. AIDOUDI AND R. E. MORRIS*

School of Chemistry, Purdie Building, North Haugh, St Andrews, KY16 9ST, UK
*Email: rem1@st-andrews.ac.uk

10.1 Introduction

Over recent years ionic liquids (ILs) have received a great deal of interest in various fields.[1] Many of the studies have focused on the 'green' chemistry[2] potential of these compounds, with particular interest to replace organic solvents in homogeneous catalysis.[3] The particular property of ionic liquids that makes them environmentally suitable for these purposes is their low vapour pressure,[4] which has significant advantages when replacing highly volatile organic solvents. However, there are many other uses of ionic liquids in diverse areas of technology from electrolytes in batteries and fuel cells[5] to the use of supported ionic liquids as catalysts.[6] In some reactions ILs act only as inert solvents and in others they play a more active role. The use of ILs in materials chemistry[7,8] has also been a great success and a wide range of materials from porous to dense and nano materials have been ionothermally-prepared. In ionothermal reactions ionic liquids have replaced water or other organic solvents and shown to be excellent solvents for the synthesis of crystalline materials.

RSC Catalysis Series No. 15
Catalysis in Ionic Liquids: From Catalyst Synthesis to Application
Edited by Chris Hardacre and Vasile Parvulescu

Published by the Royal Society of Chemistry, www.rsc.org

10.2 Ionic Liquids—A Special Class of Molten Salts

10.2.1 Definition of Ionic Liquids

An ionic liquid is classically defined as any material in the liquid state that consists predominantly of ionic species. Therefore, any ionic salt that can be made molten can be classified as an "ionic liquid", always assuming the salt does not decompose or vaporise on melting. By this definition, the well-known molten salts can also be regarded as 'ionic liquids'.[9] While traditional molten salts have been known for centuries, they have limited applications[10] due to their high melting point. Mainly "molten salts fluxes" have been used as direct replacements for traditional solid state synthesis[11] techniques in order to enhance diffusion and reduce reaction temperatures. However, the modern definition of ionic liquids tends to concentrate on those compounds that are liquid at relatively low temperatures, usually lower than 100 °C.[12,13] ILs that are liquid at or around room temperature are often called "room-temperature ionic liquids" (RTILs). For ionothermal synthesis, ILs can be broadly defined as being liquid below about 200 °C, the temperatures traditionally used in hydrothermal synthesis. In modern usage the term ionic liquid is almost exclusively reserved for liquids that contain at least one organic cation. The organic components of ionic liquids tend to be bulky and asymmetric, which results in a poorly coordinated network.

10.2.2 Modern Ionic Liquids *versus* Traditional Molten Salts

Fundamentally, there is no real difference between modern ILs and molten salts as both are liquids that contain only ions. However, the organic nature and the molecular asymmetry of the components of modern ILs dictate different structural organisation than traditional molten salts which is reflected in their lower melting point. In recent years, several studies have focused on the structural organisation of ILs, in particular those based on imidazolium cations, using various techniques including X-rays, IR, NMR and computational methods.[10,14–17] In addition to the ionic interaction between the anions and cations in ILs, these studies suggest the presence of other weak interactions *e.g.* "hydrogen bonding" that induce structural directionality. In contrast, as shown in Figure 10.1 classical molten salts form aggregates only through ionic bonds.

10.2.3 Properties of Ionic Liquids

The ionic nature of ILs greatly differentiates them from molecular solvents; when using them as the reaction media for the preparation of materials they provide a unique environment compared to other solvents. They can also be relatively polar solvents, ensuring reasonably good solubility of inorganic precursors.[18,19] Another important feature of ILs that has attracted academic

Figure 10.1 Structural features of: (a) modern ILs showing the existence of polar (dark grey) and nonpolar (light grey) nano domains; (b) traditional molten salts (structure of simple ionic compounds).
Figure 1.1 (a) reproduced by kind permission from reference 14.

interest is the possibility of designing new ILs and altering their physical and chemical properties simply by varying the anion or the cation, by introducing specific functionalities into the cation and/or anion or just by mixing two or more simple ILs. In addition to that many ILs display other interesting properties *e.g.*:

- Many but not all ILs tend to have good thermal stability and can be liquid over a range of 300 °C. This wide liquid range is a distinct advantage over traditional solvent systems that have a much narrower liquid range.
- ILs have low vapour pressure resulting from the strong ionic (Coulomb) interaction; although certain ILs can be distilled at high temperature and low pressure.[20]
- ILs have a wide range of solubilities and miscibilities with water and other solvents, for example some ILs are hydrophilic while others are hydrophobic.
- ILs are non-flammable, and recyclable.

10.2.4 Deep Eutectic Solvents

Deep eutectic solvents (DESs) are an extended class of ionic liquids, composed of a mixture which forms a eutectic with a melting point much lower than either of the individual components.[21] The first generation of eutectic solvents was based on mixtures of quaternary ammonium salts with hydrogen bond donors such as amides and carboxylic acids.[22] The deep eutectic phenomenon for a 2 to 1 mole ratio of choline chloride (2-hydroxyethyl-trimethylammonium chloride) and urea was first described by Abbot *et al.*[22] Choline chloride has a melting point of 302 °C and that of urea is 133 °C. The eutectic mixture, however, has a melting point of 12 °C.

The decrease in the melting point arises from charge delocalisation induced by the hydrogen bonding interaction between the urea molecules and the chloride ions. Compared to molecular solvents, eutectic solvents have a very low vapour pressure and are non-flammable. They have the exceptional features of ionic liquids but are a mixture of an ionic compound with a molecular compound. In addition they exhibit some more significant advantages such as relatively high polarity so they can dissolve many metal salts and metal oxides, their trivial preparation from easily available components and their relative unreactivity towards atmospheric moisture. Many are biodegradable and the toxicity of the components may be well characterised. The physical and chemical properties of DESs are dependent on the properties and functionalities of their components as well as the ratio of the mixture; judicious choice of the components will allow an appropriate solvent with specific properties. This flexibility in the composition and properties of DESs makes them useful solvents for many interesting applications.[23]

10.3 Synthetic Methods for the Preparation of Materials

Broadly speaking, the synthesis of crystalline solid state materials can be split into two main groups: those where the synthesis reaction takes place in the solid state and those which take place in solution.

10.3.1 Solid State Reactions

Solid state reactions require grinding and sometimes pelletising the reactants to encourage intimate contact between reactants. In these types of reactions usually high temperatures are required to overcome difficulties in transporting the reactants to the sites of the reaction, thus increasing the rate of diffusion. The high temperatures of solid state reactions make it difficult to incorporate ions that form volatile species, also preventing access to low temperature metastable products. Also, ions with higher oxidation states are often unstable at high temperatures. Typically solid state reactions are used for the synthesis of inorganic materials (*e.g.* solid state oxide).

Molten salt flux methods are direct replacements for traditional solid state synthesis techniques and have been considered as a good process for preparing inorganic materials where the synthetic temperatures can be relatively lowered compared to the solid state method. These synthetic procedures take place at a temperature above the melting point of the salt. For example, alkali metal hydroxide molten salts can be used as the molten phase, often contained in sealed inert (such as silver) vessels in the synthesis of many inorganic solids.

10.3.2 Synthesis in Solution

10.3.2.1 Hydrothermal and Solvothermal Synthesis

Transport in the liquid phase is obviously much easier than in solids, and syntheses require much lower temperatures (often less than 200 °C). The archetype of this type of preparative technique is hydrothermal synthesis, where the reaction solvent is water.[24] The most common method of accomplishing hydrothermal synthesis is to seal the reactants inside Teflon-lined autoclaves so that there is also significant autogeneous hydrothermal pressure produced, often up to 15 bar. The lower temperatures required for hydrothermal synthesis often lead to kinetic control of the products formed, and it is much easier to prepare metastable phases using this approach than it is using traditional solid state approaches. The important reaction and crystallisation processes in hydrothermal synthesis do not necessarily take place in solution (although of course they can) but can occur at the surfaces of gels present in the mixtures.

Solvothermal synthetic methods refer to the general class of using a solvent in the synthesis of materials. Of course water is by far the most important solvent, hence the usage of the term hydrothermal to describe its use. However, there are many other possible solvents. Alcohols, hydrocarbons, pyridine and many other organic solvents have all been used with varying degrees of success.[25] As with water these molecular solvents produce significant autogeneous pressure at elevated temperatures. The solvents used in solvothermal synthesis vary widely in their properties, from non-polar and hydrophobic to polar and hydrophilic.

10.3.2.2 Ionothermal Synthesis

The solvents used in hydrothermal and solvothermal synthesis differ fundamentally from ILs in that they are molecular in nature. The ionic nature of ILs imparts particular properties, including low vapour pressures (and so very little, if any, autogeneous pressure is produced at high temperature).[26] The ionothermal method[27] is the use of an ionic liquid or eutectic mixture as the reaction solvent and, in many cases, also as structure directing agent in the preparation of crystalline solids.

10.4 Ionothermal Synthesis of Zeolites

Many ionic liquids used today often have chemical structures that are very similar to the structures of commonly used structure directing agents (SDA), or templates in the hydrothermal synthesis of zeolites.[28] This realisation led to the first attempts to prepare zeotype frameworks using ILs as both the solvent and the template provider.[27] The potential advantage of this approach is that the competition between the solvent and template for interaction with any growing solid is removed when both solvent and template are the same species (Figure 10.2).

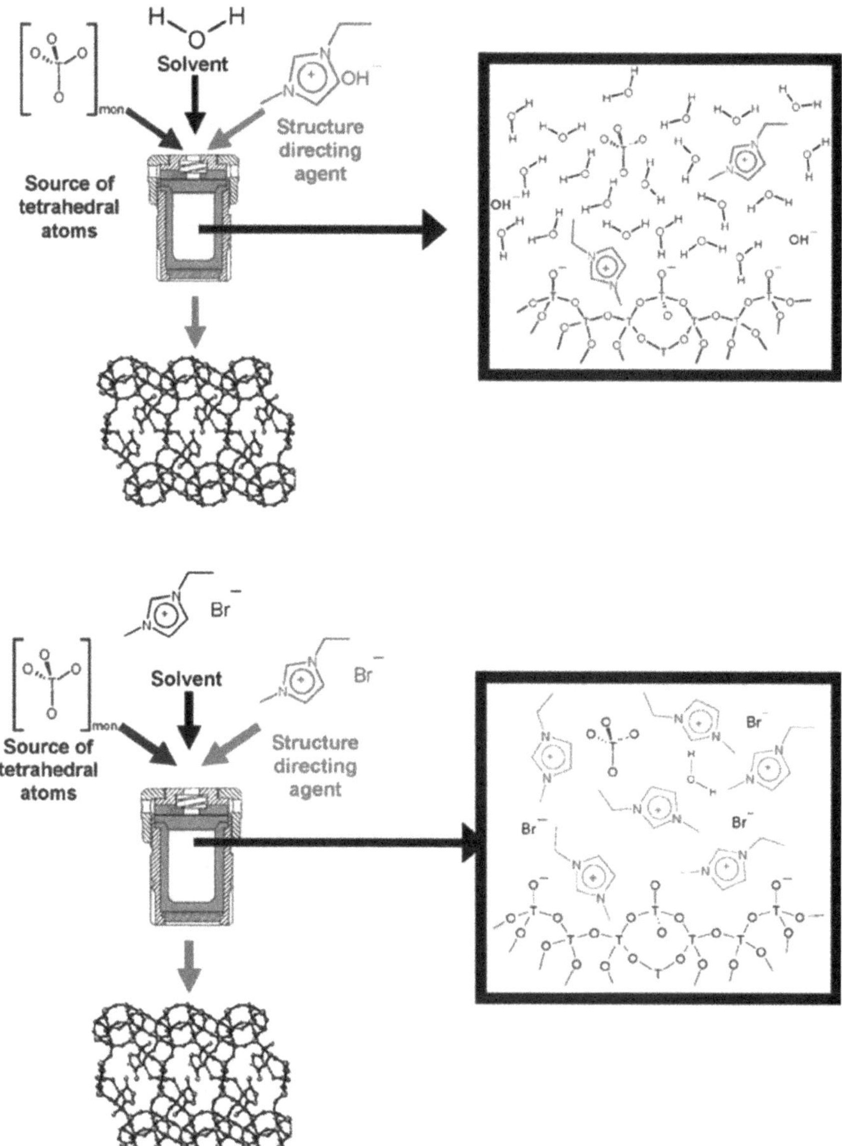

Figure 10.2 Schematic representations of the synthesis of a tetrahedral (zeotype) framework under hydrothermal (top) and ionothermal (bottom) conditions.

10.4.1 Ionothermal Synthesis of Aluminophosphates and Related Structures

The ionothermal synthesis of several aluminophosphates (AlPOs) (Figure 10.3) using 1-ethyl-3-methylimidazolium bromide and urea/quaternary ammonium salts as deep eutectic solvent was the pioneering work in this area.[27] Since then there have been many further attempts to prepare zeotype materials. The ionothermal synthesis of aluminophosphate zeolites (AlPOs) as powder or thin film forms has been by far the most successful. Many common ionic liquids proved to be suitable solvents for the preparation of AlPOs.[29–31] The ionothermal method is also suitable for incorporating the dopant metal atoms that give the frameworks their chemical activity. Silicon (to make so-called SAPOs)[32] and many different tetrahedral metals (Co, Mg *etc.*)[33,34] can all be incorporated into the ionothermally-prepared aluminophosphates.

The ionothermal synthesis of aluminophosphate provides an excellent, simple and low cost method for the ^{17}O enrichment of oxide materials that can be well characterised using ^{17}O solid state NMR.[35]

10.4.2 Ionothermal Synthesis of Silicon and Transition Metal-based Zeolites

Silicon-based zeolites have been much more of a challenge for ionothermal synthesis although there has been more success in the hydrothermal synthesis of zeolites[36,37] and mesostructured silica[38] using ILs as templates or the use of ILs as solvents in the synthesis of silica aerogels.[39,40] The problem associated with the synthesis of siliceous zeolites from ILs can be attributed to the poor solubility of the silica starting materials in commonly used ILs. Only recently one report of the synthesis of pure silica zeolites using what is called "task specific ionic liquid",[41] composed of a imidazolium cation and a mixed bromide–hydroxide anion in approximately 3 to 1 ratio, BMIM $OH_{0.65}Br_{0.35}$. It appears that silica precursors show solubility in this IL and this resulted in the crystallisation of a purely silica zeolite. Phase pure Silicate-1 (MFI) and a mixed phase product containing Silicate-1 (MFI) and Theta-1 (TON) can be prepared under the same conditions (Figure 10.4). The design of this IL is based on the recent successful synthesis of silicon-based zeolites in the presence of an added template in its hydroxide form. There is still much scope to develop the concept of task specific ILs and tailoring the IL in order to allow a full solubility of the silicate starting materials; in order to achieve this, a detailed study of the chemistry of silica in ILs would be very useful and may allow the design of a more suitable IL for this purpose.

Another option for the synthesis of purely silicate zeolites is to combine the benefits of different methods as was the case in the synthesis of zeolite MFI using a Dry-Gel Conversion (DGC) method in ionic liquids under microwave radiation (Figure 10.5).[42]

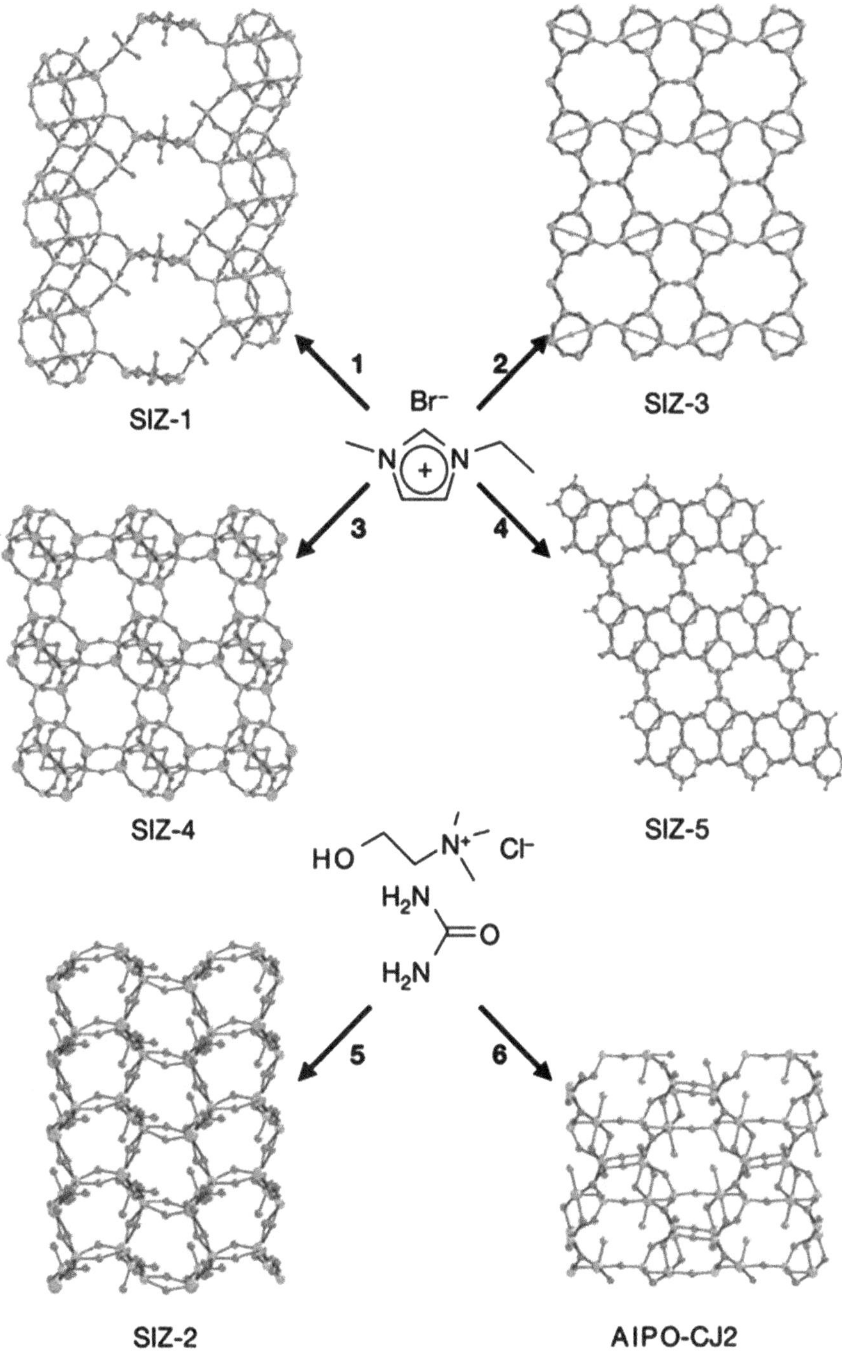

Figure 10.3 Ionothermal synthesis of AlPOs using different reaction conditions. SIZ-1, SIZ-3, SIZ-4 and SIZ-5 can be prepared using the IL EMIM Br under different reaction conditions (1–4), SIZ-2 and AlPO-CJ2 can be prepared using the deep eutectic solvent "choline chloride and urea" under different reaction conditions (1–2).

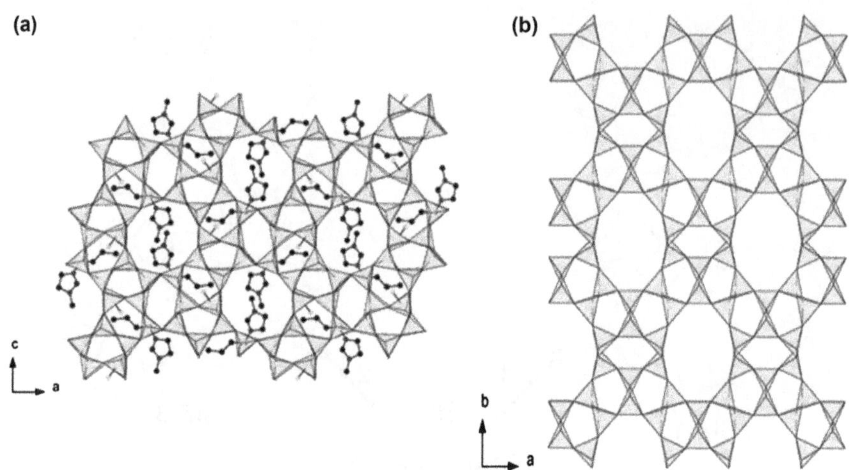

Figure 10.4 The crystal structure of: (a) [BMIM]-Silicalite-1, (b) [BMIM]-Theta-1.[41]

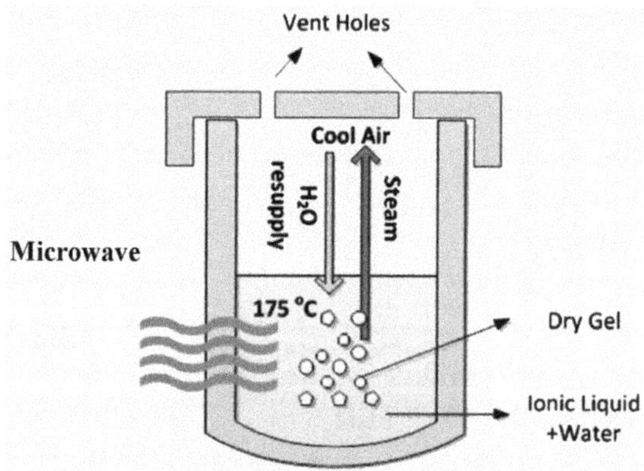

Figure 10.5 Dry-Gel Conversion method.
Figure reproduced by kind permission from reference 42.

Ionothermal synthesis was recently proved to be an efficient method for the synthesis of transition metal phosphates. Cobalt, manganese and iron phosphates can all be prepared using deep eutectic mixtures as the reaction solvent.[43] Figure 10.6 shows a series of cobalt phosphates materials that are accessible under different reaction conditions. Such materials are very interesting due to their electronic and potential catalytic properties.[44,45]

10.5 Ionothermal Synthesis of Metal–Organic Frameworks

Metal–organic frameworks (MOFs),[46,47] consisting of metal ions coordinated to organic molecules, are one class of organic–inorganic hybrid materials

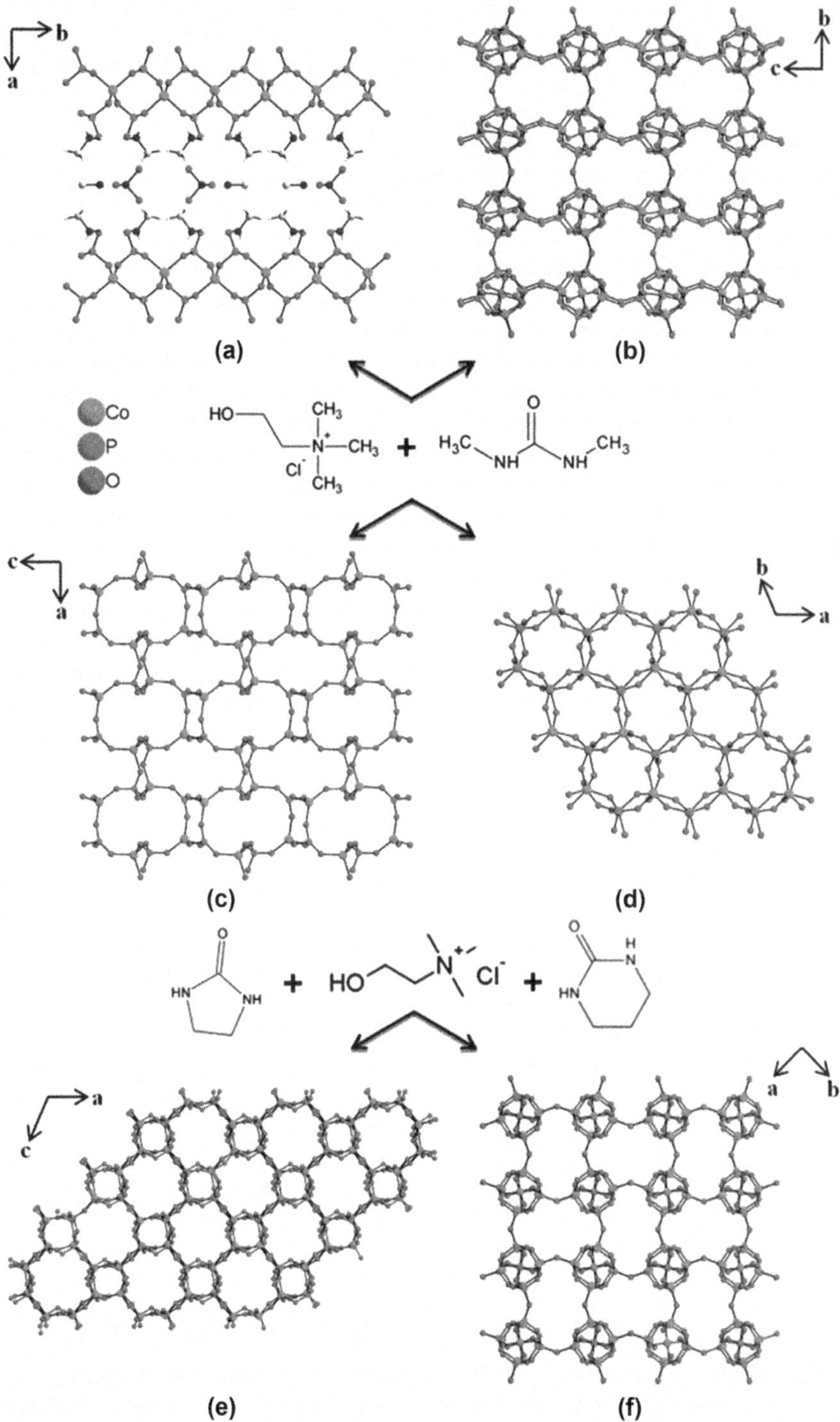

Figure 10.6 Framework structures for the ionothermally-prepared cobalt phosphates. Figure reproduced by kind permission from reference 43.

that have attracted much attention in recent years due to their potential applications in a wide variety of domains, most famously in gas adsorption and storage applications.[48–51] Recently, the potential biological and medical applications of MOFs have also been studied.[52] Normally, metal–organic frameworks are prepared using solvothermal reactions with organic solvents such as alcohols and dimethyl formamide. Over the past few years, there has been much success in the synthesis of these type of materials using ILs as the solvent and the template, and there are now many examples in the literature.[53–64] Unlike zeolites, the lower thermal stability of metal–organic frameworks leads to several issues regarding removal of the ionic templates from the ionothermally-prepared MOF. In most cases the metal–organic framework is anionic and the charge balance comes from the occluded IL cation. It was in general observed that even if a porous material can be achieved under ionothermal synthesis, the strong electrostatic interactions between the IL cation and the framework make removal of the cations to leave a true porous solid more difficult than in many other cases. Often removing the IL cation is not possible without collapsing the structure. This precludes most of the ionothermally-prepared MOFs having a proper practical interest in gas adsorption or other applications. However, porous MOFs can be prepared using deep eutectic mixtures, and a recent work by Bu and co-workers[65] has proven this very elegantly. Different types of deep eutectic mixtures based on choline chloride and urea derivatives have been used as the reaction solvent to produce different MOFs. All urea derivatives used did not decompose during the synthesis and remain intact; however the resulting materials are templated in different ways as shown in Figure 10.7. Urea or urea derivative have a strong tendency to bind to metal sites; removing the coordinated urea (or urea derivatives) led to the creation of porosity and open metal sites that might be of particular interest.

10.6 Extending Ionothermal Synthesis for the Preparation of Other Types of Materials

While the pioneering work on ionothermal synthesis targeted porous materials including zeolites and metal–organic frameworks, there is of course no reason why this technique should be limited only to the synthesis of these materials. Ionothermal synthesis can be extended and used as a general synthetic technique that can be applied to the preparation of other type of materials.

10.6.1 Ionothermal Synthesis of Hybrid Metal Fluorides

Ionothermal synthesis has been extended for the synthesis of hybrid transition metal fluorides where it shows interesting features. Using the hydrophobic IL 1-ethyl-3-methylimidazolium bis(trifluoromethylsulfonyl)imide,

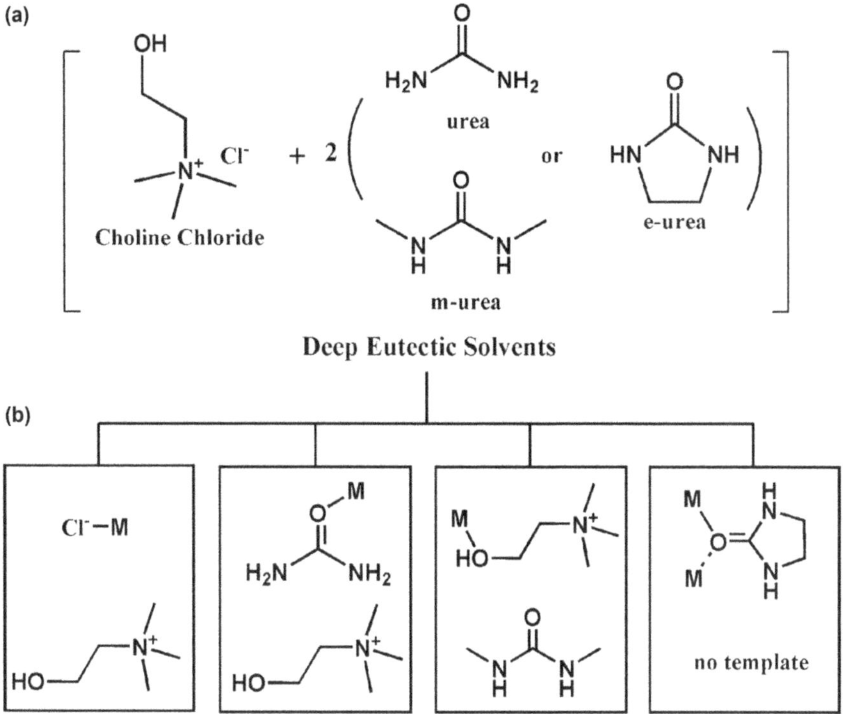

Figure 10.7 (a) Deep eutectic mixtures used, (b) their multiple roles, (M = metal). Figure reproduced by kind permission from reference 65.

[EMIM][Tf_2N], produced the first example of an organically-templated vanadium (oxy)fluoride with infinite two-dimensional V–F–V connectivity.[66] This was followed by the successful recent synthesis of a new quantum-spin-liquid candidate[67] containing a frustrated magnetic spin $\frac{1}{2}$ kagome network of d^1 V^{4+} ions. The specific properties of this IL, [EMIM][Tf_2N], make it the right medium to enable the isolation of extended solids containing V^{4+} ions that are otherwise difficult to prepare. Most V^{4+} containing solids forming low dimensional solids are prepared at low reaction temperatures; at higher temperatures in most cases V^{4+} will be further reduced to V^{3+}.[68–70] Control of the reduction of vanadium ions at higher temperatures is the key factor that may enable the preparation of extended structures and the discovery of new materials with the desired two-dimensional structures that are necessary for spin frustration to occur. The structure of $(NH_4)_2(C_7H_{14}N)(V_7O_6F_{18})$ (Figure 10.8) consists of isolated pillared double layers of stoichiometry $(V_7O_6F_{18})^{3-}$. These layers are built of two V^{4+}-containing kagome sheets that are pillared by an octahedral V^{3+} ion. Within each of these layers lie the quinuclidinium cations. The compound exhibits a high degree of magnetic frustration with significant antiferromagnetic interaction but no long range magnetic ordering or spin-freezing down to 2 K.

Figure 10.8 A view of $(NH_4)_2(C_7H_{14}N)(V_7O_6F_{18})$ (left) and the pillared double layer (right).

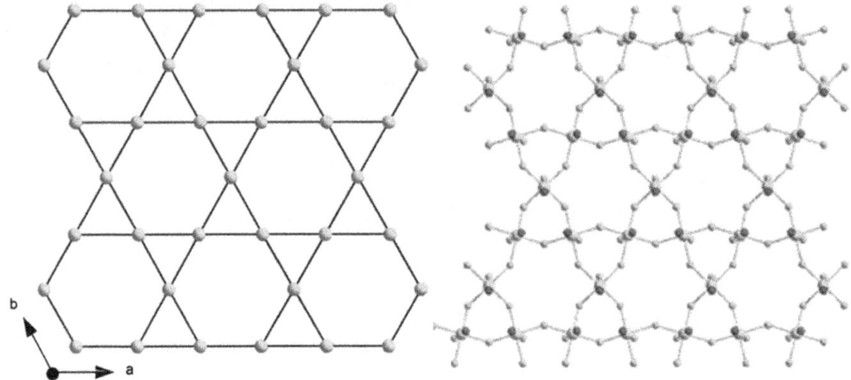

Figure 10.9 The kagome lattice (left) and the V^{4+} containing kagome lattice found in $(NH_4)_2(C_7H_{14}N)(V_7O_6F_{18})$ (right).

Kagome antiferromagnets (Figure 10.9) have been attracting considerable attention, especially those based on $S = \frac{1}{2}$ as they are the most likely candidates to display a quantum spin liquid phase (QSL).[71,72] When the frustrated network is composed of magnetic spins with spin $= \frac{1}{2}$ there is the possibility of strong quantum interactions between them, which should theoretically lead to a QSL. On the basis of this, most theoretical investigations of QSL have been concentrated on kagome lattices with $S = \frac{1}{2}$. In practice, however, only a few materials[73] can be regarded as perfect $S = \frac{1}{2}$ kagome antiferromagnets and candidates for QSL.

10.6.2 Ionothermal Synthesis of Other Solids

A wide range of other solids have been synthesised using ILs, which proves the versatility of ionothermal synthesis and its wide applicability. Most interestingly is the use of ILs in the preparation of polyoxometalates (POMs)[74–76] a family of metal oxide clusters that potentially can serve as building units for the creation of multifunctional materials, and the synthesis of polyoxometalate-based metal–organic frameworks (PMOFs).[77,78] Zeolite imidazolate frameworks (ZIFs)[79–81] and metal phosphite[82] with extra-large pores have also been successfully prepared under ionothermal conditions.

In addition, ionothermal synthesis has also been widely used for the synthesis of other materials from nanoparticles of different types,[83–89] to electrode materials[90–93] and purely organic solids including porous organic polymers (POPs).[94,95]

10.7 Ionothermal Synthesis and Microwaves

ILs are solvents with very low vapour pressure. This means that, unlike molecular solvents such as water, the IL can be heated to relatively high temperatures without the production of autogeneous pressure. High temperature reactions can therefore be carried out in simple containers such as round bottomed flasks. This important feature makes microwave heating a safe prospect as hot spots in the liquid should not cause excessive increases in pressure with their associated risk of explosion; the IL of course should be stable and not break down at the reaction conditions. The characteristic properties of ionic liquids including high ionic conductivity and polarisability make them excellent at absorbing microwaves. Combining ILs and microwave heating provides a very safe and eco-efficient method for the synthesis of materials. Several materials have been ionothermally prepared using microwave irradiation.[96–101] It was in general observed that the reaction time would be dramatically reduced from several days to minutes and products could be prepared with high levels of crystallinity and purity (Figure 10.10).

Ionothermal synthesis associated with microwave heating is an extremely useful technique for processing zeolites as thin films and coatings.[102,103] Zeolite coatings have good mechanical and thermal properties and are effective at protecting against corrosion; they are an excellent replacement for the most commonly used coatings based on chromate which are toxic and carcinogenic. The low pressure of ionothermal synthesis makes it possible to overcome the inconvenience of the previously used hydrothermal deposition process for zeolite coatings that involves autogeneous pressure, and allows the preparation of excellent coatings (Figure 10.11). However, in order to allow the microwave reactions to be carried out under ambient pressure, the IL has to be pure, dry and also stable and must not break down under the reaction conditions; as shown in Figure 10.12, a significant pressure can be generated even in the presence of a small amount of water in the system.[98]

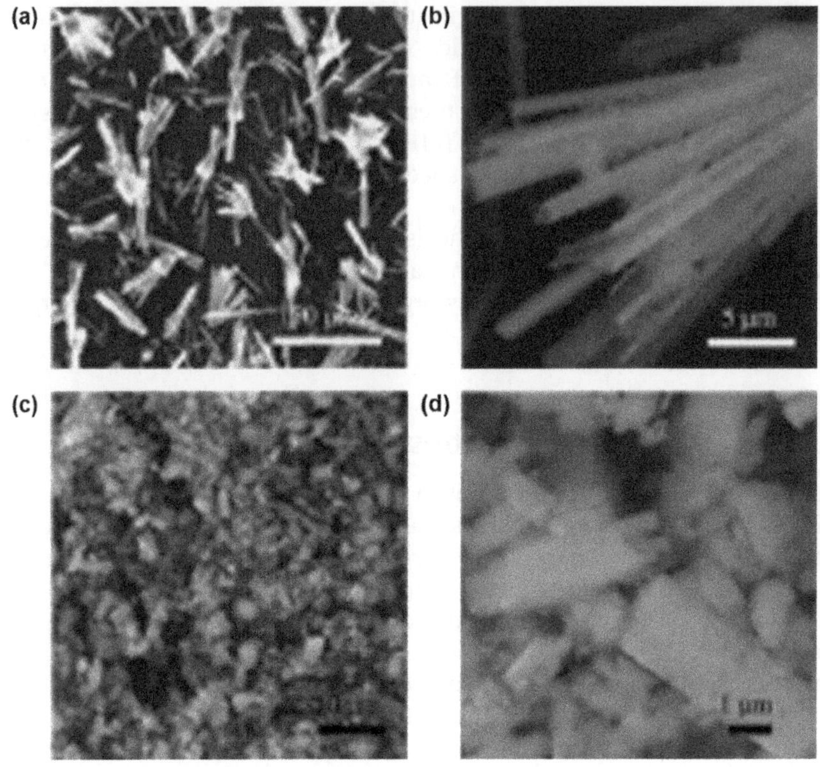

Figure 10.10 SEM images of AlPOs (AEL type): (a) and (b) samples after 68 h crystallisation with conventional heating, (c) and (d) samples after 20 minutes crystallisation with microwave heating.
Figure reproduced by kind permission from reference 99.

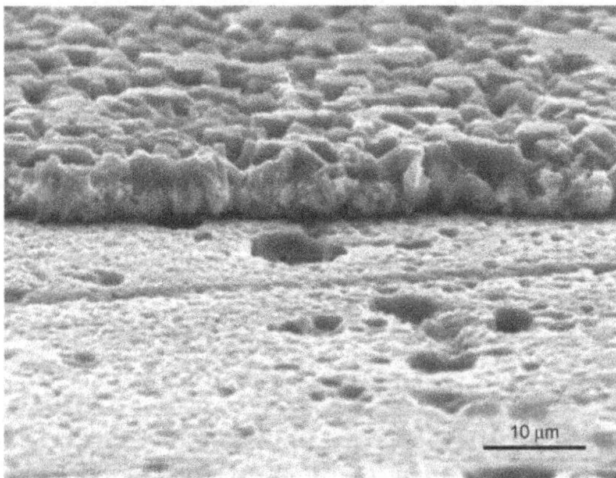

Figure 10.11 SEM micrograph showing a cross section of the SAPO-11 coating on aluminium alloy.
Figure reproduced by kind permission from reference 102.

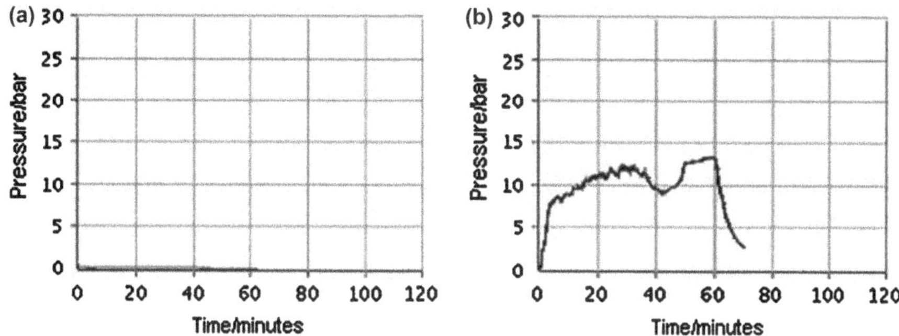

Figure 10.12 The evolution of pressure in the microwave synthesis of SIZ-4; (a) synthesis using pure ionic liquid with no added water, (b) synthesis with adding 0.018 mL of water.

10.8 The Role of ILs in Ionothermal Synthesis of Materials

In addition to its principal role as the reaction solvent, the original idea about ionothermal synthesis was to simplify the reaction for aluminophosphate synthesis by making the solvent and the template the same species. However as the range of ionothermally-prepared materials has been expanded, ILs show different behaviour in materials synthesis ranging from simply acting as a solvent to acting as a template, co-template, reactant and even as oxidation state stabiliser and catalyst. The role of an IL in ionothermal reactions can be broadly classified into two categories; where the IL plays an effective role and where the IL is just acting as solvent.

10.8.1 IL Plays an Effective Role

10.8.1.1 Cationic or Anionic Templating

In the ionothermal synthesis of zeolites and metal–organic frameworks,[96,104] an IL acts as both the solvent and the template, where the resulting material is almost always anionic and charge balancing is maintained by the organic cation. While it is not really clear if the cation plays the role of "true templating" or just as "space filling". Although changing the cation size, as shown in Figure 10.13, would influence the final structure, the larger cations form more open frameworks with extra space needed to accommodate the large template.[7]

ILs can also play the role of anionic template; where the inorganic framework is positively charged, the ionic liquid anion is occluded into the structure to balance the charge. One example can be found in copper and cadmium coordination polymers synthesised using [BMIM][BF_4] as the solvent; these materials are templated by tetrafluoroborate ([BF_4]$^-$) anion.[105,106]

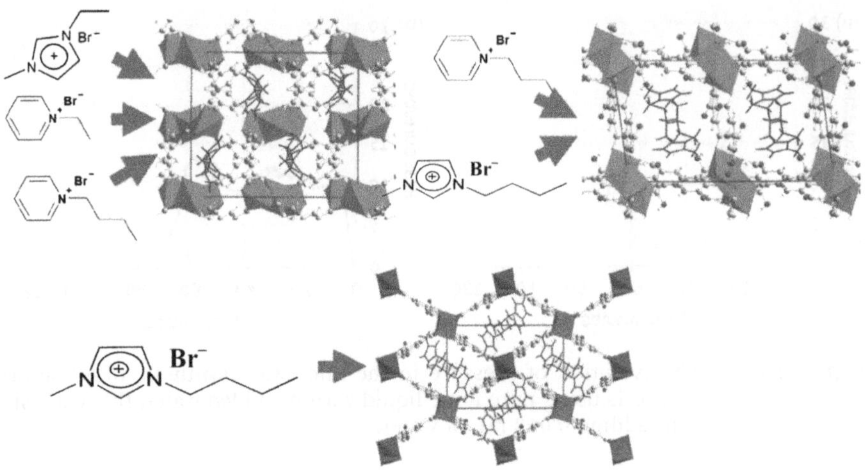

Figure 10.13 Preparation of Ni (top) and cobalt (bottom) terephthalate MOFs using ILs with different cation size.

10.8.1.2 Co-templating

Co-templating synthetic method was used in hydrothermal or solvothermal reactions where two or more organic amines or quaternary ammonium ions were added to act as the templates or structure directing agents. This has been proven to be an effective route for the preparation of many novel materials.[107,108] Of course this is no exception and the situation is exactly the same in ionothermal synthesis; the added templates offer great opportunities and open a new route to target new microporous materials. The addition of methylimidazole (MIA) to an IL 1-ethyl-3-methylimidazolium bromide ([EMIM]Br) produced a new material characterised by two distinct layers templated by both the ionic liquid cation ($[EMIM]^+$) and the added template.[109] In the absence of MIA this material could not prepared, indicating that both templates play a cooperative role for the crystallisation of this material.

Another interesting example in co-templating in ionothermal synthesis is the preparation of an aluminophosphate molecular sieve (DNL-1) analogue to cloverite[110] with 20-ring pore (Figure 10.14). This material was prepared in the IL [EMIM]Br and an added organic amine 1,6-hexanediamine (HDA) to act as the co-SDA. The Rietveld refinement of the PXRD of DNL-1 and multinuclear NMR analysis confirm that this material is isostructural to cloverite with comparable unit cell and space group. Both templates are occluded into the final structure and the HDA is essential during the synthesis. This material is characterised by higher thermal stability compared with cloverite and high surface area and micropore volume, which offer great opportunity for many potential applications in separation, catalysis and gas storage.

Figure 10.14 A view of the framework structure of DNL-1.
Figure reproduced by kind permission from reference 110.

10.8.1.3 IL Anion for Structure Induction and Phase Selection

In most ionothermally-synthesised materials including zeotypes and MOFs, the IL cations have a structure directing effect; this is not surprising since the IL cations are similar to many templates used for the hydrothermal synthesis of these types of materials. However not only the cation affects the resulting materials, the IL anion is an extremely important component of an IL and changing the anion would change the properties of the IL and therefore would potentially lead to different materials. Altering the IL anion dramatically changes the IL properties. For example, combining the $[BMIM]^+$ cation with three different anions (Br^-, $[BF_4]^-$, $[Tf_2N]^-$) produces three ILs with completely different properties, especially when it comes to their interaction with water. [BMIM]Br is a highly moisture-sensitive solid, $[BMIM][BF_4]$ is a liquid miscible with water, however, $[BMIM][Tf_2N]$ is a relatively hydrophobic liquid. From this, it is clear that changing the IL anion dramatically affects the properties of the resulting IL; this in turn would have a significant impact on the products of reactions carried out using such solvents. As stated in the previous section, the early work on ionothermal synthesis was devoted to the preparation of aluminophosphates. In such systems, the use of [EMIM]Br as the reaction solvent produced several zeotype materials[27] where the $[EMIM]^+$ cation is usually incorporated into the structure (see Figure 10.3), whereas the use of $[EMIM][Tf_2N]$ led to a new structure that has no $[EMIM]^+$ cation.[111] Another example of interest in ionothermal synthesis which shows the effect of changing the IL anion can be found in cobalt benzenetricarboxylate

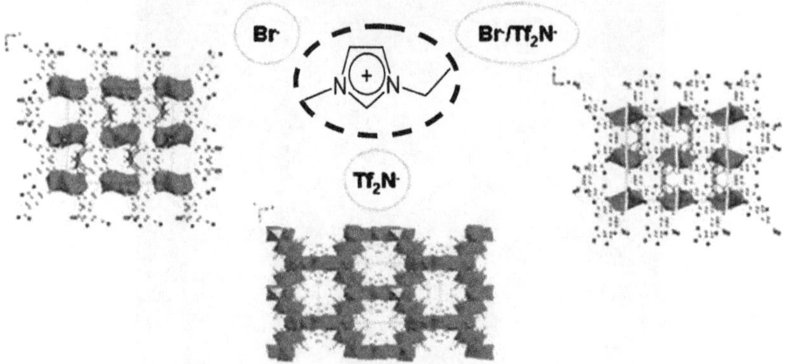

Figure 10.15 The effect of the IL anion on the resulting cobalt benzenetricarboxylate MOF structures.

MOFs,[112] again [EMIM]Br and [EMIM][Tf_2N] led to the isolation of two different types of materials, moreover a 50 : 50 mixture of both ILs led to a third structure type (Figure 10.15). These results demonstrate that changing the chemistry of the IL, by changing the anion or mixing two different ILs, has a great effect on the resulting materials.

The use of a chiral anion for the preparation of a chiral coordination polymer is another supporting example that clearly shows a direct impact of the IL anion on the structure. Morris and co-workers,[113] by combining BMIM cation with L-aspartate anion, have produced a nickel benzenetricarboxylate MOF with a chiral structure containing only achiral building blocks (Figure 10.16). While the IL anion itself is not occluded into the final structure, it has a conspicuous effect on the final product; this "induction effect" is an important feature of ionothermal synthesis and also an attractive area of research, and needs further studies and exploitations.

An important work by Kwon and co-workers[114] on nickel–base metal–organic frameworks, shows that the nature of IL halide ions governs the kinetic factors which favour one structure over the other depending on the type of halide used: Cl, Br, I. As shown in Table 10.1 five structures can be isolated using various methylimidazolium-based ILs in combination with three different halides. These structures fall into two major categories, **A** and **B**; the main difference between **A** and **B** is in the size of their cavities and also the connectivity modes of the BTC ligand. This study shows that a combination effect of the cation size and the nature of the halide used determines the final product, despite the fact that only the ionic liquid cations appear in the final structures.

10.8.1.4 Other Possible Roles

Like any other solvent, including water, there is the possibility of bonding interactions of the IL with the frameworks. However, most ILs used in ionothermal synthesis of materials are composed of dialkylimidazolium cations

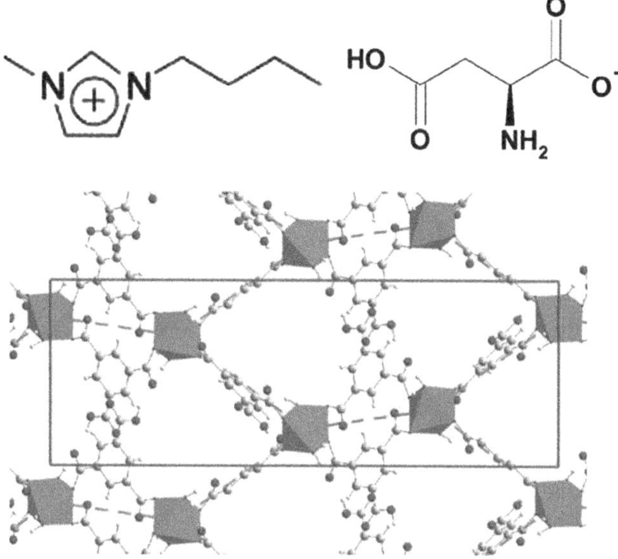

Figure 10.16 The chiral IL used by Morris and co-workers and the resulting chiral nickel metal–organic framework.

Table 10.1 Effect of the ionic liquid anion and cation on the resulting nickel benzenetricarboxylate MOF structures.[114] (**A**: $[RMIM]_2[Ni_3(BTC)_2(OAc)_2]$, **B**: $[RMIM]_2[Ni_3(HBTC)_4(H_2O)]$.

	Anion		
[RMIM]$^+$	*Cl*$^-$	*Br*$^-$	*I*$^-$
$[EMIM]^+$	$\mathbf{A_1}$	$\mathbf{A_1}$	$\mathbf{A_1}$
$[PMIM]^+$	$\mathbf{A_2}$	$\mathbf{A_2}$	$\mathbf{B_1}$
$[BMIM]^+$	$\mathbf{A_3}$	$\mathbf{B_2}$	$\mathbf{B_2}$

that lack coordinating sites. Thus, in ionothermal synthesis, dialkylimidazolium cation has acted only as a template. There is, of course, the possibility of creating coordinating sites by potential use of IL cations with functional groups. Alternatively deep eutectic solvents (DESs) that contain coordinating sites may be also used, as an example a choline chloride–dimethyurea (DMU)-based DES was used in the preparation of lanthanum-based MOFs,[115] $La(C_9O_6H_3)DMU_2$, where DMU molecules coordinate to the metal centre.

The halide-based IL anion, in some systems, may also coordinate to the metal; SIZ-13 a cobalt aluminophosphate layered material has Co–Cl bonds. Chloride ions came from choline chloride that had been used along with carboxylic acid as the reaction solvent.[33]

Another interesting role of ILs was noted in the synthesis of transition metal fluorides and it was found that the highly structured reaction environment imparted by the IL enables the stabilisation of metal ions at higher oxidation states even at elevated temperatures that allow the preparation of extended structures where the metal ions remain in higher oxidation states.[66,67]

10.8.1.5 The Hydrolysis and the In-Situ Breakdown of Unstable Ionic Liquids

While many ILs are characterised by their high chemical stability, under ionothermal conditions some of them are unstable, even those that are often relatively stable such as butylmethylimidazolium bromide. IL cations can break down especially in the presence of fluoride ions and water, *e.g.* alkylmethylimidazolium cation breaks down into dimethyimidazolium, which then templates the structure.[116] Or, under other specific conditions, the IL cation can even break down to monoalkylated imidazole species that can coordinate to metals.[117] Like the cations, some IL anions especially fluorinated anions (*e.g.* $[BF_4]^-$, $[PF_6]^-$) are also unstable and can undergo fast hydrolysis under special conditions.[118,119]

This property has been recently well exploited and used for the creation of interesting materials,[107] where [BMIM]$[BF_4]$ plays multifunctional roles, solvent, anionic template, fluoride source and catalyst, for the preparation of fluorinated cadmium metal–organic frameworks.

Deep eutectic solvents based on choline chloride and urea derivatives are also unstable under some conditions—presence of fluoride, high temperatures—and break down in the same way to release the corresponding alkylammonium ions that template the final structure. This interesting property was used for the creation of several structure types including metal phosphates[120] and metal fluorides (Figure 10.17).[121]

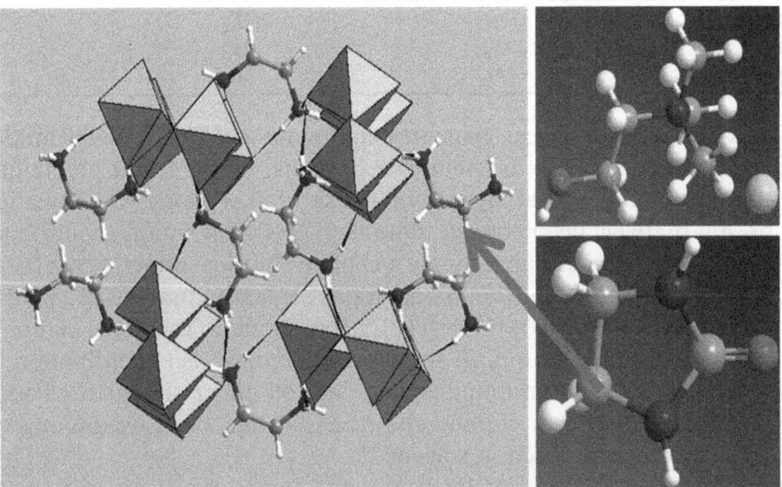

Figure 10.17 Structure of a vanadium oxyfluoride material synthesised using a choline chloride–2-imidazolidinone deep eutectic mixture; 2-imidazolidinone breaks down to ethylenediamine.

10.8.2 ILs as Solvent Only

ILs, again like any other solvent, can simply play the role of solvent only and not be occluded in the final structure at all. In the case of aluminophosphate and MOF syntheses, it was in general noted that the more hydrophobic the IL used the less likely the IL cation is to be occluded.[7]

10.9 Effect of Water and Other Mineralisers

Since the early work on ionothermal synthesis of materials, it was in general found that, while zeotypes and other materials can be synthesised without any added water, the presence of water significantly affects the crystallisation kinetics and phase selectivity. Thus, it might be possible to control the reaction products by adding specific amounts of water. Studies in this area demonstrate that it is not always easy to predict the reaction product according to the amount of water present.[99,122] However, they clearly show that too much water is detrimental to the formation of zeolites as at low concentrations of water zeolite type materials are the main products, while at increased levels of water only dense phases are produced.

The underlying reason behind this water effect is still under investigation but it is known that the microstructure of water in ILs does change with concentration.[123–125] As shown in Figure 10.18 at low concentration water molecules are isolated from each other and exist as isolated water molecules or as very small clusters dispersed in a continuous polar network formed by the IL ions. However, as the concentration of water increases larger clusters and eventually continuous hydrogen bonded water networks start to build up which dramatically change the properties of the IL–water mixture. Finally, as more and more water is added, the IL polar network in water-richer composition starts to break apart to form small ion clusters in a

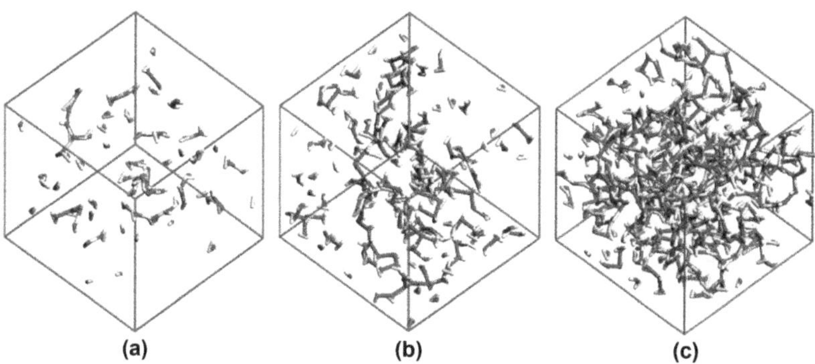

Figure 10.18 Microstructure of water in IL at different water concentrations "X": (a) $X=0.5$, (b) $X=0.8$, (c) $X=0.92$.
Figure reproduced by kind permission from reference 123.

continuous water phase and obviously the system becomes hydrothermal rather than ionothermal.

Fluorides and other mineralisers such as hydroxides are very important additives for the synthesis of microporous materials of different types. In particular, the fluoride route has been successfully applied for the hydro/solvothermal synthesis of aluminophosphates[124] and silicate.[125,126] In addition to helping solubilise the starting materials, fluoride ions can play several other roles such as framework charge balancing, catalysing the formation of some bonds like Al–O–P and they can even play the role of a structure directing agent,[127–129] where they are incorporated into the framework or occluded within small cages. The addition of fluoride in ionothermal synthesis is extremely important in determining the phase selectivity of the reaction.[27] In the synthesis of aluminophosphates, it was in general observed that the fluoride-free ionothermal synthesis produces mainly low dimensional or interrupted structures, while the addition of fluorides leads to the preparation of structures with fully connected frameworks.[7,130]

10.10 Conclusions and Perspectives

Ionic liquids have entered the scene as green alternative solvents in various fields of chemistry. In material sciences, using ionic liquids as solvents is mainly associated with eliminating the safety issues of high-pressure hydro/solvothermal reactions and simplifying the reaction. More importantly, using ionothermal as the synthesis method often results in new materials with characteristics that can be traced back to the specific IL used. One of the important and very revealing properties that differentiates the ionothermal synthesis method from the other synthesis techniques is the unique reaction environment dictated by the IL; this offers many possibilities for the preparation of important materials that are unlikely to be obtained otherwise.

Clearly, the flexibility and the possibility of designing ILs and the large number of ILs available offer great opportunities for the synthesis of other materials for specific applications and functionalities. In theory ILs can be designed to deliver almost any set of physical and chemical properties for almost any application in the chemical sciences. Hence the term "designer solvents" is often used when describing ILs. However, the tailoring or designing of ILs requires a full knowledge of patterns and trends in the characteristics of ILs, which are limited in most types. Thus, in practice ILs have been actually selected rather than designed for specific applications and this selection has often been based on trial and error. This selection approach might be very useful for high throughput methodologies.

The use of ILs as solvents has proven to be a versatile route to the preparation of many different types of materials, and this field is still wide open for further explorations and studies.

References

1. R. D. Rogers and K. R. Seddon, *Science*, 2003, **302**, 792.
2. L. A. D. Blanchard Hancu, E. J. Beckman and J. F. Brennecke, *Nature*, 1999, **399**, 28.
3. D. J. Cole-Hamilton, *Science*, 2003, **299**, 1702.
4. M. J. Earle, J. Esperanca, M. A. Gilea, J. N. C. Lopes, L. P. N. Rebelo, J. W. Magee, K. R. Seddon and J. A. Widegren, *Nature*, 2006, **439**, 831.
5. S.-L. Chou, J.-Z. Wang, J.-Z. Sun, D. Wexler, M. Forsyth, H.-K. Liu, D. R. MacFarlane and S.-X. Dou, *Chem. Mater.*, 2008, **20**, 7044.
6. W. Miao and T. H. Chan, *Acc. Chem. Res.*, 2006, **39**, 897.
7. R. E. Morris, *Chem. Commun.*, 2009, 2990.
8. R. E. Morris, *Angew. Chem., Int. Ed.*, 2008, **47**, 442.
9. R. M. Barrer, *Trans. Faraday Soc.*, 1943, **39**, 59.
10. J. Dupont, *Acc. Chem. Res.*, 2011, **44**, 1223.
11. S. J. Mugavero Iii, M. Bharathy, J. McAlum and H.-C. zur Loye, *Solid State Sci.*, 2008, **10**, 370.
12. P. Wasserscheid and T. Welton, *Ionic Liquids in Synthesis*, Wiley-VCH, Weinheim, 2003.
13. J. S. Wilkes, *Green Chem.*, 2002, **4**, 73.
14. J. N. A. Canongia Lopes and A. A. H. Pádua, *J. Phys. Chem. B*, 2006, **110**, 3330.
15. C. S. Consorti, P. A. Z. Suarez, R. F. de Souza, R. A. Burrow, D. H. Farrar, A. J. Lough, W. Loh, L. H. M. da Silva and J. Dupont, *J. Phys. Chem. B*, 2005, **109**, 4341.
16. J. Dupont, P. A. Z. Suarez, R. F. de Souza, R. A. Burrow and J.-P. Kintzinger, *Chem.–Eur. J.*, 2000, **6**, 2377.
17. K. Fumino, E. Reichert, K. Wittler, R. Hempelmann and R. Ludwig, *Angew. Chem., Int. Ed.*, 2012, **51**, 6236.
18. P. Nockemann, B. Thijs, S. Pittois, J. Thoen, C. Glorieux, K. Van Hecke, L. Van Meervelt, B. Kirchner and K. Binnemans, *J. Phys. Chem. B*, 2006, **110**, 20978.
19. W. M. Reichert, J. D. Holbrey, K. B. Vigour, T. D. Morgan, G. A. Broker and R. D. Rogers, *Chem. Commun.*, 2006, 4767.
20. M. J. Earle, J. M. S. S. Esperanca, M. A. Gilea, J. N. Canongia Lopes, L. P. N. Rebelo, J. W. Magee, K. R. Seddon and J. A. Widegren, *Nature*, 2006, **439**, 831.
21. A. P. Abbott, G. Capper, D. L. Davies, R. K. Rasheed and V. Tambyrajah, *Chem. Commun.*, 2003, 70.
22. A. P. Abbott, D. Boothby, G. Capper, D. L. Davies and R. K. Rasheed, *J. Am. Chem. Soc.*, 2004, **126**, 9142.
23. D. Carriazo, M. C. Serrano, M. C. Gutierrez, M. L. Ferrer and F. del Monte, *Chem. Soc. Rev.*, 2012, **41**, 4996.
24. C. S. Cundy and P. A. Cox, *Chem. Rev.*, 2003, **103**, 663.
25. R. E. Morris and S. J. Weigel, *Chem. Soc. Rev.*, 1997, **26**, 309.
26. H. Luo, G. A. Baker and S. Dai, *J. Phys. Chem. B*, 2008, **112**, 10077.

27. E. R. Cooper, C. D. Andrews, P. S. Wheatley, P. B. Webb, P. Wormald and R. E. Morris, *Nature*, 2004, **430**, 1012.
28. R. F. Lobo, S. I. Zones and M. E. Davis, *J. Inclusion Phenom. Mol. Recognit. Chem.*, 1995, **21**, 47.
29. L. Han, Y. Wang, C. Li, S. Zhang, X. Lu and M. Cao, *AIChE J.*, 2008, **54**, 280.
30. L. Liu, Y. Kong, H. Xu, J. P. Li, J. X. Dong and Z. Lin, *Microporous Mesoporous Mater.*, 2008, **115**, 624.
31. E. R. Parnham, P. S. Wheatley and R. E. Morris, *Chem. Commun.*, 2006, 380.
32. X. Zhao, H. Wang, C. Kang, Z. Sun, G. Li and X. Wang, *Microporous Mesoporous Mater*, 2012, **151**, 501.
33. E. A. Drylie, D. S. Wragg, E. R. Parnham, P. S. Wheatley, A. M. Z. Slawin, J. E. Warren and R. E. Morris, *Angew. Chem., Int. Ed.*, 2007, **46**, 7839.
34. E. R. Parnham and R. E. Morris, *J. Am. Chem. Soc.*, 2006, **128**, 2204.
35. J. M. Griffin, L. Clark, V. R. Seymour, D. W. Aldous, D. M. Dawson, D. Iuga, R. E. Morris and S. E. Ashbrook, *Chem. Sci.*, 2012, **3**, 2293.
36. Y. Lorgouilloux, M. Dodin, J.-L. Paillaud, P. Caullet, L. Michelin, L. Josien, O. Ersen and N. Bats, *J. Solid State Chem.*, 2009, **182**, 622.
37. X. Sun, J. King and J. L. Anthony, *Chem. Eng. J.*, 2009, **147**, 2.
38. T. Wang, H. Kaper, M. Antonietti and B. Smarsly, *Langmuir*, 2006, **23**, 1489.
39. S. Dai, Y. H. Ju, H. J. Gao, J. S. Lin, S. J. Pennycook and C. E. Barnes, *Chem. Commun.*, 2000, 243.
40. M. A. Klingshirn, S. K. Spear, J. D. Holbrey and R. D. Rogers, *J. Mater. Chem.*, 2005, **15**, 5174.
41. P. S. Wheatley, P. K. Allan, S. J. Teat, S. E. Ashbrook and R. E. Morris, *Chem. Sci.*, 2010, **1**, 483.
42. R. Cai, Y. Liu, S. Gu and Y. Yan, *J. Am. Chem. Soc.*, 2010, **132**, 12776.
43. B. T. Yonemoto, Z. Lin and F. Jiao, *Chem. Commun.*, 2012, **48**, 9132.
44. J. A. Armstrong, E. R. Williams and M. T. Weller, *J. Am. Chem. Soc.*, 2011, **133**, 8252.
45. A. K. Cheetham, G. Férey and T. Loiseau, *Angew. Chem., Int. Ed.*, 1999, **38**, 3268.
46. G. Ferey, *Chem. Soc. Rev.*, 2008, **37**, 191.
47. S. Kitagawa, R. Kitaura and S.-i. Noro, *Angew. Chem., Int. Ed.*, **43**, 2334.
48. R. Banerjee, A. Phan, B. Wang, C. Knobler, H. Furukawa, M. O'Keeffe and O. M. Yaghi, *Science*, 2008, **319**, 939.
49. R. E. Morris and P. S. Wheatley, *Angew. Chem., Int. Ed.*, 2008, **47**, 4966.
50. N. L. Rosi, J. Eckert, M. Eddaoudi, D. T. Vodak, J. Kim, M. O'Keeffe and O. M. Yaghi, *Science*, 2003, **300**, 1127.
51. B. Xiao, P. S. Wheatley, X. Zhao, A. J. Fletcher, S. Fox, A. G. Rossi, I. L. Megson, S. Bordiga, L. Regli, K. M. Thomas and R. E. Morris, *J. Am. Chem. Soc.*, 2007, **129**, 1203.

52. A. C. McKinlay, R. E. Morris, P. Horcajada, G. Férey, R. Gref, P. Couvreur and C. Serre, *Angew. Chem., Int. Ed.*, 2010, **49**, 6260.
53. S. Chen, J. Zhang and X. Bu, *Inorg. Chem.*, 2008, **47**, 5567.
54. T. Hogben, R. E. Douthwaite, L. J. Gillie and A. C. Whitwood, *CrystEngComm*, 2006, **8**, 866.
55. W.-J. Ji, Q.-G. Zhai, M.-C. Hu, S.-N. Li, Y.-C. Jiang and Y. Wang, *Inorg. Chem. Commun.*, 2008, **11**, 1455.
56. J.-H. Liao and W.-C. Huang, *Inorg. Chem. Commun.*, 2006, **9**, 1227.
57. J.-H. Liao, P.-C. Wu and Y.-H. Bai, *Inorg. Chem. Commun.*, 2005, **8**, 390.
58. J.-H. Liao, P.-C. Wu and W.-C. Huang, *Cryst. Growth Des.*, 2006, **6**, 1062.
59. Z. Lin, Y. Li, A. M. Z. Slawin and R. E. Morris, *Dalton Trans.*, 2008, 3989.
60. L. Xu, E.-Y. Choi and Y.-U. Kwon, *Inorg. Chem.*, 2007, **46**, 10670.
61. J. Zhang, S. Chen and X. Bu, *Angew. Chem., Int. Ed.*, 2008, **47**, 5434.
62. Q.-Y. Liu, Y.-L. Wang, N. Zhang, Y.-L. Jiang, J.-J. Wei and F. Luo, *Cryst. Growth Des.*, 2011, **11**, 3717.
63. J.-J. Wei, Q.-Y. Liu, Y.-L. Wang, N. Zhang and W.-F. Wang, *Inorg. Chem. Commun.*, 2012, **15**, 61.
64. Z.-F. Wu, B. Hu, M.-L. Feng, X.-Y. Huang and Y.-B. Zhao, *Inorg. Chem. Commun.*, 2011, **14**, 1132.
65. J. Zhang, T. Wu, S. Chen, P. Feng and X. Bu, *Angew. Chem., Int. Ed.*, 2009, **48**, 3486.
66. F. Himeur, P. K. Allan, S. J. Teat, R. J. Goff, R. E. Morris and P. Lightfoot, *Dalton Trans.*, 2010, **39**, 6018.
67. F. H. Aidoudi, D. W. Aldous, R. J Goff, A. M. Z. Slawin, J. P. Attfield, R. E. Morris and P. Lightfoot, *Nat. Chem.*, 2011, **3**, 801.
68. K. Adil, M. Leblanc, V. Maisonneuve and P. Lightfoot, *Dalton Trans.*, 2010, **39**, 5983.
69. D. W. Aldous, N. F. Stephens and P. Lightfoot, *Dalton Trans.*, 2007, 2271.
70. D. W. Aldous, N. F. Stephens and P. Lightfoot, *Dalton Trans.*, 2007, 4207.
71. H. Andrew, *J. Phys.: Condens. Matter*, 2004, **16**, S553.
72. A. P. Ramirez, *Annu. Rev. Mater. Sci.*, 1994, **24**, 453.
73. M. P. Shores, E. A. Nytko, B. M. Bartlett and D. G. Nocera, *J. Am. Chem. Soc.*, 2005, **127**, 13462.
74. E. Ahmed and M. Ruck, *Angew. Chem., Int. Ed.*, 2012, **51**, 308.
75. S. Lin, W. Liu, Y. Li, Q. Wu, E. Wang and Z. Zhang, *Dalton Trans.*, 2010, **39**, 1740.
76. A. S. Pakhomova and S. V. Krivovichev, *Inorg. Chem. Commun.*, 2010, **13**, 1463.
77. H. Fu, Y. Li, Y. Lu, W. Chen, Q. Wu, J. Meng, X. Wang, Z. Zhang and E. Wang, *Cryst. Growth Des.*, 2011, **11**, 458.
78. H. Fu, C. Qin, Y. Lu, Z.-M. Zhang, Y.-G. Li, Z.-M. Su, W.-L. Li and E.-B. Wang, *Angew. Chem., Int. Ed.*, 2012, **51**, 7985.

79. G. A. V. Martins, P. J. Byrne, P. Allan, S. J. Teat, A. M. Z. Slawin, Y. Li and R. E. Morris, *Dalton Trans.*, 2010, **39**, 1758.
80. L. Yang and H. Lu, *Chin. J. Chem.*, 2012, **30**, 1040.
81. L. Yang, H. Lu and S. Zhou, *Energy Technology 2012: Carbon Dioxide Management and Other Technologies*, M. D. Salazar-Villalpando, N. R. Neelameggham, D. P. Guillen, S. Pati and G. K. Krumdick (ed.), Wiley, Weinheim, 2012, 117.
82. H. Xing, W. Yang, T. Su, Y. Li, J. Xu, T. Nakano, J. Yu and R. Xu, *Angew. Chem., Int. Ed.*, 2010, **49**, 2328.
83. Y. Gao, A. Voigt, M. Zhou and K. Sundmacher, *Eur. J. Inorg. Chem.*, 2008, **2008**, 3769.
84. H. Kaper, M.-G. Willinger, I. Djerdj, S. Gross, M. Antonietti and B. M. Smarsly, *J. Mater. Chem.*, 2008, **18**, 5761.
85. L.-L. Li, W.-M. Zhang, Q. Yuan, Z.-X. Li, C.-J. Fang, L.-D. Sun, L.-J. Wan and C.-H. Yan, *Cryst. Growth Des.*, 2008, **8**, 4165.
86. M.-Y. Li, W.-S. Dong, C.-L. Liu, Z. Liu and F.-Q. Lin, *J. Cryst. Growth*, 2008, **310**, 4628.
87. Z.-X. Li, L.-L. Li, Q. Yuan, W. Feng, J. Xu, L.-D. Sun, W.-G. Song and C.-H. Yan, *J. Phys. Chem. C*, 2008, **112**, 18405.
88. X. Liu, J. Ma and W. Zheng, *Rev. Adv. Mater. Sci.*, 2011, **27**, 43.
89. J. Ma, T. Wang, X. Duan, J. Lian, Z. Liu and W. Zheng, *Nanoscale*, 2011, **3**, 4372.
90. P. Barpanda, N. Recham, J.-N. Chotard, K. Djellab, W. Walker, M. Armandn and J.-M. Tarascon, *J. Mater. Chem.*, 2010, **20**, 1659.
91. N. Recham, J. N. Chotard, J. C. Jumas, L. Laffont, M. Armand and J. M. Tarascon, *Chem. Mater.*, 2010, **22**, 1142.
92. N. Recham, J. Oro-Sole, K. Djellab, M. R. Palacin, C. Masquelier and J. M. Tarascon, *Solid State Ionics*, 2012, **220**, 47.
93. J.-M. Tarascon, N. Recham, M. Armand, J.-N. Chotard, P. Barpanda, W. Walker and L. Dupont, *Chem. Mater.*, 2010, **22**, 724.
94. S. Hug, M. E. Tauchert, S. Li, U. E. Pachmayr and B. V. Lotsch, *J. Mater. Chem.*, 2012, **22**, 13956.
95. P. Kuhn, M. Antonietti and A. Thomas, *Angew. Chem., Int. Ed.*, 2008, **47**, 3450.
96. Z. Lin, D. S. Wragg and R. E. Morris, *Chem. Commun.*, 2006, 2021.
97. L. Wang, Y.-P. Xu, Y. Wei, J. C. Duan, A.-B. Chen, B.-c. Wang, H.-j. Ma, Z.-j. Tian and L.-W. Lin, *J. Am. Chem. Soc.*, 2006, **128**, 7432.
98. D. S. Wragg, A. M. Z. Slawin and R. E. Morris, *Solid State Sci.*, 2009, **11**, 411.
99. Y.-P. Xu, Z.-J. Tian, S.-J. Wang, Y. Hu, L. Wang, B.-C. Wang, Y.-C. Ma, L. Hou, J.-Y. Yu and L.-W. Lin, *Angew. Chem., Int. Ed.*, 2006, **45**, 3965.
100. X. Zhao, C. Kang, H. Wang, C. Luo, G. Li and X. Wang, *J. Porous Mater*, 2011, **18**, 615.
101. X. Zhao, H. Wang, C. Kang, Z. Sun, G. Li and X. Wang, *Microporous Mesoporous Mater.*, 2012, **151**, 501.

102. R. Cai, M. Sun, Z. Chen, R. Munoz, C. O'Neill, D. E. Beving and Y. Yan, *Angew. Chem., Int. Ed.*, 2008, **47**, 525.
103. R. E. Morris, *Angew. Chem., Int. Ed.*, 2008, **47**, 442.
104. E. R. Parnham and R. E. Morris, *Acc. Chem. Res.*, 2007, **40**, 1005.
105. K. Jin, X. Huang, L. Pang, J. Li, A. Appel and S. Wherland, *Chem. Commun.*, 2002, 2872.
106. Z.-L. Xie, M.-L. Feng, B. Tan and X.-Y. Huang, *CrystEngComm*, 2012, **14**, 4894.
107. M. Castro, R. Garcia, S. J. Warrender, A. M. Z. Slawin, P. A. Wright, P. A. Cox, A. Fecant, C. Mellot-Draznieks and N. Bats, *Chem. Commun.*, 2007, 3470.
108. S. I. Zones, S.-J. Hwang and M. E. Davis, *Chem.–Eur. J*, 2001, **7**, 1990.
109. H. Xing, J. Li, W. Yan, P. Chen, Z. Jin, J. Yu, S. Dai and R. Xu, *Chem. Mater.*, 2008, **20**, 4179.
110. Y. Wei, Z. Tian, H. Gies, R. Xu, H. Ma, R. Pei, W. Zhang, Y. Xu, L. Wang, K. Li, B. Wang, G. Wen and L. Lin, *Angew. Chem., Int. Ed.*, 2010, **49**, 5367.
111. E. R. Parnham and R. E. Morris, *J. Mater. Chem.*, 2006, **16**, 3682.
112. Z. Lin, D. S. Wragg, J. E. Warren and R. E. Morris, *J. Am. Chem. Soc.*, 2007, **129**, 10334.
113. Z. Lin, A. M. Z. Slawin and R. E. Morris, *J. Am. Chem. Soc.*, 2007, **129**, 4880.
114. L. Xu, S. Yan, E.-Y. Choi, J. Y. Lee and Y.-U. Kwon, *Chem. Commun.*, 2009, 3431.
115. F. Himeur, I. Stein, D. S. Wragg, A. M. Z. Slawin, P. Lightfoot and R. E. Morris, *Solid State Sci.*, 2010, **12**, 418.
116. E. R. Parnham and R. E. Morris, *Chem. Mater.*, 2006, **18**, 4882.
117. P. J. Byrne, D. S. Wragg, J. E. Warren and R. E. Morris, *Dalton Trans.*, 2009, 795.
118. M. G. Freire, C. M. S. S. Neves, I. M. Marrucho, J. o. A. P. Coutinho and A. M. Fernandes, *J. Phys. Chem. A*, 2009, **114**, 3744.
119. S. Steudte, J. Neumann, U. Bottin-Weber, M. Diedenhofen, J. Arning, P. Stepnowski and S. Stolte, *Green Chem.*, 2012, **14**, 2474.
120. E. R. Parnham, E. A. Drylie, P. S. Wheatley, A. M. Z. Slawin and R. E. Morris, *Angew. Chem. Int. Ed.*, 2006, **45**, 4962.
121. F. H. Aidoudi, P. J. Byrne, P. K. Allan, S. J. Teat, P. Lightfoot and R. E. Morris, *Dalton Trans.*, 2011, **40**, 4324.
122. H. Ma, Z. Tian, R. Xu, B. Wang, Y. Wei, L. Wang, Y. Xu, W. Zhang and L. Lin, *J. Am. Chem. Soc.*, 2008, **130**, 8120.
123. C. E. S. Bernardes, M. E. Minas da Piedade and J. N. Canongia Lopes, *J. Phys. Chem. B*, 2011, **115**, 2067.
124. R. E. Morris, A. Burton, L. M. Bull and S. I. Zones, *Chem. Mater.*, 2004, **16**, 2844.
125. M. A. Camblor, L. A. Villaescusa and M. J. Díaz-Cabañas, *Top. Catal.*, 1999, **9**, 59.

126. S. I. Zones, R. J. Darton, R. Morris and S.-J. Hwang, *J. Phys. Chem. B*, 2004, **109**, 652.
127. I. Bull, L. A. Villaescusa, S. J. Teat, M. A. Camblor, P. A. Wright, P. Lightfoot and R. E. Morris, *J. Am. Chem. Soc.*, 2000, **122**, 7128.
128. L. A. Villaescusa, P. Lightfoot and R. E. Morris, *Chem. Commun.*, 2002, 2220.
129. L. A. Villaescusa, P. S. Wheatley, I. Bull, P. Lightfoot and R. E. Morris, *J. Am. Chem. Soc.*, 2001, **123**, 8797.
130. D. S. Wragg, G. M. Fullerton, P. J. Byrne, A. M. Z. Slawin, J. E. Warren, S. J. Teat and R. E. Morris, *Dalton Trans.*, 2011, **40**, 4926.

CHAPTER 11

Metal Nanoparticle Synthesis in Ionic Liquids†

CHRISTOPH JANIAK

Institut für Anorganische Chemie und Strukturchemie, Heinrich-Heine-Universität Düsseldorf, Universitätsstrasse 1, D-40225 Düsseldorf, Germany
Email: janiak@uni-duesseldorf.de

11.1 Introduction

Metal nanoparticles (M-NPs) are of increasing interest for technological applications. The area of catalysis benefits from their high surface area. The controlled and reproducible synthesis of defined and stable M-NPs with a small size distribution is very important for a range of applications.[1–5] In literature nanoparticles are also referred to as nanophase clusters, nanocrystals and colloids. In the following we primarily use the term nanoparticles for simplicity. The chemistry and physics of nanoparticles with their high surface-to-volume ratio are dominated by their surface energy.[6] Small NPs are only kinetically stable and will combine to thermodynamically favored larger particles *via* agglomeration (Figure 11.1).

This agglomeration is based on the principle of Ostwald ripening:[7] Ostwald ripening is a thermodynamically-driven spontaneous process and occurs because larger particles are more energetically favored than

†The material of this book chapter follows a published review article by the author in *Z. Naturforsch.*, 2013, **68b**, 1059–1089.

RSC Catalysis Series No. 15
Catalysis in Ionic Liquids: From Catalyst Synthesis to Application
Edited by Chris Hardacre and Vasile Parvulescu

Published by the Royal Society of Chemistry, www.rsc.org

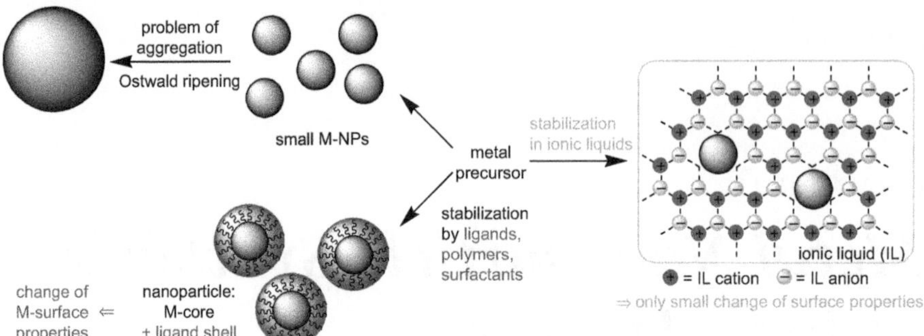

Figure 11.1 Schematic presentation of the stabilization of metal nanoparticles (M-NP) through protective stabilizers or in ionic liquids to prevent aggregation.

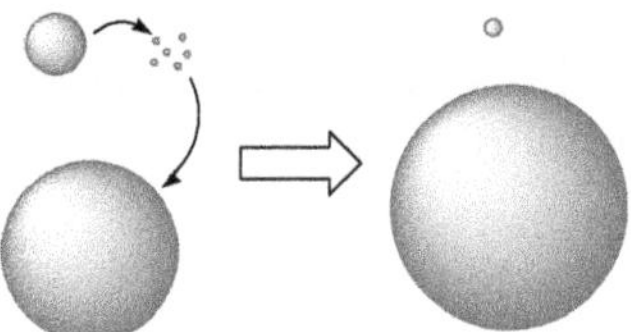

Figure 11.2 Schematic presentation of Ostwald ripening.

smaller ones. This stems from the fact that coordinatively unsaturated atoms or molecules on the surface of a particle are energetically less stable than atoms well-ordered and fully coordinated in the bulk. Large particles, with their lower surface to volume ratio, result in a lower energy state (and have a lower surface energy). As the system tries to lower its overall energy, atoms or molecules on the surface of a small (energetically unfavorable) particle will tend to detach and diffuse through solution and then attach to the surface of a larger particle. Therefore, the number of smaller particles continues to shrink, while larger particles continue to grow (Figure 11.2).

Typically, the synthesis of small metal nanoparticles requires the addition of surface capping ligands or stabilizing agents like polymers and surfactants to give a protective layer which provides electrostatic and/or steric coverage to prevent agglomeration (Figure 11.3).[8–10] In ionic liquids (ILs) metal nanoparticles can be formed without any additional stabilizers. The electrostatic and steric properties of ionic liquids allow for the stabilization of M-NPs without the need of additional stabilizers, surfactants or capping ligands (Figure 11.3). ILs may be seen as a "nanosynthetic template"[11] that, on the basis of their ionic nature,[12] high polarity, high dielectric constant and supramolecular network, stabilizes M-NPs without the need of additional protective ligands (*cf.* Figure 11.4).[13–18]

11.2 Ionic Liquids (ILs)

Ionic liquids are molten salts which consist of weakly coordinating inorganic or organic cations and anions. By definition their melting point is below 100 °C, more typically ILs are liquid at room temperature (RT-ILs).[19–22] The liquid state is thermodynamically favorable, due to the large size and conformational flexibility of the ions involved, which leads to small lattice enthalpies and large entropy changes that favor melting.[23] ILs are characterized and set apart from other solvents by their physical properties like high charge density, high polarity, high dielectric constant and supramolecular network formation.[16] Typical IL cations include 1-alkyl-3-methylimidazolium, tetraalkylammonium. Typical anions for ILs are halide anions, tetrafluoroborate ($[BF_4]^-$), hexafluorophosphate ($[PF_6]^-$), trifluoromethylsulfonate (triflate, $[TFO]^-$, $[CF_3SO_3]^-$, bis(trifluoromethylsulfonyl)amide ($[Tf_2N]^-$, $[(CF_3SO_2)_2N]^-$) (Figure 11.3).[24,25]

Ionic liquids are unique alternatives to traditional aqueous or organic solvents.[26] The preparation of advanced functional materials making use of ILs, through ionothermal synthesis, has been shown to be very promising.[27] The use of ILs and the concomitant ionothermal method is increasingly noticeable because of the excellent solvating properties of ILs, such as negligible vapor pressure, high thermal stability, high ionic conductivity, a broad liquid-state temperature range, and the ability to dissolve a variety of materials.[28]

IL-properties can be designed through judicious combination of anions and cations. For example, ILs containing $[Tf_2N]^-$ offer low viscosity and high electrochemical and thermal stability.[29] If bis(trifluoromethylsulfonyl)amide $[Tf_2N]^-$ is replaced by bis(methylsulfonyl)amide, viscosity increases and stability decreases.[30] Scattering experiments suggested that ILs are not liquids in the conventional sense, but have an organizational behavior intermediate between isotropic liquids and liquid crystals.[19] ILs have an intrinsic "nanostructure" which is caused by electrostatic, hydrogen bonding and van der Waals interactions.[14,19] The mesoscopic structure of imidazolium ionic liquids in particular can be described in part as a supramolecular three-dimensional hydrogen-bonded network of the type $\{[(RR'Im)_x(A)_{x-n})]^{n+}\ [(RR'Im)_{x-n}(A)_x)]^{n-}\}_n$ where $[RR'Im]^+$ is the 1,3-dialkylimidazolium cation and A the anion.[14,16,17] This structural pattern is not only seen in the solid phase but is also maintained to a great extent in the liquid phase. The introduction of other molecules and macromolecules proceeds with a disruption of the hydrogen bonding network and in some cases can generate nanostructures with polar and non-polar regions where inclusion-type compounds can be formed.[13,14] When mixed with other molecules or M-NPs, ILs become nanostructured materials with polar and nonpolar regions.[31–34] The combination of undirected Coulomb forces and directed hydrogen bonds leads to a high attraction of the IL building units. This is the basis for their (high) viscosity, negligible vapor pressure and

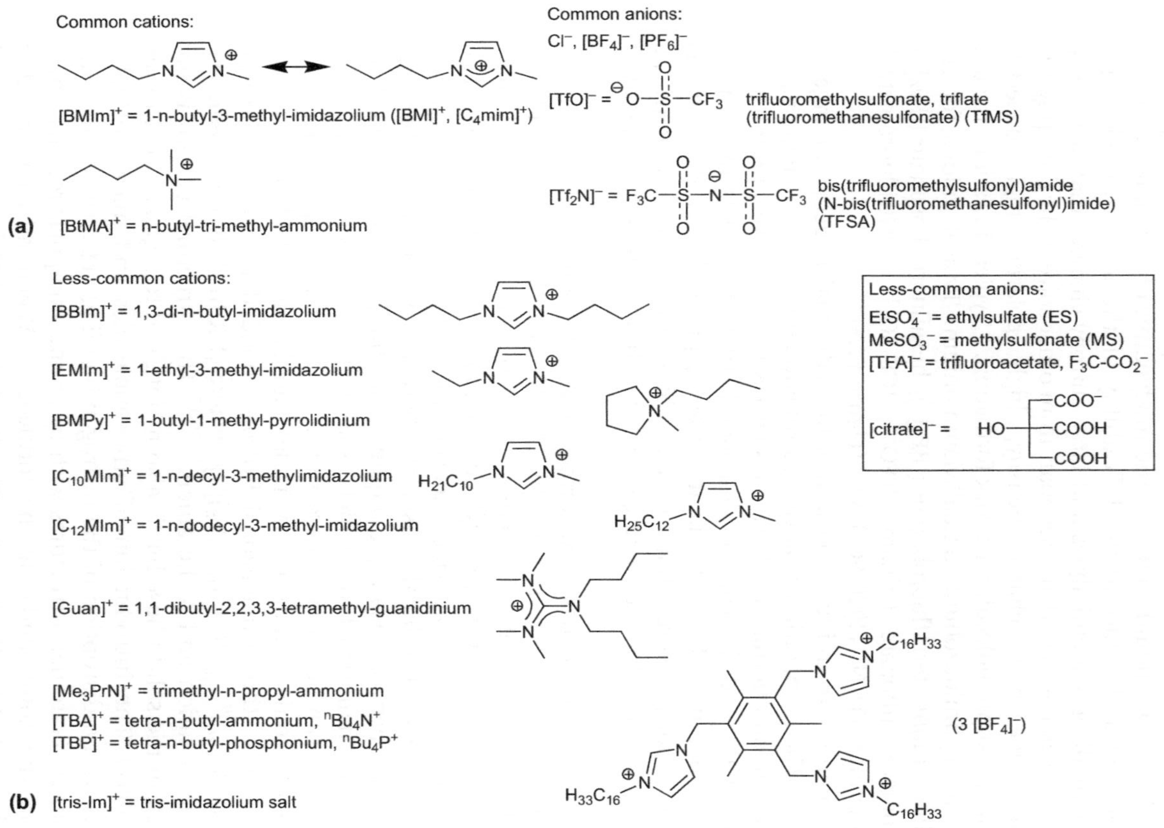

Common cations:

$[BMIm]^+$ = 1-n-butyl-3-methyl-imidazolium ($[BMI]^+$, $[C_4mim]^+$)

(a) $[BtMA]^+$ = n-butyl-tri-methyl-ammonium

Common anions:

Cl^-, $[BF_4]^-$, $[PF_6]^-$

$[TfO]^-$ = $^{\ominus}O–SO_2–CF_3$ trifluoromethylsulfonate, triflate (trifluoromethanesulfonate) (TfMS)

$[Tf_2N]^-$ = $F_3C–SO_2–N^{\ominus}–SO_2–CF_3$ bis(trifluoromethylsulfonyl)amide (N-bis(trifluoromethanesulfonyl)imide) (TFSA)

Less-common cations:

$[BBIm]^+$ = 1,3-di-n-butyl-imidazolium

$[EMIm]^+$ = 1-ethyl-3-methyl-imidazolium

$[BMPy]^+$ = 1-butyl-1-methyl-pyrrolidinium

$[C_{10}MIm]^+$ = 1-n-decyl-3-methylimidazolium

$[C_{12}MIm]^+$ = 1-n-dodecyl-3-methyl-imidazolium

$[Guan]^+$ = 1,1-dibutyl-2,2,3,3-tetramethyl-guanidinium

$[Me_3PrN]^+$ = trimethyl-n-propyl-ammonium

$[TBA]^+$ = tetra-n-butyl-ammonium, $^nBu_4N^+$

$[TBP]^+$ = tetra-n-butyl-phosphonium, $^nBu_4P^+$

(b) $[tris\text{-}Im]^+$ = tris-imidazolium salt (3 $[BF_4]^-$)

Less-common anions:

$EtSO_4^-$ = ethylsulfate (ES)

$MeSO_3^-$ = methylsulfonate (MS)

$[TFA]^-$ = trifluoroacetate, $F_3C\text{-}CO_2^-$

$[citrate]^-$ = HO–C(COO⁻)(COOH)(COOH)

Figure 11.3 Cations and anions of non-functionalized ILs, differentiated by (a) common and (b) less-common. Abbreviations and the use of capital or small letters, charges added and brackets around abbreviations vary in the literature. The abbreviations used here are given in front. Other abbreviations found in the literature follow in parentheses. For functional ILs. see Figure 11.5.

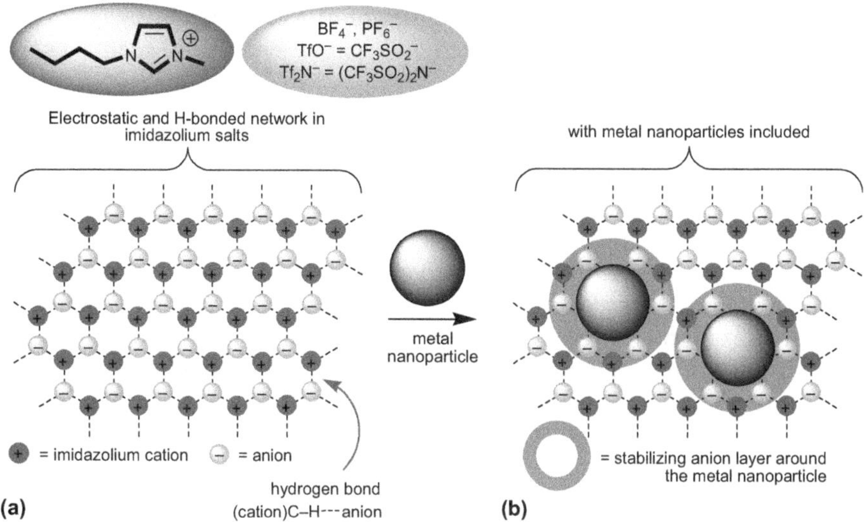

Figure 11.4 (a) Schematic network structure in 1,3-dialkylimidazolium-based ionic liquids projected in two dimensions. (b) The inclusion of metal nanoparticles in the supramolecular IL network with electrostatic and steric (= *electrosteric*) stabilization is indicated through the formation of the suggested primary anion layer forming around the M-NPs.
Adapted from ref. 18 with permission from the author; © 2011 Elsevier B.V.

three-dimensional constitution. The IL network properties should be well suited for the synthesis of defined nano-scaled metal colloid structures (see Figure 11.4).[13–15]

11.3 Metal Nanoparticles and Ionic Liquids

While some reviews note a parallel and synergistic development of both nanoparticles and ionic liquids for materials chemistry,[15] others devote only a short section to the use of ionic liquids in the synthesis of inorganic nanoparticles.[35]

The inclusion of metal nanoparticles in the supramolecular ionic liquid network brings with it the needed electrostatic and steric (= *electrosteric*) stabilization through the formation of an ion layer around the M-NPs. The type of this ion layer, hence, the mode of stabilization of metal nanoparticles in ILs is still a matter of some discussion.[15,36] Aside from the special case of thiol-, ether-, carboxylic acid-, amino-, hydroxyl- and other functionalized ILs (see Figure 11.5 and accompanying text) one could argue between IL–cation or –anion coordination to the NP surface. The electrostatic stabilization of a negatively charged surface of Au-NPs by parallel coordination mode of the imidazolium cation was proposed on the basis of surface-enhanced Raman spectroscopy (SERS) studies (*cf.* Figure 11.6a).[37] This proposal was supported by the finding of a negative zeta potential of M-NPs prepared by chemical reduction processes which indicated a negative charge of such NPs in aqueous solutions.[38]

According to DLVO (Derjaguin-Landau-Verwey-Overbeek) theory,[39] ILs provide an electrostatic protection in the form of a "protective shell" for M-NPs.[40–45] DLVO theory predicts that the first inner shell must be anionic and the anion charges should be the primary source of stabilization for the electrophilic metal nanocluster.[39] DLVO theory treats anions as ideal point charges. Real anions with a molecular volume would be better classified as "*electrosteric* stabilizers" meaning to combine both the *electro*static and the *steric* stabilization. However, the term "electrosteric" is ill-defined.[46] The stabilization of metal nanoclusters in ILs could, thus, be attributed to "extra-DLVO" forces[46] which include effects from the network properties of ILs such as hydrogen bonding, the hydrophobicity and steric interactions.[47]

Density functional theory (DFT) calculations in a gas-phase model favor interactions between IL anions, such as $[BF_4]^-$, instead of imidazolium cations, and Au_n clusters ($n = 1$, 2, 3, 6, 19, 20). This suggests a Au···F interaction and anionic Au_n stabilization in fluorous ILs. A small and Au-concentration dependent ^{19}F-NMR chemical shift difference (not seen in ^{11}B -or ^{1}H-NMR) for Au-NP–[BMIm]$[BF_4]$ supports the notion of a $[BF_4]^-$-fluorine···Au-NP contact for the NP stabilization in dynamic ILs.[48] The DFT calculations also indicate a weak covalent part in this Au···F interaction. Free imidazole bases (*e.g.* 1-methylimidazole) show similar binding energies. The Cl^- anions have the highest binding energy and can therefore be expected to bind to the NP if present in the solution. At the same time no significant binding of the $[BMIm]^+$ or $[MIm]^+$ imidazolium cations was found. These results support the model of preferred interaction between anions and Au-NPs, but also confirm the importance of considering a possible presence of Cl^- anions in the ionic liquid solution.[48,49]

The solvation of a metallic nanoparticle in the ionic liquid [BMIm]$[Tf_2N]$ was investigated by molecular simulation with a specific interaction potential. The interfacial layer of ionic liquid is only one ion thick, and thus excludes a stabilization mechanism based on an electrostatic double layer.[50]

Compared with the non-functionalized imidazolium-ILs (*cf.* Figure 11.3), functionalized imidazolium-ILs stabilize metal NPs even more efficiently because of the functional group. Thiol-,[51–53] ether-,[37] carboxylic acid-,[36] amino-,[36,54] and hydroxyl-,[52,55,56] or nitrile-[57] imidazolium-ILs (Figure 11.5) have been used to synthesize noble, primarily gold metal NPs.

The functional groups on the imidazolium cations exert an additional stabilization on M-NPs because of specific interactions of the functional group with the particle surface. The donor atom(s) of the functional group can attach to the metal nanoparticle much like an extra stabilizing capping ligand.[36] Then, the stabilization of metal nanoparticles in functionalized imidazolium-based ILs occurs through the cation with its functional group (Figure 11.6).[58,59] For both non-functionalized and functionalized ILs equally charged layers around the M-NPs lead to their separation through electrostatic repulsion and, thus, prevent their aggregation or Ostwald ripening.[7,60]

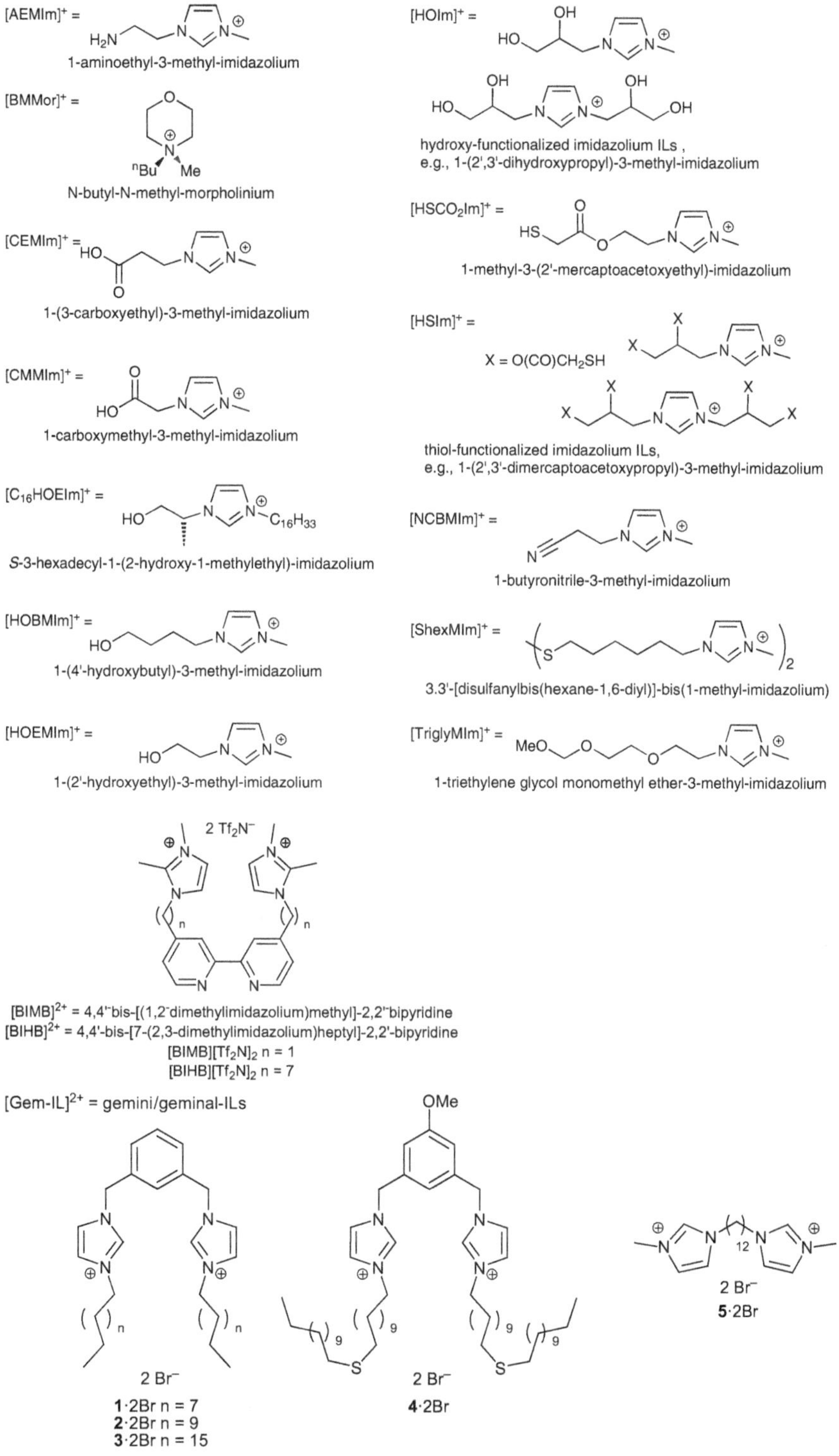

$[AEMIm]^+$ = 1-aminoethyl-3-methyl-imidazolium

$[BMMor]^+$ = N-butyl-N-methyl-morpholinium

$[CEMIm]^+$ = 1-(3-carboxyethyl)-3-methyl-imidazolium

$[CMMIm]^+$ = 1-carboxymethyl-3-methyl-imidazolium

$[C_{16}HOEIm]^+$ = *S*-3-hexadecyl-1-(2-hydroxy-1-methylethyl)-imidazolium

$[HOBMIm]^+$ = 1-(4'-hydroxybutyl)-3-methyl-imidazolium

$[HOEMIm]^+$ = 1-(2'-hydroxyethyl)-3-methyl-imidazolium

$[HOIm]^+$ = hydroxy-functionalized imidazolium ILs, e.g., 1-(2',3'-dihydroxypropyl)-3-methyl-imidazolium

$[HSCO_2Im]^+$ = 1-methyl-3-(2'-mercaptoacetoxyethyl)-imidazolium

$[HSIm]^+$ = X = O(CO)CH_2SH; thiol-functionalized imidazolium ILs, e.g., 1-(2',3'-dimercaptoacetoxypropyl)-3-methyl-imidazolium

$[NCBMIm]^+$ = 1-butyronitrile-3-methyl-imidazolium

$[ShexMIm]^+$ = 3.3'-[disulfanylbis(hexane-1,6-diyl)]-bis(1-methyl-imidazolium)

$[TriglyMIm]^+$ = 1-triethylene glycol monomethyl ether-3-methyl-imidazolium

2 Tf_2N^-

$[BIMB]^{2+}$ = 4,4'-bis-[(1,2'dimethylimidazolium)methyl]-2,2'-bipyridine
$[BIHB]^{2+}$ = 4,4'-bis-[7-(2,3-dimethylimidazolium)heptyl]-2,2'-bipyridine
$[BIMB][Tf_2N]_2$ n = 1
$[BIHB][Tf_2N]_2$ n = 7

$[Gem\text{-}IL]^{2+}$ = gemini/geminal-ILs

2 Br^-
1·2Br n = 7
2·2Br n = 9
3·2Br n = 15

2 Br^-
4·2Br

2 Br^-
5·2Br

Figure 11.5 Examples of functionalized imidazolium-ILs.[36,37,51–54]

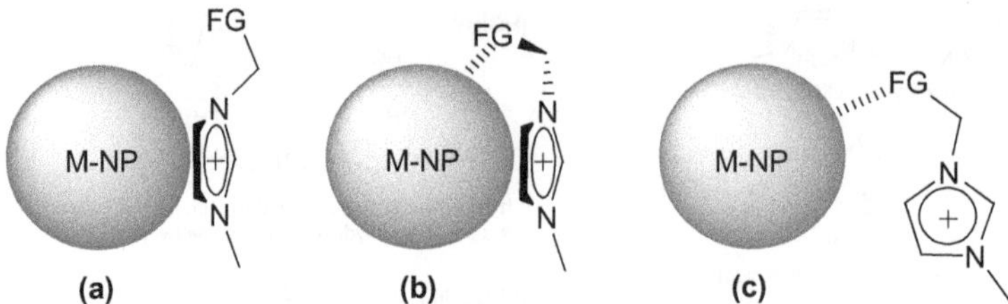

Figure 11.6 Three possible stabilization modes of metal nanoparticles by IL-imidazolium cations with functional groups (FG).[60]

11.4 Synthesis of Metal Nanoparticles in Ionic Liquids

Metal nanoparticles can be synthesized in ionic liquids[61] through chemical reduction[62–67] or decomposition,[68–71] by means of photochemical reduction[72,73] or electroreduction/electrodeposition[74–76] of metal salts where the metal atom is in a formally positive oxidation state. An elegant route is also the thermal, photolytic or chemical decomposition of compounds with zero-valent metal atoms, such as metal carbonyls $M_x(CO)_y$,[11,62,77,78] or [Ru(COD)(COT)] (see section 4.6).[79,80] Common to the synthesis of M-NPs in ILs is that no extra stabilizing molecules or organic solvents would be needed[8,13,15,45,81] even if in some cases such stabilizers are added.

Even without a chemical reaction (as exemplified below) simple dis-agglomeration of micro-sized copper flakes (1–5 μm) by stirring in ILs for 24 h at room temperature yielded copper nanoparticles of 50–100 nm diameter in [BMIm][BF_4] and of 80–100 nm diameter in [BMIm][PF_6] and [EMIm][BF_4]. A positive charge density of the Cu-NPs was deduced by XPS, owing to the strong interactions between the surface of copper nanoparticles and the anion of [BMIm][BF_4], to, therefore, give the partially positive charged surface of the Cu-NPs.[82]

11.4.1 Chemical Reduction

The reduction of metal salts is the most utilized method to generate M-NPs in ILs in general. A myriad of M-NPs have been prepared in ILs from compounds with the metal in a formally positive oxidation state M^{n+}, including M = Rh,[65] Ir,[83] Pt,[84] Ag,[63,85] Au,[51] as listed in Table 11.1. Many different types of reducing agents are used, like gases (H_2), organic (citrate, ascorbic acid, imidazolium cation of IL) and inorganic ($NaBH_4$, $SnCl_2$) agents (Table 11.1).

Molecular hydrogen (H_2) or sodium borohydride ($NaBH_4$) is often taken as reductant. The synthesis of M-NPs by reduction is not limited to conventional batch scale glass flasks. Microfluidic reactors with various continuous-flow configurations have been reported for the fabrication of

Table 11.1 Examples of M-NPs prepared in ILs by chemical reduction. Reprinted from ref. 180. Copyright Verlag der Zeitschrift für Naturforschung, Tübingen 2013.

Metal	*Metal salt precursor*	*Reducing agent*	*Ionic liquid*[a]	*M-NP average diameter ± standard deviation (nm)*	*Reference*
Mono-metallic					
Rh	$RhCl_3 \cdot 3H_2O$	H_2, 75 °C and 4 bar	$[BMIm][PF_6]$	2.0–2.5	65
	$[Rh(COD)\text{-}\mu\text{-}Cl]_2$[b]	H_2 + laser radiation	$[BMIm][PF_6]$	7.2 ± 1.3	86
	$RhCl_3$	$NaBH_4$	$[BMIm][Tf_2N]$–$[BIMB][Tf_2N]_2$ or $[BIHB][Tf_2N]_2$	1–3	87
Ir	$[Ir(COD)Cl]_2$[b]	H_2, 75 °C and 4 bar	$[BMIm][BF_4]$, $[BMIm][PF_6]$, $[BMIm]TfO^-$	2–3	88
	$[Ir(COD)_2]BF_4$, $[Ir(COD)Cl]_2$[b]	H_2	$[1\text{-alkyl-3-methyl-Im}][BF_4]$	irregular 1.9 ± 0.4, 3.6 ± 0.9	67
Pd	H_2PdCl_4	$NaBH_4$	$[HSCO_2Im][Cl]$	nanowires	53
	H_2PdCl_4	$NaBH_4$	$[Guan][Br]$–Vulcan-72 carbon	~2.8	89
	$PdCl_2$	H_2 + laser radiation	$[BMIm][PF_6]$	4.2 ± 0.8	86
	$Pd(acac)_2$	H_2	$[BMIm][PF_6]$	10 ± 0.2	69
	$Pd(acac)_2$	Imidazolium ILs, thermal, see text	$[BMIm][PF_6]$, $[HOBMIm][Tf_2N]$	5, 10, catalyst for selective acetylene hydrogenation	69
	$Pd(OAc)_2$ or $PdCl_2$	Imidazolium ILs, ultrasound, see text	$[BBIm][Br]$, $[BBIm][BF_4]$	20, catalyst for Heck reactions	70, 90
	$Pd(OAc)_2$	$[BMIm][Tf_2N]$, thermal	$[BMIm][Tf_2N]$–PPh_3	~1, catalyst for Heck reactions	71, 90
	$Pd(OAc)_2$	Imidazolium ILs, thermal, see text	$[HOEMIm][TfO]$	2.4 ± 0.5	56
			$[HOEMIm][TFA]$	2.3 ± 0.4	
			$[HOEMIm][BF_4]$	3.3 ± 0.6	
			$[HOEMIm][PF_6]$	3.1 ± 0.7	
			$[HOEMIm][Tf_2N]$	4.0 ± 0.6	
			$[BMIm][Tf_2N]$	6.2 ± 1.1	
	$Pd(OAc)_2$		$[TBA][Br]$–$[TBA][OAc]$	3.3 ± 1.2, catalyst for Heck arylations	91, 92, 90

Table 11.1 (*Continued*)

Metal	*Metal salt precursor*	*Reducing agent*	*Ionic liquid*[a]	*M-NP average diameter ± standard deviation (nm)*	*Reference*
	$Pd(OAc)_2$		[BtMA][Tf_2N]	Catalyst for Heck cross-coupling	93, 90
	$Pd(OAc)_2$	Imidazolium IL, thermal, see text	[NCBMIm][Tf_2N]	7.3 ± 2.2	94
	$Pd_2(dba)_3$[c]	H_2, 3 atm	[tris-Im][BF_4]$_3$, see Figure 11.5	Catalyst for Suzuki cross-coupling	95, 90
	Bis(benzothiazolylidene carbene)PdI_2		[TBA][Br]–[TBA][OAc]	Catalyst for Heck arylations	91
Pt	$Na_2Pt(OH)_6$	$NaBH_4$	[HSIm][A] or [HOIm][A], A = Cl^- or HS-$(CH_2)_3$-SO_3^-	3.2 ± 1.1, 2.2 ± 0.2, 2.0 ± 0.1	52
	H_2PtCl_6	$NaBH_4$	[CMMIm][Cl], [AEMIm][Br]	2.5	36
	PtO_2	H_2	[BMIm][BF_4], [BMIm][PF_6]	2–3	96
	$Pt_2(dba)_3$[c]	H_2, 75 °C, 4 atm	[BMIm][PF_6]	2.0–2.5	84
	(MeCp)$PtMe_3$	Imidazolium ILs, MWI, $h\nu$, thermal, see text	[BMIm][BF_4], [BtMA][Tf_2N]	1.5 ± 0.5, see text	97
Cu	$Cu(OAc)_2 \cdot H_2O$	$H_2NNH_2 \cdot H_2O$ (hydrazine hydrate)	[BMIm][BF_4] [BMIm][PF_6] each w. 1% PVP or PVA as stabilizer[d]	Spherical, PVP: 80–130, PVA: 260 cubic, PVP: 160 ± 14; catalyst in click reaction	98
Ag	$AgBF_4$	H_2, 85 °C, 4 atm	[BMIm][BF_4]	2.8 ± 0.8	79
		BIm as scavenger, see text	[BMIm][PF_6]	4.4 ± 1.3	
			[BMIm][TfO]	8.7 ± 3.4	
			[BtMA][Tf_2N]	26.1 ± 6.4	
	$AgBF_4$	H_2	[BMIm][BF_4] [BMpy][TfO] with TX-100/cyclohexane as reverse micellar system	~ 9 (DLS), ~ 11 (DLS), both ~ 3 from TEM	99

	$AgBF_4$	$[BMIm][BH_4]$, 1-MeIm as scavenger	$[BMIm][Tf_2N]$ in microfluidic reactor	3.73 ± 0.77	100
	Ag_2CO_3	Me_2NCHO (DMF)	$[Me_2NH_2][Me_2NCO_2]$ with small amounts of DMF	2–14	101
	$AgNO_3$	Tween 85	$[BMIm][PF_6]$	3–10	102
Au	$HAuCl_4$	Na_3citrate–$NaBH_4$, Na_3citrate, ascorbic acid	$[EMIm][EtSO_4]$	9.4, 3.9, nanorods	103
	$HAuCl_4$	Ascorbic acid	$[BMIm]$ $[C_{12}H_{25}OSO_3]$ (lauryl sulfate)	20–50	104
	$HAuCl_4$	Na_3citrate	$[CMMIm][Cl]$, $[AEMIm][Br]$	23–98	36
	$HAuCl_4 \cdot 3H_2O$	$H_2NNH_2 \cdot H_2O$ (hydrazine monohydrate)	$[TriglyMIm][MeSO_3]$	$\sim$7.5	37
	$HAuCl_4$	$NaBH_4$	$[ShexMIm][Cl]$	5.0	51
	$HAuCl_4$	$NaBH_4$	$[HSIm][A]$ or $[HOIm][A]$, $A = Cl^-$ or $HS\text{-}(CH_2)_3\text{-}SO_3^-$	3.5 ± 0.7, 3.1 ± 0.5, 2.0 ± 0.1	52
	$HAuCl_4$	$NaBH_4$	$[CMMIm][Cl]$, $[AEMIm][Br]$	3.5	36
	$[C_{16}HOEIm]AuCl_4$ from $[C_{16}HOEIm]Br$ and $HAuCl_4$	$NaBH_4$	$CHCl_3$–H_2O, $[C_{16}HOEIm][Br]$	6.0 ± 1.4	105
	$HAuCl_4$	$NaBH_4$	$[Gem\text{-}IL][Br]_2$ **1** · 2Br-**4** · 2Br, see Figure 11.5	**3:** 8.8 ± 2.2 **4:** 5.3 ± 2.4	106
	$HAuCl_4$	$NaBH_4$	$[BMIm][BF_4]$ in microfluidic reactor	4.38 ± 0.53	107
	$HAuCl_4$	$NaBH_4$	$[BMIm][PF_6]$	4.8 ± 0.7 (5.3 ± 0.8 after 2 weeks)	108
	$HAuCl_4$	$NaBH_4$	$[BMIm][PF_6]$–$[AEMIm][PF_6]$	4.3 ± 0.8	108
	$HAuCl_4$	$NaBH_4$	$[C_{12}MIm][Br]$	8.2 ± 3.5, stable for at least 8 months	109
	$HAuCl_4$	$NaBH_4$	$[Gem\text{-}IL][Br]_2$ **5** · 2Br, see Figure 11.5	10.1 ± 4.2	109
	$HAuCl_4$	$[BMIm][BH_4]$, 1-MeIm as scavenger	$[BMIm][Tf_2N]$ in microfluidic reactor	4.28 ± 0.84	100

Table 11.1 (*Continued*)

Metal	*Metal salt precursor*	*Reducing agent*	*Ionic liquid*[a]	*M-NP average diameter ± standard deviation (nm)*	*Reference*
	$HAuCl_4 \cdot 3H_2O$	$NaBH_4$, cellulose	[BMIm][Cl]	9.7 ± 2.7	110
	$HAuCl_4$	Cellulose, see text	[BMIm][Cl]	300–800	66
	$HAuCl_4 \cdot 3H_2O$	Glycerol	[EMIm][TfO],	5–7, low temp.	111
			[EMIm][$MeSO_3$],	5–7, aggregate at higher temp.	
			[EMIm][$EtSO_4$]	15–20, polydisperse	
	$HAuBr_4$	Me_2NCHO (DMF)	[Me_2NH_2][Me_2NCO_2] with small amounts of DMF	2–4	101
	Au(CO)Cl	Imidazolium ILs, thermal, MWI, hν, see text	[BMIm][BF_4]	1.8 ± 0.4, 4.1 ± 0.7	48
	$KAuCl_4$	[BMIm][BF_4] thermal, see text	[BMIm][BF_4]	1.1 ± 0.2	48
	$HAuCl_4 \cdot 4H_2O$	[$Me_3NC_2H_4OH$] [Zn_nCl_{2n+1}], thermal	[$Me_3NC_2H_4OH$] [Zn_nCl_{2n+1}]	135 °C: 35 ± 12, 140 °C: 30 ± 4, 145 °C: 24 ± 3	112
	$HAuCl_4 \cdot 3H_2O$,	[BMIm][BF_4], ultrasound, see text.	[BMIm][BF_4]–MWCNT[e]	10.3 ± 1.5	113
	$KAuCl_4$	$SnCl_2$	[BMIm][BF_4]	2.6–200	49
	$AuCl_3 \cdot 3H_2O$	[TBP][citrate]	[TBP][citrate]	15–20	114
Bi-metallic					
Pd–Au 3:1	K_2PdCl_4, $HAuCl_4$	$NaBH_4$	[BMIm][PF_6]	5.3 (± 3.0)	108
			[BMIm][PF_6]–[AEMIm][PF_6]	3.6 (± 0.7)	108

[a]For non-functional ILs see Figure 11.3, for functional ILs see Figure 11.5.
[b]COD = 1,5-cyclooctadiene, COT = 1,3,5-cyclooctatriene.
[c]dba = bis-dibenzylidene acetone.
[d]PVP = polyvinyl pyrrolidone, PVA = polyvinyl alcohol.
[e]MWCNT = multi-walled carbon nanotube.

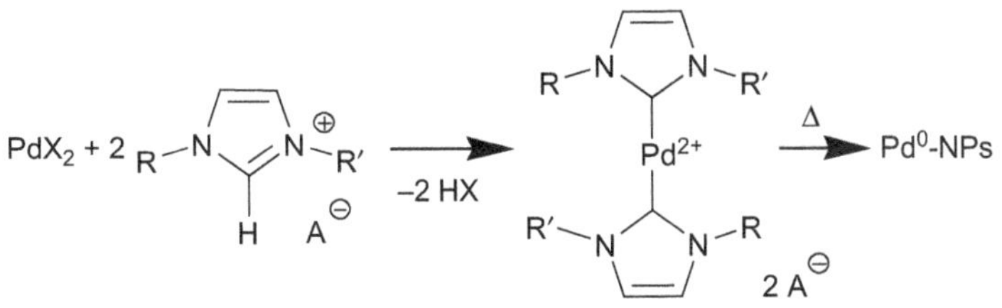

Figure 11.7 Reduction of Pd(II)-species with an imidazolium-based IL through intermediate formation of Pd-carbene complexes. Decomplexation and reduction occurs during heating.[90]
Adapted from ref. 18 with permission from the author; © 2011 Elsevier B.V.

metal nanoparticles including cobalt, copper, platinum and palladium, gold and silver, and core – shell particles.[100]

Pd-NPs from palladium(II) salts could be synthesized in the presence of imidazolium-based ILs without the need for an additional reducing agent. It is suggested that formation of Pd-*N*-heterocyclic carbene complexes precedes the formation of Pd-NPs (Figure 11.7).[70,90] The participation of carbene species in imidazolium ILs was supported by D–H exchange reactions at C2, C4 and C5 of the imidazolium cation in catalytic hydrogenation reactions promoted by classical Ir(I) colloid precursors and Ir-NPs in deuterated imidazolium ILs.[115] Imidazolium salts are also known as precursors for stable carbenes and are mild reducing agents.[116]

The thermal decomposition of $Pd(OAc)_2$ works well in various, also common organic solvents.[117] Pd-NPs with a diameter of ~1 nm formed from $Pd(OAc)_2$ in $[BMIm][Tf_2N]$ simply by heating to 80 °C in the presence of PPh_3.[71] Monodisperse Pd nanoparticles of 5 and 10 nm were obtained from $Pd(acac)_2$ dissolved in $[HOBMIm][Tf_2N]$ by heating in the absence of an additional reducing agent.[69] Heating (120 °C) of $Pd(OAc)_2$ in 1-butyronitrile-3-methylimidazolium-*N*-bis(trifluoromethane sulfonyl)imide $[NCBMIm][Tf_2N]$ under reduced pressure leads to the formation of stable and small Pd-NPs.[94]

Pd-NPs were prepared from $Pd(OAc)_2$ in hydroxyl-functionalized ILs with the 1-(2′-hydroxylethyl)-3-methylimidazolium $[HOEMIm]^+$ cation and non-functionalized control IL by thermal treatment. The influence of anions on the decomposition rate of $Pd(OAc)_2$ based on the percentage of $Pd(OAc)_2$ remaining in the sample was given the order $[Tf_2N]^- \approx [PF_6]^- > [BF_4]^- > [OTf]^- > [TFA]^-$ from investigation of a series of hydroxyl-functionalized ILs with the $[HOEMIm]^+$ cation. The OH-functionalized IL $[HOEMIm][Tf_2N]$ gave smaller Pd-NPs with diameter 4.0 ± 0.6 nm compared with Pd-NPs isolated from the non-functionalized IL $[BMIm][Tf_2N]$ with diameter 6.2 ± 1.1 nm.[56] Thermal reduction of $Pd(OAc)_2$ resulted in black NP solutions and no precipitation of the NPs was observed over a period of several months. ^1H NMR spectra recorded before and after reduction of $Pd(OAc)_2$ showed no difference and, thereby, argued against the alcohol group in the $[HOEMIm]^+$ cation as the reductant.[56]

Thermal, photolytic or microwave assisted decomposition of the air and moisture stable organometallic Pt(IV) precursor (MeCp)$PtMe_3$ in the ILs [BMIm][BF_4] and [BtMA][Tf_2N] also leads to well defined, small, crystalline and longtime stable Pt-nanoparticle (Pt-NP) dispersions without any additional reducing agents. The Pt-NP–IL dispersion was a highly active catalyst (TOF 96 000 h^{-1} at 0.0125 mol% Pt and quantitative conversion) for the biphasic hydrosilylation of phenylacetylene with triethylsilane, to the products triethyl(2- and 1-phenylvinyl)silane.[97]

When $AgBF_4$ was reduced with H_2 the AgNP particle size distribution was very broad, with a range of several 10 nm or even 100 nm in the absence of the butyl-imidazole (BIm) scavenger. This can be reasoned by proton (H^+/H_3O^+) incorporation in the IL matrix.[118] Also, the AgNP dispersion prepared without a scavenger is unstable as evidenced by clearly visible metal particle precipitation within 1–2 hours after reduction. In the presence of the BIm scavenger and soluble silver salts the distribution of the Ag nanoparticles lies largely within 10 nm and the dispersion is stable up to 3 days under argon (Figure 11.8).[63]

For Ag-NPs a correlation between the IL-anion molecular volume and the NP size was noted. The larger the volume of the IL-anion the larger is the size of the Ag-NPs. Thereby it was possible to form Ag-NPs in sizes from 2.8 to 26.1 nm with a narrow size distribution (Figure 11.9).[63]

The synthesis and characterization of Au-NPs is of great interest due to their electronic, optical, thermal and catalytic properties associated with possible applications in the fields of physics, chemistry, biology, medicine, and material science.[119] Gold particles are among the best-studied particles in nano and materials science. A well-known method to generate Au-NPs was already established by Turkevich *et al.* in 1951.[120] The reducing agent was citrate.

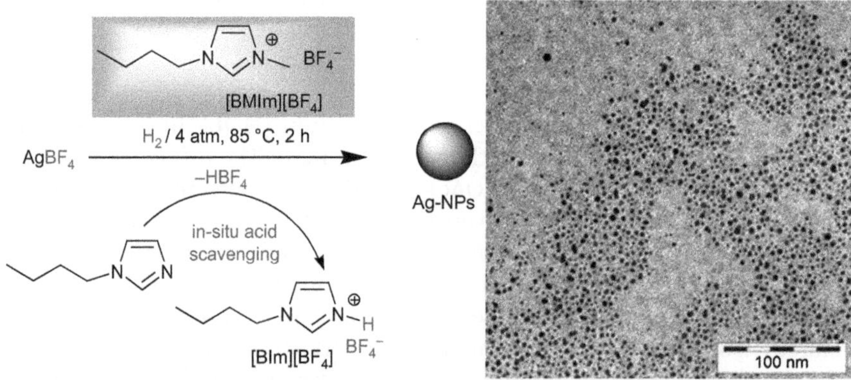

Figure 11.8 Formation of Ag-NPs (Ø 2.8 ± 0.8 nm) by hydrogen reduction of $AgBF_4$ with the imidazole scavenging process in [BMIm][BF_4] (*cf.* Figure 11.10). The ionic liquid formed from the scavenging process should be similar to the main IL solvent.[63]
TEM reprinted from ref. 63 with permission © 2008 American Chemical Society.

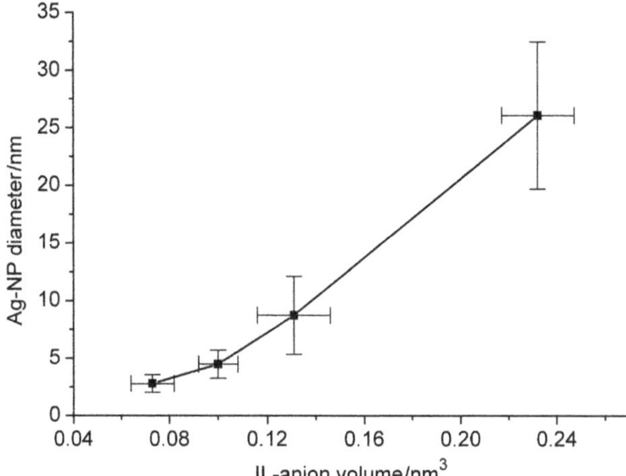

Figure 11.9 Correlation between the observed Ag nanoparticle size (from TEM) and the molecular volume of the ionic liquid anion (*cf.* Figure 11.12).

By using this method the reduction could also be carried out in the IL 1-ethyl-3-methylimidazolium ethylsulfate [EMIm][$EtSO_4$]. Afterwards it was possible to give these particles different shapes by adding a silver salt.[103]

Also, cellulose alone is a reducing agent for Au(III) in $HAuCl_4$ and at the same time acts as a morphology- and size-directing agent, which drives the crystallization towards polyhedral particles or thick plates. The gold particle morphologies and sizes mainly depend on the reaction temperature. With this route plates with a thickness from 300 nm at 110 °C to 800 nm at 200 °C were synthesized.[66]

By variation of the molar Au(III) : Sn(II) ratio it was possible to synthesize Au-NPs in different sizes in a stop-and-go, stepwise and "ligand-free" nucleation, nanocrystal growth process which can be stopped and resumed at different color steps and Au-NP sizes from 2.6 to 200 nm. This stepwise Au-NP formation was possible because the IL apparently acted as a *kinetically* stabilizing, dynamic molecular network in which the reduced Au^0 atoms and clusters can move by diffusion and cluster together, as verified by TEM analysis.[49]

Also, gold nanoparticles are reproducibly obtained by thermal, photolytic or microwave assisted decomposition/reduction under argon from Au(CO)Cl or $KAuCl_4$ in imidazolium-based ILs without an additional reducing agent. The reductive decomposition was carried out in the presence of *n*-butylimidazol as a scavenger (Figure 11.10) in the ILs [BMIm][BF_4], [BMIm][TfO] or [BtMA][Tf_2N]. The small and uniform nanoparticles of about 1–2 nm diameter in [BMIm][BF_4] increase in size with the volume of the ionic liquid anion in [BMIm][TfO] and [BtMA][Tf_2N]. Under argon the Au-NP–IL dispersion is stable without any additional stabilizers or capping molecules. From the ionic liquids the gold nanoparticles can be capped with organic thiol ligands and transferred to different polar and non-polar organic solvents. Au-NPs can also be deposited onto a polytetrafluoroethylene (PTFE, Teflon) surface.[48]

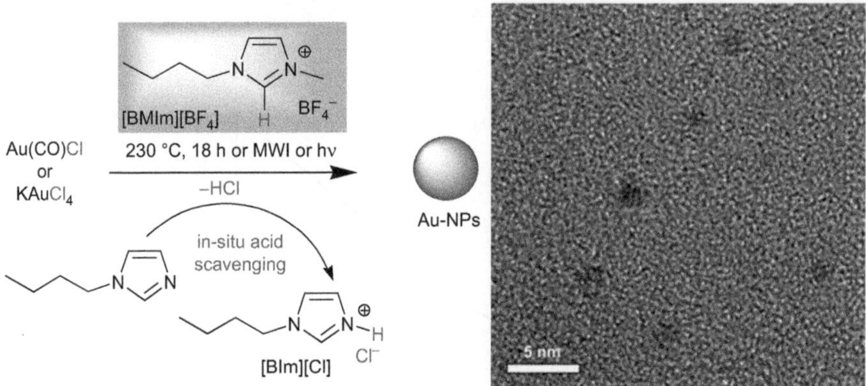

Figure 11.10 Formation of Au-NPs (Ø 1.1 ± 0.2 nm from $KAuCl_4$ in a thermal process) with the imidazole scavenging process in [BMIm][BF_4] (*cf.* Figure 11.8). The ionic liquid formed from the scavenging process should be similar to the main IL solvent.[48] The decomposition of Au(CO)Cl can also proceed by intramolecular reduction under phosgene formation according to $2\ Au(CO)Cl \rightarrow 2\ Au + CO + COCl_2$.[121] TEM reprinted from ref. 48 with permission © 2011 Wiley-VCH Verlag.

Small Au-NPs of diameter 1.1 ± 0.2 nm, generated in the IL [BMIm][BF_4], can display quantized charging at room temperature. This phenomenon is well-known for nanoparticles that are protected by a strongly bound ligand shell, but could be demonstrated for "naked" metal clusters only in the special environment of an ionic liquid. DFT methods demonstrate that the cluster charging is accompanied by a switching in the orientation of the ionic shell.[122]

Au(I) and Au(III) salts ($NaAuCl_4$ and $KAuCN_2$ respectively) in [BMIm][PF_6] underwent reductive transformation to Au(0) to afford gold nanoparticles which were found to be active catalysts for the cyclopropanation of alkenes with ethyldiazoacetate, in many cases affording high yields of cyclopropanecarboxylates. In ILs as solvents, the gold catalysts were stabilized, behaving as a metal nanoparticle reservoir, and products and catalyst separation and recycling could be achieved.[123]

Carboxylic acid- and amino-functionalized ionic liquids [CMMIm]Cl and [AEMIm]Br (*cf.* Figure 11.5) were used as the stabilizer for the synthesis of gold and platinum metal nanoparticles in aqueous solution. Smaller Au-NPs (3.5 nm) and Pt-NPs (2.5 nm) were prepared with $NaBH_4$ as the reductant. Larger gold nanospheres (23, 42, and 98 nm) were synthesized using different quantities of trisodium citrate reductant. The morphology and the surface state of the metal nanoparticles were characterized by high-resolution transmission electron microscopy, UV-visible spectroscopy, and X-ray photoelectron spectroscopy. X-ray photoelectron spectra indicated that binding energies of C 1s and N 1s from ionic liquids on the surface of metal nanoparticles shifted negatively compared with that from pure ionic liquids.

The mechanism of stabilization is proposed to be due to the interactions between imidazolium ions/functional groups in ionic liquids and metal atoms (*cf.* Figure 11.6). The metal nanoparticles could be easily assembled on multi-walled carbon nanotubes. In this case, ionic liquids acted as a linker to connect metal nanoparticles with carbon nanotubes. The imidazolium ring moiety of ionic liquids might interact with the π-electronic nanotube surface by virtue of cation-π and/or π–π interactions, and the functionalized group moiety of ionic liquids might interact with the metal NPs surface [*cf.* Figure 11.6(c)].[36]

The generation of M-NPs in ILs is also used to deposit the nanoparticles onto a support: Rhodium-NPs deposited on attapulgite (Rh-Atta) were prepared by immobilizing Rh^{3+} on Atta *via* the IL 1,1,3,3-tetramethylguanidinium lactate, followed by reduction with hydrogen at 300 °C. The loaded rhodium on Atta existed mainly in the form of Rh^0 with a small amount of its oxides and was distributed uniformly on Atta with a particle size of less than 5 nm. Atta was destroyed to some extent due to the impregnation of the IL and Rh. The activity of the composite for cyclohexene hydrogenation was investigated, which exhibited much higher efficiency compared to other catalysts, and the turnover frequencies reached 2700 (mol of cyclohexene/mol of Rh) h^{-1}.[124]

Uniform Pd nanoparticles supported on Vulcan XC-72 carbon were synthesized from H_2PdCl_4 and $NaBH_4$ using G-IL as a mediator for the nucleation and growth process.[89]

Au-NP-decorated multi-walled carbon nanotube (MWCNT) hybrids (Au-MWCNT-HBs), were prepared by the ionic liquid-assisted sonochemical method (ILASM), onto poly(ethylene terephthalate) (PET) films from $HAuCl_4 \cdot 3H_2O$, MWCNT and [BMIm][BF_4].[113] Au-NPs can also be deposited onto and stabilized by interaction with a polytetrafluoroethylene (PTFE, Teflon) surface.[48]

An example of intermediate use of the IL is the synthesis of porous supported-nanoparticle materials with the encapsulation of polyvinyl pyrrolidone (PVP)-stabilized Au-NPs into titania xerogels employing [BMIm][PF_6] as a medium followed by solvent extraction of the ionic liquid and calcination of the materials. The average Au-NP sizes increased from 5.5 ± 2.3 nm before to 8.8 ± 2.5 nm after calcination.[125]

11.4.2 Photochemical Reduction

Photochemical methods for the synthesis of M-NPs present a rather clean procedure because contaminations by reducing agents are excluded.

UV decomposition of $(MeCp)PtMe_3$ was carried out in [BMIm][BF_4] and [BtMA][Tf_2N] to yield Pt-NPs of 1.1 ± 0.5 nm diameter when fresh (1 day old) and 1.7 ± 0.1 nm (aged 67 days) or 1.2 ± 0.4 nm (aged 331 days).[97]

A high-pressure mercury lamp was used to irradiate $AgClO_4$ in a mixture of an IL, water and Tween 20 (polyoxyethylene sorbitan monolaurate). Benzoin was used as photoactivator. The average diameters of Ag-NPs prepared in water–[BMIm][BF_4] and water–[OMIm][BF_4] (1-octyl-3-methylimidazolium) microemulsions were 8.9 and 4.9 nm, respectively.[126]

$HAuCl_4 \cdot 4H_2O$ in a mixed solution of [BMIm][BF_4] and acetone (ratio 10 : 1) was irradiated for 8 h with a UV light at a wavelength of 254 nm. The UV light turns the acetone into a free radical, which then reduces the cationic Au(III) to Au-NPs. The obtained Au nanosheets were about 4 μm long and 60 nm thick.[72]

Au-NPs formed from $HAuCl_4$ in the IL 1-decyl-3-methyl-imidazolium chloride in water when irradiated with 254 nm UV light for 30 to 70 min. The obtained nanorods had different shapes and morphologies. The sizes varied between 100 and 1000 nm.[73]

Au-NPs are obtained by photolysis from Au(CO)Cl in [BMIm][BF_4] (*cf.* Figure 11.10), albeit with large diameter of 61 ± 43 nm.[48]

11.4.3 Sonochemical (Ultrasound) Reduction

$Pd(OAc)_2$ or $PdCl_2$ in the imidazolium-based ILs [BBIm]Br or [BBIm][BF_4] were irradiated with ultrasound for 1 h. The Pd-NPs were nearly spherical and a size of 20 nm was observed. The formation of a Pd–biscarbene complex as an intermediate and its subsequent sonolytic conversion to Pd nanoparticles (*cf.* Figure 11.7) has been established by NMR/MS and TEM analyses.[70]

Au-NP-decorated MWCNT hybrids, were prepared by ionic liquid-assisted sonication onto PET-films. The mixture of $HAuCl_4 \cdot 3H_2O$, MWCNT and [BMIm][BF_4] was sonicated for 60 s, resulting in the *in situ* condensation of Au-NPs with narrow size distribution of 10.3 ± 1.5 nm decorated onto the surface of ionic liquid-wrapped MWCNTs.[113] Ultrasound was also used for the synthesis of ionic liquid-functionalized multi-walled carbon nanotubes decorated with highly dispersed Au nanoparticles from $HAuCl_4 \cdot 3H_2O$ in the presence of 1-(3-aminopropyl)-3-methylimidazolium bromide and dicyclohexylcarbodiimide albeit in DMF solution.[127]

11.4.4 Electro(chemical) Reduction

Ionic liquids have high ionic conductivity, high thermal stability, negligible vapor pressure and a wide electrochemical window of up to 7 V which make them nearly inert in electrolytic processes. A clean route to prepare nanoparticles in ionic liquids is electroreduction as only electrons are used as the metal-reducing agent (see also section 11.4.5 Gas-Phase Synthesis and 11.4.5.2 Plasma Deposition Method, Glow Discharge (Plasma) Electrolysis). It should be noted, however, that the size of the metal nanoparticles from electroreduction is often above the 100 nm definition limit for nanoparticles.

For the preparation of nanocrystalline metals a successful method is pulsed electrodeposition (PED).[128] This technique was used to deposit nano-Ni[129a] nano-Pd,[129b] nano-Cu,[129c] nano-Fe,[129d] nano-Cr[129e] and other metals with $E^\circ > 0$ V as well as alloys like nano-Ni_xFe_{1-x} or nano-Ni_xCu_{1-x}[129f] from aqueous electrolytes. Nanostructured less-noble metals like Al, Mg, W and their alloys cannot be electrodeposited from aqueous electrolytes but from ionic liquids.[128] For the electrodeposition of nano-Al the electrolyte consisted of [EMIm]Cl and absolutely dry $AlCl_3$. Controlled nanostructures with crystallite sizes from 10 to 133 nm can be obtained with aromatic and aliphatic carboxylic acid additives and also influenced by temperature.[130] Nanostructured iron was deposited from [BMIm]Cl, absolutely dry $AlCl_3$ anhydrous $FeCl_3$ with benzoic acid as additive. The crystallite size of the nano-Fe deposits was adjusted from 40–160 nm by variation of the DC-current density. The alloys Al_xMn_{1-x} and Al_xIn_{1-x} were deposited with crystallite sizes of 25 nm from [BMIm]Cl–$AlCl_3$ with addition of the corresponding metal salts $MnCl_2$ and $InCl_3$, respectively.[130]

CuCl as precursor was reduced in a cavity microelectrode in [BMIm][PF_6]. The electrode potential was varied. The smallest particles had a size of 10 nm and were obtained at an electrode potential of −1.8 V.[131]

It is also possible to deposit particles on supporting material, *e.g.*, Ag-NPs from $AgBF_4$ in [BMIm][BF_4] on TiO_2. The electroreduction was performed in the high vacuum chamber of a SEM. The resulting Ag-NPs arranged themselves in a dendritic network structure.[132] The precursor Ag(TfO) was electrochemically reduced in [EMIm][TfO]. The prepared Ag nanowires were 3 μm long and 200 nm wide.[133]

Polyaniline (PANI) and Au-NPs were synthesized as a composite material by cyclic voltammetry on a modified indium–tin-oxide (ITO) glass in the IL 1-ethyl-3-methyl-imidazolium tosylate [EMIm][Tos] containing 1 mol L^{-1} trifluoroacetic acid. The Au particles were synthesized during electropolymerization of aniline and distributed in the PANI matrix. SEM showed that Au particles with diameters in the range from 500 nm to 800 nm were distributed in the PANI matrix.[134]

Graphene oxide (GO) and $HAuCl_4$ were simultaneously reduced in [BMIm][PF_6] at a potential of −2.0 V. The obtained Au-NPs on the electrochemically, reduced graphene had a size of 10 nm.[135]

Morpholinium ionic liquid, [BMMor][BF_4]-stabilized palladium nanoparticles were prepared by electrochemical reduction using a palladium foil as the anode and a platinum foil as cathode. Pd ions released from the Pd anode migrated to the Pt cathode and there Pd ions were reduced to Pd atoms forming the nanoparticles. The particle size increased with a decrease in the current density and an increase in temperature and electrolysis duration. TEM images showed an average size of 2.0 ± 0.1, 2.2 ± 0.3, 2.4 ± 0.3, 2.9 ± 0.3, 3.5 ± 0.5, 3.9 ± 0.6, and 4.5 ± 0.9 nm. Nearly a 0.5 nm-sized control of the nanoparticle was achieved. The electron diffraction patterns of the obtained nanoparticles indicated a crystalline structure.[136]

11.4.5 Gas-Phase Synthesis

Gas-phase synthesis is very effective for high purity nanoparticle products. Gas-phase synthesis methods can be discerned in gas-phase condensation and flame pyrolysis. In gas-phase condensation, the metal is vaporized from heated crucibles, by electron or laser beam evaporation or sputtering and condensed onto a liquid, here an IL. When the metal as the evaporative source is replaced with a precursor compound for decomposition, then the gas-phase condensation is termed chemical vapor decomposition or chemical vapor synthesis. For flame pyrolysis, the gaseous or liquid precursors are decomposed by a combustion reaction.[137] The negligible vapor pressure of (room temperature) ILs allows the introduction of (RT)ILs in methods requiring vacuum conditions. For metal nanoparticle synthesis, such methods are magnetron sputtering onto ILs, plasma reduction in ILs, physical vapor deposition onto ILs, and electron beam and γ-irradiation to ILs. The nanoparticles are prepared in ILs without any stabilizing agents and do not aggregate.[21]

11.4.5.1 Magnetron Sputtering

Sputtering of clusters or atoms onto ILs to yield nanoparticles therein is possible for all elements that can be ejected from a target by Ar^+ and N_2^+ plasma ion bombardment. This method has yielded various pure metal nanoparticles, such as Au, Ag, Pt and others with particle sizes less than 10 nm in diameter and without any specific stabilizing agent. Both surface tension and viscosity of the IL are important factors for the nanoparticle growth and its stabilization.[21]

Sputter deposition of indium in the ionic liquids [BMIm][BF_4], [EMIm][BF_4], [(1-allyl)MIm][BF_4] and [(1-allyl)EIm][BF_4] could produce stable indium metal nanoparticles whose surface was covered by an amorphous In_2O_3 layer to form In–In_2O_3 core–shell particles. The size of the In core was tunable from *ca.* 8 to 20 nm by selection of the IL.[138]

Pt nanoparticles were produced by Pt sputtering onto the IL trimethyl-*n*-propylammonium bis((trifluoromethyl)sulfonyl)amide [Me_3PrN][Tf_2N] without stabilizing agents. Pt nanoparticles showed mean particle diameter of *ca.* 2.3–2.4 nm independent of sputtering time.[139]

Gold nanoparticles of 1–4 nm size could be prepared by sputter deposition of the metal onto the surface of the ionic liquid [BMIm][BF_4] to generate nanoparticles with no additional stabilizing agents.[140] Likewise, Au-NPs were prepared by sputter deposition of Au metal in [BMIm][PF_6]. The size of Au nanoparticles was increased from 2.6 to 4.8 nm by heat treatment at 373 K.[141]

Au-NPs with a size of 3 to 5 nm were obtained from gold foil by sputtering deposition onto several imidazolium-based ILs.[142]

11.4.5.2 Plasma Deposition Method, Glow Discharge (Plasma) Electrolysis

When a gas is partially ionized, becomes electrically conductive and has collective behavior it is called a plasma. Plasma deposition, once called glow discharge electrolysis (GDE)[21] is an electrochemical technique in which the discharge is initiated in the gas in between the metal electrode and the

Table 11.2 Nanoparticles obtained by the plasma deposition method (glow discharge plasma electrolysis in ILs). Reprinted with permission from ref. 180. Copyright Verlag der Zeitschrift für Naturforschung, Tübingen 2013.

Metal	*Metal precursor*	*Ionic liquid*	*Average particle diameter* ± *standard deviation/nm*	*Reference*
Pd	$PdCl_2$	[BMIm][BF_4]	32.7	145
Cu	Cu(TfO)$_2$	[BMPy][TfO]	~40, deposited on gold surface	146
	Cu	[EMIm][Tf_2N]	~11	147
	Cu	[BMPy][Tf_2N]	~26	147
Ag	$AgNO_3$–Ag(TfO)	[BMIm][TfO]	~8–30	148, 137
	Ag(TfO)	[EMIm][TfO]	20	146, 137
	$AgBF_4$	[BMIm][BF_4], [BMIm][PF_6]	<100, at glassy carbon electrode	149
Au	$HAuCl_4$	[BMIm][BF_4]	~2	143
	$HAuCl_4 \cdot 4H_2O$	[BMIm][BF_4]	1.7 ± 0.8	150
Al	$AlCl_3$	[BMPy][Tf_2N]	~34, 20–64, deposited on gold surface.	146
Ge	$GeCl_2$ dioxane	[EMIm][Tf_2N]	<50	151

solution by applying high voltage. In ionic liquid glow discharge electrolysis (IL-GDE) the discharge is initiated in the gas in between the metal electrode and the ionic liquid solution. The technique is also named plasma electrochemical deposition (PECD)[128,137] at the interface of plasma and ionic liquid or gas–liquid interfacial discharge plasmas (GLIDPs).[143] The plasma is regarded as an electrode because of the deposition of the materials at the interface of ionic liquid and plasma. In IL-GDE the precursor material dissolved in IL is reduced with free electrons from the plasma.[137,144] Table 11.2 summarizes metal nanoparticles which were obtained by IL-GDE.

Gold nanoparticle-DNA encapsulated single-walled carbon nanotubes (SWNTs) were generated using GLIDP by superimposing a DC voltage to the pulse voltage, where DNA and $HAuCl_4$ dissolved in IL is used as the liquid electrode.[143]

A sub-atmospheric dielectric barrier discharge (SADBD) plasma was used for the reduction of $HAuCl_4 \cdot 4H_2O$ to Au-NPs. By introducing polyvinyl pyrrolidone (PVP) as a capping agent, the nanoparticle diameter was controlled to ~1.7 nm with a narrow size distribution.[150]

11.4.5.3 Physical Vapor Deposition Method

This method is based on a solvated metal atom dispersion technique. It offers an easy and fast method for the preparation of long-time stable metal and metal oxide particles. The use of ILs avoids the otherwise needed freezing of the solvent as well as additional stabilizers.[152]

Vaporization of elemental Cu powder under high vacuum (10^{-6} Torr) onto the surface of [BMIm][PF_6] or [BMIm][Tf_2N] generated Cu nanoparticles with an average diameter of 3 nm. Cu metal was also vaporized into an IL dispersion of ZnO to give Cu-NPs with ~3.5 nm diameter on or near the ZnO surface.[152]

Gas-phase deposition from metal vaporization on the surface of ILs gave Au-NPs whose average diameter (∅) depended on the IL: [BMIm][BF_4] ∅ 7 nm, [BMIm][Tf_2N] ∅ 4nm, [BMPy][Tf_2N] ∅ 20–40 nm, [BMIm][DCA] (DCA = dicyanamide) initial ∅ 10 nm, later 40–80 nm and [P_{66614}][DCA] (P_{66614} = trishexyltetradecylposphonium) ∅ 50 nm.[152]

11.4.5.4 Electron Beam and γ-Irradiation

Very strong electron beam and γ-irradiation of the IL-containing metal salts yield solvated electrons and/or radicals through which metal nanoparticles are then generated.[21]

$NaAuCl_4 \cdot 2H_2O$ gave Au-NPs after accelerator electron beam and γ-irradiation both at 6 kGy and 20 kGy in [BMIm][Tf_2N]. From accelerator electron beam irradation spherical Au nanoparticles were formed with a mean particle diameter of 7.6 ± 1.5 nm at 6 kGy irradiation and of 26.4 ± 3.7 nm at 20 kGy. The γ-irradiation method leads to smaller Au nanoparticles with 2.9 ± 0.3 nm and 10.7 ± 1.7 nm at 6 kGy and 20 kGy, respectively, in [BMIm][Tf_2N]. It was emphasized that the prepared Au-NPs were stable without the need for a stabilizing additive for more than three months.[153]

A low-energy electron beam irradiation was used to synthesize Au-NPs from a $NaAuCl_4 \cdot 2H_2O$ precursor in the IL [BMIm][Tf_2N]. The obtained particles had a large size of 122 nm.[154]

11.4.6 Metal Nanoparticles from Zero-Valent Metal Precursors

11.4.6.1 Metal Nanoparticles from Metal Carbonyls, $M_x(CO)_y$

Binary metal carbonyls are elegant precursors for the synthesis of metal nanoparticles. Metal carbonyls are commercially available (Table 11.3). $Fe(CO)_5$ and $Ni(CO)_4$ are industrially produced on a multi-ton scale.[155] Metal carbonyls $M_x(CO)_y$ are easily purified and handled, even if care should be exerted for the possible liberation of poisonous CO. The metal carbonyls contain the metal atoms already in the zero-valent oxidation state needed for M-NPs. No reducing agent is necessary. The side product CO is largely given off to the gas phase and removed from the dispersion. Contamination from by-products or decomposition products, which are otherwise generated during the M-NP synthesis, are greatly reduced. Thus, metal carbonyls were used early on for the preparation of M-NPs, albeit without ILs, with much of the work on Fe- or Co-NPs being devoted to magnetism.[156]

Table 11.3 Binary metal carbonyls.[a]

Group	*5*	*6*	*7*	*8*	*9*	*10*
Metal	*V, Nb, Ta*	*Cr, Mo, W*	*Mn, Tc, Re*	*Fe, Ru, Os*	*Co, Rh, Ir*	*Ni, Pd, Pt*
Mononuclear complexes	$V(CO)_6$	**$Cr(CO)_6$** **$Mo(CO)_6$** **$W(CO)_6$**		**$Fe(CO)_5$** $Ru(CO)_5$ $Os(CO)_5$		**$Ni(CO)_4$**
Polynuclear complexes			**$Mn_2(CO)_{10}$** $Tc_2(CO)_{10}$ **$Re_2(CO)_{10}$**	**$Fe_2(CO)_9$** **$Fe_3(CO)_{12}$** $Ru_2(CO)_9$ **$Ru_3(CO)_{12}$** $Os_2(CO)_9$ **$Os_3(CO)_{12}$**	**$Co_2(CO)_8$** **$Co_4(CO)_{12}$** **$Rh_4(CO)_{12}$** **$Rh_6(CO)_{16}$** **$Ir_4(CO_{12}$**	

[a]Metal carbonyls given in bold were confirmed to be commercially available, *e.g.*, from Aldrich, ABCR or Acros.

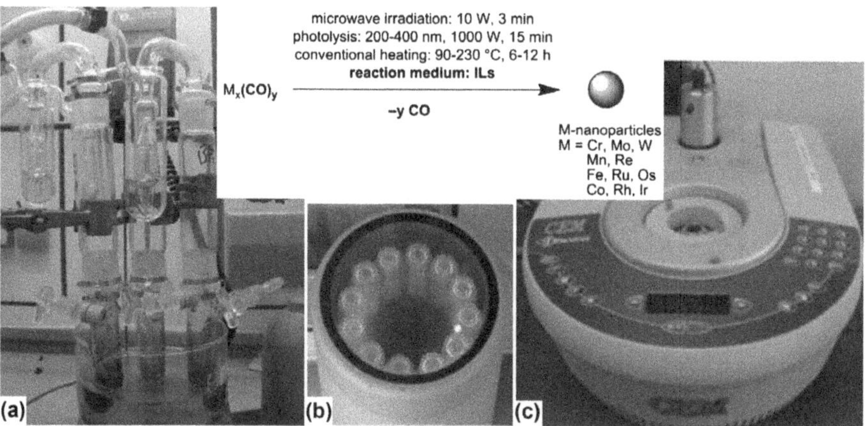

Figure 11.11 (a) Setup for conventional thermal heating of $M_x(CO)_y$–IL dispersions under argon, (b) UV reactor, (c) commercial laboratory microwave reactor.[18]
Reprinted from ref. 18 with permission from the author; © 2011 Elsevier B.V.

Metal carbonyls can be decomposed to metal nanoparticles in ILs by conventional thermal heating, UV-photolysis or microwave irradiation (MWI) (Figure 11.11a–c, Table 11.4).

ILs are especially attractive media for microwave reactions and have significant absorption efficiency for microwave energy because of their high ionic charge, high polarity and high dielectric constant.[22] Microwave heating is extremely rapid. Microwaves are a low-frequency energy source that is remarkably adaptable to many types of chemical reactions.[157] Microwave radiation can interact directly with the reaction components, the reactant mixture absorbs the microwave energy and localized superheating occurs

Table 11.4 Examples of M-NPs prepared in ILs from metal carbonyls. Reprinted from ref. 180. Copyright Verlag der Zeitschrift für Naturforschung, Tübingen 2013.

Metal	*Metal carbonyl precursor*	*Ionic liquid*	*M-NP average diameter ± standard deviation [nm]*	*Remarks*	*Reference*
Cr	$Cr(CO)_6$	$[BMIm][BF_4]$, $[BMIm][TfO]$, $[BMIm][Tf_2N]$	$\leq 1.5 \pm 0.3$, MWI, thermal;[a] 4.4 ± 1.0, $h\nu$[a]	[b,c,d,e] see Figure 11.13	78, 62
Mo	$Mp(CO)_6$	$[BMIm][BF_4]$, $[BMIm][TfO]$, $[BMIm][Tf_2N]$	$\sim 1–2$, MWI, $h\nu$;[a] $\leq 1.5 \pm 0.3$, thermal[a]	[b,c,d,e]	78, 62
W	$W(CO)_6$	$[BMIm][BF_4]$, $[BMIm][TfO]$, $[BMIm][Tf_2N]$	3.1 ± 0.8, MWI;[a] < 1, $h\nu$;[a] $\leq 1.5 \pm 0.3$, thermal[a]	[b,c,d,e] see Figure 11.12	78, 62
Mn	Mn_2CO_{10}	$[BMIm][BF_4]$	12.4 ± 3, MWI; < 1, $h\nu$ 28.6 ± 11.5, MWI	[b,c,d] see Figure 11.13	62 60
		$[CEMIm][BF_4]$	4.3 ± 1.0, MWI	see text	60
Re	Re_2CO_{10}	$[BMIm][BF_4]$	2.4 ± 0.9, MWI; < 1, $h\nu$	[b,c,d] see Figure 11.14	62
Fe	$Fe_2(CO)_9$	$[BMIm][BF_4]$	8.6 ± 3.2, MWI;[a] 7.0 ± 3.1, $h\nu$;[a] 5.2 ± 1.6, thermal[a]	[b,c,d,e]	11, 62
Ru	$Ru_3(CO)_{12}$	$[BMIm][BF_4]$	1.6 ± 0.3, MWI;[a] 2.0 ± 0.5, $h\nu$;[a] 1.6 ± 0.4, thermal[a]	[b,c,d,e]; see Figure 11.14, hydrogenation catalyst	11, 62
		$[BMIm][BF_4]$	2.2 ± 0.4, MWI	Ru-NPs deposited on TRGO, see text, Figures 11.17, 11.18	165

Os	$Os_3(CO)_{12}$	[BMIm][BF_4]	0.7 ± 0.2, MWI;[a] 2.0 ± 1.0, $h\nu$;[a] 2.5 ± 0.4, thermal[a]	[b,c,d,e]	11, 62
Co	$Co_2(CO)_8$	[BMIm][BF_4], [BMIm][TfO], [BMIm][Tf_2N]	5.1 ± 0.9, MWI;[a] 8.1 ± 2.5, $h\nu$;[a] 14 ± 8, thermal[a]	[b,c,d,e]	11, 62
		[CEMIm][BF_4]	1.6 ± 0.3, MWI	see text	60
	$Co_2(CO)_8$	[C_xMIm][Tf_2N]	7.7, thermal at 150 °C	Fischer–Tropsch catalyst giving olefins, oxygenates, and paraffins (C_7–C_{30}), reusable at least three times	166
	$Co_2(CO)_8$	[C_{10}MIm][Tf_2N]	53 ± 22, thermal at 150 °C	Co-NPs with cubic shape together with Co-NPs of irregular shape	167
Rh	$Rh_6(CO)_{16}$	[BMIm][BF_4], [BMIm][TfO], [BMIm][Tf_2N]	1.7 ± 0.3, MWI;[a] 1.9 ± 0.3, $h\nu$;[a] 3.5 ± 0.8, thermal[a]	[b,c,d,e]; see Figure 11.12, hydrogenation catalyst	11, 62
		[BMIm][BF_4]	2.8 ± 0.5	Rh-NPs deposited on TRGO, see text, Figures 11.17, 11.18	165
		[BMIm][BF_4]	2.1 ± 0.5	Rh-NPs deposited on Teflon-coated stirring bar, see text, Figure 11.20	168
Ir	$Ir_4(CO)_{12}$	[BMIm][BF_4], [BMIm][TfO], [BMIm][Tf_2N]	0.8 ± 0.2, MWI;[a] 1.4 ± 0.3, $h\nu$;[a] 1.1 ± 0.2, thermal[a]	[b,c,d,e]; hydrogenation catalyst	11, 62

[a] In [BMIm][BF_4].
[b] Median diameters and standard deviations are from TEM measurements.
[c] Microwave decomposition of metal carbonyls with 10 W for 3–10 min.
[d] Photolytic decomposition of metal carbonyls with a 1000 W Hg lamp (200–450 nm wavelength) for 15 min.
[e] Thermal decomposition of metal carbonyls from 6–12 h with 180–230 °C depending on the metal carbonyl.

resulting in a fast and efficient heating time.[158,159] Using microwaves is a fast way to heat reactants compared with conventional thermal heating. Any assumptions about abnormal "microwave effects"[160–162] have been proven wrong in the meantime.[163,164] Microwave reactions are also an "instant on"/"instant off" energy source, which reduces the risk when heating reactions.[157,158]

Metal nanoparticles were reproducibly obtained by easy, rapid (few minutes) and energy-saving (as low as 10 W) microwave irradiation under an argon atmosphere from their metal carbonyl precursors $M_x(CO)_y$ in ILs. This MWI synthesis was compared to UV-photolytic (1000 W, 15 min) or conventional thermal decomposition (180–250 °C, 6–12 h) of $M_x(CO)_y$ in ILs. The MWI-obtained nanoparticles have a very small (<5 nm) and uniform size and are prepared without any additional stabilizers or capping molecules as long-term stable M-NP–IL dispersions.[18]

For W- and Rh-NPs it was shown that the diameter increases with the molecular volume of the ionic liquid anion (Figure 11.12).[78]

Complete $M_x(CO)_y$ decomposition from the short, 3–10 min microwave irradiation was verified by Raman spectroscopy with no (metal-)carbonyl bands between 1750 and 2000 cm^{-1} being observed any more after the microwave treatment (Figure 11.13).[18,62]

Examples of TEM pictures of metal nanoparticles obtained by microwave irradiation or UV photolysis from their metal carbonyl precursors $M_x(CO)_y$ in [BMIm][BF_4] are given in Figure 11.14.[62]

The synthesis of Co-NPs and Mn-NPs by microwave-induced decomposition of the metal carbonyls $Co_2(CO)_8$ and $Mn_2(CO)_{10}$, respectively, yields smaller and better separated particles in the functionalized IL 1-(3-carboxyethyl)-3-methyl-imidazolium tetrafluoroborate [CEMIm][BF_4] (1.6 ± 0.3 nm and 4.3 ± 1.0 nm, respectively) than in the non-functionalized IL [BMIm][BF_4] (see Table 11.4). The particles are stable for more than six months although some variation in particle size could be observed by TEM.[60]

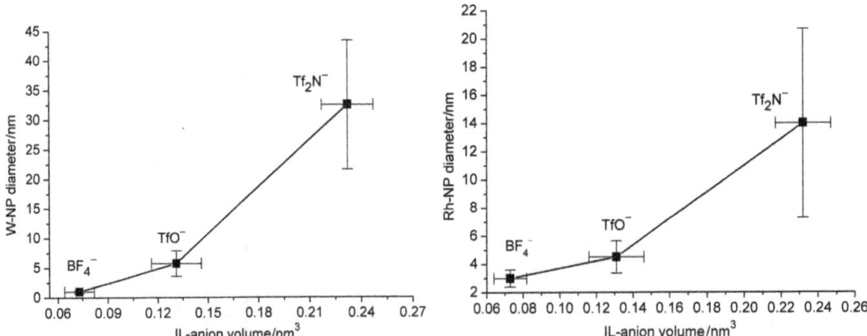

Figure 11.12 Correlation between the molecular volume of the ionic liquid anion and the observed W and Rh nanoparticle size with standard deviations as error bars (from TEM, *cf.* Figure 11.9).[78,77] Adapted from ref. 18 with permission © 2011 Elsevier B.V.

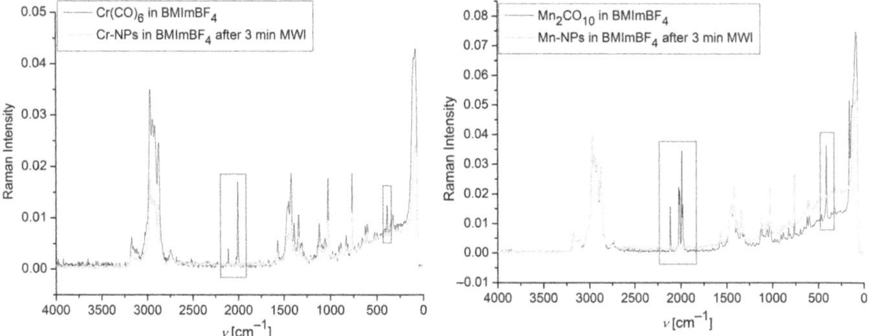

Figure 11.13 Raman-FT spectra. $Cr(CO)_6$ and $Mn_2(CO)_{10}$ in $[BMIm][BF_4]$ before and after 3 min 10 W microwave irradiation (MWI). Boxes highlight the indicative metal carbonyl bands.[62]
Reprinted from ref. 18 with permission; © 2011 Elsevier B.V.

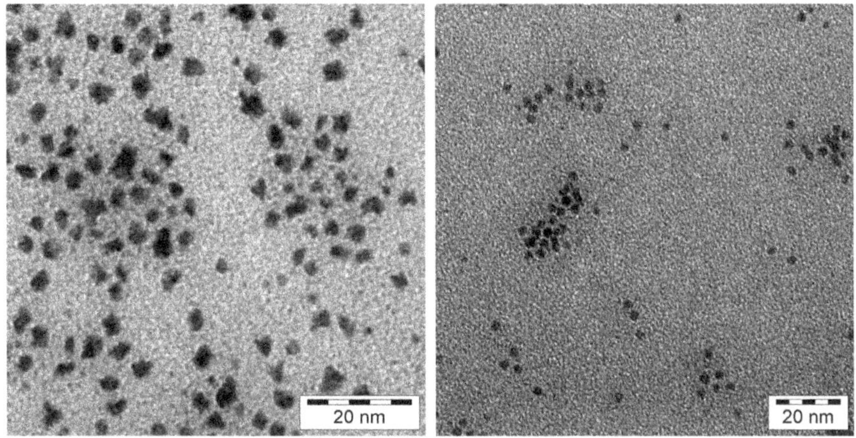

Figure 11.14 TEM photographs. Left: Re-NPs from $Re_2(CO)_{10}$ by MWI, $\varnothing$ 2.4 ± 0.9 nm. Right: Ru-NPs from $Ru_3(CO)_{12}$ by photolytic decomposition, $\varnothing$ 2.0 ± 0.5 nm (*cf.* Table 11.4).[62]
Reprinted from ref. 62 with permission; © 2010 Wiley-VCH Verlag.

11.4.6.2 *Metal Nanoparticles from Zero-Valent Metal Precursors other than $M_x(CO)_y$*

Zero-valent organometallic compounds include [Ru(COD)(COT)] and $[Ni(COD)_2]$ (COD = 1,5-cyclooctadiene, COT = 1,3,5-cyclooctatriene). Hydrogen as a reagent was used not to reduce the metal but to reduce (hydrogenate) the ligands COD and COT.[68,79,80,169] [Ru(COD)(COT)] or $[Ni(COD)_2]$ were dissolved in imidazolium-based ILs and the mixture heated under 4 bar of hydrogen under different conditions to obtain the metal nanoparticles (Table 11.5). Both organic ligands were reduced to cyclooctane and thereby

Table 11.5 Examples of M-NPs prepared in ILs from zero-valent metal precursors other than $M_x(CO)_y$.

Metal	*Metal precursor*	*IL*	*M-NP diameter ± standard deviation [nm]*	*Reference*
Ru	[Ru(COD)(COT)][a]	[BMIm][Tf_2N]	0.9–2.4	79
		[BMIm][Tf_2N] with H_2O or 1-octylamine (OA)	H_2O: ~2 nm NPs grouped in circular aggregates of 20–30 nm; OA with OA/Ru > 0.1 : 1.1 ± 0.3	80
		[BMIm][BF_4], [BMIm][PF_6], [BMIm][TfO]	2.6 ± 4	169
Ni	[$Ni(COD)_2$][a]	[C_xMIm][Tf_2N]	4.9 ± 0.9 to 5.9 ± 1.4	68

[a]COD = 1,5-cyclooctadiene, COT = 1,3,5-cyclooctatriene.

dissociated from the already zero-valent metal atom. Cyclooctane can then be removed under reduced pressure.

Ruthenium nanoparticles, Ru-NPs were obtained by decomposition, under H_2, of (η^4-1,5-cyclooctadiene)(η^6-1,3,5-cyclooctatriene)ruthenium(0), [Ru(COD)(COT)], in various 1-alkyl-3-methyl-imidazolium ionic liquids ([C_xMIm][Tf_2N] ($C_x = C_nH_{2n+1}$ where $n = 2$; 4; 6; 8; 10), and [BBIm][Tf_2N] and [BMMIm][Tf_2N] ($[BMMIm]^+$ = 1-butyl-2,3 dimethyl-imidazolium). The synthesis has been performed under 0.4 MPa of H_2, at 25 °C or at 0 °C with or without stirring. A relationship between the size of IL non-polar domains calculated by molecular dynamics simulation and the Ru-NP size measured by TEM has been found. This suggested that the phenomenon of crystal growth is probably controlled by the local concentration of [Ru(COD)(COT)] and consequently is limited to the size of the non-polar domains.[170] The size of ruthenium nanoparticles is governed by the degree of self-organization of the imidazolium-based ionic liquid in which they are generated: the more structured the ionic liquid, the smaller the size.[171] Also, the stabilization of Ru-NPs in imidazolium ILs is related to the presence of surface hydrides and their confinement in non-polar domains due to the continuous 3-D network of ionic channels.[172]

Very stable suspensions of small (*ca.* 1.2 nm) and homogeneously dispersed Ru-NPs were obtained under H_2 from [Ru(COD)(COT)], in the above imidazolium ionic liquids ([C_xMIm][Tf_2N] and in the presence of amines as ligands (1-octylamine, 1-hexadecylamine). NMR experiments (^{13}C solution and DOSY) demonstrated that the amines are coordinated to the surface of the Ru-NPs. These Ru-NPs were investigated for the hydrogenation of aromatics and have shown a high level of recyclability (up to 10 cycles) with neither loss of activity nor significant agglomeration.[173]

Also by thermal decomposition of bis(1,5-cyclooctadiene)nickel(0), [Ni(COD)$_2$] nickel nanoparticles were prepared in 1-alkyl-3-methylimidazolium bis(trifluoromethylsulfonyl)amide ionic liquids.[68]

11.4.6.3 Catalytic Applications of Metal Nanoparticles Derived from $M_x(CO)_y$

In the absence of strongly coordinating protective ligand layers, metal nanoparticles in ILs should be effective catalysts. The IL network contains only weakly coordinating cations and anions (see Figure 11.3) that bind less strongly to the metal surface and, hence, are less deactivating, than the commonly employed capping or protective ligands. The combination of M-NPs and ILs can be considered a *green catalytic system* because it can avoid the use of organic solvents. ILs are interesting in the context of green catalysis[174] which requires that catalysts be designed for easy product separation from the reaction products and multi-time efficient reuse/recycling.[22,175,176] Firstly, the very low vapor pressure of the IL and designable low miscibility of ILs with organic substrates allows for a facile separation of volatile products by distillation or removal in vacuum. Secondly, the IL is able to retain the M-NPs for catalyst reuse and recycling. M-NP–IL systems were shown to be quite easily recyclable and reusable several times without any significant changes in catalytic activity.[13] In hydrogenation reactions with Rh- or Ru-NP–IL systems the catalytic activity did not decrease upon repeated reuse.[77,62] A sizable number of catalytic reactions have successfully been carried out in ILs.[24,177] Generally, the catalytic properties (activity and selectivity) of dispersed M-NPs indicate that they possess pronounced surface-like (multi-site) rather than single-site-like character.[178,179]

The hydrogenation of internal alkynes with Pd-NPs at 25 °C and under 1 bar of hydrogen yields *Z*-alkenes with up to 98% selectivity. At higher hydrogen pressure (4 bar) alkanes were exclusively obtained without the detection of any alkenes. TOF values were up to 1282 h^{-1} with a good recyclability of the system and no loss of activity for at least 4 runs.[94]

Ru-, Rh- and Ir-NP–[BMIm][BF$_4$] dispersions were active catalysts in the biphasic liquid–liquid hydrogenation of cyclohexene or benzene to cyclohexane.[18] Even a remarkable partial hydrogenation of benzene to cyclohexene could be achieved with Ru-NP–[BMIm][PF$_6$] dispersions.[169] The low miscibility of substrates and products with the IL phase allows for easy separation by simple decantation of the hydrophobic phase.[176] The hydrogenation reaction of cyclohexene was run at 90 °C and 10 bar H_2 to 95% conversion where the reaction was intentionally stopped as thereafter the decrease in cyclohexene concentration lowered the reaction rate (Figure 11.15).[62]

Rh- and Ir-NP–IL systems function as highly effective and recyclable catalysts in the biphasic liquid–liquid hydrogenation of cyclohexene to cyclohexane with activities of up to 1900 mol cyclohexane$\times$(mol Ir)$^{-1}\times h^{-1}$

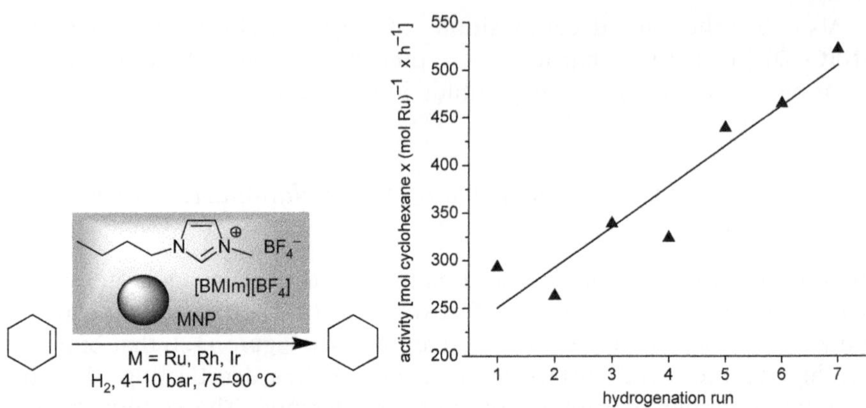

Figure 11.15 Activity for seven runs of the hydrogenation of cyclohexene with the same Ru-NP–[BMIm][BF$_4$] catalyst at 90 °C, 10 bar H_2 pressure, conversion to 95%.[62]

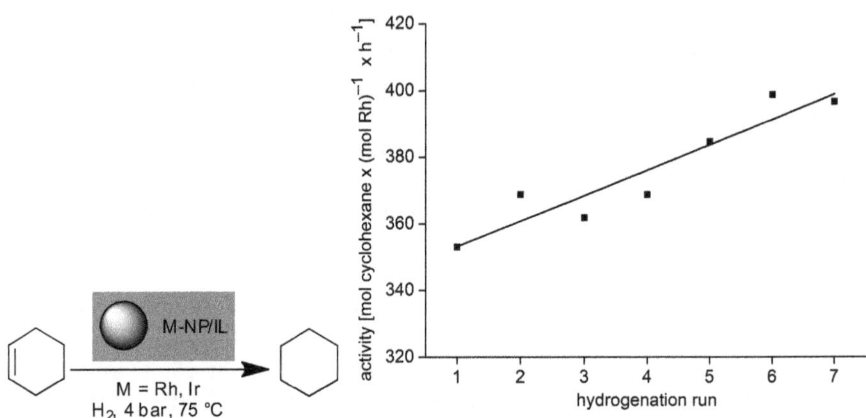

Figure 11.16 Activity over seven catalytic runs for the hydrogenation of cyclohexene with the same Rh-NP–[BMIm][BF$_4$] catalyst at 75 °C, 4 bar H_2 pressure and 2.5 h reaction time. An activity of 350 mol product$\times$(mol Rh)$^{-1}\times$h^{-1} corresponds to 88% and an activity of 400 to quantitative (100%) conversion. With the homologous Ir-NP–[BMIm][BF$_4$] catalyst even higher activities up to 1900 mol cyclohexane$\times$(mol Ir)$^{-1}\times$h^{-1} could be obtained under the same conditions, also during a shorter reaction time of 1 h for near quantitative conversion.[77]

and 380 mol cyclohexane$\times$(mol Rh)$^{-1}\times$h^{-1} for quantitative conversion at 4 bar H_2 pressure and 75 °C (Figure 11.16).[77]

Stable ruthenium or rhodium metal nanoparticles could be supported on thermally reduced graphite oxide (TRGO) (chemically derived graphene, CDG) surfaces with small and uniform particle sizes (Ru 2.2 $\pm$ 0.4 nm and Rh 2.8 $\pm$ 0.5 nm) by decomposition of their metal carbonyl precursors $Ru_3(CO)_{12}$

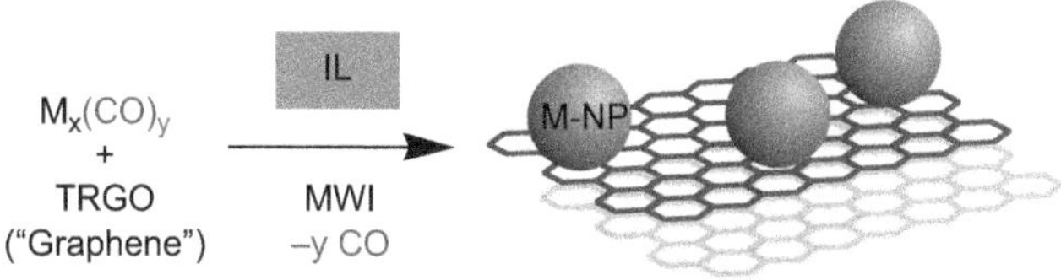

Figure 11.17 The use of microwave irradiation for the synthesis of transition metal nanoparticles supported on thermally reduced graphite oxide (TRGO) in ILs.[165]

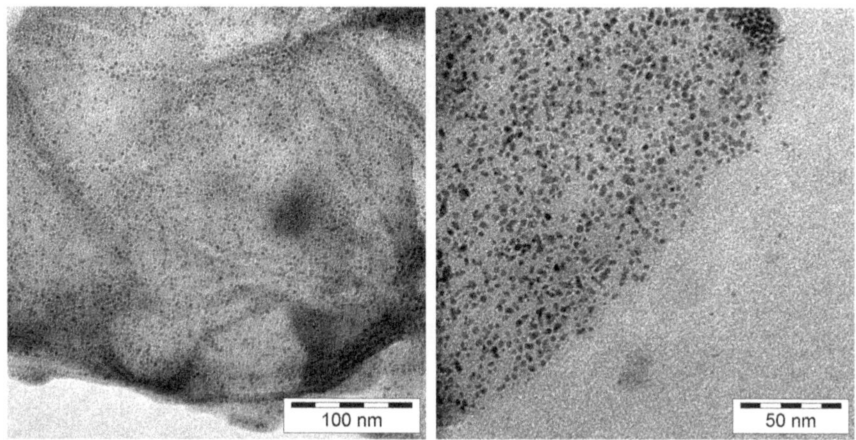

Figure 11.18 TEM photographs. Left: Ru-NP supported on thermally reduced graphite oxide (TRGO). Right: Rh-NP on TRGO, from microwave irradiation of $Ru_3(CO)_{12}$ and $Rh_6(CO)_{16}$, respectively, in a TRGO–[BMIm][BF_4] dispersion.[165]

and $Rh_6(CO)_{16}$, respectively, through microwave irradiation in a suspension of TRGO in [BMIm][BF_4] (Figures 11.17 and 11.18). The obtained hybrid nanomaterials Rh-NP–TRGO and Ru-NP–TRGO were—without further treatment—catalytically active in hydrogenation reactions yielding complete conversion of cyclohexene or benzene to cyclohexane under organic-solvent-free and mild conditions (50–75 °C, 4 bar H_2) with reproducible turnovers of 1570 mol cyclohexene$\times$(mol Ru)$^{-1}\times$h^{-1} and 310 mol benzene$\times$(mol Rh)$^{-1}\times$h^{-1}. The catalytically active M-NP–TRGO nanocomposite material could be recycled and used for several runs without any loss of activity (Figure 11.19).[165]

Rhodium nanoparticles were reproducibly deposited onto a standard, commercial Teflon-coated magnetic stirring bar by easy and rapid microwave-assisted decomposition of the metal carbonyl precursor $Rh_6(CO)_{16}$ in [BMIm][BF_4]. Such metal nanoparticle deposits are not easy to remove from the Teflon surface by simple washing procedures and represent active catalysts which one is not necessarily aware of. Such barely visible

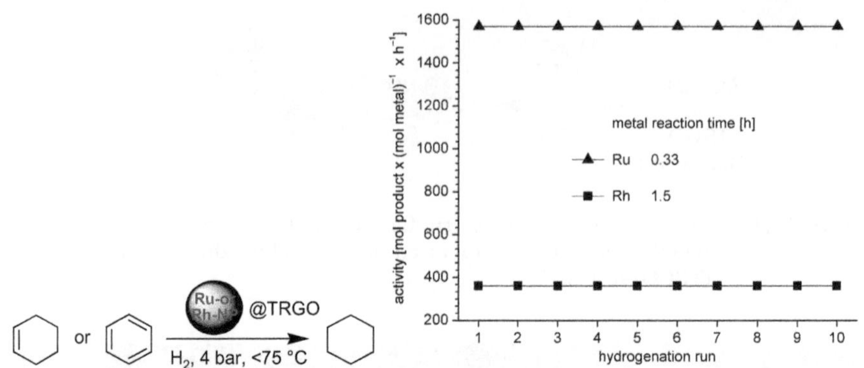

Figure 11.19 Activities for the hydrogenation of cyclohexene to cyclohexane with the same M-NP/TRGO catalyst in 10 consecutive runs.[165]
Reprinted from ref. 165 with permission © 2010 Elsevier Ltd.

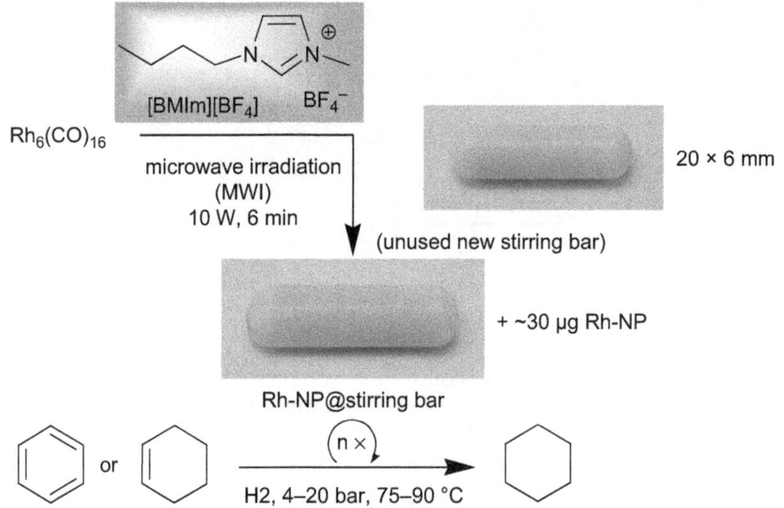

Figure 11.20 Rh-NP deposition on a Teflon-coated magnetic stirring bar from an IL dispersion and the use of Rh-NP@stirring bar in hydrogenation catalysis.[168]
Adapted from ref. 168 with permission © 2012 Elsevier B.V.

metal-nanoparticle deposits on a stirring bar can act as trace metal impurities in catalytic reactions. Rhodium nanoparticle deposits of 32 μg or less Rh metal on a 20 × 6 mm magnetic stirring bar were shown to catalyze the hydrogenation reaction of neat cyclohexene or benzene to cyclohexane with quantitative conversion. Rhodium nanoparticle-coated stirring bars were an easily handleable, separable and re-usable catalyst system for the heterogeneous hydrogenation with quantitative conversion and very high turnover frequencies of up to 32 800 mol cyclohexene $\times$ (mol Rh)$^{-1}$ $\times$ h^{-1} under organic solvent-free conditions (Figure 11.20).[168]

11.5 Conclusions

In this chapter it is shown that ionic liquids are remarkable and excellent media for the synthesis and stabilization of metal nanoparticles without the need of additional stabilizers, surfactants or capping ligands. ILs can be regarded as a supramolecular three-dimensional electrostatic and hydrogen-bonded network. The stabilization of metal nanoparticles in ILs can be further attributed to effects from the network properties of ILs such as hydrogen bonding, the hydrophobicity and steric interactions to prevent M-NPs agglomeration. The synthesis of M-NPs can proceed by chemical reduction, thermolysis, photochemical decomposition, electroreduction, microwave and sonochemical irradiation as well as gas-phase deposition methods. A microwave-induced thermal decomposition of metal carbonyls $M_x(CO)_y$ in ILs provides an especially rapid and energy-saving access to M-NPs because of the ILs' significant absorption efficiency for microwave energy due to their high ionic charge, high polarity and high dielectric constant. Metal carbonyls present attractive synthons as they are commercially available and contain the metal atoms already in the zero-valent oxidation state needed for M-NPs. No extra reducing agent is necessary and the only side product CO is given off to the gas phase and removed from the dispersion, thereby largely avoiding contaminations of the M-NP–IL dispersion.

Acknowledgements

Our work is supported by the Deutsche Forschungsgemeinschaft through grant Ja466/17-1. I thank the editor and publisher of Zeitschrift für Naturforschung for their permission to follow the article in *Z. Naturforsch. B*, 2013, **68b**, 1059–1089.[180]

References

1. A. H. Lu, E. L. Salabas and F. Schüth, *Angew. Chem., Int. Ed.*, 2007, **46**, 1222.
2. A. Gedanken, *Ultrason. Sonochem.*, 2004, **11**, 47.
3. C. N. R. Rao, S. R. C. Vivekchand, K. Biswas and A. Govindaraj, *Dalton Trans.*, 2007, 3728.
4. A. Gedanken and Y. Mastai, in *Chemistry of Nanomaterials*, ed. C. N. R. Rao, A. Müller and A. K. Cheetham, Wiley-VCH, Weinheim, 2004, p. 113.
5. J. Park, J. Joo, S. G. Kwon, Y. Jang and T. Hyeon, *Angew. Chem., Int. Ed.*, 2007, **46**, 4630.
6. M. Kim, V. N. Phan and K. Lee, *CrystEngComm*, 2012, **14**, 7535.
7. W. Ostwald, *Z. Phys. Chem.*, 1901, **37**, 385; W. Ostwald, *Lehrbuch der Allgemeinen Chemie*, Vol. 2, Part 1, Leipzig, Germany, 1896.
8. D. Astruc, F. Lu and J. R. Aranzaes, *Angew. Chem., Int. Ed.*, 2005, **44**, 7852.

9. C. Pan, K. Pelzer, K. Philippot, B. Chaudret, F. Dassenoy, P. Lecante and M.-J. Casanove, *J. Am. Chem. Soc.*, 2001, **123**, 7584.
10. J. D. Aiken III and R. G. Finke, *J. Am. Chem. Soc.*, 1999, **121**, 8803.
11. J. Krämer, E. Redel, R. Thomann and C. Janiak, *Organometallics*, 2008, **27**, 1976.
12. K. Ueno, H. Tokuda and M. Watanabe, *Phys. Chem. Chem. Phys.*, 2010, **12**, 1649.
13. J. Dupont and J. D. Scholten, *Chem. Soc. Rev.*, 2010, **39**, 1780.
14. J. Dupont, *J. Braz. Chem. Soc.*, 2004, **15**, 341.
15. M.-A. Neouze, *J. Mater. Chem.*, 2010, **20**, 9593.
16. C. S. Consorti, P. A. Z. Suarez, R. F. de Souza, R. A. Burrow, D. H. Farrar, A. J. Lough, W. Loh, L. H. M. da Silva and J. Dupont, *J. Phys. Chem. B*, 2005, **109**, 4341.
17. J. Dupont, P. A. Z. Suarez, R. F. de Souza, R. A. Burrow and J.-P. Kintzinger, *Chem.–Eur. J*, 2000, **6**, 2377.
18. C. Vollmer and C. Janiak, *Coord. Chem. Rev.*, 2011, **255**, 2039.
19. H. Weingärtner, *Angew. Chem., Int. Ed.*, 2008, **47**, 654.
20. D. Xiao, J. R. Rajian, A. Cady, S. Li, R. A. Bartsch and E. L. Quitevis, *J. Phys. Chem. B*, 2007, **111**, 4669.
21. S. Kuwabata, T. Tsuda and T. Torimoto, *J. Phys. Chem. Lett.*, 2010, **1**, 3177.
22. P. Wasserscheid and W. Keim, *Angew. Chem., Int. Ed.*, 2000, **39**, 3772.
23. I. Krossing, J. M. Slattery, C. Daguenet, P. J. Dyson, A. Oleinikova and H. Weingärtner, *J. Am. Chem. Soc.*, 2006, **128**, 13427.
24. V. I. Pârvulescu and C. Hardacre, *Chem. Rev.*, 2007, **107**, 2615.
25. N. V. Plechkova and K. R. Seddon, *Chem. Soc. Rev.*, 2008, **37**, 123.
26. T. Welton, *Chem. Rev.*, 1999, **99**, 2071.
27. R. E. Morris, *Chem. Commun.*, 2009, 2990; E. R. Parnham and R. E. Morris, *Acc. Chem. Res.*, 2007, **40**, 1005; E. R. Cooper, C. D. Andrews, P. S. Wheatley, P. B. Webb, P. Wormald and R. E. Morris, *Nature*, 2004, **430**, 1012.
28. Y. Lin and S. Dehnen, *Inorg. Chem.*, 2011, **50**, 7913; P. Lodge, *Science*, 2008, **321**, 50.
29. P. Bonhôte, A.-P. Dias, N. Papageorgiou, K. K. Kalyanasundaram and M. Grätzel, *Inorg. Chem.*, 1996, **35**, 1168.
30. J. M. Pringle, J. Golding, K. Baranyai, C. M. Forsyth, B. B. Deacon, J. L. Scott and D. R. Mc Farelane, *New J. Chem.*, 2003, **27**, 1504.
31. T. J. Gannon, G. Law, R. P. Watson, A. J. Carmichael and K. R. Seddon, *Langmuir*, 1999, **15**, 8429.
32. J. N. A. Canongia Lopes, M. F. C. Gomes and A. A. H. Padua, *J. Phys. Chem. B*, 2006, **110**, 16816.
33. G. Law, R. P. Watson, A. J. Carmichael and K. R. Seddon, *Phys. Chem. Chem. Phys.*, 2001, **3**, 2879.
34. J. N. A. Canongia Lopes and A. A. H. Pádua, *J. Phys. Chem. B*, 2006, **110**, 3330.

35. C. N. R. Rao, H. S. S. R. Matte, R. Voggu and A. Govindaraj, *Dalton Trans.*, 2012, **41**, 5089.
36. H. Zhang and H. Cui, *Langmuir*, 2009, **25**, 2604.
37. H. S. Schrekker, M. A. Gelesky, M. P. Stracke, C. M. L. Schrekker, G. Machado, S. R. Teixeira, J. C. Rubim and J. Dupont, *J. Colloid Interface Sci.*, 2007, **316**, 189.
38. R. A. Alvarez-Puebla, E. Arceo, P. J. G. Goulet, J. J. Garrido and R. F. Aroca, *J. Phys. Chem. B*, 2005, **109**, 3787.
39. E. J. W. Verwey and J. T. G. Overbeek, in *Theory of the Stability of Lyophobic Colloids*, Dover Publications Mineola, New York, 1999, p. 1.
40. E. Redel, J. Krämer, R. Thomann and C. Janiak, *GIT Labor-Fachzeitschrift*, 2008, April, 400.
41. A. N. Shipway, E. Katz and I. Willner, *ChemPhysChem*, 2000, **1**, 18.
42. T. Cassagneau and J. H. Fendler, *J. Phys. Chem. B*, 1999, **103**, 1789.
43. C. D. Keating, K. K. Kovaleski and M. J. Natan, *J. Phys. Chem. B*, 1998, **102**, 9404.
44. M. N. Kobrak and H. Li, *Phys. Chem. Chem. Phys.*, 2010, **12**, 1922.
45. G. Schmid, in *Nanoparticles: From Theory to Applications*, ed. G. Schmid, Wiley-VCH, Weinheim, 2nd edn, 2010, p. 214.
46. L. S. Ott and R. G. Finke, *Coord. Chem. Rev.*, 2007, **251**, 1075.
47. B. L. Bhargava, S. Balasubramanian and M. L. Klein, *Chem. Commun.*, 2008, 3339.
48. E. Redel, M. Walter, R. Thomann, C. Vollmer, L. Hussein, H. Scherer, M. Krüger and C. Janiak, *Chem. Eur. J.*, 2009, **15**, 10047.
49. E. Redel, M. Walter, R. Thomann, L. Hussein, M. Krüger and C. Janiak, *Chem. Commun.*, 2010, **46**, 1159.
50. A. S. Pensado and A. A. H. Pádua, *Angew. Chem., Int. Ed.*, 2011, **50**, 1.
51. H. Itoh, K. Naka and Y. Chujo, *J. Am. Chem. Soc.*, 2004, **126**, 3026.
52. K.-S. Kim, D. Demberelnyamba and H. Lee, *Langmuir*, 2004, **20**, 556.
53. S. Gao, H. Zhang, X. Wang, W. Mai, C. Peng and L. Ge, *Nanotechnology*, 2005, **16**, 1234.
54. R. Marcilla, D. Mecerreyes, I. Odriozola, J. A. Pomposo, J. Rodriguez, I. Zalakain and I. Mondragon, *Nano*, 2007, **2**, 169.
55. L. C. Branco, N. J. Rosa, J. J. M. Ramos and C. A. M. Alfonso, *Chem.-Eur. J*, 2002, **8**, 3671.
56. X. Yuan, N. Yan, S. A. Katsyuba, E. Zvereva, Y. Kou and P. J. Dyson, *Phys. Chem. Chem. Phys.*, 2012, **14**, 6026.
57. D. Zhao, Z. Fei, R. Scopelliti and P. Dyson, *Inorg. Chem.*, 2004, **43**, 2197; D. Zhao, Z. Fei, T. J. Geldbach, R. Scopeliti and P. Dyson, *J. Am. Chem. Soc.*, 2004, **126**, 15876; M. H. G. Prechtl, J. D. Scholten and J. Dupont, *J. Mol. Catal. A: Chem.*, 2009, **313**, 74.
58. A. Taubert, *Top. Curr. Chem.*, 2010, **290**, 127.
59. D.-P. Liu, G.-D. Li, Y. Su and J.-S. Chen, *Angew. Chem., Int. Ed.*, 2006, **45**, 7370.
60. D. Marquardt, Z. Xie, A. Taubert, R. Thomann and C. Janiak, *Dalton Trans.*, 2011, **40**, 8290.

61. A. Taubert and Z. Li, *Dalton Trans.*, 2007, 723.
62. C. Vollmer, E. Redel, K. Abu-Shandi, R. Thomann, H. Manyar, C. Hardacre and C. Janiak, *Chem.-Eur. J.*, 2010, **16**, 3849.
63. E. Redel, R. Thomann and C. Janiak, *Inorg. Chem.*, 2008, **47**, 14.
64. L. S. Ott and R. G. Finke, *Inorg. Chem.*, 2006, **45**, 8382.
65. G. S. Fonseca, A. P. Umpierre, P. F. P. Fichtner, S. R. Teixeira and J. Dupont, *Chem.-Eur. J*, 2003, **9**, 3263.
66. Z. Li, A. Friedrich and A. Taubert, *J. Mater. Chem.*, 2008, **18**, 1008.
67. P. Migowski, D. Zanchet, G. Machado, M. A. Gelesky, S. R. Teixeira and J. Dupont, *Phys. Chem. Chem. Phys.*, 2010, **12**, 6826.
68. P. Migowski, G. Machado, S. R. Teixeira, M. C. M. Alves, J. Morais, A. Traverse and J. Dupont, *Phys. Chem. Chem. Phys.*, 2007, **9**, 4814.
69. M. Ruta, G. Laurenczy, P. J. Dyson and L. Kiwi-Minsker, *J. Phys. Chem. C*, 2008, **112**, 17814.
70. R. R. Deshmukh, R. Rajagopal and K. V. Srinivasan, *Chem. Commun.*, 2001, 1544.
71. K. Anderson, S. C. Fernández, C. Hardacre and P. C. Marr, *Inorg. Chem. Commun.*, 2004, **7**, 73.
72. J. M. Zhu, Y. H. Shen, A. J. Xie, L.G. Qiu, Q. Zhang and X. Y. Zhang, *J. Phys. Chem. C*, 2007, **111**, 7629.
73. M. A. Firestone, M. L. Dietz, S. Seifert, S. Trasobares, D. J. Miller and N. J. Zaluzec, *Small*, 2005, **1**, 754.
74. K. Peppler, M. Polleth, S. Meiss, M. Rohnke and J. Janek, *Z. Phys. Chem.*, 2006, **220**, 1507.
75. A. Safavi, N. Maleki, F. Tajabadi and E. Farjami, *Electrochem. Commun.*, 2007, **9**, 1963.
76. K. Kim, C. Lang and P. A. Kohl, *J. Electrochem. Soc.*, 2005, **152**, E9.
77. E. Redel, J. Krämer, R. Thomann and C. Janiak, *J. Organomet. Chem.*, 2009, **694**, 1069.
78. E. Redel, R. Thomann and C. Janiak, *Chem. Commun.*, 2008, **15**, 1789.
79. T. Gutel, J. Garcia-Anton, K. Pelzer, K. Philippot, C. C. Santini, Y. Chauvin, B. Chaudret and J.-M. Basset, *J. Mater. Chem.*, 2007, **17**, 3290.
80. G. Salas, A. Podgorsek, P. S. Campbell, C. C. Santini, A. A. H. Pádua, M. F. Costa Gomes, K. Philippot, B. Chaudretd and M. Turmine, *Phys. Chem. Chem. Phys.*, 2011, **13**, 13527.
81. M. Antonietti, D. Kuang, B. Smarly and Y. Zhou, *Angew. Chem., Int. Ed.*, 2004, **43**, 4988.
82. K. I. Hana, S. W. Kang, J. Kima and Y. S. Kang, *J. Membr. Sci.*, 2011, **374**, 43.
83. J. Dupont, G. S. Fonseca, A. P. Umpierre, P. F. P. Fichtner and S. R. Teixeira, *J. Am. Chem. Soc.*, 2002, **124**, 4228.
84. C. W. Scheeren, G. Machado, J. Dupont, P. F. P. Fichtner and S. R. Texeira, *Inorg. Chem.*, 2003, **42**, 4738.
85. A. I. Bhatt, A. Mechler, L. L. Martin and A. M. Bond, *J. Mater. Chem.*, 2007, **17**, 2241.

86. M. A. Gelesky, A. P. Umpierre, G. Machado, R. R. B. Correia, W. C. Magno, J. Morais, G. Ebeling and J. Dupont, *J. Am. Chem. Soc.*, 2007, **127**, 4588.
87. R. R. Dykeman, N. Yan, R. Scopelliti and P. J. Dyson, *Inorg. Chem.*, 2011, **50**, 717.
88. G. S. Fonseca, G. Machado, S. R. Teixeira, G. H. Fecher, J. Morais, M. C. M. Alves and J. Dupont, *J. Colloid Interface Sci.*, 2006, **301**, 193.
89. X. Zhao, Y. Hua, L. Liang, C. Liu, J. Liao and W. Xing, *Int. J. Hydrogen Energy*, 2012, **37**, 51.
90. M. H. G. Prechtl, J. D. Scholten and J. Dupont, *Molecules*, 2010, **15**, 3441.
91. V. Caló, A. Nacci, A. Monopoli, S. Laera and N. Cioffi, *J. Org. Chem.*, 2003, **68**, 2929.
92. V. Caló, A. Nacci, A. Monopoli, A. Detomaso and P. Iliade, *Organometallics*, 2003, **22**, 4193.
93. F. Hassine, M. Pucheault and M. Vaultier, *C. R. Chimie*, 2011, **14**, 671.
94. R. Venkatesan, M. H. G. Prechtl, J. D. Scholten, R. P. Pezzi, G. Machadoc and J. Dupont, *J. Mater. Chem.*, 2011, **21**, 3030.
95. M. Planellas, R. Pleixats and A. Shafir, *Adv. Synth. Catal.*, 2012, **354**, 651.
96. C. W. Scheeren, J. B. Domingos, G. Machado and J. Dupont, *J. Phys. Chem. C*, 2008, **112**, 16463.
97. D. Marquardt, J. Barthel, M. Braun, C. Ganter and C. Janiak, *CrystEngComm*, 2012, **14**, 7607.
98. D. Raut, K. Wankhede, V. Vaidya, S. Bhilare, N. Darwatkar, A. Deorukhkar, G. Trivedi and M. Salunkhe, *Catal. Commun.*, 2009, **10**, 1240.
99. P. Setua, R. Pramanik, S. Sarkar, C. Ghatak, V. G. Rao, N. Sarkar and S. K. Das, *J. Mol. Liq.*, 2011, **162**, 33.
100. L. L. Lazarus, C. T. Riche, B. C. Marin, M. Gupta, N. Malmstadt and R. L. Brutchey, *ACS Appl. Mater. Interfaces*, 2012, **4**, 3077.
101. A. I. Bhatt, A. Mechler, L. L. Martin and A. M. Bond, *J. Mater. Chem.*, 2007, **17**, 2241.
102. T. Dai, L. Ge and R. Guo, *J. Mater. Res.*, 2009, **24**, 333.
103. H. R. Ryu, L. Sanchez, H. A. Keul, A. Raj and M. R. Bockstaller, *Angew. Chem., Int. Ed.*, 2008, **47**, 7639.
104. J. M. Obliosca, I. Harvey, J. Arellano, M. H. Huang and S. D. Arco, *Mater. Lett.*, 2010, **64**, 1109.
105. X. Bai, X. Li and L. Zheng, *Langmuir*, 2010, **26**, 12209.
106. L. Casal-Dujat, M. Rodrigues, A. Yagüe, A. C. Calpena, D. B. Amabilino, J. González-Linares, M. Borras and L. Pérez-García, *Langmuir*, 2012, **28**, 2368.
107. L. L. Lazarus, A. S.-J. Yang, S. Chu, R. L. Brutchey and N. Malmstadt, *Lab Chip*, 2010, **10**, 3377.

108. P. Dash, S. M. Miller and R. W. J. Scott, *J. Mol. Catal. A: Chem.*, 2010, **329**, 86.
109. A. Safavi and S. Zeinali, *Colloids Surf., A*, 2010, **362**, 121.
110. Z. Li and A. Taubert, *Molecules*, 2009, **14**, 4682.
111. V. Khare, Z. Li, A. Mantion, A. A. Ayi, S. Sonkaria, A. F. Antje Voelkl, Thünemann and A. Taubert, *J. Mater. Chem.*, 2010, **20**, 1332.
112. W. Huang, S. Chen, Y. Liu, H. Fu and G. Wu, *Nanotechnology*, 2011, **22**, 025602.
113. H. Park, J.-S. Kim, B. G. Choi, S. M. Jo, D. Y. Kim, W. H. Hong and S.-Y. Jang, *Carbon*, 2010, **48**, 1325.
114. E. Dinda, M. H. Rashid, M. Biswas and T. K. Mandal, *Langmuir*, 2010, **26**, 17568.
115. J. D. Scholten, G. Ebeling and J. Dupont, *Dalton Trans.*, 2007, 5554.
116. L. Zhao, C. Zhang, L. Zhuo, Y. Zhang and J. Y. Ying, *J. Am. Chem. Soc.*, 2008, **130**, 12586.
117. T. Tano, K. Esumi and K. Meguro, *J. Colloid Interface Sci.*, 1989, **133**, 530; K. Esumi, M. Suzuki, T. Tano, K. Torigoe and K. Meguro, *Colloids Surf.*, 1991, **55**, 9; J. S. Bradley, E. W. Hill, C. Klein, B. Chaudret and A. Duteil, *Chem. Mater.*, 1993, **5**, 254.
118. U. Schröder, J. D. Wadhawan, R. G. Compton, F. Marken, P. A. Z. Suarez, C. S. Consorti, R. F. de Souza and J. Dupont, *New J. Chem.*, 2000, **24**, 1009.
119. S. Guo and E. Wang, *Anal. Chim. Acta*, 2007, **598**, 181.
120. J. Turkevich, P. C. Stevenson and J. Hillier, *Discuss. Faraday Soc.*, 1951, **11**, 55.
121. T. A. Ryan, C. Ryan, E. A. Seddon and K. R. Seddon, in *Phosgene and Related Carbonyl Halides, Monograph 24 in Topics in Inorganic and General Chemistry*, ed. R. J. H. Clark, Elsevier, p. 242.
122. S. F. L. Mertens, C. Vollmer, A. Held, M. H. Aguirre, M. Walter, C. Janiak and T. Wandlowski, *Angew. Chem., Int. Ed.*, 2011, **50**, 9735.
123. A. Corma, I. Domínguez, T. Ródenas and M. J. Sabater, *J. Catal.*, 2008, **259**, 26.
124. S. Miao, Z. Liu, Z. Zhang, B. Han, Z. Miao, K. Ding and G. An, *J. Phys. Chem. C*, 2007, **111**, 2185.
125. P. Dash and R. W. J. Scott, *Mater. Lett.*, 2011, **65**, 7.
126. M. Harada, Y. Kimura, K. Saijo, T. Ogawa and S. Isoda, *J. Colloid Interface Sci.*, 2009, **339**, 373.
127. Z. Wang, Q. Zhang, D. Kuehner, X. Xu, A. Ivaska and L. Niu, *Carbon*, 2008, **46**, 1687.
128. F. Endres, *ChemPhysChem*, 2002, **3**, 144; F. Endres, D. MacFarlane, A. Abbott, *Electrodeposition from Ionic Liquids*, Wiley-VCH Verlag, Weinheim, 2008.
129. (*a*) U. Erb, US patent, US 5352266, 1994; (*b*) H. Natter, T. Krajewski and R. Hempelmann, *Ber. Bunsenges. Phys. Chem.*, 1996, **100**, 55; (*c*) H. Natter and R. Hempelmann, *J. Phys. Chem.*, 1996, **100**, 19525; (*d*) H. Natter, M. Schmelzer, M.-S. Löffler, C. E. Krill, A. Fitch and

R. Hempelmann, *J. Phys. Chem. B*, 2000, **104**, 2467; (*e*) R. Przenioslo, J. Wagner, H. Natter, R. Hempelmann and W. Wagner, *J. Alloys Compounds*, 2001, **328**, 259; (*f*) H. Natter, M. Schmelzer and R. Hempelmann, *J. Mater. Res.*, 1998, **13**, 1186.

130. H. Natter, M. Bukowski, R. Hempelmann, S. Zein El Abedin, E. M. Moustafa and F. Endres, *Z. Phys. Chem.*, 2006, **220**, 1275.
131. L. Yu, H. Sun, J. He, D. Wang, X. Jin, X. Hu and G. Z. Chen, *Electrochem. Commun.*, 2007, **9**, 1374.
132. P. Roy, R. Lynch and P. Schmuki, *Electrochem. Commun.*, 2009, **11**, 1567.
133. S. Zein El Abedin and F. Endres, *Electrochim. Acta*, 2009, **54**, 5673.
134. D. Wei, J. K. Baral, R. Österbacka and A. Ivaska, *J. Mater. Chem.*, 2008, **18**, 1853.
135. C. Fu, Y. Kuang, Z. Huang, X. Wang, N. Du, J. Chen and H. Zhou, *Chem. Phys. Lett.*, 2010, **499**, 250.
136. J.-H. Cha, K.-S. Kim, S. Choi, S.-H. Yeon, H. Lee, C.-S. Lee and J.-J. Shim, *Korean J. Chem. Eng.*, 2007, **24**, 1089.
137. T. A. Kareem and A. A. Kaliani, *Ionics*, 2012, **18**, 315.
138. T. Suzuki, K.-I. Okazaki, S. Suzuki, T. Shibayama, S. Kuwabata and T. Torimoto, *Chem. Mater.*, 2010, **22**, 5209.
139. T. Tsuda, K. Yoshii, T. Torimoto and S. Kuwabata, *J. Power Sources*, 2010, **195**, 5980.
140. Y. Hatakeyama, S. Takahashi and K. Nishikawa, *J. Phys. Chem. C*, 2010, **114**, 11098.
141. T. Kameyama, Y. Ohno, T. Kurimoto, K.-I. Okazaki, T. Uematsu, S. Kuwabata and T. Torimoto, *Phys. Chem. Chem. Phys.*, 2010, **12**, 1804.
142. H. Wender, L. F. de Oliveira, P. Migowski, A. F. Feil, E. Lissner, M. H. G. Prechtl, S. R. Teixeira and J. Dupont, *J. Phys. Chem. C*, 2010, **114**, 11764.
143. Q. Chen, T. Kaneko and R. Hatakeyama, *Curr. Appl. Phys.*, 2011, **11**(5), S63.
144. T. Kaneko, K. Baba and R. Hatakeyama, *J. Appl. Phys.*, 2009, **105**, 103306-1–5.
145. Y.-B. Xie and C.-J. Liu, *Plasma Processes Polym.*, 2008, **5**, 239.
146. S. Zein El Abedin, M. Pölleth, S. A. Meiss, J. Janek and F. Endres, *Green Chem.*, 2007, **9**, 549.
147. M. Brettholle, O. Höfft, L. Klarhöfer, S. Mathes, W. Maus-Friedrichs, S. Zein El Abedin, S. Krischok, J. Janek and F. Endres, *Phys. Chem. Chem. Phys.*, 2010, **12**, 1750.
148. S. A. Meiss, M. Rohnke, L. Kienle, S. Zein El Abedin, F. Endres and J. Janek, *ChemPhysChem*, 2007, **8**, 50.
149. P. He, H. Liu, Z. Li, Y. Liu, X. Xu and J. Li, *Langmuir*, 2004, **20**, 10260.
150. Z. Wei and C.-J. Liu, *Mater. Lett.*, 2011, **65**, 353.
151. A. A. Aal, R. Al-Salman, M. Al-Zoubi, N. Borissenko, F. Endres, O. Höfft, A. Prowald and S. Zein El Abedin, *Electrochim. Acta*, 2011, **56**, 10295.

152. K. Richter, A. Birkner and A.-V. Mudring, *Angew. Chem., Int. Ed.*, 2010, **49**, 2431.
153. T. Tsuda, S. Seino and S. Kuwabata, *Chem. Commun.*, 2009, 6792.
154. A. Imanishi, M. Tamura and S. Kuwabata, *Chem. Commun.*, 2009, 1775.
155. D. G. E. Kerfoot, X. Nickel, E. Wildermuth, H. Stark, G. Friedrich, F. L. Ebenhöch, B. Kühborth, J. Silver and R. Rituper, *Iron Compounds, in Ullmann's Encyclopaedia of Industrial Chemistry*, Wiley, Weinheim, 5th edn online, 2008.
156. T. Hyeon, *Chem. Commun.*, 2003, 927.
157. D. Bogdal, *Microwave-Assisted Organic Synthesis*, Elsevier, New York, 2006, p. 47.
158. A. L. Buchachenko and E. L. Frankevich, *Chemical Generation and Reception of Radio- and Microwaves*, Wiley-VCH, Weinheim, 1993, p. 41.
159. V. K. Ahluwulia, *Alternative Energy Processes in Chemical Synthesis*, Alpha Science International LTD, Oxford, 2008.
160. J. Berlan, P. Giboreau, S. Lefeuvre and C. Marchand, *Tetrahedron Lett.*, 1991, **32**, 2363.
161. F. Langa, P. de la Cruz, A. de la Hoz, A. Diaz-Ortiz and E. Diez-Barra, *Contemp. Org. Synth.*, 1997, **4**, 373.
162. L. Perreux and A. Loupy, *Tetrahedron*, 2001, **57**, 9199.
163. A. Stadler and C. O. Kappe, *J. Chem. Soc., Perkin Trans. 2*, 2000, 1363.
164. A. Stadler and C. O. Kappe, *Eur. J. Org. Chem.*, 2001, 919.
165. D. Marquardt, C. Vollmer, R. Thomann, P. Steurer, R. Mülhaupt, E. Redel and C. Janiak, *Carbon*, 2011, **49**, 1326.
166. D. O. Silva, J. D. Scholten, M. A. Gelesky, S. R. Teixeira, A. C. B. Dos Santos, E. F. Souza-Aguiar and J. Dupont, *ChemSusChem*, 2008, **1**, 291.
167. M. Scariot, D. O. Silva, J. D. Scholten, G. Machado, S. R. Teixeira, M. A. Novak, G. Ebeling and J. Dupont, *Angew. Chem., Int. Ed.*, 2008, **47**, 9075.
168. C. Vollmer, M. Schröder, Y. Thomann, R. Thomann and C. Janiak, *Appl. Catal., A*, 2012, **425–426**, 178.
169. E. T. Silveira, A. P. Umpierre, L. M. Rossi, G. Machado, J. Morais, G. V. Soares, I. J. R. Baumvol, S. R. Teixeira, R. F. P. Fichtner and J. Dupont, *Chem.–Eur. J.*, 2004, **10**, 3734.
170. T. Gutel, C. C. Santini, K. Philippot, A. Padua, K. Pelzer, B. Chaudret, Y. Chauvin and J.-M. Basset, *J. Mater. Chem.*, 2009, **19**, 3624.
171. T. Gutel, J. Garcia-Antõn, K. Pelzer, K. Philippot, C. C. Santini, Y. Chauvin, B. Chaudret and J.-M. Basset, *J. Mater. Chem.*, 2007, **17**, 3290.
172. P. S. Campbell, C. C. Santini, D. Bouchu, B. Fenet, K. Philippot, B. Chaudret, A. A. H. Pádua and Y. Chauvin, *Phys. Chem. Chem. Phys.*, 2010, **12**, 4217.
173. G. Salas, C. C. Santini, K. Philippot, V. Collière, B. Chaudret, B. Fenet and P. F. Fazzini, *Dalton Trans.*, 2011, **40**, 4660.
174. R. A. Sheldon, *Chem. Commun.*, 2008, 3352.

175. P. Wasserscheid and T. Welton in *Ionic Liquid in Synthesis*, vol. 1, Wiley-VCH, Weinheim, 2007, p. 325.
176. C. van Doorslaer, Y. Schellekens, P. Mertens, K. Binnemanns and D. De Vos, *Phys. Chem. Chem. Phys.*, 2010, **12**, 1741.
177. A. D. Sawant, D. G. Raut, N. B. Darvatkar and M. M. Salunkhe, *Green Chem. Lett. Rev.*, 2011, **4**, 41.
178. D. Astruc, *Nanoparticles and Catalysis*, Wiley-VCH, New York, 2007.
179. J. Dupont, R. F. de Souza and P. A. Z. Suarez, *Chem. Rev.*, 2002, **102**, 3667.
180. C. Janiak, *Z. Naturforsch.*, 2013, **68b**, 1059–1089.

Subject Index

References to figures are given in *italic* type. References to tables are given in **bold** type.